Lecture Notes in Computer Science 8634

Commenced Publication in 1973
Founding and Former Series Editors:
Gerhard Goos, Juris Hartmanis, and Jan van Leeuwen

Erzsébet Csuhaj-Varjú
Martin Dietzfelbinger   Zoltán Ésik (Eds.)

Mathematical Foundations of Computer Science 2014

39th International Symposium, MFCS 2014
Budapest, Hungary, August 25-29, 2014
Proceedings, Part I

 Springer

Volume Editors

Erzsébet Csuhaj-Varjú
Eötvös Loránd University
Faculty of Informatics
Budapest, Hungary
E-mail: csuhaj@inf.elte.hu

Martin Dietzfelbinger
Technische Universität Ilmenau
Fakultät für Informatik und Automatisierung
Ilmenau, Germany
E-mail: martin.dietzfelbinger@tu-ilmenau.de

Zoltán Ésik
Szeged University
Institute of Informatics
Szeged, Hungary
E-mail: ze@inf.u-szeged.hu

ISSN 0302-9743　　　　　　　　　e-ISSN 1611-3349
ISBN 978-3-662-44521-1　　　　　e-ISBN 978-3-662-44522-8
DOI 10.1007/978-3-662-44522-8
Springer Heidelberg New York Dordrecht London

Library of Congress Control Number: 2014945809

LNCS Sublibrary: SL 1 – Theoretical Computer Science and General Issues

Typesetting: Camera-ready by author, data conversion by Scientific Publishing Services, Chennai, India

Printed on acid-free paper

Springer is part of Springer Science+Business Media (www.springer.com)

Preface

The series of MFCS symposia has a long and well-established tradition. The MFCS conferences encourage high-quality research into all branches of theoretical computer science. Their broad scope provides an opportunity to bring together researchers who do not usually meet at specialized conferences. The first symposium was held in 1972. Until 2012 MFCS symposia were organized on a rotating basis in Poland, the Czech Republic, and Slovakia. The 2013 edition took place in Austria, and in 2014 Hungary joined the organizing countries. The 39th International Symposium on Mathematical Foundations of Computer Science (MFCS 2014) was held in Budapest during August 25–29, 2014.

Due to the large number of accepted papers, the proceedings of the conference were divided into two volumes on a thematical basis: Logic, Semantics, Automata and Theory of Programming (Vol. I) and Algorithms, Complexity and Games (Vol. II). The 95 contributed papers were selected by the Program Committee (PC) out of a total of 270 submissions. All submitted papers were peer reviewed and evaluated on the basis of originality, quality, significance, and presentation by at least three PC members with the help of external experts. The PC decided to give the Best Paper Award, sponsored by the European Association of Theoretical Computer Science (EATCS), to the paper "Zero Knowledge and Circuit Minimization" written by Eric Allender and Bireswar Das. In addition, the paper entitled "The Dynamic Descriptive Complexity of k-Clique" by Thomas Zeume earned the Best Student Paper Award.

The scientific program of the symposium included seven invited talks by:

- Krishnendu Chatterjee (IST Austria, Klosterneuburg, Austria)
- Achim Jung (University of Birmingham, UK)
- Dániel Marx (MTA SZTAKI, Hungary)
- Peter Bro Miltersen (Aarhus University, Denmark)
- Cyril Nicaud (Université Paris-Est Marne-la-Vallé, France)
- Alexander Sherstov (University of California, Los Angeles, USA)
- Christian Sohler (Technische Universität Dortmund, Germany)

We are grateful to all invited speakers for accepting our invitation and for their excellent presentations at the symposium. We thank all authors who submitted their work for consideration to MFCS 2014. We deeply appreciate the competent and timely handling of the submissions of all PC members and external reviewers.

The members of the Organizing Committee were Erzsébet Csuhaj-Varjú (chair, Budapest), Zsolt Gazdag (Budapest), Katalin Anna Lázár (Budapest) and Krisztián Tichler (Budapest).

The website design and maintenance were carried out by Zoltán L. Németh (University of Szeged). The publicity chair was Szabolcs Iván (University of Szeged).

The editors express their gratitude to Zsolt Gazdag, Katalin Anna Lázár, and Krisztián Tichler for their valuable contribution to the technical edition of the two volumes of proceedings.

We thank Andrej Voronkov for his EasyChair system, which facilitated the work of the PC and the editors considerably.

June 2014

Erzsébet Csuhaj-Varjú
Martin Dietzfelbinger
Zoltán Ésik

Conference Organization

The organization of the scientific part of the conference was supported by the Faculty of Informatics, Eötvös Loránd University, Budapest, and the Institute of Computer Science, Faculty of Science and Informatics, University of Szeged.

Some parts of the local arrangements were taken care of by the ELTE-Soft Non-Profit Organization and Pannonia Tourist Service.

Program Committee Chairs

Zoltán Ésik, chair	University of Szeged, Hungary
Erzsébet Csuhaj-Varjú, co-chair	Eötvös Loránd University, Hungary
Martin Dietzfelbinger, co-chair	Technische Universität Ilmenau, Germany

Program Committee

Albert Atserias	Technical University of Catalonia, Spain
Giorgio Ausiello	Sapienza University of Rome, Italy
Jos Baeten	CWI, The Netherlands
Therese Biedl	University of Waterloo, Canada
Mikołaj Bojańczyk	University of Warsaw, Poland
Gerth Stølting Brodal	Aarhus University, Denmark
Christian Choffrut	Paris Diderot University, France
Rocco De Nicola	IMT Institute for Advanced Studies Lucca, Italy
Manfred Droste	Leipzig University, Germany
Robert Elsässer	University of Salzburg, Austria
Uli Fahrenberg	IRISA/Inria Rennes, France
Fedor V. Fomin	University of Bergen, Norway
Fabio Gadducci	University of Pisa, Italy
Anna Gál	The University of Texas at Austin, USA
Dora Giammarresi	University of Tor Vergata, Rome, Italy

Roberto Grossi	University of Pisa, Italy
Anupam Gupta	Carnegie Mellon University, USA
Michel Habib	Paris Diderot University, France
Kristoffer Arnsfelt Hansen	Aarhus University, Denmark
Edith Hemaspaandra	Rochester Institute of Technology, USA
Kazuo Iwama	Kyoto University, Japan
Yoshihiko Kakutani	The University of Tokyo, Japan
Juhani Karhumäki	University of Turku, Finland
Bakhadyr Khoussainov	University of Auckland, New Zealand
Elias Koutsoupias	University of Oxford, UK
Jan Kratochvíl	Charles University, Prague, Czech Republic
Stefan Kratsch	Technical University of Berlin, Germany
Rastislav Královič	Comenius University in Bratislava, Slovakia
Amit Kumar	Indian Institute of Technology Delhi, India
Kim G. Larsen	Aalborg University, Denmark
Frédéric Magniez	Paris Diderot University, France
Ralph Matthes	IRIT (CNRS and University of Toulouse), France
Madhavan Mukund	Chennai Mathematical Institute, India
Jean-Éric Pin	LIAFA CNRS and Paris Diderot University, France
Alexander Rabinovich	Tel Aviv University, Israel
Peter Rossmanith	RWTH Aachen University, Germany
Jan Rutten	CWI and Radboud University Nijmegen, The Netherlands
Wojciech Rytter	Warsaw University and Copernicus University in Torun, Poland
Luigi Santocanale	Aix-Marseille University, France
Christian Scheideler	University of Paderborn, Germany
Thomas Schwentick	TU Dortmund University, Germany
Alex Simpson	University of Edinburgh, UK
Mohit Singh	Microsoft Research Redmond, USA
Klaus Sutner	Carnegie Mellon University, USA
Gábor Tardos	Rényi Institute, Hungary
György Turán	University of Illinois at Chicago, USA
Peter Widmayer	ETH Zürich, Switzerland
Philipp Woelfel	University of Calgary, Canada

Steering Committee

Juraj Hromkovič	ETH Zürich, Switzerland
Antonín Kučera (Chair)	Masaryk University, Czech Republic
Jerzy Marcinkowski	University of Wrocław, Poland
Damian Niwinski	University of Warsaw, Poland
Branislav Rovan	Comenius University in Bratislava, Slovakia
Jiří Sgall	Charles University, Prague, Czech Republic

Additional Reviewers

Abel, Andreas
Aceto, Luca
Acher, Mathieu
Aghazadeh, Zahra
Ahrens, Benedikt
Allender, Eric
Ambos-Spies, Klaus
Andrews, Matthew
Anselmo, Marcella
Bacci, Giorgio
Bacci, Giovanni
Baertschi, Andreas
Balbiani, Philippe
Barceló, Pablo
Bartha, Miklós
Barto, Libor
Basavaraju, Manu
Basold, Henning
Baumeister, Dorothea
Béal, Marie-Pierre
Ben-Amram, Amir
Berenbrink, Petra
Berthé, Valérie
Bilò, Davide
Bilò, Vittorio
Blanchet-Sadri, Francine
Bliznets, Ivan
Boasson, Luc
Bodlaender, Hans L.
Böhmova, Katerina
Bollig, Benedikt
Bonchi, Filippo
Bonelli, Eduardo
Bonfante, Guillaume
Bonifaci, Vincenzo
Bonsma, Paul
Boudjadar, A. Jalil
Bozzelli, Laura
Breveglieri, Luca
Brinkmann, André
Broadbent, Christopher
Brotherston, James
Brunet, Paul

Buss, Sam
Calì, Andrea
Carayol, Arnaud
Carpi, Arturo
Caucal, Didier
Cechlárová, Katarína
Chang, Richard
Christophe, Reutenauer
Chroboczek, Juliusz
Colcombet, Thomas
Colin de Verdière, Éric
Cook, Stephen
Cosme Llópez, Enric
Cranen, Sjoerd
Crescenzi, Pierluigi
Crespi Reghizzi, Stefano
Cyriac, Aiswarya
Czeizler, Elena
D'Alessandro, Flavor
D'Souza, Deepak
Dalmau, Victor
Damaschke, Peter
Datta, Samir
de Boysson, Marianne
de Mesmay, Arnaud
de Rougemont, Michel
de Wolf, Ronald
Dell, Holger
Demetrescu, Camil
Diekert, Volker
Dinneen, Michael
Drange, Pål Grønås
Dube, Simant
Dück, Stefan
Dürr, Christoph
Duparc, Jacques
Durand, Arnaud
Eikel, Martina
Elahi, Maryam
Elbassioni, Khaled
Elberfeld, Michael
Epifanio, Chiara
Espírito Santo, José

Facchini, Alessandro
Faliszewski, Piotr
Feldmann, Andreas Emil
Fernau, Henning
Ferreira, Francisco
Fertin, Guillaume
Fici, Gabriele
Finocchi, Irene
Firmani, Donatella
Fitzsimmons, Zack
Forisek, Michal
Fortier, Jérôme
Forys, Wit
Fournier, Hervé
François, Nathanaël
Fratani, Séverine
Fredriksson, Kimmo
Freivalds, Rusins
Frid, Anna
Frigioni, Daniele
Frougny, Christiane
Fusy, Éric
Gairing, Martin
Gasieniec, Leszek
Gaspers, Serge
Gastin, Paul
Gavalda, Ricard
Gawrychowski, Pawel
Gharibian, Sevag
Giustolisi, Rosario
Gliozzi, Valentina
Gmyr, Robert
Goldwurm, Massimiliano
Golovach, Petr
Grigorieff, Serge
Gualà, Luciano
Guillon, Bruno
Gupta, Sushmita
Habermehl, Peter
Harju, Tero
Hatami, Pooya
Hellerstein, Lisa
Hemaspaandra, Lane A.

Henglein, Fritz
Hertrampf, Ulrich
Hill, Cameron
Hirvensalo, Mika
Holzer, Stephan
Honda, Kentaro
Horn, Florian
Hovland, Dag
Huang, Chien-Chung
Huang, Sangxia
Huang, Zhiyi
Huber, Stefan
Hung, Ling-Ju
Hyvernat, Pierre
Imreh, Csanád
Inaba, Kazuhiro
Iván, Szabolcs
Ivanyos, Gábor
Jancar, Petr
Jansen, Bart M.P.
Jansen, Klaus
Jeřábek, Emil
Kaaser, Dominik
Kabanets, Valentine
Kaminski, Michael
Kamiński, Marcin
Kanté, Mamadou
    Moustapha
Kara, Ahmet
Kari, Jarkko
Kawamoto, Yusuke
Kerenidis, Iordanis
Kieronski, Emanuel
Kimura, Daisuke
Klasing, Ralf
Klein, Philip
Klimann, Ines
Klin, Bartek
Kling, Peter
Kniesburges, Sebastian
Kociumaka, Tomasz
Kollias, Konstantinos
Komm, Dennis
Konrad, Christian
Kontchakov, Roman

Kosowski, Adrian
Koutris, Paraschos
Koutsopoulos, Andreas
Královič, Richard
Kratsch, Dieter
Krebs, Andreas
Krokhin, Andrei
Křetinský, Jan
Kučera, Petr
Kufleitner, Manfred
Kuhnert, Sebastian
Kuperberg, Denis
Kutrib, Martin
Kuznets, Roman
La Torre, Salvatore
Labarre, Anthony
Labbé, Sébastien
Labella, Anna
Ladra, Susana
Laura, Luigi
Le Gall, François
Lecroq, Thierry
Leroy, Julien
Liaghat, Vahid
Limouzy, Vincent
Liu, Jiamou
Lluch Lafuente, Alberto
Lodaya, Kamal
Lohrey, Markus
Lokshtanov, Daniel
Lombardy, Sylvain
Lomonaco, Sam
Lonati, Violetta
Loreti, Michele
Lubiw, Anna
Luttik, Bas
Madelaine, Florent
Madonia, Maria
Malcher, Andreas
Maletti, Andreas
Mamageishvili, Akaki
Mamcarz, Antoine
Manlove, David
Markovski, Jasen
Martens, Wim

Martin, Barnaby
Marx, Dániel
Mathieson, Luke
Matoušek, Jiří
Maudet, Nicolas
Mayr, Richard
Mendler, Michael
Mihal'ák, Matúš
Mikulski, Lukasz
Mio, Matteo
Miyazaki, Shuichi
Momigliano, Alberto
Monmege, Benjamin
Montanari, Sandro
Morton, Jason
Mouawad, Amer
Mühlenthaler, Moritz
Müller, Moritz
Nagy-György, Judit
Narayanaswamy, N.S.
Nederlof, Jesper
Negri, Sara
Neven, Frank
Nichterlein, André
Niewerth, Matthias
Nishimura, Naomi
Nisse, Nicolas
Niwinski, Damian
Nonner, Tim
Nyman, Ulrik
Ochremiak, Joanna
Okhotin, Alexander
Olesen, Mads Chr.
Ouaknine, Joël
Palfrader, Peter
Pandey, Omkant
Panella, Federica
Paperman, Charles
Parberry, Ian
Pardubska, Dana
Parekh, Ojas
Paulusma, Daniel
Pavan, Aduri
Perevoshchikov, Vitaly
Perrin, Dominique

Petersen, Holger
Piątkowski, Marcin
Pignolet, Yvonne-Anne
Pinsker, Michael
Piperno, Adolfo
Plandowski, Wojciech
Poulsen, Danny Bøgsted
Pozzato, Gian Luca
Praveen, M.
Pröger, Tobias
Proietti, Guido
Quaas, Karin
Quyen, Vuong Anh
Radoszewski, Jakub
Raffinot, Mathieu
Reidl, Felix
Rettinger, Robert
Reutenauer, Christophe
Ribichini, Andrea
Ricciotti, Wilmer
Riscos-Núñez, Agustín
Riveros, Cristian
Ronchi Della Rocca,
  Simona
Rossman, Benjamin
Rubin, Sasha
Saarela, Aleksi
Sabharwal, Yogish
Sadakane, Kunihiko
Salo, Ville
Salomaa, Kai
Salvati, Sylvain
Santini, Francesco
Sau, Ignasi
Sauerwald, Thomas
Saurabh, Saket
Savicky, Petr
Schabanel, Nicolas
Schaefer, Marcus
Schmidt, Jens M.
Schmidt, Johannes

Schwartz, Roy
Schweitzer, Pascal
Segoufin, Luc
Selečéniová, Ivana
Serre, Olivier
Seto, Kazuhisa
Setzer, Alexander
Sgall, Jirí
Shallit, Jeffrey
Shen, Alexander
Siebertz, Sebastian
Silva, Alexandra
Skrzypczak, Michał
Spoerhase, Joachim
Srivathsan, B.
Stacho, Juraj
Stephan, Frank
Stiebitz, Michael
Straubing, Howard
Strozecki, Yann
Studer, Thomas
Sun, He
Suresh, S.P.
Svensson, Ola
Szörényi, Balázs
Tadaki, Kohtaro
Talebanfard, Navid
Tamaki, Suguru
Tan, Li-Yang
Tanabe, Yoshinori
Tantau, Till
Telikepalli, Kavitha
ten Cate, Balder
Thierry, Éric
Thomas, Rick
Tillich, Jean-Pierre
Törmä, Ilkka
Trinker, Horst
Tschager, Thomas
Tuosto, Emilio
Tzameret, Iddo

Umboh, Seeun
van 't Hof, Pim
van Breugel, Franck
van Hulst, Allan
van Leeuwen, Erik Jan
van Raamsdonk, Femke
Vanier, Pascal
Variyam, Vinodchandran
Vatan, Farrokh
Velner, Yaron
Vigo, Roberto
Villaret, Mateu
Volkov, Mikhail
Vrt'o, Imrich
Wagner, Uli
Waldmann, Johannes
Walen, Tomasz
Walter, Tobias
Wanka, Rolf
Warnke, Lutz
Weidner, Thomas
Weiner, Mihály
Wieder, Udi
Willemse, Tim
Williams, Ryan
Winter, Joost
Witkowski, Piotr
Wu, Zhilin
Xiao, David
Xue, Bingtian
Ye, Deshi
Zantema, Hans
Zdanowski, Konrad
Zemek, Petr
Zeume, Thomas
Zhang, Liyu
Ziegler, Martin
Zielonka, Wieslaw
Živný, Stanislav

Invited Contributions

Every Graph is Easy or Hard: Dichotomy Theorems for Graph Problems

Dániel Marx*

Institute for Computer Science and Control,
Hungarian Academy of Sciences (MTA SZTAKI),
Budapest, Hungary
dmarx@cs.bme.hu

Abstract. Given a family of algorithmic problems, a dichotomy theorem characterizes each member of the family either as "easy" or as "hard." A classical example is the result of Hell and Nešetřil classifying the complexity of H-Coloring for every fixed H: it is polynomial-time solvable if H is bipartite and NP-hard for *every* nonbipartite graph. Some dichotomy theorems characterize the complexity of a family of problems in a more general setting, where a problem in the family is defined not just by fixing a single graph H, but by fixing a (potentially infinite) *class* of graphs. For example, a result of Yannakakis characterizes the complexity of node deletion problems for *any* hereditary class of graphs, while a result of Grohe characterizes the complexity of graph homomorphisms when the left-hand side graph is restricted to be a member of a fixed class of graphs. In the talk, we survey classical and recent dichotomy theorems arising in the context of graph problems.

* Research supported by the European Research Council (ERC) grant "PARAMTIGHT: Parameterized complexity and the search for tight complexity results," reference 280152 and OTKA grant NK105645.

Computer Poker and Computational Game Theory*

Peter Bro Miltersen

Aarhus University
Department of Computer Science
Åbogade 34, 8200 Århus N
pbmiltersen@cs.au.dk

Abstract. Computationally solving two-player zero-sum games (of various kinds and with various representations) is a classical topic of computational game theory, going back at least to von Neumann's work on the relationship between linear programming and matrix games and arguably even to Zermelo's 1913 treatment of chess. Nowadays, there are at least two communities within computer science who provide an ongoing stream of interesting algorithmic problems concerning two-player zero-sum games. These are the formal methods community and the AI community. In this invited talk I survey work of a subcommunity of the latter – the computer poker community – that I think is not well known in the formal methods and algorithms communities, but should be of interest to those communities as well. I also present original joint work with Troels Bjerre Sørensen concerning the computation of asymptotically optimal push-fold strategies for heads-up no-limit Texas Hold'Em poker tournaments, as the stack sizes approach infinity.

* Work supported by The Danish National Research Foundation and The National Science Foundation of China (under the grant 61061130540) for the Sino-Danish Center for the Theory of Interactive Computation and by the Center for Research in the Foundations of Electronic Markets (CFEM), supported by the Danish Strategic Research Council.

Table of Contents – Part I

Table of Contents – Part II

Algorithms, Complexity and Games

Partial-Observation Stochastic Reachability and Parity Games[*]

Krishnendu Chatterjee

IST Austria (Institute of Science and Technology Austria)

Abstract. We consider two-player zero-sum partial-observation stochastic games on graphs. Based on the information available to the players these games can be classified as follows: (a) general partial-observation (both players have partial view of the game); (b) one-sided partial-observation (one player has partial-observation and the other player has complete-observation); and (c) perfect-observation (both players have complete view of the game). The one-sided partial-observation games subsumes the important special case of one-player partial-observation stochastic games (or partial-observation Markov decision processes (POMDPs)). Based on the randomization available for the strategies, (a) the players may not be allowed to use randomization (pure strategies), or (b) they may choose a probability distribution over actions but the actual random choice is external and not visible to the player (actions invisible), or (c) they may use full randomization. We consider all these classes of games with reachability, and parity objectives that can express all ω-regular objectives. The analysis problems are classified into the *qualitative* analysis that asks for the existence of a strategy that ensures the objective with probability 1; and the *quantitative* analysis that asks for the existence of a strategy that ensures the objective with probability at least $\lambda \in (0, 1)$.

In this talk we will cover a wide range of results: for perfect-observation games; for POMDPs; for one-sided partial-observation games; and for general partial-observation games.

1 Basic Description of Results

We present a very basic and informal description of the results to be covered in the talk. Below we discuss about pure strategies and randomized strategies with full randomization. Later we remark about randomized strategies with actions invisible.

1. *Perfect-observation stochastic games.* The decision problems for qualitative and quantitative analysis for partial-observation stochastic games with reachability and parity objectives lie in NP ∩ coNP [16,13,14,3]. Moreover, pure and memoryless optimal strategies exist for both players in such games [16,13,14,3].

---

[*] The research was partly supported by Austrian Science Fund (FWF) Grant No P 23499- N23, FWF NFN Grant No S11407-N23 (RiSE), ERC Start grant (279307: Graph Games), and Microsoft faculty fellows award.

E. Csuhaj-Varjú et al. (Eds.): MFCS 2014, Part I, LNCS 8634, pp. 1–4, 2014.

2. *POMDPs.* The quantitative analysis problem is undecidable both for infinite-memory and finite-memory strategies for POMDPs with reachability objectives [18]. The qualitative analysis problem for POMDPs with reachability objectives is EXPTIME-complete, and belief-based (subset-construction) based strategies are sufficient for randomized strategies [9,1]. The qualitative analysis problem for POMDPs with parity objectives is undecidable for infinite-memory strategies (with or without randomization) [1,8]. The qualitative analysis problem for POMDPs with parity objectives is EXPTIME-complete for finite-memory strategies (with or without randomization), and belief-based strategies are not sufficient for pure or randomized strategies [15,4].

3. *One-sided partial-observation games.* The quantitative analysis problem is undecidable both for infinite-memory and finite-memory strategies for one-sided partial-observation games with reachability objecives [18]. The qualitative analysis problem for one-sided partial-observation games with reachability objectives is EXPTIME-complete, and belief-based (subset-construction) based strategies are sufficient for randomized strategies [11], but not for pure strategies [6]. The qualitative analysis problem for one-sided partial-observation games with parity objectives is undecidable for infinite-memory strategies (with or without randomization) [1,8]. The qualitative analysis problem for one-sided partial-observation games with parity objectives is EXPTIME-complete for finite-memory strategies (with or without randomization), and belief-based strategies are not sufficient for pure or randomized strategies [17,12].

4. *General partial-observation games.* The lower bound results for POMDPs and one-sided partial-observation games carry over to the case of general partial-observation games. However the general case is more complicated. For qualitative analysis of reachability objectives with randomized strategies, the problem is 2EXPTIME-complete and belief-based strategies are sufficient [2], however, in contrast, for qualitative analysis of reachability objectives with pure strategies, the lower bound on memory requirement is non-elementary [6] and the decidability of the problem is open.

Table 1. Computational complexity results for reachability objectives

		Qualitative analysis	Quantitative analysis
	Randomized	Pure	Rand./Pure
Perfect-observation	NP ∩ coNP	NP ∩ coNP	NP ∩ coNP
POMDP	EXPTIME-c	EXPTIME-c	Undec.
One-sided	EXPTIME-c	EXPTIME-c	Undec.
Two-sided (General case)	2EXPTIME-c	Open (non-elementary mem. reqd.)	Undec.

Table 2. Computational complexity results for parity objectives

	Qualitative analysis		Quantitative analysis
	Fin. mem.	Inf. mem.	Fin./Inf. mem.
Perfect-observation	NP ∩ coNP	NP ∩ coNP	NP ∩ coNP
POMDP	EXPTIME-c	Undec.	Undec.
One-sided	EXPTIME-c	Undec.	Undec.
Two-sided (General case)	Open	Undec.	Undec.

Remark 1. For qualitative analysis with reachability objectives, finite-memory strategies are always sufficient [6], however the results are quite different for pure vs randomized strategies, and the results are summarized in Table 1. For POMDPs pure strategies are as powerful as randomized strategies, and for qualitative analysis of one-sided and general partial-observation games the results for pure strategies and randomized action-invisible strategies coincide [6,17,12]. For parity objectives, there is quite an important distinction between finite-memory and infinite-memory strategies, and the results are summarized in Table 2.

For qualitative analysis, we focus on existence of strategies that are almost-sure winning (winning with probability 1), and for other notions of qualitative analysis, such as sure winning (winning with certainty) or limit-sure winning (winning with probability arbitrarily close to 1) we refer the readers to surveys [5,10].

Acknowledgements. The talk is based on joint work with several collaborators, namely, Martin Chmelik, Laurent Doyen, Hugo Gimbert, Thomas A. Henzinger, Marcin Jurdzinski, Sumit Nain, Jean-Francois Raskin, Mathieu Tracol, and Moshe Y. Vardi.

References

1. Baier, C., Bertrand, N., Größer, M.: On decision problems for probabilistic Büchi automata. In: Amadio, R.M. (ed.) FOSSACS 2008. LNCS, vol. 4962, pp. 287–301. Springer, Heidelberg (2008)
2. Bertrand, N., Genest, B., Gimbert, H.: Qualitative determinacy and decidability of stochastic games with signals. In: LICS, pp. 319–328. IEEE Computer Society (2009)
3. Chatterjee, K.: Stochastic ω-Regular Games. PhD thesis, UC Berkeley (2007)
4. Chatterjee, K., Chmelik, M., Tracol, M.: What is decidable about partially observable Markov decision processes with omega-regular objectives. In: Proceedings of CSL 2013: Computer Science Logic (2013)
5. Chatterjee, K., Doyen, L.: The complexity of partial-observation parity games. In: Fermüller, C.G., Voronkov, A. (eds.) LPAR-17. LNCS, vol. 6397, pp. 1–14. Springer, Heidelberg (2010)

6. Chatterjee, K., Doyen, L.: Partial-observation stochastic games: How to win when belief fails. In: Proceedings of LICS 2012: Logic in Computer Science, pp. 175–184. IEEE Computer Society Press (2012)
7. Chatterjee, K., Doyen, L.: Games with a weak adversary. In: Esparza, J., Fraigniaud, P., Husfeldt, T., Koutsoupias, E. (eds.) ICALP 2014, Part II. LNCS, vol. 8573, pp. 110–121. Springer, Heidelberg (2014)
8. Chatterjee, K., Doyen, L., Gimbert, H., Henzinger, T.A.: Randomness for free. In: Hliněný, P., Kučera, A. (eds.) MFCS 2010. LNCS, vol. 6281, pp. 246–257. Springer, Heidelberg (2010)
9. Chatterjee, K., Doyen, L., Henzinger, T.A.: Qualitative analysis of partially-observable Markov decision processes. In: Hliněný, P., Kučera, A. (eds.) MFCS 2010. LNCS, vol. 6281, pp. 258–269. Springer, Heidelberg (2010)
10. Chatterjee, K., Doyen, L., Henzinger, T.A.: A survey of partial-observation stochastic parity games. Formal Methods in System Design 43(2), 268–284 (2013)
11. Chatterjee, K., Doyen, L., Henzinger, T.A., Raskin, J.-F.: Algorithms for omega-regular games of incomplete information. Logical Methods in Computer Science, 3(3:4) (2007)
12. Chatterjee, K., Doyen, L., Nain, S., Vardi, M.Y.: The complexity of partial-observation stochastic parity games with finite-memory strategies. In: Muscholl, A. (ed.) FOSSACS 2014 (ETAPS). LNCS, vol. 8412, pp. 242–257. Springer, Heidelberg (2014)
13. Chatterjee, K., Jurdziński, M., Henzinger, T.A.: Simple stochastic parity games. In: Baaz, M., Makowsky, J.A. (eds.) CSL 2003. LNCS, vol. 2803, pp. 100–113. Springer, Heidelberg (2003)
14. Chatterjee, K., Jurdziński, M., Henzinger, T.: Quantitative stochastic parity games. In: SODA 2004, pp. 121–130. SIAM (2004)
15. Chatterjee, K., Tracol, M.: Decidable problems for probabilistic automata on infinite words. In: LICS, pp. 185–194 (2012)
16. Condon, A.: The complexity of stochastic games. I&C 96(2), 203–224 (1992)
17. Nain, S., Vardi, M.Y.: Solving partial-information stochastic parity games. In: LICS, pp. 341–348 (2013)
18. Paz, A.: Introduction to probabilistic automata (Computer science and applied mathematics). Academic Press (1971)

Random Deterministic Automata[*]

Cyril Nicaud

LIGM, Université Paris-Est & CNRS, 77454 Marne-la-Vallée Cedex 2, France
cyril.nicaud@univ-mlv.fr

Abstract. In this article, we consider deterministic automata under the paradigm of average case analysis of algorithms. We present the main results obtained in the literature using this point of view, from the very beginning with Korshunov's theorem about the asymptotic number of accessible automata to the most recent advances, such as the average running time of Moore's state minimization algorithm or the estimation of the probability that an automaton is minimal. While focusing on results, we also try to give an idea of the main tools used in this field.

This article is dedicated to the memory of Philippe Flajolet.

1 Introduction

Automata theory [30] is a fundamental field of computer science, which has been especially useful in classifying formal languages, while providing efficient algorithms in several field of applications, such as text algorithms [17]. In this article, we consider deterministic automata under the paradigm of average case analysis of algorithms. We present the main results obtained in the literature with this approach, from the very beginning with Korshunov's theorem [35] to the most recent advances. We do not claim to be exhaustive, but hope to give a fair overview of the current state of this research area.

Following Knuth's seminal approach [33, Ch. 1.2.10], the area of *analysis of algorithms* aims at a more thorough analysis of algorithms by gaining insight on their running time in the best case, the worst case and the average case. Establishing that an algorithm behaves better in average than in the worst case is often done in two steps. First, one looks for properties of the inputs that makes the algorithm run faster. Then, using precise quantifications of various statistics, these properties are proved to hold with high probability[1] for random inputs.

To assist the researcher in spotting these properties, the community of analysis of algorithms also developed algorithms that randomly and uniformly generate a large variety of combinatorial structures. As we will see in the sequel, there several solutions available that are both efficient and quite generic. These random samplers are also very useful as substitutes for benchmarks for studying

---

[*] This work is supported by the French National Agency (ANR) through ANR-10-LABX-58 and through ANR-2010-BLAN-0204.

[1] i.e., with probability that tends to 1 as the size tends to infinity.

E. Csuhaj-Varjú et al. (Eds.): MFCS 2014, Part I, LNCS 8634, pp. 5–23, 2014.

algorithms experimentally. A random generator is considered good if structures of size thousand can be generated in a few seconds. A very good generator can generated objects of size one billion within the same amount of time.

The two main mathematical tools for analyzing the average running time of algorithms are discrete probabilities [23] and *analytic combinatorics* [25]. The latter is a field of computer science that aims at studying large combinatorial structures (such as permutations, trees, ...) using their combinatorial properties. The analysis starts from an enumerative description of these objects, encoded into generating series, that is, the formal series $\sum_{n\geq 0} c_n z^n$, where c_n denote the number of objects of size n. When possible, these series are interpreted as analytic functions from $\mathbb{C}$ to $\mathbb{C}$. A precise analysis of these functions then provides information on the initial combinatorial structures, especially approximate estimations of various parameters, such as the expected number of cycles in a permutation, the typical height of a tree, and so on.

Under the impulsion of Flajolet, efforts were made to systematize the approach as much as possible. It was done in two main directions. First, by providing techniques to directly characterize the generating series from a combinatorial description of the structures of interest. Then, by stating theorems that yield useful and general results, while hiding the technical complex analysis methods within their proofs. Though its main domain of application is the average case analysis of algorithms and the design of efficient random generators, analytic combinatorics has also proved useful in various fields such as statistical physics or information theory.

As an example, consider the set of total maps from $[n]$ to itself.[2] If f is such a map, its *functional graph* is the directed graph with vertex set $[n]$ and with an edge from x to y whenever $f(x) = y$. Such a graph may consist of several components, each a cycle of trees (a forest whose roots are connected by a cycle). For n and k two positive integers, let $m_{n,k}$ denote the number of maps from $[n]$ to itself having exactly k cyclic points, where x is a cyclic point of f when there exists a positive i such that $f^i(x) = x$. Intuitively, cyclic points are the vertices of the functional graph belonging to a cycle. Assume we want to estimate the proportion of cyclic points in a random map from $[n]$ to $[n]$, for large values of n. The *symbolic method* [25, Ch. II] allows to directly translate the combinatorial specification "maps = set of cycles of trees" into the equality

$$M(z,u) := \sum_{n\geq 1}\sum_{k\geq 1} \frac{m_{n,k}}{n!} z^n u^k = \frac{1}{1 - u\,T(z)} \quad \text{with} \quad T(z) = z\,e^{T(z)}\,.$$

At this point, $M(z,u)$ is viewed as an analytic function and the *singularity analysis* techniques [25, Ch. VI] yield that the average number of cyclic points is asymptotically equivalent to $\sqrt{\pi\,n/2}$.

Since a random deterministic automaton has a lot of useless states with high probability, we focus on the combinatorics of accessible automata in Section 2. We give some combinatorial bijections between automata and other combinatorial structures, as well as asymptotic results on the number of automata.

---

[2] For any positive integer n, $[n]$ denote the set $\{1,\dots,n\}$.

In Section 3, we present on a toy example a useful technique, which is widely used in the analysis of random automata. With this technique, one can export a probabilistic property of random automata to random accessible automata at low cost, avoiding the difficulty of dealing with accessible automata directly. We investigate the problem of generating accessible automata uniformly at random in Section 4. In Section 5, we briefly explained an important result, which states that a non negligible ratio of accessible automata are minimal. We then present the average case analysis of an algorithm, namely Moore's state minimization algorithm, in Section 6. In Section 7, we state some recent results about the random synchronization of automata and present some open problems that seem to be relevant for further studies.

Though not always explicitly mentioned, analytic combinatorics and discrete probabilities are used to establish most of the results presented in this article.[3]

2 Combinatorics of Automata

We start our presentation by studying the combinatorics of deterministic automata. As a guideline, we will answer the following question:[4]

Question 1: *What is the asymptotic number of accessible deterministic and complete automata with n states?*

Let us formalize the question first. Let A be a fixed alphabet with $k \geq 2$ letters.[5] For any $n \geq 1$, an *n-state transition structure* $\mathcal{T}$ on A is a pair (q_0, δ), where $q_0 \in [n]$ is the *initial state* and δ is a total map from $[n] \times A$ to $[n]$ called the *transition function* of $\mathcal{T}$. Since δ is a total map, a transition structure is a classical deterministic and complete automaton with set of states $[n]$, with q_0 as initial state, but with no final states. An *n-state deterministic and complete automaton* on A is a tuple (q_0, δ, F), where (q_0, δ) is an *n-state transition structure* and $F \subseteq [n]$ is the *set of final states*. Let $\mathfrak{T}_n$ and $\mathfrak{A}_n$ denote the set of *n-state transition structures* and *n-state automata* respectively. Obviously, we have $|\mathfrak{T}_n| = n \cdot n^{kn}$ and $|\mathfrak{A}_n| = n \cdot n^{kn} \cdot 2^n$. Since we will only consider deterministic and complete automata, we simply call them *automata* in the sequel.

Recall that a word u is *recognized* by the automaton $\mathcal{A}$ when it labels a path from the initial state to a final state. To define it formally, we extend δ to words in A^* by setting inductively that for every $p \in [n]$, $u \in A^*$ and $a \in A$, $\delta(p, \varepsilon) = p$ and $\delta(p, ua) = \delta(\delta(p, u), a)$. A word u is *recognized* by $\mathcal{A}$ when $\delta(q_0, u) \in F$. Let $\mathcal{L}(\mathcal{A})$ denote the *language recognized by $\mathcal{A}$*, i.e., the set of words it recognizes.

A state p of an automaton or a transition structure is *accessible* when there exists $u \in A^*$ such that $\delta(q_0, u) = p$. An automaton or a transition structure is *accessible* when all its states are accessible. Similarly, a state p of an automaton is

---

[3] For instance, the probabilistic study of the number of cyclic points in a random map, presented above, is one of the cornerstone of [10,19,20,42].

[4] For the exact number of automata, see Remark 7.

[5] Automata on alphabets with only one letter are very specific. See [40] for information on their typical behavior when taken uniformly at random.

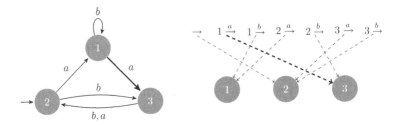

Fig. 1. An accessible automaton and its associated surjection ϕ: a transition $p \xrightarrow{a} q$ indicates that $\phi(p \xrightarrow{a}) = q$ and $\phi(\rightarrow)$ designates the initial state

co-accessible when there exists $u \in A^*$ such that $\delta(p, u) \in F$, and an automaton is *co-accessible* when all its states are co-accessible.

States that are not accessible or not co-accessible are useless, since they can be removed without changing the recognized language. We will see that from a probabilistic point of view, an accessible automaton is co-accessible with high probability (see Remark 9). The number of automata with no useless states is therefore asymptotically equivalent to the number of accessible automata. This justifies the choice of Question 1, and we now turn our attention to the set $\mathfrak{C}_n$ of n-state transition structures that are accessible.

In a transition structure, the action of each letter $a \in A$, i.e, the map $p \mapsto \delta(p, a)$, is a total map from $[n]$ to $[n]$. Using for instance techniques of analytic combinatorics [25], it is not difficult to establish that the expected number of elements with no preimage by a random map from $[n]$ to $[n]$ is $e^{-1}n$ (see [24]). Hence, roughly speaking, in a random element of $\mathfrak{T}_n$ on a two-letter alphabet, there are around $e^{-1}n$ states with no incoming transition labelled by a and $e^{-1}n$ states with no incoming transition labelled by b. "Therefore", there are around $e^{-2}n$ states with no incoming transition. This informal argument can be turned into a proof. It establishes that with high probability an element of $\mathfrak{T}_n$ is not accessible, as only the initial state can have no incoming transition in an accessible structure. Hence, $|\mathfrak{C}_n|$ is asymptotically much smaller than $|\mathfrak{T}_n|$.

The idea of Korshunov [35] is to consider elements $\mathfrak{T}_n$ whose states have at least one incoming transition, except possibly the initial state. Let $\mathfrak{T}'_n$ denote the set of such transition structures. Of course, an element of $\mathfrak{T}'_n$ is not always accessible: two strongly connected components that are totally disjoint form a non-accessible element of $\mathfrak{T}'_n$. But we will see in the sequel that it is a reasonable approximation of $\mathfrak{C}_n$. Let $E_n = [n] \times A \cup \{\rightarrow\}$ and let $\mathcal{S}_n$ denote the set of surjections from E_n onto $[n]$. To each element $\mathcal{T} = (q_0, \delta)$ of $\mathfrak{T}'_n$ one can associate bijectively an element f of $\mathcal{S}_n$, by setting $f(\rightarrow) = q_0$ and $f((p, a)) = \delta(p, a)$ (an example is depicted in Fig. 1). Hence $|\mathfrak{T}'_n|$ is equal to the number $S(kn + 1, n)$ of surjections from a set with $kn+1$ elements onto a set with n elements. Good [27] used the *saddle point method* (see [25, Ch. VIII]) to obtain an asymptotic equivalent of $S(n, m)$ when $m = \Theta(n)$. Using his result we get that the number of surjections from $[kn + 1]$ onto $[n]$ satisfies

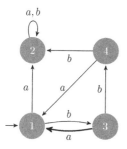
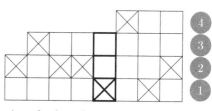

1a 1b 2a 2b **3a** 3b 4a 4b

Fig. 2. An accessible automaton and its associated diagram, as introduced in [7]. The states are labelled following a breadth-first traversal of the automaton, taking a-transitions before b-transitions. Every column of the diagram corresponds to a transition $p \xrightarrow{a} q$. The height of a column is the number of states discovered so far in the traversal, and there is a cross in the row corresponding to the target q of the transition. For example, 3 states were discovered when considering the transition $3 \xrightarrow{a} 1$. Therefore, the associated column has height 3 and there is a cross in the first row, as 1 is the target state of this transition.

$$S(kn + 1, n) \sim \alpha_k \beta_k^n n^{kn+1},$$

where $\alpha_k > 0$ and $\beta_k \in (0, 1)$ are two computable constants.

Remark 1. The quantity $\alpha_k \beta_k^n$ is exactly the probability that a mapping from $[kn + 1]$ to $[n]$ is a surjection. By Good's result, this probability is hence exponentially small.

The main result of Korshunov in [35] is that, asymptotically, the number of accessible transition structures differs from $|\mathfrak{T}_n'|$ by a multiplicative constant: if $|A| \geq 2$ then $|\mathfrak{C}_n| \sim \gamma_k |\mathfrak{T}_n'|$, where $\gamma_k > 0$ is an explicit constant. The proof is too complicated to be sketched here. It relies on a precise combinatorial study of the shape of a random element of $\mathfrak{T}_n'$. Using Good's estimation for the number of surjections, we therefore get the answer to Question 1:

$$|\mathfrak{C}_n| \sim \gamma_k \, \alpha_k \, \beta_k^n \, n^{kn+1} \, 2^n .$$

Remark 2. If ρ_k is the smallest positive solution of the equation $x = k(1 - e^{-x})$, then $\beta_k = \frac{k^k (e^{\rho_k} - 1)}{e^k \rho_k^k}$. Numerically, we have $\beta_2 \approx 0.836$ and $\beta_3 \approx 0.946$.

Remark 3. Korshunov gave a complicated formula for γ_k, using limits of converging series. Lebensztayn greatly simplified it with the theory of Lagrange inversion [36].

Remark 4. Korshunov also proved that the number of strongly connected transition structures has the same order of growth: it is asymptotically equivalent to $\delta_k \, \beta_k^n \, n^{kn+1}$, for some positive δ_k.

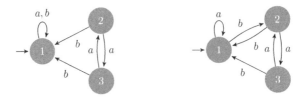

Fig. 3. On the left, an automaton with inner symmetries: there are only 3 different ways to label this shape. The automaton on the right is accessible; hence it has no inner symmetry and there are 3! different ways to label it.

Remark 5. In [7], Bassino and the author used another encoding for elements of $\mathfrak{T}'_n$. Instead of surjections, these transition structures are represented by diagrams obtained during a breadth-first traversal, as shown on Fig. 2. They have the advantage that there is a simple characterization of diagrams that correspond to accessible transition structures. These diagrams were used by Bassino, David and Sportiello to count the asymptotic number of minimal automata [6]. This result is presented in Section 5.

The answer we gave to Question 1 may seem to be unsatisfactory, since we consider two automata that only differ by their state labels as different. An *isomorphism of transition structures* is a bijection ϕ from the states of a transition structure $\mathcal{T} = (q_0, \delta)$ to those of $\mathcal{T}' = (q'_0, \delta')$ such that $\phi(q_0) = q'_0$ and for every state p and every letter a, $\delta'(\phi(p), a) = \phi(\delta(p, a))$. For the definition of *isomorphism of automata* we also require that $\phi(p)$ is final if and only if p is. It can seem more relevant to count the number of isomorphic classes rather than the number of transition structures or automata.

There is not the same number of automata in every isomorphic class, as depicted in Fig. 3. Such situations can make the counting of the number of classes quite difficult. Fortunately, the situation is easier when considering only the number of accessible structures: each state p of an n-state accessible automaton (or transition structure) is completely characterized by the smallest word u for the *radix order*,[6] also called the *length-lexicographic order*, such that $\delta(q_0, u) = p$. Hence, every bijection from the set of those words to $[n]$ define a different labelling for the automaton. Each isomorphic class of an accessible automaton or transition structure therefore contains exactly $n!$ elements. Thus, using Stirling formula, we can give the final answer to Question 1:

Answer 1: *The number of isomorphic classes of accessible automata with n states is asymptotically equivalent to $\gamma'_k \beta'^n_k n^{(k-1)n+1/2} 2^n$, where $\gamma'_k = \frac{\gamma_k \alpha_k}{\sqrt{2\pi}}$ and $\beta'_k = e \cdot \beta_k$ are two computable positive constants.*

---

[6] Compare the length first, and use the lexicographic order if they have same length.

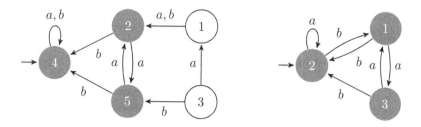

Fig. 4. On the left, a transition structure $\mathcal{T}$ with 5 states. Its states 1 and 3 are not accessible. The accessible part of $\mathcal{T}$ is depicted on the right. It is obtained by removing 1 and 3, and by normalizing the remaining state labels while preserving their relative order: $2 \mapsto 1$, $4 \mapsto 2$ and $5 \mapsto 3$.

3   From Automata to Accessible Automata

Our next goal is to obtain information on the typical properties of random accessible automata. However, we saw in the previous section that working with accessible structures can be quite complicated. In particular, we do not know any useful formula for their generating series. We therefore cannot directly use well-established techniques, such as analytic combinatorics, in order to obtain statistics on accessible automata. In this section we will explain how, in some situations, we can establish a property on random accessible automata by first proving it on random automata. We illustrate this technique on Question 2 below, which is a toy question we use to emphasize the method. More important applications will be presented in the following sections. Recall that a state p is a *sink state* if for every $a \in A$, $\delta(p, a) = p$.

Question 2: *Does a random accessible automaton often has a sink state?*

The question is easy if we remove the accessibility condition. A random automaton with n states for the uniform distribution can be seen as choosing the initial state uniformly in $[n]$, then, independently, choosing $\delta(p, a)$ uniformly in $[n]$ for every $p \in [n]$ and every $a \in A$. Assume for this section that $A = \{a, b\}$. The probability that a given state p is a sink state is therefore $\frac{1}{n^2}$. Hence, by the union bound, the probability that there is a sink state is[7] at most $\frac{1}{n}$: a uniform random automaton has no sink state with high probability. We will now show how to use this simple result to establish a similar statement for random accessible automata.

For any $\mathcal{T} \in \mathfrak{T}_n$, the *accessible part* of $\mathcal{T}$ is the accessible transition structure obtained by removing the states that are not accessible. If the accessible part has m states, we relabel them with $[m]$ while preserving their relative order, as depicted in Fig. 4.

Let m be an integer with $1 \leq m \leq n$ and let $\mathcal{T}$ be an element of $\mathfrak{C}_m$. We want to compute the number of elements of $\mathfrak{T}_n$ whose accessible part is $\mathcal{T}$.

---

[7] In fact, it is exactly $1 - (1 - \frac{1}{n^2})^n$.

To build such a transition structure, we first choose one of the $\binom{n}{m}$ possible size-m subsets of $[n]$ for labelling the states of $\mathcal{T}$. Then observe that the transitions outgoing from an accessible state are already defined by the choice of $\mathcal{T}$ and that the other transitions can end anywhere without changing the accessible part. Therefore, there are exactly $\binom{n}{m}n^{k(n-m)}$ elements of $\mathfrak{T}_n$ whose accessible part is $\mathcal{T}$. A crucial observation here is that this quantity only depends on n and m: two elements of $\mathfrak{C}_m$ are the accessible part of a random element of $\mathfrak{T}_n$ with the same probability. Using the vocabulary of probability: *conditioned by its size m, the accessible part of a uniform random element of $\mathfrak{T}_n$ is a uniform random element of $\mathfrak{C}_m$.*

Let $\#\mathtt{acc}(\mathcal{A}_n)$ be the random variable associated with the number of states of the accessible part of a random element $\mathcal{A}_n$ of $\mathfrak{T}_n$. By summing the contribution of all $\mathcal{T} \in \mathfrak{C}_m$, and since $|\mathfrak{T}_n| = n^{kn+1}$, we get that

$$\mathbb{P}(\#\mathtt{acc}(\mathcal{A}_n) = m) = \frac{1}{n}\binom{n}{m}|\mathfrak{C}_m|\,n^{-km}. \tag{1}$$

This is the starting point of the study of $\#\mathtt{acc}(\mathcal{A}_n)$ done by Carayol and the author in [13]. Thanks to Korshunov's result presented in Section 2, a fine analysis of the distribution of $\#\mathtt{acc}(\mathcal{A}_n)$ is possible. In particular, we proved that there exists a computable constant $v_k \in (0,1)$ such that $\mathbb{E}[\#\mathtt{acc}(\mathcal{A}_n)] \sim v_k n$. Numerically, $v_2 \approx 0.7968$, meaning that about 80% of the states are accessible in a typical automaton over a two-letter alphabet. The distribution is also concentrated around its mean: for any $\epsilon > 0$ the accessible part has size between $(1-\epsilon)v_k n$ and $(1+\epsilon)v_k n$ with high probability. Moreover, there exists a positive constant ω_k such that

$$\mathbb{P}\left(\#\mathtt{acc}(\mathcal{A}_n) = \lfloor v_k n \rfloor\right) \sim \frac{\omega_k}{\sqrt{n}}. \tag{2}$$

Remark 6. The main contribution of [13] is that the sequence of random variables $\#\mathtt{acc}(\mathcal{A}_n)$ is asymptotically Gaussian, of expectation and standard deviation asymptotically equivalent to $v_k n$ and $\sigma_k \sqrt{n}$ respectively, where v_k and σ_k are two computable positive constants. Hence once properly normalized, it converges in distribution to the normal law.

We can now give an answer to Question 2. Let $\mathcal{A}_n$ denote a random element of $\mathfrak{T}_n$, let $\mathtt{acc}(\mathcal{A}_n)$ denote its accessible part and let $\#\mathtt{acc}(\mathcal{A}_n)$ denote the number of states in $\mathtt{acc}(\mathcal{A}_n)$. We first use the fact that conditioned by its size, the accessible part of an automaton is a uniform accessible automaton: the probability that an element of $\mathfrak{C}_m$ has a sink state is equal to the probability that an element of $\mathfrak{T}_n$ has a sink state in its accessible part given the accessible part has size m. Therefore, we aim at studying the quantity $\mathbb{P}(\mathtt{acc}(\mathcal{A}_n)$ has a sink state $\mid \#\mathtt{acc}(\mathcal{A}_n) = m)$. But if there is a sink state in the accessible part, there is a sink state in the automaton. Hence,

$$\mathbb{P}\left(\mathtt{acc}(\mathcal{A}_n) \text{ has a sink state } \mid \#\mathtt{acc}(\mathcal{A}_n) = m\right)$$
$$\leq \mathbb{P}\left(\mathcal{A}_n \text{ has a sink state } \mid \#\mathtt{acc}(\mathcal{A}_n) = m\right).$$

Using the definition of conditional probability, we get

$$\mathbb{P}\big(\mathcal{A}_n \text{ has a sink state} \mid \#\texttt{acc}(\mathcal{A}_n) = m\big)$$
$$= \frac{\mathbb{P}\big(\mathcal{A}_n \text{ has a sink state}, \ \#\texttt{acc}(\mathcal{A}_n) = m\big)}{\mathbb{P}\big(\#\texttt{acc}(\mathcal{A}_n) = m\big)}$$
$$\leq \frac{\mathbb{P}\big(\mathcal{A}_n \text{ has a sink state}\big)}{\mathbb{P}\big(\#\texttt{acc}(\mathcal{A}_n) = m\big)}.$$

We already proved that the numerator is at most $\frac{1}{n}$ at the beginning of the section. Moreover, by Equation (2), if we choose n such that $m = \lfloor v_k\, n \rfloor$, then $n = \Theta(m)$ and the denominator is in $\Theta(\frac{1}{\sqrt{m}})$. Therefore, the probability that an element of $\mathfrak{C}_m$ has a sink state is in $\mathcal{O}(\frac{1}{\sqrt{m}})$. This gives the answer to Question 2:

Answer 2: *With high probability, a random accessible automaton has no sink state.*

Remark 7. Equation (1) rewrites in $n^{kn+1} = \sum_{m=1}^{n} \binom{n}{m} |\mathfrak{C}_m|\, n^{k(n-m)}$. This is a way to calculate the values of $|\mathfrak{C}_m|$ using a computer. The first values of the number $\frac{1}{m!}|\mathfrak{C}_m|$ of accessible transition structures up to isomorphism are,[8] for $k = 2$,

$$1, \ 12, \ 216, \ 5248, \ 160675, \ 5931540, \ 256182290, \ \ldots$$

This formula for $|\mathfrak{C}_m|$ was given by Liskovets [37]. See also Harrison [28] for the first formulas that enumerate several kind of automata.

Remark 8. The answer to Question 2 we gave follows the article [13], but the idea we used is already in Korshunov's paper [35]. To apply the method and prove that a property P does not hold with high probability for $\mathfrak{C}_n$, it is sufficient that (i) the probability P holds for $\mathfrak{T}_n$ is in $o(\frac{1}{\sqrt{n}})$ and that (ii) if $\mathcal{T} \in \mathfrak{C}_n$ satisfies P, then any transition structure whose accessible part is $\mathcal{T}$ also satisfies P.

Remark 9. By moving randomly the initial state and by using our result on the size of the accessible part, we can observe that a random automaton should have a unique stable[9] strongly connected component with high probability, which has size around $v_k\, n$. This implies that for any state p, there exists a path from p to this stable connected component. Moreover this component has a final state with high probability. Using the same technique as for sink states, this "proves" that an accessible automaton is co-accessible with high probability. By controlling the error terms, this informal argument can be turned into a full proof [20].

Remark 10. In the classical Erdős-Rényi model, a random graph with n vertices has an edge between any two vertices with probability p_n, independently. The phase transition result [11] states that there are three phases for the connectedness of such a graph: if $p_n \ll \frac{1}{n}$ then a typical graph is completely disconnected,

---

[8] This is the sequence **A006689** of the Online Encyclopedia of Integer Sequences.

[9] A set S is stable when there is no transition $p \to q$ for $p \in S$ and $q \notin S$.

with very small connected components; if $\frac{1}{n} \ll p_n \ll \frac{\log n}{n}$, there is a giant connected component of linear size and the other connected components are much smaller; if $\frac{\log n}{n} \ll p_n$, the graph is connected with high probability. This result have been extended by Karp to directed graphs [31]. One could think that by taking $p_n = \frac{k}{n}$, one would obtain a good approximation of the underlying structure of an automaton over a k-letter alphabet. In particular, the expected number of edges is kn. However, this is not really the case since random automata have a unique component with high probability according to the previous remark, whereas random graphs with $p_n = \frac{k}{n}$ do not.

4   Random Generation of Accessible Automata

As explained in the introduction, random generation is an important subfield of analysis of algorithms. Given a combinatorial class $\mathcal{C}$, the goal is to build an algorithm that efficiently generates elements of $\mathcal{C}$ of size n with the required distribution (usually the uniform distribution, in which all elements of size n have equal probability). Most basic structures, such as permutations or binary trees, have their own ad-hoc random generators. But researchers of this field also developed generic methods that translate combinatorial specifications into random samplers. They can directly be applied for objects like set partitions, partial injections, and so on. For more intricate structures, such as accessible automata, some additional work is usually required. But these general techniques form a guideline to design advanced generators, as we will see by addressing the following question:

> **Question 3:** *Is there an efficient algorithm to generate accessible automata with n states uniformly at random?*

A first idea could be to use a rejection algorithm: repeatedly generate a random automaton until it is accessible. Unfortunately, the probability p_n that an automaton is accessible is exponentially small. Since the number of iterations of this process follows a geometric law of parameter p_n, the average running time of this generator is exponential in n.

A second idea is to use the encoding into surjections of $[kn + 1]$ to $[n]$ presented in Section 2. If we can generate efficiently such a surjection, then we can successfully use the idea of a rejection algorithm. Indeed, by Korshunov's result, the probability that a random surjection corresponds to an accessible automaton tends to a positive constant. The average number of rejections is therefore in $\mathcal{O}(1)$. Moreover, using appropriate data structures, building the automaton from the surjection and testing whether it is accessible can be done in linear time. Hence, the limiting factor of this approach is the efficient generation of a random surjection from $[kn + 1]$ onto $[n]$.

Such a generator can be built using the *recursive method*, which has been introduced by Nijenhuis and Wilf [43] and developed by Flajolet, Zimmermann and Van Cutsem [26]. Let us illustrate this method on our example. Recall that $S(m, n)$ denote the number of surjections from $[m]$ onto $[n]$. In such a surjection f we distinguish two cases, depending on whether $f(m)$ has one or more preimage by f (m is of course one of these preimages). If $f(m)$ has only one preimage, then the restriction of f to $[m-1]$ is a surjection onto $[n] \setminus \{f(m)\}$; since there are n choices for $f(m)$, there are $n\,S(m-1, n-1)$ such surjections. Similarly, there are $n\,S(m-1, n)$ surjections such that $f(m)$ has more than one preimage. Hence, adding the correct initial conditions, we have the recursive formula $S(m, n) = n\,S(m-1, n-1) + n\,S(m-1, n)$. The recursive method consists of two steps. First, all the values $S(i, j)$ are computed, for $i \in [kn + 1]$ and $j \in [n]$. Then, we use the formula to build the surjection inductively: we randomly generate the image of m by f, then decide whether it has one or more preimage. It has one preimage with probability $\frac{nS(m-1,n-1)}{S(m,n)}$, in which case we remove $f(m)$ from the possible images. We then switch to $m-1$, and so on. The running time of the preprocess is $\mathcal{O}(n^2)$ and then each surjection is generated in linear time. But this analysis holds for a RAM model, where each arithmetic operation and each random choice is done in constant time. This is not realistic, since we saw that $S(kn + 1, n)$ grows extremely fast. In practice, it is hard to generate accessible automata with more than a few thousand states using this method. We have to find a better solution.

Remark 11. Following [21], it is possible to use floating point approximations to avoid the computation and the storage of $\mathcal{O}(n^2)$ big integers. Assume that we have an approximate value $p_{\approx}$ of a probability p, with $|p - p_{\approx}| \leq \epsilon$. To draw a Bernoulli law of parameter p, we generate an element x of $[0, 1]$. If $x < p_{\approx} - \epsilon$ we return 1, if $x > p_{\approx} + \epsilon$ we return 0, and if x is in the unknown zone, i.e., $|x - p_{\approx}| \leq \epsilon$, we compute a more precise estimation of p. This idea is classical in random generation but requires a careful implementation.

Remark 12. Instead of generating the surjections and rejecting those that are not *valid* (not associated to an accessible automaton), we can directly work on valid surjections. Indeed, valid surjections satisfies the same recurrence formula as surjections, but with different border conditions. This is the technique developed in [16,41], using valid diagrams of Remark 5 instead of valid surjections.

Remark 13. The algorithm designed by Almeida, Moreira and Reis [2] is another example of using the recursive method for generating accessible automata. The encoding is different, as they use string representations for accessible automata.

Boltzmann samplers were introduced in [22] and quickly became very popular in the field. This is an elegant technique that proved very efficient in many situations. Boltzmann samplers are parameterized by a positive real number x. They do not generate objects of fixed size, and the value of x has to be tuned so that the expected size of a generated object is near n. However, two objects of the same size always have the same probability to be generated. If the distribution of sizes is concentrated enough, the algorithm produces objects of size around n with high probability. The main idea of this method is thus to allow variations on the sizes in order to obtain more efficient samplers.

In our setting, the Boltzmann samplers for surjections onto $[n]$ consists of the following steps. First generates the sizes $s_1, \ldots, s_n$ of the preimage of each element of $[n]$. Each preimage size is generated independently using a non-zero Poisson law[10] of parameter x. Then fill every preimage with elements of $[m]$ where $m = \sum_{i=1}^{n} s_i$. This can be done by generating a random permutation of $[m]$ once and taking the elements in that order. The law for the s_i's guarantees that we sample surjections onto $[n]$ following a Boltzmann distribution of parameter x. For x correctly chosen, the value of m is concentrated around $kn + 1$. But we need m to be exactly $kn + 1$ for our transformation into automata to doable. This can be achieved by doing an extra rejection step: if $m \neq kn + 1$ we start the process again from the beginning. Every construction can be done in linear time, and it can be shown that the average number of rejections is in $\Theta(\sqrt{n})$. All in all, it results in an efficient random sampler for accessible automata of size n with an average running time in $\Theta(n^{3/2})$.

Remark 14. The correct value for the Boltzmann parameter x is the ρ_k of Remark 2. It also satisfies $\rho_k = k + W_0(-k e^{-k})$, where W_0 is the Lambert-W function. The series expansion of W_0 is $W_0(z) = \sum_{n \geq 1} \frac{(-n)^n}{n!} z^n$, which can be used to compute a good approximation of ρ_k, as $-k e^{-k}$ is small. This approximation is necessary for the algorithm, in order to generate the s_i's.

The third approach to random generation consists in using the result on the accessible part of a random automaton [13], which has been presented in Section 3. Recall that if $m \leq n$, then conditioned by its size m, the accessible part of a random automaton is a uniform accessible automaton with m states. Since the size of the accessible part is concentrated around $v_k n$, one can simply build a random sampler for size-m accessible automata by generating an automaton of size $\frac{n}{v_k}$ and extracting its accessible part. This is particularly efficient if we allow approximate sampling: if we use rejections until the resulting accessible automaton has size in $[(1 - \epsilon)m, (1 + \epsilon)m]$, the average running time is linear. To generate an accessible automaton of size exactly m, we use a rejection algorithm once again, and the average running time of the process is $\Theta(m^{3/2})$. It is therefore competitive with the previous method and much simpler to implement.

Remark 15. Computing v_k is not difficult, as $v_k = \frac{\rho_k}{k}$, where ρ_k can be approximated as explained in Remark 14.

The story is not over yet. In a recent preprint [8], Bassino and Sportiello presented a new method to achieve the random generation of surjections. It is based on the recursive method, mixed with the idea presented in Remark 11: probabilities are estimated more precisely only when needed. Using their method, the random generation of a surjection from $[kn + 1]$ onto $[n]$ can be done in linear expected time, yielding a linear expected time algorithm for generating accessible automata. Remark that implementing completely this technique seems

---

[10] That is, $\mathbb{P}(s = i) = \frac{x^i}{i!(e^x - 1)}$, for any $i \geq 1$.

to be quite challenging. We therefore choose to state the answer to Question 3 as follows.

Answer 3: *Using simple algorithms, one can randomly generate accessible automata with n states in expected time $\Theta(n^{3/2})$, and accessible automata having between $(1-\epsilon)n$ and $(1+\epsilon)n$ states in linear expected time. Advanced techniques under development will soon allow the generation of accessible automata with n states in linear expected time.*

Remark 16. Some implementations of these algorithms are available, such as Regal [3] for the method using Boltzmann samplers and Fado [1] that uses the recursive method on string representations (see Remark 13). The algorithm that consists in extracting the accessible part of a random automaton can be easily implemented: the random generation of an automaton is elementary, the accessible part is computed using a depth-first traversal and a good evaluation of $v_k = \frac{\rho_k}{k}$ is obtained by truncating the series of Remark 14.

5 Proportion of Minimal Automata

Let $\mathcal{A} = (q_0, \delta, F)$ be an n-state automaton. For every state $p \in [n]$, let $\mathcal{L}_p(\mathcal{A}) = \{u \in A^* : \delta(p, u) \in F\}$, i.e., the words recognized when the initial state is moved to p. Two states p and q of $\mathcal{A}$ are *equivalent* when $\mathcal{L}_p(\mathcal{A}) = \mathcal{L}_q(\mathcal{A})$. We write $p \sim q$ when p and q are equivalent. An automaton is *minimal* when its states are pairwise non-equivalent. Minimal automata are important in automata theory. In particular, up to isomorphism, there is a unique minimal automata that recognizes a given regular language $\mathcal{L}$. Moreover, it is the deterministic automaton recognizing $\mathcal{L}$ that minimizes the number of states.

Experimentations done at the end of the nineties [41] suggested that the proportion of minimal automata amongst accessible automata is not negligible. This motivate the question of this section:

Question 4: *What is the probability that a random accessible automaton is minimal?*

Bassino, David and Sportiello gave the answer to this question [6]. Their proof is complicated, we will just give the main ideas here. Two states p and q are *strongly equivalent* when both are in F or both are not in F, and for all $a \in A$, $\delta(p, a) = \delta(q, a)$. Clearly, if p and q are strongly equivalent then they are equivalent, and the automaton is not minimal.

The first step of their proof is to establish that if a random accessible automaton is not minimal, then with high probability it contains two states that are strongly equivalent. To do so, they used the technique we presented in Section 3 and proved it for random automata first. It is easier but still quite involved.

They then estimated precisely the probability of having no pair of strongly equivalent states. The critical case is for two-letter alphabets, for which this probability tends to a positive constant. Intuitively, in a random automaton, the probability that 4 given states p, p', q_a and q_b are such that $\delta(p, a) = \delta(p', a) = q_a$ and $\delta(p, b) = \delta(p', b) = q_b$ is $\frac{1}{n^4}$. Since there are $\Theta(n^4)$ choices for these 4 states, this indicates that there should be a positive probability that a random automaton is not minimal. Turning this intuition into a formal proof is difficult, especially since we have to deal with accessible automata. To do so, they use the diagram encoding depicted in Fig. 2. The main theorem of [6] is a very beautiful result, which answers Question 3:

> **Answer 4:** *If $|A| = 2$, then the probability that an accessible automaton is minimal tends to a computable constant $c_2 \in (0, 1)$. If $|A| \geq 3$, the probability that a random accessible automaton is minimal tends to 1 as the number of states tends to infinity.*

Remark 17. There is a related work of Berend and Kontorovich where they study the size of the minimal automaton of the language recognized by a random automaton [9]. They mostly rely on discrete probabilities to establish that when minimizing a random automaton, one obtains an automaton with $v_k n + \mathcal{O}(\sqrt{n} \log n)$ states with high probability.

Remark 18. Thanks to the answer to Question 4, the algorithms of Section 4 can be directly used to generate minimal automata with a given number of states. We just need to add a rejection step where we test whether the accessible automaton is minimal, and start again from the beginning if it is not. Testing minimality can be done in time $\mathcal{O}(n \log n)$ and the average number of rejections is bounded. The average running time is therefore $\mathcal{O}(n^{3/2})$ or $\mathcal{O}(n \log n)$ depending on the algorithm used for generating accessible automata.

6   Average Case Analysis of Minimization Algorithms

A minimization algorithm computes the minimal automaton of a regular language, which is usually given by an automaton. Minimal automata are important in automata theory, and there is a rich literature on the topic, with many algorithms, experimental studies, worst-case running time analysis, and so on. The best known solution to the minimization problem is Hopcroft's algorithm [29], whose worst-case running time is $\mathcal{O}(n \log n)$. This algorithm can be viewed as a tight optimization of Moore's algorithm [39], whose worst-case running time is $\mathcal{O}(n^2)$. Amongst the many other solutions, the most famous one is probably Brzozowski's algorithm [12], which is based on a different idea[11] and which also works if the input is a non-deterministic automaton. However, the running time of this elegant algorithm is exponential in the worst case, even for deterministic automata.[12]

Despite its quadratic worst-case running time, authors of libraries that implements classical algorithms for automata[13] noticed that Moore's minimization

---

[11] This is not entirely true, there are works that explain how it is related to the other minimization algorithms [15].

[12] For non-deterministic automata, the exponential blow up cannot be prevented in the worst case.

[13] Such as Vaucanson [38].

algorithm behaves well in practice. This motivates the question of this section, which is the following.

Question 5: *What is the average running time of Moore's minimization algorithm?*

Recall that two states p and q of an automaton are equivalent when the languages $\mathcal{L}_p$ and $\mathcal{L}_q$ are equal, where $\mathcal{L}_p$ is the language recognized if the initial state is moved to p. If p is a state and $\ell \geq 0$ is an integer, let $\mathcal{L}_p^{\leq \ell} = \mathcal{L}_p \cap A^{\leq \ell}$ denote the set of words of length at most ℓ that are in $\mathcal{L}_p$. Two states p and q are ℓ-*equivalent*, $p \sim_\ell q$, when $\mathcal{L}_p^{\leq \ell} = \mathcal{L}_q^{\leq \ell}$. It can be shown that if $\sim_\ell = \sim_{\ell+1}$, then $\sim_j = \sim$ for every $j \geq \ell$. Moreover, in an n-state automaton we always have $\sim_{n-2} = \sim$. Based on this facts, Moore's algorithm iteratively computes $\sim_0$, $\sim_1$, ... until $\sim_\ell = \sim_{\ell-1}$. Using appropriate data structures, each iteration can be done in linear time. Hence, the running time of the algorithm is $\mathcal{O}(n\ell)$, where ℓ is the number of iterations, that is, the smallest ℓ such that $\sim_\ell = \sim_{\ell-1}$. The minimal automaton is then built by merging states that are in the same equivalence class.

Let $\mathcal{A}$ be an automaton such that $\sim_\ell \neq \sim_{\ell-1}$ and $\sim_{\ell+1} = \sim_\ell$ for some given $\ell \geq 1$. Then there exists two states p and q that are distinguished after ℓ iterations, but not before: $p \sim_{\ell-1} q$ and there exists a word u of length ℓ such that $u \in \mathcal{L}_p$ and $u \notin \mathcal{L}_q$ (or $u \notin \mathcal{L}_p$ and $u \in \mathcal{L}_q$). Thus for every prefix v of u that is not equal to u, $\delta(p,v)$ and $\delta(q,v)$ are both final or both not final. Let G be the undirected graph whose vertices are the states of $\mathcal{A}$ and with an edge between $\delta(p,v)$ and $\delta(q,v)$ for any prefix v of u that is not equal to u. In the conference[14] paper [4], Bassino, David and the author proved that this graph is acyclic and has exactly ℓ edges. Observe that in a connect component of this graph, either all the states are final or none of them is final. But for fixed p, q and u, the graph only depends on the transition structure of $\mathcal{A}$: if we randomly choose the set of final states, the probability that the connected components satisfy the property is $2^{-\ell}$. For a good choice of $\ell \in \Theta(\log n)$, this proves that the average running time of Moore's algorithm is in $\mathcal{O}(n \log n)$.

Importantly, the proof we just sketched does not depend on the shape of the automaton: if we consider a probabilistic model where a transition structure with n states is chosen following *any* distribution and then a set of final states is added uniformly at random, then the result still holds. In particular it holds for subclasses of automata such as acyclic automata, group automata, and so on. David [18] proved that for the uniform distribution of automata, the average running time of Moore's algorithm is in $\Theta(n \log \log n)$. The proof is too involved to be presented here, but this gives the answer to Question 5:

Answer 5: *For the uniform distribution, the average running time of Moore's algorithm is in $\mathcal{O}(n \log \log n)$. For any distribution on transition structures, if final states are added uniformly at random then the average running time of Moore's algorithm is in $\mathcal{O}(n \log n)$.*

---

[14] In the journal version [5] there is no reference to this graph, the proof is done on partitions directly.

Remark 19. The proofs of [5,18] also work if each state is final with fixed probability $p \in (0, 1)$ independently. David's result, though stated for random automata, is still valid for the uniform distribution on accessible automata, using the technique presented in Section 3 (see [6,13]).

Remark 20. The proofs do not work for distributions with few final states, such as the uniform distribution on accessible automata with exactly one final state. See Section 7 for a discussion on such models.

Remark 21. Hopcroft's algorithm maintains a set of tasks to be performed. This set can be implemented in many ways (queue, stack, ...) without affecting its worst-case running time. David [18] proposed a structure for this set of tasks that guarantees that the algorithm performs at least as well as Moore's algorithm. Hence, using this implementation for the set of tasks, the average running time of Hopcroft's algorithm satisfies the same bounds as those stated in Answer 5 for Moore's algorithm.

Remark 22. De Felice and the author proved that not only Brzozowski's algorithm is inefficient in the worst case, but also that its running time is super-polynomial[15] with high probability [19]. It is thus also super-polynomial on average. This result uses a famous theorem of Erdős and Túran, which states that the order of a random permutation of $[n]$ is super-polynomial with high probability.[16]

7   Recent Results, Ongoing Works and Open Questions

An automaton is *synchronizing* when there exists a word that maps every state to the same state. Such a word is called a *synchronizing word*. In 1964, Černý [14] gave a family of synchronizing n-state automata whose smallest synchronizing word has length $(n - 1)^2$ and asked whether this bound is tight: does every synchronizing n-state automaton admit a synchronizing word of length at most $(n - 1)^2$? The question, now known as the Černý conjecture, is still open and is one of the most famous conjecture in automata theory.

The probabilistic version of this conjecture is to ask whether random automata are often synchronizing, and whether the Černý conjecture holds with high probability. In a recent preprint [10], Berlinkov proved that a random automaton is synchronizing with probability $1 - \Theta(\frac{1}{n})$ on a two-letter alphabet. This is a deep and difficult result, which was expected for quite some time, since simulations clearly shows that automata are synchronizing with high probability. Berlinkov's proof is based on classical techniques developed around the Černý conjecture and uses advanced properties of random maps from $[n]$ to $[n]$, following the approach of [34]. In another preprint [42], the author uses a different method to establish that with high probability[17] there exists a synchronizing

---

[15] i.e., grows faster that any polynomial in n.

[16] Their result is much more precise, giving a limit law for the random variable $\log O_n$, where O_n is the order of a random permutation of $[n]$.

[17] Contrarily to Berlinkov's result, there is no tight bound on the error term with this technique.

word of length $\mathcal{O}(n^{1+\epsilon})$, for any positive ϵ. Hence, the Černý conjecture holds with high probability. There is still room for sharper results in this direction, as experimentations [32] seem to indicate that the expected length of the smallest synchronizing word grows in $\sqrt{n}$.

Automata taken uniformly at random tends to have too many final states, as in practice automata with few final states are not uncommon. Unfortunately, as stated in Remark 20, most results presented in this article cannot be adapted to automata with, say, one final state. The only exception is the recent analysis of Brzozowski's algorithm for that kind of distributions [20]. One of the interesting open questions here is to confirm experimental studies that indicates that a random automaton with one final state should be minimal with non-negligible probability.

An other series of questions that is widely open is the study of the average state complexity of the classical operations on regular languages. The *state complexity* of a regular language is the number of states of its minimal automaton. A typical question in this area is "What is the average state complexity of the intersection of two languages of state complexities n?". The only known results are for unary alphabets [40] and for the reverse operation [19].

In this article, we presented different results about random deterministic automata. It is natural to try to answer the same kind of questions for non-deterministic automata too. Unfortunately, it is quite challenging to define distributions on non-deterministic automata that are both meaningful and mathematically tractable. For instance, a non-deterministic automaton taken uniformly at random recognizes all the words with high probability. The uniform model is therefore not relevant. Proving formally the experimental results such as those presented in [44], where non-deterministic automata are drawn under the Erdős-Rényi model for random graphs [11], would be an important first step in the analysis of the typical properties of random non-deterministic automata.

Acknowledgments. The author would like to thank Arnaud Carayol for his precious help during the preparation of this article.

References

1. Almeida, A., Almeida, M., Alves, J., Moreira, N., Reis, R.: Fado and guitar. In: Maneth, S. (ed.) CIAA 2009. LNCS, vol. 5642, pp. 65–74. Springer, Heidelberg (2009)
2. Almeida, M., Moreira, N., Reis, R.: Enumeration and generation with a string automata representation. Theor. Comput. Sci. 387(2), 93–102 (2007)
3. Bassino, F., David, J., Nicaud, C.: REGAL: A library to randomly and exhaustively generate automata. In: Holub, J., Žďárek, J. (eds.) CIAA 2007. LNCS, vol. 4783, pp. 303–305. Springer, Heidelberg (2007)
4. Bassino, F., David, J., Nicaud, C.: On the average complexity of Moore's state minimization algorithm. In: Albers, S., Marion, J.-Y. (eds.) STACS 2009. LIPIcs, vol. 3, pp. 123–134. Schloss Dagstuhl - Leibniz-Zentrum fuer Informatik, Germany (2009)

5. Bassino, F., David, J., Nicaud, C.: Average case analysis of Moore's state minimization algorithm. Algorithmica 63(1-2), 509–531 (2012)
6. Bassino, F., David, J., Sportiello, A.: Asymptotic enumeration of minimal automata. In: Dürr, C., Wilke, T. (eds.) STACS 2012. LIPIcs, vol. 14, pp. 88–99. Schloss Dagstuhl - Leibniz-Zentrum fuer Informatik (2012)
7. Bassino, F., Nicaud, C.: Enumeration and random generation of accessible automata. Theor. Comput. Sci. 381(1-3), 86–104 (2007)
8. Bassino, F., Sportiello, A.: Linear-time generation of specifiable combinatorial structures: General theory and first examples. arXiv, abs/1307.1728 (2013)
9. Berend, D., Kontorovich, A.: The state complexity of random DFAs. arXiv, abs/1307.0720 (2013)
10. Berlinkov, M.V.: On the probability to be synchronizable. arXiv, abs/1304.5774 (2013)
11. Bollobás, B.: Random Graphs. Cambridge University Press (2001)
12. Brzozowski, J.A.: Canonical regular expressions and minimal state graphs for definite events. In: Mathematical Theory of Automata. MRI Symposia Series, vol. 12, pp. 529–561. Polytechnic Press, Polyt. Instit. of Brooklyn, N.Y. (1962)
13. Carayol, A., Nicaud, C.: Distribution of the number of accessible states in a random deterministic automaton. In: Dürr, C., Wilke, T. (eds.) STACS 2012. LIPIcs, vol. 14, pp. 194–205. Schloss Dagstuhl - Leibniz-Zentrum fuer Informatik (2012)
14. Černý, J.: Poznámka k. homogénnym experimentom s konecnymi automatmi. Matematicko-fyzikalny Časopis Slovensk, 14 (1964)
15. Champarnaud, J.-M., Khorsi, A., Paranthoën, T.: Split and join for minimizing: Brzozowski's algorithm. In: Balík, M., Simánek, M. (eds.) Stringology 2002, pp. 96–104 (2002)
16. Champarnaud, J.-M., Paranthoën, T.: Random generation of DFAs. Theor. Comput. Sci. 330(2), 221–235 (2005)
17. Crochemore, M., Rytter, W.: Text Algorithms. Oxford University Press (1994)
18. David, J.: Average complexity of Moore's and Hopcroft's algorithms. Theor. Comput. Sci. 417, 50–65 (2012)
19. De Felice, S., Nicaud, C.: Brzozowski algorithm is generically super-polynomial for deterministic automata. In: Béal, M.-P., Carton, O. (eds.) DLT 2013. LNCS, vol. 7907, pp. 179–190. Springer, Heidelberg (2013)
20. De Felice, S., Nicaud, C.: On the average complexity of Brzozowski's algorithm for deterministic automata with a small number of final states. In: DLT 2014 (to appear in LNCS, 2014)
21. Denise, A., Zimmermann, P.: Uniform random generation of decomposable structures using floating-point arithmetic. Theor. Comput. Sci. 218(2), 233–248 (1999)
22. Duchon, P., Flajolet, P., Louchard, G., Schaeffer, G.: Boltzmann samplers for the random generation of combinatorial structures. Comb. Prob. Comp. 13(4-5), 577–625 (2004)
23. Feller, W.: An Introduction to Probability Theory and its Applications, vol. 1. Wiley (1968)
24. Flajolet, P., Odlyzko, A.M.: Random mapping statistics. In: Quisquater, J.-J., Vandewalle, J. (eds.) EUROCRYPT 1989. LNCS, vol. 434, pp. 329–354. Springer, Heidelberg (1990)
25. Flajolet, P., Sedgewick, R.: Analytic Combinatorics. Cambridge University Press (2009)

26. Flajolet, P., Zimmermann, P., Van Cutsem, B.: A calculus for the random generation of labelled combinatorial structures. Theor. Comput. Sci. 132(2), 1–35 (1994)
27. Good, I.J.: An asymptotic formula for the differences of the powers at zero. Ann. Math. Statist. 32, 249–256 (1961)
28. Harrison, M.: A census of finite automata. Canadian J. Math. 17, 100–113 (1965)
29. Hopcroft, J.E.: An $n \log n$ algorithm for minimizing the states in a finite automaton. In: Kohavi, Z. (ed.) The Theory of Machines and Computations, pp. 189–196. Academic Press (1971)
30. Hopcroft, J.E., Ullman, J.D.: Introduction to automata theory, languages, and computation. Addison-Wesley Series in Computer Science. Addison-Wesley Publishing Co., Reading (1979)
31. Karp, R.M.: The transitive closure of a random digraph. Random Struct. Algorithms 1(1), 73–94 (1990)
32. Kisielewicz, A., Kowalski, J., Szykuła, M.: A fast algorithm finding the shortest reset words. In: Du, D.-Z., Zhang, G. (eds.) COCOON 2013. LNCS, vol. 7936, pp. 182–196. Springer, Heidelberg (2013)
33. Knuth, D.E.: The Art of Computer Programming, Volume I: Fundamental Algorithms, 3rd edn. Addison-Wesley (1997)
34. Kolčin, V.: Random Mappings: Translation Series in Mathematics and Engineering. Translations series in mathematics and engineering. Springer London, Limited (1986)
35. Korshunov, A.: Enumeration of finite automata. Problemy Kibernetiki 34, 5–82 (1978) (in Russian)
36. Lebensztayn, E.: On the asymptotic enumeration of accessible automata. DMTCS 12(3) (2010)
37. Liskovets, V.A.: The number of initially connected automata. Cybernetics 4, 259–262 (1969); English translation of Kibernetika (3), 16-19 (1969)
38. Lombardy, S., Régis-Gianas, Y., Sakarovitch, J.: Introducing VAUCANSON. Theor. Comput. Sci. 328(1-2), 77–96 (2004)
39. Moore, E.F.: Gedanken experiments on sequential machines. In: Automata Studies, pp. 129–153. Princeton U. (1956)
40. Nicaud, C.: Average state complexity of operations on unary automata. In: Kutyłowski, M., Wierzbicki, T., Pacholski, L. (eds.) MFCS 1999. LNCS, vol. 1672, pp. 231–240. Springer, Heidelberg (1999)
41. Nicaud, C.: Comportement en moyenne des automates finis et des langages rationnels. PhD Thesis, Université Paris VII (2000) (in French)
42. Nicaud, C.: Fast synchronization of random automata. arXiv, abs/1404.6962 (2014)
43. Nijenhuis, A., Wilf, H.S.: Combinatorial Algorithms. Academic Press (1978)
44. Tabakov, D., Vardi, M.Y.: Experimental evaluation of classical automata constructions. In: Sutcliffe, G., Voronkov, A. (eds.) LPAR 2005. LNCS (LNAI), vol. 3835, pp. 396–411. Springer, Heidelberg (2005)

Communication Complexity Theory: Thirty-Five Years of Set Disjointness

Alexander A. Sherstov[*]

University of California, Los Angeles, USA
sherstov@cs.ucla.edu

Abstract. The set disjointness problem features k communicating parties and k subsets $S_1, S_2, \ldots, S_k \subseteq \{1, 2, \ldots, n\}$. No single party knows all k subsets, and the objective is to determine with minimal communication whether the k subsets have nonempty intersection. The important special case $k = 2$ corresponds to two parties trying to determine whether their respective sets intersect. The study of the set disjointness problem spans almost four decades and offers a unique perspective on the remarkable evolution of communication complexity theory. We discuss known results on the communication complexity of set disjointness in the deterministic, nondeterministic, randomized, unbounded-error, and multiparty models, emphasizing the variety of mathematical techniques involved.

Keywords: Set disjointness problem, communication complexity, communication lower bounds.

1 Introduction

Communication complexity theory, initiated by Andrew Yao [52] thirty-five years ago, is a central branch of theoretical computer science. The theory studies the minimum amount of communication, measured in bits, required in order to compute functions whose arguments are distributed among several parties. In addition to the basic importance of studying communication as a bottleneck resource, the theory has found a vast number of applications to other research areas, including mechanism design, streaming algorithms, machine learning, data structures, pseudorandom generators, and chip layout. Communication complexity theory is an abundant source of fascinating research questions that can be easily explained to a high school graduate but require deep mathematics and decades of collective effort to resolve. Progress in this area over the years has been truly remarkable, both in the depth and volume of research results obtained and in the diversity of techniques invented to obtain them.

Our survey focuses on a single communication problem, whose study began with the theory's inception in 1979 and actively continues to this day, with much left to discover. This problem is *set disjointness*. Its simplest version features two

---

[*] Supported by a National Science Foundation CAREER award and an Alfred P. Sloan Foundation Research Fellowship.

E. Csuhaj-Varjú et al. (Eds.): MFCS 2014, Part I, LNCS 8634, pp. 24–43, 2014.

parties who are each given a subset of $\{1, 2, \ldots, n\}$ and asked to determine with minimal communication whether the two subsets intersect. One can interpret the problem as scheduling a meeting subject to the availability of the two parties—or rather, checking whether such a meeting can be scheduled. The study of set disjointness has had a significant impact on communication complexity theory and has in many ways shaped it. First and foremost, the difficulty of determining the communication requirements of set disjointness in all but the simplest models has fueled a rapid development of the field's techniques. Moreover, set disjointness has acquired special status in communication complexity theory in that it often arises as an extremal example or as a problem separating one communication model from another. In what follows, we survey some of the highlights of this story, from basic models such as nondeterminism to advanced formalisms such as unbounded-error and multiparty communication.

2 Deterministic Communication

The simplest model of communication is the two-party deterministic model. Consider a function $f\colon X \times Y \to \{0,1\}$, where X and Y are finite sets. The model features two cooperating parties, traditionally called Alice and Bob. Alice receives an input $x \in X$, Bob receives an input $y \in Y$, and their objective is to compute $f(x,y)$. To this end, Alice and Bob communicate back and forth according to an agreed-upon protocol. The *cost* of a given communication protocol is the maximum number of bits exchanged on any input pair (x,y). The *deterministic communication complexity* of f, denoted $D(f)$, is the least cost of a communication protocol for f. In this formalism, the set disjointness problem corresponds to the function $\mathrm{DISJ}_n\colon \mathscr{P}(\{1,2,\ldots,n\}) \times \mathscr{P}(\{1,2,\ldots,n\}) \to \{0,1\}$ given by

$$\mathrm{DISJ}_n(A, B) = \begin{cases} 1 & \text{if } A \cap B = \varnothing, \\ 0 & \text{otherwise,} \end{cases}$$

where $\mathscr{P}$ refers as usual to the powerset operator.

We start by reviewing some fundamental notions, which are easiest to explain in the deterministic model and become increasingly important in more advanced models. A *combinatorial rectangle* on $X \times Y$ is any subset R of the form $R = A \times B$, where $A \subseteq X$ and $B \subseteq Y$. For brevity, we will refer to such subsets as simply *rectangles*. Given a communication problem $f\colon X \times Y \to \{0,1\}$, a rectangle R is called f-*monochromatic* if f is constant on R. Rectangles play a central role in the study of communication complexity due to the following fact [29], which shows among other things that an efficient deterministic protocol for a given function f partitions the domain into a disjoint union of a small number of f-monochromatic rectangles.

Fact 1. *Let* $\Pi\colon X \times Y \to \{0,1\}$ *be a deterministic communication protocol of cost at most c. Then there exist pairwise disjoint rectangles* $R_1, R_2, R_3, \ldots, R_{2^c}$ *such that*

$$\bigcup_{i=1}^{2^c} R_i = X \times Y$$

and Π *is constant on each* R_i.

Fooling Set Method. A straightforward technique for proving communication lower bounds is the *fooling set method* [29], which works by identifying a large set of inputs no two of which can occupy the same f-monochromatic rectangle. Formally, a *fooling set* for $f\colon X \times Y \to \{0,1\}$ is any subset $S \subseteq X \times Y$ with the following two properties: (i) f is constant on S; and (ii) if (x,y) and (x',y') are two distinct elements of S, then f is not constant on $\{(x,y),(x,y'),(x',y),(x',y')\}$. A moment's reflection reveals that an f-monochromatic rectangle can contain at most one element of S. Therefore, any partition (or even cover!) of $X \times Y$ by f-monochromatic rectangles must feature a rectangle for each point in the fooling set S, which in light of Fact 1 means that the deterministic communication complexity of f is at least $\log_2 |S|$. We summarize this discussion in the following theorem.

Theorem 2 (Fooling set method). *Let* $f\colon X \times Y \to \{0,1\}$ *be a given communication problem. If* S *is a fooling set for* f, *then*

$$D(f) \geqslant \log_2 |S|.$$

The fooling set method works perfectly for the set disjointness problem. Indeed, the set $\{(A, \{1, 2, \ldots, n\} \setminus A) : A \subseteq \{1, 2, \ldots, n\}\}$ is easily seen to be a fooling set for DISJ_n, whence $D(\mathrm{DISJ}_n) \geqslant n$. A somewhat more careful accounting yields the tight bound $D(\mathrm{DISJ}_n) = n + 1$.

Rank Bound. A more versatile technique for deterministic communication complexity was pioneered by Mehlhorn and Schmidt [33], who took an algebraic view of the question. These authors associated to every communication problem $f\colon X \times Y \to \{0,1\}$ its *characteristic matrix* $M_f = [f(x,y)]_{x \in X, y \in Y}$ and observed that any partition of $X \times Y$ into 2^c f-monochromatic rectangles gives an upper bound of 2^c on the rank of the characteristic matrix over the reals. In view of Fact 1, this gives the so-called *rank bound* on deterministic communication complexity.

Theorem 3 (Mehlhorn and Schmidt). *For any* $f\colon X \times Y \to \{0,1\}$,

$$D(f) \geqslant \log_2(\mathrm{rk}\, M_f).$$

This method, too, works well for set disjointness. Indeed, the characteristic matrix of DISJ_n is the Kronecker product of n matrices

$$\begin{bmatrix} 1 & 1 \\ 1 & 0 \end{bmatrix} \otimes \begin{bmatrix} 1 & 1 \\ 1 & 0 \end{bmatrix} \otimes \cdots \otimes \begin{bmatrix} 1 & 1 \\ 1 & 0 \end{bmatrix}$$

and therefore has full rank. As a result, $D(\mathrm{DISJ}_n) \geqslant n$ by the rank bound.

Log-Rank Conjecture. It is an instructive exercise [29] to prove that the rank bound subsumes the fooling set method, in that every communication lower bound obtained using the fooling set method can be rederived up to a constant factor using the rank bound. As a matter of fact, the *log-rank conjecture* due to Lovász and Saks [32] asserts that the rank bound is approximately tight for every function:

$$D(f) \leqslant (\log_2(\mathrm{rk}\, M_f))^{c_1} + c_2$$

for some universal constants $c_1, c_2 > 0$ and all f. This conjecture remains one of the most intriguing open questions in the area. An earlier, stronger version of the log-rank conjecture with $c_1 = 1$ has been disproved. One counterexample, due to Nisan and Wigderson [37], is a function $f\colon \{0,1\}^n \times \{0,1\}^n \to \{0,1\}$ with $D(f) = \Omega(n)$ but $\log_2(\mathrm{rk}\, M_f) = O(n^{0.631\cdots})$. As the reader might have guessed from the title of our survey, Nisan and Wigderson's construction crucially uses results [40] on the communication complexity of the set disjointness function!

3   Nondeterminism

Nondeterminism plays an important role in the study of communication, both as a natural model in its own right and as a useful intermediate formalism. In a nondeterministic protocol for a given function $f\colon X \times Y \to \{0,1\}$, Alice and Bob start by guessing a bit string, visible to them both. From then on, they communicate deterministically. A nondeterministic protocol for f must output the correct answer for *at least one* guess string when $f(x,y) = 1$ and for *all* guess strings when $f(x,y) = 0$. The cost of a nondeterministic protocol is defined as the worst-case length of the guess string, plus the worst-case cost of the deterministic phase. The *nondeterministic communication complexity* of f, denoted $N(f)$, is the least cost of a nondeterministic protocol for f. As usual, the *co-nondeterministic communication complexity* of f is the quantity $N(\neg f)$.

The nondeterministic communication complexity of a given function f is essentially characterized by the *cover number* of f, which is the smallest number of f-monochromatic rectangles whose union is $f^{-1}(1)$. Indeed, it follows easily from Fact 1 that any nondeterministic protocol of cost c gives rise to such a collection of size at most 2^c. Conversely, any size-2^c collection of f-monochromatic rectangles whose union is $f^{-1}(1)$ gives rise to a nondeterministic protocol for f of cost $c + 2$, in which Alice and Bob guess one of the rectangles and check with two bits of deterministic communication whether it contains their input pair.

Fooling Set Method. The fooling set method, reviewed in the previous section, generalizes to the nondeterministic model. Indeed, as we have already observed, no two points of a fooling set $S \subseteq f^{-1}(1)$ can reside in the same f-monochromatic rectangle, which means that any cover of $f^{-1}(1)$ by f-monochromatic rectangles must contain at least $|S|$ rectangles. In the language of nondeterministic communication complexity, we arrive at the following statement.

Theorem 4 (Fooling set method). *Let* $f\colon X \times Y \to \{0,1\}$ *be a given communication problem. If* $S \subseteq f^{-1}(1)$ *is a fooling set for* f, *then*

$$N(f) \geqslant \log_2 |S|.$$

Since the set disjointness function has a fooling set $S \subseteq \mathrm{DISJ}_n^{-1}(1)$ of size 2^n, we obtain $N(\mathrm{DISJ}_n) \geqslant n$. Set disjointness should be contrasted in this regard with its complement $\neg\mathrm{DISJ}_n$, known as *set intersection*, whose nondeterministic communication complexity is a mere $\log_2 n + O(1)$. Indeed, Alice and Bob need only guess an element $i \in \{1, 2, \ldots, n\}$ and verify with two bits of communication that it belongs to their respective sets.

Rectangle Size Bound. The most powerful method for lower bounds on nondeterministic communication complexity is the following beautiful technique, known as the *rectangle size bound* [29].

Theorem 5 (Rectangle size bound). *Let* $f\colon X \times Y \to \{0,1\}$ *be a given function. Then for every probability distribution* μ *on* $f^{-1}(1)$,

$$N(f) \geqslant \log_2 \left(\frac{1}{\max_R \mu(R)} \right),$$

where the maximum is over all rectangles $R \subseteq f^{-1}(1)$.

The rectangle size bound is a generalization of the fooling set method. Indeed, letting μ be the uniform distribution over a given fooling set $S \subseteq f^{-1}(1)$, we immediately recover Theorem 4. The proof of Theorem 5 is straightforward: any cover of $f^{-1}(1)$ by f-monochromatic rectangles must cover a set of μ-measure 1, which means that the total number of rectangles in the cover must be no less than the reciprocal of the largest μ-measure of a rectangle $R \subseteq f^{-1}(1)$. Theorem 5 is of interest in two ways. First of all, it characterizes nondeterministic communication complexity up to a small additive term [29]. Second, as we will see in the next section, ideas analogous to the rectangle size bound play a key role in the study of randomized communication complexity.

It is instructive to rederive the nondeterministic lower bound for set disjointness using the rectangle size bound. One approach is to simply consider the uniform distribution over the fooling set $\{(A, \{1, 2, \ldots, n\} \setminus A) : A \subseteq \{1, 2, \ldots, n\}\}$, which gives $N(\mathrm{DISJ}_n) \geqslant n$. A more revealing choice [29] is to let μ be the uniform distribution over $\mathrm{DISJ}_n^{-1}(1)$, so that

$$\mu(A, B) = \begin{cases} 3^{-n} & \text{if } A \cap B = \varnothing, \\ 0 & \text{otherwise.} \end{cases}$$

Now, let $R = \mathscr{A} \times \mathscr{B}$ be any rectangle in $\mathrm{DISJ}_n^{-1}(1)$. Then it is straightforward to check that the larger rectangle $\mathscr{P}(S) \times \mathscr{P}(T)$, where $S = \bigcup_{A \in \mathscr{A}} A$ and $T = \bigcup_{B \in \mathscr{B}} B$, must also be contained in $\mathrm{DISJ}_n^{-1}(1)$. It follows that $S \cap T = \varnothing$ and therefore $|R| \leqslant |\mathscr{P}(S) \times \mathscr{P}(T)| \leqslant 2^n$. In summary, we have shown that every rectangle $R \subseteq \mathrm{DISJ}_n^{-1}(1)$ contains at most 2^n inputs, whence $\mu(R) \leqslant (2/3)^n$. Applying the rectangle size bound, we arrive at $N(\mathrm{DISJ}_n) \geqslant n \log_2(3/2)$. While this bound is weaker than the previous bound $N(\mathrm{DISJ}_n) \geqslant n$, the analysis just given is more broadly applicable and is good preparation for the next section on randomized communication.

Complexity Classes. In a seminal paper, Babai, Frankl, and Simon [5] initiated a systematic study of communication from the standpoint of complexity classes. Analogous to computational complexity, the focus here is on the asymptotic communication requirements of a *family* of functions, one function for each input size. Specifically, one considers families $\{f_n\}_{n=1}^{\infty}$ where $f_n \colon \{0,1\}^n \times \{0,1\}^n \to \{0,1\}$. Among the complexity classes defined in [5] are $\mathsf{P}^{\mathsf{cc}}, \mathsf{NP}^{\mathsf{cc}}$, and $\mathsf{coNP}^{\mathsf{cc}}$, corresponding to function families with efficient deterministic, nondeterministic, and co-nondeterministic protocols, respectively. Formally, P^{cc} is the class of all families $\{f_n\}_{n=1}^{\infty}$ for which $D(f_n) \leqslant \log^c n + c$ for some constant $c > 0$ and all n. The classes $\mathsf{NP}^{\mathsf{cc}}$ and $\mathsf{coNP}^{\mathsf{cc}}$ are defined analogously with respect to the requirements $N(f_n) \leqslant \log^c n + c$ and $N(\neg f_n) \leqslant \log^c n + c$. Set disjointness is helpful in characterizing the relations among these classes. Indeed, recall that $D(\mathrm{DISJ}_n) \geqslant N(\mathrm{DISJ}_n) \geqslant n$ and $N(\neg\mathrm{DISJ}_n) \leqslant \log_2 n + O(1)$. An immediate consequence is that $\mathsf{P}^{\mathsf{cc}} \subsetneq \mathsf{NP}^{\mathsf{cc}}$, with an exponential gap between deterministic and nondeterministic complexity achieved for $\neg\mathrm{DISJ}_n$. One analogously obtains $\mathsf{NP}^{\mathsf{cc}} \neq \mathsf{coNP}^{\mathsf{cc}}$, with an exponential gap for DISJ_n. The significance of set disjointness in the study of nondeterminism is no accident: Babai et al. show that it is a complete problem for the class $\mathsf{coNP}^{\mathsf{cc}}$.

We conclude with yet another use of set disjointness. A fundamental result due to Aho, Ullman, and Yannakakis [2] states that $D(f) \leqslant cN(f)N(\neg f)$ for some absolute constant $c > 0$ and every function f. In particular, one obtains the surprising equality $\mathsf{P}^{\mathsf{cc}} = \mathsf{NP}^{\mathsf{cc}} \cap \mathsf{coNP}^{\mathsf{cc}}$. A variant of the set disjointness problem, known as *k-set disjointness* [39], shows that the upper bound of Aho et al. is tight up to a constant factor.

4   Randomized Communication

In many ways, the randomized model is the most realistic abstraction of two-party communication. As usual, consider a function $f \colon X \times Y \to \{0,1\}$, where X and Y are finite sets. Alice receives an input $x \in X$, Bob receives an input $y \in Y$, and their objective is to compute $f(x,y)$ by communicating back and forth according to an agreed-upon protocol. In addition, Alice and Bob share an unlimited supply of uniformly random bits, which they can use in deciding what messages to send. The *cost* of a randomized protocol is the maximum number of bits exchanged on any input pair (x,y). Since the random bits are shared, they do not count toward the protocol cost. A protocol is said to *compute f with error ϵ* if on every input pair (x,y), the output of the protocol is correct with probability at least $1 - \epsilon$. The *ϵ-error randomized communication complexity* of f, denoted $R_\epsilon(f)$, is the least cost of a randomized protocol that computes f

with error ϵ. The canonical quantity to study is $R_{1/3}(f)$. This setting of the error parameter is without loss of generality. Indeed, for any constants $\epsilon, \epsilon' \in (0, 1/2)$, the error of a communication protocol can be reduced from ϵ to ϵ' by running the protocol constantly many times and outputting the majority answer.

There are several other ways to formalize randomized communication, all of which turn out to be equivalent [29]. Most notably, one can consider a model where Alice and Bob each have a private source of random bits, known as the *private-coin model*. A fundamental theorem due to Newman [34] shows that whether the random bits are shared or private affects the communication complexity of any given function $f: \{0,1\}^n \times \{0,1\}^n \to \{0,1\}$ by at most an additive term of $O(\log n)$.

Corruption Bound. The randomized communication complexity of a function can be vastly smaller than its deterministic or nondeterministic complexity. For example, the problem of checking two n-bit strings for equality has randomized communication complexity $O(1)$, in contrast to its $\Omega(n)$ complexity in the deterministic and nondeterministic models. The fact that randomized protocols can be so powerful means that proving lower bounds in this model is correspondingly more difficult. The most common method for lower bounds on randomized communication the *corruption bound* due to Yao [53]. As we shall soon see, this technique is strong enough to yield the celebrated $\Omega(n)$ lower bound for set disjointness.

Theorem 6 (Corruption bound). *Let $f: X \times Y \to \{0,1\}$ be a given function, $\alpha, \beta > 0$ given parameters. Let μ be a probability distribution on $X \times Y$ such that every rectangle R obeys*

$$\mu(R \cap f^{-1}(0)) \geqslant \alpha\mu(R) - \beta.$$

Then for all $\epsilon > 0$,

$$R_\epsilon(f) \geqslant \log_2\left(\frac{\alpha\mu(f^{-1}(1)) - \epsilon}{\beta}\right).$$

The technical details of this theorem are somewhat tedious, but the intuition is entirely straightforward. Fix a probability distribution μ on the domain of the given communication problem f. The hypothesis of the theorem states that with respect to μ, the "0" entries make up at least an α fraction of any rectangle— except for particularly small rectangles, with measure on the order of β. As a result, any cover of $f^{-1}(1)$ by rectangles that are "almost" f-monochromatic requires roughly $\mu(f^{-1}(1))/\beta$ rectangles, for a communication cost of roughly $\log_2(\mu(f^{-1}(1))/\beta)$. It is not too difficult to turn this informal discussion into a rigorous proof of the corruption bound, by using Fact 1 and viewing a randomized protocol of a given cost as a probability distribution on deterministic protocols of the same cost.

An $\Omega(\sqrt{n})$ Lower Bound. The randomized communication complexity of set disjointness has been extensively studied. A variety of proof techniques have been brought to bear on this question, including Kolmogorov complexity, information theory, matrix analysis, and approximation theory. The first strong result in this line of work is an $\Omega(\sqrt{n})$ lower bound on the randomized communication complexity of DISJ_n, due to Babai, Frankl, and Simon [5]. Their proof, presented below in its entirety, uses nothing but basic combinatorics and is exceedingly elegant.

Theorem 7 (Babai, Frankl, and Simon). $R_{1/3}(\mathrm{DISJ}_n) \geqslant \Omega(\sqrt{n})$.

Proof. Without loss of generality, we may assume that n is a perfect square divisible by 12. We will work with a restriction of the set disjointness problem, in which Alice and Bob's inputs are sets of size exactly $\sqrt{n}$. Let μ denote the uniform probability distribution over all such inputs. Then

$$\mu(\mathrm{DISJ}_n^{-1}(1)) = \frac{\binom{n-\sqrt{n}}{\sqrt{n}}}{\binom{n}{\sqrt{n}}} = \Omega(1).$$

The crux of the proof is the following purely combinatorial fact:

Claim. *Let* $\mathscr{R} = \mathscr{A} \times \mathscr{B}$ *be any rectangle with* $\mathbf{P}_{(A,B)\in\mathscr{R}}[A \cap B = \varnothing] \geqslant 1 - \alpha$ *and* $|\mathscr{A}| \geqslant 2^{-\delta\sqrt{n}}\binom{n}{\sqrt{n}}$, *where* $\alpha > 0$ *and* $\delta > 0$ *are sufficiently small absolute constants. Then*

$$|\mathscr{B}| \leqslant 2^{-\delta\sqrt{n}}\binom{n}{\sqrt{n}}.$$

Let us finish the proof of the theorem before moving on to the claim itself. The claim is logically equivalent to the following statement: there exist absolute constants $\alpha > 0$ and $\delta > 0$ such that any rectangle $\mathscr{R}$ with $\mu(\mathscr{R}) \geqslant 2^{-\delta\sqrt{n}}$ satisfies

$$\mu(\mathscr{R} \cap \mathrm{DISJ}_n^{-1}(0)) > \alpha\mu(\mathscr{R}).$$

Applying the corruption bound (Theorem 6) with $\beta = 2^{-\delta\sqrt{n}}$, we conclude that $R_\epsilon(\mathrm{DISJ}_n) = \Omega(\sqrt{n})$ for sufficiently small $\epsilon = \epsilon(\alpha, \delta) > 0$. By error reduction, this implies the conclusion of the theorem. $\square$

Proof of Claim. Consider the matrix $M = [\mathrm{DISJ}_n(A, B)]_{A\in\mathscr{A}, B\in\mathscr{B}}$. By hypothesis, the "0" entries make up at most an α fraction of M. Without loss of generality, we may assume that the fraction of "0" entries is at most 2α in every row of M (if not, simply remove the offending rows, which reduces the size of $\mathscr{A}$ by at most a factor of 2). Now, abbreviate $k = \sqrt{n}/3$ and inductively find sets $A_1, A_2, \ldots, A_k \in \mathscr{A}$ that are *well separated*, in the sense that for all i,

$$|A_i \setminus (A_1 \cup A_2 \cup \cdots \cup A_{i-1})| \geqslant \frac{\sqrt{n}}{2}.$$

That such sets must exist is a straightforward exercise in counting, with $\alpha > 0$ small enough.

Recall that the "0" entries make up at most a 2α fraction of the entries in $[\mathrm{DISJ}_n(A_i, B)]_{i=1,2,\ldots,k;\ B \in \mathscr{B}}$. In particular, at least half of the sets $B \in \mathscr{B}$ must satisfy

$$\mathbf{P}_{i=1,2,\ldots,k}[A_i \cap B \neq \varnothing] \leqslant 4\alpha.$$

But by the well-separated property of $A_1, A_2, \ldots, A_k$, the number of such sets B is at most $\binom{k}{4\alpha k}\binom{n-(1-4\alpha)k\sqrt{n}/2}{\sqrt{n}}$. (Verify this!) For $\alpha > 0$ and $\delta > 0$ small enough, this estimate does not exceed $\frac{1}{2} \cdot 2^{-\delta\sqrt{n}}\binom{n}{\sqrt{n}}$, which gives the claimed upper bound on $|\mathscr{B}|$. $\square$

This lower bound on the randomized communication complexity of set disjointness has an important implication for communication complexity classes. Analogous to $\mathsf{P}^{cc}, \mathsf{NP}^{cc}, \mathsf{coNP}^{cc}$, Babai et al. [5] defined BPP^{cc} as the class of all communication problems $\{f_n\}_{n=1}^{\infty}$ for which $R_{1/3}(f_n) \leqslant \log^c n + c$ for some constant $c > 0$ and all n. Theorem 7 shows that $\{\mathrm{DISJ}_n\}_{n=1}^{\infty} \notin \mathsf{BPP}^{cc}$, thus separating the classes NP^{cc} and coNP^{cc} from BPP^{cc}.

Tight Lower Bound. The problem of determining the randomized communication complexity of set disjointness remained open for several years after the work of Babai et al. It was finally resolved by Kalyanasundaram and Schnitger [24], who used Kolmogorov complexity to obtain the tight lower bound $R_{1/3}(\mathrm{DISJ}_n) = \Omega(n)$. Shortly thereafter, Razborov [40] gave a celebrated alternate proof of the linear lower bound for set disjointness. In fact, Razborov considered an easier communication problem known as *unique set disjointness*, in which Alice and Bob's input sets $A, B \subseteq \{1, 2, \ldots, n\}$ are either disjoint or intersect in a unique element. He studied the probability distribution μ that places weight $3/4$ on disjoint pairs (A, B) of cardinality $|A| = |B| = \lfloor n/4 \rfloor$, and weight $1/4$ on uniquely intersecting pairs again of cardinality $|A| = |B| = \lfloor n/4 \rfloor$; in both cases, each such pair is equally likely. He proved that $\mu(\mathscr{R} \cap \mathrm{DISJ}^{-1}(0)) \geqslant \alpha\mu(\mathscr{R}) - 2^{-\delta n}$ for some constants $\alpha > 0$ and $\delta > 0$ and every combinatorial rectangle $\mathscr{R}$, from which the tight lower bound $R_{1/3}(\mathrm{DISJ}_n) = \Omega(n)$ follows immediately by Theorem 6. Razborov's analysis is based on an entropy argument along with an ingenious use of conditioning.

Razborov's result as well as his proof inspired much follow-up work. The fact that the lower bound holds even for unique set disjointness was a crucial ingredient in Nisan and Wigderson's counterexample to the "strong" log-rank conjecture (see Section 3). The linear lower bound on the randomized communication complexity of set disjointness has found several surprising applications, including streaming algorithms [4] and combinatorial auctions [35]. In a testament to the mathematical richness of this problem, Bar-Yossef et al. [7] discovered a simpler yet, information-theoretic proof of the linear lower bound. This line of work is still active, with a recent paper by Braverman et al. [13] determining the randomized communication complexity of set disjointness up to lower-order terms.

5   Unbounded-Error Communication

The *unbounded-error model*, due to Paturi and Simon [38], is a fascinating model of communication with applications to matrix analysis, circuit complexity, and learning theory [38,3,11,19,20,28,31,43,46,42]. Let $f\colon X \times Y \to \{0,1\}$ be a communication problem of interest. As usual, Alice and Bob receive inputs $x \in X$ and $y \in Y$, respectively, and their objective is to compute $f(x,y)$ with minimal communication. They each have an unlimited private source of random bits. Their protocol is said to compute f in the unbounded-error model if on every input (x,y), the output is correct with probability strictly greater than $1/2$. The *unbounded-error communication complexity* of f, denoted $U(f)$, is the least cost of a protocol that computes f.

Observe that the unbounded-error model is the same as the private-coin randomized model discussed in Section 4, with one exception: in the latter case the protocol must produce the correct answer with probability at least $2/3$, whereas in the former case the probability of correctness merely needs to exceed $1/2$, by an arbitrarily small amount. This difference has far-reaching implications. For example, the fact that the parties in the unbounded-error model do not have a *shared* source of random bits is crucial: it is a good exercise to check that allowing shared randomness in the unbounded-error model would make the complexity of every function a constant. This contrasts with the randomized model, where making the randomness public has almost no effect on the complexity of any given function.

There are several reasons why the unbounded-error model occupies a special place in communication complexity theory. To start with, it is vastly more powerful than the deterministic, nondeterministic, randomized, and *quantum* models [42]. Another compelling reason is that unbounded-error communication complexity is closely related to the fundamental matrix-theoretic notion of *sign-rank*, which is defined for a Boolean matrix $M = [M_{ij}]$ as the minimum rank of a real matrix $R = [R_{ij}]$ such that $\operatorname{sgn} R_{ij} = (-1)^{M_{ij}}$ for all i,j. In other words, the sign-rank of a Boolean matrix M is the minimum rank of real matrix R that sign-represents it, with negative and positive entries in R corresponding to the true and false entries in M, respectively. We let $\operatorname{rk}_\pm M$ denote the sign-rank of M. Paturi and Simon [38] proved the following beautiful theorem, which shows that unbounded-error communication and sign-rank are equivalent notions.

Fact 8 (Paturi and Simon). *For some absolute constant c and every function* $f\colon X \times Y \to \{0,1\}$,

$$U(f) - c \leqslant \log_2(\operatorname{rk}_\pm M_f) \leqslant U(f) + c.$$

Proving lower bounds on sign-rank is difficult. Indeed, obtaining a strong lower bound on the unbounded-error communication complexity of any explicit function was a longstanding problem until the breakthrough work of Forster [19] several years ago. Fortunately, the unbounded-error complexity of set disjointness is easy to analyze. The following two theorems give a complete answer, up to an additive constant.

Theorem 9 (Folklore). $U(\mathrm{DISJ}_n) \leqslant \log_2 n + O(1)$.

Proof. Consider the following randomized protocol, where A and B denote Alice and Bob's input sets, respectively. The players pick an index $i \in \{1, 2, \ldots, n\}$ uniformly at random and verify with two bits of communication whether $i \in A \cap B$. If so, they output 0. In the complementary case $i \notin A \cap B$, they output 1 with probability $n/(2n-1)$ and 0 otherwise. It is easy to verify that this protocol is correct with probability at least $n/(2n-1) > 1/2$ on every input. Moreover, it clearly has cost at most $\log_2 n + O(1)$ in the private-coin model.   □

Theorem 10 (Paturi and Simon). $U(\mathrm{DISJ}_n) \geqslant \log_2 n - O(1)$.

Proof. We will give a linear-algebraic proof of this result, as opposed to the geometric argument of Paturi and Simon [38]. By Fact 8, it suffices to show that the characteristic matrix of set disjointness has sign-rank at least n. We will actually prove the claim for the submatrix $M = [x_i]_{x \in \{0,1\}^n, i=1,2,\ldots,n}$, whose rows are the 2^n distinct Boolean vectors of length n.

For the sake of contradiction, assume that $\mathrm{rk}_\pm M \leqslant n - 1$. Then there are vectors $u_1, u_2, \ldots, u_{n-1} \in \mathbb{R}^n$ such that every $\sigma \in \{-1, +1\}^n$ is the (componentwise) sign of some linear combination of $u_1, u_2, \ldots, u_{n-1}$. Let $w \in \mathbb{R}^n$ be a nonzero vector in the orthogonal complement of $\mathrm{span}\{u_1, u_2, \ldots, u_{n-1}\}$. Define $\sigma \in \{-1, +1\}^n$ by

$$\sigma_i = \begin{cases} \mathrm{sgn}\, w_i & \text{if } w_i \neq 0, \\ 1 & \text{otherwise.} \end{cases}$$

Then $\sigma = \mathrm{sgn}(\sum_{i=1}^{n-1} \lambda_i u_i)$ for some reals $\lambda_1, \ldots, \lambda_{n-1}$, where the sign function is applied componentwise. In particular, $\langle w, \sum_{i=1}^{n-1} \lambda_i u_i \rangle > 0$. But this is impossible since w was chosen to be orthogonal to $u_1, u_2, \ldots, u_{n-1}$.   □

The above theorem was in fact the *first* lower bound on unbounded-error communication complexity.

6   Multiparty Communication

We now move on to multiparty communication, a topic that is particularly rewarding in its mathematical depth and its applications to many other areas of theoretical computer science. In this setting, k communicating parties need to compute a Boolean-valued function $f(x_1, x_2, \ldots, x_k)$ with k arguments. Each party knows one or more of the arguments to f, but not all. The more information the parties have available to them, the less communication is required. In the extreme setting known as the *number-on-the-forehead model*, each party knows exactly $k-1$ arguments, namely $x_1, \ldots, x_{i-1}, x_{i+1}, \ldots, x_k$ in the ith party's case. One can visualize this model by thinking of the k parties as seated in a circle, with $x_1, x_2, \ldots, x_k$ written on the foreheads of parties $1, 2, \ldots, k$, respectively. Any given party sees all the arguments except for the one on the party's own forehead, hence the terminology. The number-on-the-forehead model, introduced by Chandra, Furst, and Lipton [16], is the most powerful model of multiparty communication and is therefore the standard setting in which to prove communication lower bounds.

In this model, the parties communicate via a broadcast channel, with a bit sent by any party instantly reaching everyone else. They also share an unlimited supply of random bits. Analogous to the two-party case, a multiparty communication protocol *computes* f *with error* ϵ if on every input $(x_1, x_2, \ldots, x_k)$, it outputs the correct answer $f(x_1, x_2, \ldots, x_k)$ with probability at least $1 - \epsilon$. The *cost* of a protocol is the total number of broadcasts on the worst-case input; as usual, the shared randomness does not count toward the communication cost. The ϵ-*error randomized communication complexity* of f, denoted $R_\epsilon(f)$, is the least cost of an ϵ-error communication protocol for f in this model. Again, the canonical quantity to study is $R_{1/3}(F)$, where the choice of $1/3$ is largely arbitrary and can be replaced by any other constant in $(0, 1/2)$ without affecting the theory in any way.

Multiparty Set Disjointness. The multiparty set disjointness problem is by far the most studied problem in this line of work. In the k-party setting, the inputs to the problem are sets $S_1, S_2, \ldots, S_k \subseteq \{1, 2, \ldots, n\}$, and the ith party knows all the inputs except for S_i. Their goal is to determine whether the sets have empty intersection: $S_1 \cap S_2 \cap \cdots \cap S_k = \varnothing$. When specialized to $k = 2$, this definition is entirely consistent with the two-party set disjointness problem in Sections 1–4. It is common to represent the input to multiparty set disjointness as a $k \times n$ Boolean matrix $X = [x_{ij}]$, whose rows correspond to the characteristic vectors of the input sets. In this notation, set disjointness is given by the simple formula

$$\mathrm{DISJ}_{k,n}(X) = \bigwedge_{j=1}^{n} \bigvee_{i=1}^{k} \overline{x_{ij}}. \tag{1}$$

Progress on the communication complexity of set disjointness for $k \geqslant 3$ parties is summarized in Table 1. In a surprising result, Grolmusz [22] proved an upper bound of $O(\log^2 n + k^2 n/2^k)$. Proving a strong lower bound, even for $k = 3$, turned out to be difficult. Tesson [51] and Beame et al. [9] obtained a lower bound of $\Omega\big(\frac{1}{k}\log n\big)$ for randomized protocols. Four years later, Lee and Shraibman [30] and Chattopadhyay and Ada [18] gave an improved result. These authors generalized the *pattern matrix method* of [44,45] to $k \geqslant 3$ parties and thereby obtained a lower bound of $\Omega(n/2^{2^k k})^{1/(k+1)}$ on the randomized communication complexity of set disjointness. Their lower bound was strengthened by Beame and Huynh-Ngoc [8] to $(n^{\Omega(\sqrt{k/\log n})}/2^{k^2})^{1/(k+1)}$, which is an improvement for k large enough. All lower bounds listed up to this point are weaker than $\Omega(n/2^{k^3})^{1/(k+1)}$, which means that they become subpolynomial as soon as the number of parties k starts to grow. Three years later, we obtained [47] a lower bound of $\Omega(n/4^k)^{1/4}$ on the randomized communication complexity of set

Table 1. Communication complexity of k-party set disjointness

Bound	Reference
$O\left(\log^2 n + \dfrac{k^2 n}{2^k}\right)$	Grolmusz [22]
$\Omega\left(\dfrac{\log n}{k}\right)$	Tesson [51] Beame, Pitassi, Segerlind, and Wigderson [9]
$\Omega\left(\dfrac{n}{2^{2^k} k}\right)^{\frac{1}{k+1}}$	Lee and Shraibman [30] Chattopadhyay and Ada [18]
$\left(\dfrac{n^{\Omega(\sqrt{k/\log n})}}{2^{k^2}}\right)^{\frac{1}{k+1}}$	Beame and Huynh-Ngoc [8]
$\Omega\left(\dfrac{n}{4^k}\right)^{1/4}$	Sherstov [47]
$\Omega\left(\dfrac{\sqrt{n}}{2^k k}\right)$	Sherstov [49]

disjointness, which remains polynomial for up to $k \approx \frac{1}{2}\log n$ and comes close to matching Grolmusz's upper bound. Most recently [49], we improved the lower bound quadratically to $\Omega(\sqrt{n}/2^k k)$, which is the strongest bound known. This lower bound also holds for *quantum* multiparty protocols, in which case it is tight. However, it is conceivable that the *classical* randomized communication complexity of set disjointness is $\Omega(n/c^k)$ for some constant $c > 1$. Proving such a lower bound, or showing that it does not hold, is a fascinating open problem.

The lower bound from [49] is too demanding to discuss in this survey. In what follows, we will instead focus on the next best lower bound $\Omega(n/4^k)^{1/4}$.

Anatomy of Multiparty Protocols. Recall that the building blocks of two-party communication protocols are combinatorial rectangles. The corresponding objects in k-party communication are called *cylinder intersections* [6]. For a k-party problem with domain $X_1 \times X_2 \times \cdots \times X_k$, a cylinder intersection is an arbitrary function $\chi \colon X_1 \times X_2 \times \cdots \times X_k \to \{0, 1\}$ of the form

$$\chi(x_1, \ldots, x_k) = \prod_{i=1}^{k} \chi_i(x_1, \ldots, x_{i-1}, x_{i+1}, \ldots, x_k),$$

where $\chi_i \colon X_1 \times \cdots \times X_{i-1} \times X_{i+1} \times \cdots \times X_k \to \{0, 1\}$. In other words, a k-dimensional cylinder intersection is the product of k Boolean functions, where the ith function does not depend on the ith coordinate but may depend arbitrarily on the other $k-1$ coordinates. As one would expect, combinatorial rectangles are cylinder intersections for $k = 2$. The following fundamental result is the multiparty analogue of Fact 1.

Fact 11 (Babai, Nisan, and Szegedy). *Let $\Pi\colon X_1 \times X_2 \times \cdots \times X_k \to \{0,1\}$ be a deterministic k-party communication protocol with cost c. Then there exist cylinder intersections $\chi_1, \ldots, \chi_{2^c}$ with pairwise disjoint support such that*

$$\Pi = \sum_{i=1}^{2^c} \chi_i.$$

By viewing a randomized protocol with cost c as a probability distribution on deterministic protocols of cost at most c, one obtains the following corollary, where $\|\cdot\|_\infty$ denotes as usual the ℓ_∞ norm.

Corollary 12. *Let $f\colon X_1 \times X_2 \times \cdots \times X_k \to \{0,1\}$ be a given communication problem. If $R_\epsilon(f) = c$, then there exists a linear combination $\Pi = \sum_\chi a_\chi \chi$ of cylinder intersections with $\sum_\chi |a_\chi| \leqslant 2^c$ such that*

$$\|f - \Pi\|_\infty \leqslant \epsilon.$$

Analytic Preliminaries. For the past few years, analytic tools have played an increasingly important role in communication complexity theory. We will need two such tools, the Fourier transform and polynomial approximation theory. Consider the real vector space of functions $\phi\colon \{0,1\}^n \to \mathbb{R}$. For $S \subseteq \{1, 2, \ldots, n\}$, define $\chi_S\colon \{0,1\}^n \to \{-1, +1\}$ by $\chi_S(x) = \prod_{i \in S}(-1)^{x_i}$. Then every function $\phi\colon \{0,1\}^n \to \mathbb{R}$ has a unique representation of the form $\phi = \sum_S \hat{\phi}(S)\chi_S$, where $\hat{\phi}(S) = 2^{-n}\sum_{x \in \{0,1\}^n} \phi(x)\chi_S(x)$. The reals $\hat{\phi}(S)$ are called the *Fourier coefficients of ϕ*, and the mapping $\phi \mapsto \hat{\phi}$ is the *Fourier transform* of ϕ.

The *ϵ-approximate degree* of a function $\phi\colon \{0,1\}^n \to \mathbb{R}$, denoted $\deg_\epsilon(\phi)$, is the least degree of a multivariate real polynomial p that approximates ϕ within ϵ pointwise: $\|\phi - p\|_\infty \leqslant \epsilon$. We also define $E(\phi, d) = \min_p \|\phi - p\|_\infty$, where the minimum is over multivariate real polynomials p of degree at most d. Thus, $E(\phi, d)$ is the least error to which ϕ can be approximated pointwise by a polynomial of degree at most d. In this notation, $\deg_\epsilon(\phi) = \min\{d : E(\phi, d) \leqslant \epsilon\}$. The approximate degree is an extensively studied complexity measure of Boolean functions. The first result in this line of work is due to Nisan and Szegedy [36], who studied the function $\text{AND}_n(x) = \bigwedge_{i=1}^n x_i$.

Theorem 13 (Nisan and Szegedy). $\deg_{1/3}(\text{AND}_n) = \Theta(\sqrt{n})$.

The $\Omega(n/4^k)^{1/4}$ Lower Bound. We are now in a position to present the lower bound on the randomized communication complexity of multiparty set disjointness from [47]. The technical centerpiece of this result is the following lemma, which analyzes the correlation of cylinder intersections with the XOR of several independent copies of set disjointness.

Lemma 14 (Sherstov). *Let k and r be given parameters. Then there is a probability distribution μ on the domain of $\mathrm{DISJ}_{k,r}$ such that*

$$\mu(\mathrm{DISJ}_{k,r}^{-1}(0)) = \mu(\mathrm{DISJ}_{k,r}^{-1}(1))$$

and

$$\left| \underset{X_1,\ldots,X_n \sim \mu}{\mathbf{E}} \left[\chi(X_1,\ldots,X_n) \prod_{i=1}^{n} (-1)^{\mathrm{DISJ}_{k,r}(X_i)} \right] \right| \leqslant \left(\frac{2^{k-1}}{\sqrt{r}} \right)^n$$

for every n and every k-party cylinder intersection χ.

A few general remarks are in order before we delve into the proof of the communication lower bound for set disjointness. The proof is best understood by abstracting away from the set disjointness problem and considering arbitrary composed functions. Specifically, let G be a k-party communication problem, with domain $X = X_1 \times X_2 \times \cdots \times X_k$. We refer to G as a *gadget*. We are interested in the communication complexity of functions of the form $F = f(G, G, \ldots, G)$, where $f \colon \{0,1\}^n \to \{0,1\}$. Thus, F is a k-party communication problem with domain $X^n = X_1^n \times X_2^n \times \cdots \times X_k^n$. The motivation for studying such compositions is clear from the defining equation (1) for multiparty set disjointness, which shows that $\mathrm{DISJ}_{k,nr} = \mathrm{AND}_n(\mathrm{DISJ}_{k,r}, \ldots, \mathrm{DISJ}_{k,r})$. A recent line of research [45,50,30,18,8,17,47,49] gives communication lower bounds for compositions $f(G, G, \ldots, G)$ in terms of the approximate degree of f. For the purpose of proving communication lower bounds for set disjointness, the gadget G needs to be representable as $G = \mathrm{DISJ}_{k,r}$ with $r = r(n, k)$ as small as possible. This miniaturization challenge quickly becomes hard.

Theorem 15 (Sherstov). *Let $f \colon \{0,1\}^n \to \{0,1\}$ be given. Consider the k-party communication problem $F = f(\mathrm{DISJ}_{k,r}, \ldots, \mathrm{DISJ}_{k,r})$ Then for all $\epsilon, \delta \geqslant 0$,*

$$2^{R_\epsilon(F)} \geqslant (\delta - \epsilon) \left(\frac{\deg_\delta(f)\sqrt{r}}{2^k en} \right)^{\deg_\delta(f)}. \tag{2}$$

Proof. Let μ be the probability distribution from Lemma 14. Let μ_0 and μ_1 stand for the probability distributions induced by μ on $\mathrm{DISJ}_{k,r}^{-1}(0)$ and $\mathrm{DISJ}_{k,r}^{-1}(1)$, respectively. Consider the following averaging operator L, which linearly sends real functions χ on $(\{0,1\}^{k \times r})^n$ to real functions on $\{0,1\}^n$:

$$(L\chi)(z) = \underset{X_1 \sim \mu_{z_1}}{\mathbf{E}} \cdots \underset{X_n \sim \mu_{z_n}}{\mathbf{E}} [\chi(X_1, \ldots, X_n)].$$

Observe that $LF = f$. When χ is a k-party cylinder intersection, the Fourier coefficients of $L\chi$ obey

$$
\begin{aligned}
|\widehat{L\chi}(S)| &= \left| \underset{z \in \{0,1\}^n}{\mathbf{E}} \underset{X_1 \sim \mu_{z_1}}{\mathbf{E}} \cdots \underset{X_n \sim \mu_{z_n}}{\mathbf{E}} \left[\chi(X_1, \ldots, X_n) \prod_{i \in S} (-1)^{z_i} \right] \right| \\
&= \left| \underset{X_1, \ldots, X_n \sim \mu}{\mathbf{E}} \left[\chi(X_1, \ldots, X_n) \prod_{i \in S} (-1)^{\mathrm{DISJ}_{k,r}(X_i)} \right] \right| \\
&\leqslant \left(\frac{2^{k-1}}{\sqrt{r}} \right)^{|S|},
\end{aligned}
\tag{3}
$$

where the second equality uses the fact that μ places equal weight on $\mathrm{DISJ}_{k,r}^{-1}(0)$ and $\mathrm{DISJ}_{k,r}^{-1}(1)$, and the final step follows by Lemma 14.

Fix a randomized protocol for F with error ϵ and cost $c = R_\epsilon(F)$. Approximate F as in Corollary 12 by a linear combination of cylinder intersections $\Pi = \sum_\chi a_\chi \chi$, where $\sum_\chi |a_\chi| \leqslant 2^c$. For any positive integer d, the triangle inequality gives

$$
E(f, d-1) \leqslant \|f - L\Pi\|_\infty + E(L\Pi, d-1). \tag{4}
$$

We proceed to bound the two terms on the right-hand side of this inequality.

(i) By the linearity of L,

$$
\|f - L\Pi\|_\infty = \|L(F - \Pi)\|_\infty \leqslant \epsilon, \tag{5}
$$

where the last step uses the bound $\|F - \Pi\|_\infty \leqslant \epsilon$ from Corollary 12.

(ii) Discarding the Fourier coefficients of $L\Pi$ of order d and higher gives

$$
\begin{aligned}
E(L\Pi, d-1) &\leqslant \min \left\{ 1, \sum_\chi |a_\chi| \sum_{|S| \geqslant d} |\widehat{L\chi}(S)| \right\} \\
&\leqslant \min \left\{ 1, 2^c \sum_{i=d}^{n} \binom{n}{i} \left(\frac{2^{k-1}}{\sqrt{r}} \right)^i \right\} \\
&\leqslant 2^c \left(\frac{2^k e n}{d \sqrt{r}} \right)^d,
\end{aligned}
\tag{6}
$$

where the second step uses (3).

Substituting the newly obtained estimates (5) and (6) into (4),

$$
E(f, d-1) \leqslant \epsilon + 2^c \left(\frac{2^k e n}{d \sqrt{r}} \right)^d.
$$

For $d = \deg_\delta(f)$, the left-hand side must exceed δ, forcing (2). $\qquad \square$

As an immediate consequence, we obtain the claimed lower bound on the multiparty communication complexity of set disjointness [47]:

Corollary (Sherstov)**.**

$$R_{1/3}(\text{DISJ}_{k,n}) \geqslant \Omega\left(\frac{n}{4^k}\right)^{1/4}. \tag{7}$$

Proof. Recall that $\text{DISJ}_{k,nr} = \text{AND}_n(\text{DISJ}_{k,r}, \ldots, \text{DISJ}_{k,r})$ for all integers n, r. Theorem 13 guarantees that $\deg_{1/3}(\text{AND}_n) > c\sqrt{n}$ for some constant $c > 0$. Thus, letting $f = \text{AND}_n$, $\delta = 1/3$, $\epsilon = 1/4$, and $r = 4^{k+2}\lceil\sqrt{n}/c\rceil^2$ in Theorem 15 gives

$$R_{1/4}(\text{DISJ}_{k,4^{k+2}n\lceil\sqrt{n}/c\rceil^2})$$
$$= R_{1/4}(\text{AND}_n(\text{DISJ}_{k,4^{k+2}\lceil\sqrt{n}/c\rceil^2}, \ldots, \text{DISJ}_{k,4^{k+2}\lceil\sqrt{n}/c\rceil^2})) \geqslant \Omega(\sqrt{n}),$$

which is logically equivalent to (7). □

7   Other Gems

We have only focused on a small sample of results on the set disjointness problem. Prominently absent in our survey is the fascinating and influential body of work on the *quantum* communication complexity of set disjointness [14,41,1,45,50]. Much can also be said about deterministic, nondeterministic, and Merlin-Arthur multiparty protocols [25,21,47,49]. Another compelling topic is the multiparty communication complexity of the set disjointness problem in the *number-in-hand* model [7,15,12], where each party sees only one of the input sets $S_1, S_2, \ldots, S_k$ as opposed to all but one. Lower bounds for such multiparty protocols play an important role in the study of streaming algorithms. Finally, we have not discussed XOR lemmas and direct product theorems, which deal with the communication complexity of simultaneously solving several independent copies of set disjointness [27,9,10,23,26,48,47,49].

Acknowledgments. I am thankful to the organizers of MFCS 2014 and Zoltán Ésik in particular for this opportunity to share my passion for the set disjointness problem.

References

1. Aaronson, S., Ambainis, A.: Quantum search of spatial regions. Theory of Computing 1(1), 47–79 (2005)
2. Aho, A.V., Ullman, J.D., Yannakakis, M.: On notions of information transfer in VLSI circuits. In: Proceedings of the Fifteenth Annual ACM Symposium on Theory of Computing (STOC), pp. 133–139 (1983)

3. Alon, N., Frankl, P., Rödl, V.: Geometrical realization of set systems and probabilistic communication complexity. In: Proceedings of the Twenty-Sixth Annual IEEE Symposium on Foundations of Computer Science, FOCS, pp. 277–280 (1985)

4. Alon, N., Matias, Y., Szegedy, M.: The space complexity of approximating the frequency moments. J. Comput. Syst. Sci. 58(1), 137–147 (1999)

5. Babai, L., Frankl, P., Simon, J.: Complexity classes in communication complexity theory. In: Proceedings of the Twenty-Seventh Annual IEEE Symposium on Foundations of Computer Science, FOCS, pp. 337–347 (1986)

6. Babai, L., Nisan, N., Szegedy, M.: Multiparty protocols, pseudorandom generators for logspace, and time-space trade-offs. J. Comput. Syst. Sci. 45(2), 204–232 (1992)

7. Bar-Yossef, Z., Jayram, T.S., Kumar, R., Sivakumar, D.: An information statistics approach to data stream and communication complexity. J. Comput. Syst. Sci. 68(4), 702–732 (2004)

8. Beame, P., Huynh-Ngoc, D.T.: Multiparty communication complexity and threshold circuit complexity of AC^0. In: Proceedings of the Fiftieth Annual IEEE Symposium on Foundations of Computer Science, FOCS, pp. 53–62 (2009)

9. Beame, P., Pitassi, T., Segerlind, N., Wigderson, A.: A strong direct product theorem for corruption and the multiparty communication complexity of disjointness. Computational Complexity 15(4), 391–432 (2006)

10. Ben-Aroya, A., Regev, O., de Wolf, R.: A hypercontractive inequality for matrix-valued functions with applications to quantum computing and LDCs. In: Proceedings of the Forty-Ninth Annual IEEE Symposium on Foundations of Computer Science, FOCS, pp. 477–486 (2008)

11. Ben-David, S., Eiron, N., Simon, H.U.: Limitations of learning via embeddings in Euclidean half spaces. J. Mach. Learn. Res. 3, 441–461 (2003)

12. Braverman, M., Ellen, F., Oshman, R., Pitassi, T., Vaikuntanathan, V.: A tight bound for set disjointness in the message-passing model. In: Proceedings of the Fifty-Fourth Annual IEEE Symposium on Foundations of Computer Science, FOCS, pp. 668–677 (2013)

13. Braverman, M., Garg, A., Pankratov, D., Weinstein, O.: From information to exact communication. In: Proceedings of the Forty-Fifth Annual ACM Symposium on Theory of Computing, STOC, pp. 151–160 (2013)

14. Buhrman, H., Cleve, R., Wigderson, A.: Quantum vs. classical communication and computation. In: Proceedings of the Thirtieth Annual ACM Symposium on Theory of Computing, STOC, pp. 63–68 (1998)

15. Chakrabarti, A., Khot, S., Sun, X.: Near-optimal lower bounds on the multiparty communication complexity of set disjointness. In: Proceedings of the Eighteenth Annual IEEE Conference on Computational Complexity, CCC, pp. 107–117 (2003)

16. Chandra, A.K., Furst, M.L., Lipton, R.J.: Multi-party protocols. In: Proceedings of the Fifteenth Annual ACM Symposium on Theory of Computing, STOC, pp. 94–99 (1983)

17. Chattopadhyay, A.: Circuits, Communication, and Polynomials. Ph.D. thesis, McGill University (2008)

18. Chattopadhyay, A., Ada, A.: Multiparty communication complexity of disjointness. In: Electronic Colloquium on Computational Complexity (ECCC), report TR08-002 (January 2008)

19. Forster, J.: A linear lower bound on the unbounded error probabilistic communication complexity. J. Comput. Syst. Sci. 65(4), 612–625 (2002)

20. Forster, J., Krause, M., Lokam, S.V., Mubarakzjanov, R., Schmitt, N., Simon, H.U.: Relations between communication complexity, linear arrangements, and computational complexity. In: Hariharan, R., Mukund, M., Vinay, V. (eds.) FSTTCS 2001. LNCS, vol. 2245, pp. 171–182. Springer, Heidelberg (2001)

21. Gavinsky, D., Sherstov, A.A.: A separation of NP and coNP in multiparty communication complexity. Theory of Computing 6(10), 227–245 (2010)

22. Grolmusz, V.: The BNS lower bound for multi-party protocols in nearly optimal. Inf. Comput. 112(1), 51–54 (1994)

23. Jain, R., Klauck, H., Nayak, A.: Direct product theorems for classical communication complexity via subdistribution bounds. In: Proceedings of the Fortieth Annual ACM Symposium on Theory of Computing, STOC, pp. 599–608 (2008)

24. Kalyanasundaram, B., Schnitger, G.: The probabilistic communication complexity of set intersection. SIAM J. Discrete Math. 5(4), 545–557 (1992)

25. Klauck, H.: Rectangle size bounds and threshold covers in communication complexity. In: Proceedings of the Eighteenth Annual IEEE Conference on Computational Complexity, CCC, pp. 118–134 (2003)

26. Klauck, H.: A strong direct product theorem for disjointness. In: Proceedings of the Forty-Second Annual ACM Symposium on Theory of Computing, STOC, pp. 77–86 (2010)

27. Klauck, H., Špalek, R., de Wolf, R.: Quantum and classical strong direct product theorems and optimal time-space tradeoffs. SIAM J. Comput. 36(5), 1472–1493 (2007)

28. Klivans, A.R., Servedio, R.A.: Learning DNF in time $2^{\tilde{O}(n^{1/3})}$. J. Comput. Syst. Sci. 68(2), 303–318 (2004)

29. Kushilevitz, E., Nisan, N.: Communication complexity. Cambridge University Press (1997)

30. Lee, T., Shraibman, A.: Disjointness is hard in the multiparty number-on-the-forehead model. Computational Complexity 18(2), 309–336 (2009)

31. Linial, N., Mendelson, S., Schechtman, G., Shraibman, A.: Complexity measures of sign matrices. Combinatorica 27(4), 439–463 (2007)

32. Lovász, L., Saks, M.E.: Lattices, Möbius functions and communication complexity. In: Proceedings of the Twenty-Ninth Annual IEEE Symposium on Foundations of Computer Science, FOCS, pp. 81–90 (1988)

33. Mehlhorn, K., Schmidt, E.M.: Las Vegas is better than determinism in VLSI and distributed computing. In: Proceedings of the Fourteenth Annual ACM Symposium on Theory of Computing, STOC, pp. 330–337 (1982)

34. Newman, I.: Private vs. common random bits in communication complexity. Inf. Process. Lett. 39(2), 67–71 (1991)

35. Nisan, N., Segal, I.: The communication requirements of efficient allocations and supporting prices. J. Economic Theory 129(1), 192–224 (2006)

36. Nisan, N., Szegedy, M.: On the degree of Boolean functions as real polynomials. Computational Complexity 4, 301–313 (1994)

37. Nisan, N., Wigderson, A.: On rank vs. communication complexity. Combinatorica 15(4), 557–565 (1995)

38. Paturi, R., Simon, J.: Probabilistic communication complexity. J. Comput. Syst. Sci. 33(1), 106–123 (1986)

39. Razborov, A.A.: Applications of matrix methods to the theory of lower bounds in computational complexity. Combinatorica 10(1), 81–93 (1990)

40. Razborov, A.A.: On the distributional complexity of disjointness. Theor. Comput. Sci. 106(2), 385–390 (1992)

41. Razborov, A.A.: Quantum communication complexity of symmetric predicates. Izvestiya: Mathematics 67(1), 145–159 (2003)
42. Razborov, A.A., Sherstov, A.A.: The sign-rank of AC^0. SIAM J. Comput. 39(5), 1833–1855 (2010); preliminary version in Proceedings of the Forty-Ninth Annual IEEE Symposium on Foundations of Computer Science, FOCS (2008)
43. Sherstov, A.A.: Halfspace matrices. Computational Complexity 17(2), 149–178 (2008); preliminary version in Proceedings of the Twenty-Second Annual IEEE Conference on Computational Complexity, CCC (2007)
44. Sherstov, A.A.: Separating AC^0 from depth-2 majority circuits. SIAM J. Comput. 38(6), 2113–2129 (2009); preliminary version in Proceedings of the Thirty-Ninth Annual ACM Symposium on Theory of Computing, STOC (2007)
45. Sherstov, A.A.: The pattern matrix method. SIAM J. Comput. 40(6), 1969–2000 (2011); preliminary version in Proceedings of the Fortieth Annual ACM Symposium on Theory of Computing, STOC) (2008)
46. Sherstov, A.A.: The unbounded-error communication complexity of symmetric functions. Combinatorica 31(5), 583–614 (2011); preliminary version in Proceedings of the Forty-Ninth Annual IEEE Symposium on Foundations of Computer Science, FOCS (2008)
47. Sherstov, A.A.: The multiparty communication complexity of set disjointness. In: Proceedings of the Forty-Fourth Annual ACM Symposium on Theory of Computing, STOC, pp. 525–544 (2012)
48. Sherstov, A.A.: Strong direct product theorems for quantum communication and query complexity. SIAM J. Comput. 41(5), 1122–1165 (2012); preliminary version in Proceedings of the Forty-Third Annual ACM Symposium on Theory of Computing, STOC (2011)
49. Sherstov, A.A.: Communication lower bounds using directional derivatives. In: Proceedings of the Forty-Fifth Annual ACM Symposium on Theory of Computing, STOC, pp. 921–930 (2013)
50. Shi, Y., Zhu, Y.: Quantum communication complexity of block-composed functions. Quantum Information & Computation 9(5-6), 444–460 (2009)
51. Tesson, P.: Computational complexity questions related to finite monoids and semigroups. Ph.D. thesis, McGill University (2003)
52. Yao, A.C.C.: Some complexity questions related to distributive computing. In: Proceedings of the Eleventh Annual ACM Symposium on Theory of Computing, STOC, pp. 209–213 (1979)
53. Yao, A.C.C.: Lower bounds by probabilistic arguments. In: Proceedings of the Twenty-Fourth Annual IEEE Symposium on Foundations of Computer Science, FOCS, pp. 420–428 (1983)

What Does the Local Structure of a Planar Graph Tell Us About Its Global Structure?

Christian Sohler

Technische Universität Dortmund
christian.sohler@tu-dortmund.de

Abstract. The local k-neighborhood of a vertex v in an unweighted graph $G = (V, E)$ with vertex set V and edge set E is the subgraph induced by all vertices of distance at most k from v. The rooted k-neighborhood of v is also called a k-disk around vertex v. If a graph has maximum degree bounded by a constant d, and k is also constant, the number of isomorphism classes of k-disks is constant as well. We can describe the local structure of a bounded-degree graph G by counting the number of isomorphic copies in G of each possible k-disk. We can summarize this information in form of a vector that has an entry for each isomorphism class of k-disks. The value of the entry is the number of isomorphic copies of the corresponding k-disk in G. We call this vector *frequency vector* of k-disks. If we only know this vector, what does it tell us about the structure of G?

In this paper we will survey a series of papers in the area of Property Testing that leads to the following result (stated informally): There is a $k = k(\epsilon, d)$ such that for any planar graph G its local structure (described by the frequency vector of k-disks) determines G up to insertion and deletion of at most ϵdn edges (and relabelling of vertices).

1 Introduction

Very large networks like social networks, the web graph, transportation networks and road maps appear in many applications. Analyzing huge networks is already a difficult task and things become even more involved when we want to analyze a collection of very large networks or learn certain concepts from it. An illustrative example is the question, if one can learn from the Facebook graph of a country, whether it is a democracy or a totalitarian state. In order to answer this question one has to design learning algorithms that extract information from very large graphs. One possible approach is to extract features from these graphs and use standard learning methods on the extracted feature vectors. In order to make this approach work, we want to extract many features from a set of very large graphs, i.e. we have a problem that is severely time-constraint. One approach to this problem is to use random sampling to approximately extract features. Such random sampling approaches can be studied using the framework of *Property Testing*.

E. Csuhaj-Varjú et al. (Eds.): MFCS 2014, Part I, LNCS 8634, pp. 44–49, 2014.
© Springer-Verlag Berlin Heidelberg 2014

2 Property Testing

Property Testing provides a framework to study sampling approaches to approximately decide if a given object has a property or is far away from it. The notion of "far away" is parametrized by ϵ, which typically measures the fraction of the object's description that has to be modified to obtain an object that has the studied property. The concept of Property Testing has first been formulated by Rubinfeld and Sudan in the context of program checking [12]. It has then been extended to graphs by Goldreich, Goldwasser and Ron [5]. In this paper, we consider Property Testing in the bounded degree graph model, which has been introduced by Goldreich and Ron [7]. In this model, an algorithm is given oracle access to a graph $G = (V, E)$ with vertex set $V = \{1, \ldots, n\}$ and edge set E and maximum degree bounded by d. Furthermore, the algorithm is given the values n and d. It can query the oracle about the i-th neighbor of vertex j for $i \in \{1, \ldots, d\}$ and $j \in \{1, \ldots, n\}$ and the answer is either this neighbor or a special symbol that indicates that such a neighbor does not exist. Next we define the notion of ϵ-far.

Definition 1. *A graph is ϵ-far from a property Π in the bounded degree graph model, if one has to insert or delete more than ϵdn edges to obtain a graph that has property Π and maximum degree at most d.*

A Property Testing algorithm or property tester for a property Π must accept every graph with property Π with probability at least $3/4$ and reject every graph that is ϵ-far from Π with probability $3/4$. If a graph neither has Π nor is ϵ-far from it, the algorithm may answer arbitrarily.

The goal of Property Testing is to find algorithms that approximately decide a property in the above sense without looking at the whole input. In fact, there are many examples for Property Testing algorithms that make only a constant number of queries to the input graph (assuming ϵ to be constant). In order to study these properties we define testable graph properties as follows.

Definition 2. *A graph property Π is called testable, if there exists a function $q(\epsilon, d)$ such that for every n, d and ϵ there is an algorithm $A_{\epsilon,d,n}$ that gets as input a graph G with n vertices, makes at most $q(\epsilon, d)$ queries to G, accepts with probability at least $3/4$, if G has Π and rejects with probability at least $3/4$, if G is ϵ-far from Π.*

Note that the above notion allows to have different property testers for different values of ϵ, d and n. This is required if one wants to obtain results of a generality as presented later in this survey.

3 Property Testing of Planar Graphs

The first properties shown to be testable in the bounded degree graph model included connectivity, k-connectivity, cycle-freeness and being Eulerian [7]. However, no results of testable classes of properties were known. The first result in this direction was proved by Czumaj, Shapira and Sohler [3] who studied Property Testing when the input graph is restricted to be a planar graph of maximum degree at most d. Under this assumption they could prove that every hereditary property, i.e. any property that is closed under vertex removal, is testable. The property tester is very simple: It samples a set of vertices S and checks whether the subgraph induced by S has the studied property. If it does, the tester accepts and otherwise it rejects. By closedness under vertex removal, the tester accepts every hereditary graph property.

It remains to prove that the tester rejects, if the graph is ϵ-far from Π. The proof exploits the fact that every planar graph has a small vertex separator:

Theorem 1. *[10] Let $G = (V, E)$ be a planar graph with n vertices. Then V can be partitioned into three sets A, B, C such that there is no edge between the sets A and B. Furthermore, $|A|, |B| \leq \frac{2}{3} \cdot n$ and $|C| \leq 2\sqrt{2n}$.*

Repeatedly applying this theorem leads to a set of, say, at most $\epsilon n/2$ vertices whose removal partitions G into connected components of size $O(1/\epsilon^2)$. Since G has degree at most d we can achieve the same effect by removing the at most $\epsilon dn/2$ edges incident to the vertices in the separator. This implies that any graph that is ϵ-far from having a property Π is still $\epsilon/2$-far from Π after removing this set of edges. This essentially reduces testing a property of a graph G to testing the property in a graph G that only has connected components of constant size. This implies (and here we are simplifying a bit) that there are $\Omega(n)$ connected components that do not have property Π (assuming ϵ to be a constant). If we choose our value of k sufficiently large, i.e. larger than the diameter of the connected components, then sampling a vertex inside such a component will lead to the discovery of a subgraph that contains the component (recall that the sampling is done in the original graph). By closedness under vertex removal this subgraph does not have Π and the tester rejects.

4   Testing Planar Graph Properties

Given that many properties in planar graphs are testable there is the question if something similar holds for general graphs. Benjamini, Schramm and Shapira proved that this is indeed the case [1]. They showed that if the frequency vector of a graph is close to that of a planar graph, then this graph can be partitioned into small components by removing, say, $\epsilon dn/2$ edges. Furthermore, if a graph can be partitioned into such a set of small components, we can use arguments similar to that in [3] to prove that every property that is closed under insertion, deletion and contraction of edges is testable. Such a property is also called minor-closed. This implies that every planar and minor-closed graph property is testable (a graph property Π is planar, if every graph that has Π is also planar). The arguments can be generalized to prove that every minor-closed graph property is testable [1].

5   Local Algorithms to Access a Partition

Another interesting question one can ask in this context is if it is possible to get local access to a partition of a planar graph into small components that is obtained by removing at most ϵdn edges, i.e. whether for a query vertex v one can locally compute its connected component and the answer is consistent with some global partitioning. Hassidim, Kelner, Nguyen and Onak [6] introduced a remarkably simple local algorithm that gives such access. Their algorithm is a local instantiation of the following global algorithm also given in [6].

GLOBALPARTITIONING (k, ϵ)
    Compute a random permutation $v_1, \ldots, v_n$ of the vertices in V
    $P = \emptyset$
    for $i = 1$ to n do
        if v_i is still in the graph then
            if there is a connected set $S \subset V$ with $v_i \in S$ and that has at most
            $\epsilon |S|$ edges that leave S then Let S be this set
            else $S = \{v_i\}$
        $P = P \cup \{S\}$
        remove S from G
    return P

If we want to locally simulate the algorithm then each vertex computes a random value between 0 and 1 and the permutation is given by sorting the vertices increasingly according to these values. It is proved in [6] that in order to compute the component of a vertex v we typically only need to look at a constant size neighborhood of v and we only need to instantiate the random values for this neighborhood. This new algorithm can also be used to simplifiy and improve some Property Testing results.

Some more efficient constructions in terms of dependency on $1/\epsilon$ are known for more restricted classes of graphs [4,9].

6   Approximating a Planar Graph by Its Local Structure

If we consider two planar graphs that are ϵ-close to being isomorphic, i.e. one has to change at most an ϵ fraction of the edges in one graph to obtain an isomorphic copy of the other graph, then their local structures will also be similar (we here think of ϵ being a very small constant). An interesting question is, if the converse is true as well. Based on the previous results, Newman and Sohler proved that this is indeed the case [11]. They proved that if two graphs on n vertices have the same distribution of local neighborhoods, then they can be partitioned into the same set of connected componented by removing at most $\epsilon dn/2$ edges in each graph. This implies that they are ϵ-close to being isomorphic. In the formal statement of the theorem below note that $f_G(k)$ denotes the normalized frequency vector of the k-disks in G, i.e. their distribution. Also, recall that a

k-disk around v is the rooted subgraph induced by all vertices of distance at most k from v.

Theorem 2. *[11] Let G_1, G_2 be two planar graphs with maximum degree at most d on n vertices. Then for every ϵ, $0 < \epsilon \leq 1$, there exists $\eta = \eta(\epsilon, \rho, d)$, $k = k(\epsilon, \rho, d)$, such that, if $|f_{G_1}(k) - f_{G_2}(k)| \leq \eta$ then G_1 is ϵ-close to (being an isomorphic copy of) G_2.*

7   Extensions

Most of the results mentioned above extend to more general classes of graphs, i.e. to all graphs that, for every $\epsilon, 1 > \epsilon > 0$, can be partitioned into a set of small components of size $f(\epsilon, d)$ by removing at most ϵdn edges.

8   Open Problems

There are several interesting open problems in Property Testing for sparse graphs. We will mention three of the most interesting ones.

Query Complexity of Planarity Testing. Currently, the best Property Testing algorithm for planarity testing is from [9] and has a query complexity of $(1/\epsilon)^{O(\log 1/\epsilon)}$. Can this be improved to a polynomial? In order to prove such a result one may need to develop improved local partitioning algorithms or a different way to approach the problem.

Testable Properties in Expander Graphs. We do not know much about the testability of properties that contain expander graph. It would be very nice to find a characterization of a large set of testable properties that contain expander graphs.

Testable Properties in Bounded Average Degree Graphs. If we do not have a degree bound, most of the techniques presented in this survey do not work any more. It would be interesting to prove similar results as in the bounded degree graph model. So far, we know that bipartiteness is testable in arbitrary planar graphs [2]. Furthermore, it is known that one can test forst isomorphism [8].

Acknowledgements. The author acknowledges the support of ERC grant 307696.

References

1. Benjamini, I., Schramm, O., Shapira, A.: Every Minor-Closed Property of Sparse Graphs is Testable. Advances in Mathematics 223, 2200–2218 (2010)
2. Czumaj, A., Monemizadeh, M., Onak, K., Sohler, C.: Planar Graphs: Random Walks and Bipartiteness Testing. In: FOCS, pp. 423–432 (2011)

3. Czumaj, A., Shapira, A., Sohler, C.: Testing Hereditary Properties of Nonexpand-ing Bounded-Degree Graphs. SIAM Journal on Computing 38(6), 2499–2510 (2009)
4. Edelman, A., Hassidim, A., Nguyen, H.N., Onak, K.: An Efficient Partitioning Oracle for Bounded-Treewidth Graphs. In: Goldberg, L.A., Jansen, K., Ravi, R., Rolim, J.D.P. (eds.) APPROX/RANDOM 2011. LNCS, vol. 6845, pp. 530–541. Springer, Heidelberg (2011)
5. Goldreich, O., Goldwasser, S., Ron, D.: Property Testing and its Connection to Learning and Approximation. Journal of the ACM 45(4), 653–750 (1998)
6. Hassidim, A., Kelner, J.A., Nguyen, H.N., Onak, K.: Local Graph Partitions for Approximation and Testing. In: FOCS, pp. 22–31 (2009)
7. Goldreich, O., Ron, D.: Property Testing in Bounded Degree Graphs. Algorith-mica 32(2), 302–343 (2002)
8. Kusumoto, M., Yoshida, Y.: Testing Forest-Isomorphism in the Adjacency List Model. In: Esparza, J., Fraigniaud, P., Husfeldt, T., Koutsoupias, E. (eds.) ICALP 2014. LNCS, vol. 8572, pp. 763–774. Springer, Heidelberg (2014)
9. Levi, R., Ron, D.: A Quasi-Polynomial Time Partition Oracle for Graphs with an Excluded Minor. In: Fomin, F.V., Freivalds, R., Kwiatkowska, M., Peleg, D. (eds.) ICALP 2013, Part I. LNCS, vol. 7965, pp. 709–720. Springer, Heidelberg (2013)
10. Lipton, R.J., Tarjan, R.E.: A separator theorem for planar graphs. SIAM Journal on Applied Mathematics 36(2), 177–189 (1979)
11. Newman, I., Sohler, C.: Every Property of Hyperfinite Graphs Is Testable. SIAM Journal on Computing 42(3), 1095–1112 (2013)
12. Rubinfeld, R., Sudan, M.: Robust Characterizations of Polynomials with Applica-tions to Program Testing. SIAM Journal on Computing 25(2), 252–271 (1996)

Choiceless Polynomial Time on Structures with Small Abelian Colour Classes

F. Abu Zaid, E. Grädel, M. Grohe, and W. Pakusa

RWTH Aachen University, Germany
{abuzaid,graedel,pakusa}@logic.rwth-aachen.de,
grohe@informatik.rwth-aachen.de

Abstract. Choiceless Polynomial Time (CPT) is one of the candidates in the quest for a logic for polynomial time. It is a strict extension of fixed-point logic with counting (FPC) but to date it is unknown whether it expresses all polynomial-time properties of finite structures. We study the CPT-definability of the isomorphism problem for relational structures of bounded colour class size q (for short, *q-bounded structures*). Our main result gives a positive answer, and even CPT-definable canonisation procedures, for classes of q-bounded structures with small Abelian groups on the colour classes. Such classes of q-bounded structures with *Abelian colours* naturally arise in many contexts. For instance, 2-bounded structures have Abelian colours which shows that CPT captures PTIME on 2-bounded structures. In particular, this shows that the isomorphism problem of multipedes is definable in CPT, an open question posed by Blass, Gurevich, and Shelah.

1 Introduction

The quest for a logical characterisation of PTIME remains an important challenge in the field of finite model theory [10,12]. A natural logic of reference is fixed-point logic with counting (FPC) which comes rather close to capturing PTIME. It can express many fundamental graph properties and algorithmic techniques including for instance by a recent result of Anderson, Dawar and Holm, the ellipsoid method for linear programs [1]. Moreover, FPC captures PTIME on many important classes of graphs such as planar graphs and graphs of bounded tree-width, and more generally, on *every* class of graphs which excludes a fixed graph as a minor [13]. More specifically, the aforementioned classes even admit FPC-*definable canonisation* which means that FPC can define, given an input graph, an isomorphic copy of that graph over a linearly ordered universe. Clearly, if a class of structures admits FPC-definable canonisations, then FPC captures PTIME on this class, since by the Immerman-Vardi Theorem (see e.g. [10]) fixed-point logic can define every polynomial-time query on ordered structures. The technique of *definable canonisation* will also play a crucial role in this paper.

On the other hand, FPC fails to capture PTIME in general, which was shown by the CFI-construction of Cai, Fürer and Immerman [6]. Given our current knowledge, the two main sources of "hard" problems for FPC are tractable

E. Csuhaj-Varjú et al. (Eds.): MFCS 2014, Part I, LNCS 8634, pp. 50–62, 2014.
© Springer-Verlag Berlin Heidelberg 2014

cases of the graph isomorphism problem and queries from the field of linear algebra. First of all, the CFI-construction shows that FPC cannot define the graph isomorphism problem on graphs with bounded degree and with bounded colour class size. Recall that a graph of *colour class size* q is a graph coloured by an ordered set, say natural numbers, where at most q vertices get the same colour. On the other hand, the graph isomorphism problem is tractable on graphs with bounded degree or bounded colour class size [3,11,16]. Secondly, Atserias, Bulatov and Dawar [2] proved that FPC cannot express the solvability of linear equation systems over finite Abelian groups. Interestingly, also the CFI-query can be formulated using a linear equation system over $\mathbb{Z}_2$ [7].

This observation motivated Dawar, Holm, Grohe and Laubner [7] to introduce an extension of FPC by operators which compute the rank of definable matrices. The resulting logic, denoted as *rank logic* (FPR), is a strict extension of FPC and capable of defining the solvability of linear equation systems over finite fields and the CFI-query. Similar extensions of FPC by operators which solve linear equation systems over finite rings (and not only over finite fields) have been studied in [8]. It remains open whether one of these extensions suffices to capture PTIME and specifically, whether it can define the graph isomorphism problem on graphs of bounded degree and bounded colour class size.

In this paper we focus on *Choiceless Polynomial Time* (CPT), an extension of FPC which has been proposed by Blass, Gurevich and Shelah in [4]. Instead of extending the expressive power of FPC by operators for certain undefinable queries (such as the rank of a matrix), the basic idea of CPT is to combine the manipulation of higher order objects (hereditarily finite sets over the input structure) with a bounded amount of parallel computations. Technically, Choiceless Polynomial Time is based on BGS-machines (for Blass, Gurevich and Shelah), a computation model which directly works on relational input structures (and not on string encodings of those like Turing machines do). As a matter of fact, computations of BGS-machines have to respect symmetries of the input structure. Specifically, the set of states in a run of a BGS-program is closed under automorphisms of the input structure. More informally this means that BGS-computations are *choiceless*: it is impossible to implement statements like "pick an arbitrary element x and continue" which occur in many high-level descriptions of polynomial-time algorithms (e.g. Gaussian elimination, the Blossom algorithm for maximum matchings, and so on). On the other hand, BGS-machines are also very powerful which is due to their ability to construct and manipulate hereditarily finite sets built over the atoms of the input structure. If one imposes no further restriction on BGS-logic then *every* decidable class of structures can be defined in BGS-logic. Thus, to define CPT, the *polynomial-time* fragment of BGS-logic, one clearly has to restrict the amount of access a BGS-program has to the class of hereditarily finite sets.

Choiceless Polynomial Time is a strict extension of FPC [5]. More strikingly, Dawar, Richerby and Rossman [9] were able to show that CPT can define the CFI-query. Their very clever construction uses the power of CPT to avoid arbitrary choices by finding succinct (polynomial-time representable) encodings of exponential-sized sets of symmetric objects. However, to date, it is not known whether CPT suffices to capture PTIME, whether it can express the graph isomorphism problem for graphs of bounded colour class size or bounded degree, and similarly, it is open whether CPT can define the solvability of linear equation systems over finite fields. As a consequence, the relation between rank logic FPR and Choiceless Polynomial Time CPT remains unclear.

This paper is motivated by the question whether for every fixed q the isomorphism problem for relational structures of colour class size q (for short, q-*bounded structures*) can be defined in CPT. Our main result gives a positive answer for classes of q-bounded structures with *Abelian colours*, i.e. q-bounded structures where all colour classes induce substructures with Abelian automorphism groups (we give the formal definition in Section 4). More generally we establish for every class of q-bounded structures with Abelian colours a CPT-definable canonisation procedure which shows that CPT captures PTIME on such classes.

Classes of q-bounded structures with Abelian colours naturally arise in many contexts. First of all, every class of 2-bounded structures has Abelian colours which in turn shows that CPT captures PTIME on 2-bounded structures. On the other hand, FPC fails to capture PTIME on this class, since the CFI-query can easily be formulated using 2-bounded structures. Moreover, this solves an open question from [5] where the authors ask whether the isomorphism problem of *multipedes* is CPT-definable (cf. *Question 24* in [5]). Since multipedes are 2-bounded structures our result shows that the isomorphism problem for multipedes is CPT-definable.

Another important example arises from generalising the CFI-query for other Abelian groups than $\mathbb{Z}_2$. In particular, in [15] Holm uses such generalisations (called $\mathcal{C}$-*structures*) to define a query which separates certain fragments of rank logics from each other. Interestingly, $\mathcal{C}$-structures are q-bounded structures with Abelian colours which means that CPT can define the queries used by Holm which separates CPT from the fragments of FPR considered in [15].

Choiceless Polynomial Time. In this paper, we consider *finite* structures $\mathfrak{A} = (A, R_1^{\mathfrak{A}}, \ldots, R_k^{\mathfrak{A}})$ over *relational* signatures $\tau = \{R_1, \ldots, R_k\}$. To define CPT compactly, we follow Rossman [17]. For a vocabulary τ we define $\tau^{\mathsf{HF}} = \tau \uplus \{\emptyset, \mathsf{Atoms}, \mathsf{Pair}, \mathsf{Union}, \mathsf{Unique}, \mathsf{Card}\}$ as the extension of τ by the set-theoretic function symbols $\emptyset, \mathsf{Atoms}$ (constant symbols), $\mathsf{Union}, \mathsf{Unique}, \mathsf{Card}$ (unary function symbols) and Pair (binary function symbol). For a set A we denote by $\mathsf{HF}(A)$ the class of *hereditarily finite sets* over the *atoms* A, i.e. $\mathsf{HF}(A)$ is the least set with $A \subseteq \mathsf{HF}(A)$ and $x \in \mathsf{HF}(A)$ for every $x \subseteq \mathsf{HF}(A)$. A set $x \in \mathsf{HF}(A)$ is *transitive* if for all $z \in y \in x$ we have $z \in x$. The *transitive closure* of $x \in \mathsf{HF}(A)$ is the least transitive set $\mathrm{TC}(x)$ with $x \subseteq \mathrm{TC}(x)$.

For a τ-structure $\mathfrak{A}$, its *hereditarily finite expansion* $\mathsf{HF}(\mathfrak{A})$ is the following τ^{HF}-structure over the universe $\mathsf{HF}(A)$ where relations $R \in \tau$ are interpreted as in $\mathfrak{A}$ and the set theoretic functions in $\tau^{\mathsf{HF}} \setminus \tau$ are interpreted as follows:

- $\emptyset^{\mathsf{HF}(\mathfrak{A})} = \emptyset$, $\mathsf{Atoms}^{\mathsf{HF}(\mathfrak{A})} = A$, and
- $\mathsf{Pair}^{\mathsf{HF}(\mathfrak{A})}(x, y) = \{x, y\}$, $\mathsf{Union}^{\mathsf{HF}(\mathfrak{A})}(x) = \{y \in z : z \in x\}$, and

$$- \ \mathsf{Unique}^{\mathsf{HF}(\mathfrak{A})}(x) = \begin{cases} y, & \text{if } x = \{y\} \\ \emptyset, & \text{else,} \end{cases} \text{ and } \mathsf{Card}^{\mathsf{HF}(\mathfrak{A})}(x) = \begin{cases} |x|, & x \not\subseteq A \\ \emptyset, & \text{else.} \end{cases},$$

where $|x|$ is the cardinality of x encoded as a von Neumann ordinal.

A bijection $\pi : A \to A$ extends to a bijection $\pi' : \mathsf{HF}(A) \to \mathsf{HF}(A)$ in a natural way. If π is an automorphism of $\mathfrak{A}$, then π' is an automorphism of $\mathsf{HF}(\mathfrak{A})$. BGS-*logic* is evaluated over hereditarily finite expansions $\mathsf{HF}(\mathfrak{A})$ and is defined using three syntactic elements: *terms, formulas* and *programs*.

- *Terms* are built from *variables* and functions from τ^{HF} using the standard rules. For an input structure $\mathfrak{A}$, terms take values in $\mathsf{HF}(A)$. Additionally we allow *comprehension terms*: if $s(\bar{x}, y)$ and $t(\bar{x})$ are terms, and $\varphi(\bar{x}, y)$ is a formula then $r(\bar{x}) = \{s(\bar{x}, y) : y \in t(\bar{x}) : \varphi(\bar{x}, y)\}$ is a term (in which y is bound). The value $r^{\mathfrak{A}}(\bar{a})$ of the term $r(\bar{x})$ under an assignment $\bar{a} \subseteq \mathsf{HF}(A)$ is the set $r^{\mathfrak{A}}(\bar{a}) = \{s^{\mathfrak{A}}(\bar{a}, b) : b \in t^{\mathfrak{A}}(\bar{a}) : \mathsf{HF}(\mathfrak{A}) \models \varphi(\bar{a}, b)\} \in \mathsf{HF}(A)$.
- *Formulas* can be built from terms $t_1, t_2, \ldots, t_k$ as $t_1 = t_2$ and $R(t_1, \ldots, t_k)$ (for $R \in \tau$), and from other formulas using the Boolean connectives $\wedge, \vee, \neg$.
- *Programs* are triples $\Pi = (\Pi_{\mathrm{step}}, \Pi_{\mathrm{halt}}, \Pi_{\mathrm{out}})$ where $\Pi_{\mathrm{step}}(x)$ is a term, and $\Pi_{\mathrm{halt}}(x)$ and $\Pi_{\mathrm{out}}(x)$ are formulas. On an input structure $\mathfrak{A}$ a program Π induces a *run* which is the sequence $(x_i)_{i \geq 0}$ of *states* $x_i \in \mathsf{HF}(A)$ defined inductively as $x_0 = \emptyset$ and $x_{i+1} = \Pi_{\mathrm{step}}(x_i)$. Let $\rho = \rho(\mathfrak{A}) \in \mathbb{N} \cup \{\infty\}$ be minimal such that $\mathfrak{A} \models \Pi_{\mathrm{halt}}(x_\rho)$. The *output* $\Pi(\mathfrak{A})$ of the run of Π on $\mathfrak{A}$ is *undefined* $(\Pi(\mathfrak{A}) = \bot)$ if $\rho = \infty$ and is defined as the truth value of $\mathfrak{A} \models \Pi_{\mathrm{out}}(x_\rho)$ otherwise.

BGS-programs transform states (objects in $\mathsf{HF}(A)$) until a halting condition is reached, and produce their output from the ultimately constructed state. To obtain CPT-programs we put polynomial bounds on both, the complexity of states and the length of a run. To measure the complexity of objects in $\mathsf{HF}(A)$ we use the size of their transitive closure.

Definition 1. *A CPT-program is a pair $\mathcal{C} = (\Pi, p(n))$ of a BGS-program Π and a polynomial $p(n)$. The output $\mathcal{C}(\mathfrak{A})$ on an input structure $\mathfrak{A}$ is $\mathcal{C}(\mathfrak{A}) = \Pi(\mathfrak{A})$ if the following resource bounds are satisfied (otherwise we set $\mathcal{C}(\mathfrak{A}) = false$):*

- *the length $\rho(\mathfrak{A})$ of the run of Π on $\mathfrak{A}$ is at most $p(|A|)$ and*
- *for each state in the run $(x_i)_{i \leq \rho(\mathfrak{A})}$ of Π on $\mathfrak{A}$ it holds that $|TC(x_i)| \leq p(|A|)$.*

The main difference to fixed-point logics like FPC is that CPT can manipulate *higher-order* objects from $\mathsf{HF}(A)$ which have polynomial size. These objects can be, for example, clever data structures which succinctly encode exponential-sized sets, or just exhaustive search trees on small parts of the input. In contrast, FPC can access only (constant-sized) lists of elements.

Algebra and Permutation Groups. For a set V, we denote by $\mathrm{Sym}(V)$ the *symmetric group* acting on V. As usual we use *cycle notation* $(v_1 \, v_2 \cdots v_\ell)$ to specify permutations in $\mathrm{Sym}(V)$. For a *permutation group* $\Gamma \leq \mathrm{Sym}(V)$ and

$v \in V$ we write $\Gamma(v) = \{\gamma(v) : \gamma \in \Gamma\}$ to denote the *orbit* of v under the action of Γ. The set of Γ-orbits $\{\Gamma(v) : v \in V\}$ yields a partition of V. We say that Γ acts *transitively* on V if $\Gamma(v) = V$ for some (equivalently each) $v \in V$. We read group operations from *right to left* and use the notation γ^σ as a shorthand for $\sigma\gamma\sigma^{-1}$ whenever this makes sense (hence $(\gamma^\sigma)^\tau = \gamma^{\tau\sigma}$). Likewise, we let $\sigma\Gamma = \{\sigma\gamma : \gamma \in \Gamma\}$ and $\Gamma^\sigma = \{\gamma^\sigma : \gamma \in \Gamma\}$.

For a τ-structure $\mathfrak{A}$ we let $\mathrm{Aut}(\mathfrak{A}) \leq \mathrm{Sym}(A)$ denote the *automorphism group* of $\mathfrak{A}$. In this paper, $\mathrm{Aut}(\mathfrak{A})$ will often be *Abelian*. Recall that every finite Abelian group is an inner direct sum of cyclic groups of prime power order. For a group Γ and $\gamma \in \Gamma$ we denote by $\langle\gamma\rangle$ the *cyclic* subgroup of Γ generated by γ.

We define *linear equation systems* over finite rings $\mathbb{Z}_d$ where $d = p^k$ is a prime-power. Let V be a set of variables over $\mathbb{Z}_d$. By $\mathbb{Z}_d^V$ we denote the set of (unordered) $\mathbb{Z}_d$-vectors $x : V \mapsto \mathbb{Z}_d$ with indices in V. An *atomic linear term* is of the form $z \cdot v$ for $z \in \mathbb{Z}_d, v \in V$. A *linear term* is a set of atomic linear terms. An *assignment* is a map $\alpha : V \to \mathbb{Z}_d$. The *value* $t[\alpha] \in \mathbb{Z}_d$ of an atomic linear term $t = z \cdot v$ under α is $t[\alpha] = z \cdot \alpha(v)$. The value $t[\alpha] \in \mathbb{Z}_d$ of a term t under α is $t[\alpha] = \sum_{s \in t} s[\alpha]$. A *linear equation* is a pair (t, z) where t is a linear term and $z \in \mathbb{Z}_d$. An assignment $\alpha : V \to \mathbb{Z}_d$ *satisfies* $e = (t, z)$ if $t[\alpha] = z$. A *linear equation system* is a set S of linear equations. A linear equation system S is *solvable* (or *consistent*) if an assignment $\alpha : V \to \mathbb{Z}_d$ satisfies all equations in S. For more background on (linear) algebra and permutation groups see [14].

2 Relational Structures of Bounded Colour Class Size

We describe a procedure to define, given an input structure of bounded colour class size, an isomorphic copy over an ordered universe (a *canonical copy* or *canonisation*). The idea is to split the input structure into an ordered sequence of small substructures which can be canonised easily. We then combine these small canonised parts to obtain a canonisation of the full structure. To guarantee consistency, we maintain a set of isomorphisms between (the canonised part of) the input structure and its (partial) canonisation.

A *(linear) preorder* $\preceq$ of *width* $q \geq 1$ is a reflexive, transitive and total binary relation where the induced equivalence $x \sim y := (x \preceq y$ and $y \preceq x)$ only contains classes of size $\leq q$. A preorder $\preceq$ on A induces a linear order on the equivalence classes $A/_\sim$ and we write $A = A_1 \preceq \cdots \preceq A_n$ if A_i is the i-th equivalence class with respect to this linear order. A preorder $\preceq'$ *refines* $\preceq$ if $x \preceq' y$ implies $x \preceq y$.

Definition 2. *Let* $\tau = \{R_1, \ldots, R_k\}$. *A* q-*bounded* τ-*structure* $\mathcal{H}$ *is a* $\tau \uplus \{\preceq\}$-*structure* $\mathcal{H} = (H, R_1^{\mathcal{H}}, \ldots, R_k^{\mathcal{H}}, \preceq)$ *where* $\preceq$ *is a preorder on* H *of width* $\leq q$. *We write* $H = H_1 \preceq \cdots \preceq H_n$ *and denote by* $q_i := |H_i| \leq q$ *the size of the* i-*th colour class* H_i. *We set* $H_i^< = \{(i, 0), \ldots, (i, q_i - 1)\}$ *and write* $\mathcal{O}(H_i)$ *to denote the set of bijections between* H_i *and* $H_i^<$, *that is* $\mathcal{O}(H_i) = \{\pi : H_i \to H_i^<, \pi \text{ is a bijection}\}$.

For a class of q-bounded structures we always assume a *fixed* vocabulary τ. Thus the arity of all relations is bounded by a constant, say by r. Let $\mathcal{P} = \mathcal{P}(n, r)$ denote the set of non-empty subsets $I \subseteq \{1, \ldots, n\}$ of size $\leq r$. We can define $\mathcal{P}$ together with a linear order in CPT (as r is fixed). For $I \in \mathcal{P}$ we set $H_I = \biguplus_{i \in I} H_i$ and denote by $\mathcal{H}_I \subseteq \mathcal{H}$ the substructure of $\mathcal{H}$ *induced* on H_I. Since r bounds the arity of relations in τ we have $\mathcal{H} = \bigcup_{I \in \mathcal{P}} \mathcal{H}_I$.

We set $\mathcal{O}(H) = \mathcal{O}(H_1) \times \cdots \times \mathcal{O}(H_n)$ and $\mathcal{O}(H_I) = \mathcal{O}(H_{i_1}) \times \cdots \times \mathcal{O}(H_{i_\ell})$ for $I = \{i_1, \ldots, i_\ell\} \in \mathcal{P}$. Given $C \subseteq \mathcal{O}(H_I)$ the *extension* of C to $\mathcal{O}(H)$ is the set $\mathrm{ext}(C) = \{(\sigma_1, \ldots, \sigma_n) \in \mathcal{O}(H) : (\sigma_{i_1}, \ldots, \sigma_{i_\ell}) \in C\}$.

Every $\sigma \in \mathcal{O}(H)$ defines a bijection between H and the *ordered* set $H^< = \{(i,j) : 1 \le i \le n, 0 \le j < q_i\}$. The preorder $\preceq$ on H translates to the preorder $\sigma(\preceq)$ on $H^<$ which is defined as $(i,j)\sigma(\preceq)(i',j')$ if, and only if, $i \le i'$. Specifically, $\sigma \in \mathcal{O}(H)$ defines an isomorphism between the input structure $\mathcal{H}$ and the structure $\sigma(\mathcal{H}) = (H^<, \sigma(R_1^{\mathcal{H}}), \ldots, \sigma(R_k^{\mathcal{H}}), \sigma(\preceq))$. Of course we can apply $\sigma \in \mathcal{O}(H)$ also to substructures of $\mathcal{H}$. In particular for $I \in \mathcal{P}$, every $\sigma \in \mathcal{O}(H_I)$ defines an isomorphism between $\mathcal{H}_I$ and $\sigma(\mathcal{H}_I) = (H_I^<, \sigma(R_1^{\mathcal{H}_I}), \ldots, \sigma(R_k^{\mathcal{H}_I}), \sigma(\preceq^{\mathcal{H}_I}))$ where $H_I^< = \{(i,j) \in H^< : i \in I\}$. We want to construct, given $\mathcal{H}$, an isomorphic copy $\sigma(\mathcal{H})$ which we call the *canonisation* or the *canonical copy* of $\mathcal{H}$.

In general, for different $\sigma, \tau \in \mathcal{O}(H)$ we have $\sigma(\mathcal{H}) \ne \tau(\mathcal{H})$. Since the structures $\sigma(\mathcal{H})$ and $\tau(\mathcal{H})$ are defined over an ordered universe we can distinguish them in CPT. Moreover, $\sigma(\mathcal{H}) = \tau(\mathcal{H})$ holds if, and only if, $\tau^{-1}\sigma \in \mathrm{Aut}(\mathcal{H})$.

Lemma 3. $\{\tau : \tau(\mathcal{H}) = \sigma(\mathcal{H})\} = \sigma\,\mathrm{Aut}(\mathcal{H}) = \mathrm{Aut}(\sigma(\mathcal{H}))\sigma$ *for* $\sigma \in \mathcal{O}(H)$.

Let $I_1 < I_2 < \cdots < I_m$ be the enumeration of $\mathcal{P}$ according to the definable order. We denote by $\mathcal{H}[1 \cdots s] \subseteq \mathcal{H}$ the (not necessarily induced) substructure of $\mathcal{H}$ that consists of the first s components, i.e. $\mathcal{H}[1 \cdots s] = \mathcal{H}_{I_1} \cup \cdots \cup \mathcal{H}_{I_s}$.

Definition 4. *An s-canonisation is a canonisation of* $\mathcal{H}[1 \cdots s]$, *i.e. a structure* $\sigma(\mathcal{H}[1 \cdots s]) = \sigma(\mathcal{H}_{I_1}) \cup \cdots \cup \sigma(\mathcal{H}_{I_s})$ *for* $\sigma \in \mathcal{O}(H)$. *A non-empty set* $C \subseteq \mathcal{O}(H)$ *witnesses an s-canonisation if* $\tau(\mathcal{H}_{I_j}) = \sigma(\mathcal{H}_{I_j})$ *for all* $\sigma, \tau \in C$ *and* $j = 1, \ldots, s$.

Since $\mathcal{H} = \bigcup_{I \in \mathcal{P}} \mathcal{H}_I$, an m-canonisation of $\mathcal{H}$ also is a canonisation of $\mathcal{H}$. To describe our generic CPT-canonisation procedure for q-bounded structures, we assume that we have already preselected for each colour class H_i a set of linear orderings $\sigma_i \Gamma_i \subseteq \mathcal{O}(H_i)$ where $\Gamma_i \le \mathrm{Sym}(H_i)$ and $\sigma_i \in \mathcal{O}(H_i)$. The group $\Gamma = \Gamma_1 \times \cdots \times \Gamma_n$ acts on $\mathcal{O}(H)$ in the obvious way and for $\sigma = (\sigma_1, \ldots, \sigma_n) \in \mathcal{O}(H)$ we have $\sigma\Gamma = \tau\Gamma$ for every $\tau \in \sigma\Gamma$. For an index set $I = \{i_1, \ldots, i_\ell\} \in \mathcal{P}$ we write Γ_I to denote the group $\Gamma_I = \Gamma_{i_1} \times \cdots \times \Gamma_{i_\ell}$ and $(\sigma\Gamma)_I$ to denote the set $(\sigma\Gamma)_I = \sigma_{i_1}\Gamma_{i_1} \times \cdots \times \sigma_{i_\ell}\Gamma_{i_\ell} \subseteq \mathcal{O}(H_I)$. The *extension* of a set of partial orderings $C \subseteq (\sigma\Gamma)_I$ to $\sigma\Gamma$ is the set $\mathrm{ext}(C) = \{(\tau_1, \ldots, \tau_n) \in \sigma\Gamma : (\tau_{i_1}, \ldots, \tau_{i_\ell}) \in C\} \subseteq \sigma\Gamma$. The canonisation procedure for q-bounded structures is given below.

Given: q-bounded structure $\mathcal{H}$ and sets $\sigma_i \Gamma_i \subseteq \mathcal{O}(H_i)$ for $\Gamma_i \le \mathrm{Sym}(H_i)$, $\sigma_i \in \mathcal{O}(H_i)$

  $C_0 := \sigma\Gamma$ and $\mathcal{H}_0^< := \emptyset$
  for $s = 1$ **to** m **do**
    Set $I := I_s$ and define $\Delta := \mathrm{Aut}(\mathcal{H}_I) \cap \Gamma_I$ and $D := \{\tau\Delta : \tau \in (\sigma\Gamma)_I\}$
    Fix $\tau\Delta \in D$ such that $C_{s-1} \cap \mathrm{ext}(\tau\Delta) \ne \emptyset$ (possible by Lemma 3)
    Set $C_s := C_{s-1} \cap \mathrm{ext}(\tau\Delta)$ and $\mathcal{H}_s^< := \mathcal{H}_{s-1}^< \cup \tau'(\mathcal{H}_I)$ for some (all) $\tau' \in \tau\Delta$
  end for

Return: The canonisation $\mathcal{H}^< := \mathcal{H}_m^<$ of $\mathcal{H}$

To express this procedure in CPT, the difficulty is to find suitable representations for the sets C_s. Clearly, storing them explicitly is impossible as their size is exponential in the size of the input structure. In the following sections we establish suitable representations based on linear algebra. We summarise the requirements for such representations in the following definition.

Definition 5. *For explicitly given sets* $\sigma_i \Gamma_i \subseteq \mathcal{O}(H_i)$, *a CPT-definable representation of sets* $\tau \Delta$ *with* $\Delta \leq \Gamma$ *and* $\tau \in \sigma \Gamma$ *is* suitable *if the following operations are CPT-definable.*

(i) Consistency. *Given a representation of* $\tau \Delta$, *it is CPT-definable whether* $\tau \Delta \neq \emptyset$.

(ii) Intersection. *Given two representations of sets* $\tau_1 \Delta_1$ *and* $\tau_2 \Delta_2$, *a representation of the set* $\tau_1 \Delta_1 \cap \tau_2 \Delta_2$ *is CPT-definable.*

(iii) Representation of basic sets. *Given* $\tau \Delta$ *with* $\tau \in (\sigma \Gamma)_I$ *and* $\Delta \leq \Gamma_I$ *for* $I \in \mathcal{P}$, *a representation of* $ext(\tau \Delta) \subseteq \sigma \Gamma$ *can be defined in CPT.*

3   Cyclic Linear Equation Systems over Finite Rings

We proceed to show that the solvability of *cyclic linear equation systems* (CESs) over finite rings $\mathbb{Z}_d$, where $d = p^k$ is a prime power, can be defined in CPT. In Section 4, we will see that solution spaces of CESs can be used to represent sets of witnessing isomorphisms as required in Definition 5. Having this connection a *consistency check* corresponds to deciding the solvability of a linear equation system, the *intersection operation* corresponds to combining the equations of two linear systems, and the *representation of basic sets* corresponds to constructing a linear equation systems over a small set of variables.

Definition 6. *Let* V *be a set of variables over* $\mathbb{Z}_d$ *where* d *is a prime power.*

(a) *A* cyclic constraint *on* $W \subseteq V$ *is a consistent set* C *of linear equations with variables in* W *which contains for every pair* $v, w \in W$ *an equation of the form* $v - w = z$ *for* $z \in \mathbb{Z}_d$.

(b) *A* cyclic linear equation systems (CES) *over* $\mathbb{Z}_d$ *is a triple* $(V, S, \preceq)$ *where* $\preceq$ *is a preorder on* $V = V_1 \preceq \cdots \preceq V_n$ *and* S *is a linear equation system which contains for every block* V_i *a cyclic constraint* C_i.

In the definition we do not require that $\preceq$ is of bounded width. However, given the cyclic constraints $C_i \subseteq S$ we can assume that $|V_i| = d$ for all $1 \leq i \leq n$.

Lemma 7. *Given a CES* $(V, S, \preceq)$ *over* $\mathbb{Z}_d$, *we can define in CPT a CES* $(V', S', \preceq')$ *over* $\mathbb{Z}_d$ *such that* $V' = V_1' \preceq' \cdots \preceq' V_n'$ *and* $|V_i'| = d$ *for all* i, *together with a bijection between the set of assignments that satisfy the two CESs.*

For $z \in \mathbb{Z}_d$ and $v \in V_i$ we denote by $v^{+z} \in V_i$ the (unique) variable such that C_i contains the constraint $v^{+z} - v = z$. There are precisely d different assignments $\alpha : V_i \to \mathbb{Z}_d$ with $\alpha \models C_i$ and each one is determined by fixing the value of a single variable $v \in V_i$. The crucial ingredient of our CPT-procedure for solving CESs over $\mathbb{Z}_d$ is the notion of a *hyperterm* which is based on the CPT-procedure of Dawar, Richerby and Rossman for expressing the CFI-query [9].

Definition 8. *Let A be the set of assignments that satisfy all cyclic constraints C_i, i.e. $A := \{\alpha : V \to \mathbb{Z}_d : \alpha \models C_i$ for $i = 1, \ldots, n\}$. We inductively define*

(i) *hyperterms T and associated* shifted *hyperterms T^{+z} for $z \in \mathbb{Z}_d$ such that $T^{+(z_1+z_2)} = (T^{+z_1})^{+z_2}$ for $z_1, z_2 \in \mathbb{Z}_d$, and $T^{+d} = T$,*

(ii) *for assignments $\alpha \in A$ the* value *$T[\alpha] \in \mathbb{Z}_d$ such that $T^{+z}[\alpha] - T[\alpha] = z$,*

(iii) *and the* coefficient *$c(V_i, T) = c(V_i, T^{+z}) \in \mathbb{Z}_d$ of variable block V_i in the hyperterms $T, T^{+1}, \ldots, T^{+(d-1)}$.*

- *For $z \in \mathbb{Z}_d$ we define the hyperterm $T = z$ and set $T^{+y} = z + y$ for $y \in \mathbb{Z}_d$. We let $c(V_i, T) = c(V_i, T^{+y}) = 0$ for each variable block V_i and all $y \in \mathbb{Z}_d$ and let $T[\alpha] = z$ and $T^{+y}[\alpha] = z + y$ for all assignments $\alpha \in A$ and $y \in \mathbb{Z}_d$. Moreover, for $v \in V_i$, $T = v$ is a hyperterm where $T^{+y} = v^{+y}$ for $y \in \mathbb{Z}_d$. We set $c(V_j, T) = c(V_j, T^{+y}) = 1$ for $y \in \mathbb{Z}_d$ if $j = i$ and $c(V_j, T) = c(V_j, T^{+y}) = 0$ otherwise. Finally, we let $T[\alpha] = \alpha(v)$. Then $T^{+y}[\alpha] = \alpha(v^{+y}) = \alpha(v) + y$.*
- *Let Q, R be hyperterms. Then $T = Q \oplus R := \{\langle Q^{+z_1}, R^{+z_2}\rangle : z_1 + z_2 = 0\}$ is a hyperterm with shifted hyperterm $T^{+y} = \{\langle Q^{+z_1}, R^{+z_2}\rangle : z_1 + z_2 = y\}$ for $y \in \mathbb{Z}_d$. We set $c(V_i, T) = c(V_i, T^{+y}) = c(V_i, Q) + c(V_i, R)$, $T[\alpha] := Q[\alpha] + R[\alpha]$ and we have $T^{+y}[\alpha] = Q[\alpha] + R[\alpha] + y$ for $\alpha \in A$.*
- *Let Q be a hyperterm, $z \in \mathbb{Z}_d$. Then a new hyperterm $T = z \odot Q := Q \oplus \cdots \oplus Q$ results by applying the $\oplus$-operation z-times to Q (where we implicitly agree on an application from left to right). The definitions of $T^{+y}, c(V_i, T)$ and $T[\alpha]$ follow from the definition of $\oplus$.*

Definition 9. *For $\alpha \in A$, $1 \leq i \leq n$ and $z \in \mathbb{Z}_d$ we define the assignment $\alpha^{i:+z} \in A$ which results from a* semantical *z-shift of variable block V_i which means that $\alpha^{i:+z}(v) = \alpha(v) + z$ for $v \in V_i$ and $\alpha^{i:+z}(v) = \alpha(v)$ for $v \notin V_i$. Moreover we let $\pi^{i:+z} : V_i \to V_i$ be the* syntactic *z-shift on the set V_i which is defined as $\pi^{i:+z}(v) := v^{+z}$ for $v \in V_i$ lifted to a permutation acting on $\mathsf{HF}(V)$.*

There is a tight correspondence between the syntactic structure and the intended semantics for hyperterms as expressed in the following lemma.

Lemma 10. *Let $1 \leq i \leq n, z \in \mathbb{Z}_d$, let T be a hyperterm and let $c = c(V_i, T) \in \mathbb{Z}_d$ be the coefficient of variable block V_i in T.*

(a) *Then $\pi^{i:+z}(T) = T^{+c \cdot z}$. In particular if $c = 0$ then $\pi^{i:+z}(T) = T$.*

(b) *For any assignment $\alpha \in A$ we have $T[\alpha^{i:+z}] = \pi^{i:+z}(T)[\alpha]$.*

Intuitively, a hyperterm is a succinct encoding of a class of linear terms that are (given the cyclic constraints) equivalent. Using the preorder $\preceq$ it is possible to define in CPT a linearly ordered partition $S = \biguplus_{i=1}^{m} S_i$ of S, corresponding hyperterms $T_1, \ldots, T_m$ and constants $z_1, \ldots, z_m \in \mathbb{Z}_d$ such that for every equation $(t, z) \in S_i$ and $\alpha \in A$ we have $t[\alpha] = T_i[\alpha]$ and $z = z_i$ (or the CES is inconsistent). This means that the system S^* consisting of the *ordered* set of hyperequations (T_i, z_i) is equivalent to the given CES. Given the linear order on S^*, we want to use Gaussian elimination in order to determine the solvability of S^*. As a simple preparation we observe that elementary transformations can be applied to systems of hyperequations.

Lemma 11. *Let S^* be a system of hyperequations, and let $(T, z), (T', z') \in S^*$. Then S^* and $(S^* \setminus \{(T, z)\}) \cup \{(T \oplus T', z + z')\}$ have the same solutions (in A).*

We assign the $m \times n$-matrix $M[S^*] : \{1, \ldots, m\} \times \{1, \ldots, n\} \to \mathbb{Z}_d$ to the system S^* of hyperequations defined as $M[S^*](i, j) := c(V_j, T_i)$. Applying elementary operations to S^* as in Lemma 11 corresponds to applying elementary row operations to $M[S^*]$. Using a slightly adapted version of Gaussian elimination ($\mathbb{Z}_d$ is a *ring*, not a *field*) it is possible to transform S^* such that $M[S^*]$ is in *Hermite normal form*. This transformation can be expressed in CPT.

We say that a hyperterm T is *atomic* if $c(V_i, T) = 0$ for every variable block V_i. By Lemma 10 this means $T[\alpha] = T[\alpha']$ for all $\alpha, \alpha' \in A$, hence, T has a constant value $c_T = T[\alpha]$ for some (all) $\alpha \in A$. By exploiting the fact that $M[S^*]$ is in Hermite normal form it can be shown that the solvability of S^* can be characterised by determining the consistency of a set of hyperequations (T, z) with *atomic* hyperterms T, which is to check whether $c_T = z$.

It remains to express the consistency of hyperequations (T, z) for atomic hyperterms T in CPT. This is easy if T is built from constants in $\mathbb{Z}_d$.

Lemma 12. *The value of a hyperterm $T' \in \mathsf{HF}(\mathbb{Z}_d)$ can be defined in CPT.*

Given an atomic hyperterm T, it remains to construct in CPT an equivalent hyperterm $T' \in \mathsf{HF}(\mathbb{Z}_d)$. To this end, we crucially make use of the strong connection between syntax and semantics of hyperterms as stated in Lemma 10.

Lemma 13. *Let $T \in \mathsf{HF}(V)$ be an atomic hyperterm. Then we can define in CPT an equivalent hyperterm $T' \in \mathsf{HF}(\mathbb{Z}_d)$.*

Theorem 14. *The solvability of CESs over $\mathbb{Z}_d$ can be defined in CPT.*

4   Canonising q-Bounded Structures with Abelian Colours

We apply the CPT-procedure for solving CESs to show that q-bounded structures with *Abelian colours* can be canonised in CPT. Recall that we denote by $\mathcal{H}_i$ the substructure of $\mathcal{H}$ induced on the colour class H_i.

Definition 15. *A class $\mathcal{K}$ of q-bounded τ-structures has* Abelian colours *if $Aut(\mathcal{H}_i) \leq Sym(H_i)$ is Abelian for all $\mathcal{H} \in \mathcal{K}$ and colour classes $H_i \subseteq H$.*

Moreover, we say that $\mathcal{K}$ allows (CPT-)constructible transitive Abelian symmetries *if there are CPT-programs which define, given $\mathcal{H} \in \mathcal{K}$, on each colour class $H_i \subseteq H$ a transitive Abelian group $\Gamma_i \leq Sym(H_i)$ together with a linear order on $\{\sigma\Gamma_i : \sigma \in \mathcal{O}(H_i)\}$ and a linear order on Γ_i.*

We proceed to show that classes of q-bounded structures with Abelian colours can be reduced to classes with constructible transitive Abelian symmetries.

Theorem 16. *Let $\mathcal{K}$ be a class of q-bounded τ-structures with Abelian colours. There is a CPT-program which defines, given $\mathcal{H} \in \mathcal{K}$, a refinement $\preceq_r^{\mathcal{H}}$ of the preorder $\preceq^{\mathcal{H}}$ on H such that the class $\mathcal{K}'$ of structures $\mathcal{H}' = \mathcal{H}[\preceq^{\mathcal{H}} \setminus \preceq_r^{\mathcal{H}}]$ (which result from substituting $\preceq^{\mathcal{H}}$ by its refinement $\preceq_r^{\mathcal{H}}$) allows constructible transitive Abelian symmetries.*

The translation $\mathcal{H} \in \mathcal{K} \mapsto \mathcal{H}' \in \mathcal{K}'$ only refines the preorder on H, hence a canonisation of $\mathcal{H}'$ yields a canonisation of $\mathcal{H}$. Thus, CPT-definable canonisation procedures on classes of q-bounded structures with constructible transitive Abelian symmetries provide CPT-definable canonisation procedures on classes of q-bounded structures with Abelian colours.

Fix a class $\mathcal{K}$ of q-bounded structures with constructible transitive Abelian symmetries. Let $\mathcal{H} \in \mathcal{K}$ with colour classes $H = H_1 \preceq \cdots \preceq H_n$ and let $\Gamma_i \leq \mathrm{Sym}(H_i)$ denote the associated Abelian transitive groups. To express our generic canonisation procedure from Section 2 in CPT, it suffices to find CPT-definable representations of sets $\tau\Delta$ where $\Delta \leq \Gamma$ and $\tau \in \mathcal{O}(H)$ which satisfy the requirements of Definition 5. Let us first find appropriate representations for the basic sets $\sigma\Delta \subseteq \mathcal{O}(H_i)$ with $\Delta \leq \Gamma_i$ and $\sigma \in \mathcal{O}(H_i)$ for each colour class H_i.

Lemma 17. *Given a set $B \subseteq \mathsf{HF}(H)$ with $|B| \leq q$ and an Abelian transitive group $\Gamma \leq \mathrm{Sym}(B)$ which is the direct sum of k cyclic subgroups of prime-power order, i.e. $\Gamma = \langle \delta_1 \rangle \oplus \cdots \oplus \langle \delta_k \rangle$ for $\delta_1, \ldots, \delta_k \in \Gamma$ where $|\delta_i| = d_i$ is a prime-power, and given a set $\sigma\Gamma \subseteq \mathcal{O}(B)$ for $\sigma \in \mathcal{O}(B)$ we can define in CPT*

- *sets $W_1, \ldots, W_k \subseteq \mathsf{HF}(B)$ with $|W_i| = d_i$, an order $W_1 < W_2 < \cdots < W_k$,*
- *and if we set $L_i := \mathbb{Z}_{d_i}^{W_i}$ and let $e_i \in L_i$ denote the L_i-unit vector which is $e_i(w) = 1$ for all $w \in W_i$, then we can define in CPT an embedding $\varphi : \sigma\Gamma \to L_1 \times \cdots \times L_k$ which respects the action of Γ on $\sigma\Gamma$ in the following way. For all $\tau \in \sigma\Gamma$ and $\gamma = \ell_1 \cdot \delta_1 \oplus \cdots \oplus \ell_k \cdot \delta_k \in \Gamma$ we have that*

$$\varphi(\tau \circ \gamma) = \varphi(\tau) + (\ell_1 \cdot e_1, \cdots, \ell_k \cdot e_k).$$

Using the linear order on $\{\sigma\Gamma_i : \sigma \in \mathcal{O}(H_i)\}$ we fix for every colour class H_i a set $\sigma_i\Gamma_i \subseteq \mathcal{O}(H_i)$. Let $\sigma\Gamma = \sigma_1\Gamma_1 \times \cdots \times \sigma_n\Gamma_n$. Using Lemma 17 we write $\Gamma_i = \langle \delta_1^i \rangle \oplus \cdots \oplus \langle \delta_{k_i}^i \rangle$ where $|\delta_j^i| = d_j^i$ is a prime-power and define in CPT

- sets $W_1^i < W_2^i < \cdots < W_{k_i}^i$ of size $|W_j^i| = d_j^i$ and for $L_j^i := \mathbb{Z}_{d_j^i}^{W_j^i}$ embeddings

$$\varphi^i : \sigma_i\Gamma_i \to L_1^i \times \cdots \times L_{k_i}^i,$$

- such that for the L_j^i-unit vectors $e_j^i \in L_j^i$, each $\gamma = \ell_1 \cdot \delta_1^i \oplus \cdots \oplus \ell_{k_i} \cdot \delta_{k_i}^i \in \Gamma_i$ and each $\tau \in \sigma_i\Gamma_i$ it holds that $\varphi^i(\tau \circ \gamma) = \varphi^i(\tau) + (\ell_1 \cdot e_1^i, \ldots, \ell_{k_i} \cdot e_{k_i}^i)$.

We let $L = L_1^1 \times \cdots \times L_{k_1}^1 \times \cdots \times L_1^n \times \cdots \times L_{k_n}^n$ and combine the mappings φ^i to get a CPT-definable mapping $\varphi : \sigma\Gamma \to L, (\tau_1, \ldots, \tau_n) \mapsto (\varphi^1(\tau_1), \ldots, \varphi^n(\tau_n))$. Since $\Gamma = \Gamma_1 \times \cdots \times \Gamma_n = \langle \delta_1^1 \rangle \oplus \cdots \oplus \langle \delta_{k_1}^1 \rangle \times \cdots \times \langle \delta_1^n \rangle \oplus \cdots \oplus \langle \delta_{k_n}^n \rangle$ we also obtain a definable group embedding $\psi : \Gamma \to L$ as the homomorphic extension of setting $\psi(\delta_j^i) = e_j^i$. For all $\tau \in \sigma\Gamma$ and $\gamma \in \Gamma$ we have $\varphi(\tau \circ \gamma) = \varphi(\tau) + \psi(\gamma)$.

Next we analyse for $\sigma_i \Gamma_i$ the image under φ restricted to a component L_j^i, i.e. the set $(\varphi(\sigma_i \Gamma_i) \upharpoonright L_j^i) \subseteq L_j^i$. If we denote by $E_j^i := \{\ell \cdot e_j^i : 0 \leq \ell \leq d_j^i - 1\} \subseteq L_j^i$, we get $O_j^i := (\varphi(\sigma_i \Gamma_i) \upharpoonright L_j^i) = (\varphi(\sigma_i) \upharpoonright L_j^i) + E_j^i$. This means that for two vectors $x, y \in O_j^i$ it holds that $x - y \in E_j^i$. This in turn implies that for all vectors $x, y \in O_j^i$ and indices $w, w' \in W_j^i$ we have $x(w) - x(w') = y(w) - y(w')$. Hence we can define a cyclic constraint C_j^i on the set W_j^i such that O_j^i precisely corresponds to the set of assignments $\alpha : W_j^i \to \{0, \ldots, d_j^i - 1\}$ with $\alpha \models C_j^i$.

Let $P := \{p_1, \ldots, p_s\}$ be the set of all primes p_i such that Γ contains elements of order p_i. For $p \in P$ let $\Gamma_i^p \leq \Gamma_i$ denote the subgroup of Γ_i which consists of all elements $\gamma \in \Gamma_i$ whose order is a power of p. Then $\Gamma_i = \Gamma_i^{p_1} \oplus \cdots \oplus \Gamma_i^{p_s}$. In particular we have $\psi(\Gamma_i) = \psi(\Gamma_i^{p_1}) + \cdots + \psi(\Gamma_i^{p_s})$.

Similarly, for any subgroup $\Delta \leq \Gamma$ and prime $p \in P$ we let $\Delta^p \leq \Delta$ denote the subgroup of Δ which consists of elements $\delta \in \Delta$ whose order is a p-power. Then $\Delta = \Delta^{p_1} \oplus \cdots \oplus \Delta^{p_s}$ and $\Delta^p \leq \Gamma_1^p \times \Gamma_2^p \times \cdots \times \Gamma_n^p =: \Gamma^p$.

We also obtain a corresponding decomposition of L. For $p \in P$ we let $L[p] = \{(v_1^1, \ldots, v_{k_1}^1, \ldots, v_1^n, \ldots, v_{k_n}^n) \in L : \text{if } v_j^i \neq 0 \text{ then } d_j^i \text{ is a } p\text{-power}\}$. Then $\psi(\Gamma^p) \leq L[p]$ and $L = L[p_1] \oplus \cdots \oplus L[p_s]$.

For $\tau \in \mathcal{O}(H)$ and $\Delta \leq \Gamma$ we let $\varphi(\tau)^{L[p]}$ denote the projection of $\varphi(\tau) \in L$ onto the component $L[p]$. Then we have

$$\varphi(\tau \Delta) = \varphi(\tau)^{L[p_1]} + \psi(\Delta^{p_1}) \oplus \cdots \oplus \varphi(\tau)^{L[p_s]} + \psi(\Delta^{p_s}) \subseteq L[p_1] \oplus \cdots \oplus L[p_s].$$

To represent $\varphi(\tau \Delta)$ it thus suffices to represent each individual component $\varphi(\tau)^{L[p]} + \psi(\Delta^p) \subseteq L[p]$ as the set of solutions of a CES $\mathcal{S}_p$ over $\mathbb{Z}_d$ where d is a p-power. Using the cyclic constraints C_j^i from above, this is indeed possible. Altogether we represent a set $\tau \Delta$ with $\Delta \leq \Gamma$ and $\tau \in \sigma \Gamma$ as a sequence of CESs $(\mathcal{S}_{p_1}, \ldots, \mathcal{S}_{p_s})$ where the solutions of $\mathcal{S}_p$ correspond to $\varphi(\tau)^{L[p]} + \psi(\Delta^p)$. This representation is suitable with respect to Definition 5:

(i) *Consistency.* To express whether $(\mathcal{S}_{p_1}, \ldots, \mathcal{S}_{p_s})$ represents a non-empty set we check each $\mathcal{S}_p$ for consistency. This is CPT-definable by Theorem 14.

(ii) *Intersection.* Given two representations of sets $\tau_1 \Delta_1$ and $\tau_2 \Delta_2$ as sequences of CESs $(\mathcal{S}_{p_1}, \ldots, \mathcal{S}_{p_s})$ and $(\mathcal{T}_{p_1}, \ldots, \mathcal{T}_{p_s})$, we represent $\tau_1 \Delta_1 \cap \tau_2 \Delta_2$ by the sequence $(\mathcal{S}_{p_1} \cup \mathcal{T}_{p_1}, \ldots, \mathcal{S}_{p_s} \cup \mathcal{T}_{p_s})$ where $\mathcal{S}_p \cup \mathcal{T}_p$ is the CPT-definable CES which results from combining the linear equations of $\mathcal{S}_p$ and $\mathcal{T}_p$.

(iii) *Representation of basic sets.* Given a set $\rho \Delta$ with $\rho \in (\sigma \Gamma)_I$ and $\Delta \leq \Gamma_I$ for $I \in \mathcal{P}$, we get a sequence of CESs $(\mathcal{S}_{p_1}, \ldots, \mathcal{S}_{p_s})$ which represents $\text{ext}(\rho \Delta)$ of $\rho \Delta$ simply by trying all possible sequences of CESs (this is definable in CPT since the set of relevant variables is bounded by a constant).

Theorem 18. CPT *captures* PTIME *on classes of q-bounded structures with constructible transitive Abelian symmetries.*

Corollary 19. CPT *captures* PTIME *on classes of q-bounded structures with Abelian colours, and specifically, on 2-bounded structures.*

5 Discussion

We showed that CPT captures PTIME on classes of q-bounded structures with Abelian colours. It remains open whether this holds for every class of q-bounded structures. A natural way to proceed would be to allow more complex groups acting on the colour classes, for example *solvable* groups. In fact, we can modify our techniques to show that 3-bounded structures can be canonised in CPT.

Another question is whether CPT can define the solvability of linear equation systems over finite rings. A positive answer would render rank logic FPR [7] and solvability logic [8] a fragment of CPT, and otherwise, we would have a candidate for separating CPT from PTIME. It is also interesting to investigate how far our canonisation procedures for CPT can be expressed in such extensions of FPC by operators from linear algebra. For example, it is easy to see that our canonisation procedure for 2-bounded structures can be expressed in FPR.

We also want to study CPT on other classes of graphs with polynomial-time canonisation algorithms on which FPC fails to capture PTIME. Important examples are graphs of bounded degree or graphs of moderately growing treewidth.

References

1. Anderson, M., Dawar, A., Holm, B.: Maximum matching and linear programming in fixed-point logic with counting. In: LICS 2013, pp. 173–182 (2013)
2. Atserias, A., Bulatov, A., Dawar, A.: Affine systems of equations and counting infinitary logic. Theoretical Computer Science 410(18), 1666–1683 (2009)
3. Babai, L.: Monte-Carlo algorithms in graph isomorphism testing. Université de Montréal Technical Report, DMS, pp. 79–10 (1979)
4. Blass, A., Gurevich, Y., Shelah, S.: Choiceless polynomial time. Annals of Pure and Applied Logic 100(1), 141–187 (1999)
5. Blass, A., Gurevich, Y., Shelah, S.: On polynomial time computation over unordered structures. Journal of Symbolic Logic 67(3), 1093–1125 (2002)
6. Cai, J., Fürer, M., Immerman, N.: An optimal lower bound on the number of variables for graph identification. Combinatorica 12(4), 389–410 (1992)
7. Dawar, A., Grohe, M., Holm, B., Laubner, B.: Logics with rank operators. In: LICS 2009, pp. 113–122 (2009)
8. Dawar, A., Grädel, E., Holm, B., Kopczynski, E., Pakusa, W.: Definability of linear equation systems over groups and rings. LMCS 9(4) (2013)
9. Dawar, A., Richerby, D., Rossman, B.: Choiceless polynomial time, counting and the Cai–Fürer–Immerman graphs. Annals of Pure and Applied Logic 152(1-3), 31–50 (2008)
10. Grädel, E., et al.: Finite Model Theory and its Applications. Springer (2007)
11. Furst, M., Hopcroft, J.E., Luks, E.: A subexponential algorithm for trivalent graph isomorphism. Technical report (1980)
12. Grohe, M.: The quest for a logic capturing PTIME. In: LICS 2008, pp. 267–271 (2008)
13. Grohe, M.: Fixed-point definability and polynomial time on graphs with excluded minors. Journal of the ACM (JACM) 59(5), 27 (2012)
14. Hall, M.: The theory of groups. American Mathematical Soc. (1976)

15. Holm, B.: Descriptive complexity of linear algebra. PhD thesis, University of Cambridge (2010)
16. Luks, E.M.: Isomorphism of graphs of bounded valence can be tested in polynomial time. Journal of Computer and System Sciences 25(1), 42–65 (1982)
17. Rossman, B.: Choiceless computation and symmetry. In: Blass, A., Dershowitz, N., Reisig, W. (eds.) Gurevich Festschrift. LNCS, vol. 6300, pp. 565–580. Springer, Heidelberg (2010)

Sofic-Dyck Shifts

Marie-Pierre Béal[1], Michel Blockelet[2], and Cătălin Dima[2,⋆]

[1] Université Paris-Est, LIGM UMR 8049
77454 Marne-la-Vallée Cedex 2, France
marie-pierre.beal@u-pem
[2] Université Paris-Est, LACL
61 avenue du Général de Gaulle
94010 Créteil Cedex, France
{michel.blockelet,catalin.dima}@u-pec.fr

Abstract. We define the class of sofic-Dyck shifts which extends the class of Markov-Dyck shifts introduced by Krieger and Matsumoto. The class of sofic-Dyck shifts is a particular class of shifts of sequences whose finite factors are unambiguous context-free languages. We show that it corresponds exactly to shifts of sequences whose set of factors is a visibly pushdown language. We give an expression of the zeta function of a sofic-Dyck shift which has a deterministic presentation.

1 Introduction

Shifts of sequences are defined as sets of bi-infinite sequences of symbols over a finite alphabet avoiding a given set of finite blocks (or factors) called forbidden blocks. Well-known classes of shifts of sequences are shifts of finite type which avoid a finite set of forbidden factors and sofic shifts which avoid a regular set of forbidden blocks. Sofic shifts may also be defined as labels of bi-infinite paths of a finite-state labelled graph where there are no constraints of initial or infinitely repeated states. They are used to model constrained sequences in the framework of constrained coding The goal of this paper is to define and start the study of a new class of subshifts going beyond the sofic shifts, called the class of sofic-Dyck shifts.

In [11], [15], [12], Inoue, Krieger, and Matsumoto introduced and studied several classes of shifts of sequences whose set of factors are not regular but context-free. The simplest class is the class of Dyck shifts whose finite factors are factors of well-parenthesized words, also called Dyck words. Markov-Dyck shifts generalize Dyck shifts. Such shifts are accepted (or presented) by a finite-state graph equipped with a graph inverse semigroup. The graph can be considered as an automaton which operates on words over an alphabet which is partitioned into two disjoint sets, one for the left parentheses, the other one for the right parentheses. In [12], Inoue and Krieger introduced an extension of Markov-Dyck

---

⋆ This work was supported by the French National Agency (ANR) through PAI LabEx Bézout (Project ACRONYME n°ANR-10-LABX-58) and through the ANR EQINOCS.

E. Csuhaj-Varjú et al. (Eds.): MFCS 2014, Part I, LNCS 8634, pp. 63–74, 2014.
© Springer-Verlag Berlin Heidelberg 2014

shifts by constructing shifts from sofic systems and Dyck shifts. Their class contains the Motzkin shifts. In [14] (see also [10]) Krieger considers subshift presentations, called $\mathcal{R}$-graphs, with word-labelled edges partitioned into two disjoint sets of positive and negative edges equipped with a relation $\mathcal{R}$ between negative and positives edges, and such a positive edge starting at p and ending in q may be matched with a negative edge going back from q to p.

In this paper, we introduce a class of subshifts which strictly contains the above ones. We consider shifts of sequences accepted by a finite-state automaton (a labelled graph) equipped with a graph semigroup (which is no more an inverse semigroup) and a set of pairs of matched edges which may not be consecutive edges of the graph. We call such structures Dyck automata. The labelled graph does not necessarily present a Markov shift as for the case of Markov-Dyck shifts. The automaton operates on words over an alphabet which is partitioned into three disjoint sets of symbols, the call symbols, the return symbols, and internal symbols (for which no matching constraints are required).

We call the shifts presented by Dyck automata *sofic-Dyck shifts*. We prove that this class is exactly the class of shifts of sequences whose set of factors is a visibly pushdown language of finite words. So these shifts could also be called *visibly pushdown shifts*.

Visibly pushdown languages [1,2] are embeddings of context-free languages which are rich enough to model many program analysis questions. They form a natural and meaningful class in between the class of regular languages and the class of context-free languages extending the parenthesis languages [18], the bracketed languages [9], and the balanced languages [4]. Visibly pushdown languages are accepted by the so-called visibly pushdown automata. The class of these languages is moreover tractable and robust. For instance the intersection of two visibly pushdown languages is a computable visibly pushdown language.

In a second part of the paper, we compute the zeta function of a sofic-Dyck shift accepted by a deterministic Dyck automaton. The zeta function counts the number of periodic sequences of a subshift. It is a conjugacy invariant for subshifts. Two subshifts which are conjugate (or isomorphic) have the same zeta functions. The invariant is not complete and it is not known, even for shifts of finite type, whether conjugacy is a decidable property [16]. The issue of the restriction to deterministic presentations of the shift is not adressed in this paper. It is shown in [3] that the formula can be obtained also with reduced presentations and thus holds for all sofic-Dyck shifts.

The formula of the zeta function of a shift of finite type is due to Bowen and Lanford [7]. Formulas for the zeta function of a sofic shift were obtained by Manning [17] and Bowen [6]. Proofs of Bowen's formula can be found in [16] An N-rational expression of the zeta function of a sofic shift has been obtained in [19] The zeta functions of the Dyck shifts were determined by Keller in [13]. For the Motzkin shifts where some unconstrained symbols are added to the alphabets of a Dyck shift, the zeta function was determined by Inoue in [11]. In [15], Krieger and Matsumoto obtained an expression for the zeta function of a Markov-Dyck shift by applying a formula of Keller and using a clever encoding of periodic sequences of the shift.

We give an expression of the zeta function of a sofic-Dyck which has a deterministic presentation by combining techniques used to compute the zeta function of a (non Dyck) sofic shift and of a Markov-Dyck shift. We implicitly use the fact that the intersection of two visibly pushdown languages is a visibly pushdown language. With this result, sofic-Dyck shifts constitute now the largest class of shifts of sequences for which a general formula of the zeta function can be obtained.

The paper is organized as follows. A quick background on subshifts is given in Section 2.1. The notion of Dyck automata and sofic-Dyck shift is introduced in Section 2.2. In Section 3, we prove that the class of sofic-Dyck shifts is the class of visibly pushdown shifts. The definition and the computation of the zeta function of a sofic-Dyck shift is done in Section 4. The computation of the formula is given on a simple example in Section 4.4. Most proofs are omitted in this conference article.

2   Sofic-Dyck Shifts

In the section we define the class of sofic-Dyck shifts. We start with basic notions of symbolic dynamics which can be found in [16].

2.1   Subshifts

Let A be a finite alphabet. The set of finite sequences or words over A is denoted by A^* and the set of nonempty finite sequences or words over A is denoted by A^+. More generally, if L is a set of words over an alphabet A, then L^* is the set of concatenations of words of L, the empty word included. A prefix u of a word v is called *strict* if u is distinct from v and from the empty word. The set of bi-infinite sequences over A is denoted by $A^{\mathbb{Z}}$.

Let F be a set of finite words over the alphabet A. We denote by X_F the set of bi-infinite sequences of $A^{\mathbb{Z}}$ avoiding each word of F. The set X_F is called a *shift* (or also a *subshift* since it is a shift included in the full shift $A^{\mathbb{Z}} = \mathsf{X}_\emptyset$). When F can be chosen finite (resp. regular), the shift X_F is called a *shift of finite type* (resp. *sofic*).

The set of finite factors of the bi-infinite sequences belonging to a shift X is denoted $\mathcal{B}(X)$, its elements being called *blocks* (or allowed blocks) of X.

Let L be a language of finite words over a finite alphabet A. The language is *extensible* if for any $u \in L$, there are letters $a, b \in A$ such that $aub \in L$. It is *factorial* if any factor of a word of the language belongs to the language.

If X is a subshift, $\mathcal{B}(X)$ is a factorial extensible language. Conversely, if L is a factorial extensible language, then the set $\mathcal{B}^{-1}(L)$ of bi-infinite sequences x such that any finite factor of x belongs to L is a subshift [16]. The shift $\mathcal{B}^{-1}(L)$ is called the subshift *defined* by the factorial extensible language L.

2.2   Dyck Automata and Sofic-Dyck Shifts

We consider here an alphabet A which is partitioned into three disjoint sets of letters, the set A_c of *call letters*, the set A_r of *return letters*, and the set A_i of *internal letters*.

A *(finite) Dyck automaton* over A is a pair $(\mathcal{A}, M)$ of an automaton (or a directed labelled graph) $\mathcal{A} = (Q, E, A)$ over A where Q is the finite set of states or vertices, $E \subset Q \times A \times Q$ is the set of edges, and with a set M of pairs of edges $((p, a, q), (r, b, s))$ such that $a \in A_c$ and $b \in A_r$. The edges labelled by call letters (resp. return, internal) letters are also called *call* (resp. *return, internal*) *edges* and are denoted by E_r (resp. E_c, E_i). The set M is called the set of *matched edges*. We define the *graph semigroup* S associated to $(\mathcal{A}, M)$ as the free semigroup generated by the set $E \cup \{x_{pq} \mid p, q \in Q\} \cup \{0\}$ quotiented by the following relations.

$$0s = s0 = 0 \qquad\qquad s \in S, \tag{1}$$

$$x_{pq} x_{qr} = x_{pr} \qquad\qquad p, q, r \in Q, \tag{2}$$

$$x_{pq} x_{rs} = 0 \qquad\qquad p, q, r, s \in Q, q \neq r, \tag{3}$$

$$(p, \ell, q) = x_{pq} \qquad\qquad p, q, \in Q, \ell \in A_i, \tag{4}$$

$$(p, a, q) x_{qr} (r, b, s) = x_{ps} \qquad ((p, a, q), (r, b, s)) \in M, \tag{5}$$

$$(p, a, q) x_{qr} (r, b, s) = 0 \qquad ((p, a, q), (r, b, s)) \in (E_c \times E_r) \setminus M, \tag{6}$$

$$(p, a, q)(r, b, s) = 0, \qquad q \neq r, a, b \in A, \tag{7}$$

$$x_{pp}(p, a, q) = (p, a, q) \qquad p, q \in Q, a \in A, \tag{8}$$

$$(p, a, q) x_{qq} = (p, a, q) \qquad p, q \in Q, a \in A, \tag{9}$$

$$x_{pq}(r, a, s) = 0 \qquad\qquad a \in A, q \neq r. \tag{10}$$

$$(r, a, s) x_{tu} = 0 \qquad\qquad a \in A, s \neq t. \tag{11}$$

Note that we will consider here only Dyck automata with a finite number of states and thus a Dyck automaton will always be finite in the sequel. The semigroup S is in general infinite.

Informally speaking, the element 0 represents a forbidden path in the automaton and the element x_{pq} stands as a surrogate for an allowed path from p to q whose label is well-matched, as if it were a "path of zero length" from p to q. For instance, Equation 5 expresses the fact that a path made of a path with a well-matched label and extended left and right by matched edges, is again a path with a well-matched label. Equation 6 expresses the fact that a path made of a path with a well-matched label and extended left and right respectively by a call and a return edges which are not matched, is forbidden. Note that a path made of consecutive internal edges is a path with a well-matched label.

If e is an edge of $\mathcal{A}$, we denote by $f(e)$ its image in the graph semigroup S. If π is a finite path of $\mathcal{A}$, we denote by $f(\pi)$ the product in S of the images of its consecutive edges in S. A finite path π of $\mathcal{A}$ such that $f(\pi) \neq 0$ is said to be an *admissible path* of $(\mathcal{A}, M)$. A finite word is *admissible* for $(\mathcal{A}, M)$ if it is the label of some admissible path of $(\mathcal{A}, M)$. A bi-infinite path is *admissible* if all its finite factors are admissible.

The *sofic-Dyck shift presented* by $(\mathcal{A}, M)$ is the set of labels of bi-infinite admissible paths of $(\mathcal{A}, M)$ and $(\mathcal{A}, M)$ is called a *presentation* of the shift.

A finite word w labeling an admissible path π such that $f(\pi) = x_{pq}$ for some states p, q is called a *Dyck word* (or a *well-matched word*) of $(\mathcal{A}, M)$. Any Dyck word which has no Dyck word as strict prefix is called a *prime Dyck word.*

Note that an admissible word may not be a block of X since it is not always possible to extend it to a bi-infinite sequence whose all factors are admissible.

Lemma 1. *The sofic-Dyck shift accepted by a Dyck automaton $(\mathcal{A}, M)$ is exactly the set of bi-infinite sequences x such that each finite factor of x is an admissible word of $(\mathcal{A}, M)$.*

Proposition 1. *A sofic-Dyck shift is a subshift.*

Proof. Let X be a sofic-Dyck shift defined by an automaton $(\mathcal{A}, M)$ and its graph semigroup. Then $X = \mathsf{X}_F$ where F is the set of non admissible finite words for $(\mathcal{A}, M)$. Thus X is a subshift. □

The *full-Dyck shift* over the alphabet $A = (A_c, A_r, A_i)$, denoted X_A, is the set of all sequences accepted by the one-state Dyck automaton $\mathcal{A}$ containing a single state q and all loops (p, a, p) for $a \in A$, and where each edge (p, a, p) is matched with each edge (p, b, p) when $a \in A_c, b \in A_r$. Thus X_A is the set of all sequences over $A_r \cup A_c \cup A_i$.

Example 1. Consider the sofic-Dyck shift over A presented by the Dyck automaton $(\mathcal{A}, M)$ shown in the left part of Fig. 1 and with the following matched edges. The (-labelled edge is matched with the)-labelled edge and the [-labelled edge is matched with the]-labelled edge. This shift is called the *Motzkin shift.* A sequence is a block (or is allowed) if it is a factor of a well-parenthesized word, the internal letters being omitted. For instance "$(i[ii][i]((" $ is a block while the patterns "$(]$" or "$(ii]$" are forbidden. The block "$(i[i])$" is a prime Dyck word of the Motzkin shift while "$([])()$" is a Dyck word (or a well-matched block) which is not prime and "$()[[" $ is a block which is not well-matched.

Consider now the sofic-Dyck shift over A presented by the Dyck automaton $(\mathcal{B}, N)$ in the right part of Fig. 1 and with the following matched edges. The (-labelled edge is matched with the)-labelled edge and the [-labelled edge is matched with the]-labelled edge. We call this shift the *even-Motzkin shift.* The constraint is now stronger than the constraint described by the Dyck automaton $(\mathcal{A}, M)$. An even number of internal letters i is required between each pair of parenthesis symbols. The even-Motzkin shift is a subshift of the Motzkin shift and is not in any of the classes of nonsofic shifts mentioned earlier.

A Dyck automaton is *deterministic*[1] if there is at most one edge starting in a given state and with a given label. Sofic shifts (see [16]) have deterministic presentations. Although visibly pushdown languages are accepted by deterministic visibly pushdown automata [2], sofic-Dyck shifts may not be presented by any deterministic Dyck automaton.

---

[1] Deterministic presentations are also called *right-resolving* in [16].

Fig. 1. On the left, a presentation $(\mathcal{A}, M)$ of the Motzkin shift on the alphabet $A = (A_c, A_r, A_i)$ with $A_c = \{(, [\}, A_r = \{),]\}$ and $A_i = \{i\}$. The (-labelled edge is matched with the)-labelled edge and the [-labelled edge is matched with the]-labelled edge. On the right, a presentation $(\mathcal{B}, N)$ of the even-Motzkin shift on the same alphabet. Again the (-labelled edge is matched with the)-labelled edge and the [-labelled edge is matched with the]-labelled edge.

3   Characterization of Sofic-Dyck Shifts

In this section we give a characterization of sofic-Dyck shifts based on their sets of blocks (or allowed factors). We start with some background on visibly pushdown languages (see [2]).

3.1   Visibly Pushdown Languages

Visibly pushdown languages are unambiguous context-free languages of finite words accepted by visibly pushdown automata defined as follows.

Let A be an alphabet partitioned into three disjoint sets of call symbols, return symbols, and internal symbols. A *visibly pushdown automaton* on finite words over $A = (A_c, A_r, A_i)$ is a tuple $M = (Q, I, \Gamma, \Delta, F)$ where Q is a finite set of states, $I \subseteq Q$ is a set of initial states, Γ is a finite stack alphabet that contains a special bottom-of-stack symbol $\perp$, $\Delta \subseteq (Q \times A_c \times Q \times (\Gamma \setminus \{\perp\})) \cup (Q \times A_r \times \Gamma \times Q) \cup (Q \times A_i \times Q)$, and $F \subseteq Q$ is a set of final states. A transition (p, a, q, γ), where $a \in A_c$ and $\gamma \neq \perp$, is a push-transition. On reading a, the stack symbol γ is pushed onto the stack and the control changes from state p to q. A transition (p, a, γ, q) is a pop-transition. The symbol γ is read from the top of the stack and popped. If $\gamma = \perp$, the symbol is read but not popped. A transition (p, a, q) is a local action.

A stack is a nonempty finite sequence over Γ ending in $\perp$. A *run* of M labelled by $w = a_1 .. a_k$ is a sequence $(p_0, \sigma_0) \cdots (p_k, \sigma_k)$ where $p_i \in Q$, $\sigma_j \in (\Gamma \setminus \{\perp\})^* \perp$ for $0 \leq j \leq k$ and such that, for $1 \leq i \leq k$:

- If $a_i \in A_c$, then there are $\gamma_i \in \Gamma$ and $(p_{i-1}, a_i, p_i, \gamma_i) \in \Delta$ with $\sigma_i = \gamma_i \cdot \sigma_{i-1}$.
- If $a_i \in A_r$, then there are $\gamma_i \in \Gamma$ and $(p_{i-1}, a_i, \gamma_i, p_i) \in \Delta$ with either $\gamma_i \neq \perp$ and $\sigma_{i-1} = \gamma_i \cdot \sigma_i$ or $\gamma_i = \perp$ and $\sigma_i = \sigma_{i-1} = \perp$.
- If $a_i \in A_i$, then $(p_{i-1}, a_i, p_i) \in \Delta$ and $\sigma_i = \sigma_{i-1}$.

A run is *accepting* if $p_0 \in I$, $\sigma_0 = \perp$, and the last state is final, *i.e.* $p_k \in F$. A word over A is *accepted* if it is the label of an accepting run. Visibly pushdown languages have also the following grammar-based characterization (see [2]). They are accepted by *visibly pushdown grammar*.

A context-free grammar $G = (V, S, P)$ over A is a *visibly pushdown grammar* with respect to the partitioning $A = (A_c, A_r, A_i)$ if the set V of variables is partitioned into two disjoint sets V^0 and V^1, such that all the the productions in P are of one of the following forms

- $X \to \varepsilon$;
- $X \to aY$, such that if $X \in V^0$, then $a \in A_i$ and $Y \in V^0$;
- $X \to aYbZ$, such that $a \in A_c$, $b \in A_r$, $Y \in V^0$, and if $X \in V^0$, then $Z \in V^0$.

The variables in V^0 derive only well-matched words (*i.e.* where there is a one-to-one correspondence between symbols a and matching b).

3.2 Visibly Pushdown Shifts

In this section we show that the class of sofic-Dyck shifts is the class of *visibly pushdown shifts*, *i.e.* the class of subshifts X_F such that F is a visibly pushdown language.

Proposition 2. *The language of finite admissible words of a Dyck automaton is a visibly pushdown language.*

Proof. Let $(\mathcal{A} = (Q, E, A), M)$ be a Dyck automaton over A. We define a visibly pushdown automaton $V = (Q, I, \Gamma, \Delta, F)$ over A, where $I = F = Q$ and Γ is the set of edges of $\mathcal{A}$. The set of transitions Δ is obtained as follows.

- If $(p, a, q) \in E$ with $a \in A_c$, then $(p, a, q, (p, a, q)) \in \Delta$.
- If $(p, a, q) \in E$ with $a \in A_r$, then $(p, a, \gamma, q) \in \Delta$ for each call edge γ which is matched with the return edge (p, a, q).
- If $(p, a, q) \in E$ with $a \in A_i$, then $(p, a, q) \in \Delta$.

Let w be a finite word over A. There is a run $(p_0, \sigma_0) \cdots (p_k, \sigma_k)$ in V labelled by w such that $\sigma_0 = \bot$ if and only if w is the label of a path π of $(\mathcal{A}, M)$ such that $f(\pi) \neq 0$. Thus w is the label of an admissible path of $(\mathcal{A}, M)$ if and only if it is the label of an accepting run of V, which proves the proposition.   □

Let L be a language of finite words over A. We denote by $\mathcal{E}(L)$ the set of words $w \in L$ where, for any integer n, there are words u, v of length greater than n such that $uwv \in L$. Note that $\mathcal{E}(L)$ is a factorial language. This set is called in [8] the *bi-extensible* subset of L.

In order to prove that the set of blocks of a sofic-Dyck shift accepted by $(\mathcal{A}, M)$ is a visibly pushdown language, we have to prove that the bi-extensible subset of a factorial visibly pushdown language is also a factorial visibly pushdown language. It is shown in [8] that it is not true that the bi-extensible subset of a context-free language is a context-free language but the result holds for factorial context-free language. We have a similar result for factorial visibly pushdown languages which allows one to get the following characterization of the class of sofic-Dyck shifts.

Theorem 1. *The set of blocks of a sofic-Dyck shift is a visibly pushdown language. Conversely, the subshift defined by a factorial extensible visibly pushdown language is a sofic-Dyck shift.*

Since visibly pushdwon languages are closed by complementation, sofic-Dyck shifts are the class of shifts X_F such that F is a visisbly pushdown language.

4   Zeta Functions of Sofic-Dyck Shifts

Zeta functions count the periodic orbits of subshifts and constitute stronger invariants by conjugacies than the entropy (see [16] for these notions).

In this section, we give an expression of the zeta function of a sofic-Dyck shift which extends the formula obtained by Krieger and Matsumoto in [15] for Markov-Dyck shifts.

4.1   Definitions and General Formula

Every shift space is invariant by the *shift transformation* σ defined by $\sigma((x_i)_{i\in\mathbb{Z}}) = (x_{i+1})_{i\in\mathbb{Z}}$. The *zeta function* $\zeta_X(z)$ of the shift X is defined as the zeta function of its set of periodic patterns, *i.e.*

$$\zeta_X(z) = \exp \sum_{n \geq 1} p_n \frac{z^n}{n},$$

where p_n the number of sequences of X of period n, *i.e.* of sequences x such that $\sigma^n(x) = x$. This definition is extended to σ-invariant sets of bi-infinite sequences which may not be shifts (*i.e.* which may not be closed subsets of sequences).

Let X be a sofic-Dyck shift presented by a deterministic Dyck automaton $(\mathcal{A}, M)$ over A, where Q is the set of states.

We define the following matrices.

- $C = (C_{pq})_{p,q \in Q}$ where C_{pq} is the set of prime Dyck words labeling an admissible path of $(\mathcal{A}, M)$ going from p to q.
- $M_c = (M_{c,pq})_{p,q \in Q}$ (resp. $M_r = (M_{r,pq})_{p,q \in Q}$) where $M_{c,pq}$ (resp. $M_{r,pq}$) is the set of labels of paths going from p to q and made only of consecutive edges of $\mathcal{A}$ labelled by call letters (resp. return letters).
- $C_c = C M_c^*$ (resp. $C_r = M_r^* C$) .

In the sequel H will denote one of the matrices C, C_c, M_c, C_r, M_r. We denote by X_H the σ-invariant set containing all orbits (shifted sequences) of sequences $x \in A^{\mathbb{Z}}$ which are labels of bi-infinite paths $(p_i, w_i, p_{i+1})_{i\in\mathbb{Z}}$, where $w_i \in H_{p_i p_{i+1}}$.

We denote by $H(z)$ the matrix $(H_{pq}(z))_{p,q \in Q}$ where $H_{pq}(z)$ is the ordinary generating series of H_{pq}, *i.e.* $H_{pq}(z) = \sum_{n \geq 0} \mathrm{card}(H_{pq} \cap A^n) z^n$. Whenever $u \in H_{pq} \cap H_{pq'}$, $v \in H_{qr} \cap H_{q'r}$, one has $q = q'$. This unambiguous property comes from the determinism of the Dyck automaton $\mathcal{A}$. As a consequence, for any nonnegative integer k, the series $H_{pq}^k(z)$ is the ordinary generating series of the language of words $w_0 \cdots w_{k-1}$ where $w_i \in H_{p_i p_{i+1}}$ for $0 \le i \le (k-1)$, $p_0 = p$ and $p_k = q$.

We say that the matrix $(H_{pq})_{p,q \in Q}$, where each H_{pq} is a set of nonempty words over A is *circular* if for all $n, m \ge 1$ and $x_1 \in H_{p_0,p_1}, x_2 \in H_{p_1,p_2}, \ldots, x_n \in H_{p_{n-1} p_0}$, $y_1 \in H_{q_0,q_1}, y_2 \in H_{q_1,q_2}, \ldots, y_m \in H_{q_{m-1} q_0}$ and $p \in A^*$ and $s \in A^+$, the equalities

$$sx_2x_3 \cdots x_n p = y_1 y_2 \cdots y_m, \tag{12}$$

$$x_1 = ps \tag{13}$$

imply $n = m$, $p = \varepsilon$ and $x_i = y_i$ for $1 \le i \le n$.

This notion extends the classical notion of circular codes (see for instance [5]).

Proposition 3. *Let $(\mathcal{A}, M)$ be a deterministic Dyck automaton. The matrices C, C_c and C_r defined from $(\mathcal{A}, M)$ are circular matrices.*

In order to count periodic sequences of sofic-Dyck shifts, we need some machinery similar to the one used to count periodic sequences of sofic shifts (see for instance [16]).

Let $(\mathcal{A}, M)$ be a deterministic Dyck automaton where $\mathcal{A} = (Q, E)$. Let ℓ be a positive integer. We fix an ordering on the states Q. We define the Dyck automaton $(\mathcal{A}_{\otimes \ell}, M_{\otimes \ell})$ over a new alphabet A' where $\mathcal{A}_{\otimes \ell} = (Q_{\otimes \ell}, E_{\otimes \ell})$ as follows.

- We set $A' = (A'_c, A'_r, A'_i)$ with $A'_c = A_c \cup \{-a \mid a \in A_c\}$, $A'_r = A_r \cup \{-a \mid a \in A_r\}$, and $A'_i = A_i \cup \{-a \mid a \in A_i\}$.
- We denote by $Q_{\otimes \ell}$ the set of ordered ℓ-uples of distinct states of Q.
- Let $P = (p_1, \ldots, p_\ell)$, $R = (r_1, \ldots, r_\ell)$, be two elements of $Q_{\otimes \ell}$. Thus $p_1 < \cdots < p_\ell$ and $r_1 < \cdots < r_\ell$. There is an edge labelled by a from P to R in $\mathcal{A}_{\otimes \ell}$ if and only if there are edges labelled by a from p_i to p'_i for $1 \le i \le \ell$ and R is an even permutation of $(p'_1, \ldots, p'_\ell)$. If the permutation is odd we assign the label $-a$. Otherwise, there is no edge with label a or $-a$ from P to R.
- We define $M_{\otimes \ell}$ as the set of pairs of edges $((p_1, \ldots, p_\ell), a, (p'_1, \ldots, p'_\ell))$, and $((r_1, \ldots, r_\ell), \pm b, (r'_1, \ldots, r'_\ell))$ of $\mathcal{A}_{\otimes \ell}$ such that each edge (p_i, a, p'_i) is matched with (r_i, b, r'_i) for $1 \le i \le \ell$.

We say that a path of $\mathcal{A}_{\otimes \ell}$ is *admissible* if it is admissible when the signs of the labels are omitted, the sign of the label of a path being the product of the signs of the labels of the edges of the path.

We denote by $C_{\otimes \ell, PP'}$ the set of signed prime Dyck words c labeling an admissible path in $\mathcal{A}_{\otimes \ell}$ from P to P'. We denote by $C_{\otimes \ell}$ the matrix $(C_{\otimes \ell, PP'})_{P,P' \in Q_{\otimes \ell}}$ whose coefficients are sums of signed words of A^+. More generally, if H denotes one of the matrices C, C_c, M_c, C_r, M_r defined from $(\mathcal{A}, M)$ in this section, we denote by $H_{\otimes \ell}$ the matrix defined from $(\mathcal{A}_{\otimes \ell}, M_{\otimes \ell})$ similarly.

4.2   Zeta Functions of X_H

Let H be one of the matrices C, C_c, C_r, M_c, M_r. We denote by X_H the σ-invariant subset of X containing all orbits of sequences labels of bi-infinite paths $(p_i, w_i, p_{i+1})_{i \in \mathbb{Z}}$, where w_i is a prime Dyck word in $H_{p_i p_{i+1}}$.

Denoting by p_n the number of sequences of X_H of period n, the zeta function of the invariant set X_H is defined by

$$\zeta_{X_H}(z) = \exp \sum_{n>0} \frac{p_n}{n} z^n.$$

The zeta function of X_H has the following formula.

Proposition 4. *The zeta function of X_H, where H is one of the matrices C, C_c, M_c, C_r, M_r is*

$$\zeta_{X_H}(z) = \prod_{\ell=1}^{|Q|} \det(I - H_{\otimes \ell}(z))^{(-1)^\ell}.$$

4.3   Zeta Function of X

The following proposition is an extension of a similar result of Krieger and Matsumoto [15] for Markov-Dyck shifts.

Proposition 5. *The zeta function $\zeta_X(z)$ of a sofic-Dyck shift satisfies*

$$\zeta_X(z) = \frac{\zeta_{X_{C_c}}(z)\zeta_{X_{C_r}}(z)\zeta_{X_{M_c}}(z)\zeta_{X_{M_r}}(z)}{\zeta_{X_C}(z)}, \tag{14}$$

where X_C, X_{C_c}, X_{C_r}, X_{M_c}, X_{M_r} are the σ-invariants subsets of subshifts defined above.

It is based on the following proposition realizing an encoding of the periodic patterns of X.

Proposition 6. *Let $p(Y)$ denotes the periodic points of a σ-invariant set Y. We have $p(X) = p(X_{M_c}) \dot{\cup} p(X_{M_r}) \dot{\cup} (p(X_{C_c}) \cup p(X_{C_r}))$ and $p(X_{C_c}) \cap p(X_{C_r}) = p(X_C)$, where $\dot{\cup}$ denotes a disjoint union.*

The previous computations allow us to obtain the following general formula for the zeta function of X.

Theorem 2. *The zeta function of a sofic-Dyck shift accepted by a deterministic Dyck automaton $(\mathcal{A}, M)$ is given by the following formula.*

$$\zeta_X(z) = \frac{\zeta_{X_{C_c}}(z)\zeta_{X_{C_r}}(z)\zeta_{Z_c}(z)\zeta_{Z_r}(z)}{\zeta_{X_C}(z)}$$

$$= \prod_{\ell=1}^{|Q|} \det(I - C_{c,\otimes\ell}(z))^{(-1)^\ell} \det(I - C_{r,\otimes\ell}(z))^{(-1)^\ell}$$

$$\det(I - C_{\otimes\ell}(z))^{(-1)^\ell+1} \det(I - M_{r,\otimes\ell}(z))^{(-1)^\ell} \det(I - M_{c,\otimes\ell}(z))^{(-1)^\ell}.$$

Corollary 1. *The zeta function of a sofic-Dyck shift with a deterministic presentation is $\mathbb{Z}$-algebraic.*

Although sofic-Dyck shifts may not have deterministic presentations in general, it is shown in [3] that one can overcome this difficulty with reduced presentations so that the above formula holds for all sofic-Dyck shifts. It is also proved in [3] that for the subclass of sofic-Dyck shifts called finite-type-Dyck shifts, the zeta function is an $\mathbb{N}$-algebraic function, *i.e.* that it is the generating series of an unambiguous context-free language. We conjecture that the result also holds for sofic-Dyck shifts.

4.4 Example

Let X be the even-Motzkin shift of Example 1. It is presented by the Dyck automaton $(\mathcal{A}, M)$ pictured on the left part of Figure 2. The Dyck automaton $\mathcal{A}_{\otimes 1}$ is the same as $\mathcal{A}$. The Dyck automaton $\mathcal{A}_{\otimes 2}$ is pictured on the right part of Figure 2. Let us first compute the zeta function of X_C for this automaton.

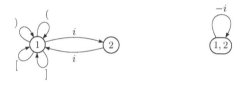

Fig. 2. The Dyck automaton $(\mathcal{A}, M)$ over $A = (\{(, [\}, \{),]\}, \{i\})$ (on the left) and the Dyck automaton $\mathcal{A}_{\otimes 2}$ on the right

We have $C = \begin{bmatrix} C_{11} & C_{12} \\ C_{21} & C_{22} \end{bmatrix}$, $C_{\otimes 2} = \begin{bmatrix} C_{(1,2),(1,2)} \end{bmatrix}$, with $C_{11} = (\, D_{11} \,) + [\, D_{11} \,]$, $C_{22} = 0, C_{12} = i, C_{21} = i$, and $D_{11} = (\, D_{11} \,)\, D_{11} + [\, D_{11} \,]\, D_{11} + ii D_{11} + \varepsilon$. We get

$$C_{11}(z) = 2z^2 D_{11}(z) = \frac{1 - z^2 - \sqrt{1 - 10z^2 + z^4}}{2}.$$

We have $C_{22}(z) = 0$, $C_{12}(z) = C_{21}(z) = z$. We also have $C_{(1,2),(1,2)} = -i$ and thus $C_{(1,2),(1,2)}(z) = -z$. We obtain

$$\zeta_{\mathsf{X}_C}(z) = \prod_{\ell=1}^{2} \det(I - C_{\otimes \ell}(z))^{(-1)^\ell} = \frac{1 + z}{1 - z^2 - \frac{1 - z^2 - \sqrt{1 - 10z^2 + z^4}}{2}}.$$

For $H = M_c, M_r$, we have $\prod_{\ell=1}^{2} \det(I - H_{\otimes \ell}(z))^{(-1)^\ell} = \frac{1}{1 - 2z}$. We also have

$$C_c = C M_c^* = \begin{bmatrix} C_{11}\{(, [\}^* \ i \\ i\{(, [\}^* \ 0 \end{bmatrix}, C_r = M_r^* C = \begin{bmatrix} \{),]\}^* C_{11} & \{),]\}^* i \\ i & 0 \end{bmatrix}.$$

$$\prod_{\ell=1}^{2} \det(I - C_{c,\otimes \ell}(z))^{(-1)^\ell} = \frac{(1 + z)(1 - 2z)}{1 - 2z - z^2 - \frac{1 - z^2 - \sqrt{1 - 10z^2 + z^4}}{2}}.$$

The same equality holds for C_r. We finally get

$$\zeta_X(z) = \frac{(1+z)(1-z^2 - \frac{1-z^2 - \sqrt{1-10z^2+z^4}}{2})}{(1-2z-z^2 - \frac{1-z^2-\sqrt{1-10z^2+z^4}}{2})^2}.$$

Acknowledgements. The authors would like to thank Arnaud Carayol and Wolfgang Krieger for helpful discussions.

References

1. Alur, R., Madhusudan, P.: Visibly pushdown languages. In: Proceedings of the 36th Annual ACM Symposium on Theory of Computing, pp. 202–211. ACM, New York (2004) (electronic)
2. Alur, R., Madhusudan, P.: Adding nesting structure to words. J. ACM 56(3) (2009)
3. Béal, M.-P., Blockelet, M., Dima, C.: Zeta function of finite-type Dyck shifts are ℕ-algebraic. In: Information Theory and Applications Workshop (2014)
4. Berstel, J., Boasson, L.: Balanced grammars and their languages. In: Brauer, W., Ehrig, H., Karhumäki, J., Salomaa, A. (eds.) Formal and Natural Computing. LNCS, vol. 2300, pp. 3–25. Springer, Heidelberg (2002)
5. Berstel, J., Perrin, D., Reutenauer, C.: Codes and automata. Encyclopedia of Mathematics and its Applications, vol. 129. Cambridge University Press, Cambridge (2010)
6. Bowen, R.: Symbolic dynamics. In: On axiom A Diffeomorphism. CBMS Reg. Conf. American Mathematical Society, vol. (35) (1978)
7. Bowen, R., Lanford, O.: Zeta functions of restrictions of the shift transformation. In: Proc. Sympos. Pure Math., vol. 14, pp. 43–50. American Mathematical Society (1970)
8. Culik II, K., Yu, S.: Cellular automata, $\omega\omega$-regular sets, and sofic systems. Discrete Appl. Math. 32(2), 85–101 (1991)
9. Ginsburg, S., Harrison, M.A.: Bracketed context-free languages. J. Comput. Syst. Sci. 1(1), 1–23 (1967)
10. Hamachi, T., Krieger, W.: On certain subshifts and their associated monoids. CoRR, abs/1202.5207 (2013)
11. Inoue, K.: The zeta function, periodic points and entropies of the Motzkin shift. CoRR, math/0602100 (2006)
12. Inoue, K., Krieger, W.: Subshifts from sofic shifts and Dyck shifts, zeta functions and topological entropy. CoRR, abs/1001 (2010)
13. Keller, G.: Circular codes, loop counting, and zeta-functions. J. Combin. Theory Ser. A 56(1), 75–83 (1991)
14. Krieger, W.: On subshift presentations. CoRR, abs/1209.2578 (2012)
15. Krieger, W., Matsumoto, K.: Zeta functions and topological entropy of the Markov-Dyck shifts. Münster J. Math. 4, 171–183 (2011)
16. Lind, D., Marcus, B.: An Introduction to Symbolic Dynamics and Coding. Cambridge University Press, Cambridge (1995)
17. Manning, A.: Axiom A diffeomorphisms have rationnal zeta fonctions. Bull. London Math. Soc. 3, 215–220 (1971)
18. McNaughton, R.: Parenthesis grammars. J. ACM 14(3), 490–500 (1967)
19. Reutenauer, C.: ℕ-rationality of zeta functions. Adv. in Appl. Math. 18(1), 1–17 (1997)

A Logical Characterization
of Timed (non-)Regular Languages*

Marcello M. Bersani[1], Matteo Rossi[1], and Pierluigi San Pietro[1,2]

[1] Dipartimento di Elettronica Informazione e Bioingegneria, Politecnico di Milano
[2] CNR IEIIT-MI, Milano, Italy
{marcellomaria.bersani,matteo.rossi,pierluigi.sanpietro}@polimi.it

Abstract. CLTLoc (Constraint LTL over clocks) is a quantifier-free extension of LTL allowing variables behaving like clocks over real numbers. CLTLoc is in PSPACE [9] and its satisfiability can polynomially be reduced to a SMT problem, allowing a feasible implementation of a decision procedure. We used CLTLoc to capture the semantics of metric temporal logics over continuous time, such as Metric Interval Temporal Logic (MITL), resulting in the first successful implementation of a tool for checking MITL satisfiability [7]. In this paper, we assess the expressive power of CLTLoc, by comparing it with various temporal formalisms over dense time. When interpreted over timed words, CLTLoc is equivalent to Timed Automata. We also define a monadic theory of orders, extending the one introduced by Kamp, which is expressively equivalent to CLTLoc. We investigate a decidable extension with an arithmetical next operator, which allows the expression of timed non-ω-regular languages.

1   Introduction

Linear Temporal Logic (LTL) is one of the most popular descriptive languages for modeling temporal behavior. Its time model is the structure $(\mathbb{N}, <)$, allowing the expression of positional orders of events, e.g., "if a query is received, then a reply will be delivered within 5 *positions* from now", but now allowing the formulation of real time constraints, that typically require a dense time domain. The absence of real time constitutes a major limitation of LTL, which has been addressed by adding variables, primitive operations or suitable modalities embedding real time, e.g., [19,12]. The reference model in this field is MTL (Metric Temporal Logic) [17,3], an extension of LTL that allows a temporal modality $\mathbf{U}_I$ (and $\mathbf{S}_I$), over a real time interval I. On a dense time domain, both satisfiability and model checking for MTL are undecidable [4], but various decidable fragments have been defined. MITL [2] restricts intervals I in $\mathbf{U}_I$ to be non punctual, e.g., a punctual eventuality such as $true\mathbf{U}_{[a,a]}\phi$ is not allowed. MITL is EXPSPACE-complete and it is closed under all Boolean operations. A smaller fragment of MTL, called QTL [15], is obtained by restricting

---

* Work supported by the Programme IDEAS-ERC, Project 227977-SMScom, and by PRIN Project 2010LYA9RH-006.

E. Csuhaj-Varjú et al. (Eds.): MFCS 2014, Part I, LNCS 8634, pp. 75–86, 2014.
© Springer-Verlag Berlin Heidelberg 2014

the temporal modalities to $\mathbf{U}_{(0,\infty)}$ and to $\mathbf{F}_{(0,1)}$, but it is actually equivalent to MITL. For many years, punctual eventualities were considered the main source of undecidability over dense time. In [10,20] this result has been revised: decidability has been shown not to solely depend on banning punctual intervals. For instance, Flat-MTL [10] and Safety MTL [20] allow both punctual eventualities and invariance properties, but they add some syntactical restrictions on the Until modality. They are not closed under negation, but their satisfiability is decidable. However, satisfiability of Safety-MTL formulae is non-elementary, while for Flat-MTL is EXPSPACE-complete. The dual version of Flat-MTL (i.e., consisting of the negation of formulae in Flat-MTL) is called coFlat-MTL: unfortunately, its satisfiability is undecidable (although model checking is EXPSPACE-complete).

The above research on MTL fragments, although also considered satisfiability, was mainly focused on devising logics that are suitable for model-checking of timed automata, while the interest in satisfiability has been quite limited. On the other hand, the need for full descriptive formalisms specifying reactive systems, and possibly with tractable complexity, is widely accepted [21].

Operational models, such as Büchi Automata (BA), are the most adopted and widely used alternative to temporal logic. Timed Automata (TA) [1] are the standard operational formalism for real time modeling. In [24], Büchi's famous result about the equivalence of Monadic Second-order Logic (MSO) and BA automata was extended to real time, by showing that the Monadic second-order logic $\mathcal{L}^d$ (augmenting MSO with a function measuring time between positions) is equivalent to TA. Because of its undecidability, MTL is not the proper logical formalism to capture TA, whose emptiness is decidable. Also, MTL does not embed explicit clocks, which are instead essential resources in TA: their absence makes the relationship between the expressiveness of clock constraints in TA and the syntactical restrictions of the various MTL fragments far from being evident.

In general, real time logics such as MTL are well suited to be interpreted with the *continuous-time semantics*, where atomic formulae are interpreted as state predicates, i.e., continuous flows or signals (i.e., mappings associating values in $\mathbb{R}_+$ with states). On the other hand, TA are naturally defined on the *pointwise semantics*, where atomic formulae are interpreted as instantaneous events associated with a timestamp, hence leading to an interpretation over timed words (sequences of timestamped events). TA can precisely be captured over the continuous-time semantics by various logic formalisms, such as quantified MITL [13] and Second-order real-time Sequential Calculus [2]. However, no temporal logic has so far been shown equivalent to TA in the pointwise semantics, where the construction of [2] cannot be applied.

In this paper, we bridge the gap between TA and temporal logic over the pointwise semantics. We consider Constraint LTL over clocks (CLTLoc), a quantifier-free extension of LTL that still considers discrete positions, but it has also a finite set of variables over a dense time domain, behaving like clocks of TA, to measure time elapsing among events occurring at discrete positions. Unlike MTL, clocks are explicit resources in CLTLoc and, as in TA, they can be compared with constants over $\mathbb{N}$ (or $\mathbb{Q}$). In [7], we prove that satisfiability of

CLTLoc is PSPACE-complete, by combining results on the decidability of CLTL [11],[5] over $\mathbb{R}$ with Region Graphs [1] capturing time behavior of variables. Moreover, the satisfiability of CLTLoc can be reduced to an instance of a SMT (satisfiability Modulo Theory) problem. A decision procedure was then easily devised and implemented (http://code.google.com/p/zot), by adopting SMT solvers such as Z3 [18]. CLTLoc has been successfully employed to reduce MITL over continuous semantics [6,8], allowing us to implement the first effective tool solving the satisfiability of MITL (http://code.google.com/p/qtlsolver).

In this paper, we prove the equivalence of CLTLoc and TA over timed sequences, and that CLTLoc is expressively complete with respect to an extension of the monadic first-order logic used by Kamp in [16], called Timed Monadic First-Order logic (T-MFO). This result extends the Kamp's equivalence between LTL and MFO to timed models. T-MFO is similar to logic $\mathcal{L}_T$ of [3], but it uses a restriction on atomic formulae and suitable *time behavior functions* to represent clocks of CLTLoc. As a consequence, as it is the case for LTL, CLTLoc with past modalities is equivalent to CLTLoc with only future operators. In analogy to TA, the number of clocks that are allowed in CLTLoc formulae determines the expressiveness of the language. We prove, in fact, that there is an infinite hierarchy of languages based on the number of clocks. An arithmetical "next" modality may also be allowed, e.g., to state formulae such as $Xx = y$, meaning that the value of clock x at the next position and the current value of clock y are equal. This modality allows the expression of properties of the length of intervals between two positions, e.g., timed non-regular behaviors where the period can be any real number, which cannot be defined by any temporal logic over dense time so far investigated. The paper is organized as follows: Sect. 2 introduces CLToc, MTL, MITL and TA; Sect. 3 shows the equivalence of CLTLoc and TA; Sect 4 defines the logic T-MFO. showing its equivalence with CLTLoc. Sect. 5 investigates CLTLoc extended with the next arithmetical modality.

2   Languages

Constraint LTL over clocks [7] (CLTLoc) is a semantic fragment of CLTL [11] where formulae are defined with respect to a finite set AP of atomic propositions, a finite set V of clocks and a pair $\mathcal{D} = (\mathbb{R}, \{<, =\})$.

Temporal terms α are defined by the syntax $\alpha := c \mid x \mid X\alpha$, where c is a constant in $\mathbb{N}$ and $x \in V$. Operator X only applies to temporal terms, with the meaning that $X\alpha$ is the *value* of temporal term α in the next position.

Formulae are defined as follows:

$$\phi := p \mid \alpha \sim \alpha \mid \phi \wedge \phi \mid \neg\phi \mid \mathbf{X}\phi \mid \mathbf{Y}\phi \mid \phi\mathbf{U}\phi \mid \phi\mathbf{S}\phi$$

where $\sim$ is a relation of $\{<, =\}$ and $\mathbf{X}$, $\mathbf{Y}$, $\mathbf{U}$ and $\mathbf{S}$ are the usual "next", "previous", "until" and "since" operators of LTL, with the same meaning.

For $n \in \mathbb{N}, n \geq 0$, let $\text{CLTLoc}(X^n)$ denote the class of CLTLoc formulae allowing atomic formulae of the form $X^h x \sim y$, $X^h x \sim c$, where x, y are clocks, c is constant and h is an integer with $0 \leq h \leq n$ and let CLTLoc^X denote the class of formulae defined as $\bigcup_{n\in\mathbb{N}}\text{CLTLoc}(X^n)$. Throughout the paper, we write CLTLoc instead of $\text{CLTLoc}(X^0)$. For convenience, one can still use in CLTLoc the operator X in formulae of the form $Xx \sim c$, as a shorthand for $\mathbf{X}(x \sim c)$. Sections 3 and 4 study the expressiveness of CLTLoc, while Section 5 deals with CLTLoc^X. The semantic definitions of CLTLoc^X in the remainder of this section also apply to its syntactic fragment CLTLoc.

The semantics of CLTLoc^X is defined with respect to a strict linear order $(\mathbb{N}, <)$ representing positions in time.

The valuation of clocks can be defined by a mapping $\sigma : \mathbb{N} \times V \to \mathbb{R}$, assigning, for every position $i \in \mathbb{N}$, a value $\sigma(i, x)$ to each clock $x \in V$. Intuitively, a clock x measures the time elapsed since the last time when $x = 0$, i.e., the last "reset" of x. To ensure that time progresses at the same rate for every clock, σ is called a *valuation* when satisfies the following condition: for every position $i \in \mathbb{N}$, there exists a "time delay" $\delta_i > 0$ such that for every clock $x \in V$:

$$\sigma(i+1, x) = \begin{cases} \sigma(i, x) + \delta_i, & \text{time progress} \\ 0 & \text{reset } x. \end{cases}$$

By definition of the sequence of δ_i, it follows that time progress is strongly monotonic and, moreover, resets in a valuation are represented by value 0, leading to a very simple definition of CLTLoc^X: there is no distinction between the action of resetting a clock x and of testing whether $x = 0$ in the current valuation. To consider monotonic time progress instead, i.e., allowing $\delta_i \geq 0$, CLTLoc^X must be enriched with a special operator to represent clock resets whose semantics is different from the one of tests $x \sim c$.

We assume that at every position there is at least one clock which is not reset: if this is not the case, just add a new clock *Now*, which is never reset. Hence, the time delay δ_i is uniquely defined in each position $i > 0$ as $\sigma(i + 1, Now) - \sigma(i, Now)$. The initial value of clocks, $\sigma(0, x)$ may be any non-negative value. When comparing CLTLoc with MTL and TA, other assumptions may be introduced to deal with some specific cases, e.g., requiring that clocks start from 0 at position 0 (which is obtained by syntactically imposing $x = 0$ at 0).

An interpretation for a CLTLoc^X formula ϕ is a pair (π, σ), where σ is a valuation and $\pi : \mathbb{N} \to \wp(AP)$ maps every position to a set of atomic propositions. The semantics of ϕ at position $i \in \mathbb{N}$ over (π, σ) is defined in Table 1. The only case requiring an explanation is the clause for interpreting $\alpha_1 \sim \alpha_2$. Given a temporal term α containing an occurrence of a variable x_α let the *depth* $|\alpha|$ of α be the total amount of temporal shift needed in evaluating α: $|x_\alpha| = 0$ and $|X\alpha| = |\alpha| + 1$. The value $\sigma(i, \alpha)$ is then defined as: $\sigma(i, \alpha) = \sigma(i + |\alpha|, x_\alpha)$. If α has an occurrence of a constant c_α then $\sigma(i, \alpha) = c_\alpha$. Hence, we can always assume that X does not appear in front of a constant. A CLTLoc^X formula ϕ is *satisfiable* if $(\pi, \sigma), 0 \models \phi$, for some (π, σ); in this case, (π, σ) is called a *model* of ϕ, and we write $(\pi, \sigma) \models \phi$.

To compare the expressiveness of CLTLoc^X with other formalisms, we introduce the satisfiability of CLTLoc^X formulae over timed ω-words (or timed ω-sequences). A timed ω-word over $\wp(AP)$ is a pair (π, τ) where $\pi : \mathbb{N} \to \wp(AP)$

Table 1. Semantics of CLTLoc^X (propositional connectives are omitted)

$$(\pi, \sigma), i \models p \Leftrightarrow p \in \pi(i) \text{ for } p \in AP$$
$$(\pi, \sigma), i \models \alpha_1 \sim \alpha_2 \Leftrightarrow \sigma(i + |\alpha_1|, x_{\alpha_1}) \sim \sigma(i + |\alpha_2|, x_{\alpha_2})$$
$$(\pi, \sigma), i \models \mathbf{X}(\phi) \Leftrightarrow (\pi, \sigma), i + 1 \models \phi$$
$$(\pi, \sigma), i \models \mathbf{Y}(\phi) \Leftrightarrow (\pi, \sigma), i - 1 \models \phi \wedge i > 0$$
$$(\pi, \sigma), i \models \phi \mathbf{U} \psi \Leftrightarrow \exists j \geq i : (\pi, \sigma), j \models \psi \wedge \forall i \leq n < j, (\pi, \sigma), n \models \phi$$
$$(\pi, \sigma), i \models \phi \mathbf{S} \psi \Leftrightarrow \exists 0 \leq j \leq i : (\pi, \sigma), j \models \psi \wedge j < n \leq i, (\pi, \sigma), n \models \phi$$

and τ is a monotonic function $\tau : \mathbb{N} \to \mathbb{R}$ such that $\forall i \; \tau(i) < \tau(i + 1)$ (strong monotonicity). Without loss of generality, to simplify some of the proofs that follow, we depart slightly from the standard definition of timed words, considering the first position 0 as "special". Given a CLTLoc^X interpretation (π', σ), let τ and π be such that: $\tau(0) - 0$, $\pi(0) = \emptyset$ and for every $i \geq 0$

$$\tau(i + 1) = \sigma(i, Now), \pi(i + 1) = \pi'(i).$$

Then, (π, τ) is called the timed ω-word associated with (π', σ) and it is denoted by $[(\pi', \sigma)]$. A relation $\models$ can be defined for every timed ω-word (π, τ) and CLTLoc^X formula ϕ as follows. Let $(\pi, \tau) \models \phi$ hold if *there exists* an interpretation (π', σ) such that $(\pi', \sigma) \models \phi$ and $(\pi, \tau) = [(\pi', \sigma)]$. A CLTLoc^X formula ϕ is satisfiable *over timed ω-words*, if $(\pi, \tau) \models \phi$, for some (π, τ).

3   Equivalence of CLTLoc and Timed Automata

This section shows that the set of timed ω-words satisfying a CLTLoc formula is timed ω-regular, i.e., it is accepted by a Timed Automaton. In particular, CLTLoc is an extension of LTL capturing exactly timed ω-regular languages.

We recall the basic definitions of Timed Automata. Let X be a finite set of clocks with values in $\mathbb{R}$. $\Gamma(X)$ is the set of clock constraints over X of the form $x \sim c \mid \neg\gamma \mid \gamma \wedge \gamma$, where $\sim \in \{<, =\}$, $x \in X$ and $c \in \mathbb{N}$. A clock valuation is a function $v : X \to \mathbb{R}$. We write $v \models \gamma$ when the clock valuation satisfies γ. For $t \in \mathbb{R}$, $v + t$ denotes the clock valuation mapping clock x to value $v(x) + t$, i.e., $(v + t)(x) = v(x) + t$.

A Timed Automaton is a tuple $\mathcal{A} = (\Sigma, Q, T, q_0, B)$ where Σ is a finite alphabet, Q is a finite set of control states, $q_0 \in Q$ is the initial state, $B \subseteq Q$ is a subset of control states (corresponding to a Büchi condition) and $T \subseteq Q \times Q \times \Gamma(X) \times \Sigma \times 2^X$ is a set of transitions.

A transition has the form $q \xrightarrow{\gamma, a, S} q'$ where $q, q' \in Q$, γ is a clock constraint of $\Gamma(X)$, a is a symbol of Σ, and S is a set of clocks to be reset. Two transitions $q \xrightarrow{\gamma, a, S} q'$ and $p \xrightarrow{\gamma', b, P} p'$ of T are *consecutive* when $q' = p$. A (finite or infinite) sequence of consecutive transitions in T is a *path* in $\mathcal{A}$. A pair (q, v), where $q \in Q$ and $v : X \to \mathbb{R}$ is a clock valuation, is a *configuration* of $\mathcal{A}$. A *run* ρ of $\mathcal{A}$ over a timed ω-word $(\pi, \tau) \in (\Sigma \times \mathbb{R})^\omega$ is an infinite sequence of configurations

$(q_0, v_0) \xrightarrow[\tau(1)]{\pi(1)} (q_1, v_1) \xrightarrow[\tau(2)]{\pi(2)} \ldots$, where $q_0 \in I$, $v_0(x) = 0$ for all $x \in X$, $v_i(x)$
is either 0 or $v_{i-1}(x) + \tau(i) - \tau(i-1)$ for all $x \in X$ and $i > 0$; moreover, the
sequence $q_0 \xrightarrow{\gamma_1, \pi(1), S_1} q_1 \xrightarrow{\gamma_2, \pi(2), S_2} q_2 \ldots$, such that $v_{i-1} + \tau(i) - \tau(i-1) \models \gamma_i$
and $x \in S_i$ iff $v_i(x) = 0$, must be a path of $\mathcal{A}$. Let $inf(\rho)$ be the set of control
states $q \in Q$ such that $q = q_i$ for infinitely many positions $i \geq 0$ in ρ. A run is
accepting when a Büchi condition holds, i.e., $inf(\rho) \cap B \neq \emptyset$.

Since we consider strictly monotonic time sequences τ, a transition with a
guard $x = 0$ can never be taken (hence, it can be replaced with *false*): such
a transition would be fired at i only when a transition at $i-1$ resets x, thus
entailing $\tau(i) = \tau(i-1)$, contradicting strict monotonicity. From now on, we
assume that guards of the form $x = 0$ are not allowed in a TA.

3.1 From Timed Automata to CLTLoc

Following a rather standard approach, we provide a CLTLoc formula which captures the semantics of a TA. To this end, we introduce a set of fresh clocks X_Q
representing the control states of $\mathcal{A}$. More precisely, a clock $c_q \in X_Q$ is associated
with each control state $q \in Q$; the value of c_q is 0 whenever $\mathcal{A}$ is in q, and it is left
to grow (i.e., $c_q > 0$) otherwise. Since in CLTLoc, unlike in TA, a clock cannot be
read and reset at the same time, following the approach of [7] for each $x \in X$ we
introduce two CLTLoc clocks, x_1 and x_2, which are alternately reset. In addition,
we introduce a third clock, x_{12}, which is used to keep track of whether $x_1 < x_2$
(x_{12} is 0 if, and only if, $x_1 < x_2$). We define a set of formulae whose conjunction $\phi_{\mathcal{A}}$
describes a given TA $\mathcal{A}$. The first formula (where $\mathbf{G}(\phi) = \neg(true\mathbf{U}\neg\phi)$) is globally
quantified and states that if x_1 is reset then it cannot be reset again, unless x_2 is
reset before it; in addition, $x_{12} = 0$ until x_2 is reset:

$$x_1 = 0 \Rightarrow \neg\mathbf{X}((x_2 > 0)\mathbf{U}(x_1 = 0)) \wedge x_{12} = 0 \wedge \mathbf{X}\left(\begin{array}{c} \mathbf{G}(x_{12} = 0 \wedge x_2 > 0) \vee \\ (x_{12} = 0)\mathbf{U}(x_2 = 0) \end{array}\right).$$

A symmetrical formula is defined also for x_2, but evaluated at position 1 rather
than at the origin. A clock constraint $x \sim c$ for a clock x of $\mathcal{A}$ is expressed by
the formula $(x_{12} = 0 \wedge x_1 \sim c) \vee (x_{12} > 0 \wedge x_2 \sim c)$. For all $q \in Q$, the following
formula translates the transition relation of $\mathcal{A}$:

$$\mathbf{G}(c_q = 0 \Rightarrow \bigvee_{q \xrightarrow{\gamma, a, S} q' \in T} \mathbf{X}(a \wedge c_{q'} = 0 \wedge \phi_\gamma \wedge \phi_S)) \tag{1}$$

where ϕ_γ is the CLTLoc formula that captures the clock constraint γ and ϕ_S is
the conjunction of formulae of the form $x_1 = 0 \vee x_2 = 0$ for each $x \in S$. The
first transition from the initial state must be dealt with separately, because at
that position all clocks in the TA are set to 0, by the following formula, to be
evaluated at the initial position:

$$\bigvee_{q_0 \xrightarrow{\gamma, a, S} q' \in T} a \wedge c_{q'} = 0 \wedge \phi_\gamma \wedge \phi_S. \tag{2}$$

To represent valid runs of $\mathcal{A}$, suitable formulae are introduced that guarantee the uniqueness of the control state and of the input symbol at each position i. The Büchi acceptance condition is obtained by enforcing that at least one final control state $q_j \in B$ is visited infinitely often. These formulae are rather trivial and are not shown here for brevity. Finally, let $\phi_{\mathcal{A}}$ be the conjunction of all previous CLTLoc formulae that capture the semantics of $\mathcal{A}$.

Theorem 1. *Let $\mathcal{A}$ be a TA with $k \geq 1$ clocks and $n \geq 1$ control states, and let (π, τ) be a timed word over alphabet Σ. Then, (π, τ) is accepted by $\mathcal{A}$ if, and only if, $(\pi, \tau) \models \phi_{\mathcal{A}}$. Moreover, ϕ_A has $3k + n$ clocks.*

3.2   From CLTLoc to Timed Automata

The timed automaton recognizing the language of timed words that are models of a given CLTLoc formula is easily obtained by exploiting the Vardi-Wolper construction [23] for LTL formulae. We take care of clock constraints that are handled as atomic formulae and, in particular, all formulae of the form $x = 0$ are converted into resets. Observe that, with the assumption of strictly monotonic time sequence, CLTLoc formulae $x = 0$ are equivalent to resets of TA.

Theorem 2. *Let ϕ be a CLTLoc formula with k clocks. Then, there exists a k-clock TA $\mathcal{A}_\phi$ recognizing the timed language defined by ϕ: for all timed words (π, τ) over alphabet $\wp(AP)$, $(\pi, \tau) \models \phi$ if, and only if, (π, τ) is recognized by $\mathcal{A}_\phi$.*

From the equivalence of CLTLoc and TA some results follow immediately. The first statement derives from the universality problem for TA.

Corollary 1. *The validity problem for CLTLoc is Π_1^1 complete.*

Let CLTLoc$_{\mathbf{X},\mathbf{U}}$ be the set CLTLoc formulae with no past operators $\mathbf{S}$ and $\mathbf{Y}$.

Corollary 2. *CLTLoc is equivalent to CLTLoc$_{\mathbf{X},\mathbf{U}}$.*

The number of clocks plays a crucial role for the expressiveness of TA. In fact, timed regular languages can be arranged in a strict hierarchy, determined by the minimum number of clocks necessary for accepting a given language.

Theorem 3. *[14] For all $k \geq 0$, the class of timed languages accepted by TA with k clocks is strictly included in the class of timed languages accepted by TA with $k + 1$ clocks.*

Example 1. Consider the family of timed languages $\{L_k\}_{k>0}$ where L_k is the set of timed words over the alphabet $\{a\}$ of the form (π, τ), such that $\pi : \mathbb{N} \to \{\emptyset, \{a\}\}$, $\pi(i) = \{a\}$ for all $i > 0$ and there exist at least k distinct pairs (i, j), $0 < i < j \in \mathbb{N}$, such that $\tau(j) - \tau(i) = 1$ (i.e., there are at least k distinct pairs of a's at exactly distance 1). In [22], it was shown that every L_k is accepted by a TA with k clocks, but it cannot be accepted by any TA with $k - 1$ clocks.

The above hierarchy result can easily be extended to CLTLoc. Let CLTLoc^k be the set of CLTLoc formulae where at most k clocks occur, i.e., $|V| \leq k$.

Theorem 4. *For all integers $k > 0$, CLTLoc^k is strictly more expressive than CLTLoc^{k-1}.*

4 Timed Monadic First Order Logic of Orders

We define the logic T-MFO, an extension of the monadic first order logic of order (MFO) that Kamp [16] showed to be equivalent to LTL. We then prove that CLTLoc is expressively complete with respect to T-MFO. T-MFO has two kinds of elements: monadic predicates whose domain is $\mathbb{N}$, and monotonic unary functions $\mathbb{N} \to \mathbb{R}$ relating positions in $\mathbb{N}$ to timestamps in $\mathbb{R}$. Similarly to the logic $\mathcal{L}_T$ of [3], T-MFO includes a special function, denoted as $t : \mathbb{N} \to \mathbb{R}$, associating each discrete position with its absolute timestamp. For simplicity and without loss of generality, in this section we assume that, given a CLTLoc formula, all clocks that appear in it are reset in position 0.

$(\mathbb{N}, <)$ is the theory of discrete positions, whereas $(\mathbb{R}, <, =, +)$ is the structure where timestamps are evaluated. The elements of T-MFO are:

- a set AP of monadic predicates over the set $\mathbb{N}$ of discrete positions;
- relation $<$ and function $+1$ on discrete positions;
- a set T of unary functions $\mathbb{N} \to \mathbb{R}$ from discrete positions to timestamps; one of them is called t;
- relations $<, =$ and function $+1$ on timestamps.

In $(\mathbb{N}, <)$, the constant 0 and the successor function $+1$ can be defined by means of $<$ and first-order quantification, but we introduce them as primitive to overcome the syntactic restrictions introduced next. Let t_x and t_y be unary functions of T from discrete positions to timestamps. We restrict the atomic formulae on timestamps to be of the form $t_x(i) \sim t_y(j)+c$, where $\sim \in \{<, =\}$ and $+c$ $(c \in \mathbb{N})$ corresponds to the application c times of function $+1$. In addition, we impose that atomic formulae only have one free variable. As a consequence, the atomic formulae that can be written on timestamps have the form

$$\beta(i) := t_x(i + h) \sim t_y(i + k) + c$$

where either t_x or t_y may be t and h, k are constants in $\mathbb{N}$ and $\sim \in \{<, =\}$ (the case where both are t is straightforward, because β reduces to either true or false based on $\sim$, h and k).

Function t of T captures the passing of time, and is similar to function f used in [3]. Hence, the following constraint holds: $\forall i \; t(i + 1) > t(i)$. The functions of set T are intended to capture the timestamps when clocks are reset – more precisely, $t_x(i)$ is the last timestamp where clock x is reset. As a consequence, the functions obey the following constraints: $t_x(0) = t(0)$ and

$$\forall i \left((t_x(i + 1) = t_x(i) \lor t_x(i + 1) = t(i + 1)) \right).$$

Finally, formulae of T-MFO are defined by the following grammar (where $p \in AP$ and i, j are variables over $\mathbb{N}$):

$$\phi := p(i) \mid \beta(i) \mid i < j \mid \neg\phi \mid \phi \land \phi \mid \forall i \phi.$$

We consider only formulae of T-MFO that do not contain free individual variables. The semantics for T-MFO formulae is defined with respect to a structure

$\mathcal{M}^{\mathcal{I}} = (\mathbb{N}, <, \mathcal{I})$ (or simply $\mathcal{I}$) where interpretation $\mathcal{I}$ specifies the sets $p^{\mathcal{I}} \subseteq \mathbb{N}$ for each $p \in \Sigma$ and the behavior of all functions of T. The satisfaction relation $\models$ is defined in the standard way. A T-MFO formula ϕ is *satisfiable* if there is $\mathcal{I}$ such that $\mathcal{I}, 0 \models \psi$; in this case, we say that $\mathcal{I}$ is a *model* of ϕ.

Given an interpretation $\mathcal{I}$, define the corresponding timed word (π, τ) as:

- For all $p \in AP$, $p \in \pi(i)$ iff $i \in p^{\mathcal{I}}$.
- For all $i \in \mathbb{N}$, $\tau(i) = t(i)$.

Relation $\models$ can be extended to timed words. Let (π, τ) be a timed word and ϕ be a T-MSO formula. We write $(\pi, \tau) \models \phi$ if there exists an interpretation $\mathcal{I}$ that is a model for ϕ such that (π, τ) is obtained from $\mathcal{I}$.

4.1   From CLTLoc to T-MFO

Every CLTLoc formula ϕ can be translated into a T-MFO formula by introducing a monadic predicate $p(i)$ for each CLTLoc proposition p, and a function $t_x(i)$ for each clock x. The following definition of translation r, mapping CLTLoc formulae to T-MFO formulae, follows [3] and is defined inductively on the structure of the CLTLoc formula. First, we introduce mapping r_i, for $i \geq 0$, which is the same as F_i of [3] for $p \in AP$, $\neg$, $\wedge$, $\mathbf{X}$ and $\mathbf{U}$, plus the following:

$$r_i(x \sim c) = t(i) \sim t_x(i) + c.$$

Let ϕ be a CLTLoc formula. Then, $r_0(\phi)$ is the corresponding T-MFO formula.

Theorem 5. *Let ϕ be a CLTLoc formula and (π, σ) be a CLTLoc interpretation. If $(\pi, \sigma) \models \phi$ then there exists an interpretation $\mathcal{I}$ such that $\mathcal{I}, 0 \models r_0(\phi)$. Conversely, let $\mathcal{I}$ be an interpretation such that $\mathcal{I}, 0 \models r_0(\phi)$. Then, there is an interpretation (π, σ) such that $(\pi, \sigma) \models \phi$.*

4.2   From T-MFO to CLTLoc

To obtain the opposite equivalence, we again exploit Kamp's results proving the equivalence between MFO and LTL. To extend the result to T-MFO, we have to show how atomic formulae $\beta(i)$ can be translated into CLTLoc, since MFO does not have formulae in this form.

Theorem 6. *Let ϕ be a T-MFO formula. There exists a CLTLoc formula ϕ' such that, for all timed words (π, τ), $(\pi, \tau) \models \phi$ if, and only if, $(\pi, \tau) \models \phi'$.*

As a consequence of Theorem 6, formulae of the form $t(i) \sim t_y(i + h) + c$ and $t(i + h) \sim t_y(i) + c$ (i.e., where one of t_x, t_y in terms β is t) are enough to characterize timed ω-languages; in fact, by exploiting the equivalence with CLTLoc, one can always remove formulae β where neither t_x nor t_y is t.

Corollary 3. *Let ϕ be a T-MFO formula. Then, there is a T-MFO formula ϕ' without instances of formula $t_x(i + h) \sim t_y(i + k) + c$, with $t_x \neq t \neq t_y$, such that for each timed word (π, τ), it is $(\pi, \tau) \models \phi$ if, and only if, $(\pi, \tau) \models \phi'$.*

5   Timed Non-regular Languages

Many extensions of the class of TA have been proposed with the goal of increasing its expressiveness. For instance, [1] introduced *diagonal constraints*, i.e., of the form $x \sim y + c$, as guards of transitions. They proved, however, that this extension does not augment the expressiveness of TA, since the construction of the region graph can be generalized to consider diagonal constraints, by refining the equivalence relation $\simeq$ on clock valuations.

In CLTLoc one can also allow diagonal constraints, but as in the case of TA, they do not augment the expressiveness of the language; the occurrence of a formula $x \sim y + c$, with $c \in \mathbb{N}$ a constant in and x, y two clocks, can equivalently be rewritten in CLTLoc as $(x > 0 \wedge y > 0)\mathbf{S}(y = 0 \wedge x \sim c)$. In fact, since time progression is the same for both clocks x and y, the formula $x \sim y + c$ is satisfied at a position $i \in \mathbb{N}$ if there exists a position $j \leq i$ such that at j both $y = 0$ and $x \sim c$ hold, and from position $j + 1$ up to i, neither x nor y are reset. Therefore, $x > 0 \wedge y > 0$ holds in every position in the interval $[j + 1, i]$.

When considering CLTLoc^{X}, i.e., when temporal terms include also X, the expressive power increases. Notice that atomic formulae of the form $\mathrm{X}^n x \sim \mathrm{X}^m y$, for $m \leq n$, may be ignored since they can equivalently be rewritten as $\mathbf{X}^m(\mathrm{X}^{n-m} x \sim y)$.

Example 2. Let L be the set of timed ω-words over the alphabet $\{a\}$ such that a is periodical. Formally, an ω-word (π, τ) is in L if, and only if, $\pi : \mathbb{N} \to \{\{a\}\}$ and for all $i \in \mathbb{N}$, $\tau(i+2) - \tau(i+1) = \tau(i+1) - \tau(i)$. L is defined in CLTLoc(X) with two clocks x, y as: $y = 0 \wedge x > 0 \wedge \mathbf{G}(a \wedge \mathrm{X}y = x \wedge \mathrm{X}x = y)$. Condition $\mathrm{X}x = y$ states that for all $i \geq 0$ the value of x at position $i + 1$ is the same of y at position i. Similarly for $\mathrm{X}y = x$. The formula imposes that at position 0 $y = 0$ and x may assume any real value $\alpha > 0$. Therefore, for all $i \in \mathbb{N}$, if i is even then $\sigma(i, y) = 0, \sigma(i, x) = \alpha$, else $\sigma(i, y) = \alpha, \sigma(i, x) = 0$. Hence, $\sigma(i + 1, Now) - \sigma(i, Now) = \alpha$. It is now obvious that (π, σ) is a model of the formula if, and only if, $[(\pi, \sigma)]$ is a timed ω-word of L.

Theorem 7. *The language of Ex. 2 is not timed regular, but it is in CLTLoc(X).*

The following immediate property of operator X is crucial in showing that the timed non-regular language L of Example 2 is in CLTLoc(X).

For a clock $z \in V$, for a position $m \in \mathbb{N}$, let $R(z, m) = \max\{j \mid 0 \leq j \leq m \wedge \sigma(j, z) = 0\}$, i.e., the largest position between 0 and m where z is reset.

Statement 1. *Let $\sigma : \mathbb{N} \times V \to \mathbb{R}$ be a valuation such that all clocks are reset at least position 0. For $i > 0, x, y \in V$, let $R_x = R(x, i+n)$, and $R_y = R(y, i)$. Then, $\sigma(i, \mathrm{X}^n x) = \sigma(i, y)$ if, and only if, $R_y < R_x$ and:*

$$\sigma(R_x, Now) - \sigma(R_y, Now) = \sigma(i + n, Now) - \sigma(i, Now)$$

In fact, $\sigma(i, \mathrm{X}^n x) = \sigma(i + n, x) = \sigma(i + n, Now) - \sigma(R_x, Now)$, and $\sigma(y, i) = \sigma(i, Now) - \sigma(R_y, Now)$: it follows that $\sigma(i, \mathrm{X}^n x) = \sigma(i, y) \Leftrightarrow \sigma(i + n, Now) - \sigma(R_x, Now) = \sigma(i, Now) - \sigma(R_y, Now)$. Hence, a formula of the form $\mathrm{X}^n x = y$

compares the time distance of positions i and $i + n$ with the time distance of positions R_y and R_x (where y and x were last reset). A special case is when $R_x = i + n$ and $R_y = i$, which entails that $\sigma(i, X^n x) = \sigma(y, i) = 0$.

For all $k \geq 0$, let CLTLoc(X^n, k) be the class of CLTLoc(X^n) formulae with at most k clocks. The number of clocks induces an infinite hierarchy over CLTLoc(X^n, k):

Theorem 8. *For all $k, n \geq 1$, the class of languages in CLTLoc($X^n, k - 1$) is strictly included in the class of languages in CLTLoc(X^n, k).*

We notice that CLTLoc(X^n, k) $\not\subseteq$ CLTLoc($k + 1$) for $k \geq 2$, since the example of Theorem 7 can be defined with just two clocks, and the language L_k in the proof of Theorem 8 is the same of Ex. 1, which obviously is in CLTLoc^k. Therefore, operator X cannot be used to replace some of the clocks:

Corollary 4. *For all $k \geq 2$, the class of languages in CLTLoc($k + 1$) is incomparable with the class of languages in CLTLoc(X^n, k).*

The operator X^n also induces an infinite hierarchy:

Theorem 9. *For all $k, n \geq 1$, the class of languages in CLTLoc(X^{n-1}, k) is strictly included in the class of languages in CLTLoc(X^n, k).*

6 Conclusions

This paper studies the expressiveness of Constraint LTL over clocks (CLTLoc), whose main interest of CLTLoc is that its decidability procedure, based on SMT solvers, has actually been implemented, allowing the verification of real time logics such as MITL or QTL.

CLTLoc is equivalent to Timed Automata in the pointwise semantics and it is expressively complete with respect to an extension of Kamp's monadic first-order logic. Its family of languages is organized in an infinite hierarchy based on the number of clocks. When an arithmetical "next" operator is allowed, CLTLoc defines also timed non-ω-regular languages, while still being decidable.

References

1. Alur, R., Dill, D.L.: A theory of timed automata. Theoretical Computer Science 126(2), 183–235 (1994)
2. Alur, R., Feder, T., Henzinger, T.A.: The benefits of relaxing punctuality. Journal of the ACM 43(1), 116–146 (1996)
3. Alur, R., Henzinger, T.A.: Real-time logics: complexity and expressiveness. Information and Computation 104, 390–401 (1993)
4. Alur, R., Henzinger, T.A.: A really temporal logic. Journal of the ACM 41(1), 181–204 (1994)
5. Bersani, M.M., Frigeri, A., Rossi, M., San Pietro, P.: Completeness of the bounded satisfiability problem for constraint LTL. In: Delzanno, G., Potapov, I. (eds.) RP 2011. LNCS, vol. 6945, pp. 58–71. Springer, Heidelberg (2011)

6. Bersani, M.M., Rossi, M., San Pietro, P.: On the satisfiability of metric temporal logics over the reals. In: Proc. of the Int. Work. on Automated Verification of Critical Systems (AVOCS), pp. 1–15 (2013)

7. Bersani, M.M., Rossi, M., San Pietro, P.: A tool for deciding the satisfiability of continuous-time metric temporal logic. In: Proc. of TIME, pp. 99–106 (2013)

8. Bersani, M.M., Rossi, M., Pietro, P.S.: Deciding continuous-time metric temporal logic with counting modalities. In: Abdulla, P.A., Potapov, I. (eds.) RP 2013. LNCS, vol. 8169, pp. 70–82. Springer, Heidelberg (2013)

9. Bersani, M.M., Rossi, M., San Pietro, P.: Deciding the satisfiability of MITL specifications. In: Proc. of the Int. Symp. on Games, Automata, Logics and Formal Verification (GandALF), pp. 64–78 (2013)

10. Bouyer, P., Markey, N., Ouaknine, J., Worrell, J.B.: On expressiveness and complexity in real-time model checking. In: Aceto, L., Damgård, I., Goldberg, L.A., Halldórsson, M.M., Ingólfsdóttir, A., Walukiewicz, I. (eds.) ICALP 2008, Part II. LNCS, vol. 5126, pp. 124–135. Springer, Heidelberg (2008)

11. Demri, S., D'Souza, D.: An automata-theoretic approach to constraint LTL. Information and Computation 205(3), 380–415 (2007)

12. Harel, E., Lichtenstein, O., Pnueli, A.: Explicit clock temporal logic. In: LICS, pp. 402–413. IEEE Computer Society (1990)

13. Henzinger, T.A., Raskin, J.-F., Schobbens, P.-Y.: The regular real-time languages. In: Larsen, K.G., Skyum, S., Winskel, G. (eds.) ICALP 1998. LNCS, vol. 1443, pp. 580–591. Springer, Heidelberg (1998)

14. Henzinger, T.A., Kopke, P.W., Wong-Toi, H.: The expressive power of clocks. In: Fülöp, Z., Gécseg, F. (eds.) ICALP 1995. LNCS, vol. 944, pp. 417–428. Springer, Heidelberg (1995)

15. Hirshfeld, Y., Rabinovich, A.: Quantitative temporal logic. In: Flum, J., Rodríguez-Artalejo, M. (eds.) CSL 1999. LNCS, vol. 1683, pp. 172–187. Springer, Heidelberg (1999)

16. Kamp, J.A.W.: Tense Logic and the Theory of Linear Order. Ph.D. thesis, University of California at Los Angeles (1968)

17. Koymans, R.: Specifying real-time properties with metric temporal logic. Real-Time Systems 2(4), 255–299 (1990)

18. Microsoft Research: Z3: An efficient SMT solver, http://z3.codeplex.com

19. Ostroff, J.S.: Temporal Logic for Real Time Systems. John Wiley & Sons, Inc., New York (1989)

20. Ouaknine, J., Worrell, J.B.: Safety metric temporal logic is fully decidable. In: Hermanns, H., Palsberg, J. (eds.) TACAS 2006. LNCS, vol. 3920, pp. 411–425. Springer, Heidelberg (2006)

21. Pradella, M., Morzenti, A., San Pietro, P.: Bounded satisfiability checking of metric temporal logic specifications. ACM Transactions on Software Engineering and Methodology (TOSEM) 22(3) (2013)

22. Suman, P.V., Pandya, P.K.: An introduction to timed automata, ch. 4, pp. 111–146. World Scientific (2012)

23. Vardi, M.Y., Wolper, P.: An automata-theoretic approach to automatic program verification. In: Proc. of LICS, pp. 332–344 (1986)

24. Wilke, T.: Specifying timed state sequences in powerful decidable logics and timed automata (extended abstract). In: Langmaack, H., de Roever, W.-P., Vytopil, J. (eds.) FTRTFT 1994 and ProCoS 1994. LNCS, vol. 863, pp. 694–715. Springer, Heidelberg (1994)

Asymptotic Monadic Second-Order Logic

Achim Blumensath[1],[*], Olivier Carton[2], and Thomas Colcombet[3],[**]

[1] TU Darmstadt, Germany
`blumensath@mathematik.tu-darmstadt.de`
[2] Liafa, Université Paris Diderot-Paris 7, France
`olivier.carton@liafa.univ-paris-diderot.fr`
[3] Liafa, Université Paris Diderot-Paris 7, France
`thomas.colcombet@liafa.univ-paris-diderot.fr`

Abstract. In this paper we introduce so-called *asymptotic logics*, logics that are meant to reason about weights of elements in a model in a way inspired by topology. Our main subject of study is Asymptotic Monadic Second-Order Logic over infinite words. This is a logic talking about ω-words labelled by integers. It contains full monadic second-order logic and can express asymptotic properties of integers labellings.

We also introduce several variants of this logic and investigate their relationship to the logic MSO+U. In particular, we compare their expressive powers by studying the topological complexity of the different models. Finally, we introduce a certain kind of tiling problems that is equivalent to the satisfiability problem of the weak fragment of asymptotic monadic second-order logic, i.e., the restriction with quantification over finite sets only.

1 Introduction

In this paper we consider logics that are able to express asymptotic properties about structures whose elements are labelled by weights. We call such logics 'asymptotic logics'. In general, these logics refer to a structure $\mathfrak{A}$ together with a labelling function d, called the 'weight map', that maps elements or tuples of elements to non-negative reals. A typical example of such an object is a metric structure, i.e., a structure $\mathfrak{A}$ equipped with a distance map $d : \mathcal{U}_{\mathfrak{A}} \times \mathcal{U}_{\mathfrak{A}} \to [0, \infty)$. In general, we refer to such structures as 'weighted structures'.

We are interested in the formalisation of properties of asymptotic nature over weighted structures. Typical examples, in the case of a metric structure, are:

- Continuity of a function f:

$$(\forall x)(\forall \varepsilon > 0)(\exists \delta > 0)(\forall y)\big[d(x, y) < \delta \to d(f(x), f(y)) < \varepsilon\big] \,.$$

---

[*] Work partially supported by DFG grant BL 1127/2-1 and the European Union's Seventh Framework Programme (FP7/2007-2013) under grant agreement no 259454.
[**] Received funding from the European Union's Seventh Framework Programme (FP7/2007-2013) under grant agreement no 259454.

E. Csuhaj-Varjú et al. (Eds.): MFCS 2014, Part I, LNCS 8634, pp. 87–98, 2014.
© Springer-Verlag Berlin Heidelberg 2014

– Uniform continuity of f:

$$(\forall \varepsilon > 0)(\exists \delta > 0)(\forall x)(\forall y)\big[d(x, y) < \delta \to d(f(x), f(y)) < \varepsilon\big] .$$

– Cauchy convergence of a sequence $(a_i)_{i \in \mathbb{N}}$:

$$(\forall \varepsilon > 0)(\exists k)(\forall i, j > k)\big[d(a_i, a_j) < \varepsilon\big] .$$

– Density of a set Y: $(\forall x)(\forall \varepsilon > 0)(\exists y \in Y)\big[d(x, y) < \varepsilon\big] .$

Inspecting the syntax of these formulae, we note the following properties. First, there are two sorts: the objects that live in the universe of the structure, such as elements, series, functions, etc...., and the objects living in $\mathbb{R}$ that are used to refer to distances. The map d is the only way to relate these two sorts, and all tests in which elements of $\mathbb{R}$ are involved are comparisons with variables ε, δ. More interesting is the remark that if a variable, say ε, ranging over $\mathbb{R}^+$ is quantified universally, it is always used as an upper bound, i.e, positively in a test of the form $d(-) < \varepsilon$ (positively in the sense that an even number of negations separate the quantifier from its use). Dually, if it is quantified existentially, it is always used as a lower bound, i.e., positively in a test of the form $d(-) \geq \varepsilon$. In particular, this is the case for the test $d(x, y) < \delta$ in the sentences expressing continuity and uniform continuity, since it occurs in the left hand-side of an implication.

This syntactic property witnessed in the above examples can be turned into a definition. An *asymptotic formula* is a formula in which it is possible to quantify over quantities $\exists \varepsilon, \forall \delta$, and the only way to use the map d is in tests of the form $d(-) \geq \varepsilon$ positively below $\exists \varepsilon$ and $d(-) < \delta$ positively below $\forall \delta$.

This restriction captures the intuition that variables ranging over $\mathbb{R}^+$ are always thought as 'tending to 0' or 'to be very small'. In other words, they are only used to state properties of a topological nature. Our objective is to understand the expressive power and the decidability status of logics to which we have added this asymptotic capability.

Link with Topological Logics. Of course, logics as described above are related to topological notions, and as such these logics are not very far from the topological logics as studied in the seventies and eighties. These were logics (variants of first-order logic) in which it is possible to quantify over open sets. There are several variants. Flum and Ziegler introduced a logic in which it is possible to quantify over open sets, but it is only allowed to test the membership in these sets under a positivity assumption with respect to the quantifier [11] (in a way very similar to our case). Rabin proved, as a consequence of the decidability of the theory of the infinite binary tree that the theory of the real line $(\mathbb{R}, <)$ with quantification over open sets is decidable [14]. On the other hand, Shelah and Gurevich showed that monadic formulas over Cantor space equipped with an 'is open' predicate is undecidable [12].

Our approach is slightly different. Our base object is not, as above, a topology of open sets, but a weight map d. Of course, if d is required to be a distance, it induces a topology. However, there is no such assumption in general (and d may even be non-binary). Nevertheless, we can consider the topology over the non-negative reals in which the open sets are the neighbourhoods of 0 (as well as $\emptyset$ of course). Then the quantifiers $\forall \varepsilon$, $\exists \delta$,... can be replaced by quantifiers ranging over open sets, and tests of the form $d(-) < \varepsilon$ by membership tests of $d(-)$ in an open set. Furthermore, these tests respect the positivity assumption as defined by Flum and Ziegler. However, this relationship of our logic with those from the literature does not seem to help with solving the questions raised in the present paper.

Monadic Second-Order Logic and Asymptotic Monadic Second-Order Logic. In this paper, we consider the asymptotic variant of monadic second-order logic, though certainly this notion of asymptoticity can be combined with other formalisms. Let us recall that monadic second-order logic is the extension of first-order logic by set quantifiers. There is a long history of works dealing with the decidability of monadic second-order logic over some classes of structures, the prominent examples being the results over ω by Büchi [7] and over the infinite binary tree by Rabin [14]. These results can be regarded as foundations for a theory of 'regular languages' of infinite words and trees. We are interested in knowing whether this logic can be 'made asymptotic' while keeping these strong decidability properties. We have good hopes that – at least some of – these results can be generalised to more general ones, in which monadic logic is extended with asymptotic capabilities.

Before continuing, let us formalise what is 'asymptotic monadic second-order logic' (AMSO for short). The first aspect is that weight maps range over the elements of the structure, and not tuples. This is a design choice, our goal being to concentrate our attention on the simplest situation. The second aspect is cosmetic: instead of considering quantities ranging over $\mathbb{R}^+$, we consider quantities ranging over $\mathbb{N}$. Essentially, this amounts for weights to exchange $d(-)$ with $\lceil 1/d(-) \rceil$ and, for quantifiers $\exists \varepsilon$, $\forall \delta$ ranging over $\mathbb{R}^+$, to exchange them for $\exists r, \forall s$ ranging over $\mathbb{N}$. As a consequence, existentially quantified numbers are used as upper bounds, while universally quantified ones are used as lower bounds. Hence, the syntax of 'asymptotic monadic second-order logic' is the one of MSO, extended by number quantifiers $\exists r, \forall s$ ranging over $\mathbb{N}$ and by predicates of the form $d(x) < r$ and $d(x) \geq s$, where x is a first-order variable, under the assumption that there is an even number of negations between the quantifier and the use.

Let us give some examples. The structure here is ω and $f : \omega \to \mathbb{N}$ is a weight map. The convention is that variables x, y, z range over elements of ω, upper case variables X, Y, Z over subsets of ω, and r, s over $\mathbb{N}$.

- f is bounded: $\exists s \forall x [f(x) \leq s]$.
- f tends to ∞: $\forall r \exists x (\forall y > x)[f(y) > r]$.
- f takes infinitely many values infinitely many times:
  $\forall r \exists s \forall x (\exists y > x)[r \leq f(y) < s]$.

The subject of this paper is to analyse the expressive power of AMSO, as well as its variants, and study its decidability status.

Link with MSO+U. A logic closely related to AMSO is MSO+U [1,3]. MSO+U is monadic second-order logic extended by a quantifier $UX\varphi(X)$ stating that 'there are arbitrarily large finite sets X satisfying $\varphi(X)$', i.e., $UX\varphi(X)$ is equivalent to $\forall s \exists X[|X| \geq s \wedge \varphi(X)]$. Thus, MSO+U can be regarded as a fragment of an asymptotic logic, if the weight map is chosen to be 'the cardinality map' that associates to each set its size (and, say, 0 for infinite sets).

So far, the precise decidability status of MSO+U is unknown. The most expressive decidable fragment over infinite words corresponds (essentially) to Boolean combination of formulas in which the U-quantifier occurs positively [3] (in fact a bit more). On the negative side, it is known that over infinite trees MSO+U is undecidable [4] under the set-theoretic assumption $V = L$. This proof is inspired from the undecidability proof of MSO over the real line by Shelah [15], and it is absolutely not adaptable to infinite words as such. Hence, there is a very large gap in our knowledge of the decidability of MSO+U. The case of the weak fragment of MSO+U, i.e., where set quantifiers do only range over finite sets, has been positively settled in [5,6] over infinite words and trees. In terms of the expressive power, this weak fragment still falls in the classes that are understood from [3].

In some sense, this paper can be seen as an attempt to better understand the logic MSO+U. This is also the subject of another branch of research: the theory of regular cost functions [8,10,9]. However, that approach concentrates on how to measure the cardinality of sets (the quantifier U involves such a computation), and does not give any asymptotic analysis of quantities.

Contributions of the Paper. In this paper, we study AMSO and some of its variants over infinite words. These variants are: BMSO in which number quantifiers are replaced by a boundedness predicate; EAMSO which extends AMSO with quantification over weight functions; and EBMSO that combines these two modifications. We also study the weak fragment WAMSO of AMSO, and its 'number prenex' fragment $AMSO^{np}$. The contributions are in several directions: expressive power, topological complexity, and decidability.

Concerning the expressive power we show that EAMSO is equivalent to EBMSO, AMSO is equivalent to BMSO, and WAMSO is equivalent to $AMSO^{np}$. All other pairs of logics can be separated. However, more interestingly, we can show that as far as the decidability of satisfiability is concerned, AMSO, BMSO, EAMSO, EBMSO and MSO+U are all equivalent, and WAMSO is equivalent to $AMSO^{np}$. We are hence confronted with only two levels of difficulty.

Concerning topological complexity, we perform an analysis in terms of descriptive set theory. We prove that AMSO reaches all levels of the projective hierarchy, while WAMSO reaches all finite levels of the Borel hierarchy. This separates the two classes. In particular, this shows that – as far as topological complexity is concerned – WAMSO is far simpler than AMSO, and at the same time far more complex than any variant of MSO known to be decidable (for instance the weak fragment of MSO+U remains at the third level of the Borel hierarchy).

On the decidability front, the case of MSO+U is notoriously open and difficult, and as explained above (in particular, it is known to be undecidable over infinite trees, though this gives no clue about the infinite word case). AMSO is not easier. In this paper, we advocate the importance of the weak fragment WAMSO as a logic of intermediate difficulty. Though we have to leave its decidability status open as well, we are able to disclose new forms of tiling problems that are equivalent to the decidability of the satisfiability of WAMSO. This provides a promising line of attack for understanding the decidability status of AMSO and MSO+U.

We believe that these numerous results perfectly describe how the asymptotic notions relate to other notions from the literature, the prominent one being MSO+U. In particular we address and answer the most important questions: expressive power, topological complexity, and – in some very preliminary form – decidability. We are finally convinced that the tiling problems that we introduce deserve to be studied on their own.

Structure of the Paper. In Section 2, asymptotic monadic second-order logic is introduced as well as several fragments. Some first results are proved: the weak fragment is introduced and it is shown to be equivalent to the number prenex-form of AMSO. The extended version of asymptotic monadic second-order logic (EAMSO) is introduced and its relation to MSO+U is established. Section 2.3 characterises our logics in terms of Borel complexity. In Section 3, we introduce certain tiling problems and we show their equivalence with the satisfiability problem for WAMSO.

2 Asymptotic Monadic Second-Order Logic and Variants

In this section, we quickly recall the definition of monadic second-order logic and we introduce the new asymptotic variant AMSO (which happens to be equivalent to another formalism, called BMSO, see below). We then introduce the weak fragment WAMSO, mention some of its basic properties. We conclude with a comparison of the expressive power of these logics.

We assume that the reader is familiar with the basic notions of logic. We consider relational structures $\mathfrak{A} = \langle \mathcal{U}, R_1, \ldots, R_k \rangle$ with universe $\mathcal{U}$ and relations $R_1, \ldots, R_k$. A *word* (finite or infinite) over the alphabet Σ is regarded as a structure whose universe is the set of positions and where the relations consist of the ordering $\leq$ of positions and unary relations a, for each $a \in \Sigma$, containing those positions carrying the letter a.

Monadic second-order logic (MSO) is the extension of first-order logic (FO) by *set variables* $X, Y, \ldots$ ranging over sets of positions, quantifiers $\exists X$, $\forall X$ over such variables, and membership tests $x \in Y$.

2.1 Weighted Structures and Asymptotic Monadic Second-Order Logic

The subject of this paper is *asymptotic monadic second-order logic*. This logic expresses properties of structures whose elements have a weight which is a natural number. Formally, a *weighted structure* is a pair $\langle \mathfrak{A}, \bar{f} \rangle$ consisting of a relational structure $\mathfrak{A}$ with universe $\mathcal{U}$ and a tuple of functions $f_i : \mathcal{U} \to \mathbb{N}$ called *weight functions*. A *weighted finite word* (*resp.* a *weighted ω-word*) corresponds to the case where $\mathfrak{A}$ is a finite word (*resp.*, an ω-word).

Asymptotic monadic second-order logic (AMSO) extends MSO with the following constructions:

- quantifiers over variables of a new type, *number variables* (written $r, s, t, \ldots$) that range over natural numbers, and
- atomic formulae $f(x) \leq r$ where x is a first-order variable and r a number variable. These formulae must appear positively inside the existential quantifier binding r, *i.e.*, the predicate and the quantifier are separated by an even number of negations. As a commodity of notation, the dual predicate $f(x) > r$ can be used positively below the universal quantifier $\forall r$.

Example 1. It is possible to express in AMSO that:

- the weights in a structure are bounded: $\exists r \forall x [f(x) \leq r]$,
- an ω-word has weights tending to infinity: $\forall s \exists x (\forall y > x)[f(y) > s]$,
- infinitely many weights occur infinitely often in a weighted ω-word:
  $\forall s \exists r \forall x (\exists y > x)[f(y) > s \wedge f(y) \leq r]$.

On the other hand, the formula $\forall r \exists x [f(x) \leq r]$ is ill-formed since it does not respect the positivity constraint separating the introduction of r and its use.

There is an alternative way to define this logic, in a spirit closer to MSO+U: the logic BMSO extends MSO with *boundedness predicates* of the form $f[X] < \infty$ where X is a set variable. Such a predicate holds if the function f restricted to the set X is bounded by some natural number. Hence $f[X] < \infty$ can be seen as a shorthand for the AMSO formula $\exists r (\forall x \in X)[f(x) \leq r]$. It follows that BMSO is a fragment of AMSO. In fact, both logics are equivalent, as shown by the following theorem.

Theorem 2. AMSO *and* BMSO *are effectively equivalent over all weighted structures.*

Finally, let us mention an important invariance of the logic AMSO. Two functions $f, g : \mathcal{U} \to \mathbb{N}$ are *equivalent*, noted $f \approx g$, if they are bounded over the same subsets of their domain (this is expressible in BMSO as $\forall X (f[X] < \infty \leftrightarrow g[X] < \infty)$). We extend this equivalence to weighted structures by $\langle \mathfrak{A}, \bar{f} \rangle \approx \langle \mathfrak{B}, \bar{g} \rangle$ if $\mathfrak{A}$ and $\mathfrak{B}$ are isomorphic, and $f_i \approx g_i$ for all i.

Proposition 3. $\approx$-*equivalent weighted structures have same* AMSO-*theory.*

This is obviously true for the logic BMSO, and hence also for AMSO according to Theorem 2. In fact, Proposition 3 also holds for the logic EAMSO introduced in Section 4 below and, more generally, for every logic that would respect syntactic constraints similar to AMSO in the use of weights. An immediate consequence is that there is no formula defining $f(x) = f(y)$ in AMSO or its variants. This rules out all classical arguments yielding undecidability in similar contexts of 'weighted logics'.

2.2   The Weak and the Number Prenex Fragments of AMSO

The use of quantifiers over infinite sets combined with number quantifiers induces intricate phenomena (the complexity analysis performed in Section 2.3 will make this obvious: AMSO reaches all levels of the projective hierarchy). There are two ways to avoid it. Either we allow only quantifiers over finite sets, thus obtaining WAMSO (which defines only Borel languages), or we prevent the nesting of monadic and number quantifiers by requiring all number quantifiers to be at the head of the formula, thereby obtaining the number-prenex fragment of AMSO, named AMSO^{np}. We will see that these two logics have the same expressive power. To avoid confusion, let us immediately point out that WAMSO and WMSO+U, the weak fragment of MSO+U, are very different logics. This is due to the fact that the syntax of AMSO is not the one of MSO+U, and as a consequence assuming the sets to be finite has dramatically different effects. In particular, we will see that WAMSO inhabits all finite levels of the Borel hierarchy, while it is known that WMSO+U is confined in the third level [13].

Weak asymptotic monadic second-order logic (WAMSO) is obtained by restricting set quantification to finite sets (the syntax remains the same). We will write $\exists^w X$ and $\forall^w X$ when we want to emphasize that the quantifiers are weak, *i.e.*, range over finite sets. Let us remark that, as usual, the weak logic is not strictly speaking a fragment of the full logic since, in general, AMSO is not able to express that a set is infinite. However, on models such as words, ω-words, or even infinite trees, the property of 'being finite' is expressible, even in MSO.

It turns out that, in a certain sense, weak quantifiers commute with number quantifiers.

Lemma 4. *There exists a WAMSO-formula $\psi(X, r)$ such that, for every sequence $\bar{Q}\bar{t}$ of number quantifiers, every WAMSO-formula $\bar{Q}\bar{t}\,\varphi(X, \bar{t})$, and all weighted ω-words w,*

$$w \models \exists^w X \bar{Q}\bar{t}\varphi(X, \bar{t}) \quad \text{iff} \quad w \models \exists r \bar{Q}\bar{t}\exists^w X \left[\varphi(X, \bar{t}) \wedge \psi(X, r)\right] .$$

By this lemma, it follows that we can transform every WAMSO-formula into *number-prenex form*, *i.e.*, into the form $\bar{Q}\bar{t}\varphi$, where $\bar{Q}\bar{t}$ is a sequence of number quantifiers while φ does not contain such quantifiers. However, this translation adds new number variables in the formula. The fragment of AMSO-formulae in number prenex form is denoted AMSO^{np}. For weak quantifiers, we obtain the logic WAMSO^{np} in the same way.

Theorem 5. *The logics* WAMSO, AMSO^{np} *and* WAMSO^{np} *effectively have the same expressive power over weighted ω-words.*

2.3   Separation Results

To separate the expressive power of the logics introduced so far, we employ topological arguments. One way to show that a logic is strictly more expressive than one of its fragments is to prove that it can define languages of a topological complexity the fragment cannot define. In our case we use the Borel hierarchy and the projective hierarchy to measure topological complexity.

Theorem 6. *Languages definable in* AMSO *strictly inhabit all levels of the projective hierarchy, and not more. Languages definable in* WAMSO *strictly inhabit all finite levels of the Borel hierarchy, and not more.*

This is proved using standard reduction techniques. We obtain the following picture:

$$
\overbrace{\mathrm{MSO} \;=\; \mathrm{WMSO}}^{\mathrm{Bool}(\boldsymbol{\Sigma}_2^0)} \;\subsetneq\; \overbrace{\mathrm{WAMSO} \;=\; \mathrm{AMSO}^{\mathrm{np}}}^{\text{all Borel levels of finite rank}} \;\subsetneq\; \overbrace{\mathrm{AMSO}}^{\text{all projective levels}} \quad .
$$

As an immediate consequence, we obtain the corollary that AMSO is strictly more expressive than WAMSO and AMSO^{np}, that AMSO^{np} is strictly more expressive than MSO, and that WAMSO is strictly more expressive than WMSO.

3   Weak Asymptotic Monadic Second-Order Logic and Tiling Problems

We have introduced in the previous sections several logics with quantitative capabilities. The analysis performed shows that WAMSO (or equivalently AMSO^{np}) offers a good compromise in difficulty in the quest for solving advanced logics like MSO+U. Indeed, in terms of Borel complexity, it is significantly simpler than other logics like AMSO and EAMSO, and hence MSO+U. Despite its relative simplicity, this logic is, still in terms of Borel complexity, significantly more complex than any other extensions of WMSO known to be decidable over infinite words, e.g., WMSO+U [1] and WMSO+R [2] [2,5]. Both of these logics can define Boolean combinations of languages at the third level of the Borel hierarchy.

In this section, we develop techniques for attacking the satisfiability problem of WAMSO over weighted ω-words, though we are not able to solve this problem itself. Our contribution in this direction is to reduce the satisfiability problem of WAMSO to a natural kind of tiling problem, new to our knowledge, the decidability of which is unknown, even in the simplest cases. As a teaser, let us show the simplest form of such tiling problems:

Open Problem 7. *Given two regular languages K and L over an alphabet Σ where K is closed under letter removal, can we decide whether, for every n, there exists a Σ-labelled picture of height n such that all rows belong to L and all columns to K?*

Note that this problem would clearly be undecidable if K was not required to be sub-word closed. In the remainder of the section, we first introduce these problems in a more general setting (a multidimensional version of it), and give the essential ideas explaining why the decidability of satisfiability for WAMSO reduces to such tiling problems.

---

[1] The weak fragment of MSO+U, where set quantifiers range over finite sets.

[2] An extension of WMSO with an unusual recurrence operator. Adding this operator to MSO yields a logic equivalent to MSO+U.

3.1   Lossy Tiling Problems

A *picture* $p : [h] \times [w] \to \Sigma$ is a rectangle labelled by a (fixed) finite alphabet Σ, where $h \in \mathbb{N}$ is the *height* and $w \in \mathbb{N}$ the *width* of the picture. For $0 \le i < w$, the ith *column* of the picture is the word $p(0, i)p(1, i)\ldots p(h - 1, i)$. A *band of height* m in a picture is obtained by erasing all but m-many rows from a picture. We regard bands of height m as words over the alphabet Σ^m. Formally, for $0 \le j_1 < j_2 < \cdots < j_m < h$, the *band for rows* $j_1, \ldots, j_m$ is the word $(p(j_1, 0), \ldots, p(j_m, 0))\ldots(p(j_1, w-1), \ldots, p(j_m, w-1))$. Our *tiling problems* have the following form. Fix an alphabet Σ and a *dimension* $m \in \mathbb{N}$.

Input: A *column language* $K \subseteq A^*$ and a *row language* $L \subseteq (A^m)^*$, both regular.
Question: Does there exist, for all $h \in \mathbb{N}$, a picture p of height h such that

- all columns in p belong to K,
- all bands of height m in p belong to L?

Such a picture is called a *solution* of the *tiling system* (K, L).

Of course, in general such problems are undecidable, even in dimension $m = 1$. Consequently, we consider two special cases of tiling systems: *monotone* and *lossy* ones. A tiling system (K, L) is *lossy* if K is closed under sub-words: for all words u, v and all letters a, $uav \in K$ implies $uv \in K$. A tiling system (K, L) is *monotone* if there exists a partial order $\le$ on the alphabet Σ (which we extend component-wise, i.e., letter-by-letter, to Σ^* and to $(\Sigma^m)^*$) such that $u \le v$ and $u \in L$ implies $v \in L$, and $uabv \in K$ implies $ucv \in K$, for some c with $c \ge a$ and $c \ge b$. Consequently, if we have a solution p of a lossy tiling system, we can obtain new solutions (of smaller height) by *removing* arbitrarily many rows of p. For a monotone tiling system, we obtain a new solution by *merging* two rows.

Example 8. (a) Consider the one-dimensional lossy tiling problem defined by $L = a^*ba^*$ and $K = a^*b^?a^*$. There are solutions of every height n: take a picture that has label a everywhere but for one b in each row, and at most one b per column (see Figure 1 (a)). The width of such a solution is at least n.

$$
\begin{array}{llll}
a\,b\,a\,a\,a\,a\,a & b\,c\,c\,c\,c\,c\,c & d|b\,a\,a|c|b\,a\,a|c & a\,1\,1\,1\,1\,1\,1 \\
b\,a\,a\,a\,a\,a\,a & a\,b\,c\,c\,c\,c\,c & c|a\,b\,a|d|a\,b\,a|c & b\,1\,1\,1\,a\,1\,1\,1 \\
a\,a\,b\,a\,a\,a\,a & a\,a\,b\,c\,c\,c\,c & c|a\,a\,b|c|a\,a\,b|d & b\,1\,a\,1\,b\,1\,a\,1 \\
a\,a\,a\,a\,a\,a\,b & a\,a\,a\,a\,b\,c\,c & & b\,a\,b\,a\,b\,a\,b\,a \\
a\,a\,a\,a\,b\,a\,a & a\,a\,a\,a\,a\,a\,b & & \\
\\
\text{(a)} & \text{(b)} & \text{(c)} & \text{(d)}
\end{array}
$$

Fig. 1. Some solutions to tiling problems

(b) A similar example uses the languages $L = a^*bc^*$ and $K = a^*b^?c^*$. Again, there exist solutions for all heights n, and the corresponding width is at least n

too. However, the solution is more constrained since it involves occurrences of b letters to describe some sort of diagonal in the solution (see Figure 1 (b)).

(c) More complex is the system with $L = (ca^*ba^*)^*d(a^*ba^*c)^*$ and $K = a^*b^?a^* + c^*d^?c^*$. There are also solutions of all heights, but this time, the minimal width for a solution is quadratic in its height (see Figure 1 (c)).

(d) Our final example is due to Paweł Parys. It consists of $L = a1^* + (b1^*a1^*)^+$ and $K = b^*a^?1^*$. All solutions of this system have exponential length (see Figure 1 (d)).

Theorem 9. *The satisfiability problem for* WAMSO *and the monotone tiling problem are equivalent. Both reduce to the lossy tiling problem.*

Conjecture 10. Monotone tiling problems and lossy ones are decidable.

This is the main open problem raised in this paper, even in dimension one. In the remainder, we will sketch some ideas on how to reduce the satisfiability of WAMSO to lossy tiling problems.

3.2   From ω-Words to Finite Words

Using Ramsey arguments in the spirit of Büchi's seminal proof [7], we can reduce WAMSO over ω-words to the following question concerning sequences of finite words. Consider a formula $\bar{Q}\bar{t}\varphi(\bar{t})$ in AMSO[np] and a sequence $\bar{u} = u_1, u_2, \ldots$ of weighted finite words. We say that $\bar{u}$ $(\bar{Q}\bar{t})$-*ultimately satisfies* $\varphi(\bar{t})$ if

$$\bar{Q}\bar{t}[u_i \models \varphi(\bar{t})] \text{ for all but finitely many } i] \ .$$

The *limit satisfiability problem* for AMSO[np] is to decide, given a formula $\bar{Q}\bar{t}\varphi(\bar{t})$, whether $\varphi(\bar{t})$ is $(\bar{Q}\bar{t})$-ultimately satisfied by some sequence $\bar{u}$.

Lemma 11. *The satisfiability problem for* AMSO[np] *and the limit satisfiability problem for* AMSO[np] *can be reduced one to the other. Furthermore, the prefix of number quantifiers is preserved by these reductions.*

Of course, the interesting reduction is from satisfiability of AMSO[np] on infinite words to limit satisfiability. We follow here an approach similar to Büchi's technique or, more precisely, its compositional variant developed by Shelah [15]. It amounts to use Ramsey's Theorem for chopping ω-words into infinitely many pieces that have the same theory. However, in this weighted situation, this kind of argument requires significantly more care.

A typical example would be to solve the satisfiability of the AMSO[np]-formula $\varphi := \forall s \exists r \forall x (\exists y > x)[s < f(y) \leq r]$ stating that there are infinitely many values that occur infinitely often (Example 1). It reduces to solving the limit satisfiability of the formula $\psi := \forall s \exists r \exists y[s < f(y) \leq r]$. A limit model for this formula would be the sequence (in which we omit the letters and only mention the weights) $\bar{u} = 0, 01, 012, 0123, \ldots$. Indeed, for all s, fixing $r = s + 1$, the formula $\exists y[s < f(y) \leq r]$ holds for almost all u_i. If we concatenate this sequence of words, we obtain the weighted ω-word $0010120123\ldots$ which satisfies φ. Conversely, every ω-word satisfying φ can be chopped into an infinite sequence of finite weighted words that satisfy ψ in the limit. In fact, this last reduction is more complex since it requires us to take care of the values contained in the finite prefixes. This is just an example, since in general the reduction is 'one-to-many' and involves regular properties of the finite prefixes.

4   Extended Asymptotic Monadic Logic

In this section we prove that the decidability problem for AMSO over ω-words is equivalent to the corresponding problem for MSO+U. To do so we introduce an extension of AMSO called *extended asymptotic monadic second-order logic* (EAMSO). This logic extends AMSO by quantifiers over weight functions. Inside a quantifier $\exists f$ we can use the function f in the usual constructions of AMSO. Note that variables for weight functions are not subject to any positivity constraint. Only number variables do have to satisfy such constraints.

Example 12. Let L_S be the language of ω-words over the alphabet $\{a, b, c\}$ such that, either there are finitely many occurrences of the letter b, or the number of a appearing between consecutive b tends to infinity. Consider the EAMSO-formula

$$\psi := \exists f \forall r \exists s \exists w \forall x \forall z \big[(w < x < z \wedge b(x) \wedge b(z)) \rightarrow$$
$$\exists y (x < y < z \wedge a(y) \wedge r < f(y) \leq s) \big] .$$

This formula defines L_S as follows. It guesses a weight function f and expresses that, for every number r, there exists a number s such that, ultimately, every two b-labeled positions $x < z$ are separated by an a-labeled position y with weight in $(r, s]$. It is easy to see that, if the number of a in an ω-word separating consecutive b tends to infinity, the weight function f defined by

$$f(x) = \begin{cases} 0 & \text{if the letter at } x \text{ is not } a \\ r & \text{if } x \text{ is the } r\text{-th occurrence of the letter } a \text{ after the last} \\ & \text{occurrence of the letter } b \text{ or the beginning of the word} \end{cases}$$

witnesses that the ω-word is a model of ψ. One can show that the converse also holds, *i.e.*, an ω-word satisfies ψ if and only if the number of a occurring between b tends to infinity (or there are finitely many occurrences of b).

The interesting point concerning EAMSO is that we can prove that, as far as satisfiability over infinite words is concerned, this logic is essentially equivalent to both AMSO and MSO+U. Let us recall that MSO+U is the extension of MSO with a new quantifier $\mathbb{U}X\varphi$ which signifies that 'there exists sets of arbitrarily large finite size such that φ holds'. For instance, it is straightforward to define the above language L_S in MSO+U.

Theorem 13. (a) *For every* MSO+U-*sentence, we can compute an EAMSO-sentence equivalent to it over ω-words. Conversely, for every EAMSO-sentence,* *there effectively exists an* MSO+U-*sentence such that the former is satisfiable over* ω-*words if, and only if, the latter is.*

(b) *For every* EAMSO-*sentence, we can compute an* AMSO-*sentence such that the former is satisfiable over* ω-*words if, and only if, the latter is.*

To compare the expressive power of EAMSO and AMSO, we again employ topological arguments. It is easy to show that, over ω-words without weights, AMSO collapses to MSO and, therefore, defines only Borel sets. However, according to Theorem 13, EAMSO is at least as expressive as MSO+U which reaches all levels of the projective hierarchy [13], even over non-weighted ω-words. Consequently, EAMSO is strictly more expressive than AMSO.

References

1. Bojańczyk, M.: The finite graph problem for two-way alternating automata. Theoret. Comput. Sci. 298(3), 511–528 (2003)
2. Bojańczyk, M.: Weak MSO with the unbounding quantifier. In: STACS. LIPIcs, vol. 3, pp. 159–170 (2009)
3. Bojańczyk, M., Colcombet, T.: Bounds in ω-regularity. In: LICS 2006, pp. 285–296 (2006)
4. Bojańczyk, M., Gogacz, T., Michalewski, H., Skrzypczak, M.: On the decidability of mso+u on infinite trees. Personal Communication
5. Bojańczyk, M., Toruńczyk, S.: Deterministic automata and extensions of weak mso. In: FSTTCS, pp. 73–84 (2009)
6. Bojańczyk, M., Toruńczyk, S.: Weak MSO+U over infinite trees. In: STACS (2012)
7. Büchi, J.R.: On a decision method in restricted second order arithmetic. In: Proceedings of the International Congress on Logic, Methodology and Philosophy of Science, pp. 1–11. Stanford Univ. Press (1962)
8. Colcombet, T.: The theory of stabilisation monoids and regular cost functions. In: Albers, S., Marchetti-Spaccamela, A., Matias, Y., Nikoletseas, S., Thomas, W. (eds.) ICALP 2009, Part II. LNCS, vol. 5556, pp. 139–150. Springer, Heidelberg (2009)
9. Colcombet, T.: Regular cost functions, part I: Logic and algebra over words. Logical Methods in Computer Science, 47 (2013)
10. Colcombet, T., Löding, C.: Regular cost functions over finite trees. In: LICS, pp. 70–79 (2010)
11. Flum, J., Ziegler, M.: Topological Model Theory, vol. 769. Springer (1980)
12. Gurevich, Y., Shelah, S.: Rabin's uniformization problem. J. Symb. Log. 48(4), 1105–1119 (1983)
13. Hummel, S., Skrzypczak, M.: The topological complexity of mso+u and related automata models. Fundam. Inform. 119(1), 87–111 (2012)
14. Rabin, M.O.: Decidability of second-order theories and automata on infinite trees. Trans. Amer. Math. Soc. 141, 1–35 (1969)
15. Shelah, S.: The monadic theory of order. Annals of Math. 102, 379–419 (1975)

Towards Efficient Reasoning Under Guarded-Based Disjunctive Existential Rules

Pierre Bourhis[1], Michael Morak[2], and Andreas Pieris[2]

[1] CNRS LIFL Université Lille 1/INRIA Lille, France
[2] Department of Computer Science, University of Oxford, UK
pierre.bourhis@univ-lille1.fr,
{michael.morak,andreas.pieris}@cs.ox.ac.uk

Abstract. The complete picture of the complexity of answering (unions of) conjunctive queries under the main guarded-based classes of disjunctive existential rules has been recently settled. It has been shown that the problem is very hard, namely 2EXPTIME-complete, even for fixed sets of rules expressed in lightweight formalisms. This gives rise to the question whether its complexity can be reduced by restricting the query language. Several subclasses of conjunctive queries have been proposed with the aim of reducing the complexity of classical database problems such as query evaluation and query containment. Three of the most prominent subclasses of this kind are queries of bounded hypertree-width, queries of bounded treewidth and acyclic queries. The central objective of the present paper is to understand whether the above query languages have a positive impact on the complexity of query answering under the main guarded-based classes of disjunctive existential rules. We show that (unions of) conjunctive queries of bounded hypertree-width and of bounded treewidth do not reduce the complexity of our problem, even if we focus on predicates of bounded arity, or on fixed sets of disjunctive existential rules. Regarding acyclic queries, although our problem remains 2EXPTIME-complete in general, in some relevant settings the complexity reduces to EXPTIME-complete; in fact, this requires to bound the arity of the predicates, and for some expressive guarded-based formalisms, to fix the set of rules.

1 Introduction

Rule-based languages lie at the core of several areas of central importance to artificial intelligence and databases, such as knowledge representation and reasoning, data exchange and integration, and web data extraction. A prominent rule-based formalism, originally intended for expressing complex recursive queries over relational databases, is Datalog, i.e., function-free first-order Horn logic. As already criticized in [28], the main weakness of this language for representing knowledge is its inability to infer the existence of new objects which are not explicitly stated in the extensional data set.

Existential rules, a.k.a. *tuple-generating dependencies (TGDs)* and *Datalog$^{\pm}$ rules*, overcome this limitation by extending Datalog with existential quantification in rule-heads; see, e.g., [6, 12–14, 26, 27]. More precisely, existential rules are

E. Csuhaj-Varjú et al. (Eds.): MFCS 2014, Part I, LNCS 8634, pp. 99–110, 2014.
© Springer-Verlag Berlin Heidelberg 2014

implications among conjunctions of atoms, and they essentially say that some tuples in a relational instance I imply the presence of some other tuples in I (hence the name tuple-generating dependencies). Unfortunately, the addition of existential quantifiers immediately leads to undecidability of conjunctive query answering [10, 12], which is the main reasoning service under existential rules. *Conjunctive queries (CQs)*, which form one of the most commonly used language for querying relational databases, are assertions of the form $\exists \mathbf{Y} \, \varphi(\mathbf{X}, \mathbf{Y})$, where φ is a conjunction of atoms, and correspond to the select-project-join fragment of relational algebra [1]. The answer to a CQ w.r.t. a database D and a set Σ of existential rules consists of all the tuples $\mathbf{t}$ of constants such that $\exists \mathbf{Y} \varphi(\mathbf{t}, \mathbf{Y})$ evaluates to *true* in every model of $(D \wedge \Sigma)$.

Several concrete languages which ensure the decidability of CQ answering have been proposed over the last five years; see, e.g., [6, 12, 14, 18, 23, 25–27]. Nevertheless, existential rules are not expressive enough for nondeterministic reasoning; for example, the statement "each parent of a father is the grandparent of a boy *or* a girl" is not expressible via existential rules. Such a statement can be expressed using the rules

$$\forall X \forall Y \, parentOf(X,Y) \wedge isfather(Y) \; \rightarrow \; \exists Z \, grandparentOf(X,Z)$$
$$\forall X \forall Y \, grandparentOf(X,Y) \; \rightarrow \; boy(Y) \vee girl(Y).$$

Obviously, to represent such kind of disjunctive knowledge, we need to extend the existing classes of existential rules with disjunction in the head of rules. Enriching existential rules with disjunction yields the formalism of *disjunctive existential rules*, a.k.a. *disjunctive TGDs (DTGDs)* [17]; henceforth, for brevity, we adopt the terms (D)TGDs.

Guarded-Based DTGDs. Guardedness is a well-known restriction which guarantees good model-theoretic and computational properties for first-order sentences [3]. Recently, inspired by guardedness, the class of guarded TGDs, that is, rules with a guard atom in the left-hand side which contains (or guards) all the universally quantified variables, has been defined [12]. Several extensions and restrictions of guarded TGDs have been proposed [6, 13]; we refer to all those formalisms by the term guarded-based TGDs, and more details will be given in Section 2. Guarded-based TGDs can be naturally extended to DTGDs. For example, the above set of rules is guarded since the atoms $parentOf(X,Y)$ and $grandparentOf(X,Y)$ are guards.

The complexity picture for query answering under the main guarded-based classes of DTGDs has been recently completed for arbitrary CQs [11]. Moreover, the complexity of answering atomic CQs, i.e., CQs consisting of a single atom, has been also investigated [2, 22]. However, the complexity picture of the problem restricted on some important subclasses of CQs, namely queries of bounded hypertree-width [20], queries of bounded treewidth [16], and acyclic queries [19], is still foggy, and there are several challenging open questions.

Research Challenges. The above subclasses of CQs have been proposed with the aim of reducing the complexity of several key decision problems on CQs such as evaluation of (Boolean) queries and query containment; in fact, those problems are NP-complete in general, but become tractable if restricted to one of the above subclasses [16, 19, 20]. The main objective of this work is to understand whether the subclasses of CQs in question have an analogous positive impact on query answering under the main guarded-based classes of DTGDs.

Although we know that our problem is unlikely to become tractable (implicit in [15]), we would like to understand whether its complexity is reduced. To achieve this, we focus on the following fundamental questions: (1) What is the exact complexity of answering queries which fall in one of the above subclasses of CQs under the main guarded-based classes of DTGDs?; (2) How is it affected if we consider predicates of bounded arity, or a fixed set of DTGDs, or a fixed set of DTGDs and a fixed query (a.k.a. the data complexity, where only the database is part of the input)?; and (3) How is it affected if we consider unions of CQs, i.e., disjunctions of a finite number of CQs? We provide answers to all these questions. This allows us to close the picture of the complexity of our problem, and come up with some general and insightful conclusions.

Our Findings. Our findings can be summarized as follows:

1. We show that (unions of) CQs of bounded hypertree-width and of bounded treewidth do not reduce the complexity of the problem under investigation. In particular, we show that for all the guarded-based classed of DTGDs in question, the problem remains 2EXPTIME-complete, even if we focus on predicates of bounded arity, or on fixed sets of DTGDs, while the data complexity remains coNP-complete. The data complexity results are inherited from existing works. However, all the other results are obtained by establishing a remarkably strong lower bound, namely query answering under a *fixed* set of DTGDs expressed in a lightweight fragment of guarded DTGDs, that is, constant-free rules with just one atom in the left-hand side without repeated variables, is 2EXPTIME-hard.

2. Regarding acyclic (unions of) CQs, we show that for all the classes of DTGDs under consideration, the problem remains 2EXPTIME-complete in general, and coNP-complete in data complexity. Again, the data complexity is inherited from existing results, while the 2EXPTIME-completeness is obtained by establishing a non-trivial lower bound. However, in some relevant cases the acyclicity of the query reduces the complexity of our problem to EXPTIME-complete. In fact, this requires to focus on predicates of bounded arity, and for some expressive classes of DTGDs, on fixed sets of DTGDs. The upper bounds are obtained by exploiting results on the guarded fragment of first-order logic, while the lower bounds required a non-trivial proof.

To sum up, queries of bounded hypertree-width and of bounded treewidth, as well as acyclic queries, do not have the expected positive impact on query answering under the main guarded-based classes of DTGDs. However, a positive impact can be observed on some relevant settings of our problem if we consider acyclic queries.

2   Preliminaries

General. Let $\mathbf{C}$, $\mathbf{N}$ and $\mathbf{V}$ be pairwise disjoint infinite countable sets of *constants, (labeled) nulls* and *variables*, respectively. We denote by $\mathbf{X}$ sequences (or sets) of variables $X_1, \ldots, X_k$. Let $[n] = \{1, \ldots, n\}$, for $n \geqslant 1$. A *term* is a constant, null or variable. An *atom* has the form $p(t_1, \ldots, t_n)$, where p is an n-ary predicate, and $t_1, \ldots, t_n$ are terms. For an atom $\underline{a}$, $dom(\underline{a})$ and $var(\underline{a})$ are the set of its terms and the set of its variables, respectively; those notations extend to sets of atoms. Usually conjunctions and disjunctions of atoms are treated as sets of atoms. An *instance* I is a (possibly infinite) set of atoms of the form $p(\mathbf{t})$, where $\mathbf{t}$ is a tuple of constants and nulls. A *database* D is a finite instance with only constants. Whenever an instance I is treated as a logical formula, is the formula $\exists \mathbf{X} (\bigwedge_{\underline{a} \in I} I)$, where $\mathbf{X}$ contains a variable for each null in I.

Conjunctive Queries. A *conjunctive query (CQ)* q is a sentence $\exists \mathbf{X} \varphi(\mathbf{X})$, where φ is a conjunction of atoms. If q does not have free variables, then it is called *Boolean*. For brevity, we consider only Boolean CQs; however, all the results of the paper can be easily extended to non-Boolean CQs. A *union of conjunctive queries (UCQ)* is a disjunction of a finite number of CQs. By abuse of notation, sometimes we consider a UCQ as set of CQs. A CQ $q = \exists \mathbf{X} \varphi(\mathbf{X})$ has a positive answer over an instance I, written $I \models q$, if there exists a homomorphism h such that $h(\varphi(\mathbf{X})) \subseteq I$. The answer to a UCQ Q over I is positive, written $I \models Q$, if there exists $q \in Q$ such that $I \models q$. A key subclass of CQs is the class of CQs of *bounded treewidth* (BTWCQs) [16], i.e., the treewidth of their hypergraph is bounded. The hypergraph of a CQ q, denoted $\mathcal{H}(q)$, is a hypergraph $\langle V, H \rangle$, where $V = dom(q)$, and, for each $\underline{a} \in q$, there exists a hyperedge $h \in H$ such that $h = dom(\underline{a})$. The treewidth of q is defined as the treewidth of its hypergraph $\mathcal{H}(q)$, that is, the treewidth of the Gaifman graph $G_{\mathcal{H}(q)}$ of $\mathcal{H}(q)$. The Gaifman graph of $\mathcal{H}(q)$ is the graph $\langle V, E \rangle$, where V is the node set of $\mathcal{H}(q)$, and $(v, u) \in E$ iff $\mathcal{H}(q)$ has a hyperedge h such that $\{v, u\} \subseteq h$. Another important subclass of CQs is the class of *acyclic* CQs (ACQs) [16]. A CQ is acyclic if $\mathcal{H}(q)$ is acyclic, i.e., it can be reduced to the empty hypergraph by iteratively eliminating some non-maximal hyperedge, or some vertex contained in at most one hyperedge.

Disjunctive Tuple-Generating Dependencies. A *disjunctive TGD* (or simply *DTGD*) σ is a first-order formula $\forall \mathbf{X} (\varphi(\mathbf{X}) \rightarrow \bigvee_{i=1}^{n} \exists \mathbf{Y}_i \, \psi_i(\mathbf{X}, \mathbf{Y}_i))$, where $n \geqslant 1$, $\mathbf{X} \cup \mathbf{Y} \subset \mathbf{V}$, and $\varphi, \psi_1, \ldots, \psi_n$ are conjunctions of atoms. The formula φ is called the *body* of σ, denoted $body(\sigma)$, while $\bigvee_{i=1}^{n} \psi_i$ is the *head* of σ, denoted $head(\sigma)$. The set of variables $var(body(\sigma)) \cap var(head(\sigma)) \subseteq \mathbf{X}$ is known as the *frontier* of σ, denoted $frontier(\sigma)$. If $n = 1$, then σ is called *tuple-generating dependency (TGD)*. The *schema* of a set Σ of DTGDs, denoted $sch(\Sigma)$, is the set of all predicates occurring in Σ. For brevity, we will omit the universal quantifiers, and use the comma (instead of $\wedge$). An instance I satisfies σ, written $I \models \sigma$, if whenever there exists a homomorphism h such that $h(\varphi(\mathbf{X})) \subseteq I$, then there exists $i \in [n]$ and $h' \supseteq h$ such that $h'(\psi_i(\mathbf{X}, \mathbf{Y}_i)) \subseteq I$; I satisfies a set Σ of DTGDs, denoted $I \models \Sigma$, if $I \models \sigma$, for each $\sigma \in \Sigma$. A *disjunctive inclusion dependency (DID)* is a constant-free DTGD with only one body-atom, the head is a disjunction of atoms, and there are no repeated variables in the body or in the head. A DTGD σ is *linear* if it has only one body-atom. A DTGD σ is *guarded* if there exists $\underline{a} \in body(\sigma)$, called *guard*, which contains all the variables in $body(\sigma)$. *Weakly-guarded* DTGDs extend guarded DTGDs by requiring only the body-variables that appear at *affected positions*, i.e., positions at which a null value may appear during the disjunctive chase (defined below), to appear in the guard; see [12]. A DTGD σ is *frontier-guarded* if there exists $\underline{a} \in body(\sigma)$ which contains all the variables of *frontier*(σ). *Weakly-frontier-guarded* DTGDs are defined analogously.

Query Answering. The *models* of a database D and a set Σ of DTGDs, denoted $mods(D, \Sigma)$, is the set of instances $\{I \mid I \supseteq D$ and $I \models \Sigma\}$. The *answer* to a CQ q w.r.t. D and Σ is *positive*, denoted $D \cup \Sigma \models q$, if $I \models q$, for each $I \in mods(D, \Sigma)$. The answer to a UCQ w.r.t. D and Σ is defined analogously. Our problem is defined as follows: Given a CQ q, a database D, and a set Σ of DTGDs, decide whether $D \cup \Sigma \models q$. If q is a BTWCQ (resp., ACQ), then the above problem is called *BTWCQ* (resp., *ACQ*) *answering*. The problem BTWUCQ (resp., AUCQ) answering is defined analogously. The *data complexity* is calculated taking only the database as input. For the *combined complexity*, the query and set of DTGDs count as part of the input as well.

Disjunctive Chase. Consider an instance I, and a DTGD $\sigma : \varphi(\mathbf{X}) \to \bigvee_{i=1}^{n} \exists \mathbf{Y} \, \psi_i(\mathbf{X}, \mathbf{Y})$. We say that σ is *applicable* to I if there exists a homomorphism h such that $h(\varphi(\mathbf{X})) \subseteq I$, and the result of applying σ to I with h is the set $\{I_1, \dots, I_n\}$, where $I_i = I \cup h'(\psi_i(\mathbf{X}, \mathbf{Y}))$, for each $i \in [n]$, and $h' \supseteq h$ is such that $h'(Y)$ is a "fresh" null not occurring in I, for each $Y \in \mathbf{Y}$. For such an application of a DTGD, which defines a single DTGD *chase step*, we write $I\langle \sigma, h \rangle \{I_1, \dots, I_n\}$. A *disjunctive chase tree* of a database D and a set Σ of DTGDs is a (possibly infinite) tree such that the root is D, and for every node I, assuming that $\{I_1, \dots, I_n\}$ are the children of I, there exists $\sigma \in \Sigma$ and a homomorphism h such that $I\langle \sigma, h \rangle \{I_1, \dots, I_n\}$. The disjunctive chase algorithm for D and Σ consists of an exhaustive application of DTGD chase steps in a fair fashion, which leads to a disjunctive chase tree T of D and Σ; we denote by $chase(D, \Sigma)$ the set $\{I \mid I$ is a leaf of $T\}$. It is well-known that, given a UCQ Q, $D \cup \Sigma \models Q$ iff $I \models Q$, for each $I \in chase(D, \Sigma)$.

The Guarded Fragment of First-Order Logic. The *guarded fragment (GFO)* has been introduced in [3]. The set of GFO formulas over a schema $\mathcal{R}$ is the smallest set (1) containing all atomic $\mathcal{R}$-formulas and equalities; (2) closed under the logical connectives $\neg, \wedge, \vee, \to$; and (3) if $\underline{a}$ is an $\mathcal{R}$-atom containing all the variables of $\mathbf{X} \cup \mathbf{Y}$, and φ is a GFO formula with free variables contained in $(\mathbf{X} \cup \mathbf{Y})$, then $\forall \mathbf{X}(\underline{a} \to \varphi)$ and $\exists \mathbf{X}(\underline{a} \wedge \varphi)$ are GFO formulas. The *loosely guarded fragment (LGFO)* is a generalization of GFO where the quantifiers are guarded by conjunctions of atomic formulas; for details see, e.g., [24].

Alternation. An *alternating Turing machine* is a tuple $M = (S, \Lambda, \delta, s_0)$, where $S = S_\forall \uplus S_\exists \uplus \{s_a\} \uplus \{s_r\}$ is a finite set of states partitioned into universal states, existential states, an accepting state and a rejecting state, Λ is the tape alphabet, $\delta \subseteq (S \times \Lambda) \times (S \times \Lambda \times \{-1, 0, +1\})$ is the transition relation, and $s_0 \in S$ is the initial state. We assume that Λ contains a special blank symbol $\sqcup$.

Table 1. Complexity of BTW(U)CQ answering. Each row corresponds to a class of DTGDs; substitute L for linear, G for guarded, F for frontier, and W for weakly. UB and LB stand for upper and lower bound. The missing references for the upper (lower) bounds are inherited from the first lower-left (upper-right) cell with a reference.

	Combined Complexity	Bounded Arity	Fixed Rules	Data Complexity
DID	2ExpTime	2ExpTime	2ExpTime	coNP
			LB: Thm. 1	LB: [15, Thm. 4.5]
L/G	2ExpTime	2ExpTime	2ExpTime	coNP
F-G	2ExpTime	2ExpTime	2ExpTime	coNP
				UB: [11, Thm. 7]
W-G	2ExpTime	2ExpTime	2ExpTime	ExpTime
				LB: [12, Thm. 4.1]
W-F-G	2ExpTime	2ExpTime	2ExpTime	ExpTime
	UB: [11, Thm. 1]			UB: [11, Thm. 7]

3   Bounded Treewidth Queries

In this section, we focus on answering (U)CQs of bounded treewidth under our respective classes of DTGDs. Table 1 gives the complete picture of the complexity of our problem. As you can observe, the data complexity for all the classes of DTGDs under consideration is obtained from existing results. More precisely, the coNP-hardness for DIDs is obtained from [15, Theorem 4.5], where it is shown that CQ answering under a single DID of the form $p_1(X) \to p_2(X) \vee p_3(X)$, is already coNP-hard, even if the input query is fixed (and thus of bounded treewidth). The coNP upper bound for frontier-guarded DTGDs has been established in [11, Theorem 7] by a reduction to UCQ answering under GFO sentences. The ExpTime-hardness for weakly-guarded DTGDs is inherited from [12, Theorem 4.1], where it is shown that CQ answering under a fixed set of weakly-guarded TGDs is ExpTime-hard, even if the input query is a single atom. The ExpTime upper bound for weakly-frontier-guarded DTGDs has be shown in [11, Theorem 7] again by a reduction to UCQ answering under GFO.

Although the data complexity of our problem can be settled by exploiting known results, the picture for all the other cases is still foggy. The best known upper bound is the 2ExpTime upper bound for answering arbitrary UCQs under weakly-frontier-guarded DTGDs [11, Theorem 1], established by a reduction to the satisfiability problem of the guarded negation fragment [9], an extension of GFO. This result, combined with the fact that CQ answering under guarded TGDs is 2ExpTime-hard in the combined complexity [12, Theorem 6.1], even for atomic queries of the form $\exists X p(X)$ (and thus of bounded treewidth), closes the combined complexity for (weakly-)(frontier-)guarded DTGDs. However, the above lower bound for guarded DTGDs is not strong enough to complete the complexity picture of our problem. We establish a strong lower bound which, together with the above 2ExpTime upper bound, gives us the complete picture of the complexity of the problem studied in this section.

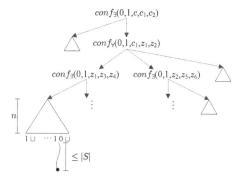

Fig. 1. Representation of the computation tree of M in the proof of Theorem 1

Theorem 1. *BTWCQ answering under fixed sets of DIDs is 2ExpTime-hard.*

Proof (sketch). The proof is by a reduction from the non-acceptance problem of an alternating exponential space Turing machine $M = (S, \Lambda, \delta, s_0)$ on the empty input. We assume that M uses exactly 2^n tape cells, $\Lambda = \{0, 1, \sqcup\}$, the initial configuration is existential, and every universal configuration is followed by two existential configurations and vice versa. Our goal is to construct a database D, a fixed set Σ of DIDs, and a BTWCQ q such that $D \cup \Sigma \models q$ iff M rejects. The general idea is to construct, by chasing D and Σ, all the trees which may encode a possible computation tree of M; in other words, each instance $I \in chase(D, \Sigma)$ will encode such a tree T_I. More precisely, the initial configuration is stored in the database D as the atom $conf_\exists(0, 1, c, c_1, c_2)$, where $\{c, c_1, c_2\} \subset \mathbf{C}$ are constants which represent the initial configuration (c), and its two subsequent configurations (c_1 and c_2) and 0 and 1 are auxiliary constants that will allow us to have access to 0 and 1 without explicitly mention them in Σ. Then, starting from the initial configuration, we construct a tree whose nodes are configurations, i.e., atoms of the form $conf_x(0, 1, t, n_1, n_2)$, where $x \in \{\exists, \forall\}$. Moreover, on each configuration node v, which represents the configuration C_v of M, we attach a configuration tree, that is, a full binary tree of depth n, and thus at its n-th level there are exactly 2^n nodes which represent the cells of the tape of M in C_v. Furthermore, for each cell we guess its content (0, 1 or $\sqcup$), and also whether the cursor of M is at this cell, and if so, we attach a chain of length at most $|S|$, which encodes the state of C_v. The above informal description is illustrated in Figure 1. Finally, we construct a BTWCQ q such that, if $I \in chase(D, \Sigma)$ entails q, then T_I is *not* a valid computation tree of M.   □

The above strong lower bound closes all the missing cases regarding the complexity of our problem, and the next result follows:

Corollary 1. *BTW(U)CQ answering under (weakly-)(frontier-)guarded DTGDs, linear DTGDs and DIDs is 2ExpTime-complete in the combined complexity. The same holds for predicates of bounded arity, and for fixed sets of DTGDs.*

Another key class of queries is the class of CQs of *bounded hypertree-width* [20]. The hypertree-width is a measure of how close to acyclic a hypergraph is, analogous

Table 2. Complexity of answering acyclic (U)CQs

	Combined Complexity	Bounded Arity	Fixed Rules	Data Complexity
DID	2ExpTime	ExpTime	ExpTime	coNP
	LB: Thm. 4		LB: Thm. 6	LB: [15, Thm. 4.5]
L/G	2ExpTime	ExpTime	ExpTime	coNP
F-G	2ExpTime	2ExpTime	ExpTime	coNP
		UB: [11, Thm. 1]		UB: [11, Thm. 7]
		LB: Thm. 5		
W-G	2ExpTime	ExpTime	ExpTime	ExpTime
		UB: Thm. 2		LB: [12, Thm. 4.1]
W-F-G	2ExpTime	2ExpTime	ExpTime	ExpTime
	UB: [11, Thm. 1]	LB: Thm. 5	UB: Thm. 3	UB: [11, Thm. 7]

to treewidth for graphs. The hypertree-width of a CQ is less than or equal to its treewidth. Since all the upper bounds in Table 1 hold for arbitrary (U)CQs, we get that arbitrary (U)CQs, (U)CQs of bounded treewidth and (U)CQs of bounded hypertree-width are indistinguishable w.r.t. to the complexity of query answering under our DTGDs.

4   Acyclic Queries

In this section, we focus on answering (unions of) acyclic queries under our respective classes of DTGDs. Table 2 gives the complete picture of the complexity of our problem. Compared with Table 1, it is immediately apparent that we can inherit from existing works the same results as for BTW(U)CQ answering, namely the data complexity in all the cases, and the 2ExpTime upper bound for answering arbitrary UCQs under weakly-frontier-guarded DTGDs. Therefore, apart from the data complexity, several non-trivial cases are still missing. As you can observe in Table 2, the combined and the data complexity do not change if we restrict our selves to acyclic queries. However, the complexity decreases from 2ExpTime to ExpTime for the non-frontier classes of DTGDs, i.e., DIDs, linear and (weakly-)guarded DTGDs, in the case of predicates of bounded arity, and also for all the classes if we consider a fixed set of DTGDs. This is an interesting finding as, in general, queries of bounded treewidth and acyclic queries behave in the same way. Let us now proceed with our results.

4.1   Upper Bounds

We start this section by showing that answering acyclic queries under weakly-guarded sets of DTGDs is in ExpTime in case of bounded arity. This is shown by a reduction to satisfiability of LGFO. Let us briefly explain the reduction via a simple example. Consider the database $D = \{s(a, a), p(a, b, c)\}$ and the set Σ consisting of

$$\sigma_1 : t(X, Y), p(Z, W, X), s(Z, V) \; \to \; \exists U \, p(Y, Z, U) \vee s(Y, Z),$$
$$\sigma_2 : p(X, Y, Z) \; \to \; \exists W \, t(Z, W).$$

Observe that the affected positions of $sch(\Sigma)$, i.e., the positions where nulls may appear during the construction of $chase(D, \Sigma)$, are $p[3]$, $t[1]$ and $t[2]$. Clearly, $t(X, Y)$ is the weak-guard for σ_1 and $p(X, Y, Z)$ the weak-guard for σ_2. Notice that σ_2 is already a GFO (and thus an LGFO) sentence. Thus, we need to convert σ_1 into an LGFO sentence. This can be done by expanding the weak-guard $t(X, Y)$ into a conjunction of atoms to guard the variables Z, V and W, and obtain the sentence Ψ_{σ_1}

$$\forall X \forall Y \forall Z \forall V \forall W \left((\hat{t}(Z, V, X, Y) \wedge \hat{t}(Z, W, X, Y) \wedge \hat{t}(V, W, X, Y)) \rightarrow \right.$$
$$\left. (p(Z, W, X) \wedge s(Z, V)) \rightarrow \exists U \, (p(Y, Z, U) \vee s(Y, Z))) \right).$$

It should not be forgotten to properly generate the atoms with the auxiliary predicate $\hat{t}$. Since the variables Z, V and W can be satisfied only with constants of $dom(D) = \{a, b, c\}$, those atoms can be generated via the GFO sentence $\Psi_{\hat{t}}$

$$\forall X \forall Y \, (t(X, Y) \rightarrow \hat{t}(a, a, X, Y) \wedge \hat{t}(a, b, X, Y) \wedge \hat{t}(a, c, X, Y) \wedge$$
$$\hat{t}(b, a, X, Y) \wedge \hat{t}(b, b, X, Y) \wedge \hat{t}(b, c, X, Y) \wedge$$
$$\hat{t}(c, a, X, Y) \wedge \hat{t}(c, b, X, Y) \wedge \hat{t}(c, c, X, Y)) \,.$$

It is not difficult to see that, for every acyclic UCQ Q, $D \cup \Sigma \models Q$ iff the sentence $\Psi = (D \wedge \Psi_{\sigma_1} \wedge \Psi_{\hat{t}} \wedge \sigma_2 \wedge \neg Q)$ is unsatisfiable. Notice that Ψ does not immediately fall into LGFO because of the query Q. However, since Q is acyclic, there exists an equivalent UCQ Q' which falls in GFO [21]. Thus, $\Psi' = (D \wedge \Psi_{\sigma_1} \wedge \Psi_{\hat{t}} \wedge \sigma_2 \wedge \neg Q')$ falls in LGFO and is equivalent to Ψ. Let us clarify that, if the head of σ_1 is a disjunction of conjunctions (instead of atoms as in the above example), then Ψ' is "almost" loosely-guarded since the existentially quantified variables are not necessarily guarded. However, as explicitly remarked in [24], the satisfiability algorithm for LGFO sentences is general enough to also treat sentences which are "almost" LGFO without increasing the complexity. From the above informal discussion we get that:

Theorem 2. *A UCQ answering under weakly-guarded sets of DTGDs is in* Ex-pTime *in case of predicates of bounded arity.*

The above machinery cannot be applied in the case of weakly-frontier-guarded DTGDs since the variables that we need to guard may appear at affected positions. However, if we focus on fixed sets of DTGDs, then the complexity is reduced to ExpTime. This is established by first reducing our problem to UCQ answering under GFO sentences, and then exploit a result in [8], where the problem of querying GFO is studied.

Theorem 3. *A UCQ answering under fixed weakly-frontier-guarded sets of DT-GDs is in* ExpTime.

We believe that the results of this section can have a practical impact on other important tasks, such as querying graph databases [5], or querying description logic ontologies [4], where the attention is usually focussed on unary and binary predicates.

4.2   Lower Bounds

We start this section by showing the following non-trivial lower bound:

Theorem 4. *ACQ answering under DIDs is* 2ExpTime-*hard in combined complexity.*

Proof (sketch). We follow the same approach as in the proof of Theorem 1. However, the way that a computation tree of the alternating Turing machine $M = (S, \Lambda, \delta, s_0)$ is represented in that proof is not useful since it will necessarily lead to a cyclic query Q. This is exactly the non-trivial part of the proof, i.e., to construct, by chasing D and Σ, all the possible trees which may encode a computation tree of M in such a way that an acyclic query Q can be employed. To this aim, the configurations of M are represented using atoms of the form $conf[s](b_1, \ldots, b_n, a, h, t, p, n_1, n_2)$, where $s \in S$ is the state of the encoded configuration (and is part of the predicate), $(b_1, \ldots, b_n) \in \{0,1\}^n$ is an integer of $\{0, \ldots, 2^n - 1\}$ in binary encoding which represents the index of the encoded cell, $h \in \{0,1\}$ and $h = 1$ means that the cursor of M is at the encoded cell, and t, p, n_1 and n_2 represent the current, the previous and the next two configurations, respectively. More precisely, using a fixed number of DIDs, one can construct a tree with nodes of the form $conf_0[s](0, 1, \sqcup, 0^{2n}, 1^n, 1, z_1, z_2, z_3, z_4)$; such an atom is associated with the configuration z_1, and contains all the auxiliary constants that will allow us to generate, via polynomially many DIDs, all the 2^n atoms of the form $conf[s](b_1, \ldots, b_n, a, h, z_1, z_2, z_3, z_4)$.   □

The above lower bound and the 2ExpTime upper bound for weakly-frontier-guarded sets of DTGDs in combined complexity [11, Theorem 1] imply that:

Corollary 2. *A(U)CQ answering under (weakly-)(frontier-)guarded DTGDs, linear DTGDs and DIDs is* 2ExpTime-*complete in combined complexity.*

Let us now focus on frontier-guarded DTGDs, and show that query answering is 2ExpTime-hard, even for ACQs and predicates of bounded arity. This is shown by exploiting the fact that a CQ $\exists \mathbf{X}\, \varphi(\mathbf{X})$ is actually the frontier-guarded TGD $\varphi(\mathbf{X}) \rightarrow p$. We thus have a reduction from CQ answering under frontier-guarded DTGDs, which is 2ExpTime-hard even for predicates of bounded arity [7].

Theorem 5. *ACQ answering under frontier-guarded DTGDs is* 2ExpTime-*hard, even for predicates of bounded arity.*

The above result and the 2ExpTime upper bound for weakly-frontier-guarded sets of DTGDs in combined complexity [11, Theorem 1] imply that:

Corollary 3. *A(U)CQ answering under (weakly-)frontier-guarded DTGDs is* 2ExpTime-*complete in case of predicates of bounded arity.*

We conclude this section by establishing the ExpTime-hardness of ACQ answering when we focus on fixed sets of DIDs. This is done by simulating an alternating linear space Turing machine M. The idea of the proof is along the lines of the proofs of Theorems 1 and 4. In particular, on each configuration node v, which represents the configuration C_v of M, we attach a cell-chain which mimics the tape in C_v, and a state-chain which encodes the state of C_v.

Theorem 6. *ACQ answering under fixed sets of DIDs is* ExpTime-*hard.*

From Theorems 2, 3 and 6 we get the that:

Corollary 4. *A(U)CQ answering under (weakly-)guarded DTGDs, linear DTGDs and DIDs is* ExpTime-*complete for predicates of bounded arity. The same problem under (weakly-)(frontier-)guarded DTGDs, linear DTGDs and DIDs is* ExpTime-*complete for fixed sets of DTGDs.*

5 Conclusions

We studied the problem of answering (U)CQs under the main guarded-based classes of DTGDs. We focussed on three key subclasses of (U)CQs, namely (U)CQs of bounded hypertree-width, (U)CQs of bounded treewidth, and acyclic (U)CQs. Our investigation shows that the above query languages do not have the expected positive impact on our problem, and in most of the cases the complexity of the problem remains 2ExpTime-complete. However, in some relevant settings, the complexity reduces to ExpTime-complete if we focus on acyclic queries. We believe that this finding can have a practical impact on crucial tasks such as querying graph databases and querying description logic ontologies, where the attention is usually focussed on unary and binary predicates.

Acknowledgements. Pierre thankfully acknowledges projet équipe associée Inria North-European Labs 2013-2016, Michael his DOC Fellowship of the Austrian Academy of Sciences, and Andreas the EPSRC grant EP/J008346/1 (ProQAW). We thank the anonymous referees for many helpful comments.

References

1. Abiteboul, S., Hull, R., Vianu, V.: Foundations of Databases. Addison-Wesley (1995)
2. Alviano, M., Faber, W., Leone, N., Manna, M.: Disjunctive Datalog with existential quantifiers: Semantics, decidability, and complexity issues. TPLP 12(4-5), 701–718 (2012)
3. Andréka, H., van Benthem, J., Németi, I.: Modal languages and bounded fragments of predicate logic. J. Philosophical Logic 27, 217–274 (1998)
4. Artale, A., Calvanese, D., Kontchakov, R., Zakharyaschev, M.: The DL-Lite family and relations. J. Artif. Intell. Res. 36, 1–69 (2009)
5. Baeza, P.B.: Querying graph databases. In: PODS, pp. 175–188 (2013)
6. Baget, J.F., Leclère, M., Mugnier, M.L., Salvat, E.: On rules with existential variables: Walking the decidability line. Artif. Intell. 175(9-10), 1620–1654 (2011)
7. Baget, J.F., Mugnier, M.L., Rudolph, S., Thomazo, M.: Walking the complexity lines for generalized guarded existential rules. In: IJCAI, pp. 712–717 (2011)

8. Bárány, V., Gottlob, G., Otto, M.: Querying the guarded fragment. In: LICS, pp. 1–10 (2010)
9. Bárány, V., ten Cate, B., Segoufin, L.: Guarded negation. In: Aceto, L., Henzinger, M., Sgall, J. (eds.) ICALP 2011, Part II. LNCS, vol. 6756, pp. 356–367. Springer, Heidelberg (2011)
10. Beeri, C., Vardi, M.Y.: The implication problem for data dependencies. In: Even, S., Kariv, O. (eds.) ICALP. LNCS, vol. 115, pp. 73–85. Springer, Heidelberg (1981)
11. Bourhis, P., Morak, M., Pieris, A.: The impact of disjunction on query answering under guarded-based existential rules. In: IJCAI (2013)
12. Calì, A., Gottlob, G., Kifer, M.: Taming the infinite chase: Query answering under expressive relational constraints. J. Artif. Intell. Res. 48, 115–174 (2013)
13. Calì, A., Gottlob, G., Lukasiewicz, T.: A general Datalog-based framework for tractable query answering over ontologies. J. Web Sem. 14, 57–83 (2012)
14. Calì, A., Gottlob, G., Pieris, A.: Towards more expressive ontology languages: The query answering problem. Artif. Intell. 193, 87–128 (2012)
15. Calvanese, D., De Giacomo, G., Lembo, D., Lenzerini, M., Rosati, R.: Data complexity of query answering in description logics. Artif. Intell. 195, 335–360 (2013)
16. Chekuri, C., Rajaraman, A.: Conjunctive query containment revisited. Theor. Comput. Sci. 239(2), 211–229 (2000)
17. Deutsch, A., Tannen, V.: Reformulation of XML queries and constraints. In: Calvanese, D., Lenzerini, M., Motwani, R. (eds.) ICDT 2003. LNCS, vol. 2572, pp. 225–241. Springer, Heidelberg (2002)
18. Fagin, R., Kolaitis, P.G., Miller, R.J., Popa, L.: Data exchange: Semantics and query answering. Theor. Comput. Sci. 336(1), 89–124 (2005)
19. Gottlob, G., Leone, N., Scarcello, F.: The complexity of acyclic conjunctive queries. J. ACM 48(3), 431–498 (2001)
20. Gottlob, G., Leone, N., Scarcello, F.: Hypertree decompositions and tractable queries. J. Comput. Syst. Sci. 64(3), 579–627 (2002)
21. Gottlob, G., Leone, N., Scarcello, F.: Robbers, marshals, and guards: game theoretic and logical characterizations of hypertree width. J. Comput. Syst. Sci. 66(4), 775–808 (2003)
22. Gottlob, G., Manna, M., Morak, M., Pieris, A.: On the complexity of ontological reasoning under disjunctive existential rules. In: Rovan, B., Sassone, V., Widmayer, P. (eds.) MFCS 2012. LNCS, vol. 7464, pp. 1–18. Springer, Heidelberg (2012)
23. Gottlob, G., Manna, M., Pieris, A.: Combining decidability paradigms for existential rules. TPLP 13(4-5), 877–892 (2013)
24. Grädel, E.: On the restraining power of guards. J. Symb. Log. 64(4), 1719–1742 (1999)
25. Grau, B.C., Horrocks, I., Krötzsch, M., Kupke, C., Magka, D., Motik, B., Wang, Z.: Acyclicity notions for existential rules and their application to query answering in ontologies. J. Artif. Intell. Res. 47, 741–808 (2013)
26. Krötzsch, M., Rudolph, S.: Extending decidable existential rules by joining acyclicity and guardedness. In: IJCAI, pp. 963–968 (2011)
27. Leone, N., Manna, M., Terracina, G., Veltri, P.: Efficiently computable Datalog$^\exists$ programs. In: KR (2012)
28. Patel-Schneider, P.F., Horrocks, I.: A comparison of two modelling paradigms in the semantic web. J. Web Sem. 5(4), 240–250 (2007)

Alternating Parity Krivine Automata

Florian Bruse*

Universität Kassel, Fachbereich Elektrotechnik und Informatik,
Wilhelmshöher Allee 73, 34121 Kassel, Germany
florian.bruse@uni-kassel.de

Abstract. Higher-Order Modal Fixpoint Logic HFL is a non-regular extension of the modal μ-calculus by a typed λ-calculus. The model-checking problem for this logic is decidable over finite structures. We present an automaton model that captures the semantics of HFL. This automaton model combines alternating parity automata with the lookup mechanisms from the Krivine machine.

Keywords: Higher-Order Modal Fixpoint Logic, Fixpoint Logic, Krivine Machine, λ-Calculus.

1 Introduction

Automata and temporal logics enjoy a tight connection as, for many logics, there is a class of automata that captures precisely the expressive power of the logic. Such automata provide operational semantics to the denotational semantics of their logical counterpart. Exploring and exploiting this connection has a long tradition in mathematical logic: The works of Rabin [1], Büchi [2] and Kupferman, Vardi and Wolper [3] have used this technique to advance research considerably. Passing to the operational point of view is particularly useful for finding efficient algorithms related to the logic.

A common technique in the area of software verification is to reduce a problem to the model-checking problem of a temporal logic over a labeled transition system. Many temporal logics are well understood, but can only express regular properties. Since interesting properties are often non-regular, non-regular logics have been getting increased attention. Classical examples of such properties are context-free ones such as "every request is answered exactly once", counting-related questions such as "the number of responses never exceeds the number of requests" or questions related to process equivalence relations [4].

Formalisms developed to express non-regular properties are, e.g., visibly pushdown languages and automata [5,6], Fixpoint Logic with Chop (FLC) [7] or the Higher-Order Modal Fixpoint Logic HFL [8], which was created to express assume-guarantee properties for modularized programs. HFL is a very expressive logic that extends the modal μ-calculus with a typed λ-calculus. Its

* This work was supported by the European Research Council under the European Community's Seventh Framework Programme [ERC grant agreement no 259267].

E. Csuhaj-Varjú et al. (Eds.): MFCS 2014, Part I, LNCS 8634, pp. 111–122, 2014.

model-checking problem for formulae of order k over finite structures is k-EXPTIME-complete [9]. The technique used for finite structures, full enumeration of function tables, does not extend to infinite structures, even if these are finitely presented (e.g., pushdown structures). Since model-checking visibly pushdown systems against a rather weak fragment of HFL is already undecidable [10], decidability results for large classes of infinite systems are not possible, but an automaton model can help to establish the precise boundary of undecidability. For finite structures, the high complexity of the model-checking problem asks for local techniques that, in practice, could mitigate some of this complexity. An automaton model for HFL could help with that.

This paper presents such a computation model for HFL and its fragments. HFL's definition as a fixpoint logic suggests a parity automaton. In order to do the bookkeeping caused by higher-order λ-expressions, we use the Krivine machine [11] that was recently used by Salvati and Walukiewicz in the context of higher-order recursion schemes [12]. The Krivine machine mimics call-by-name evaluation to do weak head reduction of λ-expressions, an evaluation strategy that was chosen to keep the space of automaton states finite. We call the resulting automaton model *Alternating Parity Krivine Automata* (APKA). While it extends nicely the machinery developed for FLC in [13], the extension is not quite straightforward, due to the higher-order features: For example, one cannot, without loss of generality, assume that negation occurs only in front of atomic formulae, nor is the nesting and alternation of fixpoints as easily treated as in the modal μ-calculus, in particular with respect to the parity condition.

The paper is structured as follows: In Section 2, we introduce the syntax and semantics of HFL. In Section 3, we define APKA, prove that they capture the semantics of HFL, and discuss some peculiarities with respect to the behavior of the parity conditions of HFL and some of its fragments. In Section 4 we outline open questions warranting further research. Some proofs have been omitted due to space constraints.

2   Syntax and Semantics of HFL

The definitions of syntax and semantics of HFL are more complex than e.g., the μ-calculus, because the incorporation of the λ-calculus demands a type system. This system also keeps track of negation to guarantee a proper definition of fixpoints. HFL's original presentation [8] treats this in greater detail.

Labeled Transition Systems. A *labeled transition system* (LTS) $\mathcal{T}$ over a set $\mathcal{P}$ of *atomic propositions* and a set $\mathcal{R}$ of *action names* is a set S of states together with interpretations $P^{\mathcal{T}} \subseteq S$ for all $P \in \mathcal{P}$ and $R^{\mathcal{T}} \subseteq S \times S$ for all $R \in \mathcal{R}$. With $\mathcal{T}, s_0$ we denote the pointed transitions system with distinguished state s_0, and with $R[s]$ we denote the set of R-successors of s.

Syntax. For the remainder of the paper, fix sets $\mathcal{P}$ of propositions, $\mathcal{R}$ of binary relations and infinite sets of variables $\mathcal{V}$ and $\mathcal{F}$ that denote variables bound by a λ-expression, respectively a fixpoint quantifier. Separating the latter is usually not done for HFL, but facilitates technical exposition of APKA. Lower case letters $x, y, \ldots$ denote variables in $\mathcal{V}$, upper case letters $X, Y, \ldots$ those in $\mathcal{F}$.

The set of types is defined inductively: Pr is the ground type. If τ, τ' are types, then so is $\tau \to \tau'$ and $\tau^0 \to \tau', \tau^+ \to \tau'$ and $\tau^- \to \tau'$ denote the functions of this type that are monotone, antitone or constant in the argument. The annotations $+, -, 0$ are called variances, v denotes an unspecified variance. The operator $\to$ is right-associative, so any type can be written as $\tau_1 \to \cdots \to \tau_n \to \mathrm{Pr}$. The order ord is defined inductively via $\mathrm{ord}(\mathrm{Pr}) = 0$ and $\mathrm{ord}(\tau_1 \to \cdots \tau_n \to \mathrm{Pr}) = \max(\mathrm{ord}(\tau_1), \ldots, \mathrm{ord}(\tau_n)) + 1$. The set of types is partially ordered via $\tau, \tau' < \tau \to \tau'$.

Definition 1. HFL-*formulae φ are defined by the grammar*

$$\varphi ::= P \mid \Diamond_R \varphi \mid \varphi \vee \varphi \mid \neg\varphi \mid x \mid X \mid \lambda(x^v : \tau).\varphi \mid (\varphi\,\varphi) \mid \mu(X : \tau).\varphi \mid \nu(X : \tau).\varphi$$

where $P \in \mathcal{P}$, $R \in \mathcal{R}$, $x \in \mathcal{V}$ and $X \in \mathcal{F}$.

The connectives $\top, \bot, \wedge, \Box_R$ can be added as derived connectives as usual.

The binder $\lambda(x^v : \tau).\varphi$ binds x in φ, the binder $\sigma(X : \tau).\varphi$ with $\sigma \in \{\mu, \nu\}$ binds X in φ. Let $\mathrm{sub}(\varphi)$ denote the set of subformulae of φ. An HFL-formula is *well-named* if, for each $X \in \mathcal{F}$, there is at most one subformula of the form $\sigma(X : \tau).\psi$. We do not insist in unique occurrences for variables in $\mathcal{V}$ but this can be obtained via renaming, i.e, α-conversion, if desired. A variable from $\mathcal{V}$ or $\mathcal{F}$ in a formula φ is *bound* if it is bound by a binder of the respective type, and *free* otherwise. A formula is called *closed* if it has no free variables and *open* otherwise. For a well-named formula φ and $X \in \mathcal{F} \cap \mathrm{sub}(\varphi)$, define $\mathrm{fp}_\varphi(X)$ as the unique subformula ψ of φ such that $\psi = \sigma X.\psi'$ for $\sigma \in \{\mu, \nu\}$. We have a partial order $<_{\mathrm{fp}_\varphi}$ on the fixpoint variables of φ via $X <_{\mathrm{fp}_\varphi} Y$ if Y appears freely in $\mathrm{fp}_\varphi(X)$. We say that Y is *outermore* than X. A variable is *outermost* among a set of variables if it is maximal in this set with respect to $<_{\mathrm{fp}_\varphi}$.

A finite set $\Sigma = \{X_1^{v_1} : \tau_1, \ldots, X_n^{v_n} : \tau_n, x_1^{v_1'} : \tau_1', \ldots, x_m^{v_m'} : \tau_m'\}$ where the $X_i \in \mathcal{F}$, $x_i \in \mathcal{V}$, the τ_i, τ_i' are types and the v_i, v_i' are variances is called a *context*. The context Σ^- is obtained from Σ by reversing the variance in each type: $x : \tau^+$ changes to $x : \tau^-$ and vice versa, $x : \tau^0$ stays fixed. For example, if $\Sigma = \{x : (\mathrm{Pr}^v \to \mathrm{Pr})^+\}$ then $\Sigma^- = \{x : (\mathrm{Pr}^v \to \mathrm{Pr})^-\}$. Given a context Σ and a formula φ, we say that φ has type τ in context Σ if $\Sigma \vdash \varphi : \tau$ can be derived via the typing rules in Figure 1. If $\Sigma \vdash \varphi : \tau$ for some Σ and τ then φ is *well-typed*. In particular, a closed formula is well typed if $\emptyset \vdash \varphi : \tau$. Typing judgments are unique if formulae are annotated with the correct types [8]. From now on, we will often omit the type annotations and tacitly assume that all formulae are well-typed if the type of the formula in question can be derived from context.

Semantics. Fix an LTS with underlying set S. The semantics of types are partially ordered sets defined inductively via

- $[\![\mathrm{Pr}]\!] = (2^S, \subseteq)$

$$\frac{}{\Sigma \vdash P \colon \mathrm{Pr}} \qquad \frac{\Sigma \vdash \varphi \colon \mathrm{Pr}}{\Sigma \vdash \Diamond_R \varphi \colon \mathrm{Pr}} \qquad \frac{\Sigma^- \vdash \varphi \colon \mathrm{Pr}}{\Sigma \vdash \neg \varphi \colon \mathrm{Pr}} \qquad \frac{\Sigma \vdash \varphi_1 \colon \mathrm{Pr} \quad \Sigma \vdash \varphi_2 \colon \mathrm{Pr}}{\Sigma \vdash \varphi_1 \vee \varphi_2 \colon \mathrm{Pr}}$$

$$\frac{}{\Sigma, x^+ \colon \tau \vdash x \colon \tau} \qquad \frac{}{\Sigma, X^+ \colon \tau \vdash X \colon \tau} \qquad \frac{\Sigma, x^v \colon \tau \vdash \varphi \colon \tau'}{\Sigma \vdash \lambda(x^v \colon \tau).\varphi \colon \tau^v \to \tau'}$$

$$\frac{\Sigma, X^+ \colon \tau \vdash \varphi \colon \tau}{\Sigma \vdash \mu(X \colon \tau).\varphi \colon \tau} \qquad \frac{\Sigma, X^+ \colon \tau \vdash \varphi \colon \tau}{\Sigma \vdash \nu(X \colon \tau).\varphi \colon \tau} \qquad \frac{\Sigma \vdash \varphi \colon \tau^+ \to \tau' \quad \Sigma \vdash \psi \colon \tau}{\Sigma \vdash (\varphi \psi) \colon \tau'}$$

$$\frac{\Sigma \vdash \varphi \colon \tau^- \to \tau' \quad \Sigma^- \vdash \psi \colon \tau}{\Sigma \vdash (\varphi \psi) \colon \tau'} \qquad \frac{\Sigma \vdash \varphi \colon \tau^0 \to \tau' \quad \Sigma \vdash \psi \colon \tau \quad \Sigma^- \vdash \psi \colon \tau}{\Sigma \vdash (\varphi \psi) \colon \tau'}$$

Fig. 1. Typing Rules for HFL

- $[\![\tau^v \to \tau']\!] = ([\![\tau']\!]^{([\![\tau]\!]^v)}, \sqsubseteq_{\tau^v \to \tau'})$

where $[\![\tau']\!]^{([\![\tau]\!]^v)}$ is the set of monotone, antitone or constant functions from $[\![\tau]\!]$ to $[\![\tau']\!]$, depending on v. The partial order $\sqsubseteq_{\tau \to \tau'}$ is defined via pointwise comparison: For $f, g \in [\![\tau']\!]^{([\![\tau]\!])}$ let $f \sqsubseteq_{\tau \to \tau'} g$ if and only if for all $x \in [\![\tau]\!]$ we have $f(x) \sqsubseteq_{\tau'} g(x)$. Then $\sqsubseteq_{\tau^+ \to \tau'}$ is the order induced by $\sqsubseteq_{\tau \to \tau'}$ and $\sqsubseteq_{\tau^- \to \tau'}$ is $\{(a,b) : (b,a) \in \{\sqsubseteq_{\tau \to \tau'}\}\}$ and $\sqsubseteq_{\tau^0 \to \tau'}$ is $\sqsubseteq_{\tau^+ \to \tau'} \cap \sqsubseteq_{\tau^- \to \tau'}$.

Note that $[\![\mathrm{Pr}]\!]$ is a boolean algebra and, hence, also a complete lattice. This makes $[\![\tau^v \to \tau']\!]$ also a complete lattice for all τ, τ', v. Let $\bigsqcup_\tau M$ and $\bigsqcap_\tau M$ denote the join and meet, respectively, of the set $M \subseteq [\![\tau]\!]$, and let $\top_\tau$ and $\bot_\tau$ denote the maximal and minimal elements of $[\![\tau]\!]$.

Let $\Sigma = X_1 \colon \tau_1, \ldots, X_n \colon \tau_n, x_1 \colon \tau_1', \ldots, x_m \colon \tau_m'$ be a context. An interpretation η is a partial map from the sets of variables $\mathcal{V}$ and $\mathcal{F}$ such that $\eta(X_i) \in [\![\tau_i]\!]$ for all $i \leq n$ and $\eta(x_j) \in [\![\tau_j']\!]$ for all $j \leq m$. Then $\eta[X \mapsto f]$ is the interpretation that maps X to f and agrees with η otherwise, similar for $\eta[x \to f]$.

We define the semantics of HFL over $\mathcal{T}$ inductively as in Figure 2. For well-typed, well-named φ, we write $\mathcal{T}, s \models_\eta \varphi$ if $s \in [\![\emptyset \vdash \varphi \colon \mathrm{Pr}]\!]_\eta$. We write $\mathcal{T}, s \models \varphi$ if φ is closed and η is the empty interpretation. Two formulae are *equivalent*, written $\varphi \equiv \varphi'$, if $[\![\Sigma \vdash \varphi]\!]_\eta = [\![\Sigma \vdash \varphi']\!]_\eta$ for all η, Σ.

Example 2. The formula

$$\varphi = \mu(X \colon (\mathrm{Pr} \to \mathrm{Pr}) \to \mathrm{Pr}).\lambda(f^+ \colon (\mathrm{Pr} \to \mathrm{Pr})).P \vee \big(X\,(\lambda(y^+ \colon \mathrm{Pr}).f(f\,y))\big)$$

says that a state where P holds can be reached after 2^n applications of f for some n. When applied to $\psi = \lambda(x^+ \colon \mathrm{Pr}).\Diamond_R x$ the term $(\varphi\,\psi)$ has type Pr and says that a state where P holds can be reached in exactly 2^n R-transitions.

Lemma 3. *For all well-named, well-typed $\varphi, \psi \in \mathrm{HFL}$ and for all environments η, we have that $[\![\varphi[\psi/x]]\!]_\eta = [\![\varphi]\!]_{\eta[x \mapsto [\![\psi]\!]_\eta]}$.*

Proof. By induction on the syntax of HFL-formulae.

$$\llbracket \Sigma \vdash P \colon \mathrm{Pr} \rrbracket_\eta = P^{\mathcal{T}}$$

$$\llbracket \Sigma \vdash \varphi \vee \psi \colon \mathrm{Pr} \rrbracket_\eta = \llbracket \Sigma \vdash \varphi \colon \mathrm{Pr} \rrbracket_\eta \cup \llbracket \Sigma \vdash \psi \colon \mathrm{Pr} \rrbracket_\eta$$

$$\llbracket \Sigma \vdash \neg \varphi \colon \mathrm{Pr} \rrbracket_\eta = S \setminus \llbracket \Sigma \vdash \varphi \rrbracket_\eta$$

$$\llbracket \Sigma \vdash \Diamond_R \varphi \colon \mathrm{Pr} \rrbracket_\eta = \{ s \in S \colon \exists s' \in R[s] \colon s' \in \llbracket \Sigma \vdash \varphi \colon \mathrm{Pr} \rrbracket_\eta \}$$

$$\llbracket \Sigma \vdash \lambda(x^v \colon \tau).\varphi \colon \tau^v \to \tau' \rrbracket_\eta = \{ f \in \llbracket \tau^v \to \tau' \rrbracket \colon \forall y \in \llbracket \tau \rrbracket .$$
$$f(y) = \llbracket \Sigma, x^v \colon \tau \vdash \varphi \colon \tau' \rrbracket_{\eta[x \mapsto y]} \}$$

$$\llbracket \Sigma \vdash X \colon \tau' \rrbracket_\eta = \eta(X)$$

$$\llbracket \Sigma \vdash x \colon \tau' \rrbracket_\eta = \eta(x)$$

$$\llbracket \Sigma \vdash \mu(X \colon \tau).\varphi \colon \tau \rrbracket_\eta = \bigsqcap \{ d \in \llbracket \tau \rrbracket \colon \llbracket \Sigma, (X \colon \tau^+) \vdash \varphi \colon \tau \rrbracket_{\eta[X \mapsto d]} \sqsubseteq_\tau d \}$$

$$\llbracket \Sigma \vdash \nu(X \colon \tau).\varphi \colon \tau \rrbracket_\eta = \bigsqcup \{ d \in \llbracket \tau \rrbracket \colon \llbracket \Sigma, (X \colon \tau^+) \vdash \varphi \colon \tau \rrbracket_{\eta[X \mapsto d]} \sqsupseteq_\tau d \}$$

$$\llbracket \Sigma \vdash (\varphi \, \psi) \colon \tau' \rrbracket_\eta = \llbracket \Sigma \vdash \varphi \colon \tau^v \to \tau' \rrbracket_\eta (\llbracket \Sigma \vdash \psi \colon \tau \rrbracket_\eta)$$

Fig. 2. Semantics of HFL

3   Alternating Parity Krivine Automata

Fix a well-named, well-typed and closed formula $\varphi \colon \mathrm{Pr}$ and an LTS $\mathcal{T}, s_0$ with state set S. We define an Alternating Parity Krivine Automaton (APKA) for φ.

Closures and Environments. Closures and environments are defined recursively. An environment maps variables to closures, a closure is a subformula of φ together with an environment that stipulates how free variables of φ are to be resolved. For convenience of proofs, environments are consecutively numbered. The first environment e_0 is used as a constant for the empty environment.

Definition 4. *The constant e_0 is the empty environment. If $\psi \in \mathrm{Sub}(\varphi)$ and e is an environment, then (ψ, e) is a closure. If e_i and e_j are environments, (ψ, e_i) is a closure, $i, j < k$, and $x \in \mathcal{V}$ then $e_k = (x \mapsto (\psi, e_i), e_j)$ is an environment.*

We write Clos for the set of closures, and Env for the set of environments. The intuition of a closure (ψ, e) is that ψ is to be evaluated in the context of e, i.e., free variables of ψ are resolved using e. An environment of the form $e_k = (x \mapsto (\psi, e_i), e_j)$ means that in the context of e_k, the variable x is bound to ψ, interpreted in context e_i. Other free variables are resolved in e_j, with both $i, j < k$. The root environment e_0 does not bind any variables. A variable lookup that ends here fails, but invariants of the APKA prevent that this happens.

Let (ψ, e_i) be a closure and let $\mathcal{E} = \{e_0, \ldots, e_n\}$ be a set of environments such that $i \leq n$. Then $e_i(\psi)$ denotes the expression where free variables in (ψ, e_i) are replaced by their value in the environment recursively via

$$e_0(\psi) = \psi$$
$$(x \mapsto (\psi', e_k), e_j)(\psi) = e_j(\psi[e_k(\psi')/x]).$$

Configurations. A configuration of an alternating parity Krivine automaton $\mathcal{A}$ for φ over $\mathcal{T}$ is of the form $C = (s, (\psi, e), \Gamma, \mathcal{E}, p)$, where $s \in S$, (ψ, e) is a closure, Γ is a finite stack of closures, $\mathcal{E}$ is a set of environments and $p \in \{\exists, \forall\}$ denotes the player that handles branching in the acceptance game (see below). $\bar{p}$ is the player opposite to p. We call a configuration *well-formed* if the following hold:

- $\mathcal{E} = \{e_1, \ldots, e_n\}$ for some $n \in \mathbb{N}$.
- No free variables: $e(\psi)$ does not have free variables from $\mathcal{V}$, nor does $e_i(\psi_i)$ for every environment $e_i = (x \mapsto (\psi_i, e_j), e_k)$. If $\Gamma = (\psi_m, e_{i_m}) \cdots (\psi_1, e_{i_1})$, then for all $1 \leq k \leq m$ we have $i_k \leq n$ and $e_{i_k}(\psi_k)$ has no free $\mathcal{V}$-variables.
- Variable types agree with their bound values: If x has type τ, and $e_i = (x \mapsto (\psi_i, e_j), e_k)$, then $\Sigma \vdash e_j(\psi_i) : \tau$ where Σ only consists of hypotheses for variables in $\mathcal{F}$ uniquely obtained from their annotations in φ.
- $\Sigma \vdash ((\cdots ((\psi\, \psi_m)\, \psi_{m-1}) \cdots)\, \psi_1) : \text{Pr}$ for Σ as before.

We write $\text{Conf}(\mathcal{A}, \mathcal{T})$ for the set of well-formed configurations of $\mathcal{A}$ over $\mathcal{T}$.

Let $C = (s, (\psi, e), \Gamma, \mathcal{E}, p)$ be a well-formed configuration such that $\Gamma = (\psi_n, e_{i_n}) \cdots (\psi_1, e_{i_1})$. Let $e_{i_k}(\psi_k) = \psi'_k$ for all $1 \leq k \leq n$ and let $e(\psi) = \psi'$. We write $[C]$ for the expression $((\cdots ((\psi'\, \psi'_n)\, \psi'_{n-1}) \cdots)\, \psi'_1)$ of type Pr. The configuration C is called *positive* if $\mathcal{T}, s \models [C]$ and $p = \exists$ or if $\mathcal{T}, s \not\models [C]$ and $p = \forall$. Otherwise, it is called *negative*.

The function $\text{lookup} : \mathcal{V} \times \text{Env} \to \text{Clos}$ is defined as follows:

$$\text{lookup}(x, e_0) = \text{undefined}$$
$$\text{lookup}(x, (x \mapsto (\varphi, e_j), e_k)) = (\varphi, e_j)$$
$$\text{lookup}(x, (y \mapsto (\varphi, e_j), e_k)) = \text{lookup}(x, e_k) \text{ if } y \neq x$$

Given a stack $\Gamma = (\psi_n, e_n), \cdots (\psi_1, e_1)$, define $\text{top}(\Gamma)$ as (ψ_n, e_n) and $\text{pop}(\Gamma)$ as $(\psi_{n-1}, e_{n-1}) \cdots (\psi_1, e_1)$ and $\text{push}(\Gamma, (\psi, e))$ as $(\psi, e)(\psi_n, e_n) \cdots (\psi_1, e_1)$.

Transition Relation. Acceptance for the APKA $\mathcal{A}$ for φ over $\mathcal{T}, s_0$ is defined via a two-player *acceptance game* between players $\exists$ and $\forall$ playing on $\text{Conf}(\mathcal{A}, \mathcal{T})$, starting from the well-formed configuration $(s_0, (\varphi, e_0), \epsilon, \emptyset, \exists)$. The automaton accepts if and only if $\exists$ has a winning strategy. The parameter p in a configuration denotes which player handles branching. We call branching by $\exists$ *nondeterministic* and branching by $\forall$ *universal*. Moves that never branch are called *deterministic*. If the play encounters a negation, p is changed to $\bar{p}$. The transition relation from configuration $C = (s, (\psi, e), \Gamma, \mathcal{E}, p)$ depends on the form of ψ as follows:

- If ψ is P, then $\exists$ wins if C is positive, otherwise $\forall$ wins.
- If ψ is $\psi_1 \vee \psi_2$, then p chooses $i \in \{1, 2\}$ and $(s, (\psi_i, e), \Gamma, \mathcal{E}, p)$ is the successor.
- If $\psi \neg \psi'$, then $(s, (\psi', e), \Gamma, \mathcal{E}, \bar{p})$ is the successor.
- If ψ is $\Diamond_R \psi_1$, then p chooses a successor $(t, (\psi_1, e), \Gamma, \mathcal{E}, p)$ such that $t \in R[s]$.
- If ψ is $\sigma X . \psi_1$, then $(s, (\psi_1, e), \Gamma, \mathcal{E}, p)$ is the successor.
- If ψ is $(\psi_1\, \psi_2)$, then $(s, (\psi_1, e), \text{push}(\Gamma, (\psi_2, e)), \mathcal{E}, p)$ is the successor.
- If ψ is $\lambda x . \psi_1$, then the configuration $(s, (\psi_1, e'), \text{pop}(\Gamma), \mathcal{E} \cup \{e' = (x \to (\text{top}(\Gamma), e))\}, p)$ is the successor.

- If ψ is x, then $(s, \mathrm{lookup}(x, e), \Gamma, \mathcal{E}, p)$ is the successor.
- If ψ is X, then $(s, (\mathrm{fp}_\varphi(X), e), \Gamma, \mathcal{E}, p)$ is the successor.

The automaton performs weak head reduction. It is not hard to see that individual transitions transform well-formed configurations into well-formed configurations. Moreover, if the automaton transitions deterministically then the successor configuration is positive if and only if the original configuration is so. If $\forall$ moves in C and C is positive then all successors are positive and if C is negative then so is at least one successor. Conversely, if $\exists$ moves in C and C is positive then there is at least one positive successor and if C is negative then no successor is positive. Because $\exists$ can enforce that play stays in positive configurations and because she wins finite plays that end in a positive configuration, $\exists$ wins all finite plays if the initial configuration is positive, i.e., if $\mathcal{T}, s_0 \models \varphi$.

The purpose of the call-by-name strategy, as opposed to, e.g., explicit syntactic β-reduction, is to keep the number of automaton states finite when projected to the formula component, which stays within the finite set of $\mathrm{Sub}(\varphi)$. This makes the transition relation finitely presentable. The set of environments is not finite, and, since it is tree-like, it is much more complicated than, e.g., the stack of a pushdown automaton. This is unavoidable with such an expressive logic.

Unfolding Trees. For the winning condition on infinite plays, we need additional machinery. The goal is to decide acceptance by a parity condition. Unfortunately, simply observing which fixpoints occur during the play is not sufficient. Already in the fragment FLC [13], some fixpoint expansions represent function calls that return after a finite number of steps. Such expansions are finite approximations at different arguments and should not be considered for the winning condition, since we are only interested in fixpoints with infinitely many approximations at the same argument. In order to filter out these finite approximations, we present the sequence of configurations of a given play in a tree-like fashion. This *unfolding tree* is an infinite, not necessarily finitely branching tree labeled by the formula component of configurations in the play, i.e., subformulae of φ. It is not to be confused with the game graph itself. This extends results from [13].

From an infinite play $(C_i)_{i \in \mathbb{N}} = (s_i, (\psi_i, e_i), \Gamma_i, \mathcal{E}_i, p_i)_{i \in \mathbb{N}}$ in the acceptance game for φ we inductively generate an unfolding tree via the formulae of configurations in the path. During the induction, there is always an active node that is added somewhere in the tree. Each step consists in applying a label to the active node and then adding a new active node. The induction starts with only one node labeled by $\varphi = \psi_0$. The active node is added as per below. In the ith step, the active node is labeled ψ_i and a new active node is added:

- If ψ_i is of the form $\varphi_1' \vee \varphi_2'$, $\Diamond_R \varphi'$, $\neg\varphi'$, $\sigma X.\varphi'$, $\lambda x.\varphi'$, X or $(\varphi_1' \ \varphi_2')$, then the active node is added as a left son of the previous active node.
- If $\psi_i = x$, then (ψ_{i+1}, e') equals $\mathrm{lookup}(x, e_i)$, so there is a node labeled $\psi_j = (\psi' \ \psi_{i+1})$ such that $e' = e_j$, because ψ_{i+1} was bound to a variable, and expressions bound to a variable occur as the operand of an application. There may already be right sons of the node labeled ψ_j, add the active node as a new rightmost son.

The unfolding tree is the limit of this inductive definition.[1] We tacitly identify a node in the tree labeled ψ_i with the configuration C_i. We say that a formula *occurs* on a node in the tree if that node is labeled by the formula. We say that a fixpoint variable occurs *under an even number of negations* if the play so far has passed an even number of negations, i.e., the parameter p for the configuration is $\exists$. If p is $\forall$, the variable occurs *under an odd number of negations*.

Finite paths in the unfolding tree represent function calls that return after finitely many steps. Hence, fixpoint expansions on such paths are not relevant to the winning condition. An unfolding tree can be infinitely branching in nodes labeled by an application, but we show that it contains exactly one infinite path, and only fixpoint expansions on that path are relevant for the winning condition.

Example 5. Consider the formula[2]

$$\Big(\big(\lambda f.(\mu X.\lambda x.(\Diamond_R x \vee \Box_{R'}((\nu Y.\lambda y.f(X(Y\,y)))x)))\big)\,(\lambda z.\Diamond_R z)\Big)\,\top,$$

which holds on the root of a tree such that from the root to the second level, the transitions are labeled with R' and labeled with R otherwise. A winning strategy for $\exists$ is to choose the second disjunct the first time X is evaluated and the first disjunct from then on. This strategy induces the unfolding tree in Figure 3, with the nodes labeled by the number of their configuration. The node labeled 2 is infinitely branching, and the rightmost branch of its left subtree is an infinite path. The only fixpoint variable that occurs on the infinite path is Y, while both X and Y are unfolded infinitely often. Since X occurs only on finite paths, the variable relevant for the winning condition is the greatest fixpoint variable Y.

Lemma 6. *Every infinite unfolding tree contains exactly one infinite path.*

This may seem obvious, because an unfolding tree is just a reorganization of an infinite path. However, we have to rule out the possibility that the tree is infinitely branching, but does not have a finite path. The proof of this lemma spans several pages and needs to be omitted due to space reasons.

Lemma 7. *On the infinite path in an unfolding tree, there is a unique outermost fixpoint variable that occurs infinitely often. Moreover it eventually occurs only under an even or odd number of negations.*

Proof. A path in an unfolding tree corresponds to expanding fixpoints and passing to subformulae. Since the path is infinite, there must be infinitely many occurrences of fixpoint variables on it, and one variable must occur infinitely often. For two variables that occur infinitely often, but are incomparable with

---

[1] A referee pointed out that Salvati and Walukiewicz use a similar technique [12] that can be traced back to [14]. They include the node of creation into the definition of a closure to facilitate construction of the tree. Since the unfolding tree is not part of ongoing plays, we do not incorporate this into the definition of closure.

[2] This is a modified version of an example from [13].

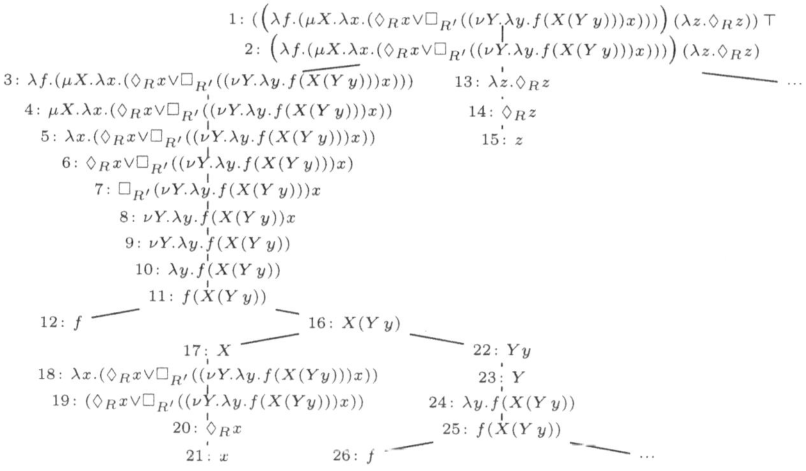

1: $\left(\left(\lambda f.(\mu X.\lambda x.(\Diamond_R x\vee\Box_{R'}((\nu Y.\lambda y.f(X(Y y)))x)))\right)(\lambda z.\Diamond_R z)\right)\top$

2: $\left(\lambda f.(\mu X.\lambda x.(\Diamond_R x\vee\Box_{R'}((\nu Y.\lambda y.f(X(Y y)))x)))\right)(\lambda z.\Diamond_R z))$

3: $\lambda f.(\mu X.\lambda x.(\Diamond_R x\vee\Box_{R'}((\nu Y.\lambda y.f(\overline{X(Y y)}))x)))$ 13: $\lambda z.\Diamond_R z$

4: $\mu X.\lambda x.(\Diamond_R x\vee\Box_{R'}((\nu Y.\lambda y.f(X(Y y)))x))$ 14: $\Diamond_R z$

5: $\lambda x.(\Diamond_R x\vee\Box_{R'}((\nu Y.\lambda y.f(X(Y y)))x))$ 15: z

6: $\Diamond_R x\vee\Box_{R'}((\nu Y.\lambda y.f(X(Y y)))x)$

7: $\Box_{R'}(\nu Y.\lambda y.f(X(Y y)))x$

8: $\nu Y.\lambda y.f(X(Y y))x$

9: $\nu Y.\lambda y.f(X(Y y))$

10: $\lambda y.f(X(Y y))$

11: $f(X(Y y))$

12: f 16: $X(Y y)$

17: X 22: $Y y$

18: $\lambda x.(\Diamond_R x\vee\Box_{R'}((\nu Y.\lambda y.f(X(Yy)))x))$ 23: Y

19: $(\Diamond_R x\vee\Box_{R'}((\nu Y.\lambda y.f(X(Yy)))x))$ 24: $\lambda y.f(X(Y y))$

20: $\Diamond_R x$ 25: $f(X(Y y))$

21: x 26: f

Fig. 3. The upper part of an infinite unfolding tree

respect to $<_{\mathrm{fp}_\varphi}$, there must be a common superformula that also occurs infinitely often. In particular, there must be a common fixpoint term superformula that occurs infinitely often. This fixpoint is outermore than the first two.

By the typing rules, an even number of negations occurs between two expansions of the same instance of a fixpoint variable. Hence, all expansions of a variable happen with the same p, until the play sees the formula $\sigma X.\psi$ again. For the outermost fixpoint variable, this happens only finitely often and eventually all occurrences of this variable on the infinite path are with the same p.

Acceptance Condition. $\exists$ wins an infinite play in the acceptance game if and only if the outermost variable that occurs infinitely often on the infinite path in the play's unfolding tree is a greatest fixpoint variable occuring with $p = \exists$ or if it is a least fixpoint variable occurring infinitely often with $p = \forall$.

Theorem 8. *The* APKA *for* φ *accepts* $\mathcal{T}, s_0$ *if and only if* $\mathcal{T}, s_0 \models \varphi$.

The difficult part of the proof is to establish that there is exactly one infinite path on the unfolding tree of an infinite play (Lemma 6). Together with the preceding lemma, this allows to adapt the proof of correctness for the FLC-model-checking game from [13]. Since the correctness proof here proceeds roughly alongside the same lines, we omit it. The proof idea is that $\exists$ moves such that the play always is locally consistent, i.e. the semantics of the current position are positive. In particular, whenever a play encounters a fixpoint variable that is in $\exists$'s domain (e.g., a least fixpoint under an even number of negations), $\exists$ can restrict herself to an approximation of the respective fixpoint up to a certain ordinal. She can play with that approximation in mind, staying in configurations that are positive even if restricted to the approximations, which is a winning strategy.

Corollary 9. HFL-*model-checking is decidable over finite LTS.*

Proof (Sketch). For finite transition systems, the game graph of the acceptance game for any APKA can be truncated at finite depth: Because all type lattices are finite (k-fold exponential height[3] for types of order k), nested unfolding of fixpoint variables can be limited to the height of the lattice in question.

The known k-EXPTIME completeness result for HFL-formulae of order at most k does not follow directly from Corollary 9. Consider the formula $\varphi = (\mu X.\lambda f.P \vee (X(\lambda y.f(f\,y)))) (\lambda x.\Diamond_R x)$ from Example 2, which says that a state where P holds can be reached in exactly 2^n R-transitions for some n. Since X is of order 2, the maximal number of unfoldings of X in a play is doubly exponential. This leaves the play with a triply exponential number of Diamonds after fixpoint unfolding is finished, a result not compatible with k-fold exponential time. Criteria to abort iteration early to avoid such problems are subject of ongoing research.

On Parity Conditions. The acceptance condition for APKA does not consider all fixpoint variables that are unfolded during a play of the acceptance game, because this would yield incorrect semantics (cf. Example 5). Consider the μ-calculus and FLC [7], both fragments of HFL. A computation model for the μ-calculus are alternating parity automata. FLC does not have an associated automaton model, but the author of [13] presents a parity game that captures the semantics of FLC. Since both logics are fragments of HFL, APKA are also an automaton model for these logics. Clearly, an alternating parity automaton is just an APKA that does use neither the stack nor environment features. Consequently, the unfolding tree for a run is degenerate, i.e., a straight path. If one considers the FLC-model-checking game as the acceptance game of an APKA, such an APKA does not use the environment feature either, so function parameters are used in a last-in-first-out fashion due to the stack mechanics.

These differences are reflected in the parity conditions for the respective automata. For the μ-calculus, if a play in the acceptance game unfolds a fixpoint variable, this variable will count against the player responsible for this type of variable, e.g., against $\exists$ for least fixpoints under an even number of negations. In the FLC-model-checking game, an occurrence of a fixpoint variable only counts towards the winning condition if it is *stack-increasing*, i.e., if the elements on the stack at the moment of occurrence are never read (cf. also the so-called *stair-parity* conditions [6]). For FLC, being a stack-increasing variable coincides with being on the infinite path of the unfolding tree. Since for FLC, the unfolding tree has at most binary branching and the infinite path can be obtained by always choosing the rightmost branch, it is not a priori known whether an occurrence of a variable does count towards the winning condition. If it does not count, the players will know after finitely many steps, the occurrence can be discarded, and the winner of the play can be obtained from the remaining occurrences without explicitly constructing the unfolding tree. For full HFL, however, it is not possible to tell whether an occurrence of a fixpoint counts towards the winning condition until after the game is finished. This is because here, the location of the

---

[3] Precise bounds can be found in Lemma 3.5 in [9].

infinite path in the unfolding tree is not subject to any constraints. In particular, it cannot be obtained by always choosing the rightmost branch, (cf. Example 5). Hence, the tree has to be explicitly constructed to find the winner of a play.

4 Further Work

Further research is needed into conditions to stop to explore a given path in the game graph of the APKA-acceptance game over finite LTS. As seen in the discussion after Corollary 9, just limiting consecutive fixpoint unfoldings via the height of the respective type lattice is suboptimal. In the μ-calculus model-checking game, a play is decided once the same combination of subformula and state occurs twice. Already for FLC, the presence of a stack makes things harder. For HFL, conditions on the environment structure will also need to be included.

Questions that are settled for the modal μ-calculus, and, to a certain extent, for FLC, are still open for HFL. Consider syntactic negation: Sometimes negation is only allowed at the ground level (e.g., [8]), sometimes negation may occur in front of expressions of arbitrary types, subject to monotonicity conditions (e.g., [9]). In both cases, negation cannot be pushed inwards to occur only in front of atomic expressions. Based on preliminary research we conjecture that it is possible to obtain a negation normal form at the cost of blowup in formula size. The exact semantics of higher-order negation are also not clear. The game-theoretic formulation in the context of APKA may help to understand this better.

HFL also differs from the μ-calculus and FLC with respect to alternation. For both we know how to properly define alternation of fixpoints, and the alternation hierarchy is strict [15,13]. For HFL, it is not even clear how exactly to define alternation. In particular, it is possible to syntactically hide alternation by λ-abstracting away outer fixpoints, trading syntactic alternation for a higher order. Even if a meaningful alternation hierarchy exists for HFL, it is possible that the alternation hierarchy of the μ-calculus collapses to a finite level inside HFL.

For both questions, it seems useful to relax the way an APKA is defined very closely alongside the syntax tree of an HFL-formula. Consider alternating parity automata, which first were conceived as an alternative presentation of the syntax tree of μ-calculus- or CTL-formulae. Now they are considered in their own right, without a specific formula in mind, because they proved to be a useful tool for a number of problems. While this may not necessarily be the case for APKA, it might be useful to do the same for APKA to learn properties of APKA not visible at first glance, and then to apply them back to HFL-formulae.

Finally, the formalism of a Krivine machine is used in the investigation of higher-order recursion schemes and collapsible pushdown automata. Salvati and Walukiewicz use the Krivine machine for this in [12], and Carayol and Serre expand on their work in [16]. The difference to our work is that their work on higher-order recursion schemes and collapsible pushdown automata concerns model-checking of a regular logic (e.g., MSO) over higher-order structures, while we explored model-checking a higher-order logic on finite and, hence, regular, structures. The relation between the two problems warrants further investigation, given that the Krivine machine appears on both sides.

Acknowledgments. I thank Martin Lange and Étienne Lozes for discussing the topic with me at length and for many helpful suggestions.

References

1. Rabin, M.O.: Decidability of second-order theories and automata on infinite trees. Transactions of the American Mathematical Society 141, 1–35 (1969)
2. Büchi, J.R.: Weak second-order arithmetic and finite automata. Mathematical Logic Quarterly 6(1-6), 66–92 (1960)
3. Kupferman, O., Vardi, M.Y., Wolper, P.: An automata-theoretic approach to branching-time model checking. J. ACM 47(2), 312–360 (2000)
4. Lange, M., Lozes, É., Guzmán, M.V.: Model-checking process equivalences. In: Faella, M., Murano, A. (eds.) GandALF. EPTCS, vol. 96, pp. 43–56 (2012)
5. Alur, R., Madhusudan, P.: Visibly pushdown languages. In: Babai, L. (ed.) STOC, pp. 202–211. ACM (2004)
6. Löding, C., Madhusudan, P., Serre, O.: Visibly pushdown games. In: Lodaya, K., Mahajan, M. (eds.) FSTTCS 2004. LNCS, vol. 3328, pp. 408–420. Springer, Heidelberg (2004)
7. Müller-Olm, M.: A modal fixpoint logic with chop. In: Meinel, C., Tison, S. (eds.) STACS 1999. LNCS, vol. 1563, pp. 510–520. Springer, Heidelberg (1999)
8. Viswanathan, M., Viswanathan, R.: A higher order modal fixed point logic. In: Gardner, P., Yoshida, N. (eds.) CONCUR 2004. LNCS, vol. 3170, pp. 512–528. Springer, Heidelberg (2004)
9. Axelsson, R., Lange, M., Somla, R.: The complexity of model checking higher-order fixpoint logic. Logical Methods in Computer Science 3(2) (2007)
10. Axelsson, R., Hague, M., Kreutzer, S., Lange, M., Latte, M.: Extended computation tree logic. In: Fermüller, C.G., Voronkov, A. (eds.) LPAR-17. LNCS, vol. 6397, pp. 67–81. Springer, Heidelberg (2010)
11. Krivine, J.L.: A call-by-name lambda-calculus machine. Higher-Order and Symbolic Computation 20(3), 199–207 (2007)
12. Salvati, S., Walukiewicz, I.: Krivine machines and higher-order schemes. In: Aceto, L., Henzinger, M., Sgall, J. (eds.) ICALP 2011, Part II. LNCS, vol. 6756, pp. 162–173. Springer, Heidelberg (2011)
13. Lange, M.: The alternation hierarchy in fixpoint logic with chop is strict too. Inf. Comput. 204(9), 1346–1367 (2006)
14. Walukiewicz, I.: Pushdown processes: Games and model-checking. Inf. Comput. 164(2), 234–263 (2001)
15. Bradfield, J.C.: The modal mu-calculus alternation hierarchy is strict. In: Sassone, V., Montanari, U. (eds.) CONCUR 1996. LNCS, vol. 1119, pp. 233–246. Springer, Heidelberg (1996)
16. Carayol, A., Serre, O.: Collapsible pushdown automata and labeled recursion schemes: Equivalence, safety and effective selection. In: LICS, pp. 165–174. IEEE (2012)

Advances in Parametric Real-Time Reasoning*

Daniel Bundala and Joël Ouaknine

Department of Computer Science, University of Oxford
Wolfson Building, Parks Road, Oxford, OX1 3QD, UK

Abstract. We study the decidability and complexity of the reachability problem in parametric timed automata. The problem was introduced 20 years ago by Alur, Henzinger, and Vardi in [1], where they showed decidability in the case of a single parametric clock, and undecidability for timed automata with three or more parametric clocks.

By translating such problems as reachability questions in certain extensions of parametric one-counter machines, we show that, in the case of two parametric clocks (and arbitrarily many nonparametric clocks), reachability is decidable for parametric timed automata with a single parameter, and is moreover $\mathsf{PSACE}^{\mathsf{NEXP}}$-hard. In addition, in the case of a single parametric clock (with arbitrarily many nonparametric clocks and arbitrarily many parameters), we show that the reachability problem is NEXP-complete, improving the nonelementary decision procedure of Alur *et al*.

1 Introduction

The problem of reachability in parametric timed automata (PTA) was introduced over two decades ago in a seminal paper of Alur, Henzinger, and Vardi [1]: given a timed automaton in which some of the constants appearing within guards on transitions are parameters, is there some assignment of integers to the parameters such that an accepting location of the resulting concrete timed automaton becomes reachable?

In this framework, a clock is said to be *nonparametric* if it is never compared with a parameter, and *parametric* otherwise. Alur *et al*. [1] showed that, for timed automata with a single parametric clock, reachability is decidable (irrespective of the number of nonparametric clocks). The decision procedure given in [1] however has provably nonelementary complexity. In addition, [1] showed that reachability becomes undecidable for timed automata with at least three parametric clocks.

The decidability of reachability for PTAs with *two* parametric clocks (and arbitrarily many nonparametric clocks) was left open in [1], with hardly any progress (partial or otherwise) that we are aware of in the intervening period. The problem was shown in [1] to subsume the question of reachability in Ibarra *et al*.'s "simple programs" [9], also open for over 20 years, as well as a decision problem for a fragment of Presburger arithmetic with divisibility.

---

* Full version of the paper is available at: `http://www.cs.ox.ac.uk/people/joel.ouaknine/publications/advances_parametric14abs.html`

E. Csuhaj-Varjú et al. (Eds.): MFCS 2014, Part I, LNCS 8634, pp. 123–134, 2014.

Our main results are as follows: (i) We show that, in the case of two parametric clocks (and arbitrarily many nonparametric clocks), reachability is decidable for PTAs with a *single* parameter. Furthermore, we establish a $\mathsf{PSPACE}^{\mathsf{NEXP}}$ lower bound on the complexity of this problem. (ii) In the case of a single parametric clock (with arbitrarily many nonparametric clocks and arbitrarily many parameters), we show that the reachability problem is NEXP-complete, improving the nonelementary decision procedure of Alur *et al.*

Our results rest in part on new developments in the theory of one-counter machines [5], their encodings in Presburger arithmetic [4], and their application to reachability in (ordinary) timed automata [6,3]. We achieve this by restricting our attention to PTAs with *closed* (i.e. *non-strict*) clock constraints. As parameters are restricted to ranging over integers[1], standard digitisation techniques apply [7,15], reducing the reachability problem over dense time to discrete (integer) time. (Alternatively, our results also apply directly to timed automata interpreted over discrete time, regardless of the type of constraints used.) The restriction to integer time enables us, among others, to keep track of the values of two parametric clocks using a single counter, in effect reducing the reachability problem for timed automata with two parametric clocks to a halting problem for parametric one-counter machines.

Related Work. The decidability of reachability for PTAs can be achieved in certain restricted settings, for instance by bounding the allowed range of the parameters [10] or by requiring that parameters only ever appear either as upper or lower bounds, but never as both [8]: in the latter case, if there is a solution at all then there is one in which parameters are set either to zero or infinity. The primary concern in such restricted settings is usually the development of practical verification tools, and indeed the resulting algorithms tend to have comparatively good complexity.

Miller [14] observed that over dense time and with parameters allowed to range over rational numbers, reachability for PTA becomes undecidable already with a single parametric clock. In the same setting, Doyen [2] showed undecidability of reachability for two parametric clocks even when using exclusively open (i.e. strict) time constraints.

A connection between timed automata and counter machines was previously established in nonparametric settings [6], and used to show that reachability for (ordinary) two-clock timed automata is polynomial-time equivalent to the halting problem for one-counter machines, even when constants are encoded in binary. Unfortunately, it is not obvious how to extend and generalise this construction to PTA, specifically in the case of two parametric clocks and an arbitrary number of nonparametric clocks, as we handle in the present paper. The reduction of [6] was used in [3] to show that halting for bounded one-counter machines, and

---

[1] Other researchers have considered variations in which parameters are allowed to range over rationals, yielding different outcomes as regards the decidability of reachability; see, e.g., [14,2], discussed further below.

hence reachability for two-clock timed automata, is PSPACE-complete, solving what had been a longstanding open problem.

Finally, parametric one-counter machines without upper bounds imposed on the value of the counter were studied in [5], where reachability was shown to be decidable. The techniques used in [5] make crucial use of the unboundedness of the counter and therefore do not appear applicable in the present setting.

2 Preliminaries

We now give definitions used throughout the rest of the paper. A *timed automaton* is a finite automaton extended with clocks; each clock measuring the time since it was last reset. A parametric timed automaton is obtained by replacing the known constants in the guards by parameters.

Formally, let P be a finite set of *parameters*. An *assignment* for P is a function $\gamma : P \to \mathbb{N}$ assigning a *natural number* to each parameter. A *parametric timed automaton (PTA)* $A = (S, s_0, C, P, F, E)$ is a tuple where S is the set of states, $s_0 \in S$ is the initial state, C is the set of clocks, P is the set of parameters, $F \subseteq S$ is the set of final states and $E \subseteq S \times S \times 2^C \times G(C, P)$ is the set of edges where $G(C, P)$ is the set of guards of the form $x \leq v, x \geq v$ where x is a clock and $v \in \mathbb{N} \cup P$. An edge (s, s', R, G) is from state s to state s'. Set R specifies which clocks are reset. A clock is *parametrically constrained* if it is compared to a parameter in some guard. The class of PTAs with k parametrically constrained clocks is denoted k-PTA. If γ is an assignment to parameters then A^γ denotes the automaton obtained by setting each parameter $p \in P$ to $\gamma(p)$.

A *configuration* (s, ν) of A^γ consists of state s and function $\nu : C \to \mathbb{N}$ assigning a value to each clock. A transition exists from configuration (s, ν) to (s', ν') in A^γ, written $(s, \nu) \to (s', \nu')$, if either there is $t \in \mathbb{N}$ such that $\nu(c) + t = \nu'(c)$ for every clock $c \in C$ or there is an edge $e = (s, s', R, G) \in E$ such that G is satisfied for current clock values and if $c \in R$ then $\nu'(c) = 0$ and if $c \notin R$ then $\nu'(c) = \nu(c)$.

The initial clock valuation ν_0 assigns 0 to every clock. A *run* of a machine is a sequence $\pi = c_1, c_2, \ldots, c_k$ of configurations such that $c_i \to c_{i+1}$ for each i. A run is called *accepting* if c_1 is the initial configuration (s_0, ν_0) and c_k is in a final state. The *existential halting problem*, also known as parametric reachability or the emptiness problem, asks whether there is some parameter valuation γ such that A^γ has an accepting run. From here onwards, we omit "existential" and write simply "halting problem". We say that two automata A_1 and A_2 have *equivalent halting problem* if A_1 halts if and only if A_2 halts.

Given a run π, we use $start(\pi) = c_1$ and $end(\pi) = c_k$ to denote the first and the last configuration of the run, respectively. If τ is a run, we write $\pi \to \tau$ if the runs can be connected by a transition, i.e. $end(\pi) \to start(\tau)$.

A *parametric timed 0/1 automaton* [1] $A = (S, s_0, C, P, F, E)$ is a timed automaton such that each edge $e \in E$ is labeled by a time increment $t \in \{0, 1\}$. A transition from (s, ν) to (s', ν') is valid only if $\nu'(c) - \nu(c) = t$ for each $c \in C$ not reset by the edge giving rise to the transition.

A *one-counter machine* is a finite-state machine equipped with a single counter. Each edge is labelled by an integer, which is added to the counter whenever that edge is taken. The counter is required to be nonnegative at all times. E.g., subtracting $c \in \mathbb{N}$ in one transition and adding c in the next transition leaves the counter unchanged but can be performed only if the counter is at least c.

A *bounded one-counter machine* also allows $\leq x$ edges. Such an edge can be taken only when the counter is at most x. Reachability in these two classes of counter machines are respectively known to be NP-complete [5] and PSPACE-complete [3] if the numbers are encoded in binary.

Parametric machines are obtained by replacing the known constants by parameters. A *parametric bounded one-counter machine (PBOCA)* $C = (S, s_0, F, P, E, \lambda)$ is a tuple where S is the set of states, s_0 is the initial state, $F \subseteq S$ are the final states, P is the set of parameters, $E \subseteq S \times S$ is the set of edges and $\lambda : E \to Op$ assigns an operation to each edge and has codomain Op: $\{+c, -c, +p, -p, \leq c, = c, \geq c, \leq p, = p, \geq p, +[0, p], \equiv 0 \bmod c \ : \ c \in \mathbb{N}, p \in P\}$.

A *parametric one-counter machine* allows only operations: $\pm c, \pm p, \geq c, \geq p, = 0$. Note that parametric one-counter machines are a subclass of parametric bounded one-counter machines.

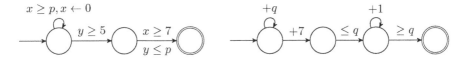

Fig. 1. A parametric timed automaton (left) and a parametric bounded one-counter machine (right). The final states are reachable if, for example, $p = 10$ and $q = 11$.

A *configuration* (s, x) of C consists of a state $s \in S$ and counter value $x \in \mathbb{N}$. Thus, the counter is always nonnegative. Machine C starts in state s_0 and counter equal to 0 and then takes individual edges updating the counter. We use $counter(s, x) = x$ to denote the counter value in a configuration. We extend the definition to runs componentwise and write $counter(\pi) \leq C$ (resp. $counter(\pi) \geq C$) if the comparison holds for every element: $\forall i \ . \ counter(\pi(i)) \leq C$ (resp. $\forall i \ . \ counter(\pi(i)) \geq C$).

Let Z be a (nonparametric) one-counter machine. For configurations c, d of Z and numbers $x, y \in \mathbb{N}$, we write $(c, d) \in Z(x, y)$ if there is a run $\pi : c \to d$ such that the counter stays between x and y, i.e. $x < counter(\pi) < y$.

For a parameter assignment γ, configuration (s', x') is directly reachable from (s, x) (written $(s, x) \to (s', x')$) in C^γ if an edge $e = (s, s') \in E$ exists such that
- if $\lambda(e) = \pm c, c \in \mathbb{N}$ then $x \pm c = x'$
- if $\lambda(e) = \pm p, p \in P$ then $x \pm \gamma(p) = x'$
- if $\lambda(e) = \sim c, c \in \mathbb{N}$ then $x = x'$ and $x \sim c$ where $\sim \in \{\leq, \geq\}$
- if $\lambda(e) = \sim p, p \in P$ then $x = x'$ and $x \sim \gamma(p)$ where $\sim \in \{\leq, \geq\}$
- if $\lambda(e) = +[0, p], p \in P$ then $x \leq x' \leq x + \gamma(p)$
- if $\lambda(e) = \equiv 0 \bmod c, x \in \mathbb{Z}$ then $x = x'$ and $x \equiv 0 \bmod c$

The *existential halting problem* asks whether there is a parameter valuation γ such that C^γ has an accepting run.

2.1   Presburger Arithmetic

Presburger Arithmetic with Divisibility is the first-order logical theory of $\langle \mathbb{N}, <, +, |, 0, 1 \rangle$. The existential fragment (formulae of the form $\exists x_1, x_2, \ldots, x_k.\varphi$ where φ has no quantifiers) is denoted as $\exists$PAD . The satisfiability of $\exists$PAD formulae was shown decidable in [12] and in NP [13]. Given a set $S \subseteq \mathbb{N}^k$ we say that S is $\exists$PAD **definable** if there is a finite set R of $\exists$PAD formulae, each formula with free variables $x_1, \ldots x_k$ such that $(n_1, \ldots, n_k) \in S \iff \bigvee_{\varphi \in R} \varphi(n_1, \ldots, n_k)$. Note that $\exists$PAD sets are closed under finite union, intersection and projection. It was shown in [4,5] that the reachability relation of parametric one-counter machines is $\exists$PAD definable.

Lemma 1 ([4], Lemma 4.2.2). *Given a parametric one-counter machine B and states s, t, the relation $Reach(B, s, t) = \{(x, y, n_1, \ldots, n_k) \mid (s, x) \to^* (t, y)$ in B^γ where $\gamma(p_i) = n_i\}$ is $\exists$PAD definable.*

2.2   Nonparametric Clock Elimination

Let A be a PTA. By modifying the region construction, we show how to build a PTA with equivalent halting problem without nonparametric clocks.

Once the value of a nonparametric clock c is above the largest constant appearing in A, the precise value of c does not affect any comparison. Now, the value of c is always a natural number. Hence, we eliminate nonparametric clocks by storing in the state space of C the values of the clocks up to the largest constant. However, we must ensure that the eliminated clocks progress simultaneously with the remaining parametric ones. This motivates $0/1$ timed automata where the $+1$ updates correspond to the progress of time whereas the $+0$ updates correspond to taking an edge in A. Formally:

Lemma 2 ([1]). *Let $A = (S, s_0, C, P, F, E)$ be a PTA. Then there is a parametric $0/1$ timed automaton $A' = (S', s_0', C', P', F', E')$ such that $C' \subseteq C$ contains only parametrically constrained clocks of C and A and A' have equivalent halting problem. Moreover, $|A'| = O(2^{|A|})$.*

3   One Parametric Clock

For the rest of the section, fix a 1-PTA A. We show how to decide the halting problem for A. By Lemma 2, there is an exponentially larger parametric $0/1$ automaton B with one (parametrically constrained) clock and equivalent halting problem. In Lemma 4 we show how to eliminate clock resets from B by introducing -1 edges, thereby turning B into a PBOCA. Hence, to decide the halting problem for A it suffices to decide the halting problem for a PBOCA with only $-1, 0, +1$ counter updates. We establish such a result in Theorem 5. Hence:

Theorem 3. *The halting problem for 1-PTAs is decidable in* NEXP.

Decidability of the halting problem 1-PTAs originally appeared in [1], albeit with nonelementary complexity. We give a completely different proof using one-counter machines yielding a NEXP algorithm. Later we show that the problem is also NEXP-hard. In the full version of the paper we prove the technical lemma:

Lemma 4. *Let B be a parametric 0/1 timed one-clock automaton. Then there is a PBOCA C such that B and C have equivalent halting problem. Further, all updates in C are either $-1, 0$ or $+1$ and $|C| = O(|B|)$.*

3.1   Decidability for Counter Machines with Constant Updates

We now show how to decide the halting problem for PBOCAs with all counter updates either $-1, 0$ or $+1$. Fix such a machine C. To show that C halts, we have to find an assignment γ and an accepting run π in C^γ. Even without knowing γ, we show that π splits into subruns of a simple form independent of γ the existence of which is reducible to satisfiability of certain $\exists$PAD formulae.

Let γ be a parameter assignment and assume that we guessed the order of parameters, let's say, $\gamma(p_1) < \gamma(p_2) < \ldots < \gamma(p_k)$, but not their precise values. Let c_1 and c_2 be arbitrary configurations of C^γ such that $c_1 \to^* c_2$ in C^γ and consider a shortest run $\pi : c_1 \to c_2$. There is a constant $M \in \mathbb{N}$, determined in Lemma 7, such that the run π can be factored into subruns between successive parameters and subruns around individual parameters. Formally, $\pi = \pi_0 \to \pi_1 \to \pi_2 \to \cdots \to \pi_l$ such that (π_0 can be possible empty)
  – Even-indexed runs: $\gamma(p) - M \leq counter(\pi_{2i}) \leq \gamma(p) + M$ for a parameter p,
  – Odd-indexed runs: $\gamma(p_r) + M < counter(\pi_{2i+1}) < \gamma(p_{r+1}) - M$ for some consecutive parameters $\gamma(p_r) < \gamma(p_{r+1})$,
  – For every i, the runs π_i and π_{i+1} are joined by an edge $end(\pi_i) \to start(\pi_{i+1})$.
Notice that every edge in C changes the counter by at most 1. Hence, we have $counter(start(\pi_{2i+1})) = p_r + M + 1$ or $counter(start(\pi_{2i+1})) = p_{r+1} - M - 1$. Thus, $start(\pi_i)$ is always of the form $start(\pi_i) = (s_i, p_{f(i)} + x_i)$ for some state s_i, some $|x_i| \in \{M, M+1\}$ and parameter $p_{f(i)}$. Hence, $start(\pi_i)$ is uniquely determined by the triple $(s_i, f(i), x_i)$. Similarly, $end(\pi_i)$ is uniquely determined by some triple $(t_i, g(i), y_i)$ with $|y_i| \in \{M, M+1\}$.

By minimality, π visits every configuration only once. Hence an odd-indexed run can start in only one of $2nk$ configurations (n states, k parameters). Hence, the number of odd-indexed runs, and hence the total number of runs is $O(nk)$.

To show that there is a run from c_1 to c_2 we guess a factoring of the above form. We shall show (justifying the choice of M) in Lemma 8 that the odd-indexed runs π_{2i+1} correspond to runs in some one-counter machine $C_{h(2i+1)}$. By Lemma 1, the existence of a run in $C_{h(2i+1)}$ is $\exists$PAD expressible as: $\varphi_{2i+1} = Reach(C_{h(2i+1)}, s_{2i+1}, t_{2i+1})(n_{f(2i+1)} + x_{2i+1}, n_{g(2i+1)} + y_{2i+1}, n_1, \ldots, n_k)$.

In Lemma 9, we show that the even-indexed runs are independent of γ, can be precomputed and the reachability relation can be hardwired into the formula. Thus, we express the existence of a particular factoring from c_1 to c_2 as $\varphi =$

$\bigwedge_i \varphi_{2i+1} \wedge \psi(f, g, h, \overrightarrow{s}, \overrightarrow{t}, \overrightarrow{x}, \overrightarrow{y}) \wedge \bigwedge_i (n_i + M < n_{i+1})$ where the middle term encodes that the odd- and even-indexed runs are adjacent (directly computable) and that the even-indexed runs are valid (Lemma 9). The last conjunct encodes the technical restriction $\gamma(p_i) + M < \gamma(p_{i+1})$ imposed in Lemmas 8 and 9.

The restriction is relaxed as follows. First, if the parameters are not in the increasing order $\gamma(p_i) < \gamma(p_{i+1})$ then we relabel the parameters and build the appropriate formula. If $\gamma(p_i) \leq \gamma(p_{i+1}) < \gamma(p_i) + M$ then M depends only on $|C|$ (Lemma 7) and so only finitely many possibilities exist for $\gamma(p_{i+1}) - \gamma(p_i)$. Hence we replace each occurrence of p_{i+1} in C by $p_i + w$ for the appropriate $w < M$.

Theorem 5. *Given states* $s, t \in C$ *the set* $G(C, s, t) = \{(x, y, n_1, \ldots, n_k) \mid (s, x) \to^* (t, y)$ *in* C^γ *where* $\gamma(p_i) = n_i\}$ *is* $\exists$PAD *definable.*

Recall that satisfiability of $\exists$PAD formulae is in NP [13] and that $|C|$ is exponential in $|A|$ (Lemmas 2 and 4). Hence, Theorem 3 follows. We have also proved the corresponding lower bound, in fact, already for a single parameter.

Theorem 6. *The halting problem for* 1-*PTAs with one parameter is* NEXP-*hard.*

The proof of $\exists$PAD definability relied on two lemmas that we prove now. First, we show how to calculate the odd-indexed runs. Let c_1, c_2 be configurations of C^γ between two successive parameters: $\gamma(p_i) < counter(c_1), counter(c_2) < \gamma(p_{i+1})$.

Consider the counter machine C_i obtained from C by evaluating all comparisons as if the counter was between $\gamma(p_i)$ and $\gamma(p_{i+1})$. Formally, C_i is obtained from C by removing all $\geq p_j$ and $\leq p_k$ edges for $k \leq i < j$ and all $\leq p_j$ and $\geq p_k$ edges for $k \leq i < j$ are replaced by $+0$ edges. Further, for $i > 0$ and $c \in \mathbb{N}$ we also remove all $\leq c$ edges from C_i. Note that the definition of C_i's depends only on the order of parameters in γ.

During a run $\pi : c_1 \to c_2$ in C_i, the counter can become less than $\gamma(p_i)$ or greater than $\gamma(p_{i+1})$. So π does not necessarily correspond to a run in C. However, notice that C_i is a one-counter machine without parameters or $\leq x$ comparisons, i.e. an ordinary one-counter machine and thus has the following property [11]: If there is a run between two configurations then there is a run where the counter does not deviate much from the initial and the final counter value:

Lemma 7 ([11], Lemma 42). *Let* C_i *be as above. There is a constant* M *(polynomial in* $|C_i|$*) s.t. for any configurations* c_1 *and* c_2 *of* C_i *if* $c_1 \to^* c_2$ *then there is a run* $\pi : c_1 \to c_2$ *such that* $U - M \leq counter(\pi) \leq V + M$ *where* $U = \min(counter(c_1), counter(c_2))$ *and* $V = \max(counter(c_1), counter(c_2))$.

So as long as $\gamma(p_1) + M < counter(c_1), counter(c_2) < \gamma(p_2) - M$, the runs $c_1 \to c_2$ in C_i correspond to runs in C. See the full version for the proof:

Lemma 8. *Let* γ *be an assignment with* $\gamma(p_i) + M < \gamma(p_{i+1})$ *for all* i. *Let* c, d *be configurations with* $\gamma(p_i) + M < counter(c), counter(d) < \gamma(p_{i+1}) - M$. *Then* $(c, d) \in C^\gamma(\gamma(p_i), \gamma(p_{i+1})) \iff c \to^* d$ *in* C_i^γ.

For the even-indexed runs, the reachability around individual parameters, i.e. in intervals $(\gamma(p_i) - M, \gamma(p_i) + M)$, can be precomputed. Suppose that $\gamma(p_{i-1}) < \gamma(p_i) - M < \gamma(p_i) + M < \gamma(p_{i+1})$ so that the interval $(\gamma(p_i) - M, \gamma(p_i) + M)$ does not contain $\gamma(p_{i-1})$ or $\gamma(p_{i+1})$. Let $-M < x, y < M$ and let π be a run from $(s, \gamma(p_i) + x)$ to $(t, \gamma(p_i) + y)$ such that $\gamma(p_i) - M \leq counter(\pi) \leq \gamma(p_i) + M$. Then for every component $\pi(i)$, we can write $counter(\pi(j)) = \gamma(p_i) + z_j$ for some $-M \leq z_j \leq M$. But now, the run π is valid for any specific value of $\gamma(p_i)$ as only z_j determines which edges are enabled in C^γ. (See the full version)

Lemma 9. *Let γ, δ be parameter assignments with $\gamma(p_i) + M < \gamma(p_{i+1}), \delta(p_i) + M < \delta(p_{i+1})$ for all i. Let $s, t \in c$ be states and $-M < x, y < M$ integers. Then*

$$((s, \gamma(p_i) + x), (t, \gamma(p_i) + y)) \in C^\gamma(\gamma(p_i) - M, \gamma(p_i) + M) \iff$$
$$((s, \delta(p_i) + x), (t, \delta(p_i) + y)) \in C^\delta(\delta(p_i) - M, \delta(p_i) + M)$$

Furthermore, it is decidable in polynomial time whether $((s, \gamma(p_i) + x), (t, \gamma(p_i) + y)) \in C^\gamma(\gamma(p_i) - M, \gamma(p_i) + M)$ for any (and all) such assignment γ.

4   Two Parametric Clocks

We now show that the halting problem for 2-PTAs is equivalent to the halting problem for PBOCAs. The equivalence is used in Section 4.2 to show decidability of the halting problem in certain classes of 2-PTAs.

First, observe that a counter can be stored as a difference of two clocks, which can be used (see the full version) to show the easier direction of the equivalence.

Theorem 10. *Let C be a PBOCA. Then there is a 2-PTA A such that A and C have equivalent halting problem. Moreover, if C has no '$\equiv 0 \bmod c$' edges then A has no nonparametric clocks. Otherwise, A has one nonparametric clock.*

4.1   Reduction to Parametric Bounded One-Counter Machines

For the converse, fix A to be a 2-PTA. We reduce A to a PBOCA C. To begin, we construct (Lemma 2) a parametric 0/1 timed automaton B with two parametrically constrained clocks, denoted x and y, with the halting problem equivalent to A. We then transform B to C. Denote the counter of C by z.

For the time being, we need to relax the assumption that z stays nonnegative. That is, subtracting 5 when the counter is 2 results in the counter being -3. In Remark 12 we later show how to restore the nonnegativity of the counter.

The idea of the reduction is that, after a clock of B is reset, that clock equals zero, so we use z to store the value of the other clock. We construct C in such a way that after a reset of y, counter z stores the value of x and after a reset of x, counter z stores $-y$. Initially C starts with the counter equal to 0.

Machine C then operates in phases. Each phase corresponds to a run of B between two consecutive resets of some (possibly different) clock.

Suppose y was the last clock to reset. After the reset, the configuration of B is $(s, (z, 0))$ for some state $s \in B$ and the counter $z = x$. We show how C calculates the configuration after the next clock reset in B.

After time Δ, the clocks go from configuration $(z, 0)$ to $(z + \Delta, \Delta)$. Based on the guards, different edges in B^γ are enabled as time progresses. Precisely, suppose we know the order of the parameters $p_1 < p_2 < \ldots < p_k$. Then let **region** $R_{(i,j)}$ be the set of clock valuations $[p_i, p_{i+1}] \times [p_j, p_{j+1}]$. Then the set of enabled edges depends only on the region $R_{(i,j)}$ the clocks (x, y) lie in.[2]

Therefore, machine C guesses the regions $R_{(i_0, j_0)}, R_{(i_1, j_1)}, \ldots, R_{(i_m, j_m)}$ in the order in which they are visited by the clocks (x, y) and it also guesses the states $s_0, s_1, \ldots, s_m$ of B when each region R_l is visited for the first time, the state t in which the next reset occurs and which clock is reset next (see Fig 2).

Machine C checks that the sequence is valid as follows. First, C checks, that $(z, 0)$ lies in R_0. Second, it checks that the regions are adjacent: $i_{l+1} - i_l = 1 \wedge j_{l+1} = j_l$ or $i_{l+1} = i_l \wedge j_{l+1} - j_l = 1$ or $i_{l+1} - i_l = j_{l+1} - j_l = 1$. The last case corresponds to the clocks hitting a corner of a region. Then, C checks that starting in clock configuration $(z, 0)$, the regions can be visited in the guessed order.

Consider region $R_{(u,v)}$ for some u, v. When the region is visited for the first time, then either clock x equals p_u or clock y equals p_v. In the former case, the clock configuration is $(p_u, p_u - z)$, in the latter case, it is $(p_v + z, p_v)$. The configuration depends on the direction in which $R_{(u,v)}$ is visited. See Fig. 2.

- If $i_{l+1} - i_l = 1$ then C checks that clock x reaches $p_{i_{l+1}}$ before clock y reaches p_{j_l+1}. That is, $p_{i_{l+1}} - z \le p_{j_l+1}$. Equivalently, $p_{i_{l+1}} \le z + p_{j_l+1}$, which can be easily tested by a PBOCA. In Fig. 2 this corresponds to region $R_{(1,0)}$, which is visited before $R_{(2,0)}$.
- Similarly, if $j_{l+1} - j_l = 1$. E.g, in Fig. 2 region $R_{(2,1)}$ is visited before $R_{(2,2)}$.

We say $R_{(u,v)}$ was **reached from left** in the first and that $R_{(u,v)}$ was **reached from bottom** in the second case. See Fig. 2 for the intuition behind the names.

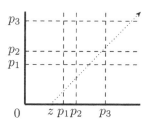

Finally, C checks reachability within individual regions. For $l = (u, v)$, let c_l be the configuration in which the region R_l is visited for the first time. Then C checks that a run from c_l to c_{l+1} exists in R_l.

Now, with each $R_{(i,j)}$, we introduce a one-counter machine $B_{(i,j)}$ obtained from B assuming clock $x \in [p_i, p_{i+1}]$ and clock $y \in [p_j, p_{j+1}]$, instantiating all comparisons accordingly and by removing all edges resetting a clock. Each $B_{(i,j)}$ corresponds to the region $R_{(i,j)}$ in the same way automata C_i corresponded to one-dimensional regions in Section 3.

Fig. 2. Regions for parameters $p_1 < p_2 < p_3$. The dotted line shows an evolution of clock configuration, which visits $R_{(0,0)}, R_{(1,0)}, R_{(2,0)}, R_{(2,1)}, R_{(2,2)}, R_{(3,2)}, R_{(3,3)}$.

---

[2] Our definition of rectangular regions differs slightly from the one usually given in the literature. However, as all inequalities are nonstrict the regions are sufficient. For ease of presentation, we also use the convention $p_0 = 0$ and $p_{k+1} = \infty$.

Notice that $B_{(i,j)}$'s are $0/1$ automata without resets or comparisons, i.e. one-counter machines. In particular, the reachability relation for $B_{(i,j)}$'s is semilinear. For a pair of states s and t of a one-counter machine X define $\Pi(X, s, t)$ to be the set of counter values reachable at t by a run starting in state s and counter equal to 0: $\Pi(X, s, t) = \{v \mid \exists \pi \in X. \, start(\pi) = (s, 0) \wedge end(\pi) = (t, v)\}$.

Lemma 11. *Let X be a one-counter machine with $0/1$ updates. Then for any states $s, t \in X$ the set $\Pi(X, s, t)$ is effectively semilinear: $\Pi(X, s, t) = N \cup \bigcup_{j=1}^{j=r} \{a_j + b_j \mathbb{N}\}$ where $N \subseteq \mathbb{N}$ is finite and $a_j, b_j \in \mathbb{N}$.*

Now, to check that a run from c_l to c_{l+1} exists in R_l, machine C distinguishes whether R_l and R_{l+1} are reached from bottom or from left and uses the semilinearity of the reachability relation of the corresponding $B_{(i,j)}$.

The translation is mundane and is given in the full version of the paper. For example, suppose $R_l = R_{(p_x, p_y)}$ for parameters p_x and p_y. Then $c_l = (s_l, (p_x, p_x - z))$ or $c_l = (s_l, (p_y + z, p_y))$ depending on the direction. If R_l was reached from left and R_{l+1} from bottom then C checks that $(s_{l+1}, (p_{y+1} + z, p_{y+1}))$ is reachable from $(s_l, (p_x, p_x - z))$. That is, that $z + p_{y+1} - p_x \in \Pi(B_l, s_l, s_{l+1})$. All such constraints can be checked using '$\equiv 0 \bmod c$' edges (see Fig. 3).

Finally note that once the value of a clock becomes larger than p_k its exact value is irrelevant to any future comparison. Hence, C tracks x and y only up to p_k and remembers which clocks exceed it. Hence, we can assume that the counter of C is always inside $[-p_k, p_k]$.

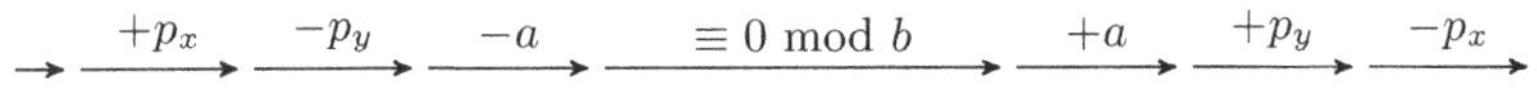

Fig. 3. Gadget testing that for given $a, b \in \mathbb{N}$ there is $k \in \mathbb{N}$ such that $z + p_x - p_y = a + kb$, i.e. $z + p_x - p_y - a \equiv 0 \bmod b$. Letter z denotes the current counter value.

Next, we modify C to ensure that the counter is always nonnegative. Let C' be obtained from C by adding a new initial state and a $+p_k$ edge from the new to the original initial state. Further, any comparison edge (s, G, t) (e.g., where G is $\le p_i$) is replaced by a gadget of three edges $(s, -p_k, q), (q, G, q')$ and $(q', +p_k, t)$ which subtract p_k from the counter, perform the original check and then add p_k to the counter thereby offsetting the counter by p_k.

Remark 12. We can assume that the counter of C is always inside $[0, 2p_k]$.

Note that the construction depends on the order of parameters. However, we can build an automaton for each possible order, check the order of parameters and then transition into the automaton for the appropriate order.

Theorem 13. *Given a 2-PTA there is a PBOCA with equivalent halting problem.*

The reduction was inspired by [6] (see Related Work). Unlike [6], we exploit semilinearity in individual regions and perform one phase in a single stage of C.

4.2   The One-Parameter Case

Suppose that the 2-PTA A uses only a single parameter p and consider the corresponding PBOCA C. We show that all '$\equiv 0 \bmod c$' and '$+[0, p]$' edges can be eliminated from C. Using Remark 12, we show in Lemma 15 how to decide the halting problem in the resulting class of PBOCAs.

Inspecting the detailed proof of the reduction from A to C (as found in the full version), observe that '$+[0, p]$' edges are introduced only when two successive regions are both visited from left or both visited from bottom. For a single parameter, only regions $[0, p] \times [0, p], [0, p] \times [p, \infty], [p, \infty] \times [0, p], [p, \infty] \times [p, \infty]$ exist. Simple case analysis shows that this can occur only when the counter starts at 0—this can be treated separately thereby eliminating '$+[0, p]$' edges from C.

Next, we also eliminate '$\equiv 0 \bmod c$' edges from C. Intuitively, C shall store in its state space the counter modulo c_i for each $c_1, \ldots, c_r$ appearing as '$\equiv 0 \bmod c_i$' in C. The construction depends on the value of $p \bmod c_i$ for each i.

Given $D = (d_1, \ldots, d_r)$, let C_D be the one-counter machine obtained from C which tracks the counter modulo each c_i assuming $p \equiv d_i \bmod c_i$. Formally, the states of C_D are $S \times \mathbb{Z}_{c_1} \times \ldots \times \mathbb{Z}_{c_r}$ where S are the states of C and $\mathbb{Z}_{c_i}$ denotes the ring of integers modulo c_i. The machine C_D contains all comparison edges of C. Further, let $(v_1, \ldots, v_r) \in \mathbb{Z}_{c_1} \times \ldots \times \mathbb{Z}_{c_r}$. Let E be the edges of C, then C_D also contains the following edges:

- $((q, v_1, \ldots, v_r), \pm c, (q', v_1 \pm c, \ldots, v_r \pm c)$ if $(q, \pm c, q') \in E$,
- $((q, v_1, \ldots, v_r), \pm p, (q', v_1 \pm d_1, \ldots, v_r \pm d_r)$ if $(q, \pm p, q') \in E$,
- $((q, v_1, \ldots, v_r), +0, (q', v_1, \ldots, v_r))$ if $v_i = 0$ and $(q, \equiv 0 \bmod c_i, q') \in E$.

Notice that there are no '$\equiv 0 \bmod c$' edges in C_D. By construction, runs in C_D^γ are equivalent to runs C^γ provided $d_i \equiv \gamma(p) \bmod c_i$. That is:

Lemma 14. *Let γ be an assignment such that $\gamma(p) = d_i \bmod c_i$ for each i. Let $(s, x), (t, y)$ be configurations of C. Then $(s, x) \to^* (t, y)$ in C^γ if and only if $((s, x \bmod c_1, \ldots, x \bmod c_r), x) \to^* ((t, y \bmod c_1, \ldots, y \bmod c_r), y)$ in C_D^γ.*

For given D, finding an accepting run π such that $counter(\pi) \leq 2 \cdot \gamma(p)$ suffices (Remark 12) to decide the halting problem for C_D. For any such run π and index i we can write $counter(\pi(i)) = a\gamma(p) + b$ where $a \leq 2$ and $b < \gamma(p)$.

Since a is bounded, we can build a one-counter machine G keeping a in the state space and b in the counter. We do not enforce $b < \gamma(p)$ (or any other $\leq x$ constraint) in G. Instead, we use Lemma 7 on G and split π into subruns close to and far from a multiple of $\gamma(p)$. We write $\pi = \tau_0 \to \pi_1 \to \tau_1 \ldots \pi_l \to \tau_l$ such that for every τ_i the value $counter(\tau_i) \bmod \gamma(p) \in [0, \ldots, M] \cup [\gamma(p) - M, \gamma(p))$. For every π_i we have $counter(\pi_i) \bmod \gamma(p) \in (M, \gamma(p) - M)$. Then we use techniques on factoring of runs analogous to those used for one 1-PTAs (Section 3.1). In general, we have: (See the full version)

Lemma 15. *Given C with one parameter p, no '$\equiv 0 \bmod c$' and no '$+[0, p]$' edges, $k \in \mathbb{N}$ and states $s, t \in C$ the set $G(C, s, t, k) = \{(x, y, q) \mid \exists \pi : (s, x) \to (t, y) \in C^\gamma \text{ s.t. } counter(\pi) \leq k \cdot q \text{ where } q = \gamma(p)\}$ is $\exists$PAD definable.*

Theorem 16. *The halting problem is decidable for 2-PTAs with one parameter.*

This settles the case of 2-PTAs with a single parameter. However, even the case of only two parameters is open. On the other hand, already for a single parameter, we have the following lower bound. (See the full version)

Theorem 17. *The decidability of the halting problem for 2-PTAs with a single parameter is* $\mathsf{PSPACE}^{\mathsf{NEXP}}$-*hard.*

Acknowledgments. This research was financially supported by EPSRC.

References

1. Alur, R., Henzinger, T.A., Vardi, M.Y.: Parametric real-time reasoning. In: Proceedings of the 25th Annual Symposium on Theory of Computing. ACM Press (1993)
2. Doyen, L.: Robust parametric reachability for timed automata. Information Processing Letters 102(5), 208–213 (2007)
3. Fearnley, J., Jurdziński, M.: Reachability in two-clock timed automata is PSPACE-Complete. In: Fomin, F.V., Freivalds, R., Kwiatkowska, M., Peleg, D. (eds.) ICALP 2013, Part II. LNCS, vol. 7966, pp. 212–223. Springer, Heidelberg (2013)
4. Haase, C.: On the Complexity of Model Checking Counter Automata. PhD thesis, University of Oxford (2012)
5. Haase, C., Kreutzer, S., Ouaknine, J., Worrell, J.: Reachability in succinct and parametric one-counter automata. In: Bravetti, M., Zavattaro, G. (eds.) CONCUR 2009. LNCS, vol. 5710, pp. 369–383. Springer, Heidelberg (2009)
6. Haase, C., Ouaknine, J., Worrell, J.: On the relationship between reachability problems in timed and counter automata. In: Finkel, A., Leroux, J., Potapov, I. (eds.) RP 2012. LNCS, vol. 7550, pp. 54–65. Springer, Heidelberg (2012)
7. Henzinger, T.A., Manna, Z., Pnueli, A.: What good are digital clocks? In: Kuich, W. (ed.) ICALP 1992. LNCS, vol. 623, pp. 545–558. Springer, Heidelberg (1992)
8. Hune, T., Romijn, J., Stoelinga, M., Vaandrager, F.: Linear parametric model checking of timed automata. In: Margaria, T., Yi, W. (eds.) TACAS 2001. LNCS, vol. 2031; pp. 189–203. Springer, Heidelberg (2001)
9. Ibarra, O.H., Jiang, T., Trân, N., Wang, H.: New decidability results concerning two-way counter machines and applications. In: Lingas, A., Carlsson, S., Karlsson, R. (eds.) ICALP 1993. LNCS, vol. 700, pp. 313–324. Springer, Heidelberg (1993)
10. Jovanović, A., Lime, D., Roux, O.H.: Integer parameter synthesis for timed automata. In: Piterman, N., Smolka, S.A. (eds.) TACAS 2013. LNCS, vol. 7795, pp. 401–415. Springer, Heidelberg (2013)
11. Lafourcade, P., Lugiez, D., Treinen, R.: Intruder deduction for AC-like equational theories with homomorphisms. In: Research Report LSV-04-16, LSV, ENS de Cachan (2004)
12. Lipshitz, L.: The Diophantine Problem for Addition and Divisibility. Transactions of the American Mathematical Society, 235 (1978)
13. Lipshitz, L.: Some remarks on the diophantine problem for addition and divisibility, vol. 33 (1981)
14. Miller, J.S.: Decidability and complexity results for timed automata and semi-linear hybrid automata. In: Lynch, N., Krogh, B.H. (eds.) HSCC 2000. LNCS, vol. 1790, pp. 296–310. Springer, Heidelberg (2000)
15. Ouaknine, J., Worrell, J.B.: Universality and language inclusion for open and closed timed automata. In: Maler, O., Pnueli, A. (eds.) HSCC 2003. LNCS, vol. 2623, pp. 375–388. Springer, Heidelberg (2003)

Universal Lyndon Words

Arturo Carpi[1,*], Gabriele Fici[2,*], Štěpán Holub[3,**],
Jakub Opršal[3,**], and Marinella Sciortino[2,*]

[1] Dipartimento di Matematica e Informatica, Università di Perugia, Italy
carpi@dmi.unipg.it
[2] Dipartimento di Matematica e Informatica, Università di Palermo, Italy
{fici,mari}@math.unipa.it
[3] Department of Algebra, Univerzita Karlova, Czech Republic
{holub,oprsal}@karlin.mff.cuni.cz

Abstract. A word w over an alphabet Σ is a Lyndon word if there exists an order defined on Σ for which w is lexicographically smaller than all of its conjugates (other than itself). We introduce and study *universal Lyndon words*, which are words over an n-letter alphabet that have length $n!$ and such that all the conjugates are Lyndon words. We show that universal Lyndon words exist for every n and exhibit combinatorial and structural properties of these words. We then define particular prefix codes, which we call Hamiltonian lex-codes, and show that every Hamiltonian lex-code is in bijection with the set of the shortest unrepeated prefixes of the conjugates of a universal Lyndon word. This allows us to give an algorithm for constructing all the universal Lyndon words.

Keywords: Lyndon word, Universal cycle, Universal Lyndon word, Lex-code.

1 Introduction

A word is called Lyndon if it is lexicographically smaller than all of its conjugate words (other than itself). Lyndon words are an important and well studied object in Combinatorics. Recall, for example, the fact that every Lyndon word is unbordered, or the existence of a unique factorization of any word into a nondecreasing sequence of Lyndon words [5]. The definition of Lyndon word implicitly assumes a lexicographic order. Therefore, for different orders, we typically obtain several distinct Lyndon conjugates of the same word. The motivation of this paper is to push the idea to its limits, and ask whether there is a *universal Lyndon word*, that is, a word of length $n!$ over n letters such that for each of its conjugates there exists an order with respect to which this conjugate is Lyndon.

Such a word resembles similar objects known in the literature as universal cycles. A *universal cycle* [2] is a circular word containing every object of a particular type exactly once as a factor. Probably the most prominent example

---

* Partially supported by Italian MIUR Project PRIN 2010LYA9RH, "Automi e Linguaggi Formali: Aspetti Matematici e Applicativi".
** Partially supported by the Czech Science Foundation grant number 13-01832S.

E. Csuhaj-Varjú et al. (Eds.): MFCS 2014, Part I, LNCS 8634, pp. 135–146, 2014.
© Springer-Verlag Berlin Heidelberg 2014

of universal cycles are de Bruijn cycles, which are circular words of length 2^n containing every binary word of length n exactly once.

The set represented by a universal Lyndon word is the set of all total orders on n letters or, equivalently, all permutations of n letters. The most convenient way is to represent the order $a_1 < a_2 < \cdots < a_n$ by its "shorthand encoding", which is the word $a_1 a_2 \cdots a_{n-1}$. Jackson [4] showed that the corresponding universal cycles exist for every n and can be obtained from an Eulerian graph in a manner similar to the generation of de Bruijn cycles. Ruskey and Williams [6] gave efficient algorithms for constructing shorthand universal cycles for permutations. Our paper can be seen as a generalization of this concept. Indeed, it is easy to note that every shorthand universal cycle for permutations is a universal Lyndon word (see [3] for more details), but the opposite is not true—that is, there exist universal Lyndon words such that the Lyndon conjugate for some order $a_1 < a_2 < \cdots < a_n$ does not start with $a_1 a_2 \cdots a_{n-1}$.

We study the structural properties of universal Lyndon words and give combinatorial characterizations. We then develop a method for generating all the universal Lyndon words. This method is based on the notion of Hamiltonian lex-code, which we introduce in this paper.

2   Notation

Given a finite non-empty ordered set Σ (called the *alphabet*), we let Σ^* denote the set of words over the alphabet Σ. Given a finite word $w = a_1 a_2 \cdots a_n$, with $n \geq 1$ and $a_i \in \Sigma$, the length n of w is denoted by $|w|$. The *empty word* will be denoted by ε and we set $|\varepsilon| = 0$. We let Σ^n denote the set of words of length n and by Σ^+ the set of non-empty words. For $u, v \in \Sigma^+$ we let $|u|_v$ denote the number of (possibly overlapping) occurrences of v in u. For instance, $|011100|_{00} = 1$ and $|011100|_{11} = 2$.

Given a word $w = a_1 a_2 \cdots a_n$, $a_i \in \Sigma$, we say a word $v \in \Sigma^+$ is a *factor* of w if $v = a_i a_{i+1} \cdots a_j$ for some integers i, j with $1 \leq i \leq j \leq n$. We let $\mathrm{Fact}(w)$ denote the set of all factors of w and $\mathrm{Alph}(w)$ the set of all factors of w of length 1. If $i = 1$ (resp., $j = n$), we say that the factor v is a *prefix* (resp., a *suffix*) of w. We let $\mathrm{Pref}(w)$ (resp., $\mathrm{Suff}(w)$) denote the set of prefixes (resp., suffixes) of the word w. The empty word ε is a factor, a prefix and a suffix of any word. A factor (resp., a prefix, resp., a suffix) of a word w is *proper* if it is different from ε and from w itself.

A *border* of w is a proper prefix of w that is also a suffix of w. A word is said to be *unbordered* if it does not have borders. A word u is a *cyclic factor* of w if $u \in \mathrm{Fact}(ww)$ and $|u| \leq |w|$. We let $|w|_u^c$ denote the number of (possibly overlapping) occurrences of u as a cyclic factor of w. For instance, $|011100|_{00}^c = 2$. We say that a word u is *conjugate* to a word v if there exist words w_1, w_2 such that $u = w_1 w_2$ and $v = w_2 w_1$. The conjugate is *proper* if both w_1 and w_2 are non-empty. The conjugacy being an equivalence relation, we can define a *cyclic word* as a conjugacy class of words. Note that u is a cyclic factor of a word w if and only if u is a factor of a conjugate of w.

Every total order on the alphabet Σ induces a different lexicographic (or dictionary) order on Σ^*. Recall that the lexicographic order $\lhd$ on Σ^* induced by the order $<$ on the alphabet Σ is defined as follows: $u \lhd v$ if u is a prefix of v or za is a prefix of u and zb is a prefix of v, with $a < b$. We say that a word w over Σ is a *Lyndon word* if there exists a total order on Σ such that, with respect to this order w is lexicographically smaller than all of its proper conjugates (or, equivalently, proper suffixes). For example, the word $w = abcabb$ is a Lyndon word, because for the order $a < c < b$ it is the smallest word in its conjugacy class. Note that a Lyndon word must be primitive (i.e., it cannot be written as a concatenation of two or more copies of a shorter word), and therefore its conjugates are all distinct.

A set of words $X \subset \Sigma^+$ is a *code* if for every $x_1, x_2, \ldots, x_h, x_1', x_2', \ldots, x_k' \in X$, if $x_1 x_2 \cdots x_h = x_1' x_2' \cdots x_k'$, then $h = k$ and $x_i = x_i'$ for every $1 \leq i \leq h$. For example, $X = \{ab, abb\}$ is a code. Every set $X \subset \Sigma^+$ with the property that no word in X is a prefix of another word in X is a code, and is called a *prefix code*.

A *directed graph* (or *digraph*) is a pair $G = (V, E)$, where V is a set, whose elements are called *vertices*, and E is a binary relation on V (i.e., a set of ordered pairs of elements of V) whose elements are called *edges*. The *indegree* (resp., *outdegree*) of a vertex v in a digraph G is the number of edges incoming to v (resp., outgoing from v). A *walk* in a digraph G is a non-empty alternating sequence $v_0 e_0 v_1 e_1 \cdots e_{k-1} v_k$ of vertices and edges of G such that $e_i = (v_i, v_{i+1})$ for every $i < k$. If $v_0 = v_k$ the walk is *closed*. A closed walk in a digraph G is an *Eulerian cycle* if it traverses every edge of G exactly once. A digraph is *Eulerian* if it admits an Eulerian cycle. A fundamental property of graphs is that a connected digraph is Eulerian if and only if the indegree of each vertex is equal to its outdegree. A closed walk in a digraph G is a *Hamiltonian cycle* if it contains every vertex of G exactly once. A digraph is *Hamiltonian* if it admits a Hamiltonian cycle.

In the rest of the paper, we let Σ_n denote the alphabet $\{1, 2, \ldots, n\}$, $n > 0$.

3   Universal Lyndon Words

Definition 1. *A universal Lyndon word (ULW) of degree n is a word over Σ_n that has length $n!$ and such that all its conjugates are Lyndon words.*

Remark 1. Since there exist $n!$ possible orders on Σ_n, a universal Lyndon word w of degree n has the property that for every order on Σ_n, there is exactly one conjugate of w that is Lyndon with respect to this order; on the other hand, from the definition it follows that a conjugate of a universal Lyndon word cannot be Lyndon for more than one order.

We consider universal Lyndon words up to rotation, i.e., as cyclic words.

Example 1. The only universal Lyndon word of degree 1 is 1, and the only universal Lyndon word of degree 2 is 12. There are three universal Lyndon words of degree 3, namely 212313, 323121 and 131232. Note that these words are pairwise isomorphic (i.e., one can be obtained from another by renaming letters). There are 492 universal Lyndon words of degree 4. There are 41 if we consider them up to isomorphism, and are presented in Tables 1 and 2.

Remark 2. It is worth noticing that a universal Lyndon word cannot contain a square (i.e., a concatenation of two copies of the same word) as a cyclic factor. That is, a universal Lyndon word is cyclically square-free. Indeed, if uu is a factor of w, then there is a conjugate of w that has u as a border, and it is easily shown that every Lyndon word must be unbordered, and therefore no conjugate of a universal Lyndon word can have a border.

The following proposition gives a sufficient condition for a word being a ULW.

Proposition 1. *Let $n \geq 2$, and w be a word over Σ_n such that every permutation of $n - 1$ elements of Σ_n appears as a cyclic factor in w exactly once. Then w is a universal Lyndon word.*

Proof. Suppose that every permutation of $n - 1$ elements of Σ_n appears as a cyclic factor in w exactly once. Since there are $n!$ such words, this implies that w has length $n!$. Now, for any order $a_1 < a_2 < \ldots < a_{n-1} < a_n$ over Σ_n, there is exactly one conjugate of w beginning with $a_1 a_2 \cdots a_{n-1}$, and this conjugate is Lyndon with respect to this order. So w has exactly $n!$ distinct Lyndon conjugates and therefore is a universal Lyndon word. $\qquad\qquad\square$

Remark 3. One might wonder whether it is sufficient to suppose that *each* of w's factors of length $n - 1$ appears exactly once in the word w to guarantee that w is a ULW. This is not the case. For example, let $n = 4$; the word $w = 123412431324134214231432$ has $n!$ distinct factors of length $n - 1$ but is not a universal Lyndon word, since its conjugate 314321234124313241342142 is not Lyndon for any order (in fact this is a consequence of the fact that the conjugate 313241342142314321234124 is Lyndon both for the orders $3 < 1 < 2 < 4$ and $3 < 1 < 4 < 2$).

We now use the result of Proposition 1 to show that there exist universal Lyndon words for each degree.

Given an integer $n > 2$, the *Jackson graph of degree n*, denoted $J(n)$, is a directed graph in which the nodes are the words over Σ_n that are permutations of $n - 2$ letters, and there is an edge from node u to node v if and only if the suffix of length $n - 3$ of u is equal to the prefix of length $n - 3$ of v and the first letter of u is different from the last letter of v. The label of such an edge is set to the first letter of u. In Fig. 1, the Jackson graph $J(4)$ is depicted.

Proposition 2. *There exist universal Lyndon words of degree n for every $n > 0$.*

Proof. We can suppose $n > 2$. Take the Jackson graph $J(n)$. By construction, this graph is connected and the indegree and outdegree of each vertex are both equal to 2. Therefore, it contains an Eulerian cycle. Let w denote the word obtained by concatenating the labels of such an Eulerian cycle. Note that every

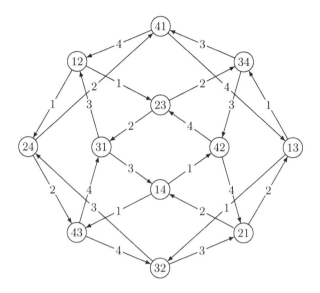

Fig. 1. The Jackson graph $J(4)$ of degree 4. Every Eulerian cycle of $J(4)$ is a universal Lyndon word.

word that is the permutation of $n - 2$ letters appears as a cyclic factor in w exactly twice and the two occurrences are followed by the two letters that do not appear in the factor. By Proposition 1, w is then a universal Lyndon word of degree n.   □

A universal Lyndon word that is an Eulerian cycle of a Jackson graph is called a *universal cycle* [4], or *shorthand universal cycle for permutations* [6], but in this paper we will call it a *universal Lyndon word of Jackson type*, or simply a *Jackson universal Lyndon word*.

The Jackson universal Lyndon words of degree 4 are presented in Table 1 (the list contains only pairwise non-isomorphic words, in their representation starting with 1231).

However, there are universal Lyndon words that are not of Jackson type. In fact, the converse of Proposition 1 is not true. For instance, $w = 12343124231413242134321414$ is a universal Lyndon word of degree 4 but it does not contain any of 142, 143, 241, 243, 341, 342 as a factor.

4   Order-Defining Words

In this section, we give combinatorial results on the structure of universal Lyndon words.

Table 1. The 20 Jackson universal Lyndon words of degree 4, up to isomorphisms

123124132431432142342134	123124132134214324314234
123124314214321324134234	123124134213243214314234
123124132431421432134234	123124314234213214324134
123124314213214324134234	123124321431423421324134
123124132431423421432134	123124134213243143214234
123124314214324132134234	123124132431432134214234
123124132432143142342134	123124321342143142341324
123124132431432142134234	123124132143243142134234
123124314234132134214324	123124132432143142134234
123124314234134214321324	123124134214321324314234

Let $w = a_1 a_2 \cdots a_{n!}$ be a universal Lyndon word of degree n. Let w_i denote the conjugate of w starting at position i, that is,

$$w_i = a_i a_{i+1} \cdots a_{n!} a_1 a_2 \cdots a_{i-1} \, .$$

Definition 2. *We say that a partial order $\lhd$ on Σ_n is a partial alphabet order with respect to $I \subseteq \Sigma_n$ if $\lhd$ is a total order on I, $i \lhd j$ for each $i \in I$ and $j \in \Sigma_n \setminus I$, and all $j, k \in \Sigma_n \setminus I$ are incomparable. The size of $\lhd$ is set to $|I|$.*

Note that a partial alphabet order of size $n - 1$ is a total order on Σ_n.

Every word $u \in \Sigma_n^+$ defines a partial alphabet order $\lhd_u$ with respect to $\mathrm{Alph}(u)$, defined as follows: $i \lhd_u j$ if and only if the first occurrence of i in u precedes the first occurrence of j in u.

The following proposition shows that in a universal Lyndon word, every conjugate is Lyndon with respect to the order it defines. This is an important structural property of universal Lyndon words, which is not true in general. Take, for example, the word $w = 123122$. It is Lyndon with respect to the order $1 < 3 < 2$, but it is not Lyndon with respect to the order it defines, $1 < 2 < 3$.

We let $\lhd_i$ denote the order defined by w_i and by $\blacktriangleleft_i$ the order with respect to which w_i is Lyndon.

Theorem 1. *Let w be a word of length $n!$ over Σ_n. Then w is a ULW if and only if every conjugate of w is Lyndon with respect to the order it defines. That is, $\lhd_i = \blacktriangleleft_i$ for every i.*

Proof. If every conjugate of w is Lyndon, then w is ULW by definition. So we only have to prove the "only if" part of the statement.

Suppose that $\lhd_j \neq \blacktriangleleft_j$ for some j, and let k be such that $\blacktriangleleft_k = \lhd_j$. Let z be the longest common prefix of w_j and w_k. Then za is a prefix of w_j and zb a prefix of w_k, where $a \neq b$ are letters. We have $a \blacktriangleleft_j b$ and $b \blacktriangleleft_k a$. Therefore, also $b \vartriangleleft_j a$, which implies that there exists $u \in \Sigma_n^*$ such that bua is a suffix of za and bub is a suffix of zb. Let w_ℓ be the conjugate starting with bua. Obviously, $b \blacktriangleleft_\ell a$, since b is the first letter of w_ℓ. But then we have that $bub \blacktriangleleft_\ell bua$, and therefore w_ℓ has a conjugate smaller than itself for the order $\blacktriangleleft_\ell$, a contradiction. $\square$

Proposition 3. *Let w be a universal Lyndon word, and u a cyclic factor of w. Then for every conjugate w_i of w, we have that u is a prefix of w_i if and only if $\vartriangleleft_u \subseteq \blacktriangleleft_i$.*

Proof. By Theorem 1, we have $\vartriangleleft_u \subseteq \blacktriangleleft_i$ for each i such that u is a prefix of w_i. Choose one such w_i (which exists since u is a cyclic factor of w). Let $\vartriangleleft_u \subseteq \blacktriangleleft_j$ and suppose that za is a prefix of u and zb a prefix of w_j for two distinct letters a and b and some $z \in \Sigma_n^*$. Then $a \blacktriangleleft_i b$, and, since $a \in \text{Alph}(u)$, we deduce that $a \vartriangleleft_u b$. This implies that $a \blacktriangleleft_j b$, since $\vartriangleleft_u \subseteq \blacktriangleleft_j$. Therefore, $w_i \blacktriangleleft_j w_j$, a contradiction. $\square$

Proposition 3 states that the cyclic factors of a ULW are in one-to-one correspondence with the orders they define. As an example, if $1 < 2$ and, say, 212 is a cyclic factor of a universal Lyndon word w, then every other occurrence of 21 in w must be followed by 2.

Corollary 1. *Let w be a universal Lyndon word of degree n, and u a cyclic factor of w of length $k > 0$. Then u is the lexicographically smallest cyclic factor of w of length k with respect to any total order $\blacktriangleleft$ on Σ_n such that $\vartriangleleft_u \subseteq \blacktriangleleft$.*

We now give a combinatorial characterization of universal Lyndon words.

Theorem 2. *Let w be a word over Σ_n. Then w is a universal Lyndon word if and only if for every cyclic factor u of w, one has*

$$|w|_u^c = (n - |\text{Alph}(u)|)! \tag{1}$$

Proof. Suppose that w is a ULW. There are $(n - |\text{Alph}(u)|)!$ many total orders $\blacktriangleleft$ on Σ_n such that $\vartriangleleft_u \subseteq \blacktriangleleft$. Hence, (1) follows from Corollary 1.

Suppose now that (1) holds for every cyclic factor u of w and let us prove that w is a ULW. For every letter $a \in \Sigma_n$, one has $|w|_a = |w|_a^c = (n-1)!$, so that $|w| = \sum_{a \in \Sigma_n} |w|_a = n!$. Moreover, w is primitive, since $|w|_w^c = 1$. We show that w is a Lyndon word with respect to $\vartriangleleft_w$. Let v be a proper conjugate of w and let z be the longest common prefix of w and v. Let a and b be the letters that follow the prefix z in w and v respectively. Since both za and zb occur in w, we have $|w|_z^c > |w|_{zb}^c$ which implies $b \notin \text{Alph}(z)$ by (1). Because za is a prefix of w and $b \notin \text{Alph}(z)$, one has $a \vartriangleleft_w b$, and therefore $w \vartriangleleft_w v$. This proves that w is a Lyndon word. By a similar argument, all conjugates of w are Lyndon words, so that w is a ULW. $\square$

Corollary 2. *The reversal of a ULW is a ULW.*

Note that the fact that the set of universal Lyndon words is closed under reversal is not an immediate consequence of the definition. This property is not true for Lyndon words, e.g. the word 112212 is Lyndon but its reversal is not.

Definition 3. *We say that u is a* minimal order-defining word *if no proper factor of u defines $\lhd_u$.*

Proposition 4. *Given a universal Lyndon word w of degree n, for each partial alphabet order $\lhd$ on Σ_n there is a unique minimal order-defining word with respect to $\lhd$ that is a cyclic factor of w.*

Proof. Let $\lhd$ be a partial alphabet order with respect to I. Let w_i be such that $\lhd \subseteq \blacktriangleleft_i$, and let u be the shortest prefix of w_i such that $\mathrm{Alph}(u) = I$. Note that $\lhd_u = \lhd$ by Theorem 1. Clearly, u is a minimal order-defining word, and the uniqueness is a consequence of Proposition 3. $\square$

Let w be a universal Lyndon word. We let $MT(w)$ denote the minimal total order-defining words of w, i.e., the set of cyclic factors of w that are minimal order-defining words with respect to a total order on Σ_n. The next proposition is a direct consequence of the definitions and of the previous results.

Proposition 5. *Let w be a universal Lyndon word of degree n, and u a cyclic factor of w. The following conditions are equivalent:*

1. *$u \in MT(w)$;*
2. *$|\mathrm{Alph}(u)| = n - 1$, and $|\mathrm{Alph}(u')| < n - 1$ for each proper prefix u' of u;*
3. *there exists a unique conjugate w_i of w such that u is the shortest unrepeated prefix of w_i.*

The shortest unrepeated prefix of a word is also called its *initial box* [1].

In what follows, we exhibit a structural property of ULW.

Definition 4. *We say that a cyclic factor v of a word w is a* stretch *if w contains a cyclic factor avb with $a, b \in \Sigma_n \setminus \mathrm{Alph}(v)$. Let u be a cyclic factor of w. We say that a cyclic factor v of w is a* stretch extension *of u in w if u is a factor of v, $\mathrm{Alph}(u) = \mathrm{Alph}(v)$, and v is a stretch.*

Of course, a stretch is always a stretch extension of itself.

Example 2. Let $w = 12341243132413421423 1432$. Then 31 has two stretch extensions in w, namely 313 and itself.

Lemma 1. *Each cyclic factor u of a ULW w has a unique stretch extension in w. Moreover, it has a unique occurrence in its stretch extension.*

Proof. Let v be a stretch extension of u in w. Then u and v have the same number of cyclic occurrences in w by Theorem 2. $\square$

Theorem 3. *If asa is a cyclic factor of a ULW w, with $a \in \Sigma_n \setminus \mathrm{Alph}(s)$, then bsb is a cyclic factor of w for each $b \in \Sigma_n \setminus \mathrm{Alph}(s)$.*

Proof. Proceed by induction on $|s|$. The claim trivially holds for $|s| = 0$, since aa is not a cyclic factor of w. Let now $|s| > 0$. We first show that if bs is a cyclic factor of w, then also bsb is a cyclic factor of w. Let therefore bs be a cyclic factor of w, where $b \neq a$ is a letter, and let j be such that $\triangleleft_{bsa} \subseteq \blacktriangleleft_j$. By Lemma 1, the word bsa is not a prefix of w_j. Let therefore $bs'e$ be a prefix of w_j and $bs'f$ a prefix of bsa where e and f are distinct letters. Suppose first that $e = b$. If $s' = s$, then bsb is a cyclic factor of w as required. If, on the other hand, the word s' is a proper prefix of s, then the induction assumption for the word $bs'b$ implies that $as'a$ is a cyclic factor of w. This is a contradiction with Proposition 3 since $\triangleleft_{as'a} \subseteq \triangleleft_{asa}$. Let now $e \neq b$. Note that then $\triangleleft_{s'e} \subseteq \triangleleft_{sa}$ since $\triangleleft_{bsa} \subseteq \blacktriangleleft_j$. But we have also $\triangleleft_{s'f} \subseteq \triangleleft_{sa}$, a contradiction with Proposition 3.

The proof is concluded by a counting argument. Theorem 2 implies that, for any $b \notin \mathrm{Alph}(s)$, the word s has m times more cyclic occurrences in w than bsb, where m is the cardinality of $\Sigma_n \setminus \mathrm{Alph}(s)$. □

The previous result shows the combinatorial structure of universal Lyndon words. Note that the factors of the form asa, $a \in \Sigma_n \setminus \mathrm{Alph}(s)$, with $|\mathrm{Alph}(s)| < n - 2$, only appear in non-Jackson universal Lyndon words. In fact, they can be viewed as premature repetitions of the letter a.

5   Universal Lyndon Words and Lex-Codes

Proposition 1 implies that an Eulerian cycle in a Jackson graph is a universal Lyndon word. However, there exist universal Lyndon words that do not arise from a Jackson graph, as we showed at the end of Section 3.

The non-Jackson universal Lyndon words of degree 4 are presented in Table 2 (the list contains only pairwise non-isomorphic words, in their representation starting with 2123).

We now exhibit a method for constructing all the universal Lyndon words. This method is based on particular prefix codes, whose definition is given below.

Definition 5. *A set $X \subseteq \Sigma_n^*$ is a* lex-code *of degree n if:*

1. *for any $x \in X$, there exists a unique ordering of Σ_n such that x is the lexicographical minimum of X;*
2. *if u is a proper prefix of some word of X, then u is a prefix of at least two distinct words of X.*

A lex-code X of degree n is Hamiltonian *if the relation*

$$S_X = \{(x, y) \in X \times X \mid \exists a \in \Sigma, x \text{ is a prefix of } ay\}$$

has a Hamiltonian digraph.

Notice that Condition 1 in the previous definition ensures that a lex-code is a prefix code.

The following theorem shows the relationships between Hamiltonian lex-codes and universal Lyndon words.

Table 2. The 21 non-Jackson universal Lyndon words of degree 4, up to isomorphisms

212313243134212414234143	212313241432124313414234
212313241423414321243134	212313241432124313423414
212313421243132414234143	212313414234212431324143
212313241421243134234143	212341423132414321243134
212313241423414313421243	212313212414324313414234
212313243134142342124143	212313212414324313423414
212313212432414234143134	212313414234212414313243
212313212414234143243134	212313212432414313423414
212313212433342341432414	212313243321241431342414
212313241431342341421243	212313212432414313414234
212313212431341432414234	

Theorem 4. *Let w be a ULW. Then the set $MT(w)$ is a Hamiltonian lex-code. Conversely, if $X \subseteq \Sigma_n^*$ is a Hamiltonian lex-code, then there exists a ULW w such that $X = MT(w)$.*

Proof. We assume that w is a ULW and show that $MT(w)$ verifies the definition of lex-code. Since there is a bijection between the elements of $MT(w)$, the conjugates of w (Proposition 5) and the total orders on Σ_n (Theorem 1), Condition 1 is a direct consequence of Corollary 1. Always from Proposition 5, any proper prefix x' of a word x in $MT(w)$ contains less than $n-1$ distinct letters. From Theorem 2, x' has at least two occurrences as a cyclic factor of w. Therefore, there exist at least two distinct conjugates w_i and w_j of w beginning with x'. Then x' is a proper prefix of the shortest unrepeated prefixes of w_i and w_j respectively. By Proposition 5, we conclude that Condition 2 holds.

Now, we show that the lex-code X is Hamiltonian. For every $0 \leq i \leq n!-1$, let a_i be the first letter of the conjugate w_i of w. Notice that for every $0 \leq i \leq n!-2$ one has $a_i w_{i+1} = w_i a_i$. By Proposition 5, every word in $MT(w)$ is the shortest unrepeated prefix x_i of a conjugate w_i. As x_{i+1} is an unrepeated prefix of w_{i+1}, the word $v = a_i x_{i+1}$ is an unrepeated prefix of $a_i w_{i+1} = w_i a_i$. Thus, either $v = w_i a_i$ or v is an unrepeated prefix of w_i. In both cases, x_i is a prefix of v and therefore $(x_i, x_{i+1}) \in S_X$. Similarly, one has $(x_{n!-1}, x_0) \in S_X$. We conclude that $(x_0, x_1, \ldots, x_{n!-1}, x_0)$ is a Hamiltonian cycle in the digraph of S_X.

Conversely, we assume that X is a Hamiltonian lex-code and show that $X = MT(w)$ for a suitable ULW w. Let $(x_0, x_1, \ldots, x_{k-1}, x_0)$ be a Hamiltonian cycle in the digraph of S_X. By Condition 1, one has $k = n!$ and X is a prefix code. Since $(x_i, x_{i+1}) \in S_X$, $0 \leq i < k$ (where $x_k = x_0$) one has

$$x_i u_i = a_i x_{i+1} \tag{2}$$

for suitable $a_i \in \Sigma_n$, $u_i \in \Sigma_n^*$, $0 \le i < k$.

Set $w_i = a_i \cdots a_{k-1} a_0 \cdots a_{i-1}$, $0 \le i < k$. By iterated application of (2), one obtains that x_i is a prefix of a power of w_i. Now let $0 \le i, j < k$ and $i \ne j$. For a sufficiently large m, x_i, x_j are prefixes of w_i^m, w_j^m, respectively. Thus, taking into account that X is a prefix code, for every total order $\triangleleft$ on Σ_n, one has $w_i \triangleleft w_j$ if and only if $x_i \triangleleft x_j$. From this remark, in view of Condition 1, one derives that $w = w_0$ is a ULW.

To complete the proof, it is sufficient to show that x_i is the shortest unrepeated prefix of w_i, $0 \le i < k$. In fact, this implies that $X = MT(w)$. Suppose that the shortest unrepeated prefix h_i of w_i is a proper prefix of x_i. Then by Condition 2, h_i is also prefix of x_j and consequently of w_j, for some $j \ne i$. But this contradicts Proposition 5. Thus x_i is a prefix of h_i. Now, suppose $x_i \ne h_i$. Since by Proposition 5, h_i is a shortest word containing $n - 1$ distinct letters, $|\mathrm{Alph}(x_i)| < n-1$ and, by Theorem 2, x_i has at least another occurrence starting at a position $j \ne i$. So we have that the words x_i and x_j are one a prefix of the other, against the fact that X is a prefix code. $\qquad \square$

From Theorem 4, in order to produce a ULW, one can construct a lex-code and check whether it is Hamiltonian. Let S_n be the set of the total orders on Σ_n. All lex-codes of degree n can be obtained by a construction based on iterated refinements of a partition of S_n as follows:

1. set $X = \{\varepsilon\}$ and $C_\varepsilon = S_n$;
2. repeat the following steps until C_x is a singleton for all $x \in X$:
   (a) select $x \in X$ such that C_x contains at least two elements;
   (b) choose $\Gamma \subseteq \Sigma_n$;
   (c) for any $a \in \Gamma$, let C_{xa} be the set of the orders of C_x such that $a = \min \Gamma$;
   (d) replace X by $(X \setminus \{x\}) \cup \{xa \mid a \in \Gamma, \ C_{xa} \ne \emptyset\}$.

An example of execution of the previous algorithm is presented in Ex. 3.

One can verify that after each iteration of loop 2, X is a prefix code, $(C_x)_{x \in X}$ is a partition of S_n, and any $x \in X$ is the lexicographic minimum of X for all orders of C_x. It follows that the procedure halts when X is a lex-code. Moreover, one can prove that any lex-code X may be obtained by the procedure above, choosing conveniently Γ at step (b) of each iteration.

Clearly, not all lex-codes are Hamiltonian. Thus, the main problem is to understand which limitations the Hamiltonianicity of the lex-code imposes to the construction above. For example, the words in a lex-code can be arbitrarily long. But by Theorem 4, if X is a lex-code of degree n and $u \in X$ is longer than $n!$, then X cannot be Hamiltonian.

Example 3. Let $n = 3$. At the beginning of the algorithm, $X = \{\varepsilon\}$ and $C_\varepsilon = S_3 = \{123, 132, 213, 231, 312, 321\}$. The first choice of a word x in X is forced, we must take $x = \varepsilon$. Let us choose $\Gamma = \{1, 2\}$. We then get $C_1 = \{123, 132, 312\}$, $C_2 = \{213, 231, 321\}$ and X becomes $\{1, 2\}$. Let us now choose $x = 1$ and $\Gamma = \{1, 3\}$. We get $C_{11} = \{123, 132\}, C_{13} = \{312\}$ and therefore $X = \{2, 11, 13\}$. Next, take $x = 2$ and $\Gamma = \{2, 3\}$; now $C_{22} = \{213, 231\}, C_{23} = \{321\}$ and

$X = \{11, 13, 22, 23\}$. Then pick $x = 11$ and $\Gamma = \{2, 3\}$, so that $C_{112} = \{123\}$, $C_{113} = \{132\}$ and $X = \{13, 22, 23, 112, 113\}$. Finally, the last choice of a word in X is forced, $x = 22$ (since C_{22} is the only set of cardinality greater than 1 left). We choose $\Gamma = \{1, 3\}$ and get $C_{221} = \{213\}$ and $C_{223} = \{231\}$. The lex-code obtained is thus $X = \{13, 23, 112, 113, 221, 223\}$. The reader can verify that this lex-code is not Hamiltonian.

The following choices of x and Γ lead to the Hamiltonian lex-code $X = \{12, 13, 21, 23, 31, 32\}$: $x = \varepsilon$, $\Gamma = \{1, 2, 3\}$; $x = 1$, $\Gamma = \{2, 3\}$; $x = 2$, $\Gamma = \{1, 3\}$; $x = 3$, $\Gamma = \{1, 2\}$.

6 Conclusion and Open Problems

We introduced universal Lyndon words, which are words over an n-letter alphabet having $n!$ Lyndon conjugates. We showed that this class of words properly contains the class of shorthand universal cycles for permutations. We gave combinatorial characterizations and constructions for universal Lyndon words. We leave open the problem of finding an explicit formula for the number of ULW of a given degree.

We exhibited an algorithm for constructing all the universal Lyndon words of a given degree. The algorithm is based on the search for a Hamiltonian cycle in a digraph defined by a particular code, called Hamiltonian lex-code, that we introduced in this paper. It would be natural to find efficient algorithms for generating (or even only counting) universal Lyndon words.

Finally, universal Lyndon words have the property that every conjugate defines a different order, with respect to which it is Lyndon. We can define a *universal order word* as a word of length $n!$ over Σ_n such that every conjugate defines a different order. Universal Lyndon words are therefore universal order words, but the converse is not true, e.g. the word 123421323121424314324134 is a universal order word but is not ULW. Thus, it would be interesting to investigate which properties of universal Lyndon words still hold for this more general class.

References

1. Carpi, A., de Luca, A.: Words and special factors. Theoret. Comput. Sci. 259(1-2), 145–182 (2001)
2. Chung, F., Diaconis, P., Graham, R.: Universal cycles for combinatorial structures. Discrete Math. 110, 43–59 (1992)
3. Holroyd, A.E., Ruskey, F., Williams, A.: Shorthand universal cycles for permutations. Algorithmica 64(2), 215–245 (2012)
4. Jackson, B.: Universal cycles of k-subsets and k-permutations. Discrete Math. 117(1-3), 141–150 (1993)
5. Lothaire, M.: Combinatorics on words. Cambridge University Press, Cambridge (1997)
6. Ruskey, F., Williams, A.: An explicit universal cycle for the $(n - 1)$-permutations of an n-set. ACM Trans. Algorithms 6(3), article 45 (2010)

Subword Complexity and Decomposition of the Set of Factors

Julien Cassaigne[1], Anna E. Frid[1,3],
Svetlana Puzynina[2,3], and Luca Q. Zamboni[2,4]

[1] Aix-Marseille Université, France
cassaigne@iml.univ-mrs.fr, anna.e.frid@gmail.com
[2] Department of Mathematics and Statistics, University of Turku, Finland
svepuz@utu.fi
[3] Sobolev Institute of Mathematics, Russia
[4] Université de Lyon 1, France
zamboni@math.univ-lyon1.fr

Abstract. In this paper we explore a new hierarchy of classes of languages and infinite words and its connection with complexity classes. Namely, we say that a language belongs to the class $\mathcal{L}_k$ if it is a subset of the catenation of k languages $S_1 \cdots S_k$, where the number of words of length n in each of S_i is bounded by a constant. The class of infinite words whose set of factors is in $\mathcal{L}_k$ is denoted by $\mathcal{W}_k$. In this paper we focus on the relations between the classes $\mathcal{W}_k$ and the subword complexity of infinite words, which is as usual defined as the number of factors of the word of length n. In particular, we prove that the class $\mathcal{W}_2$ coincides with the class of infinite words of linear complexity. On the other hand, although the class $\mathcal{W}_k$ is included in the class of words of complexity $O(n^{k-1})$, this inclusion is strict for $k > 2$.

1 Preliminaries

The complexities of infinite words and languages is a widely studied area in formal languages theory. We follow the general approach where the complexity is measured as the number of fragments of a given size. Applied to words, it means that the complexity of a language L (or an infinite word u) is the function $p_L(n)$ (resp., $p_u(n)$) counting the number of elements of L (resp., factors of u) of length n. This function was introduced by Morse and Hedlund in 1938 [9] under the name *block growth* as a tool to study symbolic dynamical systems. The name *subword complexity* was given by Ehrenfeucht, Lee, and Rozenberg [4]; as the term "factor" replaces "subword", the term "factor complexity" is more and more popular [3].

An infinite word is ultimately periodic if and only if its complexity is ultimately constant, and it is a classical result that the smallest complexity of aperiodic words is $p(n) = n + 1$ [9]. The words of this complexity are called Sturmian and form a very interesting and well-explored family (see, e.g., Chapter 2 in [8]). Results on the complexity usually belong to one of the two families: they give either conditions or formulas on the complexity of words from given families

E. Csuhaj-Varjú et al. (Eds.): MFCS 2014, Part I, LNCS 8634, pp. 147–158, 2014.

(see, e.g., [10]), or conditions on words with given restrictions on the complexity. As an example of a complicated problem of that kind, we mention the S-adic conjecture on words of linear complexity (see [7] and references therein). For a recent survey and deep results on subword complexity, see [3].

In the paper we relate the subword complexity to local conditions of factorization type. Namely, we are interested in the following question: What is the relation between the complexity of the word and the condition that each its factor can be decomposed into a product of a finite number k of words belonging to a language of a bounded complexity? In a related paper [5] instead of languages of bounded complexity we considered the language of palindromes. Note that in both cases we need the language of factors to be a subset of the concatenation of these languages and not the concatenation itself. For another family of problems where the equality to the concatenation is needed, see e.g. [1,6].

2   Classes and Basic Hierarchy

We consider finite and infinite words over a finite alphabet Σ, i.e., finite or infinite sequences of elements from the set Σ. A *factor* or a *subword* of an infinite word is any sequence of its consecutive letters. The factor $u_i \cdots u_j$ of an infinite word $u = u_1 \cdots u_n \cdots$, with $u_k \in \Sigma$, is denoted by $u[i..j]$. As usual, the set of factors of a finite or infinite word u is denoted by $\mathrm{Fac}(u)$. A factor s of a right infinite word u is called *right* (resp., *left*) *special* if $sa, sb \in \mathrm{Fac}(u)$ (resp., $as, bs \in \mathrm{Fac}(u)$) for distinct letters $a, b \in \Sigma$. The length of a finite word s is denoted by $|s|$, and the number of occurrences of a letter a in s is denoted by $|s|_a$. The empty word is denoted ε and we define $|\varepsilon| = 0$. An infinite word $u = vwwww \cdots = vw^\omega$ for some non-empty word w is called ultimately ($|w|$-)periodic. In the paper we mostly follow the terminology and notation from [8].

Denote by $\mathcal{P}(\alpha)$ the set of infinite words of complexity $O(n^\alpha)$.

Let us introduce the classes $\mathcal{L}_k$ of languages and $\mathcal{W}_k$ of infinite words as follows: a language L (infinite word u) belongs to the class $\mathcal{L}_k$ (resp., $\mathcal{W}_k$) if

$$L \subseteq S_1 \cdots S_k$$

(resp., $\mathrm{Fac}(u) \subseteq S_1 \cdots S_k$) for some languages S_i with $p_{S_i}(n) = O(1)$. In other words, $u \in \mathcal{W}_k$ if and only if $\mathrm{Fac}(u) \in \mathcal{L}_k$, and the condition $p_{S_i}(n) = O(1)$ means exactly that for some constant C we have $p_{S_i}(n) \leq C$ for all n. We also have $\mathcal{P}(0) = \mathcal{W}_1$.

By a simple cardinality argument, we have the following inclusion:

Lemma 2.1. *For each integer $k > 0$, we have $\mathcal{W}_{k+1} \subseteq \mathcal{P}(k)$.*

PROOF. Suppose a word u is in $\mathcal{W}_{k+1}$ and consider the factors of length n of u. There is $\binom{n+k}{k} = O(n^k)$ ways to decompose a positive integer n to $k + 1$ non-negative summands in a given order: $n = n_1 + n_2 + \ldots + n_{k+1}$. If the summand n_i is the length of the ith factor in a decomposition of a word of length n to $k+1$ factors, and there are at most C words of length n_i in the set S_i, it means that in total, there are not more than C^{k+1} decompositions of words corresponding to a given decomposition of n. Taking all the factors of u of length n together, we see that they are not more than $C^{k+1}\binom{n+k}{k} = O(n^k)$, which means exactly that $u \in \mathcal{P}(k)$.   □

Example 2.2. Now we are going to show that the Thue-Morse word $t = 01101001\cdots$, defined as the fixed point starting with 0 of the morphism $\varphi : 0 \to 01, 1 \to 10$, belongs to $\mathcal{W}_2$. For each n the Thue-Morse word consists of words $t_n = \varphi^n(0)$ and $\overline{t_n} = \varphi^n(1)$, both of them of length 2^n: $t = t_n\overline{t_n}\overline{t_n}t_n\overline{t_n}t_nt_n\overline{t_n}\cdots$. Defining S_1 to be the set of suffixes of all t_n and $\overline{t_n}$, and S_2 to be the set of their prefixes, we see that S_1 and S_2 contain exactly two words of length k each. To cut each factor w of t, we just choose any of its occurrences and a position m in it divided by the maximal power n of 2: $w = t[i..j] = t[i..m]t[m+1..j]$. By the definition of m, $t[i..m]$ is a suffix of t_n or $\overline{t_n}$, and $t[m+1..j]$ is a prefix of one of them, and thus, $w \in S_1S_2$. So, $t \in \mathcal{W}_2$. This construction can be generalized to any fixed point of a primitive morphism but obviously not to fixed points whose complexity is higher than linear (see [10] for examples).

Example 2.3. Sturmian words, which can be defined as infinite words with complexity $n + 1$ for each n, also belong to $\mathcal{W}_2$. These words have exactly one right and one left special factor of each length. One of the ways to construct the sets S_1 and S_2 for a Sturmian word s is the following:

$$S_1 = \{va|a \in \{0,1\}, v \text{ is a right special factor of } s\} \cup \{\varepsilon\},$$
$$S_2 = \{av|a \in \{0,1\}, v \text{ is a left special factor of } s\} \cup \{\varepsilon\}.$$

Remark that in fact the set S_2 is the set of reversals of factors from S_1, and $\#S_1(n) = \#S_2(n) = 2$ for each $n > 0$. The fact that every factor of s belongs to S_1S_2 follows from the properties of Sturmian words: it can be proved that every factor w of s has an occurrence $[i..j]$ with $i \le 0, j \ge 0$ in the biinfinite characteristic Sturmian word u of s, where either $u = c^R 01c$ or $u = c^R 10c$, with c the right infinite characteristic word (i.e., the infinite left special word).

Now let us introduce the *accumulative complexity* function $g_L(n)$ (resp., $g_u(n)$) of a language L (resp., a word u) as

$$g_L(n) = \sum_{i=1}^{n} p_L(i) \qquad (\text{resp., } g_u(n) = \sum_{i=1}^{n} p_u(i)).$$

As above, we introduce the classes $\mathcal{L}'_k$ of languages and $\mathcal{W}'_k$ of infinite words as follows: a language L (resp., infinite word u) belongs to the class $\mathcal{L}'_k$ (resp., $\mathcal{W}'_k$) if

$$L \subseteq S_1 \cdots S_k$$

(resp., $\mathrm{Fac}(u) \subseteq S_1 \cdots S_k$) for some languages S_i with $g_{S_i}(n) = O(n)$.

As above, $u \in \mathcal{W}'_k$ if and only if $\mathrm{Fac}(u) \in \mathcal{L}'_k$. The condition $g_{S_i}(n) = O(n)$ means exactly that for all n we have $g_{S_i}(n) \le Kn$ for some constant K.

Clearly, $\mathcal{L}_k \subseteq \mathcal{L}'_k$, since $p_{S_i}(n) \le C$ for all n implies $g_{S_i}(n) \le Cn$. As for an opposite inclusion, we can only can prove the following theorem and its corollary.

Theorem 2.4. $\mathcal{L}'_1 \subseteq \mathcal{L}_2$.

PROOF. Consider a language $L \in \mathcal{L}'_1$, by definition this means that $g_L(n) \leq Kn$ for some K. We shall construct inductively the sets S and T of complexity $p_S(n), p_T(n) \leq 2K + 1$ such that $L \subseteq ST$.

Let us order the elements of L according to their length: $L = \{v_1, \ldots, v_n, \ldots\}$ with $|v_n| \leq |v_{n+1}|$. The sets S and T are constructed inductively: we choose any $S_1 = \{s_1\}$ and $T_1 = \{t_1\}$ so that $v_1 = s_1 t_1$ and then do as follows. Suppose that we constructed the sets S_{n-1} and T_{n-1} of cardinality less than or equal to $n-1$ each so that $\{v_1, \ldots, v_{n-1}\} \subseteq S_{n-1}T_{n-1}$ and the number of words of each length l in each of S_{n-1}, T_{n-1} is bounded by $2K + 1$.

Consider the word v_n and denote its length by m. It admits $m+1$ factorizations $v_n = st$. If for a given factorization we have $s \in S_{n-1}$ and $t \in T_{n-1}$, we do not need to add anything to these sets and can take $S_n = S_{n-1}, T_n = T_{n-1}$. If for example $s \notin S_{n-1}$, we can construct S_n by adding s to S_{n-1}: $S_n = S_{n-1} \cup \{s\}$ if the words of length $|s|$ in S_{n-1} are at most $2K$ (and symmetrically for T_{n-1}). But the number N of lengths l such that $p_{S_{n-1}}(l) > 2K$ (resp., $p_{T_{n-1}}(l) > 2K$) and thus no more of words of length l can be added to S_{n-1} (resp., T_{n-1}) is bounded by $N \leq (n-1)/(2K)$, since the total number of words in S_{n-1} (resp., T_{n-1}) is at most $(n-1)$.

So, to assure that at least one of $m+1$ factorizations is admitted and we (if necessary) can add new words s_n and t_n: $S_n = S_{n-1} \cup \{s_n\}$, $T_n = T_{n-1} \cup \{t_n\}$ such that $v_n = s_n t_n$, we should check that $m + 1 > 2(n-1)/(2K)$. But since m is the length of the word number n in L, we have $n \leq g_L(m) \leq Km$ and thus $2(n-1)/(2K) \leq (2Km - 2)/(2K) < m + 1$, which was to be proved. $\quad\square$

Corollary 2.5. *For each $k > 0$, we have $\mathcal{L}'_k \subseteq \mathcal{L}_{2k}$.*

PROOF. Take a language $L \in \mathcal{L}'_k$: by the definition, $L \subseteq S_1 \ldots S_k$ with $S_i \in \mathcal{L}'_1$ for all i. Due to the theorem above, all $S_i \in \mathcal{L}_2$, that is, $S_i \subseteq S_i^{(l)} S_i^{(r)}$ where the complexities of $S_i^{(l)}, S_i^{(r)}$ are bounded. Clearly, we have $L \subseteq S_1^{(l)} S_1^{(r)} \ldots S_k^{(l)} S_k^{(r)}$, which proves the corollary. $\quad\square$

So, for all $k > 0$ we have $\mathcal{L}_k \subseteq \mathcal{L}'_k \subseteq \mathcal{L}_{2k}$ and thus $\mathcal{W}_k \subseteq \mathcal{W}'_k \subseteq \mathcal{W}_{2k}$.

3 Linear Complexity and $\mathcal{W}_2$

In this section, we prove the main result of this paper, namely,

Theorem 3.1. *An infinite word is of linear complexity if and only if its language of factors is a subset of the catenation of two languages of bounded complexity: $\mathcal{W}_2 = \mathcal{P}(1)$.*

The $\subseteq$ inclusion has been proven in Lemma 2.1. Since for periodic words the statement is obvious, it remains to find the languages S, T of bounded complexity for a given infinite word u of linear complexity $p_u(n) \leq Cn$ such that the set of factors of u is a subset of ST.

The construction of the sets S and T is based on so-called *markers* which we define below.

3.1   Markers and Classification of Occurrences

Let u be an infinite word. Given a length n, we say that a subset M of the set of factors of u of length n is a set of *markers*, or, more precisely, of *D-markers* for a constant D, if each factor of u of length Dn contains at least one word $m \in M$ as a factor.

Recall that a factor v of u is called *right special* if $va, vb \in \text{Fac}(u)$ for at least two different symbols a, b.

Lemma 3.2. *The set of right special factors of u of length n is a set of $(C+1)$-markers, where $p_u(n) \leq Cn$.*

PROOF. Consider a factor v of u of length $(C+1)n$ and suppose that none of its factors of length n is right special. It means that each factor of v of length n, whenever it occurs in u, uniquely determines the next factor of length n, shifted by one letter. But there are $Cn+1$ occurrences of factors of length n in v. So, at least two of them correspond to the same factor, and what happens after its second occurrence repeats what happens after the first one. So, the word u is ultimately periodic, a contradiction. $\square$

The number of right special factors of u of length n is uniformly bounded by a constant R which is a polynomial of C, where $p_u(n) \leq Cn$, due to a result of Cassaigne [2,3]. Thus, we have the following

Corollary 3.3. *For each length n, there exists a set of cardinality R of $(C+1)$-markers of length n in u.*

Remark that the set of right special factors is just one the possible ways to build the set of markers. For the proof below it does not matter how the set of markers was constructed, the only thing we use is that the set of markers of each length is bounded.

Consider a factor $w = w_1 \cdots w_n$ of u and denote by $p(w)$ its minimal period, that is, the minimal positive integer such that $w_i = w_{i+p(w)}$ for all $i > 0$ and $i+p(w) \leq n$. The word $w[1..p(w)]$, also called the minimal period of w, is denoted by $P(w)$; each time it will be clear from the context whether the period means the word or the number.

An occurrence $w = u[j+1..j+n]$ of w in u is called *internal* if two conditions hold. First, $u_{j+p} = u_{j+p-p(w)}$ for all p such that $1 \leq p \leq p(w)$ and $j+p-p(w) \geq 1$; second, symmetrically, $u_{j+p} = u_{j+p+p(w)}$ for all p such that $n - p(w) + 1 \leq p \leq n$. In other words, due to the definition of $p(w)$, for an internal occurrence of w in the infinite word u we have $u[j+p(w)+1..j+p(w)+n] = w$ and, provided that $j \geq p(w)$, $u[j-p(w)+1..j-p(w)+n] = w$.

An occurrence which is not internal is called *extreme*. More precisely, if $u_{j+i} \neq u_{j+i-p(w)}$ for some i such that $\max(1, p(w)-j+1) \leq i \leq p(w)$, it is called *initial*, and if $u_{j+i} \neq u_{j+i+p(w)}$ for some i such that $n - p(w) + 1 \leq i \leq n$, it is called *final*. Clearly, an occurrence of a word in u can be initial and final at the same time.

Since u is not ultimately periodic, each its factor w admits a final occurrence, otherwise u would be ultimately $p(w)$-periodic.

3.2 Construction and Proof

For each $k \geq 1$, consider the set of D-markers of length 2^k whose cardinality is bounded by R. Due to Corollary 3.3, such a set exists and we shall call its elements markers of order k.

Consider a factor v of length $n \geq 2D$ of u. Our goal is to construct two words $s \in S$ and $t \in T$ such that $u = st$. By the definition of markers, v contains a marker of order one; now consider the largest k such that it contains a marker m of order k. Choose an occurrence of v in u: $v = u[i+1..i+n]$. If all occurrences of m in $u[i+1..i+n]$ are internal, take one of them (say, the first one). If not, choose an extreme occurrence of m in $u[i+1..i+n]$ (again, the first of them if they are several). In both cases, we denote the chosen occurrence $m = u[j+1..j+2^k]$; here $j \geq i$ and $j + 2^k \leq i + n$.

Now we define $s = s(v) = u[i+1..j+2^{k-1}]$ and $t = t(v) = u[j+2^{k-1}+1..i+n]$. Clearly, $v = st$. Note that the marker m is cut exactly in the middle of an occurrence: $m = m_l m_r$ with $|m_l| = |m_r| = 2^{k-1}$. Here s ends by m_l and t starts with m_r.

At last, let us define

$$S = (\text{Fac}(u) \cap \Sigma^{<2D}) \cup \{s(v)|v \in (\text{Fac}(u) \cap \Sigma^{\geq 2D})\},$$
$$T = \{\varepsilon\} \cup \{t(v)|v \in (\text{Fac}(u) \cap \Sigma^{\geq 2D})\},$$

where ε is the empty word, $\Sigma^{<n} = \bigcup_{k=0}^{n-1} \Sigma^k$ and $\Sigma^{\geq n} = \Sigma^* \setminus \Sigma^{<n}$.

It follows immediately from the definitions that $\text{Fac}(u) \subseteq ST$. It remains to prove that the cardinalities of $S \cap \Sigma^n$ and $T \cap \Sigma^n$ are uniformly bounded.

Consider a length $l \geq 2D$. Let us count the words from $T \cap \Sigma^l$.

What can be the length of a marker m used to construct a word $t \in T \cap \Sigma^l$? It is equal to 2^k, where the word m_r of length 2^{k-1} is a prefix of t and thus $2^{k-1} \leq l$. On the other hand, since k was chosen to be maximal and by the definition of D, we have $l < D2^{k+1}$. These two inequalities can be rewritten as

$$\frac{l}{2D} < 2^k \leq 2l, \tag{1}$$

which means that k can take at most $\log_2 D + 2$ values for a given l.

Since we use a construction with at most R markers of each order k, in total there are at most $R(\log_2 D + 2)$ markers which are used to construct the words from $T \cap \Sigma^l$. Exactly the same counting works for the words from $S \cap \Sigma^l$. They can be a bit shorter with respect to k in average, since we choose the first occurrence of a longest marker whenever we have a choice, and since the factor which we decompose can be close to the beginning of u. However, the same bounds hold, and the same $R(\log_2 D + 2)$ (or less) markers can be used to construct the words from $S \cap \Sigma^l$.

Now let us consider separately the cases when the occurrence of a marker used for a decomposition is internal, initial or final.

Lemma 3.4. *Consider an occurrence of a factor v of length $n \geq 2D$ in u and a longest marker m in it. If all the occurrences of m to the chosen occurrence of v are internal, then v is $p(m)$-periodic.*

PROOF. Follows from the definition of an internal occurrence.   □

Let us fix a length $l \geq 2D$. Clearly, for a given marker m of a suitable length 2^k, there is exactly one possible word in Σ^l which can belong to T because of internal occurrences of m: It is $p(m)$-periodic and obtained from the prefix of length $l + 2^{k-1}$ of $P(m)^\omega$ by deleting the first 2^{k-1} symbols. Symmetrically, there is exactly one possible word in Σ^l which can belong to S because of internal occurrences of m.

It follows that for each $l \geq 2D$, each of at most $R(\log_2 D + 2)$ possible markers for this length, its internal occurrences can give at most one word of length l in T and at most one word in S. Now let us consider words arising from extreme occurrences.

For the sake of convenience, define a new symbol $z \notin \Sigma$ and fix $u_n = z$ for $n \leq 0$. So, instead of u, we can now consider a bi-infinite word $u' = \cdots zzzu_1u_2u_3\cdots$.

Let us fix a marker m of length 2^k and a length l satisfying (1) and consider the set $T_f(m, l)$ of words from T of length l arising from final occurrences of m to u. For any word $t \in T_f(m, l)$ consider a place in u which gives rise to it, that is, fix a position $j \geq 0$ such that $m = u[j+1..j+2^k]$ and $t = u[j+2^{k-1}+1..j+2^{k-1}+l]$. Now for each i such that $0 \leq i < 2^{k-1}$ define the word $e_f(m, t, j, i)$ of length $l + 2^k$ as $e_f(m, t, j, i) = u[j + 1 - i..j + l + 2^k - i]$ (see Fig. 1). Note that if $j + 1 < 2^k$, the word $e_f(m, t, j, i)$ for sufficiently large i-s starts with one or several (but not more than $2^{k-1} - 1$) symbols z.

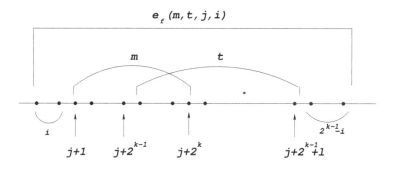

Fig. 1. Construction of $e_f(m, t, j, i)$

Lemma 3.5. *If* $e_f(m, t, j, i) = e_f(m, t', j', i')$ *with* $|t| = |t'| = l$, *then* $t = t'$ *and* $i = i'$.

PROOF. Denote $e_f(m, t, j, i) = e_f(m, t', j', i') = e$. Note also that k can be uniquely reconstructed from m.

Suppose that $i = i'$; then $t = t' = e[i + 2^{k-1} + 1..i + 2^{k-1} + l]$.

Suppose that $i < i'$. Then the word $e[i + 1..i' + 2^k]$ has m as a prefix and a suffix and thus is $(i' - i)$-periodic. In particular, m is $(i' - i)$-periodic. Since $p(m)$ is the minimal period of m, we have $p(m) \leq i' - i < 2^{k-1} = |m|/2$. So, for each $h = 1, \ldots, 2^k - p(m) + i' - i$ both symbols e_{i+h} and $e_{i+h+p(m)}$ belong to either the prefix copy of m or to the suffix copy of m (or to both).

So, $e_{i+h} = e_{i+h+p(m)}$ for all h from 1 to $2^k - p(m) + i' - i \geq 2^k$, and in particular for all h such that $2^k - p(m) + 1 \leq h \leq 2^k$. This contradicts to the fact that $u[j + 1..j + 2^k] = e[i + 1..i + 2^k]$ is a final occurrence of m to u.   □

So, the number of possible words $e_f(m, t, j, i)$ for a given marker m and a given length l of t is minorized by the number of pairs (t, i); here t is a word from $T \cap \Sigma^l$ arising from a final occurrence of a marker m, and for each m, t and j, the parameter i takes exactly 2^{k-1} values. On the other hand, all $e_f(m, t, j, i)$ are words of length $l + 2^k$, which are either factors of u or its prefixes preceded by at most 2^{k-1} new symbols z: the number of factors of u of length $l + 2^k$ is $p_u(l + 2^k)$, the number of words with z is at most 2^{k-1}, and the number of words $e_f(m, t, j, i)$ is majorized by $p_u(l + 2^k) + 2^{k-1} \leq C(l + 2^k) + 2^{k-1}$. So, we have

$$2^{k-1} t_f(m, l) \leq C(l + 2^k) + 2^{k-1},$$

where $t_f(m, l)$ is the contribution to $T \cap \Sigma^l$ of all the final occurrences of a marker m of length 2^k.

Since $l < 2^{k+1}D$, the latter inequality can be rewritten as

$$t_f(m, l) < \frac{C(2D + 1)2^k + 2^{k-1}}{2^{k-1}} = 2C(2D + 1) + 1.$$

In other words,

$$t_f(m, l) \leq 2C(2D + 1).$$

Exactly the same upper bound can be symmetrically proved for the contribution to $T \cap \Sigma^l$ of initial occurrences of a marker m: $t_i(m, l) \leq 2C(2D + 1)$. So, each of $R(\log_2 D + 2)$ possible markers for the length l can contribute at most for the following number of words to $T \cap \Sigma^l$: one word arising from its internal occurrences, plus $2C(2D + 1)$ words arising from final occurrences, plus $2C(2D + 1)$ words arising from initial occurrences. This gives the desired upper bound: the total number of words in the set $T \cap \Sigma^l$ is bounded by the constant

$$R(\log_2 D + 2)[1 + 4C(2D + 1)].$$

The proof for $S \cap \Sigma^l$ is similar and gives the same constant as the upper bound. □

Note that the analogous fact for general languages is not true: there exists a language of linear complexity not belonging to any $\mathcal{L}_k$. However, this language (which we do not describe here because of the lack of space) is not closed under taking a factor.

4   Word of Quadratic Complexity

Lemma 2.1 and Theorem 3.1 imply that $\mathcal{W}_2 = \mathcal{P}(1)$, and in general $\mathcal{W}_{k+1} \subseteq \mathcal{P}(k)$ for all k. So, the following natural question arises: is it true that $\mathcal{W}_{k+1} = \mathcal{P}(k)$ for all k?

The answer is negative, and, since $\mathcal{W}_k \subseteq \mathcal{W}'_k$, to show it we just point an example of a word of quadratic complexity which does not belong to $\mathcal{W}'_3$.

Consider the word $u = ababbabbb \cdots = \prod_{i=1}^{\infty} ab^i$. Its complexity $p_u(n) = \Theta(n^2)$: this can be either proved directly or derived from the famous paper by Pansiot [10], since u is obtained by erasing the first letter c from the fixed point starting with c of the morphism $c \mapsto cab, a \mapsto ab, b \mapsto b$.

Lemma 4.1. *The word u does not belong to $\mathcal{W}_3'$.*

PROOF. Suppose the opposite: $\mathrm{Fac}(u) \subseteq XYZ$ with $g_X(n), g_Y(n), g_Z(n) = O(n)$. Now for each word $v \in \mathrm{Fac}(u)$ of length at most n fix some its decomposition $v = x(v)y(v)z(v) = xyz$ with $x \in X$, $y \in Y$, $z \in Z$. We shall estimate the number of words v which can be decomposed like that.

Now for each $k, l > 0$ define the word $w_{k,l} = ab^l ab^{l+1} \cdots ab^{l+k-1}a$. Clearly, $w_{k,l}$ is a factor of u of length $k(l + (k+1)/2) + 1$.

Claim. Let $E(n)$ be the set of pairs (k, l) such that $|w_{k,l}| \leq n$, $k \geq 3$ and $l \geq \sqrt{n}$. Then $\#E(n) = \Theta(n \log n)$.

PROOF. Note that the condition $|w_{k,l}| = k(l + (k+1)/2) + 1 \leq n$ implies the inequality $l \leq \dfrac{n-1}{k} - \dfrac{k+1}{2}$. So,

$$\#E(n) = \sum_{k=3}^{\infty} \# \left\{ l \in \mathbb{N} : \sqrt{n} \leq l \leq \frac{n-1}{k} - \frac{k+1}{2} \right\}.$$

Observe that this set is empty for $k \geq \sqrt{2n}$: indeed, if $k \geq \sqrt{2n}$, then $\dfrac{n-1}{k} - \dfrac{k+1}{2} \leq \dfrac{n}{\sqrt{2n}} - \dfrac{\sqrt{2n}+1}{2} < 0$. So,

$$\#E(n) = \sum_{k=3}^{\lfloor \sqrt{2n} \rfloor} \left(\frac{n-1}{k} - \frac{k+1}{2} - \sqrt{n} + 1 \right).$$

Here

$$\sum_{k=3}^{\lfloor \sqrt{2n} \rfloor} \frac{n-1}{k} = (n-1) \sum_{k=3}^{\lfloor \sqrt{2n} \rfloor} \frac{1}{k} = \Theta(n \ln n)$$

and

$$\sum_{k=3}^{\lfloor \sqrt{2n} \rfloor} \left(\frac{k+1}{2} + \sqrt{n} - 1 \right) = \Theta(n).$$

The claim follows. □

Let us say that a factor v of u is *of type* (k, l) if $v = b^i w_{k,l} b^j$ for some i and j. Clearly, each factor of u either is of some type (k, l), or contains at most one letter a.

Denote by $F(n)$ the set of pairs (k, l) with $k \geq 3$ and $l \geq \sqrt{n}$ such that there exists a factor v of u of length at most n and of type (k, l) whose decomposition is xyz with $|x|_a \leq 1$, $|z|_a \leq 1$. There were $k+1 \geq 4$ letters a in v, and at least

$k - 1 \geq 2$ of them stay in the word y. The type of y is thus one of the four following: (k, l), $(k - 1, l + 1)$, $(k - 1, l)$, $(k - 2, l + 1)$. But the total number of words in Y of length at most n is $g_Y(n) = O(n)$, and each word y can give rise to at most four types from $F(n)$. So, $\#F(n) \leq 4g_Y(n) = O(n)$, and due to the previous claim, there are still $\#E(n) \backslash F(n) = \Theta(n \log n)$ pairs (k, l) with $k \geq 3$ and $l \geq \sqrt{n}$ such that each word v of type (k, l) and of length at most n is decomposed so that its middle part $y(v)$ contains at most one letter a. Since there are $k + 1 \geq 4$ letters a in v, we see that either $x(v)$ or $z(v)$ contains at least two letters a.

We denote this set of pairs by $H(n) = E(n) \backslash F(n)$. The number of all factors v of u whose types are in $H(n)$ is denoted by $s(n)$.

Consider a factor v of u of length at most n whose type is in $H(n)$. Suppose first that the word $x(v)$ contains more than one letter a. Then the word v is uniquely determined by $x(v)$ and the length $|v| \leq n$. So, the number of words v of length $\leq n$ admitting such a decomposition is bounded by $ng_X(n) = O(n^2)$.

Symmetrically, the number of words v such that $z(v)$ contains more than one letter a is bounded by $ng_Z(n) = O(n^2)$.

So, the number $s(n)$ of words whose types are in $H(n)$ is $O(n^2)$. But on the other hand, the number of types in $H(n)$ is $\Theta(n \log n)$, and for each type (k, l), the number of words of this type is $l(l + k + 1)$: indeed, such a word is of the form $b^i w_{k,l} b^j$, where i can take l values from 0 to $l - 1$ and j can take $l + k + 1$ values from 0 to $l + k$. Since we restricted ourselves to the case of $l \geq \sqrt{n}$, the number of words of each type is $l(l + k + 1) > n$. In total, we have that $s(n) \geq n\Theta(n \log n)$, that is,

$$s(n) = \Omega(n^2 \log n).$$

A contradiction to the previous condition $s(n) = O(n^2)$.   □

Since $\mathcal{W}_3 \subseteq \mathcal{W}'_3$, we get also the following

Corollary 4.2. *There exists a word of quadratic complexity which does not belong to $\mathcal{W}_3$.*

5   Belonging to Some $\mathcal{W}_k$

The word u of quadratic complexity considered in the previous section does not belong to $\mathcal{W}'_3$, but it can be proved that it belongs to $\mathcal{W}'_4$. We omit this proof here since it does not add much to the theory. However, this result suggests the following question: given a word of complexity majorated by a polynomial, is it true that it belongs to $\mathcal{W}_k$ for some k?

As we show in the next proposition, the answer to this question is negative.

Proposition 5.1. *For any growing integer function $f(n)$ such that $f(1) \geq 1$, $f(n) \leq n$ and $f(n) \to \infty$, there exists an infinite word w of complexity $O(n^2 f(n))$ which does not belong to $\mathcal{W}_k$ for any k.*

PROOF. First we describe the construction of the word w, then we prove that w does not belong to $\mathcal{W}_k$ for any k, and after that we prove that the word has complexity $O(n^2 f(n))$.

Define the infinite word w as follows:

$$w = \prod_{p=1}^{\infty} \prod_{q=1}^{f(p)} (a^p b^q)^{k(p,q)},$$

where $k(p,q)$ is a growing function: $k(p,q) \leq k(p,q+1)$ and $k(p,f(p)) \leq k(p+1,1)$ for all p and q.

Let us prove that $w \notin \mathcal{W}_k$ for any k. Suppose by contrary that $w \in \mathcal{W}_k$: $\mathrm{Fac}(w) \subseteq S_1 \cdots S_k$ with $p_{S_i}(n) \leq M_i$ for all i. Define $S = \cup_i S_i$; then $p_S(n) \leq \sum_i p_{S_i}(n) \leq \sum_i M_i = M$ for an appropriate constant M. Consequently, $g_S(n) \leq Mn$ for all n.

Claim. For every pair of integers (p,q), such that $p + q < \frac{n-2}{2k-1}$, $q \leq f(p)$ and $k(p,q) \geq 2k - 1$, there exists a word $s_{p,q} \in S$, $|s_{p,q}| \leq n$, such that $s_{p,q}$ contains $ba^p b^q a$ as a factor, and all those words $s_{p,q}$ are distinct.

PROOF. Consider the word $b(a^p b^q)^{2k-1} a$. Since $k(p,q) \geq 2k - 1$ and $q \leq f(p)$, it is a factor of w, and since $p + q < \frac{n-2}{2k-1}$, its length is at most n. However we cut the word $b(a^p b^q)^{2k-1} a$ into at most k pieces, at least one piece will contain $ba^p b^q a$ as a factor. The claim follows. □

Let us estimate the number of words $ba^p b^q a$ for $p + q < \frac{n-2}{2k-1}$, $q \leq f(p)$ and $k(p,q) \geq 2k - 1$. Since the function $k(p,q)$ is growing, there exists a constant p_k such that $k(p,q) \geq 2k - 1$ for all $p \geq p_k$ and all $q \leq f(p)$. Since $f(p) \leq p$ for all p, we have $p + q \leq p + f(p) \leq 2p$, and thus the number of pairs (p,q) is bounded from below by the sum $\sum_{p=p_k}^{\frac{n-2}{2(2k-1)}} f(p)$. Since $f(p) \to \infty$ as $p-> \infty$, and since $g_S(n)$ is bounded from below by the number of pairs (p,q) due to Claim 5, we have

$$g_S(n) \geq \sum_{p=p_k}^{\frac{n-2}{2(2k-1)}} f(p) > Mn$$

for some sufficiently large n. A contradiction to the fact that $g_S(n) \leq Mn$.

Now let us check that the complexity of the word w is $O(n^2 f(n))$. The word w contains factors of the following types:

1. Factors of a block $(a^p b^q)^k$ for some p, q and k.
2. Factors of a concatenation of blocks $(a^p b^q)^{k_1} (a^p b^{q+1})^{k_2}$.
3. Factors of a concatenation of blocks $(a^p b^{f(p)})^{k_1} (a^{p+1} b)^{k_2}$.
4. Factors containing some complete block $(a^p b^q)^{k_{p,q}}$ as a factor.

Remark that some of these families intersect, but this is not a problem since we only need a bound. So, let us estimate the number of words of length n in each family.

In the family 1, we have $O(n)$ words of the form $a^i b^{n-i}$ or $b^i a^{n-i}$, plus $O(n^2)$ words of the form $a^i b^q a^{n-q-i}$ (uniquely determined by $0 < i, q < n$) or $b^i a^p b^{n-p-i}$ (uniquely determined by $0 < i, p < n$), plus words containing a factor $b a^p b^q a$ or $a b^q a^p b$. The latter words are uniquely determined by $p < n$, $q \le f(p)$ and the position of the first occurrence of a^p, which takes values from 0 to $p+q < n$. So, the number of such words (and thus of all the words in family 1) is $O(n^2 f(n))$.

Treating the other three families analogously, we see that the complexity of each of them is at most $O(n^2 f(n))$ too. So, the complexity $p_w(n) = O(n^2 f(n))$, which completes the proof. $\square$

6   Conclusion

We finalize this paper by suggesting the following open problem: What is the minimal possible complexity of a word which does not belong to any $\mathcal{W}_k$?

Remark that Theorem 3.1 and Proposition 5.1 imply that this complexity is strictly bigger than linear and is at most quadratic.

Supported in part by RFBR grants 12-01-00089 and 12-01-00448, as well as by the Academy of Finland under a FiDiPro grant and under grant 251371.

References

1. Avgustinovich, S.V., Frid, A.E.: A unique decomposition theorem for factorial languages. Int. J. Alg. Comput. 15, 149–160 (2005)
2. Cassaigne, J.: Special factors of sequences with linear subword complexity. DLT 1995, 25–34. World Sci. Publishing, Singapore (1996)
3. Cassaigne, J., Nicolas, F.: Factor complexity. Combinatorics, automata and number theory. Encyclopedia Math. Appl., vol. 135, pp. 163–247. Cambridge Univ. Press (2010)
4. Ehrenfeucht, A., Lee, K.P., Rozenberg, G.: Subword complexities of various deterministic developmental languages without interactions. Theoret. Comput. Sci. 1, 59–76 (1975)
5. Frid, A., Puzynina, S., Zamboni, L.Q.: On palindromic factorization of words. Advances in Applied Mathematics 50, 737–748 (2013)
6. Han, Y.-S., Salomaa, K., Wood, D.: Prime Decompositions of Regular Languages. In: Ibarra, O.H., Dang, Z. (eds.) DLT 2006. LNCS, vol. 4036, pp. 145–155. Springer, Heidelberg (2006)
7. Leroy, J.: Some improvements of the S -adic conjecture. Adv. Appl. Math. 48(1), 79–98 (2012)
8. Lothaire, M.: Algebraic combinatorics on words. Cambridge University Press (2002)
9. Morse, M., Hedlund, G.: Symbolic dynamics. Amer. J. Math. 60, 815–866 (1938)
10. Pansiot, J.-J.: Complexité des facteurs des mots infinis engendrés par morphismes itérés. In: Paredaens, J. (ed.) ICALP 1984. LNCS, vol. 172, pp. 380–389. Springer, Heidelberg (1984)

Cyclic Complexity of Words

Julien Cassaigne[1], Gabriele Fici[2,*],
Marinella Sciortino[2,*], and Luca Q. Zamboni[3,4]

[1] Institut de Mathématiques de Luminy, Marseille, France
`cassaigne@iml.univ-mrs.fr`
[2] Dipartimento di Matematica e Informatica, University of Palermo, Italy
`{fici,mari}@math.unipa.it`
[3] Université Claude Bernard Lyon 1, France
[4] Department of Mathematics, University of Turku, Finland
`lupastis@gmail.com`

Abstract. We introduce and study a new complexity function on words, which we call *cyclic complexity*, which counts the number of conjugacy classes of factors of each given length. We extend the famous Morse-Hedlund theorem to the setting of cyclic complexity by showing that a word is ultimately periodic if and only if it has bounded cyclic complexity. Unlike most complexity functions, cyclic complexity distinguishes between Sturmian words having different slopes. More precisely, we prove that if x is a Sturmian word and y is a word having the same cyclic complexity of x then y is Sturmian and, up to renaming letters, it has the same language of factors of x.

Keywords: Cyclic complexity, factor complexity, Sturmian words, minimal forbidden factor.

1 Introduction

The usual notion of complexity of a discrete system counts the number of distinct patterns of the same size that the system can generate. In the case of sequences (words), this is the number of distinct blocks (factors) of each given length. This measure of complexity is usually called *factor complexity* (or *block complexity*). The words with the "simplest" structure are the periodic ones. They are of the form $x = u^\omega$ (i.e., an infinite concatenation of a same finite block u) called *purely periodic*, or of the form $x = vu^\omega$, called *ultimately periodic*. The non-periodic words are called *aperiodic*. The factor complexity distinguishes between periodic and aperiodic words. In fact, a fundamental result dating back to the late 30's is the famous theorem of Morse and Hedlund [14] stating that a word is aperiodic if and only if it has at least $n+1$ factors of length n for every n. From this result, it is natural to consider those aperiodic words which have minimal factor complexity, i.e., those having exactly $n + 1$ distinct factors of length n for every n. These

---

* Partially supported by Italian MIUR Project PRIN 2010LYA9RH, "Automi e Linguaggi Formali: Aspetti Matematici e Applicativi".

E. Csuhaj-Varjú et al. (Eds.): MFCS 2014, Part I, LNCS 8634, pp. 159–170, 2014.

are called *Sturmian words* and a vast bibliography exists showing their interest both from the theoretical viewpoint and in applications. For example, Sturmian words code the digital approximations in the plane of Euclidean straight lines with irrational slope, with the property that two Sturmian words have the same slope if and only if they have the same language of factors.

There exist many other measures of complexity of words in literature. For example, a lot of attention has recently been given (see for instance [3,10,16,18,19]) to the *abelian complexity*, which is the function counting the number of factors of each given length up to permutations. Other new measures of complexity of words have been introduced over the time, which are intermediate between factor and abelian complexity (e.g. maximal pattern complexity [7], k-abelian complexity [8], binomial complexity [17]) or involve different definitions that appear naturally in the study of sequences (e.g. periodicity complexity [12], minimal forbidden factor complexity [13], palindromic complexity [5], etc.) For most of these measures, Sturmian words are those aperiodic words of lowest complexity. However, they do not distinguish between two Sturmian words having different slopes.

In this paper we propose a new measure of complexity, *cyclic complexity*, which consists in counting the factors of each given length of a word up to conjugacy. The notion of conjugacy is a basic notion in Combinatorics on Words. Two words are said conjugate if they are equal when read on a circle[1]. That is, the cyclic complexity of a word is the function counting the number of conjugacy classes of factors of each given length.

One of the main results of this paper is that cyclic complexity distinguishes between periodic and aperiodic words. In fact, we prove the following theorem.

Theorem 1. *A word is ultimately periodic if and only if it has bounded cyclic complexity.*

That is, the Morse-Hedlund theorem can be extended to the setting of cyclic complexity. Note that a word is (purely) periodic if and only if there exists an integer n such that all the factors of length n are conjugate. Therefore, the minimum value that the cyclic complexity of an aperiodic word can take is 2. We will prove that Sturmian words have the property that the cyclic complexity takes value 2 infinitely often.

Since the Sturmian words are characterized by having $n + 1$ factors of length n for every n, the factor complexity does not distinguish between two Sturmian words with different languages of factors. In contrast, for cyclic complexity, two Sturmian words with different languages of factors have different cyclic complexity. Indeed, we prove something stronger:

Theorem 2. *Let x be a Sturmian word. If a word y has the same cyclic complexity as x then, up to renaming letters, y is a Sturmian word having the same slope of x.*

---

[1] More formally, u and v are conjugate if and only if one can write $u = w_1 w_2$ and $v = w_2 w_1$ for some words w_1, w_2.

That is, not only two Sturmian words with different languages of factors cannot have the same cyclic complexity, but the only words which have the same cyclic complexity of a Sturmian word x are those Sturmian words with the same slope of x.

These two results suggest that cyclic complexity can be considered as an interesting refinement of the classical notion of factor complexity and can open new perspectives in the study of complexity of discrete systems.

Note that factor complexity, abelian complexity and cyclic complexity can all be viewed as actions of different subgroups of the symmetric group on the indices of a finite word (respectively, the trivial subgroup, the whole symmetric group and the cyclic subgroup). Since factor and abelian complexity are very well studied, looking at other subgroups of the symmetric group seems a very natural way of investigation.

2 Basics

Given a finite non-empty ordered set A (called the *alphabet*), we let A^* and $A^{\mathbb{N}}$ denote respectively the set of finite words and the set of (right) infinite words over the alphabet A. The order on the alphabet A can be extended to the usual lexicographic order on the set A^*.

For a finite word $w = w_1 w_2 \cdots w_n$ with $n \geq 1$ and $w_i \in A$, the length n of w is denoted by $|w|$. The *empty word* is denoted by ε and we set $|\varepsilon| = 0$. We let A^n denote the set of words of length n and A^+ the set of non-empty words. For $u, v \in A^+$, $|u|_v$ is the number of occurrences of v in u. For instance $|0110010|_{01} = 2$. The *Parikh vector* of w is the vector whose components are the number of occurrences of the letters of A in w. For example, if $A = \{a, b, c\}$, then the Parikh vector of $w = abb$ is $(1, 2, 0)$. The *reverse* (or *mirror image*) of a finite word w is the word obtained by reading w in the reverse order.

Given a finite or infinite word $\omega = \omega_1 \omega_2 \cdots$ with $\omega_i \in A$, we say a word $u \in A^+$ is a *factor* of ω if $u = \omega_i \omega_{i+1} \cdots \omega_{i+n-1}$ for some positive numbers i and n. We let $\mathrm{Fact}(\omega)$ denote the set of all factors of ω, and $\mathrm{Alph}(\omega)$ the set of all factors of ω of length 1. If $\omega = u\nu$, we say that u is a *prefix* of ω, while ν is a *suffix* of ω. A factor u of ω is called *right special* (resp. *left special*) if both ua and ub (resp. au and bu) are factors of ω for distinct letters $a, b \in A$. The factor u is called *bispecial* if it is both right special and left special.

For each factor u of ω, we set

$$\omega\big|_u = \{n \in \mathbb{N} \mid \omega_n \omega_{n+1} \cdots \omega_{n+|u|-1} = u\}.$$

We say ω is *recurrent* if for every $u \in \mathrm{Fact}(\omega)$ the set $\omega\big|_u$ is infinite. We say ω is *uniformly recurrent* if for every $u \in \mathrm{Fact}(\omega)$ the set $\omega\big|_u$ is syndetic, i.e., of bounded gap. A word $\omega \in A^{\mathbb{N}}$ is *(purely) periodic* if there exists a positive integer p such that $\omega_{i+p} = \omega_i$ for all indices i, while it is *ultimately periodic* if $\omega_{i+p} = \omega_i$ for all sufficiently large i. Finally, a word $\omega \in A^{\mathbb{N}}$ is called *aperiodic* if it is not ultimately periodic. For a finite word $w = w_1 w_2 \cdots w_n$, we call p a *period* of u if $w_{i+p} = w_i$ for every $1 \leq i \leq n - p$.

Two finite or infinite words are said to be *isomorphic* if the two words are equal up to a renaming of the letters.

A (finite or infinite) word w over A is *balanced* if and only if for any u, v factors of w of the same length and for every letter $a \in A$, one has $||u|_a - |v|_a| \leq 1$. More generally, w is *C-balanced* if there exists a constant $C > 0$ such that for any u, v factors of w of the same length and for every letter $a \in A$, one has $||u|_a - |v|_a| \leq C$. For example, the word 010111 is not balanced, but it is 2-balanced. Note that if w is C-balanced, then it is C'-balanced for any $C' \geq C$.

The *factor complexity* of an infinite word ω is the function

$$p_\omega(n) = |\operatorname{Fact}(\omega) \cap A^n|,$$

i.e., the function that counts the number of distinct factors of length n of ω, for every $n \geq 0$. The factor complexity is a standard measure of the complexity of an infinite word. By Morse-Hedlund theorem, words with bounded factor complexity are precisely ultimately periodic words and aperiodic words with minimal factor complexity have linear factor complexity. In the binary case, aperiodic words with minimal factor complexity have factor complexity equal to $n+1$, i.e., they are Sturmian words. An example of word achieving maximal factor complexity over an alphabet of size $k > 1$ can be obtained by concatenating the k-ary expansions of non-negative integers. For example, if $k = 2$, one obtains the so called Champernown word 0110111001011101111000$\cdots$

The factor complexity counts the factors appearing in the word. A dual point of view consists in counting the shortest factors that *do not* appear in the word. This leads to another measure of complexity, described below.

Let w be a (finite or infinite) word over an alphabet A. A finite non-empty word v is a *minimal forbidden factor* for w if v does not belong to $\operatorname{Fact}(w)$ but every proper factor of v does. We denote by $\operatorname{MF}(w)$ the set of all minimal forbidden words for w. The *minimal forbidden factor complexity* of an infinite word ω is the function

$$mf_\omega(n) = |\operatorname{MF}(\omega) \cap A^n|,$$

i.e., the function that counts the number of distinct minimal forbidden factors of length n of ω, for every $n \geq 0$.

We now introduce a new measure of complexity. The idea is to count the factors of each given length that are different up to a rotation. Recall that two finite words u, v are *conjugate* if there exist words w_1, w_2 such that $u = w_1 w_2$ and $v = w_2 w_1$. The conjugacy relation is an equivalence over A^*, which is denoted by $\sim$, whose classes are called *conjugacy classes*. Note that two words belonging to the same conjugacy class must have the same Parikh vector.

The *cyclic complexity* of an infinite word ω is the function

$$c_\omega(n) = \left| \frac{\operatorname{Fact}(\omega) \cap A^n}{\sim} \right|,$$

i.e., the function that counts the number of distinct conjugacy classes of factors of length n of ω, for every $n \geq 0$.

Observe that, by the definition, $c_\omega(n) \leq p_\omega(n)$ for every n. Moreover, if a word ω has maximal cyclic complexity, then it has maximal factor complexity. In fact, let $w \in A^*$ be any word. We want to show that $w \in \mathrm{Fact}(\omega)$. Consider the word ww. From the maximality of the cyclic complexity of ω, some conjugate of ww is an element of $\mathrm{Fact}(\omega)$. But every conjugate of ww contains w as a factor, hence $w \in \mathrm{Fact}(\omega)$.

Since a word having maximal factor complexity clearly also has maximal cyclic complexity, we have the following proposition.

Proposition 1. *An infinite word has maximal cyclic complexity if and only if it has maximal factor complexity.*

The cyclic complexity, as well as the other mentioned complexity functions, can be naturally extended to any factorial language. Recall that a language is any subset of A^*. A language L is called *factorial* if it contains all the factors of its words, i.e., if $uv \in L \Rightarrow u, v \in L$. The cyclic complexity of L is defined by

$$c_L(n) = \left| \frac{L \cap A^n}{\sim} \right|.$$

The cyclic complexity is an invariant for several operations on languages. For example, it is clear that if two languages are isomorphic (i.e., one can be obtained from the other by renaming letters), then they have the same cyclic complexity. Furthermore, if L is a language and $\tilde{L}$ is obtained from L by reversing (mirror image) each word in L, then L and $\tilde{L}$ have the same cyclic complexity.

3 Cyclic Complexity Distinguishes between Periodic and Aperiodic Words

In this section we give a proof of Theorem 1. The following lemma connects cyclic complexity to balancedness.

Lemma 1. *Let $\omega \in A^{\mathbb{N}}$ and suppose that there exists a constant C such that $c_\omega(n) \leq C$ for every n. Then ω is C-balanced.*

Proof. For every n, there are at most C conjugacy classes of factors of length n in ω. This implies that there are at most C different Parikh vectors for the factors of ω of length n, that is, ω has abelian complexity bounded by C. It can be proved (see [16]) that this implies that the word ω is C-balanced. $\square$

Lemma 2. *Let $\omega \in A^{\mathbb{N}}$ be aperiodic and let $v \in A^+$ be a factor of ω which occurs in ω an infinite number of times. Then, for each positive integer K there exists a positive integer n such that ω contains at least $K + 1$ distinct factors of length n beginning in v.*

Proof. Suppose to the contrary that for some K, ω has at most K distinct factors of each length n which begin in v. Since ω is aperiodic and v occurs infinitely often in ω, there exist $K + 1$ distinct suffixes of ω (say $y_0, y_1, \ldots, y_K$) beginning in v. By the pigeonhole principle, for each positive integer n there exist $0 \le i < j \le K$ such that y_i and y_j begin in the same prefix of length n. Again by the pigeonhole principle, there exist $0 \le i < j \le K$ such that y_i and y_j begin in the same prefix of length n for infinitely many distinct values of n. Hence, $y_i = y_j$, a contradiction. $\square$

Proof of Theorem 1. If ω is ultimately periodic, then it has bounded factor complexity by Morse-Hedlund theorem, hence it must have bounded cyclic complexity.

Let us now prove that if ω is aperiodic, then for any fixed positive integer M, $c_\omega(n) \ge M$ for some n. Short of replacing ω by a suffix of ω, we can suppose that each letter occurring in ω occurs infinitely often in ω. First, suppose that for each positive integer C, ω is not C-balanced. Then, by Lemma 1, the cyclic complexity of ω is unbounded and we are done. Next, suppose that each $u \in A^+$ is a factor of ω. In this case, ω would have full complexity, whence the cyclic complexity of ω is again unbounded. Thus, we can suppose that ω is C-balanced for some positive integer C, and that some $u \in A^+$ is not a factor of ω. Since ω is C-balanced, there exists a positive integer N such that each factor of ω of length N contains an occurrence of each $a \in \mathrm{Alph}(\omega)$. As u is a forbidden factor of ω, it follows that u is a forbidden factor of each suffix of ω. Since each letter occurring in ω occurs infinitely often, it follows there exist a suffix ω' of ω, a letter $a \in A$ and a word $v \in A^+$ such that av is a forbidden factor of ω' and v occurs in ω' infinitely often. By Lemma 2, there exists a positive integer $n_0 \ge 2|v|$ such ω' contains at least MN distinct factors of length n_0 beginning in v. We denote these factors by $u_1, u_2, \ldots, u_{MN}$. There exist $v_1, v_2, \ldots, v_{MN}$, each in A^N, such that $u_i v_i$ are factors of ω' (of length $n_0 + N$) for each $1 \le i \le MN$. Since each v_i contains an occurrence of a, it follows there exists $n \ge n_0$ such that ω' contains at least M distinct factors of length n beginning in v and terminating in a. Since av is a forbidden factor of ω', no two of these factors are conjugate to one another. Hence, $c_{\omega'}(n) \ge M$ and thus $c_\omega(n) \ge M$. $\square$

4 Cyclic Complexity Distinguishes between Sturmian Words with Different Languages

In this section we exhibit results on the cyclic complexity of Sturmian words and give a sketch of the proof of Theorem 2.

There exists a vast bibliography on Sturmian words (see for instance the survey papers [1,2], [9, Chap. 2], [15, Chap. 6] and references therein).

A Sturmian word is an infinite word having exactly $n + 1$ distinct factors of length n, for every $n \ge 0$. That is, a word x is Sturmian if and only if $p_x(n) = n+1$ for every $n \ge 0$. Note that an immediate consequence of the definition is that $|\mathrm{Alph}(x)| = 2$, so a Sturmian word is a binary word. In this section we fix the alphabet $A = \{0,1\}$.

A very well known instance of Sturmian words is the Fibonacci word $F = 010010100100101001\cdots$, obtained as the limit of the substitution $0 \mapsto 01$, $1 \mapsto 0$.

Sturmian words have a multitude of combinatorial properties that make them fundamental objects in the field of Combinatorics on Words. By Morse-Hedlund Theorem, Sturmian words are those aperiodic words with minimal factor complexity. We recall some other characterizations in the next proposition.

Proposition 2. *Let $x \in A^{\mathbb{N}}$. The following conditions are equivalent:*

1. *x is Sturmian;*
2. *x is balanced and aperiodic;*
3. *x has exactly one right (resp. left) special factor for each length.*

Recall that the *slope* of a finite binary word w over the alphabet A is defined as $s(w) = \frac{|w|_1}{|w|}$. The slope of an infinite binary word, when it exists, is the limit of the slopes of its prefixes. A Sturmian word can also be defined by considering the intersections with a squared-lattice of a semi-line having a slope which is an irrational number. A horizontal intersection is denoted by the letter 0, while a vertical intersection is denoted by the letter 1. Note that the slope of the Sturmian word is exactly the slope of such a semi-line. For example, the slope of the Fibonacci word is $(1 + \phi)^{-1}$, where $\phi = (1 + \sqrt{5})/2$ is the golden ratio.

An important property of Sturmian words is that their factors depend on their slope only, i.e., we have the following result (see [14]).

Proposition 3. *Let x, y be two Sturmian words. Then $\mathrm{Fact}(x) = \mathrm{Fact}(y)$ if and only if x and y have the same slope.*

A fundamental role in the study of factors of Sturmian words is played by the *central words*. A word is central if it has coprime periods p and q and length $p + q - 2$. There are several characterizations of central words (see [1] for a survey). Here we recall the following ones.

Proposition 4. *Let w be a word over A. The following conditions are equivalent:*

1. *w is a central word;*
2. *$0w1$ and $1w0$ are conjugate;*
3. *w is a bispecial factor of some Sturmian word;*
4. *w is a palindrome and the words $w0$ and $w1$ are balanced;*
5. *$0w1$ is balanced and is the least element (w.r.t. the lexicographic order) in its conjugacy class;*
6. *w is a power of a letter or there exist central words p_1, p_2 such that $w = p_1 01 p_2 = p_2 10 p_1$. Moreover, in this latter case $|p_1| + 2$ and $|p_2| + 2$ are coprime periods of w and $\min(|p_1| + 2, |p_2| + 2)$ is the minimal period of w.*

Let w be a central word with coprime periods p and q and length $p + q - 2$. The words $0w1$ and $1w0$, which, by Proposition 4, are conjugate, are called *Christoffel words*. Let $r = |0w1|_0$ and $s = |0w1|_1$. It can be proved that r and s are the multiplicative inverses of p and q modulo $p + q$, respectively. Moreover, the conjugacy class of $0w1$ and $1w0$ contains exactly $|w| + 2$ words. If we sort these words lexicographically and arrange them as rows of a matrix, we obtain a square matrix with remarkable combinatorial properties (see [4, 6, 11]). We call this matrix the (r, s)-*Christoffel array* and denote it by $\mathcal{A}_{r,s}$ (see Figure 1 for an example). Two consecutive rows of $\mathcal{A}_{r,s}$ differ only by a swap of two consecutive positions. Moreover, the columns are also conjugate and in particular the first one is $0^r 1^s$, while the last one is $1^s 0^r$.

Every aperiodic word (and therefore, in particular, every Sturmian word) contains infinitely many bispecial factors. If w is a bispecial factor of a Sturmian word x, then w is central by Proposition 4 and there exists a unique letter $a \in A$ such that w', the shortest palindrome beginning with wa, is a bispecial factor of x. Moreover, if p and q are the coprime periods of w such that $|w| = p+q-2$, then the word w' is central and its coprime periods p' and q' verifying $|w'| = p'+q'-2$ satisfy either $p' = p + q$ and $q' = p$, or $p' = p + q$ and $q' = q$, depending on the letter a. For example, 010 is a bispecial factor of the Fibonacci word F and has coprime periods 3 and 2 (and length $3 + 2 - 2$). The successive (in length order) bispecial factor of F is 010010, which is the shortest palindrome beginning in $010 \cdot 0$ and has coprime periods 5 and 3 (and length $5 + 3 - 2$). There exist other Sturmian words having 010 as a bispecial factor and for which the successive bispecial factor is 01010 (i.e., the shortest palindrome beginning with $010 \cdot 1$) whose coprime periods are 5 and 2.

These combinatorial properties of central words and the bispecial factors of a Sturmian word will be needed in our proof of Theorem 2.

Sturmian words have unbounded cyclic complexity (by Theorem 1) but their cyclic complexity takes value 2 for infinitely many n. More precisely, we have the following result.

Lemma 3. *Let x be a Sturmian word. Then $c_x(n) = 2$ if and only if $n = 1$ or there exists a bispecial factor of x of length $n - 2$.*

The value 2 is the minimal possible for an aperiodic word. In fact, it is well known that a word ω is (purely) periodic if and only if there exists $n \geq 1$ such that all the factors of length n of ω are conjugate.

Since a Sturmian word contains infinitely many bispecial factors, the previous result implies that for a Sturmian word x one has that $\liminf c_x(n) = 2$. However, this is not a characterization of Sturmian words. In fact, there exist non-Sturmian aperiodic words with minimal cyclic complexity (in the sense of having limit inferior of the cyclic complexity equal to 2). Consider for example the morphism $\mu : 0 \mapsto 00, 1 \mapsto 01$. It is possible to prove that in the word $\mu(F) = 0001000010001000001 \cdots$, image of the Fibonacci word F under μ, there are exactly 2 conjugacy classes of factors of length n for every n that is the double of a Fibonacci number[2], so that $\liminf c_{\mu(F)}(n) = 2$. However, the word $\mu(F)$ is not Sturmian (it contains the factors 00000 and 10001 and therefore is not balanced). We show in Table 1 the first values for the cyclic complexity of F and $\mu(F)$.

---

[2] Recall that Fibonacci numbers are defined by: $F_0 = 1$, $F_1 = 1$, and $F_n = F_{n-1} + F_{n-2}$ for every $n > 1$.

Table 1. The initial values of the cyclic complexity for the Fibonacci word F and its morphic image $\mu(F)$

n	1	2	3	4	5	6	7	8	9	10	11	12	13	14	15	16	17	18	19	20	21	22
$c_F(n)$	2	2	2	3	2	4	4	2	7	4	5	8	2	9	9	4	13	5	9	14	2	16
$c_{\mu(F)}(n)$	2	2	2	2	3	2	3	3	3	2	5	4	5	4	6	2	7	7	7	4	9	5

We now give a sketch of the proof of Theorem 2.

Proof of Theorem 2 (Sketch). Since y has the same cyclic complexity of x, we have that in particular $2 = c_x(1) = c_y(1)$, so y is a binary word. Since x is aperiodic, by Theorem 1 c_x is unbounded. Since x and y have the same cyclic complexity we have, always by Theorem 1, that y is aperiodic.

We want to prove that y and x have the same factors. By contradiction, let $n + 2$ be the least length for which x and y have different factors. This implies that x and y have a same bispecial factor w of length n. Let p' and q', with $p' > q'$, be the two coprime periods of w such that $n = |w| = p' + q' - 2$. Let w_x (resp. w_y) be the successive (in length order) bispecial factor of x (resp. of y). It can be proved that $\{|w_x|, |w_y|\} = \{2p' + q' - 2, p' + 2q' - 2\}$ and that w_x and w_y cannot have the same length.

Suppose $|w_x| < |w_y|$. Then, by Lemma 3, y would have cyclic complexity equal to 2 at length $|w_x| + 2$, which is impossible since between $|w|$ and $|w_y|$ the word y behaves as a Sturmian word and so by Lemma 3 it should have a bispecial factor of length $|w_x| + 2$. Hence, we can suppose that $|w_x| > |w_y|$, so that w_x has periods $p' + q'$ and p' and length $2p' + q' - 2$, while w_y has periods $p' + q'$ and q' and length $p' + 2q' - 2$.

To ease notation, we set $p = p' + q'$ and $q = p'$, so that $|w_y| = 2p - q - 2$ and $|w_x| = p + q - 2$. Let us consider the set of factors of y of length $2p - q$. Since $|w| + 2 < 2p - q < |w_x| + 2$, we know by Lemma 3 that $c_x(2p - q) > 2$. So, $c_y(2p - q) > 2$.

If there was a Sturmian word y' such that $\mathrm{Fact}(y') \cap A^{2p-q} = \mathrm{Fact}(y) \cap A^{2p-q}$, then $2p - q$ would be the length of a bispecial factor plus 2 of a Sturmian word and then, by Lemma 3, we would have $c_y(2p - q) = 2$, a contradiction. This implies that w_y is a bispecial factor of y that behaves differently from a bispecial factor of a Sturmian word. More precisely, we must have that $0w_y$ and $1w_y$ are both right special factors of y. Therefore, $0w_y0$ and $1w_y1$ are in two different conjugacy classes and all the other factors of y of length $2p - q$ are in a third conjugacy class. In other words, we have $c_y(2p - q) = 3$. Thus, in order to have a contradiction we are left to prove that $c_x(2p - q) \geq 4$.

It is known that among the $p + q + 1$ factors of x of length $p + q$, there is one factor with a Parikh vector and the remaining $p + q$ factors with the other Parikh vector, these latter being in a same conjugacy class, which is in fact the conjugacy class of the Christoffel word $0w_x1$.

Let $r = |0w_x1|_0$ and $s = |0w_x1|_1$. Without loss of generality, we can suppose that $r > s$, i.e., we can suppose that 11 does not appear as a factor in x. Therefore, we can build the (r, s)-Christoffel array $\mathcal{A}_{r,s}$. The factors of length $2p - q$ of x can be obtained by removing the last $2q - p$ columns from $\mathcal{A}_{r,s}$ (of course, in this way some rows can be equal and therefore some factors appear more than once). We refer to the matrix so obtained as to $\mathcal{A}'_{r,s}$.

The cases $s = 1, 2, 3$ can be proved separately. Here we give the sketch of the proof when $s > 3$. Recall that $\{r, s\} = \{p^{-1}, q^{-1}\} \mod (p + q)$. Suppose that $s = p^{-1} < q^{-1}$. In this case, one can prove that the last three rows in $\mathcal{A}'_{r,s}$ are distinct and start and end with 1. Therefore, each of these rows is unique in its conjugacy class. Since any other row correspond to a factor with a different Parikh vector, this implies that there are at least 4 conjugacy classes and we are done.

The other case is when $s = q^{-1} < p^{-1}$. This case can be proved analogously by considering the first four rows of the matrix $\mathcal{A}'_{r,s}$. In fact, one can prove that the factors appearing in the first four rows of the matrix $\mathcal{A}'_{r,s}$ are pairwise distinct and neither is conjugate to another. $\qquad \square$

Example 1. Consider the Fibonacci word F and its bispecial factor $w = 010010$, which has periods $p = 5$ and $q = 3$. We have $s = q^{-1} = 3 < 5 = r = p^{-1}$. In Figure 1 we show the $(5, 3)$-Christoffel array $\mathcal{A}_{5,3}$. The rows are the lexicographically sorted factors of F with Parikh vector $(5, 3)$. The other factor of length 8 of F is 10100101. The factors of F of length $2p - q = 7$ can be obtained by removing the last column of the matrix. Notice that the first 4 rows (once the last character has been removed) are pairwise distinct and neither is conjugate to another.

To end this section, we compare the cyclic complexity to the minimal forbidden factor complexity for the special case of Sturmian words.

In [13] the authors proved the following result.

Theorem 3. *Let x be a Sturmian word and let y be an infinite word such that for every n one has $p_x(n) = p_y(n)$ and $mf_x(n) = mf_y(n)$, i.e., y is a word having the same factor complexity and the same minimal forbidden factor complexity as x. Then, up to isomorphism, y is a Sturmian word having the same slope as x.*

$$\mathcal{A}_{5,3} = \begin{pmatrix} 0\,0\,1\,0\,0\,1\,0\,1 \\ 0\,0\,1\,0\,1\,0\,0\,1 \\ 0\,1\,0\,0\,1\,0\,0\,1 \\ 0\,1\,0\,0\,1\,0\,1\,0 \\ 0\,1\,0\,1\,0\,0\,1\,0 \\ 1\,0\,0\,1\,0\,0\,1\,0 \\ 1\,0\,0\,1\,0\,1\,0\,0 \\ 1\,0\,1\,0\,0\,1\,0\,0 \end{pmatrix}$$

Fig. 1. The matrix $\mathcal{A}_{5,3}$ for the Fibonacci word F for $p = 5$ and $q = 3$

Note that Theorem 2 is much stronger than Theorem 3, because in this latter the fact that y is a Sturmian word follows directly from the hypothesis that y has the same factor complexity as x.

Indeed, the cyclic complexity is more fine than the minimal factor complexity. Let x be an infinite binary word such that $MF(x) = \{11, 000\}$ and y an infinite binary word such that $MF(y) = \{11, 101\}$. Then x and y have the same minimal forbidden factor complexity, but it is readily checked that $c_x(5) = 3$ while $c_y(5) = 4$. Note that x contains 7 factors of length 5 corresponding to 3 cyclic classes $(00100, 00101, 01001, 10010, 10100, 10101)$ while y contains the factors $00000, 10000, 10010, 10001$ no two of which are cyclically conjugate.

5   Conclusions and Further Developments

We introduced a new measure of complexity of words, cyclic complexity. We showed that for this measure of complexity the Morse-Hedlund theorem can be extended, that is, a word is ultimately periodic if and only if it has bounded cyclic complexity (Theorem 1). The aperiodic words with minimal cyclic complexity can be defined as those having exactly 2 conjugacy classes of factors of length n for infinitely many values of n. Among these we have Sturmian words (which are the aperiodic words with minimal factor complexity), but we also exhibited a non-Sturmian example which, however, is a morphic image of a Sturmian word. We leave as an open problem that of characterizing the aperiodic words with minimal cyclic complexity.

Contrarily to other measures of complexity, cyclic complexity characterizes the language of a Sturmian word, in the sense that two Sturmian words with different languages of factors have different cyclic complexities. More precisely, we proved that a word having the same cyclic complexity as a Sturmian word must be Sturmian and have the same slope (Theorem 2). A natural question is therefore the following: Given two infinite words x and y with the same cyclic complexity, what can we say about their languages of factors?

First, there exist two periodic words having same cyclic complexity but whose languages of factors are not isomorphic nor related by mirror image. For example, let τ be the morphism: $0 \mapsto 010$, $1 \mapsto 011$ and consider the words $x = \tau((010011)^\omega)$ and $x' = \tau((101100)^\omega)$. One can verify that x and x' have the same cyclic complexity up to length 17 and, since each has period 18, the cyclic complexities agree.

Furthermore, it is easy to show that even two aperiodic words can have same cyclic complexity but different languages of factors. For example, let x be an infinite binary word such that $MF(x) = \{000111\}$ and y an infinite binary word such that $MF(y) = \{001111\}$. Then the languages of factors of x and y are not isomorphic, nor related by mirror image, yet the two words have the same cyclic complexity. However, we do not know if this can still happen with the additional hypothesis of linear complexity, for example. In every case, these examples show that cyclic complexity does not determine, in general, the language of factors. So, our Theorem 2 is very special to Sturmian words.

In conclusion, we believe that the new notion of complexity we introduced in this paper, cyclic complexity, can open new perspectives in the study of complexity of words and languages.

References

1. Berstel, J.: Sturmian and episturmian words (a survey of some recent results). In: Bozapalidis, S., Rahonis, G. (eds.) CAI 2007. LNCS, vol. 4728, pp. 23–47. Springer, Heidelberg (2007)
2. Berstel, J., Lauve, A., Reutenauer, C., Saliola, F.: Combinatorics on Words: Christoffel Words and Repetitions in Words. CRM monograph series, vol. 27. American Mathematical Society (2008)
3. Blanchet-Sadri, F., Fox, N.: On the asymptotic abelian complexity of morphic words. In: Béal, M.-P., Carton, O. (eds.) DLT 2013. LNCS, vol. 7907, pp. 94–105. Springer, Heidelberg (2013)
4. Borel, J.-P., Reutenauer, C.: On Christoffel classes. RAIRO Theor. Inform. Appl. 40(1), 15–27 (2006)
5. Brlek, S., Hamel, S., Nivat, M., Reutenauer, C.: On the palindromic complexity of infinite words. Internat. J. Found. Comput. Sci. 15(2), 293–306 (2004)
6. Jenkinson, O., Zamboni, L.Q.: Characterisations of balanced words via orderings. Theoret. Comput. Sci. 310(1-3), 247–271 (2004)
7. Kamae, T., Zamboni, L.: Maximal pattern complexity for discrete systems. Ergodic Theory Dynam. Systems 22(4), 1201–1214 (2002)
8. Karhumäki, J., Saarela, A., Zamboni, L.Q.: On a generalization of abelian equivalence and complexity of infinite words. J. Comb. Theory, Ser. A 120(8), 2189–2206 (2013)
9. Lothaire, M.: Algebraic Combinatorics on Words. Cambridge University Press, Cambridge (2002)
10. Madill, B., Rampersad, N.: The abelian complexity of the paperfolding word. Discrete Math. 313(7), 831–838 (2013)
11. Mantaci, S., Restivo, A., Sciortino, M.: Burrows-Wheeler transform and Sturmian words. Inform. Process. Lett. 86(5), 241–246 (2003)
12. Mignosi, F., Restivo, A.: A new complexity function for words based on periodicity. Internat. J. Algebra Comput. 23(4), 963–988 (2013)
13. Mignosi, F., Restivo, A., Sciortino, M.: Words and forbidden factors. Theoret. Comput. Sci. 273(1-2), 99–117 (2002)
14. Morse, M., Hedlund, G.A.: Symbolic dynamics. Amer. J. Math. 60, 1–42 (1938)
15. Fogg, N.P.: Substitutions in Dynamics, Arithmetics and Combinatorics. Lecture Notes in Math., vol. 1794. Springer (2002)
16. Richomme, G., Saari, K., Zamboni, L.: Abelian complexity of minimal subshifts. J. Lond. Math. Soc. 83(1), 79–95 (2011)
17. Rigo, M., Salimov, P.: Another generalization of abelian equivalence: Binomial complexity of infinite words. In: Karhumäki, J., Lepistö, A., Zamboni, L. (eds.) WORDS 2013. LNCS, vol. 8079, pp. 217–228. Springer, Heidelberg (2013)
18. Saarela, A.: Ultimately constant abelian complexity of infinite words. J. Autom. Lang. Comb. 14(3-4), 255–258 (2010)
19. Turek, O.: Abelian complexity and abelian co-decomposition. Theoret. Comput. Sci. 469, 77–91 (2013)

Classifying Recognizable Infinitary Trace Languages Using Word Automata

Namit Chaturvedi* and Marcus Gelderie**

RWTH Aachen Univeristy, Lehrstuhl für Informatik 7, Aachen, Germany
{chaturvedi,gelderie}@automata.rwth-aachen.de

Abstract. We address the problem of providing a Borel-like classification of languages of infinite Mazurkiewicz traces, and provide a solution in the framework of ω-automata over infinite words – which is invoked via the sets of linearizations of infinitary trace languages. We identify trace languages whose linearizations are recognized by deterministic weak or deterministic Büchi (word) automata. We present a characterization of the class of linearizations of all recognizable ω-trace languages in terms of Muller (word) automata. Finally, we show that the linearization of any recognizable ω-trace language can be expressed as a Boolean combination of languages recognized by our class of deterministic Büchi automata.

1 Introduction

Traces were introduced as models of concurrent behaviors of distributed systems by Mazurkiewicz, who later also provided an explicit definition of infinite traces [7]. Zielonka demonstrated the close relation between traces and trace-closed sets of words, which can be viewed as "linearizations" of traces, and established automata-theoretic results regarding recognizability of languages of finite traces [11] (cf. also [3,6]). Later, [4,2,8] enriched the theory of recognizable languages of infinite traces (*recognizable ω-trace languages*), by introducing models of computations viz. asynchronous Büchi automata and deterministic asynchronous Muller automata. Being closely related to word languages, a set of infinite traces is recognizable iff the corresponding trace-closed set of infinite words is.

In the case of ω-regular word languages, there exists a straightforward characterization of languages recognized by deterministic Büchi automata, and a result due to Landweber states that it is decidable whether a given ω-regular language is deterministically Büchi recognizable [9, Chapter 1]. However, analogous results over recognizable ω-trace languages have only recently been established in terms of "synchronization-aware" asynchronous automata [1].

While asynchronous automata are useful in implementing distributed monitors and distributed controllers, their constructions are prohibitively expensive even by automata-theoretic standards. On the other hand, for applications like

---

* Supported by the DFG Research Training Group-1298 AlgoSyn, and the CASSTING Project funded by the European Commission's 7th Framework Programme.
** Supported by the DFG Research Training Group-1298 AlgoSyn.

E. Csuhaj-Varjú et al. (Eds.): MFCS 2014, Part I, LNCS 8634, pp. 171–182, 2014.

model-checking and formal verification, word automata recognizing trace-closed languages would already allow for analysis of most of the interesting properties pertaining to distributed computations.

Therefore, in this paper, we study classes of ω-regular word languages that allow us to "transfer" interesting results to the corresponding classes of recognizable ω-trace languages. In particular, motivated by the Borel hierarchy for regular languages of infinite words, our main contribution is a new setup for a classification theory for recognizable ω-trace languages in terms of trace-closed, ω-regular word languages.

Recall that in the sequential setting, reachability languages and deterministically Büchi recognizable languages – constituting the lowest levels of the Borel hierarchy – can be obtained via natural operations over regular languages $K \subseteq \Sigma^*$ in the following ways:

- $\mathsf{ext}(K) = K \cdot \Sigma^\omega = \{\alpha \in \Sigma^\omega \mid \alpha \text{ has a prefix in } K\}$
- $\mathsf{lim}(K) = \{\alpha \in \Sigma^\omega \mid \alpha \text{ has infinitely many prefixes in } K\}$

These operations, which we call the *infinitary extension* and the *infinitary limit* of K, can be generalized to obtain infinitary extensions $\mathsf{ext}(T)$ and infinitary limits $\mathsf{lim}(T)$ of regular trace languages T.

In this paper, given the trace-closed word language K corresponding to a regular trace language T, we firstly show that K can be modified to K_I such that $\mathsf{ext}(K_I)$ is also trace-closed and corresponds to the linearization of $\mathsf{ext}(T)$ (here I denotes the independence relation over the alphabet Σ). Building on this, we are able to characterize the class of Boolean combinations of languages $\mathsf{ext}(T)$ as precisely those whose linearizations are recognized by the class of "I-diamond" deterministic weak automata (DWAs).

Next, we consider infinitary limits. Here the situation is different, in that there exist regular trace languages T such that although the trace-closed word language L corresponding to $\mathsf{lim}(T)$ is ω-regular, it is not recognized by any I-diamond deterministic Büchi automaton (DBA). We therefore introduce the class of *limit-stable* word languages K – and by extension limit-stable trace languages T – such that the correspondence of Fig. 1b holds, and $\mathsf{lim}(K)$ can be characterized in terms of I-diamond DBA.

It is well known that every trace-closed ω-regular language is recognized by an I-diamond Muller automata [2]. We characterize these languages in terms of a well defined class of I-diamond Muller automata. And lastly, justifying our

(a) The correspondence of infinitary extensions for all regular trace languages T.   (b) The correspondence of infinitary limits for all *limit-stable* trace languages T.

Fig. 1. From trace-closed regular languages to trace-closed ω-regular languages

definitions, we show that every trace-closed ω-regular word language (that is, the linearization of any recognizable ω-trace language) can be expressed as a finite Boolean combination of languages $\lim(K)$, with K limit-stable.

In related work, Diekert & Muscholl [2] consider a form of "deterministic" trace languages. It is shown that every recognizable language of infinite traces is a Boolean combination of these deterministic languages. However, in the attempt to characterize the corresponding "deterministic" trace-closed word languages in terms of I-diamond automata, it is necessary to extend the Büchi acceptance condition beyond what we know from standard definitions [8]. It had been left open in [8] whether there exists a class of deterministic asynchronous Büchi automata for deterministic trace languages.

We begin with presenting the basic definitions and notations. In Sec. 3, we present the operations ext and lim that allow for construction of recognizable ω-trace languages from regular trace languages. In particular, we exhibit recognizable ω-trace languages whose linearizations are recognized by I-diamond DWA, and those whose linearizations are I-diamond DBA recognizable. Finally, we establish our main result demonstrating the expressiveness of I-diamond DBA recognizable trace-closed languages.

2   Preliminaries

We denote a recognizable language of finite words, or simply a *regular language*, with the upper case letter K and a class of such languages with $\mathcal{K}$. Finite words are denoted with lower case letters u, v, w etc. Infinite words are denoted by lower case Greek letters α and β, and a recognizable language of infinite words, or simply an *ω-regular language*, by upper case L. For a word u or α, we denote its infix starting at position i and ending at position j by $u[i,j]$ or $\alpha[i,j]$, and the ith letter with $u[i]$ or $\alpha[i]$. For a language K, we define $\overline{K} := \Sigma^* \setminus K$.

We assume the reader is familiar with the notions of Deterministic Finite Automata (DFAs) and Deterministic Büchi Automata (DBAs). We say that a language is *DBA recognizable* if it is recognized by a DBA. For the class REG of regular languages, the class $\lim(\text{REG})$ coincides with the DBA recognizable languages. Further, the class $\text{BC}(\lim(\text{REG}))$ of finite Boolean combinations of languages from $\lim(\text{REG})$ is also the class of ω-regular languages, and it coincides with the class of languages recognized by nondeterministic Büchi or deterministic Muller automata.

Recall that a Deterministic Weak Automaton (DWA) is a DBA where every strongly connected component of the transition graph has only accepting states or only rejecting states. For a regular language K, the minimal DFA recognizing K also recognizes $\lim(K)$ as a DBA. Given the minimal DFA $\mathfrak{A} = (Q, \Sigma, q_0, \delta, F)$ recognizing K, a DWA $\mathfrak{A}' := (Q', \Sigma, q_0, \delta', F')$ recognizing $\text{ext}(K)$, respectively $\text{ext}(\overline{K})$, can be constructed as follows:

1. For a symbol $\perp \notin Q$ and define $Q' := (Q \setminus F) \cup \{\perp\}$.

2. For each $q \in Q', a \in \Sigma$, define $\delta'(q,a) := \begin{cases} \delta(q,a) & \text{if } q \neq \perp \text{ and } \delta(q,a) \notin F, \\ \perp & \text{otherwise.} \end{cases}$

3. Define $F' := \{\bot\}$, respectively $F' := Q' \setminus \{\bot\}$.

The family of DWAs is closed under Boolean operations. For an ω-language L, define a congruence $\sim_L \subseteq \Sigma^* \times \Sigma^*$ where $u \sim_L v \Leftrightarrow \forall \alpha \in \Sigma^\omega, u\alpha \in L$ iff $v\alpha \in L$. If L is recognized by a DWA then this congruence has a finite index. We say that an ω-language is *weakly recognizable* if it is recognized by a DWA. The class BC(ext(REG)) of finite Boolean combinations of languages in ext(REG) is exactly the set of weakly recognizable languages [10].

Remark 1 (The minimal DWA [5]). For a weakly recognizable language L, if M is the index of the congruence defined above, then L is recognized by a DWA $\mathfrak{A} = (Q, \Sigma, q_0, \delta, F)$ with $|Q| = M$. Also, for every $q \in Q$ there exists a word $u_q \in \Sigma^*$ such that for each $u \in \Sigma^*, \delta(q_0, u) = q$ iff $u \in [u_q]_{\sim_L}$.   ⊠

Turning to traces, let $I \subseteq \Sigma \times \Sigma$ denote an irreflexive[1], symmetric *independence* relation over an alphabet Σ, then $D := \Sigma^2 \setminus I$ is the reflexive, symmetric *dependence* relation over Σ. We refer to the pair (Σ, I) as the *independence alphabet*. For any letter $a \in \Sigma$, we define $I_a := \{b \in \Sigma \mid aIb\}$ and $D_a := \{b \in \Sigma \mid aDb\}$. A *trace* can be identified with a labeled, acyclic, directed *dependence graph* $[V, E, \lambda]$ where V is a set of countably many vertices, $\lambda: V \to \Sigma$ is a labeling function, and E is a countable set of edges such that, firstly, for every $v_1, v_2 \in V: \lambda(v_1)D\lambda(v_2) \Leftrightarrow (v_1, v_2) \in E \vee (v_2, v_1) \in E$; secondly, every vertex has only finitely many predecessors. $\mathbb{M}(\Sigma, I)$ and $\mathbb{R}(\Sigma, I)$ represent the sets of all finite and infinite traces whose dependence graphs satisfy the two conditions above. We denote finite traces with the letter t, and an infinite trace with θ; the corresponding languages with T and Θ respectively. For a trace $t = [V, E, \lambda]$, define alph$(t) := \{a \in \Sigma \mid \emptyset \neq \lambda^{-1}(a) \subseteq V\}$, and similarly for a trace θ. For an infinite trace, define alphinf$(\theta) := \{a \in \Sigma \mid |\lambda^{-1}(a)| = \infty\}$.

For two traces t_1, t_2, $t_1 \sqsubseteq t_2$ (or $t_1 \sqsubset t_2$) denotes that t_1 is a (proper) prefix of t_2. We denote the prefix relation between words similarly. The least upper bound of two finite traces, whenever it exists, denoted $t_1 \sqcup t_2$ is the smallest trace s such that $t_1 \sqsubseteq s$ and $t_2 \sqsubseteq s$. Whenever it exists, one can similarly refer to the least upper bound $\bigsqcup S$ of a finite or an infinite set S of traces. The concatenation of two traces is denoted as $t_1 \odot t_2$. Note that for any t, θ the concatenation $t \odot \theta \in \mathbb{R}(\Sigma, I)$. However, $\theta \odot t \in \mathbb{R}(\Sigma, I)$ iff alphinf$(\theta)I$alph(t).

The canonical morphism $\Gamma: \Sigma^* \to \mathbb{M}(\Sigma, I)$ associates finite words with finite traces, and the inverse mapping $\Gamma^{-1}: \mathbb{M}(\Sigma, I) \to 2^{\Sigma^*}$ associates finite traces with equivalence classes of words. The morphism Γ can also be extended to a mapping $\Gamma: \Sigma^\omega \to \mathbb{R}(\Sigma, I)$. For a (finite or infinite) trace t, the set $\Gamma^{-1}(t)$ represents the *linearizations* of t. Two words u, v are equivalent, denoted $u \sim_I v$, iff $\Gamma(u) = \Gamma(v)$. We note that for finite traces the relation $\sim_I$ coincides with the reflexive, transitive closure of the relation $\{(uabv, ubav) \mid u, v \in \Sigma^* \wedge aIb\}$. For a word w, define the set $[w]_{\sim_I} := \Gamma^{-1}(\Gamma(w))$. Finally, we say that a word language K is *trace-closed* iff $K = [K]_{\sim_I}$, where $[K]_{\sim_I} := \bigcup_{u \in K} [u]_{\sim_I}$.

---

[1] A relation R is *irreflexive* if for no x we have xRx.

Definition 2. *A trace language $T \subseteq \mathbb{M}(\Sigma, I)$ (resp. $\Theta \subseteq \mathbb{R}(\Sigma, I)$) is called recognizable iff $\Gamma^{-1}(T)$ (resp. $\Gamma^{-1}(\Theta)$) is a recognizable word language.*

We denote the classes of recognizable languages of finite and infinite traces with $\mathsf{Rec}(\mathbb{M}(\Sigma, I))$ and $\mathsf{Rec}(\mathbb{R}(\Sigma, I))$ respectively.

Asynchronous cellular automata have been introduced [2,4] as acceptors of recognizable ω-trace languages. However, a global view of their (local) transition relations yields a notion of automata that recognize trace-closed word languages. Throughout this paper, we take this global view of asynchronous automata. Formally, a *deterministic asynchronous cellular automaton (DACA)* over (Σ, I) is a 4-tuple $\mathfrak{a} = (\prod_{a \in \Sigma} Q_a, (\delta_a)_{a \in \Sigma}, q_0, F)$, consisting of sets Q_a of *local states* for each letter $a \in \Sigma$, and where $q_0 \in \prod_{a \in \Sigma} Q_a$, $\delta_a \colon \prod_{b \in D_a} Q_b \to Q_a$ and $F \subseteq \prod_{a \in \Sigma} Q_a$. Given a state $q \in \prod_{a \in \Sigma} Q_a$ and a letter $b \in \Sigma$, the unique b-successor $\delta(q, b) = q' = (q'_a)_{a \in \Sigma} \in \prod_{a \in \Sigma} Q_a$ is given by $q'_b = \delta_b((q_a)_{a \in D_b})$ and $q'_a = q_a$ for all $a \neq b$. That is, the only component that changes its state is the component corresponding to b. Given a word $u \in \Sigma^*$ the run ρ_u of $\mathfrak{a}$ on u is given as usual by $\rho_u(0) = q_0$ and $\rho_u(i + 1) = \delta(\rho_u(i), u[i])$. This definition extends naturally to infinite runs ρ_α on infinite $\alpha \in \Sigma^\omega$. Define $\mathsf{occ}_a(\rho)$ of (a finite or an infinite) run ρ to be the set $\{\rho(0)_a, \rho(1)_a, \ldots\} \subseteq Q_a$. Likewise, $\mathsf{inf}_a(\rho) = \{q \in Q_a \mid \exists^\infty n \colon \rho(n)_a = q\}$.

A *deterministic asynchronous cellular Muller automaton* [2] (a *DACMA*) is an asynchronous automaton $\mathfrak{a} = (\prod_{a \in \Sigma} Q_a, (\delta_a)_{a \in \Sigma}, q_0, \mathcal{F})$ with the acceptance table $\mathcal{F} \subseteq \prod_{a \in \Sigma} \mathscr{P}(Q_a)$, where $\mathscr{P}(Q_a)$ denotes the power set of Q_a. A DACMA *accepts* $\alpha \in \Sigma^\omega$ if for some $F = (F_a)_{a \in \Sigma} \in \mathcal{F}$ we have $\forall a \in \Sigma \colon \mathsf{inf}_a(\rho_\alpha) = F_a$. A *deterministic asynchronous cellular Büchi automaton* (a *DACBA*) is a tuple $\mathfrak{a} = (\prod_{a \in \Sigma} Q_a, (\delta_a)_{a \in \Sigma}, q_0, \mathcal{F})$, $F \subseteq \prod_{a \in \Sigma} \mathscr{P}(Q_a)$. A DACBA *accepts* $\alpha \in \Sigma^\omega$ if for some $F = (F_a)_{a \in \Sigma} \in \mathcal{F}$ we have $F_a \subseteq \mathsf{inf}_a(\rho_\alpha)$.

While it is known that the class of DACMAs characterize precisely the class of recognizable ω-trace languages [2], no such correspondence is known for the class of languages recognized by DACBAs [8].

A word automaton $\mathfrak{A} = (Q, \Sigma, q_0, \delta)$ is called *I-diamond* if for every $(a, b) \in I$ and every state $q \in Q$, $\delta(q, ab) = \delta(q, ba)$. Every $T \in \mathsf{Rec}(\mathbb{M}(\Sigma, I))$ (resp. $\Theta \in \mathsf{Rec}(\mathbb{R}(\Sigma, I))$) is recognized by a DACA [3] (resp. a DACMA). Via their global behaviors, asynchronous automata accept the corresponding trace-closed languages, and in particular, every regular trace-closed language (resp. trace-closed ω-regular language) is recognized by an I-diamond DFA (resp. I-diamond Muller automaton). In fact for every trace-closed $K \in \mathsf{REG}$, the minimal DFA $\mathfrak{A}_K$ accepting K is I-diamond.

3 From Regular Trace Languages to ω-Regular Trace Languages

We wish to extend the well-studied relations between regular and ω-regular languages to trace languages. We first look at reachability languages and their Boolean combinations, i.e. the weakly recognizable languages, and study how they can be obtained as a result of infinitary operations on regular trace languages. After this, we observe that the case of Büchi recognizability is not as straightforward and provide a resolution.

3.1 Infinitary Extensions of Regular Trace Languages

In the classification hierarchy of ω-regular languages, reachability and safety languages occupy the lowest levels. For trace languages we have the following.

Definition 3. *Let* $T \in \mathsf{Rec}(\mathbb{M}(\Sigma, I))$. *The* infinitary extension *of* T *is the* ω-*trace language given by* $\mathsf{ext}(T) := T \odot \mathbb{R}(\Sigma, I)$.

Extrapolating the definition of E-automata for word languages, we define E-automata for trace languages where a run is accepting if for each $a \in \Sigma$ some predefined local states from Q_a are reached. Formally, a *deterministic asynchronous E-automaton* (a *DAEA*) is a tuple $\mathfrak{a} = (\prod_{a \in \Sigma} Q_a, (\delta_a)_{a \in \Sigma}, q_0, \mathcal{F})$ with $\mathcal{F} \subseteq \prod_{a \in \Sigma} \mathscr{P}(Q_a)$. The DAEA $\mathfrak{a}$ accepts $\alpha \in \Sigma^\omega$ if for some $F = (F_a)_{a \in \Sigma} \in \mathcal{F}$ we have that $\mathsf{occ}_a(\rho_\alpha) \cap F_a \neq \emptyset$. Note that given a DACA $\mathfrak{A}$ with $L(\mathfrak{A}) = T$, in order to accept $\mathsf{ext}(T)$ any DAEA $\mathfrak{a}$ must infer the "global-state reachability" of $\mathfrak{A}$ by referring only to "local-state reachability" in $\mathfrak{a}$. A simple counterexample suffices to show that this is a difficult task.

Proposition 4. *There exist languages* $T \subseteq \mathsf{Rec}(\mathbb{M}(\Sigma, I))$ *such that* $\mathsf{ext}(T)$ *is not recognized by any DAEA.*

A similar argument can be drawn against a possible definition of deterministic asynchronous weak automata, defined in terms of SCCs that occur locally within Q_a for each $a \in \Sigma$. This means that the class of reachability languages resists characterization in terms of deterministic asynchronous cellular automata. We therefore concentrate on the classes of I-diamond automata and trace-closed reachability languages in the hope of finding reasonable characterizations.

First we note that the definition of infinitary extensions of a trace-closed languages is not sound with respect to trace equivalence of ω-words; i.e. if $T \in \mathsf{Rec}(\mathbb{M}(\Sigma, I))$ and $K = \Gamma^{-1}(T)$, then, in general, $\mathsf{ext}(K) \neq \Gamma^{-1}(\mathsf{ext}(T))$.

Example 5. Let $\Sigma = \{a, b, c\}$, and bIc. Define $K := [ab]_{\sim_I}$. Clearly K is trace-closed and, moreover, $acb \notin K$. Let $T = \Gamma(K)$. Clearly $abc^\omega, acbc^\omega, accbc^\omega, \ldots$ are equivalent words since they induce the same infinite trace which belongs to $\mathsf{ext}(T)$. However, while $abc^\omega \in \mathsf{ext}(K)$, $ac^+bc^\omega \nsubseteq \mathsf{ext}(K)$.   ⊠

Definition 6. *Let* $K \subseteq \Sigma^*$ *be trace-closed. Define the* I-*suffix extended trace-closed language (or* I-*suffix extension) of* K *as* $K_I := K \cup \bigcup_{a \in \Sigma} [K a^{-1} a I_a^*]_{\sim_I}$.

Due to the closure of $\mathsf{Rec}(\mathbb{M}(\Sigma, I))$ under concatenation and finite union [3], we know that K_I is regular whenever K is regular.

Proposition 7. *For a language* $T \in \mathsf{Rec}(\mathbb{M}(\Sigma, I))$, *let* $K = \Gamma^{-1}(T)$, *and let* K_I *be the* I-*suffix extension of* K. *Then it holds that* $\Gamma^{-1}(\mathsf{ext}(T)) = \mathsf{ext}(K_I)$.

Remark 8. In general $K_I \neq (K_I)_I$. However, iterated I-suffix extensions preserve the infinitary extension languages, i.e. $\mathsf{ext}(K) \subseteq \mathsf{ext}(K_I) = \mathsf{ext}((K_I)_I) = \mathsf{ext}(((K_I)_I)_I) \ldots$ and so on.   ⊠

Proposition 7 provides us the basis for generating the class of weakly recognizable trace-closed languages corresponding to the recognizable subset of $\mathrm{BC}(\mathsf{ext}(\mathbb{M}(\Sigma, I)))$. Henceforth, whenever we speak of the language $\Gamma^{-1}(\mathsf{ext}(T))$ we refer to $\mathsf{ext}(\Gamma^{-1}(T)_I)$. Similarly, for a trace-closed language K we always mean $\mathsf{ext}(K_I)$ whenever we say $\mathsf{ext}(K)$.

Theorem 9. *A trace-closed language $L \subseteq \Sigma^\omega$ is recognized by an I-diamond DWA iff $L \in \mathrm{BC}(\mathsf{ext}(\mathcal{K}))$ for a set $\mathcal{K} \subseteq 2^{\Sigma^*}$ of trace-closed regular languages.*

3.2   Infinitary Limits of Regular Trace Languages

We now consider the *infinitary limit* operator. In the case of word languages, this operator extends regular languages to the family ω-regular languages that are DBA recognizable. This is not straight forward for traces, and here we seek an effective characterization of languages $T \in \mathsf{Rec}(\mathbb{M}(\Sigma, I))$, such that $\Gamma^{-1}(\mathsf{lim}(T))$ is recognized by an I-diamond DBA.

Definition 10 ([2]). *Let $T \in \mathsf{Rec}(\mathbb{M}(\Sigma, I))$, the infinitary limit $\mathsf{lim}(T)$ is the ω-trace language containing all $\theta \in \mathbb{R}(\Sigma, I)$ such that there exists a sequence $(t_i)_{i \in \mathbb{N}}, t_i \in T$ satisfying $t_i \sqsubset t_{i+1}$ and $\bigsqcup_{i \in \mathbb{N}} t_i = \theta$.*

It is open whether there exists any characterization for the class of languages recognized by the family of DACBAs, however there do exist regular languages $T \subseteq \mathbb{M}(\Sigma, I)$ such that $\mathsf{lim}(T)$ is not recognized by any DACBA [8]. In fact, even when relying on trace-closed word languages and I-diamond automata, we cannot hope to characterize these languages in the manner of infinitary extensions as demonstrated previously in Section 3.1.

Example 11. Let $\Sigma = \{a, b\}$, and aIb. Define $K := [(aa)^+(bb)^+]_{\sim_I}$ as the trace-closed language with even number of occurrences of a's and b's. The minimal DFA accepting this language is shown in Figure 2. If $T = \Gamma(K)$, then

$$\mathsf{lim}(T) = \Theta := \left\{ \theta \in \mathbb{R}(\Sigma, I) \,\middle|\, \begin{array}{l} |\theta|_a \text{ even, } |\theta|_b = \infty, \text{ or} \\ |\theta|_a = \infty, |\theta|_b \text{ even, or} \\ |\theta|_a = |\theta|_b = \infty \end{array} \right\}$$

The trace-closed language $L = \Gamma^{-1}(\Theta)$ consists of all infinite words $\alpha \in \Sigma^\omega$ that satisfy the same conditions as $\theta \in \Theta$ above.   ⊠

It is easy to verify that the DFA of Figure 2 does not accept L when equipped with a Büchi acceptance condition. For instance, the automaton can loop forever in states 4, 6, and 7, thereby witnessing infinitely many a's and b's, without ever visiting state 8.

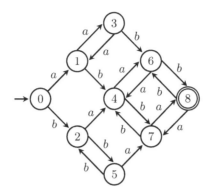

Fig. 2. The minimal DFA recognizing language K of Example 11

Proposition 12. *There exists no I-diamond deterministic parity automaton, and therefore no I-diamond deterministic Büchi automaton, that recognizes the language $L \subseteq \Sigma^\omega$ as described in Example 11.*

Corollary 13. *There exists a family $\mathcal{K}$ of regular trace-closed languages, namely $\mathcal{K} := \{[(a^m)^+(b^n)^+]_{\sim_I} \mid m, n \geq 2\}$ over $\Sigma = \{a, b\}$, such that given $T = \Gamma(K)$ for any $K \in \mathcal{K}$, there exists no I-diamond DBA recognizing $\Gamma^{-1}(\lim(T))$.*

Definition 14. *A trace-closed language $K \subseteq \Sigma^*$ is I-limit-stable (or simply limit-stable) if $\lim(K)$ is also trace-closed. By extension, $T \subseteq \mathbb{M}(\Sigma, I)$ is limit-stable if $\Gamma^{-1}(T)$ is.*

Given an automaton and its states p, q, we write $p \xrightarrow{u} q$ if some $u \in \Sigma^*$ leads from p to q, and $p \overset{u}{\Rightarrow} q$ if a final state is also visited.

Definition 15. *Given (Σ, I), let $\mathfrak{A} = (Q, \Sigma, q_0, \delta, F)$ be an I-diamond automaton. $\mathfrak{A}$ is F, I-cycle closed, if for all $u \sim_I v$ and all q we have $q \overset{u}{\Rightarrow} q$ iff $q \overset{v}{\Rightarrow} q$.*

We can now give an effective characterization of limit-stable languages.

Theorem 16. *For any $T \in \mathsf{Rec}(\mathbb{M}(\Sigma, I))$ and $K = \Gamma^{-1}(T)$, the following are equivalent:*

(a) K, and therefore T, is limit-stable.
(b) For all sequences $(t_i)_i = t_0 \sqsubset t_1 \sqsubset t_2 \cdots \subseteq T$ and all sequences $(u_i)_i$ with $u_i \in \Gamma^{-1}(t_i)$, there exists a subsequence $(u_{j_i})_i$ and a sequence $(v_{j_i})_i$ of proper prefixes $v_{j_i} \sqsubset u_{j_i}$ with $|v_{j_i}| < |v_{j_{i+1}}|$ and $v_{j_i} \in K$ for all $i \in \mathbb{N}$.
(c) Any DFA $\mathfrak{A}$ recognizing K is F, I-cycle closed.

Proof. (a) $\Longrightarrow$ (b): If (b) is false, then we may choose a sequence $(t_i)_i$ of traces in T with the property that for some sequence $(u_i)_i$ of linearizations of $(t_i)_i$, every subsequence $(u_{n_i})_i$, and every sequence $(v_{n_i})_i$ of proper prefixes $v_{n_i} \sqsubset u_{n_i}, v_{n_i} \in K$, we have $\sup_i |v_{n_i}| < \infty$. Since $|\Sigma| < \infty$ we have that Σ^∞ is a compact space. Hence $(u_i)_i$ has a converging subsequence $(u_{m_i})_i$. Because every subsequence of $(u_i)_i$ has the properties given in the previous sentence, so does $(u_{m_i})_i$. Let $\alpha = \lim_{i \to \infty} u_{m_i}$. Then $\alpha \sim_I \beta$ for some $\beta = x \cdot y_1 \cdot y_2 \cdots$ with $x \cdot y_1 \cdots y_i \in \Gamma^{-1}(t_{m_i})$. Hence, $\beta \in \lim(L)$. But, by construction, $\alpha \notin \lim(K)$ because for some $n \in \mathbb{N}$ no prefix of length $> n$ of α is in K.

(b) $\Longrightarrow$ (a): Let $\theta = \bigsqcup_i t_i$ for traces $t_i \in T$. We may assume that $t_i \sqsubset t \sqsubset t_{i+1}$ implies $t \notin T$. Let $\alpha \in \Gamma^{-1}(\theta)$. Then we pick prefixes $(w_i)_i$ of α, such that w_i is of minimal length with $t_i \sqsubseteq \Gamma(w_i)$. Consider the subsequence $(t_{2i})_i$ of $(t_i)_i$. Each w_{2i+1} is a prefix of some linearization of $t_{2(i+1)}$, say $u_{2(i+1)}$. We apply (b) to the sequence $(t_{2i})_i$ and get a sequence $(v_{2i})_i$ of proper prefixes of the u_{2i}, such that $\sup_i |v_{2i}| = \infty$ and $v_{2i} \in K$. We now have to show that v_{2i} is already a prefix of w_{2i-1}. Suppose not, i.e. $w_{2i-1} \sqsubset v_{2i} \sqsubset u_{2i}$. Then this would give a trace $t \in T$ with $t_{2i-1} \sqsubset t \sqsubset t_{2i}$.

(a) $\Longrightarrow$ (c): Suppose $\mathfrak{A}$ is not F, I-cycle closed. Then there exists $q \in Q$ and $u \sim_I v$ with $q \overset{v}{\Rightarrow} q$ but not $q \overset{v}{\Rightarrow} q$. Since $\mathfrak{A}$ is I-diamond, this means that the run $q \overset{v}{\to} q$ exists, but does not visit a final state. Now pick $x \in \Sigma^*$ with $q_0 \overset{x}{\to} q$. Then $\alpha = x \cdot u^\omega \in \lim(K)$ and $\beta = x \cdot v^\omega \notin \lim(L)$. But clearly $\alpha \sim_I \beta$ implies that $\lim(K)$ is not trace-closed.

(c) $\Longrightarrow$ (a): Let $\alpha \sim_I \beta$ and let $\alpha \in \lim(K)$. Take $\mathfrak{A} = \mathfrak{A}_K$ and consider extended transition profiles $\tau_w \subseteq Q \times \{0,1\} \times Q$ for $w \in \Sigma^*$ defined by $(p,1,q) \in \tau_w$ iff $p \overset{w}{\Rightarrow} q$ and $(p,0,q) \in \tau_w$ iff $p \overset{w}{\to} q$ but not $p \overset{w}{\Rightarrow} q$. Then we can factorize $\alpha = uv_0v_1v_2 \cdots$ for finite words $u, v_0, v_1, \ldots$ with $\tau_u \cdot \tau_{v_i} = \tau_u$ and $\tau_{v_i} \cdot \tau_{v_{i+1}} = \tau_{v_i}$. Likewise we can factorize $\beta = u'v_0'v_1' \cdots$.

Next, we observe that we find $r \in \mathbb{N}$ with $\Gamma(u'v_0') \sqsubseteq \Gamma(uv_0 \cdots v_r)$. This gives $x \in \Sigma^*$ with $u'v_0' \cdot x \sim_I uv_0 \cdots v_r$. Conversely, there exists $m \in \mathbb{N}$ with $\Gamma(uv_0 \cdots v_{r+1}) \sqsubseteq \Gamma(u'v_0' \cdots v_m')$ and therefore $y \in \Sigma^*$ with $u'v_0' \cdots v_m' \sim_I uv_0 \cdots v_r v_{r+1} y \sim_I u'v_0' x v_{r+1} y$, which implies $x v_{r+1} y \sim_I v_1' \cdots v_m'$.

Notice that if $q_0 \overset{u}{\to} q$ and $q_0 \overset{u'}{\to} q'$, then (by trace equivalence and the fact that $\mathfrak{A}$ is I-diamond) we have $q' \overset{x}{\to} q$. Likewise we have $q \overset{y}{\to} q'$ and $q' \overset{x v_{r+1} y}{\longrightarrow} q'$. Now we can apply (c) to see that $q' \overset{x v_{r+1} y}{\Longrightarrow} q'$ iff $q' \overset{v_1' \cdots v_m'}{\Longrightarrow} q'$. However, since $\alpha \in \lim(K)$, since $\tau_{v_{r+1}} = \tau_{v_i}$ for all i, and since $q \overset{v_{r+1}}{\Longrightarrow} q$, we have $q' \overset{x v_{r+1} y}{\Longrightarrow} q'$. Hence, $q' \overset{v_1' \cdots v_m'}{\Longrightarrow} q'$. Since furthermore $\tau_{v_1' \cdots v_m'} = \tau_{v_i'}$, we have for all i, $q' \overset{v_i'}{\underset{F}{\Rightarrow}} q'$ whence $\beta \in \lim(K)$. $\blacksquare$

Corollary 17. *Let $K = \Gamma^{-1}(T)$ for some $T \in \mathsf{Rec}(\mathbb{M}(\Sigma, I))$. Given the automaton $\mathfrak{A}_K$, it is decidable in time $\mathcal{O}(|Q|^2 \cdot |\Sigma|(|\Sigma| + \log |Q|))$ whether or not K is limit-stable.*

3.3   Characterization of Regular Infinitary Trace-Closed Languages

In [2], it was shown that for every recognizable ω-trace language $\Theta \subseteq \mathbb{R}(\Sigma, I)$ the corresponding ω-regular trace-closed language $L = \Gamma^{-1}(\Theta)$ is recognized by an I-diamond deterministic Muller automaton (DMA). On the other hand, it is not the case that every I-diamond DMA recognizes a trace-closed language. Similar to the property of F, I-cycle closure for DBAs, we present a condition over the acceptance component $\mathcal{F}$ of I-diamond DMAs to enable a characterization.

Given an automaton, two of its states p, q, and a word $u \in \Sigma^*$, we denote with $\mathsf{occ}(p \xrightarrow{u} q)$ the set of states occurring in the run from p to q over u.

Definition 18. *Given* (Σ, I), *an* I-*diamond DMA* $\mathfrak{A} = (Q, \Sigma, q_0, \delta, \mathcal{F})$ *is said to be* $\mathcal{F}, I$-*cycle closed if for all* $u, v \in \Sigma^*$ *such that* $u \sim_I v$, *and all* $q \in Q$, *we have* $\mathsf{occ}(q \xrightarrow{u} q) \in \mathcal{F}$ *iff* $\mathsf{occ}(q \xrightarrow{v} q) \in \mathcal{F}$.

$\mathcal{F}, I$-cycle closure was mentioned in [8, Chapter 7] under a different term. We obtain an independent proof of the following result by using an approach very similar to that we used to show the equivalence Theorem 16:(a) $\Leftrightarrow$ (c).

Theorem 19 (also cf. [8]). *For any language* $\Theta \subseteq \mathbb{R}(\Sigma, I)$ *of infinite traces,* Θ *is recognized by a DACMA if and only if the trace-closed language* $L = \Gamma^{-1}(\Theta)$ *is recognized by an* $\mathcal{F}, I$-*cycle closed DMA.*

4   A Borel-Like Classification

Any I-diamond DWA recognizing a trace-closed language is trivially F, I-cycle closed since for any word $u \in \Sigma^*$ and any $q \in Q$, it holds that $q \xrightarrow{u} q$ if and if all states in the path taken by u are accepting. This is because a path from q back to itself also implies an SCC, and therefore any $v \sim_I u$ will also remain in the same SCC which comprises solely of accepting states.

It is also straightforward that for a limit-stable language K, the complement language $\overline{\lim(K)}$ of K's infinitary extension is also recognized by an F, I-cycle closed deterministic co-Büchi automaton (DcBA). The following result is then a consequence of Theorem 9, Theorem 16, and the definitions.

Theorem 20. *A trace-closed language* $L \subseteq \Sigma^\omega$ *is recognized by an* I-*diamond DWA if and only if it is recognized by both an* F, I-*cycle closed DBA and an* F, I-*cycle closed DcBA.*

This result is in nice correspondence with the classical Borel level where weakly recognizable languages are precisely those that lie in the intersection of deterministic Büchi and deterministic co-Büchi recognizable languages. Finally, we now demonstrate that the class of limit-stable languages is expressive enough to generate all ω-regular trace-closed languages.

In [2], it was shown using algebraic arguments that every recognizable ω-trace language can be expressed as a finite Boolean combination of "restricted" lim-languages. This result also extends to the corresponding trace-closed linearization languages. Our characterization of limits of limit-stable languages allows for a first automata-theoretic equivalence result.

Theorem 21. *Let* L *be a trace-closed* ω-*language.* L *is* ω-*regular iff* L *is a finite Boolean combination of infinitary limits of limit-stable languages, i.e. a finite Boolean combination of* F, I-*cycle closed DBA recognizable languages.*

Proof. Recall the definition of DACMAs (cf. Sec. 2), and the result that every recognizable ω-trace language is recognized by a DACMA [2].

Let $L \subseteq \Sigma^\omega$ be recognizable, trace-closed. Pick a DACMA $\mathfrak{a}$ recognizing L. Recall that the global transition behavior of $\mathfrak{a}$ gives an I-diamond DFA, and we denote this DFA by $\mathfrak{A} = (\prod_{a \in \Sigma} Q_a, \Sigma, q_0, \delta)$. Given $q \in Q_a$ we obtain a DBA $\mathfrak{A}_q = (\prod_{a \in \Sigma} Q_a, \Sigma, q_0, \delta, F_q)$, where $F_q = \{q\} \times \prod_{b \neq a} Q_b$. Note that $\mathfrak{A}_q$ is F_q, I-cycle closed, because for any $q' \in \prod_{a \in \Sigma} Q_a$ and all $u \sim_I v$ with $q' \xrightarrow{u} q'$ and $q' \xrightarrow{v} q'$ we have[2] $\mathsf{occ}_a(q' \xrightarrow{u} q') = \mathsf{occ}_a(q' \xrightarrow{v} q')$. Now:

$$L = \bigcup_{(F_a)_{a \in \Sigma} \in \mathcal{F}} \bigcap_{a \in \Sigma} \bigcap_{q \in F_a} L(\mathfrak{A}_q) \cap \bigcap_{q \notin F_a} \overline{L(\mathfrak{A}_q)}$$

∎

We therefore obtain a Borel-like classification for recognizable ω-trace languages where the lowest level is occupied by reachability and safety languages. At the next level, we have infinitary limits of limit-stable languages and their complements. And the Boolean combinations of these languages generate the class of all recognizable ω-trace languages.

5 Conclusion

The infinitary extension operator ext and the infinitary limit operator lim offer natural mechanisms for obtaining ω-languages by expressing reachability and liveness conditions over regular languages. While in the case of word languages, these ω-languages have well-known characterizations in terms of specific classes of ω-automata, it is not easy to generalize these observations to trace languages. Analogous characterizations of recognizable ω-trace languages in terms of classes of deterministic asynchronous automata either do not exist (reachability conditions) or impose a high technical complexity (liveness conditions).

The results of this paper demonstrate that a classification of recognizable ω-trace languages in terms of trace-closed word languages is both meaningful and efficient. Once in the realm of words, for any trace language $T \in \mathsf{Rec}(\mathbb{M}(\Sigma, I))$ we investigated the relationship between its infinitary extension $\mathsf{ext}(T)$ and the infinitary extension $\mathsf{ext}(K)$, where $K = \Gamma^{-1}(T)$. We showed that any such K can be modified to K_I such that $\mathsf{ext}(K_I)$ is also trace-closed and thus corresponds to the linearizations of $\mathsf{ext}(T)$. We also showed that Boolean combinations of trace-closed languages $\Gamma^{-1}(\mathsf{ext}(T)), T \in \mathsf{Rec}(\mathbb{M}(\Sigma, I))$, are precisely the languages recognized by the class of I-diamond DWAs. In a similar vein, a trace-closed language $K \in \mathsf{Rec}(\Sigma^*)$ is limit-stable precisely when $\mathsf{lim}(K)$ is also trace-closed and recognized by F, I-cycle closed DBA, which can be obtained from any DFA recognizing K. Moreover, we showed that it is efficiently decidable whether or not a trace-closed word language K, and therefore $T = \Gamma(K)$, is limit-stable.

---

[2] This can be proven by an induction on the number of swapping operations needed to obtain v from u.

It must be noted that if the independence relation I over the alphabet is empty, then we obtain the well-known theorems for ω-regular word languages as special cases of our results. In this manner, our characterizations have two interesting consequences. First, that I-diamond DWA recognizable languages are precisely those that are both I-diamond det. Büchi and I-diamond det. co-Büchi recognizable. Second, that every recognizable language of infinite traces can be expressed as a Boolean combination of languages $\lim(T)$ for limit-stable languages T. This, in turn, gives rise to a Borel-like classification hierarchy for trace languages in terms of trace-closed word languages.

As a next step, we would like to investigate whether these classes of languages can also be characterized in terms of logic. Such a characterization will allow for a direct comparison with Borel levels. Also, in the manner of Landweber's result for ω-regular word languages in general, we would like to have the ability to decide whether or not a given trace-closed word language is recognized by an F, I-cycle closed deterministic Büchi automaton.

Acknowledgement. We would like to thank Wolfgang Thomas and Christof Löding for encouragement and numerous fruitful discussions.

References

1. Chaturvedi, N.: Toward a Structure Theory of Regular Infinitary Trace Languages. In: Esparza, J., Fraigniaud, P., Husfeldt, T., Koutsoupias, E. (eds.) ICALP 2014, Part II. LNCS, vol. 8573, pp. 134–145. Springer, Heidelberg (2014)
2. Diekert, V., Muscholl, A.: Deterministic Asynchronous Automata for Infinite Traces. Acta Informatica 31(4), 379–397 (1994)
3. Diekert, V., Rozenberg, G. (eds.): The Book of Traces. World Scientific (1995)
4. Gastin, P., Petit, A.: Asynchronous Cellular Automata for Infinite Traces. In: Kuich, W. (ed.) ICALP 1992. LNCS, vol. 623, pp. 583–594. Springer, Heidelberg (1992)
5. Löding, C.: Efficient minimization of deterministic weak ω-automata. Information Processing Letters 79(3), 105–109 (2001)
6. Madhavan, M.: Automata on Distributed Alphabets. In: D'Souza, D., Shankar, P. (eds.) Modern Applications of Automata Theory. IISc Research Monographs Series, vol. 2, pp. 257–288. World Scientific (May 2012)
7. Mazurkiewicz, A.: Trace Theory. In: Brauer, W., Reisig, W., Rozenberg, G. (eds.) Petri Nets: Applications and Relationships to Other Models of Concurrency. LNCS, vol. 255, pp. 278–324. Springer, Heidelberg (1987)
8. Muscholl, A.: Über die Erkennbarkeit unendlicher Spuren. PhD thesis (1994)
9. Perrinand., D., Pin, J.-É.: Automata and Infinite Words. In: Infinite Words: Automata, Semigroups, Logic and Games. Pure and Applied Mathematics, vol. 141. Elsevier (2004)
10. Staiger, L.: Subspaces of $GF(q)^\omega$ and Convolutional Codes. Information and Control 59(1-3), 148–183 (1983)
11. Zielonka, W.: Notes on Finite Asynchronous Automata. R.A.I.R.O. – Informatique Théorique et Applications 21, 99–135 (1987)

Bounded Variable Logic, Parameterized Logarithmic Space, and Savitch's Theorem

Yijia Chen[1] and Moritz Müller[2]

[1] Department of Computer Science, Shanghai Jiaotong University, China
yijia.chen@cs.sjtu.edu.cn
[2] Kurt Gödel Research Center, University of Vienna, Austria
moritz.mueller@univie.ac.at

Abstract. We study the parameterized space complexity of model-checking first-order logic with a bounded number of variables. By restricting the number of the quantifier alternations we obtain problems complete for a natural hierarchy between parameterized logarithmic space and FPT. We call this hierarchy the *tree hierarchy*, provide a machine characterization, and link it to the recently introduced classes PATH and TREE. We show that the lowest class PATH collapses to parameterized logarithmic space only if Savitch's theorem can be improved. Finally, we settle the complexity with respect to the tree-hierarchy of finding short undirected paths and small undirected trees.

1 Introduction

The model-checking problem for first-order logic FO asks whether a given first-order sentence φ holds true in a given relational structure $\mathbf{A}$. The problem is PSPACE-complete in general and even its restriction to primitive positive sentences and two-element structures stays NP-hard. However, Vardi [24] showed in 1995 that the problem is solvable in polynomial time when restricted to a constant number of variables.

In database theory, a typical application of the model-checking problem, we are asked to evaluate a relatively short query φ against a large database $\mathbf{A}$. Thus, it has repeatedly been argued in the literature (e.g. [23]), that measuring in such situations the computational resources needed to solve the problem by functions depending only on the length of the input $(\varphi, \mathbf{A})$ is unsatisfactory. Parameterized complexity theory measures computational resources by functions taking as an additional argument a *parameter* associated to the problem instance. For the parameterized model-checking problem $p\text{-MC(FO)}$ one takes the length of φ as parameter and asks for algorithms running in fpt time, that is, in time $f(|\varphi|) \cdot |\mathbf{A}|^{O(1)}$ for some computable function f. Sometimes, this relaxed tractability notion allows to translate (by an effective but often inefficient procedure) the formula into a form, for which the model-checking can be solved efficiently (see [14] for a survey). For example, algorithms exploiting Gaifman's locality theorem solve $p\text{-MC(FO)}$ on structures of bounded local treewidth [12] in fpt time, and on bounded degree graphs even in parameterized logarithmic

E. Csuhaj-Varjú et al. (Eds.): MFCS 2014, Part I, LNCS 8634, pp. 183–195, 2014.
© Springer-Verlag Berlin Heidelberg 2014

space [10]. Parameterized logarithmic space, para-L, relaxes logarithmic space in much the same way as FPT relaxes polynomial time [4,10].

Note that Vardi's result mentioned above implies that p-$\mathrm{MC}(\mathrm{FO}^s)$ can be solved in fpt time, where FO^s denotes the class of first-order sentences using at most s variables. The starting point of this paper is the question whether one can solve p-$\mathrm{MC}(\mathrm{FO}^s)$ in parameterized logarithmic space. We show that both a negative as well as a positive answer would imply certain breakthrough results in classical complexity theory. We now describe our results in some more details.

A first guess could be that when s increases, so does the space complexity of p-$\mathrm{MC}(\mathrm{FO}^s)$. But it turns out that there is a parameterized logarithmic space reduction from p-$\mathrm{MC}(\mathrm{FO}^s)$ to p-$\mathrm{MC}(\mathrm{FO}^2)$ (implicit in Theorem 8 below). On the other hand, one can naturally stratify p-$\mathrm{MC}(\mathrm{FO}^s)$ into subproblems p-$\mathrm{MC}(\Sigma_1^s)$, p-$\mathrm{MC}(\Sigma_2^s)$, ..., according to the number of quantifier alternations allowed in the input sentences. It leads to a hierarchy of classes TREE[t] consisting of the problems reducible to p-$\mathrm{MC}(\Sigma_t^s)$.

The lowest TREE[1] coincides with the class TREE introduced in [5]. The class TREE stems from the complexity classification of homomorphism problems under parameterized logarithmic space reductions, which refines Grohe's famous characterization of those homomorphism problems that are in FPT [13]. As shown in [5], they are either in para-L, or PATH-complete, or TREE-complete. The class PATH here had already been introduced by Elberfeld et al. [9].

All mentioned classes line up in the *tree hierarchy*:

$$\text{para-L} \subseteq \text{PATH} \subseteq \text{TREE}[1] \subseteq \text{TREE}[2] \subseteq \cdots \subseteq \text{TREE}[*] \subseteq \text{FPT}, \quad (1)$$

with p-$\mathrm{MC}(\mathrm{FO}^s)$ being complete for TREE[*].

The classes PATH and TREE deserve some special interest. They can be viewed as parameterized analogues of NL and LOGCFL (cf. [25]) respectively, and capture the complexity of some parameterized problems of central importance. For example,

---

p-DiPATH
Instance: A directed graph $\mathbf{G}$ and $k \in \mathbb{N}$.
Parameter: k.
Problem: Is there a directed path of length k in $\mathbf{G}$?

---

is complete for PATH [9,5], and here we show (Proposition 17) that

---

p-DiTREE
Instance: A directed graph $\mathbf{G}$ and a directed tree $\mathbf{T}$.
Parameter: $|\mathbf{T}|$.
Problem: Is there an embedding of $\mathbf{T}$ into $\mathbf{G}$?

---

is complete for TREE. We always assume that paths are simple, i.e. without repeated vertices. And by a *directed tree* we mean a directed graph obtained from a tree by directing all edges away from the root.

A negative answer to our question whether p-MC(FO^s) $\in$ para-L is equivalent to para-L $\neq$ TREE[*] and, in particular, implies L $\neq$ P.[1] In contrast, a positive answer would imply para-L = PATH, a hypothesis we study in some detail here. Recall that Savitch's seminal result from 1969 can be equivalently stated as NL $\subseteq$ DSPACE(log$^2 n$). In Lipton's words [18] "one of the biggest embarrassments of complexity theory [...] is the fact that Savitch's theorem has not been improved [...]. Nor has anyone proved that it is tight." Hemaspaandra et al. [16, Corollary 2.8] showed that Savitch's theorem could be improved if there were problems of sublogarithmic density $o(\log n)$ and Turing hard for NL. We refer to [20] for more on this problem. Here we show:

Theorem 1. *If* para-L = PATH, *then* NL $\subseteq$ DSPACE$\big(o(\log^2 n)\big)$.

The hypothesis para-L $\neq$ PATH is hence implied by the hypothesis that Savitch's Theorem is optimal, and in turn implies L $\neq$ NL (see the discussion before Proposition 6).

Finally, we settle the complexity of two more problems with respect to the tree-hierarchy. First we show that the undirected version

---

p-PATH
   Instance: An (undirected) graph **G** and $k \in \mathbb{N}$.
Parameter: k.
   Problem: Is there a path of length k in **G**?

---

of p-DIPATH is in para-L. To the best of our knowledge this has not been known before despite the considerable attention p-PATH has gained in parameterized complexity theory (e.g. [6,1,7]). It answers a question of [5]. Second, and in contrast to the just mentioned result, we prove that the undirected version

---

p-TREE
   Instance: An (undirected) graph **G** and a tree **T**.
Parameter: $|\mathbf{T}|$.
   Problem: Is there an embedding of **T** into **G**?

---

of p-DITREE stays TREE-complete.

2 Preliminaries

Structures and Logic. A *vocabulary* τ is a finite set of relation, function and constant symbols. Relation and function symbols have an associated *arity*, a positive natural number. A τ-*term* is a variable, a constant or of the form $f(t_1, \cdots, t_r)$ where f is an r-ary function symbol and $t_1, \ldots, t_r$ are again τ-terms. A τ-*atom* has the form $t = t'$ or $R(t_1, \ldots, t_r)$ where R is an r-ary relation symbol and $t, t', t_1, \ldots, t_r$ are τ-terms. τ-*formulas* are built from atoms by means of $\wedge, \vee, \neg$ and existential and universal quantification $\exists x, \forall x$. The vocabulary τ

---

[1] In fact, general results from [10] imply that the hypotheses para-L $\neq$ FPT, TREE[*] $\neq$ FPT and L $\neq$ P are all equivalent.

is called *relational* if it contains only relation symbols. A *(finite) τ-structure* **A** consists in a finite nonempty set A, its *universe*, and for each r-ary relation symbol $R \in \tau$ an *interpretation* $R^{\mathbf{A}} \subseteq A^r$ and for each r-ary function symbol $f \in \tau$ an *interpretation* $f^{\mathbf{A}} : A^r \to A$ and for each constant symbol $c \in \tau$ an *interpretation* $c^{\mathbf{A}} \in A$. We view *digraphs* as $\{E\}$-structures **G** for a binary relation symbol E such that $E^{\mathbf{G}}$ is irreflexive. A *graph* is a digraph **G** with symmetric $E^{\mathbf{G}}$. If **G** is a (di)graph, we refer to elements of G as *vertices* and to elements of $E^{\mathbf{G}}$ as *(directed) edges*.

Let τ be a relational vocabulary and **A**, **B** two τ-structures. A *homomorphism from* **A** *to* **B** is a function $h : A \to B$ such that for every r-ary $R \in \tau$ we have $\big(h(a_1), \ldots, h(a_r)\big) \in R^{\mathbf{B}}$ whenever $(a_1, \ldots, a_r) \in R^{\mathbf{A}}$. We understand that there do not exist homomorphisms between structures interpreting different (relational) vocabularies. As has become usual in our setting, we understand that an *embedding* is an injective homomorphism.

Parameterized Complexity. A *(classical) problem* is a subset $Q \subseteq \{0, 1\}^*$, where $\{0, 1\}^*$ is the set of finite binary strings; the length of a binary string x is denoted by $|x|$. As model of computation we use Turing machines $\mathbb{A}$ with a (read-only) input tape and several worktapes. For definiteness, let us agree that a *nondeterministic* Turing machine has special states $c_\exists, c_0, c_1$ and can nondeterministically move from state $c_\exists$ to state c_b with $b \in \{0, 1\}$, and we say $\mathbb{A}$ *existentially guesses the bit* b. An *alternating* Turing machine additionally has a state $c_\forall$ allowing to *universally guess* a bit b. For $c : \{0, 1\}^* \to \mathbb{N}$, the machine is said to *use c many nondeterministic (co-nondeterministic) bits* if for every $x \in \{0, 1\}^*$ every run of $\mathbb{A}$ on x contains at most $c(x)$ many configurations with state $c_\exists$ (resp. $c_\forall$).

A *parameterized problem* is a pair (Q, κ) of a classical problem Q and a logarithmic space computable *parameterization* $\kappa : \{0, 1\}^* \to \mathbb{N}$, mapping any instance $x \in \{0, 1\}^*$ to its *parameter* $\kappa(x) \in \mathbb{N}$. For a class $\mathcal{A}$ of structures we consider the parameterized *homomorphism problem*

---

p-HOM($\mathcal{A}$)
  Instance: A structure $\mathbf{A} \in \mathcal{A}$ and a structure **B**.
  Parameter: $|\mathbf{A}|$.
  Problem: Is there a homomorphism from **A** to **B**?

Here, $|\mathbf{A}|$ denotes the size of a reasonable encoding of **A**. Similarly, the parameterized *embedding problem* p-EMB($\mathcal{A}$) asks for an embedding instead of a homomorphism.

The class FPT contains those parameterized problems (Q, κ) that can be decided in *fpt time (with respect to κ)*, i.e. in time $f(\kappa(x)) \cdot |x|^{O(1)}$ for some computable function $f : \mathbb{N} \to \mathbb{N}$. The class para-L (para-NL) contains those parameterized problems (Q, κ) such that Q is decided (accepted) by some (nondeterministic) Turing machine $\mathbb{A}$ that runs in *parameterized logarithmic space* $f(\kappa(x)) + O(\log |x|)$ for some computable $f : \mathbb{N} \to \mathbb{N}$. A *pl-reduction* from (Q, κ) to (Q', κ') is a reduction $R : \{0, 1\}^* \to \{0, 1\}^*$ from Q to Q' such that

$\kappa'(R(x)) \leq f(\kappa(x))$ and $|R(x)| \leq f(\kappa(x)) \cdot |x|^{O(1)}$ for some computable $f : \mathbb{N} \to \mathbb{N}$, and R is *implicitly pl-computable*, that is, the following problem is in para-L:

---

p-BITGRAPH(R)
   Instance: (x, i, b) with $x \in \{0, 1\}^*$, $i \geq 1$, and $b \in \{0, 1\}$.
Parameter: $\kappa(x)$.
   Problem: Does $R(x)$ have length $|R(x)| \geq i$ and ith bit b?

PATH and TREE. The class PATH (resp. TREE) contains those parameterized problems (Q, κ) such that Q is accepted by a nondeterministic Turing machine $\mathbb{A}$ which runs in parameterized logarithmic space, and for some computable function $f : \mathbb{N} \to \mathbb{N}$ uses $f(\kappa(x)) \cdot \log |x|$ many nondeterministic bits (and additionally $f(\kappa(x))$ many co-nondeterministic bits).

The class PATH has been discovered by Elberfeld et al. [9]. It captures the complexity of the fundamental problem:

---

p-REACHABILITY
   Instance: A directed graph $\mathbf{G}$, two vertices $s, t \in G$, and $k \in \mathbb{N}$.
Parameter: k.
   Problem: Is there a (directed) path of length k from s to t in $\mathbf{G}$?

Theorem 2 ([9,5]). p-REACHABILITY *is PATH-complete (under pl-reductions).*

The class TREE has been introduced in [5] for the purpose of a classification of the complexities of homomorphism problems up to pl-reductions:

Theorem 3 ([15,13,5]). *Let $\mathcal{A}$ be a decidable class of relational structures of bounded arity. Then*

1. *if the cores of $\mathcal{A}$ have bounded tree-depth, then p-HOM($\mathcal{A}$) $\in$ para-L;*
2. *if the cores of $\mathcal{A}$ have unbounded tree-depth but bounded pathwidth, then p-HOM($\mathcal{A}$) is PATH-complete;*
3. *if the cores of $\mathcal{A}$ have unbounded pathwidth but bounded treewidth, then p-HOM($\mathcal{A}$) is TREE-complete;*
4. *if the cores of $\mathcal{A}$ have unbounded treewidth, then p-HOM($\mathcal{A}$) is not in FPT unless $W[1] = FPT$.*

Here, *bounded arity* means that there is a constant bounding the arities of symbols interpreted in structures from $\mathcal{A}$.[2] Understanding in a similar way the complexities of the embedding problems p-EMB($\mathcal{A}$) is wide open (see e.g. [11, page 355]). We know:

Theorem 4 ([5]). *For $\mathcal{A}$ as in Theorem 3 we have p-EMB($\mathcal{A}$) $\in$ para-L in case (1), p-EMB($\mathcal{A}$) $\in$ PATH in case (2), and p-EMB($\mathcal{A}$) $\in$ TREE in case (3).*

Note p-PATH and p-TREE are roughly the same as p-EMB($\mathcal{P}$) and p-EMB($\mathcal{T}$), where $\mathcal{P}$ and $\mathcal{T}$ denote the classes of paths and trees, respectively. The complexities of these important problems are left open by Theorems 3 and 4.

---

[2] We do not recall the notion of core nor the width notions here because we do not need them.

3   Model-Checking Bounded Variable First-Order Logic

The tree hierarchy. Following [5] we consider machines $\mathbb{A}$ with *mixed nondeterminism*. Additionally to the binary nondeterminism embodied in the states $c_\exists, c_\forall, c_0, c_1$ from Section 2 they use *jumps* explained as follows. Recall our Turing machines have an input tape. During a computation on an input x of length $n := |x|$ the cells numbered 1 to n of the input tape contain the n bits of x. The machine has an *existential* and a *universal jump state* $j_\exists$ resp. $j_\forall$. A successor configuration in a jump state is obtained by changing the state to the initial state and placing the input head on an arbitrary cell holding an input bit; the machine is said to *existentially resp. universally jump to* the cell.

Observe that of the number of the cell to which the machine jumps can be computed in logarithmic space by moving the input head stepwise to the left. Intuitively, a jump should be thought as a guess of a number in $[n] := \{1, \ldots, n\}$. Acceptance is defined as usual for alternating machines. Call a configuration *universal* if it has state $j_\forall$ or $c_\forall$, and otherwise *existential*. The machine $\mathbb{A}$ accepts $x \in \{0, 1\}^*$ if its initial configuration on x is *accepting*. The set of accepting configurations is the smallest set that contains all accepting halting configurations, that contains an existential configuration if it contains some of its successors, and that contains a universal configuration if it contains all of its successors.

Each run of $\mathbb{A}$ on some input x contains a subsequence of jump configurations (i.e. with state $j_\exists$ or $j_\forall$). For a natural number $t \geq 1$ the run is t-*alternating* if this subsequence consists in t blocks, the first consisting in existential configurations, the second in universal configurations, and so on. The machine $\mathbb{A}$ is t-*alternating* if for every input $x \in \{0, 1\}^*$ every run of $\mathbb{A}$ on x is t-alternating.

Let $f : \{0, 1\}^* \to \mathbb{N}$. The machine $\mathbb{A}$ *uses f jumps (bits)* if for every input $x \in \{0, 1\}^*$ every run of $\mathbb{A}$ on x contains at most $f(x)$ many jump configurations (resp. configurations with state $c_\exists$ or $c_\forall$).

As for a more general notation, note that every run of $\mathbb{A}$ on x contains a sequence of *nondeterministic configurations*, i.e. with state in $\{j_\exists, j_\forall, c_\exists, c_\forall\}$. The *nondeterminism type* of the run is the corresponding word over the alphabet $\{j_\exists, j_\forall, c_\exists, c_\forall\}$. For example, being $2t$-alternating means having nondeterminism type in $(\{j_\exists, c_\exists, c_\forall\}^* \{j_\forall, c_\exists, c_\forall\}^*)^t$. Here, we use regular expressions to denote languages over $\{j_\exists, j_\forall, c_\exists, c_\forall\}$.

Definition 5. A parameterized problem (Q, κ) is in TREE[*] if there are a computable $f : \mathbb{N} \to \mathbb{N}$ and a machine $\mathbb{A}$ with mixed nondeterminism that accepts Q, runs in parameterized logarithmic space, and uses $f \circ \kappa$ jumps and $f \circ \kappa$ bits. Furthermore, if $\mathbb{A}$ is t-alternating for some $t \geq 1$, then (Q, κ) is in TREE[t].

The definition of TREE[t] is due to Hubie Chen.

It is straightforward to verify PATH $\subseteq$ TREE = TREE[1] (cf. [5, Lemmas 4.5, 5.4]). Obviously, para-L $\subseteq$ PATH $\subseteq$ para-NL, and all classes are equal if L = NL (see [10]). Conversely, Elberfeld et al. [9] observed that L = NL if PATH = para-NL. In fact, using general results from [10] one can show:

Proposition 6. *1.* para-NL $\subseteq$ TREE[*] *if and only if* NL = L.

2. FPT $\subseteq$ TREE[*] *if and only if* P = L.

We shall need the following technical lemma.

Lemma 7 (Normalization). *Let $t \geq 1$. A parameterized problem (Q, κ) is in* TREE[t] *if and only if there are a computable $f : \mathbb{N} \to \mathbb{N}$ and a t-alternating machine $\mathbb{A}$ with mixed nondeterminism that accepts Q, runs in parameterized logarithmic space (with respect to κ) and such that for all $x \in \{0, 1\}^*$ every run of $\mathbb{A}$ on x has nondeterminism type:*

$$\left((j_\exists c_\forall)^{f(\kappa(x))} (j_\forall c_\exists)^{f(\kappa(x))} \right)^{\lfloor t/2 \rfloor} (j_\exists c_\forall)^{f(\kappa(x)) \cdot (t \bmod 2)}. \tag{2}$$

Model-Checking. For $s \in \mathbb{N}$ let FO^s denote the class of (first-order) formulas over a relational vocabulary containing at most s variables (free or bound). For $t \in \mathbb{N}$ we define the classes Σ_t and Π_t as follows. Both Σ_0 and Π_0 are the class of quantifier free formulas; Σ_{t+1} (resp. Π_{t+1}) is the closure of Π_t (resp. Σ_t) under positive Boolean combinations and existential (resp. universal) quantification. We use Σ_t^s and Π_t^s to denote $\mathrm{FO}^s \cap \Sigma_t$ and $\mathrm{FO}^s \cap \Pi_t$ respectively.

For a class of formulas Φ we consider the parameterized problem:

p-MC(Φ)
  Instance: A sentence $\varphi \in \Phi$ and a structure $\mathbf{A}$.
Parameter: $|\varphi|$.
  Problem: $\mathbf{A} \models \varphi$?

It is well known [24] that for all $s \in \mathbb{N}$ the problem p-MC(FO^s) is in FPT, indeed, the underlying classical problem is in P.

Theorem 8. *Let $t \geq 1$ and $s \geq 2$. Then p-MC(Σ_t^s) is* TREE[t]*-complete.*

Proof. The containment p-MC(Σ_t^s) $\in$ TREE[t] is straightforward. To show that p-MC(Σ_t^2) is hard for TREE[t], let $(Q, \kappa) \in$ TREE[t] be given and choose a computable f and a t-alternating machine $\mathbb{B}$ with $f \circ \kappa$ jumps and $f \circ \kappa$ bits that accepts Q and runs in space $f(\kappa(x)) + O(\log |x|)$.

Given $x \in \{0, 1\}^*$ compute an upper bound $s = f(\kappa(x)) + O(\log |x|)$ on the space needed by $\mathbb{B}$ on x; since κ is computable in logarithmic space, the number $f(\kappa(x))$ and hence s can be computed in parameterized logarithmic space. We can assume that $\mathbb{B}$ on x always halts after at most $m = 2^{f(\kappa(x))} \cdot |x|^{O(1)}$ steps. Note that the binary representation of m can be computed in parameterized logarithmic space. For two space s configurations c, c' of $\mathbb{B}$ on x, we say that $\mathbb{B}$ *reaches c' from c* if there is a length $\leq m$ computation of $\mathbb{B}$ leading from c to c' that neither passes through a nondeterministic configuration nor through a configuration of space $> s$. We assume $\mathbb{B}$ reaches a nondeterministic configuration from the initial configuration, i.e. the computation of $\mathbb{B}$ on x is not deterministic.

We define a structure $\mathbf{A}$ whose universe A comprises all (length $O(s)$ binary codes of) nondeterministic space s configurations of $\mathbb{B}$ on x. The structure $\mathbf{A}$ interprets a binary relation symbol E, unary function symbols s_0, s_1 and unary relation symbols $S, F, J_\exists, J_\forall, C_\exists, C_\forall$ as follows. A pair $(c, c') \in A^2$ is in $E^{\mathbf{A}}$ if there exists a successor configuration c'' of c such that $\mathbb{B}$ reaches c' from c''. The symbol S is interpreted by $S^{\mathbf{A}} = \{c_{\text{first}}\}$ where c_{first} is the (unique) first configuration in A reached by $\mathbb{B}$ from the initial configuration of $\mathbb{B}$ on x. The symbols $J_\exists, J_\forall, C_\exists$ and $C_\forall$ are interpreted by the sets of configurations in A with states $j_\exists, j_\forall, c_\exists$ and $c_\forall$ respectively. Obviously these sets partition A. The symbol F is interpreted by the set of those $c \in A$ such that

- $c \in C_\exists^{\mathbf{A}} \cup J_\exists^{\mathbf{A}}$ and $\mathbb{B}$ reaches a space s accepting halting configuration from at least one successor configuration of c.
- $c \in C_\forall^{\mathbf{A}} \cup J_\forall^{\mathbf{A}}$ and $\mathbb{B}$ reaches a space s accepting halting configuration from all successor configurations of c.

The function symbols s_0 and s_1 are interpreted by any functions $s_0^{\mathbf{A}}, s_1^{\mathbf{A}} : A \to A$ such that for every $c \in C_\exists^{\mathbf{A}} \cup C_\forall^{\mathbf{A}}$ with $\{d \in A \mid (c,d) \in E^{\mathbf{A}}\} \neq \emptyset$ we have:

$$\{s_0^{\mathbf{A}}(c), s_1^{\mathbf{A}}(c)\} = \{d \in A \mid (c,d) \in E^{\mathbf{A}}\}.$$

It is easy to check that $\mathbf{A}$ is implicitly pl-computable from x. For example, to check whether a given pair $(c, c') \in A^2$ is in $E^{\mathbf{A}}$ we simulate $\mathbb{B}$ starting from c for at most m steps; if the simulation wants to visit a configuration of space $> s$ or a nondeterministic configuration $\neq c'$, then we stop the simulation and reject.

For a word w of length $|w| \geq 1$ over the alphabet $\{j_\exists, j_\forall, c_\exists, c_\forall\}$ we define a formula $\varphi_w(x)$ with (free or bound) variables x, y as follows. We proceed by induction on $|w|$. If $|w| = 1$, define $\varphi_w(x) := Fx$. For $|w| \geq 1$ define:

$$\varphi_{c_\forall w}(x) := C_\forall x \wedge \big(\varphi_w(s_0(x)) \wedge \varphi_w(s_1(x))\big),$$
$$\varphi_{c_\exists w}(x) := C_\exists x \wedge \big(\varphi_w(s_0(x)) \vee \varphi_w(s_1(x))\big),$$
$$\varphi_{j_\exists w}(x) := J_\exists x \wedge \exists y \big(E(x,y) \wedge \exists x(x = y \wedge \varphi_w(x))\big),$$
$$\varphi_{j_\forall w}(x) := J_\forall x \wedge \forall y \big(\neg E(x,y) \vee \forall x(\neg x = y \vee \varphi_w(x))\big).$$

Let $|w| \geq 1$ and assume that $c \in A$ is a configuration such that every run of $\mathbb{B}$ on x starting at c has nondeterminism type w; then (recall the definition of an accepting configuration from page 188)

$$c \text{ is accepting} \iff \mathbf{A} \models \varphi_w(c). \tag{3}$$

This follows by a straightforward induction on $|w|$. Now we look for $\mathbf{A}'$ and φ'_w with this property but in a relational vocabulary.

By the Normalization Lemma 7 we can assume that all runs of $\mathbb{B}$ on x have nondeterminism type w of the form (2). For such a w we observe that $\varphi_w(x)$ is in Σ_t^2 and all its atomic subformulas containing some function symbol are of the form $E(s_b(x), y)$, $J_\exists(s_b(x))$, or $J_\forall(s_b(x))$. For $b \in \{0, 1\}$ we introduce binary relation symbols E_b and unary relation symbols $J_{\forall b}$ and $J_{\exists b}$, and then replace the atomic subformulas $E(s_b(x), y)$, $J_\exists(s_b(x))$, $J_\forall(s_b(x))$ in $\varphi_w(x)$ by $E_b(x, y)$, $J_{\exists b}(x)$, $J_{\forall b}(x)$ respectively. This defines the formula $\varphi'_w(x)$. Note that $\varphi'_w(x) \in \Sigma_t^2$.

To define $\mathbf{A}'$ we expand $\mathbf{A}$ setting $E_b^{\mathbf{A}'} := \{(c, d) \mid (s_b^{\mathbf{A}}(c), d) \in E^{\mathbf{A}}\}$, $J_{\exists b}^{\mathbf{A}'} := \{c \mid s_b^{\mathbf{A}}(c) \in J_\exists^{\mathbf{A}}\}$, and $J_{\forall b}^{\mathbf{A}'} := \{c \mid s_b^{\mathbf{A}}(c) \in J_\forall^{\mathbf{A}}\}$. Then we have for all $c \in A$:

$$\mathbf{A} \models \varphi_w(c) \iff \mathbf{A}' \models \varphi'_w(c).$$

As the assumption of (3) is satisfied for c_{first}, and c_{first} is accepting if and only if $\mathbb{B}$ accepts x, that is, if and only if $x \in Q$, we get

$$x \in Q \Longleftrightarrow \mathbf{A}' \models \varphi'_w(c_{\text{first}})$$

Then $x \mapsto \left(\exists x (Sx \wedge \varphi'_w(x)), \mathbf{A}' \right)$ is a reduction as desired. $\qquad \square$

Now, the following are derived by standard means.

Corollary 9. *Let* $t' > t \geq 1$. *If* TREE[t] *is closed under complementation, then* TREE[t'] = TREE[t].

Corollary 10. *Let* $s \geq 2$. *Then* p-MC(FO^s) *is* TREE[$*$]-*complete. In particular,* TREE[$*$] $\subseteq$ FPT.

Remark 11. It is not known whether PATH or TREE are closed under complementation (cf. [5]). Their classical counterparts NL and LOGCFL are (cf. [2]), but both proofs break under the severe restrictions on nondeterminism in the parameterized setting.

4   PATH and Classical Complexity Theory

Savitch's Theorem [22] is a milestone result linking nondeterministic space to deterministic space. It states that the problem

REACHABILITY
Instance: A directed graph $\mathbf{G}$ and two vertices $s, t \in G$.
Problem: Is there a (directed) path from s to t in $\mathbf{G}$?

is in DSPACE($\log^2 n$). It is a long-standing open problem whether this can be improved. We prove the following stronger version of Theorem 1:

Theorem 12. *Assume whether* $(\mathbf{G}, s, t, k) \in p$-REACHABILITY *can be decided in deterministic space* $f(k) + O(\log |G|)$ *for a function* $f : \mathbb{N} \to \mathbb{N}$ *(which is not necessarily computable). Then* REACHABILITY $\in$ DSPACE$\left(o(\log^2 n) \right)$.

Proof. (Sketch) Let $\mathbb{A}$ be an algorithm deciding whether $(\mathbf{G}, s, t, k) \in p$-REACHA-BILITY in space $f(k) + O(\log |G|)$. We can assume that $f(k) \geq k$ for every $k \in \mathbb{N}$. Then let $\iota : \mathbb{N} \to \mathbb{N}$ be nondecreasing and unbounded such that

$$f(\iota(n)) \leq \log n, \text{ and hence } \iota(n) \leq \log n \tag{4}$$

for all $n \in \mathbb{N}$. Note that we might not know how to compute $\iota(n)$. Now let $\mathbf{G} = (G, E^{\mathbf{G}})$ be a directed graph, $s, t \in G$, $n := |G|$, and $k \geq 2$. We compute in space $O(\log k + \log n)$ the *minimum* $\ell := \ell(k)$ such that

$$k^\ell \geq n - 1, \text{ and hence } \ell = O(\log n / \log k). \tag{5}$$

We define a sequence of directed graphs $(\mathbf{G}_i^k)_{i \leq \ell}$ with self-loops. Each $\mathbf{G}_i^k$ has vertices $G_i^k := G$ and a directed edge $(u, v) \in E^{\mathbf{G}_i^k}$ if there is a directed path from u to v in $\mathbf{G}$ of length at most k^i. In particular, $E^{\mathbf{G}_0^k}$ is the reflexive closure of $E^{\mathbf{G}}$; and by (5) there is a path from s to t in $\mathbf{G}$ if and only if there is an edge from s to t in $\mathbf{G}_\ell^k$. Furthermore, for every $i \in [\ell]$ and $u, v \in G_i^k = G_{i-1}^k = G$ there is an edge from u to v in $\mathbf{G}_i^k$ if and only if there is a path from u to v in $\mathbf{G}_{i-1}^k$ of length at most k. The following recursive algorithm $\mathbb{C}$ decides, given a directed graph $\mathbf{G}$, $k, i \in \mathbb{N}$, and $u, v \in G$, whether $(u, v) \in E^{\mathbf{G}_i^k}$.

---

1. **if** $i = 0$ **then** output whether $\left(u = v \text{ or } (u, v) \in E^{\mathbf{G}}\right)$ and return

2. simulate $\mathbb{A}$ on $\left(\mathbf{G}_{i-1}^k, u, v, k\right)$

3.   **if** $\mathbb{A}$ queries "$(u', v') \in E^{\mathbf{G}_{i-1}^k}$?" **then** call $\mathbb{C}(\mathbf{G}, k, i-1, u', v')$.

For every $k \geq 2$ let $\mathbb{C}^k$ be the algorithm which, given a directed graph $\mathbf{G}$ and $s, t \in G$, first computes $\ell = \ell(k)$ as in (5) and then simulates $\mathbb{C}(\mathbf{G}, k, \ell, s, t)$. Thus, $\mathbb{C}^k$ decides whether there is a path from s to t in $\mathbf{G}$. We analyse its space complexity. First, the depth of the recursion tree is ℓ, as $\mathbb{C}^k$ recurses on $i = \ell, \ell - 1, \ldots, 0$. As usual, $\mathbb{C}^k$ has to maintain a stack of intermediate configurations for the simulations of

$$\mathbb{A}(\mathbf{G}_\ell^k, \text{-}, \text{-}, k), \mathbb{A}(\mathbf{G}_{\ell-1}^k, \text{-}, \text{-}, k), \ldots, \mathbb{A}(\mathbf{G}_0^k, \text{-}, \text{-}, k).$$

These are space $f(k) + O(\log n)$ configurations, so by (5) $\mathbb{C}^k$ runs in space

$$O\left(\log k + \log n + \ell \cdot \left(f(k) + \log n\right)\right) = O\left(\log k + \frac{f(k) \cdot \log n + \log^2 n}{\log k}\right).$$

By (4) this is $o(\log^2 n)$ for $k := \iota(n)$. We would thus be done if we could compute $\iota(n)$, say, in space $O(\log n)$. In particular, this can be ensured under the hypothesis para-L = PATH of Theorem 1 which allows to choose f space-constructible. The general case needs some additional efforts. It can be handled using the strategy underlying Levin's optimal inverters [17,8], namely to simulate all $\mathbb{C}^2, \mathbb{C}^3, \ldots$ in a diagonal fashion. $\square$

The trivial brute-force algorithm (cf. [5, Lemma 3.11]) decides $p\text{-MC}(\Sigma_1^2)$ (indeed, the whole $p\text{-MC}(\text{FO})$) in space $O\left(|\varphi|^2 \cdot \log |\mathbf{A}|\right)$. Assuming the optimality of Savitch's Theorem, this is space-optimal in the following sense:

Corollary 13. *If* REACHABILITY $\notin$ DSPACE$(o(\log^2 n))$, *then whether* $(\varphi, \mathbf{A}) \in p\text{-MC}(\Sigma_1^2)$ *cannot be decided in deterministic space* $o\left(f(|\varphi|) \cdot \log |\mathbf{A}|\right)$ *for any* f.

We close this section by characterizing the collapse of PATH to para-L similarly as analogous characterizations of W[P] = FPT [3], or BPFPT = FPT [19].

Definition 14. *Let* $c : \mathbb{N} \to \mathbb{N}$ *be a function. The class* NL$[c]$ *contains all classical problems* Q *that are accepted by some nondeterministic Turing machine which uses* $c(|x|)$ *many nondeterministic bits and runs in logarithmic space.*

Theorem 15. *para-L = PATH if and only if there exists a space-constructible function* $c(n) = \omega(\log(n))$ *such that* NL$[c]$ = L.

5   Embedding Undirected Paths and Trees

As mentioned in the Introduction it is known that (see [5, Theorem 4.7]):

Proposition 16. *p*-DIPATH *is* PATH-*complete.*

A straightforward but somewhat tedious argument shows:

Proposition 17. *p*-DITREE *is* TREE-*complete.*

The following two results determine the complexities of the undirected versions of these two problems. Somewhat surprisingly, the complexity of the former drops to para-L while the latter stays TREE-complete:

Theorem 18. *p*-PATH $\in$ para-L.

Theorem 19. *p*-TREE *is* TREE-*complete.*

Theorem 18 answers a question posed in [5, Section 7]. Its proof is based on the well-known color-coding technique. Specifically, we shall use the following lemma from [11, page 349]:

Lemma 20. *For every sufficiently large $n \in \mathbb{N}$, it holds that for all $k \leq n$ and for every k-element subset X of $[n]$, there exists a prime $p < k^2 \cdot \log n$ and $q < p$ such that the function $h_{p,q} : [n] \to \{0, \ldots, k^2 - 1\}$ given by $h_{p,q}(m) := (q \cdot m \mod p) \mod k^2$ is injective on X.*

Proof of Theorem 18. Let $\mathbf{G} = ([n], E^{\mathbf{G}})$ be a graph and $0 < k < n$. Assume n is large enough for Lemma 20 to apply. Using its notation we set

$$F := \left\{ g \circ h_{p,q} \mid g : \{0, \ldots, (k+1)^2 - 1\} \to [k+1] \text{ and } q < p < (k+1)^2 \log n \right\}.$$

For $f \in F$ let $\mathbf{G}(f)$ be the graph obtained from $\mathbf{G}$ by deleting all edges $(u, v) \in E^{\mathbf{G}}$ with $|f(u) - f(v)| \neq 1$. By Lemma 20 one readily verifies that $\mathbf{G}$ contains a path of length k if and only if there are $f \in F$ and $u, v \in [n]$ such that $f(u) = 1$, $f(v) = k + 1$, and there is a path from u to v in $\mathbf{G}(f)$.

To decide whether $(\mathbf{G}, k) \in$ *p*-PATH we cycle through all tuples (g, p, q, u, v) with $g : \{0, \ldots, (k+1)^2 - 1\} \to [k+1], q < p < (k+1)^2 \log n$, and $u, v \in [n]$, and test whether $g(h_{p,q}(u)) = 1$, $g(h_{p,q}(v)) = k + 1$, and there is a path from u to v in $\mathbf{G}(g \circ h_{p,q})$. For every such test we simulate Reingold's logarithmic space algorithm [21] for REACHABILITY on (undirected) graphs. The simulation relies on the fact that $\mathbf{G}(g \circ h_{p,q})$ is implicitly pl-computable from (g, p, q) and $\mathbf{G}$. $\square$

Acknowledgements. We are indepted to Hubie Chen who proposed the classes TREE[*t*] for study. We thank him and Jörg Flum for their comments on earlier drafts of this paper. Further we want to thank the referees for their comments. The research of the first author is partially supported by National Nature Science Foundation of China via Projects 61373029 and 61033002. The second author has been supported by the FWF (Austrian Science Fund) Project P 24654 N25.

References

1. Bodlaender, H.L., Downey, R.G., Fellows, M.R., Hermelin, D.: On problems without polynomial kernels. J. of Comp. and Syst. Sciences 75(8), 423–434 (2009)
2. Borodin, A., Cook, S.A., Dymond, P.W., Ruzzo, W.L., Tompa, M.: Two applications of inductive counting for complementation problems. SIAM J. on Computing 18(3), 559–578 (1989)
3. Cai, L., Chen, J., Downey, R.G., Fellows, M.R.: On the structure of parameterized problems in NP. Information and Computation 123, 38–49 (1995)
4. Cai, L., Chen, J., Downey, R.G., Fellows, M.R.: Advice classes of parameterized tractability. Annals of Pure and Applied Logic 84(1), 119–138 (1997)
5. Chen, H., Müller, M.: The fine classification of conjunctive queries and parameterized logarithmic space complexity. In: Proc. of PODS 2013, pp. 309–320 (2013); full version arXiv:1306.5424 [cs.CC]
6. Chen, J., Kneis, J., Lu, S., Mölle, D., Richter, S., Rossmanith, P., Sze, S.-H., Zhang, F.: Randomized divide-and-conquer: Improved path, matching, and packing algorithms. SIAM J. on Computing 38(6), 2526–2547 (2009)
7. Chen, Y., Flum, J.: On parameterized path and chordless path problems. In: Proc. of CCC 2007, pp. 250–263 (2007)
8. Chen, Y., Flum, J.: On optimal inverters. Bull. Symb. Logic (to appear, 2014)
9. Elberfeld, M., Stockhusen, C., Tantau, T.: On the space complexity of parameterized problems. In: Thilikos, D.M., Woeginger, G.J. (eds.) IPEC 2012. LNCS, vol. 7535, pp. 206–217. Springer, Heidelberg (2012)
10. Flum, J., Grohe, M.: Describing parameterized complexity classes. Information and Computation 187(2), 291–319 (2003)
11. Flum, J., Grohe, M.: Parameterized Complexity Theory. Springer (2006)
12. M. Frick and M. Grohe. Deciding first-order properties of locally tree-decomposable structures. *J. of the ACM*, 48(6):1184–1206, 2001.
13. Grohe, M.: The complexity of homomorphism and constraint satisfaction problems seen from the other side. J. of the ACM 54(1), 1:1–1:24 (2007)
14. Grohe, M., Kreutzer, S.: Methods for algorithmic meta theorems. In: Model Theoretic Meth. in Finite Combinatorics, Cont. Math., vol. 558, pp. 181–206. AMS (2011)
15. Grohe, M., Schwentick, T., Segoufin, L.: When is the evaluation of conjunctive queries tractable? In: Proc. of STOC 2001 (2001)
16. Hemaspaandra, L.A., Ogihara, M., Toda, S.: Space-efficient recognition of sparse self-reducible languages. Computational Complexity 4, 262–296 (1994)
17. Levin, L.: Universal sequential search problems. Problems of Information Transmission 9(3), 265–266 (1973)
18. Lipton, J.R.: The P = NP Question and Gödel's Lost Letter. Springer (2010)
19. Montoya, J.-A., Müller, M.: Parameterized random complexity. Theory of Computing Systems 52(2), 221–270 (2013)
20. Potechin, A.: Bounds on monotone switching networks for directed connectivity. In: Proc. of FOCS 2010, pp. 553–562 (2010)
21. Reingold, O.: Undirected connectivity in log-space. J. of the ACM 55(4) (2008)
22. Savitch, W.J.: Relationships between nondeterministic and deterministic tape complexities. J. of Comp. and Syst. Sciences 4(2), 177–192 (1970)

23. Schweikardt, N., Schwentick, T., Segoufin, L.: Database theory: Query languages. In: Algorithms and Theory of Computation, pp. 19.1–19.34. Chapman & Hall/CRC (2010)
24. Vardi, M.Y.: On the complexity of bounded-variable queries. In: Proc. of PODS 1995, pp. 266–276. ACM Press (1995)
25. Venkateswaran, H.: Properties that characterize LOGCFL. J. of Comp. and Syst. Sciences 43(2), 380–404 (1991)

An Algebraic Characterization
of Unary Two-Way Transducers

Christian Choffrut and Bruno Guillon

LIAFA, CNRS and Université Paris 7 Denis Diderot, France

Abstract. Two-way transducers are ordinary finite two-way automata that are provided with a one-way write-only tape. They perform a word to word transformation. Unlike one-way transducers, no characterization of these objects as such exists so far except for the deterministic case. We study the other particular case where the input and output alphabets are both unary but when the transducer is not necessarily deterministic. This yields a family which extends properly the rational relations in a very natural manner. We show that deterministic two-way unary transducers are no more powerful than one-way transducers.

1 Introduction

In the theory of words, two different terms are more or less indifferently used to describe the same objects: transductions and binary relations. The former term distinguishes an input and an output, even when the input does not uniquely determine the output. In certain contexts it is a synonym for translation where one source and one target are understood. The latter term is meant to suggest pairs of words playing a symmetric role.

Transducers and two-tape automata are the devices that implement the transductions and relations respectively. The concept of multitape- and thus in particular two-tape automata was introduced by Rabin and Scott [8] and also by Elgot and Mezei [3] almost fifty years ago. Most closure and structural properties were published in the next couple of years. As an alternative to a definition via automata it was shown that these relations were exactly the rational subsets of the direct product of free monoids. On the other hand, transductions, which are a generalization of (possibly partial) functions, is a more suitable term when the intention is that the input preexist the output. The present work deals with two-way transducers which are such a model of machine using two tapes. An input tape is read-only and is scanned in both directions. An output tape is write-only, initially empty and is explored in one direction only. The first mention of two-way transducers is traditionally credited to Shepherdson.

Our purpose is to define a structural characterization of these relations in the same way that the relations defined by multi-tape automata are precisely the rational relations. However we limit our investigation to the case where the input and output are words over a one letter alphabet, i.e., to the case where they both belong to the free monoid a^* generated by the unique letter a. Our

E. Csuhaj-Varjú et al. (Eds.): MFCS 2014, Part I, LNCS 8634, pp. 196–207, 2014.

technique does not apply to non-unary alphabets. The input is written over one tape and is delimited by a left and a right endmarker which prevents the reading head to fall off the input. An output is written on a second write-only tape.

We now state our main result more precisely. Given a binary relation $R \subseteq X \times Y$ where X and Y are two arbitrary subsets and an element $x \in X$ we put $R(x) = \{y \mid (x, y) \in R\}$. Now assume that X and Y are two monoids and recall that R is rational if it belongs to the smallest family of subsets of $X \times Y$ which contains the finite subsets and which is closed under set union, componentwise concatenation and Kleene star. We are able to prove the following

Theorem 1. *A relation of the monoid $a^* \times a^*$ is defined by a two-way transducer if and only if it is a finite union of relations R satisfying the following condition: there exist two rational relations $S, T \subseteq a^* \times a^*$ such that for all $x \in a^*$ we have*

$$R(x) = S(x)T(x)^*$$

The relation $\{(a^n, a^{kn}) \mid n, k \geq 0\}$ is a simple example. It is of the previous form, however it is not rational. Indeed, identifying a^* with the additive monoid of integers $\mathbb{N}$ this relation defines the relation "being a multiple of". However rational subsets of $\mathbb{N}$ are first-order definable in Presburger arithmetics, i.e., arithmetics with addition only.

We quickly review the few results which to the best of our knowledge are published on two-way transducers when considering them for their own sake. Engelfriet and Hoogeboom showed that a function on the free monoid is defined by a deterministic two-way transducer if and only if it is the set of models of an MSO formula, [4]. In [5] the authors show that given a transducer accepting a function, it is decidable whether or not it is equivalent to a one-way transducer, and when it is, an equivalent one-way transducer is computable.

We now turn to a short presentation of the content of our manuscript. In the next section we recall all the basic concepts concerning rational subsets, rational relations, the different types of automata and transducers. We also introduce a few notions which are indispensable for guaranteeing rigorous proofs of the results. Binary relations over free monoids can be viewed as functions of the first component into the semiring of subsets of the second component. In section 3 we revisit the notion of formal series and show how binary relations can fit into this setting. Important closure properties are established that are instrumental for the proof of our result. The actual proof of our result is done in section 4 by decomposing general two-way transductions into simpler ones. The conclusion contains remarks on interesting consequences of our results that, were it not for the space restriction, could be drawn using our approach.

2   Preliminaries

We assume the reader is familiar with language and automata theory. For the sake of completeness we recall some notions and fix some notation.

An *alphabet* Σ is a finite set of symbol. The free monoid it generates is denoted by Σ^*, and its elements are *words* over Σ including the *empty word* ϵ. The *length* of a word u is $|u|$. The *concatenation* of two words u and v is denoted uv. A *language* is a set of words *i.e.*, a subset of Σ^*.

Given a monoid M, the family of rational subsets denoted $\mathrm{Rat}(M)$ is the least family $\mathcal{F}$ of subsets, containing the finite sets and closed under set union: $X, Y \in \mathcal{F} \Rightarrow X \cup Y \in \mathcal{F}$, set concatenation: $X, Y \in \mathcal{F} \Rightarrow XY \in \mathcal{F}$ and Kleene star: $X \in \mathcal{F} \Rightarrow X^* \in \mathcal{F}$ (recall that $XY = \{xy \mid x \in X, y \in Y\}$ and that $X^* = \{1\} \cup X \cup \cdots X^i \cdots$ where 1 is the unit of the monoid). Here we are mainly interested in the case where M is a free monoid or a direct product of free monoids.

2.1   Finite Automata

We fix an alphabet Σ, called *input alphabet*, and let $\triangleright$ and $\triangleleft$ be two special symbols which do not belong to Σ, called respectively *left* and *right endmarkers*. The set $\Sigma \cup \{\triangleright, \triangleleft\}$ is denoted by $\overline{\Sigma}$.

Definition 1. *A* finite automaton *over Σ is a tuple (Q, q_-, Q_+, δ), where Q is a finite set whose elements are called* states, *q_- is the* initial state, *Q_+ is the set of* accepting states *and δ is the set of* transitions, *included in $Q \times \overline{\Sigma} \times \{-1, 0, +1\} \times Q$, with the restriction that it does not contain any transition of the form $(\_, \triangleright, -1, \_)$, $(\_, \triangleleft, +1, \_)$ and $(q, \triangleleft, \_, \_)$ for $q \in Q_+$.*

We recall the dynamics of the device. Given an input word $u = u_1 \cdots u_n$ on Σ we augment it to $\tilde{u} = u_0 \cdot u_1 \cdots u_n \cdot u_{n+1}$ where $u_0 = \triangleright$ and $u_{n+1} = \triangleleft$. The automaton starts the computation with the word $\tilde{u}$ written on the tape, the input head positioned on the leftmost cell scanning u_0, and in state q_-. At each step, the automaton reads the input symbol $a \in \overline{\Sigma}$ scanned by the head, and according to its current state q chooses a direction d and a state q' with $(q, a, d, q') \in \delta$. Then it enters the state q' and moves its head according to d. The automaton *accepts* the input word u if it eventually enters an accepting state at the rightmost position. Because of the restrictions on transition set, the input head cannot move out of $\tilde{u}$. The set of all words accepted by the automaton is the *language accepted*. Two automata are *equivalent* if they accept the same language.

Now we consider some restricted versions of finite automata. An automaton is *one-way* (resp. *restless*) if no transition is of the form $(\_, \_, -1, \_)$ (resp. $(\_, \_, 0, \_)$). It is *deterministic* if for each pair (q, a) in $Q \times \overline{\Sigma}$, there exists at most one pair (d, q') in $\{-1, 0, +1\} \times Q$ with $(q, a, d, q') \in \delta$, in other words δ is a (partial) function from $Q \times \overline{\Sigma}$ into $\{-1, 0, +1\} \times Q$. It is well-known that all versions accept the same family, that of regular languages, [8,10].

2.2   Configurations, Runs, Traces

The description of the system at a fixed time is given by the current state and the input head position: a *configuration* of A over a word u of length n is a pair (q, p) where q is a state and p is a *position* of $\tilde{u}$ *i.e.*, an integer such that $0 \leq p \leq n+1$.

The *initial configuration* is the configuration $(q_-, 0)$. An *accepting configuration* is any configuration $(q, n + 1)$ with $q \in Q_+$. We call *border configuration*, any configuration whose position is equal to 0 or $n + 1$. We speak of *left-* and *right-border* configurations respectively.

From the transition set follows the successor relation on configurations on u. A pair of configurations $((q, p), (q', p'))$ belongs to the *successor relation*, written $(q, p) \to (q', p')$, if the automaton may enter (q', p') from (q, p) in one step that is, $(q, u_p, (p' - p), q')$ belongs to δ. In particular $(p' - p)$ has to be equal to -1, 0 or 1. Observe that the relation depends on the input word. Due to the last restriction on transition sets in Definition 1, an accepting configuration has no successor.

Definition 2. *A run of A on u is a sequence $c_0, c_1, \ldots, c_\ell$ of successive configurations of A on u i.e., for each $0 \le i < \ell$, $c_i \to c_{i+1}$.*

A run is accepting *if it starts from the initial configuration and halts in some accepting configuration.*

An input word u is accepted by an automaton A if there exists an accepting run of A on u.

The following notion is probably superfluous when dealing with automata but it is instrumental when working with transducers.

Definition 3. *The trace of a run $\mathbf{r} = (q_0, p_0), (q_1, p_1), \ldots, (q_\ell, p_\ell)$ of A on u is the sequence $\mathbf{t_r} = t_1, \ldots, t_\ell$ of transitions such that for each $0 < i \le \ell$, t_i is the witness of $(q_{i-1}, p_{i-1}) \to (q_i, p_i)$ i.e., $t_i = (q_{i-1}, u_{p_{i-1}}, p_i - p_{i-1}, q_i)$.*

We are now interested in some particular runs. We view the set of configurations on u as an alphabet, the runs as words and the sets of runs as languages. Similarly, traces of runs are viewed as words over the transition set of A. Consequently, commas between successive letters in runs and traces are no longer necessary.

Given two runs $\mathbf{r} = c_0 \cdots c_\ell$ and $\mathbf{r}' = d_0 \cdots d_k$ with $c_\ell = d_0$ we consider their *composition* $\mathbf{r}@\mathbf{r}' = c_0 \cdots c_\ell d_1 \cdots d_k$ by deleting d_0. Its trace is the word $\mathbf{t_r}\mathbf{t_{r'}}$ where $\mathbf{t_r}$ and $\mathbf{t_{r'}}$ are the traces of $\mathbf{r}$ and $\mathbf{r}'$. The composition is a partial operation and is thus different from the concatenation.

A run $\mathbf{r}$ is *central* if none of its configurations is border. A *hit* is a run $c_0 c_1 \cdots c_\ell$ such that c_0 and c_ℓ are border configurations and $c_1 \cdots c_{\ell-1}$ is central, i.e., it is a run between two visits of endmarkers. Because the initial and accepting configurations are border, every accepting run is a composition of finitely many hits. We distinguish a hit by a pair of *border points* which are of the form $(q, \triangleright)$ or $(q, \triangleleft)$ for some $q \in Q$ with the natural meaning. For two border points b_0 and b_1, we speak of a b_0 *to* b_1 *hit* on u.

A *loop* of A on u is a run $c_0 \ldots c_\ell$ where $c_0 = c_\ell$. When the first configuration (or its state component) is fixed and equal to c (resp. to q), we speak of a c- (or q-) loop. Trivially if c is a configuration of A on u, and $\mathbf{r}$ and $\mathbf{r}'$ are two c-loops then $\mathbf{r}@\mathbf{r}'$ exists and is a c-loop. We consider a run reduced to a single configuration as a trivial loop.

We need a last particular type of run which is crucial in our work: a run of A on u is *loop-free* if it contains no nontrivial loop. Loop-free runs have in particular bounded length (depending on the length of u), since the set of configurations on u is finite. The following technical result is more or less trivial but happens to be useful.

Lemma 1. *Every run on u can be factored into $\lambda(c_0)\lambda(c_1)\cdots\lambda(c_\ell)$ such that $c_0c_1\cdots c_\ell$ is a loop-free run on u and $\lambda(c)$ is a c-loop.*

2.3 Transducers

In this section, Σ and Δ are two fixed *input* and *output* alphabets. Transducers are finite automata, which are provided with the ability to output symbols during the computation. A natural way to define such machines, is to add a function that maps every transition into some kind of output. At each step, the machine performs a transition, and produces an output.

Definition 4. *A transducer is a pair $T = (A, \phi)$ where A is an automaton over Σ with transition set δ and where ϕ is an output function which is a mapping of δ into the set of nonempty rational subsets of Δ^*.*

Let u be a word in Σ^* and let $\mathbf{r}$ be a run on u of trace $t_1 \cdots t_\ell$. The word $v \in \Delta^*$ is produced by $\mathbf{r}$ if it belongs to the subset $\phi(t_1)\cdots\phi(t_\ell)$. We will also use the notation $\Phi_T(\mathbf{r}) = \phi(t_1)\cdots\phi(t_\ell)$ or simply $\Phi(\mathbf{r})$ when the transducer T is understood.

A pair $(u, v) \in \Sigma^* \times \Delta^*$ is accepted by the transducer if v is produced by an accepting run on u. The relation accepted by T is the set of all such (u, v) and is denoted by $\|T\|$.

The transducer T is *deterministic*, if A is deterministic and ϕ is single valued. It is *one-way* (resp. *restless*) if A is. Each transducer is equivalent to a transducer of some simpler forms:

Lemma 2. *Let T be a transducer. There exists a computable equivalent transducer which is restless and whose output function maps δ into $\Delta \cup \{\epsilon\}$ (in particular δ is single valued).*

It is well known that the family of relations accepted by one-way transducers is the family of rational relations, e.g., [1, Theorem III. 7.1] [2,9].

Theorem 2. *One-way transducers accept exactly the family of rational relations.*

The family of rational relations is strictly smaller than the family of relations accepted by general transducers, even when the alphabets are unary.

Example 1. The relation $MULT = \bigcup_{k \in \mathbb{N}} MULT_k = \{(a^n, a^{kn}) \mid n, k \in \mathbb{N}\}$ is accepted by the three-state two-way restless sweeping transducer:

$$(\{\overrightarrow{q}, \overleftarrow{q}, q_+\}, \overleftarrow{q}, \{q_+\}, \delta), \phi)$$

where:
$$\delta = \left\{ \begin{array}{l} (\overrightarrow{q}, a, +1, \overrightarrow{q})\ (\overleftarrow{q}, a, -1, \overleftarrow{q})\ (q_+, a, +1, q_+) \\ (\overrightarrow{q}, \lhd, -1, \overleftarrow{q})\ (\overleftarrow{q}, \rhd, +1, \overrightarrow{q})\ (\overleftarrow{q}, \rhd, +1, q_+) \end{array} \right\}$$

and ϕ maps $(\overrightarrow{q}, a, +1, \overrightarrow{q})$ to a and all other transitions to ϵ (see Figure 1).

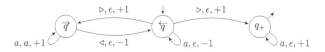

Fig. 1. A two-way transducer accepting the relation $MULT$ (an edge (q, q') is labeled (s, w, d) if ϕ maps the transition (q, s, d, q') to w.)

This relation is not rational as observed in the introduction. It follows from Theorem 2 that no one-way transducer accept it. As a consequence of Corollary 1, no deterministic transducer can accept the relation.

3   Formal Series

Here $\mathbb{K}$ denotes a semiring of subsets of a free monoid such as $\mathrm{Rat}(\Delta^*)$ or $\mathcal{P}(\Delta^*)$. Many of the properties stated below can be extended but this is not the place to make a general theory.

3.1   General Definitions

A *(formal) series* over Σ^* with coefficients in $\mathbb{K}$ is a function from Σ^* to $\mathbb{K}$. The set of series over Σ^* with coefficients in $\mathbb{K}$ is denoted by $\mathbb{K}\langle\langle\Sigma^*\rangle\rangle$. For a series s in $\mathbb{K}\langle\langle\Sigma^*\rangle\rangle$ and a given element $u \in \Sigma^*$, the *image* of u by s is denoted by $\langle s, u\rangle$. A series is a *polynomial* if there are finitely many words with non-zero image, *i.e.*, the set $\{u \mid \langle s, u\rangle \neq 0\}$ is finite.

The set $\mathbb{K}\langle\langle\Sigma^*\rangle\rangle$ is provided with the usual binary operations of restriction to a subset ($\langle s_X, u\rangle = \langle s, u\rangle$ if $u \in X$ and $= 0$ otherwise), sum ($\langle s + t, u\rangle = \langle s, u\rangle + \langle t, u\rangle$), Cauchy product ($\langle st, u\rangle = \sum_{u=vw} \langle s, v\rangle \langle t, w\rangle$) and Kleene star ($\langle s^*, u\rangle = \sum_{u=u_1 \cdots u_n, n \geq 0} \langle s, u_1\rangle \cdots \langle s, u_n\rangle$). We need two operations which we called *Hadamard*- or simply *H-operations*. The first one is usual the second happens to be new.

– the *Hadamard product* (or H-product): $s \otimes t : \forall u \in \Sigma^*,\ \langle s \otimes t, u\rangle = \langle s, u\rangle \langle t, u\rangle$

– the *Hadamard star* (or H-star): $s^{H\star} : \forall u \in \Sigma^*,\ \langle s^{H\star}, u\rangle = \langle s, u\rangle^*$.

Convention. We can take advantage of the properties of formal series in order to study the relations $R \subseteq \Sigma^* \times \Delta^*$. Such a relation can be viewed as the function $x \to f_R(x) = \{y \mid (x, y) \in R\}$ of Σ^* into the power set of Δ^*, or equivalently as a formal series with coefficients in the semiring $\mathbb{K} = \mathcal{P}(\Delta^*)$. We will thus identify R with f_R by speaking of the *formal series associated with the relation*. In the same spirit we will speak of *formal series accepted by a transducer*.

3.2   Rational Series and Beyond

The family of *rational series* over the serimiring $\mathbb{K}$, denoted $\mathrm{Rat}_{\mathbb{K}}\langle\langle\Sigma^*\rangle\rangle$, is the smallest family of series over Σ^* with coefficients in the semiring $\mathbb{K}$ which contains the polynomials and which is closed under sum, Cauchy product and star. The following theorem is a reformulation of Theorem 2.

Theorem 3. *The family of series in $\mathcal{P}(\Delta^*)\langle\langle\Sigma^*\rangle\rangle$ accepted by one way transducers is equal to the family of rational series over Σ^* with coefficients in $\mathrm{Rat}(\Delta^*)$.*

The family $\mathrm{Rat}_{\mathbb{K}}\langle\langle\Sigma^*\rangle\rangle$ is not closed under H-operations for an arbitray semiring. However when $\mathbb{K}$ is commutative the following holds, [9, Thm III. 3.1]

Theorem 4. *If $\mathbb{K}$ is commutative then $\mathrm{Rat}_{\mathbb{K}}\langle\langle\Sigma^*\rangle\rangle$ is closed under H-product.*

The H-star of a rational series is not necessarily rational, even when Σ is unary. Therefore the following defines a broader family.

Definition 5. *The family of* Hadamard series, *denoted $\mathrm{Had}_{\mathbb{K}}\langle\langle\Sigma^*\rangle\rangle$ is the set of finite sums of Hadamard products of the form $\alpha_{\textcircled{\tiny H}}\beta^{H\star}$ with $\alpha,\beta\in\mathrm{Rat}_{\mathbb{K}}\langle\langle\Sigma^*\rangle\rangle$.*

This family enjoys nice closure properties whose routine proof is left to the reader.

Proposition 1. *If $\mathbb{K}$ is commutative, the family $\mathrm{Had}_{\mathbb{K}}\langle\langle\Sigma^*\rangle\rangle$ is closed under finite sum, H-product and H-star.*

The following is general but provides, when restricted to the case where Δ is unary, one direction of our main theorem 5.

Proposition 2. *If s and t are series accepted by two-way transducers, so are the series the $s_{\textcircled{\tiny H}}t$ and $s^{H\star}$.*

4   Unary Two-Way Transductions

From now on we concentrate on unary two-way transducers, *i.e.*, on those with input and output alphabets reduced to the letter a and characterize the relations they define. We fix a transducer (A,ϕ).

The following is a reformulation of Theorem 1 in terms of series (remember that binary relations in $a^*\times a^*$ are identified with formal series in $\mathcal{P}(a^*)\langle\langle a^*\rangle\rangle$ as observed by the convention of paragraph 3.1).

Theorem 5. *Let $\mathbb{K}$ denote the semiring $\mathrm{Rat}(a^*)$. A series $s\in\mathcal{P}(a^*)\langle\langle a^*\rangle\rangle$ is accepted by some two-way finite transducer if and only if $s\in\mathrm{Had}_{\mathbb{K}}\langle\langle a^*\rangle\rangle$, i.e., there exist a finite collection of rational series $\alpha_i,\beta_i\in\mathrm{Rat}_{\mathbb{K}}\langle\langle a^*\rangle\rangle$ such that:*

$$s=\sum_i \alpha_i{}_{\textcircled{\tiny H}}\beta_i^{H\star}$$

The fact that the condition is sufficient is a direct consequence of Theorem 3 and Proposition 2. The other direction is more involved. We proceed as follows. We first show that if the transducer performs a unique hit, $i.e.$, it never visits endmarkers except at the beginning and at the end of the computation, it defines a rational relation. Then we use the closure properties of paragraph 3.2 to prove that the full binary relation with the possibility of performing an arbitrary number of hits, belongs to $\mathrm{Had}_{\mathbb{K}}\langle\langle a^*\rangle\rangle$.

4.1 One-Way Simulation of Hits

We fix two border points b_0 and b_1 and we consider the set of pairs (u, v) such that v is produced by some b_0 to b_1 hit on u. The idea of the proof can be explained as follows where some technicalities are ignored.

We adapt Lemma 1 by saying that a b_0 to b_1 hit over an input u is of the form $c_0\lambda(c_1)\cdots\lambda(c_{\ell-1})c_\ell$ where $c_0c_1\cdots c_\ell$ is a loop-free b_0 to b_1 hit and $\lambda(c_i)$ is a central c_i-loop for $i = 1,\ldots,\ell-1$. Then the set produced on all possible b_0 to b_1 hits on u is the union over all loop-free b_0 to b_1 hits $c_0c_1\cdots c_\ell$ of the following subset (recall the notation of Definition 4).:

$$\Phi(c_0c_1)\Phi(\lambda(c_1))\cdots\Phi(\lambda(c_{\ell-1}))\Phi(c_{\ell-1}c_\ell) \tag{1}$$

Since the output alphabet is unary, the above terms commute and may be rewritten as as many products as there are positions in u. For each position $0 \le p \le n+1$ we group 1) the $\Phi(\lambda(c_i))$ around all the configurations in position p and 2) all $\Phi(c_ic_{i+1})$ involving a transition occurring at position p. The former product leads us to investigate all outputs of central loops and the latter product leads us to adapt the notion of crossing sequences for loop free runs.

Loop-Free Hits. We adapt the traditional notion of crossing sequences, e.g., [6, pages 36-42], to our purpose.

Fix a position $0 \le p \le n$ on the input u. The *crossing sequence* of a run at position p is the record, in the chronological order, of all the destination states in the transition performed between positions p and $p + 1$, i.e., the states at position $p+1$ in a left to right move and the states at position p in a right to left move, see Figure 2. The following is general and does not assume any condition on the run:

Definition 6. *Let* $\mathbf{r} = ((q_i, p_i))_{0 \le i \le \ell}$ *be a run over some input word u of length n and let $0 \le p \le n$ be a position on u. The* crossing sequence *of $\mathbf{r}$ at p, denoted $X_{\mathbf{r}}(p)$, is the ordered state sequence extracted from the sequence $q_0,\ldots,q_\ell$ as follows: for each $1 \le i \le \ell$, q_i is selected if and only if:*

$$\left(p_{i-1} = p \text{ and } p_i = p+1\right) \quad or \quad \left(p_{i-1} = p+1 \text{ and } p_i = p\right)$$

In the run $\mathbf{r}$ there exist two types of states, those which are entered from the left and those which are entered from the right. These two types alternate. Which type occurs first depends on whether the initial border is left or right.

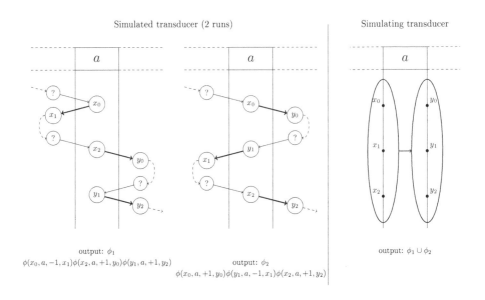

Simulated transducer (2 runs)   Simulating transducer

output: ϕ_1

$\phi(x_0, a, -1, x_1)\phi(x_2, a, +1, y_0)\phi(y_1, a, +1, y_2)$

output: ϕ_2

$\phi(x_0, a, +1, y_0)\phi(y_1, a, -1, x_1)\phi(x_2, a, +1, y_2)$

output: $\phi_1 \cup \phi_2$

Fig. 2. Two different runs producing the same crossing sequence

In particular, if $\mathbf{r}$ is an accepting run, then for each position p, the first and last state of $X_{\mathbf{r}}(p)$ correspond to left to right moves. If $\mathbf{r}$ is loop-free, then for all crossing sequences $\mathbf{q} = q_0, q_1, \ldots$ and for all integers $0 \le i < j$ of the same parity we have $q_i \ne q_j$.

The following technical result implies that the set of all pairs $(u, v) \in \Sigma^* \times \Delta^*$ such that v is the output of a b_0 to b_1 loop-free hit on u is a rational relation. It works because the output alphabet is unary.

Lemma 3. *Given a restless transducer* $T = (A, \phi)$ *and two border points* b_0 *and* b_1 *of* A, *there exists a computable one-way transducer* $T' = (A', \phi')$ *satisfying the following condition:*

Let $\mathbf{r} = c_0 \cdots c_\ell$ *be a* b_0 *and* b_1 *loop free hit on* u *in* A, *let* $v \in \Phi_T(\mathbf{r})$ *and let* $\mathbf{r}' = X_{\mathbf{r}}(0), X_{\mathbf{r}}(1) \ldots, X_{\mathbf{r}}(|u|)$ *the associated crossing sequences. Then* $\mathbf{r}'$ *is an accepting run in* T' *with* $v \in \Phi_{T'}(\mathbf{r}')$.

Conversely, if $\mathbf{r}' = \mathbf{r}'_0, \ldots, \mathbf{r}'_{|u|}$ *is a run on* u *in* T' *with* $v \in \Phi_{T'}(\mathbf{r}')$, *then there exists a* b_0 *and* b_1 *loop free hit* $\mathbf{r}$ *on* u *in* T *with crossing sequences* $\mathbf{r}'$ *such that* $v \in \Phi_T(\mathbf{r})$.

Computing the Outputs of Central Loops. As said previously we now turn to the investigation of central loops. We show that there are only finitely many different sets of words produced by central loops. Intuitively for an initial configuration (q, p) the set of outputs does not depend on p provided it is sufficiently far away from each endmarker.

Given a central loop $\mathbf{r} = (q_0, p_0), \ldots, (q_k, p_k)$, and two elements ℓ and r in $\mathbb{N} \cup \{\infty\}$, we say that $\mathbf{r}$ is (ℓ, r)-*limited* if for all $0 \le i \le k$ we have $p_0 - \ell \le p_i \le$

$p_0 + r$. Observe that if $\ell' \geq \ell$ and $r' \geq r$, every (ℓ, r)-limited loop is (ℓ', r')-limited. Any central loop is (∞, ∞)-limited.

We denote by $\mathcal{O}_{\ell,r}^{(q)}$ the union of all $\Phi(\mathbf{r})$ where $\mathbf{r}$ is a (ℓ, r)-limited q-central loop (observe that it does not depend on the initial position since no endmarker is visited). In particular, for each $\ell' \geq \ell$ and each $r' \geq r$, $\mathcal{O}_{\ell,r}^{(q)}$ is included in $\mathcal{O}_{\ell',r'}^{(q)}$. The language $\mathcal{O}_{\infty,\infty}^{(q)}$ is the set of all outputs of q-central loops. For any ℓ, r and q, $\mathcal{O}_{\ell,r}^{(q)}$ contains in particular the empty word ϵ, since for any central configuration (q, p), the run reduced to (q, p) is a trivial central-loop. The following shows that each language $\mathcal{O}_{\ell,r}^{(q)}$ is rational via Parikh's Theorem [7] and that there exist finitely many different such languages:

Lemma 4. *Let (A, ϕ) be a transducer. For any ℓ and r in $\mathbb{N} \cup \{\infty\}$, and any state q of A there exists a computable one-way automaton accepting the language $\mathcal{O}_{\ell,r}^{(q)}$. It follows that the language is rational.*

There exists a computable N, such that for each $r \geq N$, each $\ell \geq N$ and each state q, $\mathcal{O}_{\ell,r}^{(q)} = \mathcal{O}_{N,N}^{(q)} = \mathcal{O}_{\infty,\infty}^{(q)}$.

Putting Things Together. We are now able to simulate all hits:

Proposition 3. *Given two border points b_0 and b_1, we can compute a one-way transducer accepting the set of pairs (u, v) such that v is produced by a b_0 to b_1 hit on u. The relation accepted is thus rational.*

Proof. Let $T = (A, \phi)$ be a transducer that we suppose thanks to Lemma 2 restless. We denote by $N + 1$ the integer computed from Lemma 4.

We define the following two one-way restless automata which share the same state set $Q_B = \{0, 1, \ldots, N, \infty\}$. The first one $B_L = (Q_B, 0, Q_B, \delta_L)$ counts the distance to the left endmarker up to N. The second one $B_R = (Q_B, \infty, \{0\}, \delta_B)$ guesses whether the distance to the right endmarker is greater than N or is equal to some integer less than or equal to N and checks this guess by counting down until reaching the right endmarker. The transition sets δ_L and δ_R are defined as follows with the convention $N + 1 = \infty + 1 = \infty$ and $\infty - 1 = \infty$

- $\delta_L = \{(q, s, +1, q + 1)\}$ where $s \in \overline{\Sigma}$
- $\delta_R = \{(\infty, \triangleright, +1, i) \mid i \in Q_B\} \cup \{(\infty, a, +1, N)\} \cup \{(q, a, +1, q - 1) \mid q \neq 0\}$

We build a one-way restless transducer (A', ϕ') from T as in Lemma 3. Because A', B_L and B_R are one-way restless, we can take the product automaton $A'' = A' \times B_L \times B_R$. Let us now consider a transition $t = ((\mathbf{q}, \ell, r), a, +1, (\mathbf{q}', \ell', r'))$ of A'' performed at position p. The states appearing in $\mathbf{q}$ (*resp.* $\mathbf{q}'$) correspond to states entered by T in position $p - 1$ or p (*resp.* p or $p + 1$). In any case, this difference can be determined from the initial border point and the parity of the index of the state in the crossing sequence. In particular, we may extract from both crossing sequences $\mathbf{q}$ and $\mathbf{q}'$ the set S_t of states that are entered by T in position p. Hence, we can add for each such state q, the possible output of (ℓ, r)-limited central q-loops. Formally, the image of t by ϕ'' is defined by:

$$\phi''(t) = \phi'(t) \left(\bigcup_{q \in S_t} \mathcal{O}_{\ell,r}^{(q)} \right)$$

By construction, the one-way restless transducer (A'', ϕ'') simulates any b_0 to b_1 hit of T. Hence, the relation of pairs (u, v) such that v is produced by some b_0 to b_1 hit of T on u is rational. □

4.2   Unlimited Number Hits

We prove Theorem 5 by considering an unlimited number of hits. We show that the series associated with the relation defined by a two-way transducer can be expressed via a transitive closure of a square matrix with entries in $\mathrm{Had}_{\mathbb{K}} \langle\langle \Sigma^* \rangle\rangle$. More precisely, if a run $\mathbf{r}$ is accepting, then there exists a sequence of composable hits $\mathbf{r}_0, \mathbf{r}_1, \ldots, \mathbf{r}_k$ such that $\mathbf{r} = \mathbf{r}_0 @ \mathbf{r}_1 @ \ldots @ \mathbf{r}_k$.

We first adapt the matrix multiplication to the Hadamard product. Given an integer N and two matrices $X, Y \in (\mathbb{K} \langle\langle \Sigma^* \rangle\rangle)^{N \times N}$ we define their H-product as the matrix $Z = X \oplus Y \in (\mathbb{K} \langle\langle \Sigma^* \rangle\rangle)^{N \times N}$ where

$$Z_{i,j} = \sum_{k=1}^{N} X_{i,k} \oplus Y_{k,j}$$

Also, the H-star of the matrix X is defined as the infinite sum

$$(X)^{H\star} = \sum_{k=0}^{\infty} \overbrace{X \oplus \cdots \oplus X}^{k \text{ times}}$$

Proposition 4. *If the matrix X is in $(\mathrm{Had}_{\mathbb{K}} \langle\langle \Sigma^* \rangle\rangle)^{N \times N}$ then so is $(X)^{H\star}$.*

Proof. Let Ξ be the alphabet consisting of the letters $x_{i,j}$ for $1 \leq i, j \leq N$ and let $\mathcal{X}$ be the $N \times N$-matrix whose (i, j)-entry is the symbol $x_{i,j}$. Each (i, j)-entry of $\mathcal{X}^*$ is a rational subset over the alphabet Ξ: $(\mathcal{X}^*)_{i,j} \in \mathrm{Rat}(\Xi^*)$. Consider the morphism $h : \Xi^* \to \mathbb{K} \langle\langle M \rangle\rangle$ defined by $h(x_{i,j}) = X_{i,j}$. Then $h((\mathcal{X}^*)_{i,j})$ is an expression involving the entries of X and the three operations of set union, H-sum and H-product. We conclude by Proposition 1. □

Proof (Theorem 5). In one direction this is an immediate consequence of the fact that the family of series associated with a two-way transducer is closed under sum, Hadamard product and Hadamard star, see Proposition 2.

It remains to prove the converse. Let T be transducer. Consider a matrix X whose rows and columns are indexed by the pairs $Q \times \{\triangleright, \triangleleft\}$ of border points. For all pairs of border points b_0 and b_1, its (b_0, b_1) entry is, by Proposition 3, the rational series associated to b_0 to b_1 hits. The series accepted by T is the sum of the entries of $X^{H\star}$ in positions $((q_-, \triangleright), (q, \triangleleft))$ for $q \in Q_+$. Since all rational series are also Hadamard series, we conclude by Proposition 4. □

5   Conclusion

Our main result of Theorem 1 gives a characterization of relations (series) accepted by two-way unary transducers. A key point is that crossing sequences of loop-free runs have bounded size. In consequence, any loop-free run can be simulated by a one-way transducer as done in Lemma 3. We point that this simulation does not require any hypothesis on the size of the input alphabet.

We fix a transducer $T = (A, \phi)$ accepting a relation $R \subseteq \Sigma^* \times \Gamma^*$, with $|\Gamma| = 1$. If A is deterministic or *unambiguous* (*i.e.*, for each input word u, there exists at most one accepting run of A on u), then every accepting run is loop-free. Therefore, by Lemma 3, T is equivalent to some constructible one-way transducer. Another interesting case is when R is a function. Then for each u, all the accepting runs on u produce the same output word. Hence, considering only loop-free runs preserves the acceptance of T.

Corollary 1. *Let $R \subseteq \Sigma \times \Delta$ with $|\Delta| = 1$ be accepted by some two-way transducer $T = (A, \phi)$. If A is unambiguous or if R is a function then R is rational.*

A *rational uniformization* of a relation $R \subseteq \Sigma^* \times \Gamma^*$, is a rational function $F \subseteq R$, such that the domain of F is equal to the one of R. Under the hypothesis $|\Gamma| = 1$, it is possible to build, from Lemma 3, a one-way transducer accepting such a F. Since the transducer obtained from Lemma 3 is not necessarily functional, the construction involves a result of Eilenberg [2, Prop. IX 8. 2] solving the rational uniformization problem for rational relation.

Corollary 2. *There exists a computable one-way transducer accepting a rational uniformization of R.*

As a consequence of Proposition 3, in the case of unary transducers, the change of direction of the input head can be restricted to occur at the endmarkers only. In the literature such machines are known as *sweeping* machines.

References

1. Berstel, J.: Transductions and context-free languages. B. G. Teubner (1979)
2. Eilenberg, S.: Automata, Languages and Machines, vol. A. Academic Press (1974)
3. Elgot, C.C., Mezei, J.E.: On Relations Defined by Finite Automata. IBM Journal 10, 47–68 (1965)
4. Engelfriet, J., Hoogeboom, H.J.: MSO definable string transductions and two-way finite-state transducers. ACM Trans. Comput. Log. 2(2), 216–254 (2001)
5. Filiot, E., Gauwin, O., Reynier, P.A., Servais, F.: From two-way to one-way finite state transducers. In: LICS, pp. 468–477 (2013)
6. Hopcroft, J.E., Ullman, J.D.: Introduction to Automata Theory, Languages and Computation. Addison-Wesley (1979)
7. Parikh, R.: On context-free languages. J. ACM 13(4), 570–581 (1966)
8. Rabin, M.O., Scott, D.: Finite automata and their decision problems. IBM Journal of Research and Development 3, 114–125 (1959)
9. Sakarovitch, J.: Elements of Automata Theory. Cambridge University Press (2009)
10. Shepherdson, J.C.: The reduction of two-way automata to one-way automata. IBM Journal of Research and Development Archive 3, 198–200 (1959)

Size-Change Abstraction and Max-Plus Automata[*]

Thomas Colcombet[1], Laure Daviaud[1], and Florian Zuleger[2]

[1] LIAFA, CNRS, Université Paris Diderot, Paris, France
[2] Vienna University of Technology, Vienna, Austria

Abstract. Max-plus automata (over $\mathbb{N} \cup \{-\infty\}$) are finite devices that map input words to non-negative integers or $-\infty$. In this paper we present (a) an algorithm allowing to compute the asymptotic behaviour of max-plus automata, and (b) an application of this technique to the evaluation of the computational time complexity of programs.

1 Introduction

The contributions of this paper are two-fold. First, we provide an algorithm that given a function computed by a max-plus automaton over $\mathbb{N} \cup \{-\infty\}$ computes the asymptotic minimal behaviour of the automaton as a function of the length of the input. We then apply this result for characterizing the asymptotic complexity bounds that can be obtained by the size-change abstraction, which is a widely used technique in automated termination analysis. These two contributions are of independent interest. Let us introduce them successively.

Weighted Automata, and the Main Theorem

Max-plus automata belong to the wider family of weighted automata, as introduced by Schützenberger [8]. The principle of weighted automata is to consider non-deterministic automata that produce values in a semiring $(S, \oplus, \otimes, 0, 1)$ (i.e., a ring in which the addition is not required to have an inverse). Weighted automata interpret the non-determinism of the automaton as the sum in the semiring and the sequence as the product. Standard non deterministic automata correspond to the case of the Boolean semiring $(\{0, 1\}, \vee, \wedge, 0, 1)$. Probabilistic automata correspond to the case $([0, 1], +, \times, 0, 1)$ (with a stochasticity restriction). Distance automata (or min-plus automata) correspond to the case $(\mathbb{N} \cup \{\infty\}, \min, +, 0, \infty)$.

In this paper, we concentrate our attention to max-plus automata, which correspond to the semiring $(\mathbb{N} \cup \{-\infty\}, \max, +, 0, -\infty)$. Such automata have

---

[*] The research leading to these results has received funding from the European Union's Seventh Framework Programme (FP7/2007-2013) under grant agreement n°259454 and from the Vienna Science and Technology Fund (WWTF) through grant ICT12-059.

E. Csuhaj-Varjú et al. (Eds.): MFCS 2014, Part I, LNCS 8634, pp. 208–219, 2014.
© Springer-Verlag Berlin Heidelberg 2014

transition with weights in $\mathbb{N}$. Over a given input, they output the maximum over all accepting runs of the sum of the weights of transitions (and $-\infty$ if there is no accepting run). Such automata are natural candidates for modelling worst case behaviours of systems, as shown in the subsequent application. Remark that max-plus automata share a lot of common points with min-plus automata, and indeed, many results for max-plus automata can be converted into results for min-plus automata and vice-versa[1].

We seek to analyse the asymptotic behaviour of such automata. More precisely, fix a max-plus automaton computing a function f from the words in $\mathbb{A}^*$ to $\mathbb{N} \cup \{-\infty\}$. We study the asymptotic evolution of $c(n)$ defined for $n \in \mathbb{N}$ as:

$$c(n) = \inf\{f(w) \ : \ w \in \mathbb{A}^*, |w| \geq n\} \ .$$

We show that this quantity either is $-\infty$ for all n, or it is in $\Theta(n^\beta)$ for a computable rational $\beta \in [0,1]$. Our main theorem, Theorem 2, expresses this property in a dual, yet equivalent, way as the asymptotic behaviour of the longest word that happens to have a value smaller than n.

From a logical perspective, it has to do with a quantifier alternation since the quantity studied is computed as a minimum (inf) of a function which, itself, is defined as a maximum (as a max-plus-automaton). In particular, in our case, it is immediately PSPACE hard (using reduction of the universality problem for non-deterministic automata). Such quantifier alternations are often even more complex when weighted automata are considered. For instance, a natural question involving such an alternation is to test whether $f(u) < |u|$ for some u, and it turns out to be undecidable [5]. On the other side, the boundedness question for min-plus automata (determining if there exists n such that $f(u) \leq n$ for all words u), which also has a similar quantifier alternation flavour, turns out to be decidable [4]. The work of Simon [10] has the most similarities with our contribution. It shows that, for a min-plus automaton computing a function g, the dual quantity $d(n) = \sup\{g(w) \ : \ w \in \mathbb{A}^*, |w| \leq n\}$ has a behaviour that is asymptotically between $n^{1/(k+1)}$ and $n^{1/k}$ for some non-negative integer k. Our result differs in two ways. First, the results for min-plus automata and for max-plus automata cannot be converted directly into results over the other form of automata. Second, our main result is significantly more precise since it provides the exact asymptotic coefficient. The proof of this theorem is the subject of the first part of this paper.

Program Analysis and Size Change Abstraction

The second contribution in this work consists in applying Theorem 2 for characterizing the asymptotic complexity bounds that can be obtained by the size-change abstraction, which is a popular program abstraction for automated

---

[1] Indeed, if we allow negative weights, then negating all weights turns max-plus automata into min-plus automata and vice-versa, while preserving the semantics. However, such kind of reductions can get more complicated, if not impossible, when negative values are forbidden, as it is in our case.

termination analysis (e.g. [6,7]). This question was the primary reason for this investigation.

We start with definitions needed to precisely state our contribution. We fix some finite set of *size-change variables* Var. We denote by Var' the set of primed versions of the variables Var. A *size-change predicate* (SCP) is a formula $x \triangleright y'$ with $x, y \in Var$, where $\triangleright$ is either $>$ or $\geq$. A *size-change transition* (SCT) T is a set of SCPs. A *size-change system* (SCS) $\mathcal{S}$ is a set of SCTs.

We define the semantics of size-change systems by *valuations* $\sigma : Var \to [0..N]$ of the size-change variables to natural numbers in the interval $[0..N]$, where N is a (symbolic) natural number. We write $\sigma, \tau' \models x \triangleright y'$ for two valuations σ, τ, if $\sigma(x) \triangleright \tau(y)$ holds over the natural numbers. We write $\sigma, \tau' \models T$, if $\sigma, \tau' \models x \triangleright y'$ holds for all $x \triangleright y' \in T$. A *trace* of an SCS $\mathcal{S}$ is a sequence $\sigma_1 \xrightarrow{T_1} \sigma_2 \xrightarrow{T_2} \cdots$ such that $T_i \in \mathcal{S}$ and $\sigma_i, \sigma'_{i+1} \models T_i$ for all i. The *length* of a trace is the number of SCTs that the trace uses, counting multiple SCTs multiple times. An SCS $\mathcal{S}$ is *terminating*, if $\mathcal{S}$ does not have a trace of infinite length.

We note that in earlier papers, e.g. [6], the definition of a size-change system includes a control flow graph that restricts the set of possible traces. For the ease of development we restrain from adding control structure but our result also holds when we add control structure. Moreover, earlier papers, e.g. [6], consider SCSs semantics over the natural numbers, i.e., valuations $\sigma : Var \to \mathbb{N}$. In contrast, we restrict values to the interval $[0, N]$ in order to guarantee that the length of traces is bounded for terminating SCSs: no valuation $\sigma \in Var \to [0..N]$ can appear twice in a trace (otherwise we would have a cycle, which could be pumped to an infinite trace); thus the length of traces is bounded by $(N + 1)^k$ for SCSs with k variables.

Problem Statement: Our goal is to determine a function $h_{\mathcal{S}} : \mathbb{N} \to \mathbb{N}$ such that the length of the longest trace of a terminating SCS $\mathcal{S}$ is of asymptotic order $\Theta(h_{\mathcal{S}}(N))$. This question has also been of interest in a recent report [1], which claims that SCSs always have a polynomial bound, i.e., a bound $\Theta(N^k)$ for some $k \in \mathbb{N}$. However, this is not the case (see example below). We believe that the development in [1] either contains a gap or that the results of [1] have to be stated differently.

Example 1. The length of the longest trace of the SCS $\mathcal{S} = \{T_1, T_2, T_3\}$ with
$T_1 = \{x_1 > x'_1, x_2 \geq x'_2, x_3 > x'_3, x_4 \geq x'_4\}$,
$T_2 = \{x_1 > x'_1, x_2 \geq x'_2, x_2 \geq x'_3, x_2 > x'_4, x_3 > x'_4, x_4 > x'_4\}$ and
$T_3 = \{x_2 > x'_2, x_2 > x'_3, x_2 > x'_4, x_3 > x'_2, x_3 > x'_3, x_3 > x'_4, x_4 > x'_2, x_4 > x'_3, x_4 > x'_4\}$ is of asymptotic order $\Theta(N^{\frac{3}{2}})$. For comparison, [1] considers SCSs bounded in terms of the initial state; we can make $\mathcal{S}$ bounded in terms of the initial state by adding a new variable x_N to $\mathcal{S}$, and adding the constraints $\{x_N \geq x'_N, x_N \geq x'_1, x_N \geq x'_2, x_N \geq x'_3, x_N \geq x'_4\}$ to each of T_1, T_2, T_3.

The asymptotic order $\Theta(N^{\frac{3}{2}})$ of $\mathcal{S}$ can be established by Theorem 1 stated below (a corresponding max-plus automaton is stated in Example 2). For illustration purposes, we sketch here an elementary proof. For the lower bound we consider the sequence $s_N = ((T_1^{\frac{\sqrt{N}}{2}-1}T_2)^{\frac{\sqrt{N}}{2}-1}T_3)^{\frac{\sqrt{N}}{2}-1}$. For example, for

$N = 36$ we have $s_N = T_1 T_1 T_2 T_1 T_1 T_2 T_3 T_1 T_1 T_2 T_1 T_1 T_2 T_3$. Note that s_N is of length $l_N = \frac{\sqrt{N}}{2} \cdot \frac{\sqrt{N}}{2} \cdot (\frac{\sqrt{N}}{2} - 1) = \Omega(N^{\frac{3}{2}})$. We define valuations σ_i, with $0 \le i \le l_N$, that demonstrate that s_N belongs to a trace of $\mathcal{S}$: given some index $0 \le i \le l_N$, let t_3 denote the number of T_3 before index i in the sequence s_N, let t_2 denote the number of T_2 before index i since the last T_3, and let t_1 denote the number of T_1 before index i since the last T_2 (note that we have $0 \le t_1, t_2, t_3 < \frac{\sqrt{N}}{2}$ by the shape of s_N); we set $\sigma_i(x_1) = N - t_2 \cdot \frac{\sqrt{N}}{2} - t_1, \sigma_i(x_2) = N - t_3 \cdot \sqrt{N}, \sigma_i(x_3) = N - t_3 \cdot \sqrt{N} - t_1, \sigma_i(x_4) = N - t_3 \cdot \sqrt{N} - \frac{\sqrt{N}}{2} - t_2$. It is easy to verify that the valuations σ_i satisfy all constraints of s_N.

We move to the upper bound. Let S be a sequence of SCTs that belongs to a trace of $\mathcal{S}$. We decompose $S = S_1 T_3 S_2 T_3 \cdots$ into subsequences S_i that do not contain any occurrence of T_3. We define a_i to be the maximal number of consecutive T_1 in S_i, and b_i to be the total number of T_2 in S_i. We set $c_i = \max\{a_i, b_i\}$. We start with some observations: We have $|S_i| \le c_i(c_i+1)+c_i = c_i(c_i + 2)$ (i) by the definition of the c_i. We have $|S_i| \le N$ (ii) because the inequality $x_1 > x_1'$ is contained in T_1 as well as in T_2 and the value of x_1 can only decrease N times in S_i. Combining (i) and (ii) we get $|S_i| \le \min\{c_i(c_i + 2), N\}$ (iii). We have $\sum_i c_i \le N$ (iv); this holds because there is a chain of inequalities from the beginning to the end of S that for every i either uses all inequalities $x_3 > x_3'$ of the consecutive T_1 or all inequalities $x_4 > x_4'$ of the T_2 in S_i, and this chain can only contain N strict inequalities. Finally, by the definition of the S_i we have $|S| \le \sum_i |S_i| + 1$. With (iii) we get $|S| \le \sum_i \min\{c_i(c_i + 2), N\} + 1 \le 5 \sum_i \min\{c_i^2, N\}$ (v). Using associativity and commutativity we rearrange the sum $\sum_i c_i = \sum_i d_i + \sum_i e_i + r$, where the d_i are summands $c_i > \sqrt{N}$ and the e_i and r are the sum of summands $c_i \le \sqrt{N}$ with $\frac{\sqrt{N}}{2} \le e_i \le \sqrt{N}$ and $r < \frac{\sqrt{N}}{2}$; we denote $e_i = \sum_j c_{ij}$ for some c_{ij}. By (iv) there are at most $\sqrt{N}$ of the d_i and at most $2\sqrt{N}$ of the e_i. Using these definitions in (v) we get $|S| \le 5(\sum_i \min\{d_i^2, N\} + \sum_{i,j} \min\{c_{ij}^2, N\} + \min\{r^2, N\}) \le 5(\sqrt{N} \cdot N + \sum_{i,j} c_{ij}^2 + N) \le 5(\sqrt{N} \cdot N + \sum_i e_i^2 + N) \le 5(\sqrt{N} \cdot N + 2\sqrt{N} \cdot N + N) = O(N^{\frac{3}{2}})$.

In this paper we establish the fundamental result that the *computational time complexity* of terminating SCA instances is decidable:

Theorem 1. *Let $\mathcal{S}$ be a terminating SCS. The length of the longest trace of $\mathcal{S}$ is of order $\Theta(N^\alpha)$, where $\alpha \ge 1$ is a rational number; moreover, α is computable.*

We highlight that our result provides a *complete characterization* of the complexity bounds arising from SCA and gives means for determining the exact asymptotic bound of a given abstract program. Our investigation was motivated by previous work [11], where we introduced a practical program analysis based on SCA for computing resource bounds of imperative programs; in contrast to this paper, [11] does not study the completeness of the proposed algorithms and does not contain any result on the expressivity of SCA.

Organization of the Paper. In Section 2, we give the automata definitions and sketch the proof of Theorem 2. In Section 3 we provide a reduction from

size-change systems to max-plus automata that allows to prove Theorem 1 from
Theorem 2.

2   Max-Plus Automata

In this section, we first define max-plus automata (section 2.1), and then sketch
the proof of Theorem 2 (section 2.2).

2.1   Definition of Max-Plus Automata

A **semigroup** $(S, \cdot)$ is a set S equipped with an associative binary operation '$\cdot$'.
If the product has furthermore a neutral element 1, $(S, \cdot, 1)$ is called a **monoid**.
The monoid is said to be **commutative** if $\cdot$ is commutative. An **idempotent** in
a semigroup is an element e such that $e \cdot e = e$. Given a subset A of a semigroup,
$\langle A \rangle$ denotes the closure of A under product, *i.e.*, the least sub-semigroup that
contains A. Given $X, Y \subseteq S$, $X \cdot Y$ denotes $\{a \cdot b \ : \ a \in X, \ b \in Y\}$.

A **semiring** $(S, \oplus, \otimes, 0_S, 1_S)$ is a set S equipped with two binary operations
$\oplus$ and $\otimes$ such that $(S, \oplus, 0_S)$ is a commutative monoid, $(S, \otimes, 1_S)$ is a monoid,
0_S is absorbing for $\otimes$ (for all $x \in S$, $x \otimes 0_S = 0_S \otimes x = 0_S$) and $\otimes$ distributes over
$\oplus$. We shall use the **max-plus semiring** $(\{-\infty\} \cup \mathbb{N}, \max, +, -\infty, 0)$, denoted
$\overline{\mathbb{N}}$, and its extension $\overline{\mathbb{R}^+} = \{-\infty, 0\} \cup \{x \ : \ x \in \mathbb{R}, \ x \geq 1\}$, that we name the
real semiring. This semiring will be used instead of $\overline{\mathbb{N}}$ during the computations.
The operation over matrices induced by this semiring is denoted $\otimes$. Remark that
$0_{\overline{\mathbb{N}}} = -\infty$, and $1_{\overline{\mathbb{N}}} = 0$.

Let S be a semiring. The set of matrices with m rows and n columns over S is
denoted $\mathcal{M}_{m,n}(S)$, or simply $\mathcal{M}_n(S)$ if $m = n$. As usual, $A \otimes B$ for two matrices
A, B (provided the width of A and the height of B coincide) is defined as:

$$(A \otimes B)_{i,j} = \bigoplus_{0 < k \leq n} (A_{i,k} \otimes B_{k,j}) \quad \left(= \max_{0 < k \leq n} (A_{i,k} + B_{k,j}) \text{ for } S = \overline{\mathbb{N}} \text{ or } \overline{\mathbb{R}^+} \right).$$

It is standard that $(\mathcal{M}_n(S), \otimes, I_n)$ is a monoid, whose neutral element is the
diagonal matrix I_n with 1_S (i.e., 0 for $\overline{\mathbb{N}}$) on the diagonal, and 0_S (i.e., $-\infty$ for
$\overline{\mathbb{N}}$) elsewhere. For a positive integer k, we set $M^0 = I_n$, and $M^k = M^{k-1} \otimes M$.
For $\lambda \in \mathbb{R}^+$, we denote by λA the matrix such that $(\lambda A)_{i,j} = \lambda A_{i,j}$ for all i, j
(this matrix has non-negative real coefficients, which might not be over $\overline{\mathbb{R}^+}$ if
$\lambda \leq 1$). Finally, we write $A \leq B$ if for all i, j, $A_{i,j} \leq B_{i,j}$.

A **max-plus automaton** over the alphabet $\mathbb{A}$ (with k states) is a map δ
from $\mathbb{A}$ to $\mathcal{M}_k(\overline{\mathbb{N}})$ together with initial and final vectors $I, F \in \mathcal{M}_{1,k}(\{0, -\infty\})$.
The map δ is uniquely extended into a morphism from $\mathbb{A}^*$ to $\mathcal{M}_k(\overline{\mathbb{N}})$, that we
also denote δ. The **function computed by the automaton** maps each word
$u \in \mathbb{A}^*$ to ${}^t I \otimes \delta(u) \otimes F \in \overline{\mathbb{N}}$ where ${}^t I$ denotes the transpose of I.

Example 2. We consider the following automaton, over the alphabet $\{a, b, c\}$, for $k = 6$ and defined by (where $-\infty$ is not written for readability):

$$\delta(a) = \begin{pmatrix} 0\,0\,0\,0\,0\,0 \\ 1\quad\ 0 \\ 0\quad\ 0 \\ 1\ 0 \\ 0\,0 \\ 0 \end{pmatrix}, \qquad \delta(b) = \begin{pmatrix} 0\,0\,0\,0\,0\,0 \\ 1\quad\ 0 \\ 0\,0\,1\,0 \\ 1\,0 \\ 1\,0 \\ 0 \end{pmatrix},$$

$$\delta(c) = \begin{pmatrix} 0\,0\,0\,0\,0\,0 \\ 0 \\ 1\,1\,1\,0 \\ 1\,1\,1\,0 \\ 1\,1\,1\,0 \\ 0 \end{pmatrix}, \qquad \text{and } I = F = \begin{pmatrix} 0 \\ 0 \\ 0 \\ 0 \\ 0 \\ 0 \end{pmatrix}.$$

It is sometimes convenient to see such matrices as a weighted automaton [8]. Such a presentation is provided in Figure 1. The states of the automaton are $q_1, \ldots, q_6$ and correspond respectively to the lines and the columns 1 to 6 of the matrices. There is a transition from q_i to q_j corresponding to letter $x = a, b, c$ if the entry i, j of the matrix $\delta(x)$ is $z \neq -\infty$. In this case, the transition is weighted by z. The initial states are the states q_i such that $I_i = 0$. The final states are the states q_j such that $F_j = 0$. A run over the word w is a path (a sequence of compatible transitions) in the graph labelled by w. Its weight is the sum of the weights of the transitions. Finally the weight of a given word w is the maximum of the weights of the runs labelled by w and going from an initial state to a final state. The weight of w, given by the graph representation is exactly the value ${}^tI \otimes \delta(w) \otimes F$, given by the matrix presentation.

More details about weighted automata can be found in [3].

2.2   Main Theorem

Theorem 2. *Given a max-plus automaton computing* $f : \mathbb{A}^* \to \mathbb{N} \cup \{-\infty\}$, *there exists an algorithm that computes the value* $\alpha \in \{+\infty\} \cup \{\beta \in \mathbb{Q} \;:\; \beta \geq 1\}$ *such that*

$$g(n) = \Theta(n^\alpha)$$

where $g(n) = \sup\{|w| \;:\; f(w) \leq n\}$, *with the convention that* $n^{+\infty} = +\infty$.

Example 3. The algorithm applied on the automaton given in exemple 2 outputs value $2/3$. A sequence of words that witness this growth is $((a^n b)^n)c^n)_{n \in \mathbb{N}}$.

The Semigroup of Weighted Matrices. Our goal is to analyse the relationship between the output of the automaton and the length of the input. Thus

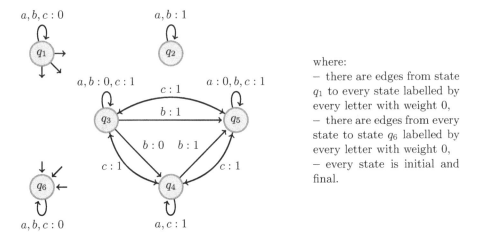

where:
- there are edges from state q_1 to every state labelled by every letter with weight 0,
- there are edges from every state to state q_6 labelled by every letter with weight 0,
- every state is initial and final.

Fig. 1. A weighted automaton over the semiring $(\overline{\mathbb{N}}, \max, +)$

we use weighted matrices that are pairs of a matrix representing the behaviour of the automaton with a value standing for the length of the input. Formally, a **weighted matrix** is an ordered pair (M, x) where $M \in \mathcal{M}_k(\overline{\mathbb{R}^+})$ and $x \geq 1$ is a real number called the **weight** of the weighted matrix. They are useful to represent pairs $(\delta(w), |w|)$. The set of weighted matrices is denoted by $\mathcal{W}_k$. Weighted matrices have a semigroup structure $(\mathcal{W}_k, \otimes)$, where $(M, x) \otimes (N, y)$ stands for $(M \otimes N, x + y)$. By definition, the function $w \mapsto (\delta(w), |w|)$ is a morphism of semigroups. As in the general case, we use $\otimes$ over subsets of $\mathcal{W}_k$. Given $A \subseteq \mathcal{W}_k$, $\langle A \rangle$ is the closure under $\otimes$ of A. Our goal is to study the set

$$\{(\delta(w), |w|) \mid w \in \mathbb{A}^*\} = \langle\{(\delta(a), 1) \mid a \in \mathbb{A}\}\rangle$$

and more precisely to give a finite representation of it up to some approximation. The key to our algorithm is the ability to (a) finitely represent infinite sets of weighted matrices and (b) define a notion of approximation between such sets. Then our algorithm computes using such sets, and guarantees that, up to the approximation, it is consistent with the behaviour of the automaton. We present these notions below. From now we fix a max-plus automaton with k states computing a function f and defined by the morphism δ. Let us first introduce another semiring useful for defining finite representation.

The $\overline{\mathbb{R}^+_\odot}$ and Small Semirings, and the Semigroup of Weighted Matrices.

We have seen the semirings $\overline{\mathbb{N}}$ and $\overline{\mathbb{R}^+}$. We use another semiring over the same ground set $\overline{\mathbb{R}^+}$ but with a different product, $\odot$. For all $x, y \in \overline{\mathbb{R}^+}$ set $x \odot y$ to be:

$$x \odot y = \begin{cases} -\infty & \text{if either } x = -\infty \text{ or } y = -\infty, \\ \max(x, y) & \text{otherwise.} \end{cases}$$

Again, $(\overline{\mathbb{R}^+}, \max, \odot, -\infty, 0)$ is a semiring, denoted $\overline{\mathbb{R}^+_\odot}$. As before, this induces a product operation $\odot$ for matrices. The product operation $\odot$ is a good approximation of $\otimes$ as shown by the following key lemma that follows from the similar property for real number and monotonicity of max and plus.

Lemma 1. *Given matrices $M_1, \ldots, M_q$, $q \geq 1$ over $\overline{\mathbb{R}^+}$, then*

$$M_1 \odot \cdots \odot M_q \leq M_1 \otimes \cdots \otimes M_q \leq q(M_1 \odot \cdots \odot M_q) .$$

The last semiring we use is the **small semiring** $(\mathbb{S}, \max, \odot, -\infty, 0)$, simply denoted $\mathbb{S}$, which is the restriction of $\overline{\mathbb{R}^+_\odot}$ to $\{-\infty, 0, 1\}$. There is a natural map φ from $\overline{\mathbb{R}^+}$ to $\mathbb{S}$ obtained by collapsing all elements above or equal to 1 to 1. It happens that φ is at the same time a **morphism** of semirings from $\overline{\mathbb{R}^+}$ to $\mathbb{S}$ and from $\overline{\mathbb{R}^+_\odot}$ to $\mathbb{S}$. Matrices over the small semiring are called **small matrices**.

The morphism φ is also extended to weighted matrices by $\varphi((M, x)) = \varphi(M)$.

Our goal is, given a finite set of weighted matrices A, to compute a presentation of $\langle A \rangle$ up to approximation (Lemma 7). The notion of presentation of sets of weighted matrices and the notion of approximation are the subject of the two subsequent sections.

Presentable Sets of Weighted Matrices. We introduce now the notion of presentable sets of matrices, i.e., sets of matrices that we can manipulate via their finite presentation. Our sets of weighted matrices are presented in 'exponential form', i.e., given a weight $x \geq 1$, an entry of the matrix will be of the form x^α. In fact, some special cases have to be treated, that results in the use of $\alpha = \bot$ or $-\infty$.

Exponents and exponentiations The semiring of **exponents** (the choice of this name will be explained when defining exponentiation in the next paragraph) is $(Exps, \max, \max_\odot, \bot, -\infty)$ where

$$Exps = \{\bot, -\infty\} \cup [0, 1] ,$$

where max is defined with respect to the order $\bot < -\infty < x < y$ for all $x < y \in [0, 1]$, and where $\max_\odot(\alpha, \beta)$ for $\alpha, \beta \in Exps$ is defined by:

$$\max_\odot(\alpha, \beta) = \begin{cases} \bot & \text{if } \alpha = \bot \text{ or } \beta = \bot, \\ \max(\alpha, \beta) & \text{otherwise.} \end{cases}$$

This semiring will be simply denoted $Exps$, and the induced operation over matrices $\odot$ (we will see that this notation is not ambiguous). We take the convention to denote by α, β exponents, and by X, Y, Z vectors and matrices of exponents.

We define now the exponentiation operation. For $x \geq 1$ and $\alpha \in Exps$, set

$$x^\alpha = \begin{cases} -\infty & \text{if } \alpha = \bot, \\ 0 & \text{if } \alpha = -\infty, \\ x^\alpha & \text{otherwise, i.e., if } \alpha \in [0, 1], \text{ for the usual exponent.} \end{cases}$$

Lemma 2. *For all $x \geq 1$, $\alpha \mapsto x^\alpha$ is a semiring morphism from Exps to $\overline{\mathbb{R}_\odot^+}$.*

Note that this morphism can be applied to vectors (or matrices). In this case, given a matrix $Y \subseteq \mathit{Exps}^{k \times k}$, and some $x \geq 1$, we denote by $Y[x] \in \overline{\mathbb{R}^+}^{k \times k}$ the matrix such that $(Y[x])_{i,j} = x^{Y_{i,j}}$ for all $i, j = 1 \ldots k$. According to the previous lemma, the map $Y \mapsto Y[x]$ is a morphism from matrices over *Exps* to matrices over $\overline{\mathbb{R}_\odot^+}$.

It is also sometimes convenient to send the small semiring to the exponent semiring. It is done using the following straightforward lemma.

Lemma 3. *The function γ that maps $-\infty$ to $\perp$, 0 to $-\infty$, and 1 to 0 is a semiring morphism from $\mathbb{S}$ to Exps such that $x^{\gamma(a)} = a$ for all $a \in \mathbb{S}$ and $x \geq 1$.*

Polytopes and presentable sets. Our goal it to describe finitely some infinite sets of matrices over $\overline{\mathbb{R}^+}$. We start from the notion of polytope. For this, we rely on the definition of polytopes in $\mathbb{R}^k$: a **polytope** (in $\mathbb{R}^k$) is a convex hull of finitely many points of $\mathbb{R}^k$. We would like to use this definition for subsets of Exps^k. For that we send *Exps* to $\mathbb{R}$ by $t(\perp) = -2$, $t(-\infty) = -1$ and $t(s) = s$ if s is real.

A subset of Exps^k is called a polytope if its image under t is a polytope in $\mathbb{R}^k$. In particular, we can use this definition for matrices of exponents, yielding polytopes of matrices.

We can now define presentable sets of matrices over $\overline{\mathbb{R}^+}$. Essentially, a set of matrices over $\overline{\mathbb{R}^+}$ is presentable if it is the image under exponentiation of a finite union of polytopes of exponent matrices. Let us define precisely how this is defined. A set of weighted matrices $A \subseteq \mathcal{W}_k$ is **presentable** if it is of the form:

$$A = \{(M, 1) \ : \ M \in S\} \cup \{(Y[x], x) \ : \ Y \in P, \ x \geq 1\} ,$$

where S is a set of small matrices of dimension $k \times k$, and P is a finite union of polytopes of $\mathit{Exps}^{k \times k}$. The pair (S, P) is called the **presentation** of A. A presentation is said **small** if $P = \emptyset$. It is said **asymptotic** if $S = \emptyset$. Obviously, any presentable set is the union of a set of small presentation with a set of asymptotic presentation. Of course presentable sets are closed under union.

The Approximation and Simulation Scheme. We describe now the notion of approximation that we use. Indeed, our goal is to compute the set of weighted matrices $\{(\delta(w), |w|)\}$. We cannot expect to do it in general, and, at any rate, presentable sets of matrices cannot capture exactly the behaviour of the automaton. That is why we reason about sets of matrices up to some approximation relation that is sufficiently precise for our purpose, and at the same time is sufficiently relaxed for allowing to approximate the behaviour of the automaton by a presentable set of weighted matrices.

Given some $a \geq 1$ and two weighted matrices (M, x) and (N, y), we write

$$(M, x) \preccurlyeq_a (N, y) \quad \text{if} \quad M \leq aN, \ y \leq ax \ \text{and} \ \varphi(M) = \varphi(N) .$$

This definition extends to sets of weighted matrices as follows. Given two such sets A, B, $A \preccurlyeq_a B$ if for all $(N, y) \in B$, there exists $(M, x) \in A$ such that

$(M, x) \preccurlyeq_a (N, y)$. We write $A \approx_a B$ if $A \preccurlyeq_a B$ and $B \preccurlyeq_a A$ and say that A is a-**equivalent** to B. We drop the a parameter when not necessary, and simply write $A \approx B$ if $A \approx_a B$ for some a.

A first consequence of this definition is that every weighted matrix (M, x) is a-equivalent to the weighted matrix $(\varphi(M), 1)$ where a is the maximum of the entries of M and x. This justifies that, in the definition of a presentable set, the weighted matrices of the finite part are of this form.

Let us give some intuition why this approximation may help. For instance consider some exponent matrix M, and let us show:

$$\{(M[x], x) \; : \; x \geq 1\} \approx_2 \{(M[y], y) \; : \; y \in \mathbb{N}, \; y \geq 1\} \; .$$

Indeed, one inclusion is obvious, yielding $\preccurlyeq_1$. For the other direction, consider some $x \geq 1$, and take $y = \lfloor x \rfloor$, then $2y \geq x$ and $M[y] \leq M[x]$, thus $(M[y], y) \preccurlyeq_2 (M[x], x)$. More generally imagine the y's would be further constrained to be multiples of some value, say 2, then the same arguments would work. Hence this equivalence relation allows to absorb a certain number of phenomena that can occur in an automaton and are irrelevant for our specific problem. In particular, if the least growing rate is achieved for words of length n for n even only, then this 'computing modulo 2' can be 'hidden' thanks to the $\approx$-approximation.

The following lemma establishes some essential properties of the $\preccurlyeq_a$ relations (as a consequence, the same properties hold for $\approx_a$).

Lemma 4. *Given A, A', B, B', C sets of weighted matrices and $a, b \geq 1$,*

1. *if $A \preccurlyeq_a B$ and $b \geq a$, then $A \preccurlyeq_b B$,*
2. *if $A \preccurlyeq_a A'$ and $B \preccurlyeq_a B'$, then $A \cup B \preccurlyeq_a A' \cup B'$,*
3. *if $A \preccurlyeq_a B$ and $B \preccurlyeq_b C$ then $A \preccurlyeq_{ab} C$,*
4. *if $A \preccurlyeq_a A'$ and $B \preccurlyeq_a B'$ then $A \otimes B \preccurlyeq_a A' \otimes B'$,*
5. *if $A \preccurlyeq_a B$ then $\langle A \rangle \preccurlyeq_a \langle B \rangle$.*

The Main Induction: The Forest Factorization Theorem of Simon. The forest factorization theorem of Simon [9] is a powerful combinatorial tool for understanding the structure of finite semigroups. In this short abstract, we will not describe the original statement of this theorem, in terms of trees of factorizations, but rather a direct consequence of it which is central in our proof (the presentation of the theorem was used in a similar way in [2]).

Theorem 3 (equivalent to the forest factorization theorem [9]). *Given a semigroup morphism φ from $(S, \otimes)$ (possibly infinite) to a finite semigroup $(T, \odot)$, and some $A \subseteq S$, set $B_0 = A$ and for all $n \geq 0$,*

$$B_{n+1} = B_n \cup B_n \otimes B_n \cup \bigcup_{\substack{e \in T \\ \text{is idempotent}}} \langle B_n \cap \varphi^{-1}(e) \rangle \; ,$$

then $\langle A \rangle = B_N$ for $N = 3|T| - 1$.

This theorem teaches us that, for computing the closure under product in the semigroup S, it is sufficient to be able to know how to compute (a) the union of sets, (b) the product of sets, and (c) the restriction of a set to the inverse image of an idempotent by φ, and (d) the closure under product of sets of elements that all have the same idempotent image under φ. Of course, this proposition is only interesting when the semigroup T is cleverly chosen.

In our case, we are going to use the above proposition with $(S, \otimes) = (\mathcal{W}_k, \otimes)$, and $(T, \odot) = (\mathcal{M}_k(\mathbb{S}), \odot)$, and φ the morphism which maps each weighted matrix (M, x) to $\varphi(M)$. Our algorithm will compute, given a presentation of a set of weighted matrices A, an approximation of $\langle A \rangle$ using the inductive principle of the factorization forest theorem. This is justified by the two following lemmas.

Lemma 5. *For all presentable sets of weighted matrices A, A', there exists effectively a presentable set of weighted matrices $\mathtt{product}(A, A')$ such that*

$$A \otimes A' \approx \mathtt{product}(A, A') \ .$$

Lemma 6. *For all presentable sets A such that $\varphi(A) = \{E\}$ for E an idempotent, there is effectively a presentable set $\mathtt{idempotent}(A)$ such that*

$$\langle A \rangle \approx \mathtt{idempotent}(A) \ .$$

Assuming that Lemmas 5 and 6 hold, it is easy to provide an algorithm which, given a presentable set A computes a presentable set $\mathtt{closure}(A)$ as follows:

– Set $A_0 = A$ and for all $n = 0 \ldots N - 1$ (N taken from Theorem 3), set

$$A_{n+1} = A_n \cup \mathtt{product}(A_n, A_n) \cup \bigcup_{\substack{E \in \mathcal{M}_k(\mathbb{S}) \\ \text{idempotent}}} \mathtt{idempotent}(A_n \cap \varphi^{-1}(E)) \ .$$

– Output $\mathtt{closure}(A) = A_N$.

The correctness of this algorithm is given by the following lemma. It derives from the good properties of $\approx$ given in Lemma 4.

Lemma 7. *For all presentable sets of weighted matrices $\mathtt{closure}(A) \approx \langle A \rangle$.*

This allows us to conclude the proof of Theorem 2. The algorithm takes an automaton δ, I, F as input, then it computes thanks to the above Lemma 7 a presentable set B that is $\approx$-equivalent to $\langle A \rangle$ where A is the set of weighted matrices corresponding to basic letters (i.e., $\{(\delta(a), 1) \ : \ a \text{ letter}\}$). Set (S, P) a presentation of B. Then the algorithm outputs $\inf\{{}^t I \odot X \odot F \mid X \in P\}$ that is computable since P is a finite union of polytopes. This coefficient is the answer of the algorithm: the minimal exponent such that the presentable set witnesses the existence of a behaviour of the automaton that has this growth-rate.

3   From Size-Change Systems to Max-Plus Automata

For proving Theorem 1, we define a translation of SCSs to max-plus automata. Let S be an SCS with k variables, which we assume to be numbered $x_1, \ldots, x_k$. We define an max-plus automaton $\phi(S)$ with $k+2$ states as follows: The alphabet A_S of $\phi(S)$ contains a letter a_T for every SCT $T \in S$. We define the mapping δ of A_S to $\mathcal{M}_{k+2}(\overline{\mathbb{N}})$ as follows:

$$\delta(a_T)_{i,j} = \begin{cases} 0, & i = 1 \text{ or } j = k + 2 \\ 1, & x_{i-1} > x'_{j-1} \in T \\ 0, & x_{i-1} \geq x'_{j-1} \in T \\ -\infty, \text{ otherwise} \end{cases}$$

Further, $\phi(S)$ has the initial and final vector $I = F = \mathbf{0} \in \mathcal{M}_{1,k+2}(\overline{\mathbb{N}})$. For example, the SCS from Example 1 is translated to the max-plus-automaton in Example 2.

The following lemmata relate SCSs and their translations; they allow us to derive Theorem 1 from Theorem 2.

Lemma 8. *Let u be a word of $\phi(S)$ with $^tI \otimes \delta(u) \otimes F = N$. Then S has a trace with valuations over $[0, N]$ of length $|u|$.*

Lemma 9. *Assume S has a trace with valuations over $[0, N]$ of length l. Then there is a word u of $\phi(S)$ with $^tI \otimes \delta(u) \otimes F \leq N$ and $|u| = l$.*

References

1. Ben-Amram, A.M., Vainer, M.: Bounded termination of monotonicity-constraint transition systems. CoRR, abs/1202.4281 (2012)
2. Colcombet, T., Daviaud, L.: Approximate comparison of distance automata. In: STACS, pp. 574–585 (2013)
3. Droste, M., Kuich, W., Vogler, H. (eds.): Handbook of Weighted Automata. Springer (2009)
4. Hashiguchi, K.: Limitedness theorem on finite automata with distance functions. J. Comput. Syst. Sci. 24(2), 233–244 (1982)
5. Krob, D.: The equality problem for rational series with multiplicities in the tropical semiring is undecidable. Internat. J. Algebra Comput. 4(3), 405–425 (1994)
6. Lee, C.S., Jones, N.D., Ben-Amram, A.M.: The size-change principle for program termination. In: POPL, pp. 81–92 (2001)
7. Manolios, P., Vroon, D.: Termination analysis with calling context graphs. In: Ball, T., Jones, R.B. (eds.) CAV 2006. LNCS, vol. 4144, pp. 401–414. Springer, Heidelberg (2006)
8. Schützenberger, M.P.: On the definition of a family of automata. Information and Control 4, 245–270 (1961)
9. Simon, I.: Factorization forests of finite height. Theoretical Computer Science 72, 65–94 (1990)
10. Simon, I.: The nondeterministic complexity of a finite automaton. In: Mots. Lang. Raison. Calc., pp. 384–400. Hermès, Paris (1990)
11. Zuleger, F., Gulwani, S., Sinn, M., Veith, H.: Bound analysis of imperative programs with the size-change abstraction. In: Yahav, E. (ed.) SAS 2011. LNCS, vol. 6887, pp. 280–297. Springer, Heidelberg (2011)

Alternating Vector Addition Systems with States*

Jean-Baptiste Courtois and Sylvain Schmitz

LSV, ENS Cachan & CNRS & INRIA, France

Abstract. Alternating vector addition systems are obtained by equipping vector addition systems with states (VASS) with 'fork' rules, and provide a natural setting for infinite-arena games played over a VASS. Initially introduced in the study of propositional linear logic, they have more recently gathered attention in the guise of *multi-dimensional energy games* for quantitative verification and synthesis.

We show that establishing who is the winner in such a game with a state reachability objective is 2-EXPTIME-complete. As a further application, we show that the same complexity result applies to the problem of whether a VASS is simulated by a finite-state system.

1 Introduction

Vector addition systems with states (VASS) allow to model systems manipulating multiple discrete resources, for instance bank accounts balances or numbers of processes running concurrently. Extending their definition to two-players games is both a very natural endeavour and a tricky problem: the most immediate definition, where both players can freely update the vector values, leads to an undecidable game even with the simplest winning condition, namely (control) state reachability [2].

Facing this difficulty, one might expect to see a flurry of competing definitions for VASS games that would retain decidability through various restrictions. Surprisingly, this is not really the case: if there is indeed a large number of denominations (e.g. *B*-VASS games [16], Z-reachability games [5], multi-dimensional energy games [7]), Abdulla, Mayr, Sangnier, and Sproston [3] noted last year that they all defined essentially the same *asymmetric* class of games, where one player is restricted and cannot update the vector values.

Our contention in this paper is that so many different people coming up independently with the same model is not a coincidence, but a sure sign of a fundamental idea deserving investigation in its own right. We find further arguments in our own initial interest in such games, which comes from the study of simulation problems between Petri nets and finite-state systems [9, 12] where they arise naturally—Abdulla et al. [1] recently made a similar observation. Furthermore the model was already explicitly defined in the '90s in the study of substructural logics [13, 10], and appears implicitly in recent proofs of complexity lower bounds in [8, 4]. We show in this paper that determining the winner of

---

* Work funded in part by the ANR grant 11-BS02-001-01 REACHARD.

E. Csuhaj-Varjú et al. (Eds.): MFCS 2014, Part I, LNCS 8634, pp. 220–231, 2014.
© Springer-Verlag Berlin Heidelberg 2014

an asymmetric VASS game with a state reachability objective is 2-EXPTIME-complete. We extend for this well-known techniques by Rackoff [15] and Lipton [14] used to establish the complexity of VASS problems, see sections 3 and 4. We also provide refined bounds when the dimension of the problem is fixed, and show how to compute the *Pareto frontier* for such games.

Perhaps more importantly than those technical contributions, we single out in Sec. 2 a simple definition for alternation in VASS by way of 'fork' rules (following [13]), for which the complexity analyses of sections 3 and 4 are relatively easy, and establish it as a pivotal definition for VASS games. Indeed, we relate it to energy games in Sec. 5 (following [3]) and to regular simulation problems for VASS in Sec. 6. Our lower bound improves on all the published bounds for those problems, including the EXPSPACE-hardness of simulations between *basic parallel processes* and finite-state processes due to Lasota [12]. Our upper bound applies to the simulation of Petri nets by finite-state systems, for which only decidability was known [9].

Due to page limits, some material is omitted, but can be found in the full version of the paper at http://hal.inria.fr/hal-00980878.

2   Alternating VASS

VASS were originally called 'and-branching' counter machines by Lincoln, Mitchell, Scedrov, and Shankar [13], and were introduced to prove the undecidability of propositional linear logic. Kanovich [10] later identified a fragment of linear logic, called the $(!, \oplus)$-Horn fragment, that captures exactly alternation in VASS, and adopted a game viewpoint. As discussed in sections 5 and 6, this class of systems has since reappeared in other contexts, which motivates its study in earnest.

2.1   Basic Definitions

An *alternating vector addition system with states* (AVASS) is syntactically a tuple $\mathcal{A} = \langle Q, d, T_u, T_f \rangle$ where Q is a finite set of *states*, d is a *dimension* in $\mathbb{N}$, and $T_u \subseteq Q \times \mathbb{Z}^d \times Q$ and $T_f \subseteq Q^3$ are respectively finite sets of *unary* and *fork* rules. We denote unary rules $(q, \mathbf{u}, q_1)$ in T_u with $\mathbf{u}$ in $\mathbb{Z}^d$ by '$q \xrightarrow{\mathbf{u}} q_1$' and fork rules (q, q_1, q_2) in T_f by '$q \rightarrow q_1 \wedge q_2$.' A *vector addition system with states* (VASS) is an AVASS with $T_f = \emptyset$.

Deduction Semantics. Given an AVASS, its semantics is defined by a deduction system over *configurations* $(q, \mathbf{v})$ in $Q \times \mathbb{N}^d$. For rules $q \xrightarrow{\mathbf{u}} q_1$ and $q \rightarrow q_1 \wedge q_2$,

$$\frac{q, \mathbf{v}}{q_1, \mathbf{v} + \mathbf{u}} \text{ (unary)} \qquad \frac{q, \mathbf{v}}{q_1, \mathbf{v} \quad q_2, \mathbf{v}} \text{ (fork)}$$

where '$+$' denotes component-wise addition in $\mathbb{N}^d$, and implicitly $\mathbf{v} + \mathbf{u}$ has no negative component, i.e. is in $\mathbb{N}^d$. When working with finite deduction trees t, we define the *height* $h(t)$ of t as the maximal length among all its branches. A *(multi)-context* C is a finite tree with n distinguished leaves labelled with

n distinct variables $x_1, \ldots, x_n$; $C[t_1, \ldots, t_n]$ then denotes the tree obtained by substituting for each $1 \leq j \leq n$ the tree t_j for the variable x_j.

Game Semantics. The top-down direction of the deduction semantics allows for potentially infinite deduction trees, and defines in a natural way an *asymmetric VASS game* as defined by Kanovich [10] and later by Raskin et al. [16]. Two players, 'Controller' and 'Environment', play over the infinite arena $Q \times \mathbb{N}^d$. In a current configuration $(q, \mathbf{v})$, Controller chooses among the applicable rules in $T_u \cup T_f$. In case of a unary rule $q \xrightarrow{\mathbf{u}} q'$, the next configuration is $(q', \mathbf{v} + \mathbf{u})$, where by assumption $\mathbf{v} + \mathbf{u} \geq \mathbf{0}$ where '$\mathbf{0}$' denotes the null vector in $\mathbb{N}^d$. In case of a fork rule $q \to q_1 \wedge q_2$, Environment then chooses which branch of the deduction tree to explore, i.e. chooses between $(q_1, \mathbf{v})$ and $(q_2, \mathbf{v})$ as the next configuration. Various winning conditions on such plays $(q_0, \mathbf{v}_0), (q_1, \mathbf{v}_1), \ldots$ can then be envisioned, and correspond to conditions that must be satisfied by all the branches of a deduction tree. As shown by Abdulla et al. [3], such asymmetric games are closely related to *multi-dimensional energy games* [7, 5], see Sec. 5.

2.2   Decision Problems and Complexity

We assume when deriving complexity bounds a binary encoding of vectors in $\mathbb{Z}^d$. That is, letting $\|\mathbf{u}\|_\infty \overset{\text{def}}{=} \max_{1 \leq i \leq d} |\mathbf{u}(i)|$ denote the norm of the vector $\mathbf{u}$ and defining $\|T_u\|_\infty \overset{\text{def}}{=} \max_{(q, \mathbf{u}, q') \in T_u} \|\mathbf{u}\|_\infty$, then the size of an AVASS $\langle Q, d, T_u, T_f \rangle$ depends polynomially on the *bitsize* $\log(\|T_u\|_\infty + 1)$. Note that we can reduce by standard techniques all our decision problems to work with a set of unary rules T_u' with effects $\mathbf{u} = \mathbf{e}_i$ or $\mathbf{u} = -\mathbf{e}_i$—where '$\mathbf{e}_i$' is the unit vector with '1' in coordinate i and '0' everywhere else—, but this comes at the expense of an increase in the dimension by a factor of $\log(\|T_u\|_\infty + 1)$.

Reachability. The decision problem that originally motivated the definition of AVASS by Lincoln et al. [13] is *reachability*: given an AVASS $\langle Q, d, T_u, T_f \rangle$ and two states q_r and q_ℓ in Q, does there exist a deduction tree with root labelled by $(q_r, \mathbf{0})$ and every leaf labelled by $(q_\ell, \mathbf{0})$? Equivalently, does Controller have a strategy that ensures that a play starting in $(q_r, \mathbf{0})$ eventually visits $(q_\ell, \mathbf{0})$?

Fact 2.1 (Lincoln et al. **[13]).** *Reachability in AVASS is undecidable.*

State Reachability. Our main problem of interest in this paper is *(control) state reachability* (aka *leaf coverability*): given as before an AVASS $\langle Q, d, T_u, T_f \rangle$ and two states q_r and q_ℓ in Q, we ask now whether there exists a deduction tree with root labelled by $(q_r, \mathbf{0})$ and every leaf label in $\{q_\ell\} \times \mathbb{N}^d$. Equivalently, does Controller have a strategy that ensures that a play starting in $(q_r, \mathbf{0})$ eventually visits $(q_\ell, \mathbf{v})$ for some $\mathbf{v}$ in $\mathbb{N}^d$? We prove in this paper that state reachability is 2-ExpTime-complete, see Thm. 3.1 and Thm. 4.1.

Non-termination. A second problem of interest is *non-termination*: given an AVASS $\langle Q, d, T_u, T_f \rangle$ and an initial state q_r in Q, does there exist a deduction tree where every branch is infinite? Equivalently, does Controller have a strategy to ensure that a play starting in $(q_r, \mathbf{0})$ never stops?

Brázdil, Jančar, and Kučera [5] show in the context of Z-reachability games that this problem is ExpSpace-hard, and in $(d-1)$-ExpTime when the dimension d is fixed. Our 2-ExpTime lower bound in Thm. 4.1 is the best known lower bound for this problem, leaving a large complexity gap.

We discuss a few other decision problems related to energy games in Sec. 5 and to regular VASS simulations in Sec. 6.

3   Complexity Upper Bounds

The state reachability problem asks about the existence of a deduction tree with root $(q_r, \mathbf{0})$ and leaves labels in $\{q_\ell\} \times \mathbb{N}^d$, which describes when using the game semantics a winning strategy for Controller. More generally, we are interested in deduction trees with root label $(q, \mathbf{v})$ and leaves in $\{q_\ell\} \times \mathbb{N}^d$, which we call *witnesses* for $(q, \mathbf{v})$. Let us write $\mathcal{A}, q_\ell \triangleright q, \mathbf{v}$ if such a witness exists in an AVASS $\mathcal{A}$; then the state reachability problem asks whether $\mathcal{A}, q_\ell \triangleright q_r, \mathbf{0}$.

Following Rackoff [15], the main idea to prove a 2-ExpTime upper bound on the state reachability problem is to prove a doubly exponential upper bound on the height of witnesses, by induction on the dimension d; see Sec. 3.1. But let us first make a useful observation: if $\mathcal{A}, q_\ell \triangleright q, \mathbf{v}$ and $(q', \mathbf{v}') \geq (q, \mathbf{v})$ for the product ordering over $Q \times \mathbb{N}^d$, i.e. if $q = q'$ and $\mathbf{v}'(i) \geq \mathbf{v}(i)$ for all $1 \leq i \leq d$, then $\mathcal{A}, q_\ell \triangleright q', \mathbf{v}'$. This means that the set of root labels that ensure reaching q_ℓ is *upward-closed*, and since $(Q \times \mathbb{N}^d, \leq)$ is a well partial order, it has a finite set of minimal elements called its *Pareto frontier*:

$$\text{Pareto}(\mathcal{A}, q_\ell) \stackrel{\text{def}}{=} \min\{(q, \mathbf{v}) \in Q \times \mathbb{N}^d \mid \mathcal{A}, q_\ell \triangleright q, \mathbf{v}\} . \tag{1}$$

We use in Sec. 3.2 the bounds on the size of witnesses to show that Pareto frontiers can be computed in doubly exponential time, which in turn proves:

Theorem 3.1. *State reachability in AVASS is in* 2-ExpTime. *It is in* ExpTime *when the dimension is fixed, and in* PTime *when furthermore the bitsize is fixed.*

Note that the PTime bound in the case of a fixed dimension and fixed bitsize, is not trivial, since it still allows for infinite arenas. In essence it shows one can add a fixed number of counters to a reachability game 'for free.'

3.1   Small Witnesses

Let us fix an instance $\langle \mathcal{A}, q_r, q_\ell \rangle$ of the state reachability problem with $\mathcal{A} = \langle Q, d, T_u, T_f \rangle$ and write $[d] \stackrel{\text{def}}{=} \{1, \ldots, d\}$ for its set of components. For a subset $I \subseteq [d]$ of the components of $\mathcal{A}$, we write $\mathbf{u}_{\restriction I}$ for the projection of a vector $\mathbf{u}$ on I, and define the *projection* $\mathcal{A}_{\restriction I} \stackrel{\text{def}}{=} \langle Q, |I|, T_{u \restriction I}, T_f \rangle$ of $\mathcal{A}$ on I as the AVASS with unary rules $T_{u \restriction I} \stackrel{\text{def}}{=} \{(q, \mathbf{u}_{\restriction I}, q') \mid (q, \mathbf{u}, q') \in T_u\}$. Let $W_I \stackrel{\text{def}}{=} \{(q, \mathbf{v}) \in Q \times \mathbb{N}^{|I|} \mid \mathcal{A}_{\restriction I}, q_\ell \triangleright q, \mathbf{v}\}$ be the set of witness roots in $\mathcal{A}_{\restriction I}$. We are interested in bounding the height $h(t)$ of *minimal* witnesses t in $\mathcal{A}_{\restriction I}$:

$$H_I \stackrel{\text{def}}{=} \sup_{(q, \mathbf{v}) \in W_I} \min\{h(t) \mid t \text{ witnesses } (q, \mathbf{v})\} , \tag{2}$$

where implicitly $H_I = 0$ if no witness exists.

A last remark before we proceed is that, if a label $(q, \mathbf{v})$ appears twice along a branch of a witness t, i.e. if $t = C[C'[t']]$ for some context C, some non-empty context C' with root label $(q, \mathbf{v})$, and tree t' with root label $(q, \mathbf{v})$, then the *shortening* $C[t']$ of t, obtained by replacing $C'[t']$ by t' in t, is also a witness.

Assume that there exists a witness for some root label $(q, \mathbf{v})$. We bound H_I by induction on $|I|$: for the base case where $I = \emptyset$, by repeated shortenings we see that no branch of a minimal witness can have the same state twice, thus

$$H_\emptyset \leq |Q| . \tag{3}$$

Consider now some non-empty set I and a minimal witness t for $(q, \mathbf{v})$. We would like to bound H_I, assuming by induction hypothesis that we are able to bound H_J for all $J \subsetneq I$ by some value $H_{\subsetneq I} = \max_{J \subsetneq I} H_J$. Define for this a large value $B_I \overset{\text{def}}{=} \|T_u\|_\infty \cdot H_{\subsetneq I}$ and consider along each branch of t the first occurrence (starting from the root) of a node with some vector value $\geq B_I$ if one exists. Let n be the number of such first occurrences in t; then t can be written as $C[t_1, \ldots, t_n]$ where C is a context where all the vector values are $< B_I$, and each t_j witnesses $\mathcal{A}_I, q_\ell \rhd q_j, \mathbf{v}_j$ where $\mathbf{v}_j(i_j) \geq B_I$ for some i_j in I.

1. By repeated shortenings, we can bound the height of C by $|Q| \cdot B_I^{|I|}$.
2. For each j, let $I_j \overset{\text{def}}{=} I \setminus \{i_j\}$. Then t_j is also a witness for $\mathcal{A}_{\upharpoonright I_j}, q_\ell \rhd q_j, \mathbf{v}_{j \upharpoonright I_j}$, and we can replace it by a witness t'_j of height at most H_{I_j}. Then t'_j also witnesses $\mathcal{A}_I, q_\ell \rhd q_j, \mathbf{v}_j$ because B_I bounds the maximal total decrease that can occur along a branch of a deduction tree of height H_{I_j}.

Hence $t' \overset{\text{def}}{=} C[t'_1, \ldots, t'_n]$ is a witness for $(q, \mathbf{v})$ and

$$H_I \leq h(t') \leq |Q| \cdot B_I^{|I|} + H_{\subsetneq I} = |Q| \cdot (\|T_u\|_\infty \cdot H_{\subsetneq I})^{|I|} + H_{\subsetneq I} . \tag{4}$$

Combining (3) with (4), we obtain by induction over d that

$$H_{[d]} \leq (|Q| \cdot (\|T_u\|_\infty + 1) + 1)^{(3d)!} . \tag{5}$$

Observe that this bound is doubly exponential in d, but only exponential in the bitsize $\log(\|T_u\|_\infty + 1)$, and polynomial in the number of states $|Q|$.

3.2   Pareto Frontier

Equation (5) yields an algorithm in $\text{AExpSpace} = \text{2-ExpTime}$ to decide given $(q, \mathbf{v})$ in $Q \times \mathbb{N}^d$ whether $\mathcal{A}, q_\ell \rhd q, \mathbf{v}$: it suffices to look for a minimal witness of height at most $H_{[d]}$, and the vector values in such a witness are bounded by $H_{[d]} \cdot \|T_u\|_\infty$.

Furthermore, as observed by Yen and Chen [18], a bound like (5) that does not depend on the initial configuration $(q, \mathbf{v})$ can be exploited to compute the Pareto frontier: if $(q, \mathbf{v})$ belongs to $\text{Pareto}(\mathcal{A}, q_\ell)$, then $\|\mathbf{v}\|_\infty \leq H_{[d]} \cdot \|T_u\|_\infty$. Thus the Pareto frontier can be computed by running the previous algorithm on at most $|Q| \cdot (1 + H_{[d]} \cdot \|T_u\|_\infty)^d$ candidates $(q, \mathbf{v})$:

Proposition 3.2. *Let $\mathcal{A} = \langle Q, d, T_u, T_f \rangle$ be an AVASS and q_ℓ be a state in Q. Then the Pareto frontier $\mathrm{Pareto}(\mathcal{A}, q_\ell)$ can be computed in doubly exponential time. If d is fixed it can be computed in exponential time, and if $\|T_u\|_\infty$ is also fixed it can be computed in polynomial time.*

4 Complexity Lower Bounds

In this section, we match the 2-EXPTIME upper bound of Thm. 3.1 for state reachability in AVASS (Sec. 4.1). Regarding the fixed dimensional cases, we also show in Sec. 4.2 that our EXPTIME upper bound is optimal—note that the case where both the dimension and the bitsize are fixed is trivially PTIME-hard by reduction from the emptiness problem for tree automata. These lower bounds on decision problems also entail that our bounds in Thm. 3.2 for the complexity of computing Pareto frontiers are optimal.

4.1 A General Lower Bound

We extend the classical EXPSPACE-hardness proof of Lipton [14] for state reachability in VASS to the AVASS case. Instead of reducing from the halting problem for Minsky machines with counter valuations bounded by 2^{2^n}, we reduce instead from the same problem for *alternating* Minsky machines.

More precisely, a Minsky machine can be defined as a VASS with additional *zero-test* rules T_z of the form $q \xrightarrow{i?=0} q'$ for $1 \le i \le d$ with deduction semantics

$$\frac{q, \mathbf{v} \quad \mathbf{v}(i) = 0}{q', \mathbf{v}} \text{ (zero-test)}$$

An *alternating* Minsky machine $\langle Q, d, T_u, T_f, T_z \rangle$ can similarly be defined by allowing fork rules. By adapting the usual encoding of Turing machines into Minsky machines to the alternating case, the halting problem for alternating Minsky machines with counter values bounded by 2^{2^n} is hard for AEXPSPACE = 2-EXPTIME. With this in mind, the necessary adaptations of Lipton's reduction are straightforward; see the full paper for details.

Proposition 4.1. *State reachability and non-termination in AVASS are hard for 2-EXPTIME.*

Proposition 4.1 was implicit in the 2-EXPTIME lower bound proofs of [8, 4] for similar questions. Reducing instead from AVASS would simplify these proofs by separating the extension of Lipton's arguments from the actual reduction.

4.2 Fixed Dimension

Similarly to Thm. 4.1, proving an EXPTIME lower bound in the case where the dimension d is fixed is rather easy: Rosier and Yen [17, Thm. 3.1] show indeed that the *boundedness* problem for VASS of dimension $d \ge 4$ is PSPACE-hard by reducing from the acceptance problem in linear bounded automata (LBA). Their proof easily extends to the state reachability and non-termination problems for

VASS, and for AVASS by reducing instead from alternating LBA; see the full paper for details.

Proposition 4.2. *State reachability and non-termination in AVASS of fixed dimension $d \geq 4$ are* ExpTime-*hard.*

5   Energy Games

The asymmetric game semantics described in Sec. 2.1 is easily seen to be equivalent to one-sided VASS games as defined in [16, 3]. Such a game is played on a VASS with a partitioned state space $Q = Q_\Diamond \uplus Q_\Box$, where Controller owns the states in $Q_\Diamond$ and can freely manipulate the current vector value, while Environment owns the states in $Q_\Box$ and can only change the current state: if $q_\Box \xrightarrow{\mathbf{u}} q'$ is a rule in T_u and $q_\Box \in Q_\Box$, then $\mathbf{u} = \mathbf{0}$; these restricted Environment rules correspond to AVASS fork rules.

Abdulla et al. [3] have shown the equivalence of AVASS games with the *(multidimensional) energy games* of Brázdil et al. [5] and Chatterjee et al. [7], where the asymmetry between Controller and Environment is not enforced in the structure of the AVASS or in restricted unary rules for Environment: in such a game, Environment can use arbitrary unary rules. This would lead to an undecidable state reachability game when played on the $Q \times \mathbb{N}^d$ arena [2], but energy games are played instead over $Q \times \mathbb{Z}^d$—which means that unary rules can be applied even if they yield some negative vector components.

Asymmetry appears instead in the winning conditions for Controller. In addition to a winning condition $Win \subseteq Q^\omega \cup Q^*$ on the sequence of states $q_0, q_1, \ldots$ appearing during the play, Controller must also ensure that all the components of the vectors $\mathbf{v}_0, \mathbf{v}_1, \ldots$ remain non-negative (positive in [5]). Such games are motivated by the synthesis of controllers able to ensure that quantitative values (represented by the integer vectors) are maintained above some critical values.

Various regular winning conditions Win can be employed in this setting: the simplest one is (state) reachability, i.e. $Win = Q^*\{q_\ell\}$, which is in 2-ExpTime by Thm. 3.1. Non-termination, i.e. $Win = Q^\omega$, is shown to be in Tower, i.e. iterated exponential time, by Brázdil et al. [5]. Finally, parity is shown decidable by Abdulla et al. [3]. Theorem 4.1 furthermore entails that state reachability and non-termination (and thus parity) multi-dimensional energy games are 2-ExpTime-hard.

6   Regular Simulations

Jančar and Moller [9] proved in 1995 that the two *regular VASS simulation problems* VASS $\preceq$ FS and FS $\preceq$ VASS, which ask whether a VASS is simulated by a finite-state system (FS) and vice versa, are decidable. They relied however on well quasi orders in their proofs and no complexity upper bounds have been published since. Regarding lower bounds, no improvement has appeared in the general case over the easy ExpSpace-hardness one can derive by reductions from the state reachability and non-termination problems for VASS and the lower bounds of Lipton [14] for these. In the particular case where we restrict ourselves to *basic parallel processes* (BPP) instead of VASS, Kučera and Mayr [11] proved that FS $\preceq$ BPP is PSPACE-hard and BPP $\preceq$ FS is co-NPTIME-hard, and both bounds were later improved to ExpSPACE-hardness by Lasota [12].

By presenting reductions to and from the state reachability and non termination problems in AVASS, we improve on all these results:

- BPP $\preceq$ FS and VASS $\preceq$ FS are both 2-ExpTIME-complete by Thm. 4.1 and Thm. 3.1, and
- FS $\preceq$ BPP and FS $\preceq$ VASS are both 2-ExpTIME-hard by Thm. 4.1 and in TOWER by the results of Brázdil et al. [5].

Abdulla et al. [1] independently showed similar connections between on the one hand the (undecidable) simulation problem PDS $\preceq$ VASS between pushdown systems (PDS) and VASS, and on the other hand energy games played on infinite pushdown graphs. They show that these problems become decidable when the PDS has a singleton stack alphabet and the VASS is 1-dimensional.

6.1 Transition Systems and Simulations

Labelled Transition Systems. Operational semantics are often defined through *labelled transition systems* (LTS) $\mathcal{S} = \langle S, \Sigma, \rightarrow \rangle$ where S is a set of states, Σ is a set of actions, and $\rightarrow \subseteq S \times \Sigma \times S$ is a labelled transition relation, with elements denoted by '$s_1 \xrightarrow{a} s_2$.' When S is finite we call $\mathcal{S}$ a *finite-state system* (FS).

For instance, the operational semantics of a VASS $\mathcal{V} = \langle Q, d, T_u \rangle$ along with a labelling $\sigma : T_u \rightarrow \Sigma$ using a set of actions Σ is the LTS $\mathcal{S}_\mathcal{V} \stackrel{\text{def}}{=} \langle Q \times \mathbb{N}^d, \Sigma, \rightarrow \rangle$ with transitions $(q, \mathbf{v}) \xrightarrow{a} (q', \mathbf{v} + \mathbf{u})$ whenever $r = q \xrightarrow{u} q'$ is a unary rule in T_u with label $\sigma(r) = a$ (which we write more simply $q \xrightarrow{u,a} q'$ in the following).

Simulations. Given two LTS $\langle S_1, \Sigma, \rightarrow_1 \rangle$ and $\langle S_2, \Sigma, \rightarrow_2 \rangle$, a *simulation* is a relation $R \subseteq S_1 \times S_2$ such that, whenever (s_1, s_2) belongs to R then for each action a in Σ, if there exists s_1' in S_1 with $s_1 \xrightarrow{a}_1 s_1'$, then there also exists s_2' in S_2 such that $s_2 \xrightarrow{a}_2 s_2'$ and (s_1', s_2') is also in R. A state s_1 is *simulated* by a state s_2, written $s_1 \preceq s_2$, if there exists a simulation R such that (s_1, s_2) is in R.

Simulations can also be characterised by two-player turn-based *simulation games* between 'Spoiler', who wishes to disprove simulation, and 'Duplicator', who aims to establish its existence, played over the arena $S_1 \times S_2$. In a position (s_1, s_2), Spoiler first chooses a transition $s_1 \xrightarrow{a}_1 s_1'$ in $\mathcal{S}_1$, and Duplicator must answer with a transition $s_2 \xrightarrow{a}_2 s_2'$ with the same label a, and the game then proceeds from (s_1', s_2'). A player loses if during one of its turns no suitable transition can be found, otherwise the play is infinite and Duplicator wins. Then $s_1 \preceq s_2$ if and only if Duplicator has a winning strategy starting from (s_1, s_2).

Given two classes of (finitely-presented) systems $\mathbf{A}$ and $\mathbf{B}$, the *simulation problem* $\mathbf{A} \preceq \mathbf{B}$ takes as input two systems A in $\mathbf{A}$ and B in $\mathbf{B}$ with operational semantics $\mathcal{S}_A$ and $\mathcal{S}_B$, and two initial states s_A from $\mathcal{S}_A$ and s_B from $\mathcal{S}_B$, and asks whether $s_A \preceq s_B$. In the following we focus on *regular VASS simulations*, where one of $\mathbf{A}$ and $\mathbf{B}$ is the class of labelled VASS and the other the class FS.

6.2 From Regular VASS Simulations to AVASS

Our two reductions from regular VASS simulations essentially implement the simulation game as an AVASS game. Given a finite set of actions Σ, a labelled VASS defined by $\mathcal{V} = \langle Q, d, T_u \rangle$ and $\sigma: T_u \to \Sigma$, a finite-state system $\mathcal{A} = \langle S, \Sigma, \to_{\mathcal{A}} \rangle$, and a pair of states (q_0, s_0) from $Q \times S$, we construct in both cases a state space $Q' \stackrel{\text{def}}{=} (Q \times S) \uplus (Q \times S \times \Sigma)$ for our AVASS. For convenience we allow forks of arbitrary finite arity $q \to q_1 \wedge \cdots \wedge q_r$.

VASS $\preceq$ FS. We actually reduce in this case from the complement problem VASS $\not\preceq$ FS to AVASS state reachability from (q_0, s_0). Controller plays the role of Spoiler, owns the states in $Q \times S$, and tries to reach the distinguished state q_ℓ. Environment plays the role of Duplicator and owns the states in $Q \times S \times \Sigma$. The rules of the AVASS are then:

$$(q, s) \xrightarrow{\mathbf{u}} (q', s, a) \qquad\qquad \text{whenever } q \xrightarrow{\mathbf{u}, a} q' \in T_u \qquad (6)$$

$$(q', s, a) \to q_\ell \wedge \bigwedge_{s \xrightarrow{a}_{\mathcal{A}} s'} (q', s') . \qquad\qquad (7)$$

Observe that Spoiler has a winning strategy from (q_0, s_0) in the simulation game if and only if it can force Duplicator into a deadlock, i.e. a state s and an action a where no transition $s \xrightarrow{a}_{\mathcal{A}} s'$ exists. This occurs if and only if Environment can be forced into going to q_ℓ in (7) in the AVASS game starting from (q_0, s_0).

Proposition 6.1. *There is a logarithmic space reduction from VASS $\not\preceq$ FS to AVASS state reachability.*

FS $\preceq$ VASS. This direction is actually a particular case of [3, Thm. 5], who show the decidability of *weak simulation* by reducing it to a parity energy game. Environment now plays the role of Spoiler and owns the states in $S \times Q$. Controller now plays the role of Duplicator, owns the states in $S \times Q \times \Sigma$, and attempts to force an infinite play. The rules of the AVASS are then:

$$(s, q) \to \bigwedge_{s \xrightarrow{a}_{\mathcal{A}} s'} (s', q, a) , \qquad\qquad (8)$$

$$(s', q, a) \xrightarrow{\mathbf{u}} (s', q') \qquad\qquad \text{whenever } q \xrightarrow{\mathbf{u}, a} q' \in T_u . \qquad (9)$$

Then, Duplicator has a winning strategy in the simulation game from (q_0, s_0) if and only if Controller has a winning strategy for non-termination in the AVASS game starting in (q_0, s_0):

Proposition 6.2. *There is a logarithmic space reduction from FS $\preceq$ VASS to AVASS non-termination.*

6.3 From AVASS to Regular VASS Simulations

Basic Parallel Processes. As announced at the beginning of the section, we prove our lower bounds on the more restricted BPP rather than VASS. Formally, a *BPP net* is a Petri net $\mathcal{N} = \langle P, T, W \rangle$ where P and T are finite sets of places and

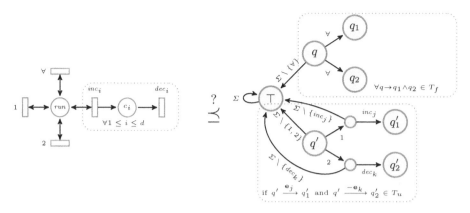

Fig. 1. Reducing AVASS state reachability to a simulation BPP $\npreceq$ FS

transitions and $W: (P \times T) \cup (T \times P) \to \mathbb{N}$ is the weighted flow, where additionally for all transitions t in T there is exactly one place p in P with $W(p, t) = 1$ and for all $p' \neq p$, $W(p', t) = 0$. Given a labelling function $\sigma: T \to \Sigma$, its semantics is defined by the LTS $\mathcal{S}_{\mathcal{N}} \stackrel{\text{def}}{=} \langle \mathbb{N}^{|P|}, \Sigma, \to_{\mathcal{N}} \rangle$ where $m \stackrel{a}{\to}_{\mathcal{N}} m - W(P, t) + W(t, P)$ if and only if there exists t with $\sigma(t) = a$ and $m \geq W(P, t)$. In figures we represent places as circles, transitions as rectangles, and positive flows as arrows.

In both our reductions, we want to implement an AVASS game as a simulation game where the FS is in charge of maintaining the state information and the BPP is in charge of maintaining the vector values. We assume we are given an AVASS $\langle Q, d, T_u, T_f \rangle$ in *ordinary form*, i.e. where the only updates vectors in T_u are unit vectors, and in *binary form*, i.e. for each state q of Q, either there is a fork $q \to q_1 \wedge q_2$ (and we call q an *universal state*), or there are exactly two unary rules $q \stackrel{u_1}{\longrightarrow} q_1$ and $q \stackrel{u_2}{\longrightarrow} q_2$ with origin q (and we call it an *existential state*), or there are no applicable rules at all (and we call it a *deadlock state*). We can ensure these two conditions through logarithmic space reductions. Our action alphabet is then defined as $\Sigma \stackrel{\text{def}}{=} \{\forall, \exists, 1, 2\} \cup \{inc_i, dec_i \mid 1 \leq i \leq d\}$.

BPP $\preceq$ FS. We actually reduce AVASS state reachability to BPP $\npreceq$ FS and assume wlog. that the target state q_ℓ is a deadlock state, and even the only deadlock state by adding rules $q_d \to q_d \wedge q_d$ for the other deadlock states q_d. We construct a BPP net for Spoiler with places $P \stackrel{\text{def}}{=} \{run\} \cup \{c_i \mid 1 \leq i \leq d\}$ where run contains a single token at all times and the c_i's encode the current vector value of the AVASS. Its transitions, labels and flows are depicted on the left of Fig. 1. Its purpose is to force Duplicator, which is playing on the FS depicted on the right of Fig. 1, into state q_ℓ. Because q_ℓ is a deadlock state and Spoiler can always fire transitions (e.g. $\forall$), it then wins the simulation game.

Duplicator plays the role of Environment in the original AVASS game and maintains the AVASS state using its state space, which contains Q. When in a universal state it can choose the following state, but when in an existential state Spoiler chooses instead the branch by firing transition 1 or 2. Duplicator

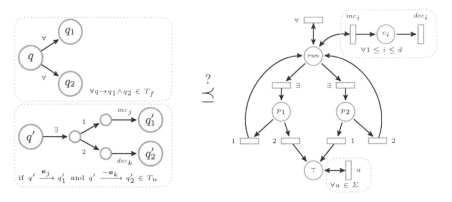

Fig. 2. Reducing AVASS non-termination to a simulation FS $\preceq$ BPP

ensures that the sequence of transitions of Spoiler is indeed valid in the original AVASS, by punishing invalid transitions by entering state '$\top$,' where it can play any symbol and thus win the simulation game.

Proposition 6.3. *There is a logarithmic space reduction from AVASS state reachability to* BPP $\npreceq$ FS.

FS $\preceq$ BPP. In this direction we reduce from the non-termination problem. Spoiler now plays in an FS depicted on the left of Fig. 1 and plays for Environment in the original AVASS game. It still maintains the current state of the AVASS in its state space.

Duplicator now plays on a BPP depicted on the right of Fig. 1. It plays the role of Controller in the original VASS game and maintains the vector values in its places c_i as before. We rely on *Duplicator's choice*: using the '$\exists$' label in existential states, Spoiler leaves the choice to Duplicator, who can punish Spoiler—if it does not comply with its choice between actions '1' and '2'—by putting a token in place '$\top$', from where it wins.

Proposition 6.4. *There is a logarithmic space reduction from AVASS non-termination to* FS $\preceq$ BPP.

7   Concluding Remarks

Alternating VASS provide a unified formalism to reason about VASS games, along with simple complexity arguments for state reachability objectives. This allows us to improve on all the previously known complexity bounds for regular VASS simulations, and show in particular that VASS $\preceq$ FS is 2-ExpTime-complete.

The main open question at this point is whether the upper bounds for non-termination and parity objectives on AVASS could be lowered to 2-ExpTime, and thus to close the gap between 2-ExpTime-hardness and Tower for FS $\preceq$ VASS. A first step to this end could be to extend the PTime upper bound of Chaloupka [6] for the fixed bitsize and unknown initial credit case from dimension two to arbitrary fixed dimensions. However, quoting Chaloupka, 'since the presented results about 2-dimensional VASS are relatively complicated, we suspect this problem is difficult.'

Acknowledgements. The authors thank Stefan Göller who drew our attention to [9, 12] and to the fact that the exact complexities of the two regular simulation problems were unknown. This work also benefited from discussions with Sławomir Lasota, Ranko Lazić, Arnaud Sangnier, and Patrick Totzke.

References

1. Abdulla, P.A., Atig, M.F., Hofman, P., Mayr, R., Kumar, K.N., Totzke, P.: Infinite-state energy games. CSL-LICS 2014. ACM (2014), to appear
2. Abdulla, P.A., Bouajjani, A., d'Orso, J.: Deciding monotonic games. CSL 2003. LNCS, vol. 2803, pp. 1–14. Springer (2003)
3. Abdulla, P.A., Mayr, R., Sangnier, A., Sproston, J.: Solving parity games on integer vectors. Concur 2013. LNCS, vol. 8052, pp. 106–120. Springer (2013)
4. Bérard, B., Haddad, S., Sassolas, M., Sznajder, N.: Concurrent games on VASS with inhibition. Concur 2012. LNCS, vol. 7454, pp. 39–52. Springer (2012)
5. Brázdil, T., Jančar, P., Kučera, A.: Reachability games on extended vector addition systems with states. ICALP 2010. LNCS, vol. 6199, pp. 478–489. Springer (2010)
6. Chaloupka, J.: Z-reachability problem for games on 2-dimensional vector addition systems with states is in P. Fund. Inform. 123(1), 15–42 (2013)
7. Chatterjee, K., Doyen, L., Henzinger, T.A., Raskin, J.F.: Generalized mean-payoff and energy games. FSTTCS 2010. LIPIcs, vol. 8, pp. 505–516. LZI (2010)
8. Demri, S., Jurdziński, M., Lachish, O., Lazić, R.: The covering and boundedness problems for branching vector addition systems. J. Comput. Syst. Sci. 79(1), 23–38 (2012)
9. Jančar, P., Moller, F.: Checking regular properties of Petri nets. Concur '95. LNCS, vol. 962, pp. 348–362. Springer (1995)
10. Kanovich, M.I.: Petri nets, Horn programs, linear logic and vector games. Ann. Pure App. Logic 75(1–2), 107–135 (1995)
11. Kučera, A., Mayr, R.: Simulation preorder over simple process algebras. Inform. and Comput. 173(2), 184–198 (2002)
12. Lasota, S.: EXPSPACE lower bounds for the simulation preorder between a communication-free Petri net and a finite-state system. Inf. Process. Lett. 109(15), 850–855 (2009)
13. Lincoln, P., Mitchell, J., Scedrov, A., Shankar, N.: Decision problems for propositional linear logic. Ann. Pure App. Logic 56(1–3), 239–311 (1992)
14. Lipton, R.: The reachability problem requires exponential space. Tech. Rep. 62, Yale University (1976)
15. Rackoff, C.: The covering and boundedness problems for vector addition systems. Theor. Comput. Sci. 6(2), 223–231 (1978)
16. Raskin, J.F., Samuelides, M., Begin, L.V.: Games for counting abstractions. AVoCS 2004. ENTCS, vol. 128(6), pp. 69–85. Elsevier (2005)
17. Rosier, L., Yen, H.C.: A multiparameter analysis of the boundedness problem for vector addition systems. J. Comput. Syst. Sci. 32(1), 105–135 (1986)
18. Yen, H.C., Chen, C.L.: On minimal elements of upward-closed sets. Theor. Comput. Sci. 410(24–25), 2442–2452 (2009)

Information Rate of Some Classes of Non-regular Languages: An Automata-Theoretic Approach[*]

(Extended Abstract)

Cewei Cui[1], Zhe Dang[1], Thomas R. Fischer[1], and Oscar H. Ibarra[2]

[1] School of Electrical Engineering and Computer Science
Washington State University, Pullman, WA 99164, USA
{ccui,zdang,fischer}@eecs.wsu.edu
[2] Department of Computer Science
University of California, Santa Barbara, CA 93106, USA
ibarra@cs.ucsb.edu

Abstract. We show that the information rate of the language accepted by a reversal-bounded deterministic counter machine is computable. For the nondeterministic case, we provide computable upper bounds. For the class of languages accepted by multi-tape deterministic finite automata, the information rate is computable as well.

1 Introduction

A software system often interacts with its environment. The complexity of an average observable event sequence, or behavior, can be a good indicator of how difficult it is to understand its semantics, test its functionality, etc. This is particularly true considering the fact that a modern software system is often too complex to analyze algorithmically by looking at the source code, line by line. Instead, the system is treated as a black-box whose behaviors can be observed by running it (with provided inputs), i.e., testing. One source to obtain all of the system's intended behaviors is from the design, though whether an intended behavior is the system's actual behavior must still be confirmed through testing. Despite this, the problem of computing the complexity of an average intended behavior from the design is important; in particular, the complexity can be used to estimate the cost of testing, even at the design stage where the code is not available yet.

In principle, a behavior is a word and the design specifies a set of words, i.e., a language L. There has already been a fundamental notion shown below, proposed by Shannon [15] and later Chomsky and Miller [4], that we have evaluated through experiments over C programs [18], fitting our need for the aforementioned

---

[*] The full version of this paper is at http://www.eecs.wsu.edu/~zdang/papers/rateFULL.pdf

E. Csuhaj-Varjú et al. (Eds.): MFCS 2014, Part I, LNCS 8634, pp. 232–243, 2014.
© Springer-Verlag Berlin Heidelberg 2014

complexity. For a number n, we use $S_n(L)$ to denote the number of words in L whose length is n. The *information rate* λ_L of L is defined as

$$\lambda_L = \lim \frac{\log S_n(L)}{n}.$$

When the limit does not exist, we take the upper limit, which always exists for a finite alphabet. Throughout this paper, the logarithm is base 2. The rate is closely related to data compression ratio [7] and hence has immediate practical applications. Information rate is a real number. Hence, as usual, when we say that the rate is computable, it means that we have an algorithm to compute the rate up to any given precision (i.e., first n digits, for any n). A fundamental result is in the following.

Theorem 1. *The information rate of a regular language is computable [4].*

The case when L is non-regular (e.g., L is the external behavior set of a software system containing (unbounded) integer variables like counters and clocks) is more interesting, considering the fact that a complex software system nowadays is almost always of infinite-state and the notion of information rate has been used in software testing [18,19]. However, in such a case, computing the information rate is difficult (sometimes even not computable [10]) in general. Existing results, such as unambiguous context-free languages [11], Lukasiewicz-languages [16], and regular timed languages [2], are limited and mostly rely on Mandelbrot generating functions and the theory of complex/real functions, which are also difficult to generalize.

In this paper, instead of taking the path of Mandelbrot generating functions, we use automata-theoretic approaches to compute the information rate for some classes of non-regular languages, including languages accepted by machines equipped with restricted counters. Our approaches make use of the rich body of techniques in automata theory developed in the past several decades and, as we believe, the approaches themselves can also be applied to more general classes of languages.

We first investigate languages accepted by reversal-bounded nondeterministic counter machines [8]. A counter is a nonnegative integer variable that can be incremented by 1, decremented by 1, or stay unchanged. In addition, a counter can be tested against 0. Let k be a nonnegative integer. A *nondeterministic k-counter machine (NCM)* is a one-way nondeterministic finite automaton, with input alphabet Σ, augmented with k counters. For a nonnegative integer r, we use NCM(k,r) to denote the class of k-counter machines where each counter is r-*reversal-bounded*; i.e., it makes at most r alternations between nondecreasing and nonincreasing modes in any computation; e.g., the following counter value sequence

$$0\ 0\ 1\ 2\ 2\ 3\ \underline{3}\ \underline{2}\ 1\ 0\ \underline{0}\ \underline{1}\ 1$$

is of 2-reversal, where the reversals are underlined. For convenience, we sometimes refer to a machine M in the class as an NCM(k,r). In particular, when k and r are implicitly given, we call M as a *reversal-bounded NCM*. When M is deterministic, we use 'D' in place of 'N'; e.g., DCM. As usual, $L(M)$ denotes the language that M accepts.

Reversal-bounded NCMs have been extensively studied since their introduction in 1978 [8]; many generalizations are identified; e.g., ones equipped with multiple tapes, with two-way tapes, with a stack, etc. In particular, reversal-bounded NCMs have found applications in areas like Alur and Dill's [1] time-automata [6,5], Paun's [14] membrane computing systems [9], and Diophantine equations [17].

In this paper, we show that the information rate of the language L accepted by a reversal-bounded DCM is computable. The proof is quite complex. We first, using automata-theoretic techniques, modify the language into essentially a regular language, specified by an unambiguous regular expression that is without nested Kleene stars, further constrained by a Presburger formula on the symbol counts in the words of the regular language. We show that the information rate of L can be computed through the information rate of the constrained language, where the latter can be reduced to a simple and solvable convex minimization problem. Unfortunately, we are not able to generalize the technique to reversal-bounded NCM. However, it is known [3] that a reversal-bounded NCM can be made to be one with counter values linearly bounded (in input size). Using this fact, we are able to obtain a computable upper bound on the rate when a reversal-bounded NCM is considered. We also consider the case when the reversal-bounded NCM does not make a lot of nondeterministic choices (i.e., sublinear-choice). In this case, the rate is shown computable. The result leads us to study a class of languages accepted by multi-tape DFAs. The information rate of such a multi-tape language is computable as well.

2 Information Rate of Languages Accepted by Reversal-Bounded Counter Machines

We now recall a number of definitions that will be used later. Let N be the set of nonnegative integers and k be a positive number. A subset S of N^k is a linear set if there are vectors $\mathbf{v}_0, \mathbf{v}_1, \cdots, \mathbf{v}_t$, for some t, in N^k such that $S = \{\mathbf{v}|\mathbf{v} = \mathbf{v}_0 + b_1\mathbf{v}_1 + \cdots + b_t\mathbf{v}_t, b_i \in N\}$. S is a semilinear set if it is a finite union of linear sets. Let $\Sigma = \{a_1, \cdots, a_k\}$ be an alphabet. For each word $\alpha \in \Sigma^*$, define the Parikh map [13] of α to be the vector $\#(\alpha) = (\#_{a_1}(\alpha), \cdots, \#_{a_k}(\alpha))$, where each symbol count $\#_{a_i}(\alpha)$ denotes the number of symbol a_i's in α. For a language $L \subseteq \Sigma^*$, the Parikh map of L is the set $\#(L) = \{\#(\alpha) : \alpha \in L\}$. The language L is semilinear if $\#(L)$ is a semilinear set. There is a classic result needed in the paper:

Theorem 2. *Let M be a reversal-bounded NCM. Then $\#(L(M))$ is a semilinear set effectively computable from M [8].*

Let Y be a finite set of integer variables. An atomic Presburger formula on Y is either a linear constraint $\sum_{y \in Y} a_y y < b$, or a mod constraint $x \equiv_d c$, where a_y, b, c and d are integers with $0 \le c < d$. A Presburger formula can always be constructed from atomic Presburger formulas using $\neg$ and $\wedge$. It is known that

Presburger formulas are closed under quantification. Let S be a set of k-tuples in N^k. S is Presburger definable if there is a Presburger formula $P(y_1, \cdots, y_k)$ such that the set of nonnegative integer solutions is exactly S. It is well-known that S is a semilinear set iff S is Presburger definable.

Let M be a reversal-bounded deterministic counter machine. The main result of this paper shows that the information rate of $L(M)$ is computable. The proof has four steps. First, we show that the information rate of $L(M)$ can be computed through the information rate of a counting language (defined in a moment) effectively constructed from M. Second, we show that the information rate of a counting language can be computed through the information rate of a counting replacement language (also defined in a moment) effectively constructed from the counting language. Third, we show that the information rate of a counting replacement language can be computed through the information rate of a simple counting replacement language. Finally, we show that the information rate of a simple counting replacement language is computable.

NOTE: Because of space limitation, we only provide the full proof of Lemma 1 and give proof sketches of other lemmas. Complete proofs will be given in the full paper at `http://www.eecs.wsu.edu/~zdang/papers/rateFULL.pdf`

A *counting* language L is specified by a regular language and a Presburger formula such that L is exactly the set of all words w in the regular language with the Parikh map $\#(w)$ satisfying the Presburger formula. For instance, $\{a^n b^{2n} c^{3n}\}$ and $\{w : \#_a(w) = 2\#_b(w) = 3\#_c(w), w \in (a+b+c)^*\}$ are counting languages.

Lemma 1. *Suppose that M is a reversal-bounded deterministic counter machine. There is a counting language L, effectively constructed from M, such that L and $L(M)$ have the same information rate; i.e., $\lambda_L = \lambda_{L(M)}$.*

Proof. Suppose that M is a DCM(k,r), for some k and r. That is, the counters in M are, say, $x_1, \cdots, x_k$. Without loss of generality, we assume that M starts and accepts with counter values being zero and every counter increments at least once in between. Furthermore, we need only consider the case when $r = 1$ (and every counter makes exactly one reversal). This is because an r-reversal-bounded counter can be easily simulated by $\lceil \frac{r}{2} \rceil$ 1-reversal-bounded counters [8].

We first show that M can be "simulated" by a finite automaton M' as follows. When M runs on an input, every 1-reversal counter x_i in M is simulated by two monotonic (i.e., 0-reversal) counters x_i^+ and x_i^- in M' that counts the number of increments and the number of decrements, respectively, made to x_i. During the run, before the reversal of x_i (M' "knows" the point of reversal), a counter test of "$x_i = 0$?" in M is simulated using M''s finite memory. After the reversal, a counter test of "$x_i = 0$?" in M is always simulated as "no" in M' until M' reads a special symbol $\clubsuit_i$ from the input. After reading this special symbol, the counter test is simulated as "yes" in M' and M' makes sure that there are no further counter increments made to x_i^-. Hence, the special symbol $\clubsuit_i$ resembles the first "time" when x_i decrements to 0 in M. At the end of input, M' accepts when M accepts. Up to now, since the monotonic counters in M' does not contribute to the state transitions in M', M' is indeed a finite automaton.

Notice that M' runs on input w' obtained by inserting k (which is a constant) special symbols $\clubsuit_1, \cdots, \clubsuit_k$ into w. Clearly, if w is accepted by M, then there is a way to insert the k special symbols into w such that the resulting w' is accepted by M'. However, the inverse is not necessarily true. This is because it requires the following Presburger test

$$\bigwedge_{1 \le i \le k} x_i^+ = x_i^- \tag{1}$$

to be successful when M' is at the end of w'. We now use a technique that removes the monotonic counters from M'.

When M' runs on an input, on every input symbol b, M' runs from a state s, reading 0 or more ϵ symbols, and then actually reading the b and, after this, entering a state s'. This is called a round. We use a *round symbol* $[s, b, s']$ to denote the round and use $P_{[s,b,s']}$ to denote the set of vectors of net increments made to the monotonic counters during the round. The proof of the following claim is an exercise which constructs a reversal-bounded NCM to accept unary encoding of vectors in the set and uses Theorem 2:

(Claim 1) $P_{[s,b,s']}$ is a Presburger definable set.

Consider an input word $w' = b_0 \cdots b_{n-1}$ in $L(M')$ for some n. Suppose that, when M' runs on the input, a sequence $[w']$ of rounds is as follows:

$$[s_0, b_0, s_1][s_1, b_1, s_2] \cdots [s_{n-1}, b_{n-1}, s_n] \tag{2}$$

where s_0 is the initial state and s_n is an accepting state. The Presburger test in (1), denoted by $Q(Y)$, where Y is the vector of the $2k$ monotonic counters in M', is equivalent to $Q(\Delta)$, for some

$$\Delta = \Delta_{[s_0,b_0,s_1]} + \Delta_{[s_1,b_1,s_2]} \cdots + \Delta_{[s_{n-1},b_{n-1},s_n]}, \tag{3}$$

where each

$$\Delta_{[s_{j-1},b_{j-1},s_j]} \in P_{[s_{j-1},b_{j-1},s_j]}. \tag{4}$$

That is, the counter values in (1) are the accumulated counter net increments Δ in all the rounds as shown in (3) and hence, the Presburger test in (1) can be performed directly over the Δ.

We now use $\#_{[s,b,s']}$ to denote the number of appearances of the round symbol $[s, b, s']$ in (2) and introduce the notation $\#_{[s,b,s']} \cdot P_{[s,b,s']}$ to denote the set of all the summations of $\#_{[s,b,s']}$ number of (not necessarily distinct) vectors in $P_{[s,b,s']}$. Clearly, the Δ in (3) can be re-written as

$$\Delta = \sum_{\substack{[s,b,s'] \text{ appearing in } (2), \\ \Delta_{[s,b,s']} \in \#_{[s,b,s']} \cdot P_{[s,b,s']}}} \Delta_{[s,b,s']}. \tag{5}$$

We now claim that

(Claim 2) The formula $\Delta_{[s,b,s']} \in \#_{[s,b,s']} \cdot P_{[s,b,s']}$ is Presburger in $\Delta_{[s,b,s']}$ and $\#_{[s,b,s']}$.

The proof of the claim will be shown in a moment. We use $\#$ to denote the vector of the counts $\#_{[s,b,s']}$, noticing that there are totally $|S| \times |\Sigma| \times |S|$ round symbols, where $|S|$ is the number of states in M' and $|\Sigma|$ is the size of its input alphabet. From the claim, the equation in (5) and hence (3) is a Presburger formula in Δ and $\#$, after eliminating quantified variables $\Delta_{[s,b,s']}$'s. We use $\hat{Q}(\Delta, \#)$ to denote the formula. Therefore, the Presburger test is equivalent to the following Presburger formula

$$\exists \Delta . Q(\Delta) \wedge \hat{Q}(\Delta, \#),$$

which is denoted by $\dot{Q}(\#)$.

We now define a language L' to be the set of all round symbol sequences $[w']$ in (2) for all w' accepted by M'. L' is regular. Let L'' be the counting language obtained from words $[w']$ in L' satisfying Presburger formula $\dot{Q}(\#([w']))$.

In summary, we have the following: w is accepted by M iff there is a w' (after inserting the aforementioned k special symbols $\clubsuit_i$'s into w) accepted by M' and the acceptance is witnessed by the round symbol sequence $[w'] \in L''$ shown in (2). Since M is deterministic, the mapping from $w \in L(M)$ to $[w'] \in L''$ is one-to-one (while it is not necessarily true for the mapping from $w \in L(M)$ to $[w'] \in L'$). Notice also that the length of $[w']$ in (2) equals $|w'| = |w| + k$. Because k is a constant, directly from definition, $\lambda_{L(M)} = \lambda_{L''}$. The result follows.

To complete the proof, we still need show Claim 2. We first assume a unary encoding $[\delta]$ for integers δ; e.g., 00000 for -5 and 11111 for +5, where the 0 and 1 are the basis. From Claim 1, $P_{[s,b,s']}$ is therefore a semilinear set. It is known that, for every semilinear set, one can construct a regular language whose Parikh map is exactly the semilinear set. This can be shown directly using the definition of semilinear set; e.g., the semilinear set $\{(1+t,t) : t \geq 0\} \cup \{(2+2t,3t) : t \geq 0\}$ corresponds to the regular language $a(ab)^* + aa(aabbb)^*$. Let $L_{[s,b,s']}$ be a regular language (on alphabet, say, $\{c_1, \cdots, c_{2k}\}$), accepted by an NFA $M_{[s,b,s']}$, corresponding to the semilinear set $P_{[s,b,s']}$. We construct a reversal-bounded NCM $\dot{M}$ as follows. Working on an input

$$[\#_{[s,b,s']}] \Diamond [\delta_1] \Diamond [\delta_2] \Diamond \cdots [\delta_{2k}], \tag{6}$$

$\dot{M}$ uses reversal-bounded counters $y_0, \cdots, y_{2k}$, initially being zero. Again, notice that, on the input, the $2k + 1$ unary encoding blocks use distinct basis. $\dot{M}$ repeatedly simulates the NFA $M_{[s,b,s']}$ from the NFA's initial state to accepting state. On each simulation, $\dot{M}$ guesses an input (one symbol by one symbol) for the NFA. Along with the simulation, $\dot{M}$ uses monotonic counters $y_1, \cdots, y_{2k}$ to count, respectively, the number of $c_1, \cdots, c_{2k}$ that the NFA has read so far. When a simulation ends, $\dot{M}$ increments the counter y_0. After a number of rounds of simulations, $\dot{M}$ nondeterministically decides to shut down the simulation. At this moment, $\dot{M}$ checks that the values $\#_{[s,b,s']}$, $\delta_1, \cdots, \delta_{2k}$ on the input (6) are exactly the same as the values stored in counters $y_0, \cdots, y_{2k}$, respectively. The checking can be done by reading through each unary block of the input while decrementing the counter corresponding to the block to zero.

In this case, $\dot{M}$ accepts the input in (6). Clearly, $\dot{M}$ is indeed reversal-bounded and, $\Delta_{[s,b,s']} \in \#_{[s,b,s']} \cdot P_{[s,b,s']}$ iff the input in (6) is accepted by $\dot{M}$ with $\Delta_{[s,b,s']} = (\delta_1, \cdots, \delta_k)$. Since $L(\dot{M})$ is a semilinear language (Theorem 2), there is a Presburger formula $Q_{[s,b,s']}$ such that the input in (6) is accepted by $\dot{M}$ iff $Q_{[s,b,s']}(\#_{[s,b,s']}, \delta_1, \cdots, \delta_{2k})$. The claim follows since the desired Presburger formula is $Q_{[s,b,s']}(\#_{[s,b,s']}, \Delta_{[s,b,s']})$. $\qquad\square$

The proof of Lemma 1 cannot be generalized to the case where M is a reversal-bounded NCM. This is because, in establishing the one-to-one correspondence in the proof, one requires that M is deterministic.

The second step of the proof for the main theorem is to establish that a counting language can be "converted" into a counting replacement language, which is defined as follows. A *replacement system* G is specified by k levels, for some $k > 0$, where

- for each $1 \le i \le k$, the i-level has a distinct alphabet Σ_i (i.e., $\Sigma_i \cap \Sigma_j = \emptyset$ if $i \ne j$);
- the first level contains a finite set of *base words* w on alphabet Σ_1;
- for each $1 < i \le k$, the i-level contains a finite set of replacement rules in the form of

$$a \leftarrow awa, \tag{7}$$

  where $a \in \Sigma_{i-1}$ and w is a word on alphabet Σ_i;
- all of the words w's mentioned above satisfy the following property: there is no symbol appearing simultaneously in any two such words and, for any symbol, if it appears in w, it appears only once.

When a replacement rule in (7) is applied on a word u, the result is to replace an appearance a in u with awa. A word u is *generated by* G if it is the result of the following sequence of replacements (in this order): starting from a basic word in the first level, we apply replacement rules in the second level for zero or more times, replacement rules in the third level for zero or more times, $\cdots$, replacement rules in the k-level for zero or more times. We use $L(G)$ to denote the set of all u's generated by G. Clearly, $L(G)$ is a regular language. A *regular replacement language* is $L(G)$ for some replacement system G.

For instance, consider a replacement system G with basic word abc, and the second level replacement rules $a \leftarrow adea, a \leftarrow afa, b \leftarrow bgb$, and the third level replacement rules $e \leftarrow ehe$. After applying the second level replacement rules for zero or more times, we obtain words in $a(dea + fa)^*b(gb)^*c$. Subsequently, after applying the third level replacement rules, we finally get $L(G) = a(de(he)^*a + fa)^*b(gb)^*c$.

A *counting replacement language* is a counting language specified by a regular replacement language and a Presburger formula.

Lemma 2. *Suppose that L is a counting language. There is a counting replacement language L', effectively constructed from the specification of L, such that $\lambda_L = 2\lambda_{L'}$.*

The proof of Lemma 2 is intuitively not difficult. Suppose that the counting language L is specified by a regular language L_R and a Presburger formula P. The proof works on the set of accepting runs (i.e., state-symbol sequences) of a DFA M accepting the regular language L_R. The runs have an unambiguous way to decompose back to loops, as specified in a replacement system G. This idea is actually classic, e.g., in the textbook construction from finite automata to regular expressions. Consequently, the Presburger formula needs to be modified after the decomposition.

A replacement system is *simple* if it has at most two levels. A regular replacement language is *simple* if it is generated by a simple replacement system. A counting replacement language is *simple* if it is specified by a simple regular replacement language and a Presburger formula. The third step in the proof for the main theorem establishes that the information rate of a counting replacement language can be computed through the information rate of a simple counting replacement language.

Lemma 3. *Suppose that L is a counting replacement language. There are a constant K and a simple counting replacement language L', constructed from the specification of L, such that $\lambda_L = K \cdot \lambda_{L'}$.*

The proof of Lemma 3 is difficult. We sketch the ideas used in the proof. In the lemma, L is a given counting replacement language. Hence, L can be specified by a regular expression with nested Kleene stars. In order to obtain the desired simple counting replacement language L', we must "collapse" the nested Kleene stars. There is a straightforward way for the collapsing. For instance, $((ab)^*c)^*$ can be converted into $(ab)^*c^+ + \Lambda$, where the latter one does not have any nested Kleene star and the two expressions have the same Parikh map. However, such a straightforward approach, used in computing the Parikh map of a regular language, has a problem in establishing Lemma 3. Using the approach, the resulting simple counting replacement language L' may not preserve the information rate. Therefore, we need a more sophisticated approach that, roughly speaking, keeps both the information rate (up to a constant ratio of K) and the Parikh map. The proof uses a one-to-one encoding that stretches a word in L (this is where the constant K in the lemma comes from) and moves around nested loops in the word so that the resulting word contains at most one level of loops.

The last step in the proof of the main theorem establishes the following.

Lemma 4. *The information rate of a simple counting replacement language is computable.*

The proof of Lemma 4 is a complex reduction from computing the information rate of a simple counting replacement language to solving a convex minimization problem, which is well-known computable. (It really says that the information rate of a simple counting replacement language is the solution of a convex minimization problem.)

Directly from Lemmas 1, 2, 3, and 4, we have the following main theorem.

Theorem 3. *The information rate of the language accepted by a reversal-bounded deterministic counter machine is computable.*

We currently do not have a precise time complexity of computing $\lambda_{L(M)}$ where M is the reversal-bounded DCM in Theorem 3. However, the lower bound of the complexity is $\Omega(2^m)$, where m is the number of states in M. The reason is as follows. Consider an M with no counters and with $m-1$ nested loops, each with length 2. That is, the replacement system G obtained in Lemma 2 is of m levels. Hence, in Lemma 3, after $m-2$ rounds of collapsing, G becomes a replacement system of two levels, where the length of each basic word is at least 2^{m-2}.

Currently, we are not able to generalize the approach used in the above proof to the case when M is nondeterministic. The reason was already mentioned right after Lemma 1. However, when M is nondeterministic, a word in $L(M)$ may correspond to multiple accepting runs. Therefore, λ_L in the statement of Lemma 1 now satisfies

$$\lambda_L \geq \lambda_{L(M)}, \tag{8}$$

and hence, the information rate computed throughout the proof serves as a computable upper bound of $\lambda_{L(M)}$.

Suppose that M is nondeterministic. Consider the one-to-many mapping from input word $w = a_0 \cdots a_{n-1}$ in M to the "run" w' (also of length n). We use $g(n)$ to denote the maximal number of distinct w' one could possibly obtain. Clearly, when M is highly nondeterministic, the number $g(n)$ should be large. Clearly, we have $g(n) \cdot S_n(L(M)) \geq S_n(L)$. Hence, when

$$\lim \frac{\log g(n)}{n} = 0, \tag{9}$$

we have $\lambda_L \leq \lambda_{L(M)}$. Combining (8), we have $\lambda_{L(M)} = \lambda_L$ and hence it is computable when condition (9) holds.

We say that M is $f(n)$-choice if, during every execution of M over input word of length n, M makes at most $f(n)$ nondeterministic choices. M is *sublinear-choice* if M is $f(n)$-choice for some f satisfying $\lim \frac{f(n)}{n} = 0$. Suppose that M contains K instructions (which is a constant). For an input word $w \in L(M)$ of length n, there are at most $K^{f(n)}$ number of accepting executions that witness the fact $w \in L(M)$. Recalling the definition $g(n)$, we have $g(n) \leq K^{f(n)}$, and therefore, when $\lim \frac{f(n)}{n} = 0$, condition (9) holds. Hence,

Theorem 4. *The information rate of the language accepted by a sublinear-choice reversal-bounded nondeterministic counter machine is computable.*

Kuich and Maurer [12] investigate computation of information rate of tuple languages from pseudolinear tuple grammars that, intuitively, cannot generate more than one copy of a nonterminal symbol. It is worth studying the relationship between the notion of "pseudolinear" and our notion of "sublinear-choice", noticing that our notion essentially limits the nondeterminism in a nondeterministic machine.

Though currently it is open whether the information rate of a reversal-bounded nondeterministic counter machine is computable, we can compute one more upper bound as follows. It is known [3] that if M is a reversal-bounded nondeterministic counter machine, we can effectively construct an equivalent reversal-bounded nondeterministic counter machine M' that runs in dn time for some effectively computable constant d. Without loss of generality, we assume that M does not stay; i.e., on a move of M', it either changes a counter value, or reads an input symbol. Again, we can also assume that each counter in M' makes exactly one reversal and M' accepts with all counters being 0. Let w be an input word of length n accepted by M'. That is, there is an accepting run $s_1 b_1 \cdots s_t b_t$ over w, with $t = dn$. In above, each s_i is a state, and each b_i is either a symbol that M' reads from w, or a counter increment or decrement symbol. Even though M' is a nondeterministic machine, the accepting runs can be accepted by a reversal-bounded deterministic counter machine M''. Hence, the information rate $\lambda_{L(M'')}$ is computable. Notice that the accepting run (of length $2dn$), after dropping all states and counter increment or decrement symbols, becomes the input word w (of length n). Immediately, we have $S_{2dn}(L(M'')) \geq S_n(L(M'))$. Hence, $\lambda_{L(M')} \leq 2d\lambda_{L(M'')}$. Recalling that M and M' are equivalent (i.e., accepting the same language), we have $\lambda_{L(M)} \leq 2d\lambda_{L(M'')}$ and the latter is computable.

Currently, it is unclear whether this upper bound or the one obtained in (8) is better.

A 2-tape NFA M is an NFA with two input tapes. If M is a 2-tape NFA, let $T(M) = \{(x,y) : M \text{ on input } (x,y) \text{ accepts }\}$. At each step, the transition of M is of the form $q : (a,b) \to p$, where $a,b \in \Sigma \cup \{\epsilon\}$ (ϵ is the null symbol). The transition means that M in state q reading a and b on the two tapes enters state p. A deterministic 2-tape NFA is denoted by 2-tape DFA.

If M is a 2-tape DFA, let $R(M) = \{xy : (x,y) \in T(M)\}$ and $S(M) = \{x\bar{y} : (x,y) \in T(M)\}$, where $\bar{y}$ is the reverse of string y. We first show that if M is a 2-tape DFA, the information rate of $R(M)$ is computable. We describe an NFA M' accepting a regular language that simulates the 2-tape DFA M. If M uses transition $q : (a,b) \to p$, then M' in state q reads a and enters state s (a new intermediate state), then reads b' (a marked version of b) in state s and enters state p (and thus the symbols in the second tape of M are marked). Clearly, the number of words of length n accepted by M' is "almost" the number of words in $R(M)$ of length n, where "almost" here refers to a ratio of $n+1$ (there are at most $(n+1)$ ways to decode an xy in $R(M)$ back to a pair (x,y) in $T(M)$); i.e.,

$$S_n(R(M)) \leq S_n(L(M')) \leq (n+1) \cdot S_n(R(M)).$$

This immediately gives $\lambda_{L(M')} = \lambda_{R(M)}$. Similarly, one can obtain $\lambda_{L(M')} = \lambda_{S(M)}$. Since the information rate of the language accepted by the NFA M' is computable, we have

Theorem 5. *For 2-tape DFA M, the information rates of $R(M)$ and $S(M)$ are computable.*

The proof ideas can be generalized. Let M be a 2-tape DFA with reversal-bounded counters. As before, $R(M) = \{xy : (x,y) \in T(M)\}$ and $S(M) = \{x\bar{y} : (x,y) \in T(M)\}$. Clearly, we can construct a deterministic reversal-bounded counter machine M' simulating M by interleaving the symbols in the two tapes as before. Since the information rate of the language accepted by a deterministic reversal-bounded machine is computable (Theorem 3), it follows that the information rate of $R(M)$ as well as $S(M)$ is computable.

The above can further be generalized to k-tape DFA with reversal-bounded counters $(k \geq 2)$, where

$$T(M) = \{(x_1, ..., x_k) : M \text{ on input } (x_1, ..., x_k) \text{ accepts}\},$$

and

$$R(M) = \{(x_1^{\mathrm{op}_1} ... x_k^{\mathrm{op}_k}) : (x_1, ..., x_k) \in T(M)\}.$$

where each op_i is either "does nothing" or "reverse x_i" (the choice depends only on R).

From this we can see that there are rather complicated examples of languages for which the information rate is computable. For example, the language $L = \{x\#y\#xy : x, y \in (a + b)^*\}$ has computable information rate, since the set of triples $T = \{(x\#, y\#, xy) : x, y \in (a + b)^*\}$ can be accepted by a 3-tape DFA (even without reversal-bounded counters). Note that L is not even a context-free language.

As we have shown above, the information rate of the language accepted by a multi-tape DFA is computable. In contrast, for multi-head DFAs, we have the following.

A 2-head DFA is a DFA with two one-way heads. The move of the machine depends on the state and the symbols scanned by the two heads. In a move, the machine changes state and moves each head (independently) at most one cell to the right.

Proposition 1. *The information rate of the language accepted by a 2-head DFA is not computable.*

Proof. The proof idea follows Kaminger [10]. It is known (using the undecidability of the halting problem for Turing machines) that the emptiness problem (i.e., is the language accepted empty?) for 2-head DFAs is undecidable. Given a 2-head DFA M, we modify it to a 2-head DFA M' such that $L(M') = \{xw : x \in L(M), w \in (a + b)^*\}$, where a, b are new symbols. M' simply simulates M on x and when M accepts, M' reads w and accepts. Hence $L(M')$ is empty if and only if $L(M)$ is empty, and moreover, $L(M')$ is infinite (and with rate 1) if and only if $L(M)$ is not empty. Then information rate $\lambda_{L(M')} = 0$ if and only if $L(M')$ is empty, which is undecidable. □

3   Conclusions and Discussions

We have shown that the information rate of a language accepted by a reversal-bounded deterministic counter machine is computable. For the nondeterministic case, we have provided computable upper bounds. We also considered the cases when the reversal-bounded NCM is sublinear-choice. For the class of languages accepted by multi-tape DFAs, the information rate is computable as well, as we have shown.

References

1. Alur, R., Dill, D.L.: A theory of timed automata. Theoretical Computer Science 126(2), 183–235 (1994)
2. Asarin, E., Degorre, A.: Volume and entropy of regular timed languages: Discretization approach. In: Bravetti, M., Zavattaro, G. (eds.) CONCUR 2009. LNCS, vol. 5710, pp. 69–83. Springer, Heidelberg (2009)
3. Baker, B.S., Book, R.V.: Reversal-bounded multipushdown machines. J. Comput. Syst. Sci. 8(3), 315–332 (1974)
4. Chomsky, N., Miller, G.A.: Finite state languages. Information and Control 1, 91–112 (1958)
5. Dang, Z.: Pushdown timed automata: A binary reachability characterization and safety verification. Theor. Comput. Sci. 1-3(302), 93–121 (2003)
6. Dang, Z., Ibarra, O.H., Bultan, T., Kemmerer, R.A., Su, J.: Binary reachability analysis of discrete pushdown timed automata. In: Emerson, E.A., Sistla, A.P. (eds.) CAV 2000. LNCS, vol. 1855, pp. 69–84. Springer, Heidelberg (2000)
7. Hansel, G., Perrin, D., Simon, I.: Compression and Entropy. In: Finkel, A., Jantzen, M. (eds.) STACS 1992. LNCS, vol. 577, pp. 513–528. Springer, Heidelberg (1992)
8. Ibarra, O.H.: Reversal-bounded multicounter machines and their decision problems. Journal of the ACM 25(1), 116–133 (1978)
9. Ibarra, O.H., Dang, Z., Egecioglu, O., Saxena, G.: Characterizations of Catalytic Membrane Computing Systems. In: Rovan, B., Vojtáš, P. (eds.) MFCS 2003. LNCS, vol. 2747, pp. 480–489. Springer, Heidelberg (2003)
10. Kaminger, F.P.: The noncomputability of the channel capacity of context-senstitive languages. Inf. Comput. 17(2), 175–182 (1970)
11. Kuich, W.: On the entropy of context-free languages. Information and Control 16(2), 173–200 (1970)
12. Kuich, W., Maurer, H.: The structure generating function and entropy of tuple languages. Information and Control 19(3), 195–203 (1971)
13. Parikh, R.J.: On Context-Free Languages. J. ACM 13(4), 570–581 (1966)
14. Paun, G.: Membrane Computing, An Introduction. Springer (2002)
15. Shannon, C.E., Weaver, W.: The Mathematical Theory of Communication. University of Illinois Press (1949)
16. Staiger, L.: The entropy of Lukasiewicz-languages. In: Kuich, W., Rozenberg, G., Salomaa, A. (eds.) DLT 2001. LNCS, vol. 2295, pp. 155–165. Springer, Heidelberg (2002)
17. Xie, G., Dang, Z., Ibarra, O.H.: A solvable class of quadratic Diophantine equations with applications to verification of infinite state systems. In: Baeten, J.C.M., Lenstra, J.K., Parrow, J., Woeginger, G.J. (eds.) ICALP 2003. LNCS, vol. 2719, pp. 668–680. Springer, Heidelberg (2003)
18. Yang, L., Cui, C., Dang, Z., Fischer, T.R.: An information-theoretic complexity metric for labeled graphs (2011) (in review)
19. Yang, L., Dang, Z., Fischer, T.R.: Information gain of black-box testing. Form. Asp. Comput. 23(4), 513–539 (2011)

Relating Nominal and Higher-Order Rewriting

Jesús Domínguez and Maribel Fernández

Department of Informatics, King's College London, Strand WC2R 2LS, UK
`jesus.dominguez_alvarez@kcl.ac.uk`

Abstract. We present a translation function from nominal rewriting systems (NRSs) to combinatory reduction systems (CRSs), transforming closed nominal rules and ground nominal terms to CRSs rules and terms while preserving the rewriting relation. This result, together with previous translations from CRSs to NRSs and between CRSs and other higher-order rewriting formalisms, opens up the path for a transfer of results between higher-order and nominal rewriting. In particular, techniques and properties of the rewriting relation, such as termination, can be exported from one formalism to the other.

1 Introduction

Programs and logical systems often include binding operators. First-order term rewriting systems [1,2] do not provide support for reasoning on binding structures. This motivated the study of combinations of term rewriting systems with the λ-calculus [3], which offers a notion of variable binding and substitution. Combinatory reduction systems (CRSs) [4,5] are well-known examples of *higher-order rewriting* formalisms, where a meta-language based on the untyped λ-calculus was incorporated to a first-order rewriting framework. Other approaches followed, such as HRSs [6] and ERSs [7,8], for example.

Techniques to prove confluence and termination of higher-order rewriting systems were studied in [9,5,10] amongst others. However, the syntax and type restrictions imposed on rules in these systems have prevented the design of completion procedures for higher-order rewriting systems [11].

More recently, the nominal approach [12,13] has been used to design rewriting systems with support for binding [14]. Nominal rewriting systems do not rely on the λ-calculus, instead, two kinds of variables are used: *atoms*, which can be abstracted but behave similarly to constants, and metalevel variables or just *variables*, which are first-order in that they cannot be abstracted and substitution does not avoid capture of unabstracted atoms. On nominal terms [15,14] α-equivalence is axiomatised using a *freshness relation* between atoms and terms. Nominal syntax enjoys many useful properties, for instance, unification modulo α-equivalence is decidable and unitary [15] and nominal matching is linear [16]. Nominal rewriting can be implemented efficiently if we use *closed* rules.

The availability of efficient algorithms to solve unification problems on nominal terms motivated the study of the relationship between higher-order and nominal syntax in a series of papers [17,18,19]. In this paper, we focus on the relationship

E. Csuhaj-Varjú et al. (Eds.): MFCS 2014, Part I, LNCS 8634, pp. 244–255, 2014.

between nominal and higher-order *rewriting*, specifically, we define a translation function from *standard* NRS rules and ground nominal terms to CRS rules and terms, preserving the rewriting relation between NRSs and CRSs, key to the translation of properties such as confluence and termination. Together with a previous translation from CRSs to NRSs [18], we now have reduction-preserving translations in both directions[1] so that properties and techniques developed for one formalism can be exported to the other (e.g., termination techniques based on the construction of a well-founded reduction ordering). A Haskell implementation of the translation functions and a tool to prove termination using the nominal recursive path ordering [20] are available from [21,22].

Related work. In [23,8] CRSs are compared with HRSs and ERSs respectively, and in [24] CRSs are expressed in terms of the ρ-calculus [25,26]. In [27] a context calculus is described to represent distinct kinds of meta-variables. A comparison of various higher-order formalisms, with many interesting examples, is provided in [28]. In [29] a termination-preserving translation between *Algebraic Functional Systems* and other higher-order formalisms is presented. See also [30] for a concise presentation of higher-order rewrite systems. Although in this paper we focus on the relationship between NRSs and CRSs, thanks to the existing translations between CRSs and other higher-order rewriting formalisms, this is sufficient to obtain a bridge between nominal and higher-order rewriting.

Our work is closely related to the work reported in [17,19]: Cheney [17] represented higher-order unification as nominal unification, and Levy and Villaret [19] transformed nominal unification into higher-order unification, providing a translation that preserves unifiers. Our translation differs from [17,19] since our requirement is to have a mapping of NRS terms and rules to CRS meta-terms and rules in such a way that reductions are preserved.

The rest of the paper is organised as follows. In section 2 we recall CRSs and NRS. In section 3 we describe the translation from nominal terms to CRSs meta-terms, while in section 4 we extend it to take into account rules and substitution. In Section 5 we prove that nominal rewrite steps can be simulated in CRSs via the translation function. Section 6 concludes and discusses future work.

2 Preliminaries

We briefly recall the main concepts of NRSs and CRSs — two rewriting formalisms that extend the syntax of first-order terms and the notion of first-order rewriting, to facilitate the specification of systems with binding operators. We refer the reader to [5,14] for more details and examples.

Nominal Rewriting. A *nominal signature* Σ is a set of term-formers, or *function symbols*, $f, g, \ldots$, each with a fixed arity. We fix a countably infinite set $\mathcal{X}$ of

---

[1] The completeness theorem in [18] states that if $s \xrightarrow[CRS]{} t$ then $s' \xrightarrow[NRS]{*} t'$, where s', t' are the translated terms, but in fact the proof shows $s' \xrightarrow[NRS]{+} t'$.

variables ranged over by $X, Y, Z, \ldots$, and a countably infinite set $\mathcal{A}$ of *atoms* ranged over by $a, b, c, \ldots$, and assume that Σ, $\mathcal{X}$, and $\mathcal{A}$ are pairwise disjoint. A *swapping* is a pair of atoms, written $(a\ b)$. *Permutations* π are bijections on $\mathcal{A}$ such that the set of atoms for which $a \neq \pi(a)$ is finite; this is called the *support* of π, written as $support(\pi)$. Permutations are represented by lists of swappings, Id denotes the *identity permutation*. We write π^{-1} for the inverse of π and $\pi \circ \pi'$ for the composition of π' and π. For example, if $\pi = (a\ b)(b\ c)$ then $\pi^{-1} = (b\ c)(a\ b)$.

Nominal terms, or just *terms*, are generated by the grammar

$$s, t ::= a \mid \pi{\cdot}X \mid [a]s \mid f s \mid (s_1, \ldots, s_n)$$

and called, respectively, atoms, moderated variables or simply variables, abstractions, function applications (which must respect the arity of the function symbol) and tuples (if $n = 0$ or $n = 1$ we may omit the parentheses). We abbreviate $\text{Id}{\cdot}X$ as X if there is no ambiguity. An abstraction $[a]t$ is intended to represent t with a bound. We call occurrences of a *abstracted* if they are in the scope of an abstraction, and *unabstracted* (or free) otherwise. For example, $f([a]X, [b]b)$ is a nominal term, and so is $f(X, (a\ b){\cdot}X)$. For more examples, we refer the reader to [15,14].

We write $V(t)$ (resp. $A(t)$) for the set of variables (resp. atoms) in t (we use the same notation for rules, contexts, etc.). The set $A(t)$ includes the atoms occurring as subterms of t and, in addition, the atoms in abstractions and in the support of permutations occurring in t. In particular, $A((a\ b){\cdot}X) = \{a, b\}$. *Ground terms* have no variables: $V(t) = \emptyset$ if t is ground.

The *action of a permutation* π on a term t, written $\pi{\cdot}t$, is defined by induction: $\text{Id}{\cdot}t = t$ and $(a\ b)\pi{\cdot}t = (a\ b){\cdot}(\pi{\cdot}t)$, where a swapping acts on terms as follows:

$$(a\ b){\cdot}a = b \quad (a\ b){\cdot}b = a \quad (a\ b){\cdot}c = c \quad (c \notin \{a, b\})$$
$$(a\ b){\cdot}(\pi{\cdot}X) = ((a\ b) \circ \pi){\cdot}X \qquad (a\ b){\cdot}[c]t = [(a\ b){\cdot}c](a\ b){\cdot}t$$
$$(a\ b){\cdot}f t = f(a\ b){\cdot}t \qquad (a\ b){\cdot}(t_1, \ldots, t_n) = ((a\ b){\cdot}t_1, \ldots, (a\ b){\cdot}t_n).$$

Substitutions are generated by the grammar: $\sigma ::= \text{Id} \mid [X{\mapsto}s]\sigma$. We use the same notation for the identity substitution and permutation, and also for composition, since there will be no ambiguity. Substitutions act on variables, without avoiding capture of atoms. We write $t\sigma$ for the application of σ on t, defined as follows: $t\text{Id} = t$, $t[X{\mapsto}s]\sigma = (t[X{\mapsto}s])\sigma$, and

$$a[X{\mapsto}s] = a \quad (\pi{\cdot}X)[X{\mapsto}s] = \pi{\cdot}s \quad (\pi{\cdot}Y)[X{\mapsto}s] = \pi{\cdot}Y \ (X \neq Y)$$
$$([a]t)[X{\mapsto}s] = [a](t[X{\mapsto}s]) \quad (f t)[X{\mapsto}s] = f t[X{\mapsto}s]$$
$$(t_1, \ldots, t_n)[X{\mapsto}s] = (t_1[X{\mapsto}s], \ldots, t_n[X{\mapsto}s])$$

Definition 1. *A* freshness *(resp. α-equivalence) constraint is a pair $a\#t$ (resp. $s \approx_\alpha t$) of an atom and a term (resp. terms). A* freshness context *(ranged over by Δ, ∇, Γ), is a set of constraints of the form $a\#X$. Freshness and α-equivalence judgements, written $\Delta \vdash a\#t$ and $\Delta \vdash s \approx_\alpha t$ respectively, are derived using the rules below, where a, b denote different atoms and $ds(\pi, \pi') = \{a \mid \pi{\cdot}a \neq \pi'{\cdot}a\}$ (difference set).*

$$\frac{}{\Delta \vdash a\#b}\;(\#\mathbf{ab}) \qquad \frac{\pi^{-1}\cdot a\#X \in \Delta}{\Delta \vdash a\#\pi\cdot X}\;(\#\mathbf{X}) \qquad \frac{\Delta \vdash a\#s}{\Delta \vdash a\#fs}\;(\#\mathbf{f})$$

$$\frac{\Delta \vdash a\#s_1 \;\cdots\; \Delta \vdash a\#s_n}{\Delta \vdash a\#(s_1,\ldots,s_n)}\;(\#\mathbf{tupl}) \qquad \frac{}{\Delta \vdash a\#[a]s}\;(\#[\mathbf{a}]) \qquad \frac{\Delta \vdash a\#s}{\Delta \vdash a\#[b]s}\;(\#[\mathbf{b}])$$

$$\frac{}{\Delta \vdash a \approx_\alpha a}\;(\approx_\alpha\mathbf{a}) \qquad \frac{\forall a \in ds(\pi,\pi') : a\#X \in \Delta}{\Delta \vdash \pi\cdot X \approx_\alpha \pi'\cdot X}\;(\approx_\alpha\mathbf{X})$$

$$\frac{\Delta \vdash s \approx_\alpha t}{\Delta \vdash fs \approx_\alpha ft}\;(\approx_\alpha\mathbf{f}) \qquad \frac{\Delta \vdash s_1 \approx_\alpha t_1 \;\cdots\; \Delta \vdash s_n \approx_\alpha t_n}{\Delta \vdash (s_1,\ldots,s_n) \approx_\alpha (t_1,\ldots,t_n)}\;(\approx_\alpha\mathbf{tupl})$$

$$\frac{\Delta \vdash s \approx_\alpha t}{\Delta \vdash [a]s \approx_\alpha [a]t}\;(\approx_\alpha[\mathbf{a}]) \qquad \frac{\Delta \vdash (b\ a)\cdot s \approx_\alpha t \quad \Delta \vdash b\#s}{\Delta \vdash [a]s \approx_\alpha [b]t}\;(\approx_\alpha[\mathbf{b}])$$

Let P_i be a freshness or α-equality constraint (for $1 \le i \le n$). We write $\Delta \vdash P_1,\ldots,P_n$ when proofs of $\Delta \vdash P_i$ exist (for $1 \le i \le n$), using the derivation rules above.

Definition 2. A nominal rewrite rule $R = \nabla \vdash l \to r$ is a tuple of a freshness context ∇ and terms l and r such that $V(r) \cup V(\nabla) \subseteq V(l)$.

A nominal rewrite system *(NRS)* is an equivariant set $\mathcal{R}$ of nominal rewrite rules, that is, a set of nominal rules that is closed under permutations. We shall generally equate a set of rewrite rules with its equivariant closure.

Example 1. The following rules are used to compute prenex normal forms in first-order logic. The signature has term-formers forall, exists, not, and. We show only two rules due to space constraints. Intuitively, equivariance means that the choice of atoms in rules is not important (see [14] for more details), therefore we could change a to b (i.e., $(a\ b)$) for instance.

$$a\#P \vdash \mathsf{and}(P, \mathsf{forall}([a]Q)) \to \mathsf{forall}([a]\mathsf{and}(P,Q))$$
$$\vdash \mathsf{not}(\mathsf{exists}([a]Q)) \to \mathsf{forall}([a]\mathsf{not}(Q)).$$

Nominal rewriting [14] operates on 'terms-in-contexts', written $\Delta \vdash s$ or just s when $\Delta = \emptyset$. Below, $C[\ ]$ varies over terms with exactly one occurrence of a distinguished variable Id-, or just -. We write $C[s]$ for $C[-\mapsto s]$, and $\Delta \vdash \nabla\theta$ for $\{\Delta \vdash a\#X\theta \mid a\#X \in \nabla\}$.

Definition 3. A *term* s rewrites with $R = \nabla \vdash l \to r$ to t in Δ, written $\Delta \vdash s \to_R t$ *(as usual, we assume $V(R) \cap V(\Delta,s) = \emptyset$), if $s = C[s']$ and there exists θ such that $\Delta \vdash \nabla\theta$, $\Delta \vdash l\theta \approx_\alpha s'$ and $\Delta \vdash C[r\theta] \approx_\alpha t$. Since Δ does not change during rewriting, a rewriting derivation is written $\Delta \vdash s_1 \to_R s_2 \to_R \cdots \to_R s_n$, abbreviated as $\Delta \vdash s_1 \to^* s_n$.*

When rules are closed, nominal rewriting can be efficiently implemented using nominal matching (there is no need to consider equivariance).

Closed terms are, roughly speaking, terms without unabstracted atoms and such that all the occurrences of any given variable are under compatible abstractions.

Definition 4. *A term-in-context $\Delta \vdash t$ is closed if*

1. *every atom $a \in A(t)$ that occurs as a subterm of t occurs in the scope of an abstraction of a;*
2. *if $\pi \cdot X$ occurs in the scope of an abstraction of $\pi \cdot a$ then any occurrence of $\pi' \cdot X$ is in the scope of an abstraction of $\pi' \cdot a$ or $a \# X \in \Delta$ (that is, if one occurrence of a variable $\pi \cdot X$ captures an atom then either this atom is captured in all occurrences of $\pi' \cdot X$ or it is fresh for X);*
3. *for any pair $\pi_1 \cdot X, \pi_2 \cdot X$ occurring in t, where $a \in ds(\pi_1, \pi_2)$, if a is not in the scope of an abstraction in one of the occurrences (that is, if a is not captured in either $\pi_1 \cdot X$ or $\pi_2 \cdot X$) then $a \# X \in \Delta$.*

A rewrite rule $\nabla \vdash l \to r$ is closed if $\nabla \vdash (l, r)$ is a closed term.

For example, $[a]f(X, a)$ is closed, but $f(X, a)$ and $f(X, [a]X)$ are not, however $a \# X \vdash f(X, [a]X)$ is closed. All the rewrite rules in Example 1 are closed.

Closedness can be easily checked using the nominal matching algorithm [16], as follows. First, given a term in context $\nabla \vdash t$, or more generally, a pair $P = \nabla \vdash (l, r)$ (this could be a rule $R = \nabla \vdash l \to r$), let us write $P^{\shortmid} = \nabla^{\shortmid} \vdash (l^{\shortmid}, r^{\shortmid})$ to denote a *freshened variant* of P, i.e., a version where the atoms and variables have been replaced by 'fresh' ones. We shall always explicitly say what $P^{\shortmid}$ is freshened for when this is not obvious. For example, a freshened version of $(a \# X \vdash f(X) \to X)$ with respect to itself and to $a' \# X \vdash a'$ is $(a'' \# X' \vdash f(X') \to X')$. We will write $A(P') \# V(P)$ to mean that all atoms mentioned in P' are fresh for each of the variables occurring in P. Let $\nabla^{\shortmid} \vdash t^{\shortmid}$ be a freshened version of $\nabla \vdash t$. Then $\nabla \vdash t$ is *closed* if there exists a substitution σ such that $\nabla, A(\nabla^{\shortmid} \vdash t^{\shortmid}) \# V(\nabla \vdash t) \vdash \nabla^{\shortmid} \sigma$ and $\nabla, A(\nabla^{\shortmid} \vdash t^{\shortmid}) \# V(\nabla \vdash t) \vdash t^{\shortmid} \sigma \approx_\alpha t$. A similar check can be done for nominal rewrite rules, or, in general, for tuples $\nabla \vdash (t_1, \ldots, t_n)$.

Combinatory Reduction Systems. A combinatory reduction system [4,5] is a pair consisting of an alphabet $\mathcal{A}$ and a set of rewrite rules.

The *alphabet* consists of: variables $a, b, c, \ldots$; meta-variables with fixed arities, written as Z_i^n where n is the arity of Z_i^n (n is omitted when there is no ambiguity); function symbols $f, g, \ldots$ with fixed arities; and an abstraction operator $[\cdot] \cdot$. Only variables can be abstracted. We write $MV(t)$ (resp. $Var(t)$) for the set of meta-variables (resp. variables) occurring in a term t (the same notation is used for rules, etc.).

In CRSs a distinction is made between *meta-terms* and *terms*. Meta-terms are the expressions built from the symbols in the alphabet, in the usual way. Variables that are in the scope of the abstraction operator are *bound*, and *free* otherwise. Meta-terms are defined modulo renaming of bound variables (i.e., a meta-term represents an α-equivalence class). Terms are meta-terms that do not contain meta-variables, and are also defined modulo α-equivalence.

A (meta-)term is closed if every variable occurrence is bound. CRSs adopt the usual naming conventions (also known as Barendregt's variable conventions): in particular, all bound variables are chosen to be different from the free variables.

A *rewrite rule* is a pair of meta-terms, written $l \Rightarrow r$, where l, r are closed, l has the form $f(s_1, \ldots, s_n)$ where $n \geq 0$ (when $n = 0$ we omit the parentheses), $MV(r) \subseteq MV(l)$, and $MV(l)$ occur only in the form $Z_i^n(a_1, \ldots, a_n)$, where $a_1, \ldots, a_n$ are pairwise distinct bound variables. We call this form a *meta-application*.

Example 2. The β-reduction rule for the λ-calculus is written:

$$\mathsf{app}(\mathsf{lam}([a]Z(a)), Z') \Rightarrow Z(Z')$$

where Z is a unary meta-variable and Z' is 0-ary.

The reduction relation is defined on terms. To generate the rewrite relation, each meta-variable in a rule is replaced by a special kind of λ-term, and in the obtained term all β-redexes and the residuals of these β-redexes are reduced (i.e. a complete development is performed). Formally, the rewrite relation is defined using *substitutes* and *valuations*. An n-ary substitute is an expression of the form $\underline{\lambda}a_1 \ldots a_n.t$, where t is a term and $a_1, \ldots, a_n$ are different variables. An n-ary substitute can be applied to a n-tuple $s_1, \ldots, s_n$ of terms, and the result is the term t where $a_1, \ldots, a_n$ are simultaneously replaced by $s_1, \ldots, s_n$. A valuation σ is a map that assigns an n-ary substitute to each n-ary meta-variable. It is extended to a mapping from meta-terms to terms: given a valuation σ and a meta-term t, first we replace in t all meta-variables by their images in σ and then we perform the developments of the β-redexes created.

A context is a term with an occurrence of a special symbol $[\,]$ called hole. A rewrite step is now defined in the usual way: if $l \Rightarrow r$ is a rewrite rule, σ a valuation and $C[\,]$ a context, then $C[l\sigma] \Rightarrow C[r\sigma]$.

Example 3. The following is a rewrite step using as example the β-rule given in Example 2:

$$\mathsf{app}(\mathsf{lam}([a]f(a,a)), t) \Rightarrow_\beta f(t,t)$$

To generate it we use the valuation σ that maps Z to $\underline{\lambda}(b).f(b,b)$ and Z' to the term t. Then $\mathsf{app}(\mathsf{lam}([a]Z(a)), Z')\sigma$ is the term $\mathsf{app}(\mathsf{lam}([a]f(a,a)), t)$ obtained by first replacing Z and Z' as indicated by σ and then reducing the β-redex $(\underline{\lambda}(b).f(b,b))(a)$. Also, $Z(Z')\sigma$ is the term $f(t,t)$ obtained by replacing Z and Z', which gives $(\underline{\lambda}(b).f(b,b))(t)$, and then β-reducing for $\underline{\lambda}$-.

3   Translating Nominal Terms

We start by designing a translation function for nominal terms; we deal with nominal rewrite rules in the next section.

For each nominal signature Σ, and sets $\mathcal{A}$ and $\mathcal{X}$ of atoms and variables, we consider a CRS alphabet containing Σ, variables $\mathcal{A}$ and meta-variables $\mathcal{X}$.

First we define an auxiliary function: Λ, to compute, for each variable occurring in a nominal term, the set of atoms that may be captured when a variable is instantiated.

Definition 5 (Mapping Λ_t). *For each nominal term t, we define $\Lambda_t : \mathcal{X} \to \mathcal{P}(\mathcal{A})$ such that $\Lambda_t(X) = \{a_1, \ldots, a_n\}$ if $X \in V(t)$ has k occurrences in t, A_i is the set of atoms abstracted above the ith occurrence of X, and $\{a_1, \ldots, a_n\} = A_1 \cup \ldots \cup A_k$. In other words, $\Lambda_t(X)$ is the set of all the atoms abstracted above occurrences of X in t. We omit the inductive definition.*

Definition 6 (Term Translation). *Let $\Delta \vdash t$ be a nominal term-in-context and Λ_t as in Definition 5. Then $\mathcal{T}(\Delta, t) = [\![t]\!]^{\Delta}_{\Lambda_t}$, where $[\![\cdot]\!]^{\Delta}_{\Lambda_t}$ is an auxiliary function defined by induction over the structure of nominal terms as follows:*

$$
\begin{aligned}
&\textbf{(atom)} \quad [\![a]\!]^{\Delta}_{\Lambda_t} &&= a, \\
&\textbf{(var)} \quad [\![\pi \cdot X]\!]^{\Delta}_{\Lambda_t} &&= X(\overline{xs}) \text{ where} \\
& && \overline{xs} \triangleq \pi \cdot xs \text{ (we omit } (\overline{xs}) \text{ if empty)} \\
& && xs \triangleq toAscList^2 \left([\pi^{-1} \cdot \Lambda_t(X)] - \{a \mid a \# X \in \Delta\}\right) \\
&\textbf{(abs)} \quad [\![[a]s]\!]^{\Delta}_{\Lambda_t} &&= [a][\![s]\!]^{\Delta}_{\Lambda_t}, \\
&\textbf{(fun)} \quad [\![f s]\!]^{\Delta}_{\Lambda_t} &&= f[\![s]\!]^{\Delta}_{\Lambda_t}, \\
&\textbf{(tuple)} \quad [\![(s_1, \ldots, s_n)]\!]^{\Delta}_{\Lambda_t} &&= ([\![s_1]\!]^{\Delta}_{\Lambda_t}, \ldots, [\![s_n]\!]^{\Delta}_{\Lambda_t}).
\end{aligned}
$$

Example 4. The nominal term $\vdash [a][b]X$ is translated as the CRS meta-term $[a][b]X(a, b)$ including both variables in the meta-application as they may appear free in a substitution $\sigma(X)$. Failure to include them could lead to a renaming of the bound variables in the meta-term to avoid variable-capture, as defined in [5]. This would subsequently lead to a translation disassociated from its original input. Freshness constraints also have to be taken into account. For example, the term $a \# X \vdash [a][b]X$ is translated as the meta-term $[a][b]X(b)$, where the atom a is not included in the meta-application. However, a freshness constraint does not always discard an atom from inclusion in the list of variable arguments. This depends on permutations. We adjust our example to show this. Consider $a \# X \vdash [a][b](a\ b) \cdot X$. In this case we want to translate the mapping $b \mapsto a$ but not $a \mapsto b$ since $a \# X \in \Delta$. Our translation outputs the meta-term $[a][b]X(a)$, which seems to suggest that a may occur free in $\sigma(X)$ contradicting the constraint $a \# X \in \Delta$. However, since $\sigma(X)$ satisfies Δ, the atom a cannot be free in $\sigma(X)$ or in its CRS translation (see Definition 9). Hence the mapping $a \mapsto b$ is discarded.

Example 5. The nominal term $\vdash [a][b](a\ c) \cdot X$ includes a mapping $a \mapsto c \in \pi$ from an abstracted atom a to an unabstracted atom c. Our algorithm is designed to translate NRSs into CRSs, which are closed by definition. Accordingly, this particular kind of mapping cannot be explicitly represented at term level thus its application is shifted to the substitute that instantiates X, if any. This method is shown in detail in section 4 when describing the translation of rules and substitutions. Our translation function for terms produces the meta-term $[a][b]X(b, a)$ which effectively takes into account the rest of the mappings in π, in this case $b \mapsto b, c \mapsto a$, generating a closed CRS meta-term.

*Property 1 (**Equivalence Relation**).* Let $\Delta \vdash t$ be a closed term-in-context and $\mathcal{T}(\Delta, t) = \hat{t}$ its CRS translation. If $\pi_1 \cdot X$ and $\pi_2 \cdot X$ are two occurrences of

---

² List of atoms in ascending lexical order. Any total ordering can be used.

the same variable X in t, and $X(\overline{xs_1})$, $X(\overline{xs_2})$ are their translations in $\hat{t}$, then $\pi_1^{-1} \cdot \overline{xs_1} \equiv \pi_2^{-1} \cdot \overline{xs_2}$.

Proof. It is sufficient to prove that, for any atom a if $a \in xs_1$ at position i then $a \in xs_2$ at position j such that $i = j$ by application of *toAscList*, else $a\#X \in \Delta$ by definition of a closed term (see Definition 4). We omit the proof. $\square$

Next we prove that the translation function produces CRS (meta-)terms where the arity of each variable is correctly enforced.

Property 2 (**Preserving Closedness of Translated Terms**).

(a) If $\Delta \vdash t$ is a closed nominal term then its CRS translation according to Definition 6 is a closed CRS meta-term.
(b) If the nominal term t is ground, then its translation is a CRS term.

Proof. This is due to the translation respecting the structure of t. It follows from Property 1 that every meta-application $X^n(\overline{xs})$ respects the arity $n = |\overline{xs}|$ for all occurrences of X in its translation $\hat{t}$. Since every variable $a \in \overline{xs}$ also exists in $\Lambda_t(X)$, thus bound, the term is closed. $\square$

4   Transforming NRS Rules

NRS rules are more general than CRS rules in that free atoms may occur in rules. In this section, we impose some conditions on NRS rules to obtain a class of rules that can be translated to CRS rules.

Definition 7 (Standard Nominal Rule). *A nominal rule is called* standard *if it is closed and the left-hand side has the form $f s$.*

Definition 8 (Rule Translation). *Let $R \equiv \nabla \vdash l \to r$ be a standard nominal rule. We define the translation of R as $\mathcal{T}^{\mathcal{R}}(\nabla, l, r) = \mathcal{T}(\nabla, l) \Rightarrow \mathcal{T}(\nabla, r)$ where $\mathcal{T}(\Delta, t)$ is given in Definition 6.*

Note that if a nominal rule $\nabla \vdash l \to r$ is closed (i.e., $\nabla \vdash (l, r)$ is closed), then $\nabla \vdash l$ and $\nabla \vdash r$ are both closed terms.

Lemma 1 (Well-Defined Rule Translation). *Let $R \equiv \nabla \vdash l \to r$ be a standard nominal rule. If $R' \equiv \hat{l} \Rightarrow \hat{r}$ is its translation according to Definition 8, then R' is a CRS rule.*

Proof. The proof is straight-forward. $\square$

Example 6. The (closed) nominal rules to compute prenex normal forms (see Example 1) can be translated to CRS rules by application of our algorithm. We show the CRS translation for the first rule, computed by our Haskell implementation (see [21]): $\mathsf{and}(P, \mathsf{forall}([a]Q(a))) \Rightarrow \mathsf{forall}([a]\mathsf{and}(P, Q(a)))$.

5   Simulating Nominal Rewrite Steps

Translation of a rewrite relation is not as straight-forward as one could expect. The rewriting relation generated by a set of CRS rules is defined on terms, not on meta-terms. In order to preserve the rewriting relation, we need to consider only ground nominal substitutions. Moreover permutations will be applied to substitutions in the translation, in order to preserve the meaning of the term. For this reason, we will define a translation function for pairs of a term-in-context $\Delta \vdash t$ and a substitution σ. In other words, some permutations will be dealt with by applying them directly to the substitution before translation. These correspond to mappings from atoms to free atoms occurring in the term. Take for instance the example ($\vdash (a\ b)\cdot X$, $[X \mapsto f(a, b)]$) The term $\vdash (a\ b)\cdot X$ is trivially closed (no free atoms occur in the term and there is only one variable). The CRS translation given in Definition 6 for nominal terms and Definition 9 for nominal substitution, given below, produce the pair ($X, [X \mapsto f(b, a)]$) where the permutation $(a\ b)$ has been applied directly to the substitute. Accordingly, the list of bindings added to σ (possibly empty, as in the above example) must also be modified to preserve the binding structure after application of π to σ. Further examples are considered after Definition 9.

Definition 9 (Substitution Translation). *Let $\Delta \vdash t$ be a closed nominal term-in-context, Λ_t as in Definition 5, and σ a nominal substitution satisfying Δ, such that $\sigma = [X_i \mapsto t_i]$, $1 \leq i \leq n$ where $dom(\sigma) \subseteq V(t)$ and $t\sigma$ is ground.*
Then $\mathcal{T}^f(\Delta, t, \sigma) = [X_i \mapsto \underline{\lambda}(\overline{xs}_i).s_i]$ where $\overline{xs}_i$ and s_i are defined as follows. Let π_i be the permutation suspended in the leftmost occurrence of X_i in t. Then

- $\overline{xs}_i \stackrel{\Delta}{=} \pi_i \cdot xs_i,$
- $xs_i \stackrel{\Delta}{=} toAscList([\pi_i^{-1} \cdot \Lambda_t(X_i)] - \{a \mid a\#X_i \in \Delta\}),$
- $s_i \stackrel{\Delta}{=} \mathcal{T}(\Delta, \pi_i \cdot t_i).$

We denote by $(\hat{t}, \hat{\sigma})$ the result of $(\mathcal{T}(\Delta, t), \mathcal{T}^f(\Delta, t, \sigma))$.

Lemma 2 justifies the use of the leftmost occurrence of $\pi \cdot X$ in t in Definition 9.

Example 7. Consider the nominal term-in-context and substitution:
$(a, c\#X \ \vdash \ g([a][b][c](a\ d)(e\ f)\cdot X, [a][b][c]((c\ d)(e\ f)\cdot X), [X \mapsto f(b, d, e, f)])$.
The term translation function produces a CRS meta-term $g([a][b][c]X(b, a),$ $[a][b][c]X(b, c))$ and the substitution translation produces the corresponding substitutes $[X \mapsto \underline{\lambda}(b, a).f(b, a, f, e)]$ and $[X \mapsto \underline{\lambda}(b, c).f(b, c, f, e)]$ associated to each occurrence of X. Note that these substitutes are α-equivalent. Hence the algorithm opts for the leftmost $[X \mapsto \underline{\lambda}(b, a).f(b, a, f, e)]$.
Moreover, applying the lexical ordering directly to $\overline{xs}$ instead of xs would produce substitutes which are no longer α-equivalent, providing incorrect instantiations.

Lemma 2 (α-equivalence of Substitutes). *Let $\Delta \vdash t$ be a closed nominal term-in-context, Λ_t as defined in Definition 5, and σ a nominal substitution satisfying Δ such that $dom(\sigma) \subseteq V(t)$ and $t\sigma$ is ground. Let $\pi_i \cdot X$, $\pi_j \cdot X$ be two occurrences of the same variable in t, and let $[X \mapsto \underline{\lambda}(\overline{xs}_i).s_i]$ and $[X \mapsto \underline{\lambda}(\overline{xs}_j).s_j]$*

be translations according to Definition 9 but using π_i and π_j respectively. Then $[X \mapsto \underline{\lambda}(\overline{xs}_i).s_i] \approx_\alpha [X \mapsto \underline{\lambda}(\overline{xs}_j).s_j]$.

Proof. It easily follows from Property 1 and Definition 4. □

Lemma 3 (Instantiation). *Let* $\Delta \vdash t$ *be a closed nominal term-in-context,* Λ_t *as defined in Definition 5, and* σ *a substitution satisfying* Δ *such that* $dom(\sigma) \subseteq V(t)$ *and* $t\sigma$ *is ground.*
   Assume $(\llbracket t' \rrbracket^\Delta_{\Lambda_t}, \mathcal{T}^f(\Delta, t, \sigma)) = (\hat{t}', \hat{\sigma})$, *where* t' *is any subterm of* t *(e.g.* $t' \equiv t$*). Then* $\llbracket t'\sigma \rrbracket^\Delta_{\Lambda_t} = \hat{t}'\hat{\sigma}$.

Proof. The result follows from Lemma 2 and Property 1. □

Note that $C[\,]$ is a term, as explained in the paragraph above Definition 3, and is translated to $\hat{C}[\,]$ using Definition 6.
   We can now derive the main result of the paper: the preservation of the rewrite relation under the translation.

Theorem 1 (Rewrite Step Translation). *Let* $R \equiv \nabla \vdash l \to r$ *be a standard nominal rule. Let* t *be a ground nominal term and* $\hat{t} = \mathcal{T}(\emptyset, t)$.
   If $t \to_R u$ *then there exists* $\hat{u}$ *such that* $\hat{t} \Rightarrow_{R'} \hat{u}$ *using* $R' \equiv \mathcal{T}^\mathcal{R}(\nabla, l, r)$, *and* $\hat{u} = \mathcal{T}(\emptyset, u)$.

Proof. If $t \to_R u$ then there exists C, σ such that $t \approx_\alpha C[l\sigma]$ with σ a ground nominal substitution satisfying ∇ such that $dom(\sigma) \subseteq V(l)$.
   Also $R' \equiv \mathcal{T}^\mathcal{R}(\nabla, l, r) \equiv \llbracket l \rrbracket^\nabla_{\Lambda_l} \Rightarrow \llbracket r \rrbracket^\nabla_{\Lambda_r} = \hat{l} \Rightarrow \hat{r}$ by Definition 8, where Lemma 1 asserts that the translation is a CRS rule.
If we have, by application of Definition 9, $\mathcal{T}^f(\nabla, l, \sigma) = \hat{\sigma}_l$ then, by Lemma 3 $\llbracket l\sigma \rrbracket^\nabla_{\Lambda_l} = \hat{l}\hat{\sigma}_l$. Hence we have $\hat{t} = \hat{C}[\hat{l}\hat{\sigma}_l]$.
   Similarly, since $u \equiv C[r\sigma]$ we have $\mathcal{T}^f(\nabla, r, \sigma) = \hat{\sigma}_r$, leading to $\hat{u} = \hat{C}[\hat{r}\hat{\sigma}_r]$ by application of Definition 9, followed by Lemma 3. Notice that $dom(\hat{\sigma}_r) \subseteq dom(\hat{\sigma}_l)$ and $\hat{\sigma}_r(\hat{X}) \approx_\alpha \hat{\sigma}_l(\hat{X})$ by Lemma 2.
   Hence we conclude by stating that if $l\sigma \to_R r\sigma$ then $\hat{t} \Rightarrow_{R'} \hat{u}$ as expected. □

Corollary 1 (Termination). [3] *Termination of the translated CRS implies termination of the NRS.* □

6 Conclusions and Future Work

We have shown two extensions of first-order rewriting, CRSs and NRSs, to be closely related. We have shown that despite their differences in the meta-language, it is possible to translate between these formalisms. We have given a translation function which transforms the class of closed NRSs into CRSs. We have shown some non-trivial examples to support our work, as well as an implementation in Haskell for our translation function. Although previous work has

---

[3] The corollary also holds in the other direction (see [18]).

been done on translating nominal syntax to higher-order syntax [19] and back to NRSs [17], our work differs from [19] by focusing on a syntax-directed mapping of NRS terms to meta-terms, extended to rules and preserving the rewriting relation, which is key to the translation of properties such as confluence and termination. Since there is also a translation from CRSs to NRSs [18], we now we have a mechanism to export results on termination of rewriting from one framework to the other. Nominal terms have good algorithmic properties, which suggests that we could translate CRSs to NRSs in order to take advantage of existing nominal procedures (i.e. orderings, completion) then transfer back results. This could lead to procedures of nominal systems being adapted to suit CRSs or creation of new procedures by combination of existing ones from both formalisms. Nominal typing systems could also be adapted to the (untyped) CRSs. This is left for future work.

Acknowledgements. We thank Elliot Fairweather and Christian Urban for many helpful discussions, and Jamie Gabbay for providing the macro for Ⅵ.

References

1. Baader, F., Nipkow, T.: Term Rewriting and all that. CUP (1988)
2. Dershowitz, N.: Terese: Term Rewrite Systems. CAMTCS, vol. 55. CUP (2003)
3. Barendregt, H.P.: The Lambda Calculus, its Syntax and Semantics, Revised edn. Studies in Log. and the Found. of Math. North-Holland, vol. 103 (1984)
4. Klop, J.W.: Combinatory reduction systems. PhD thesis, Utrecht University (1980)
5. Klop, J.W., van Oostrom, V., van Raamsdonk, F.: Combinatory reduction systems: Introduction and survey. Theor. Comp. Sci. 121, 279–308 (1993)
6. Nipkow, T.: Higher-order critical pairs. In: LICS, pp, 342–349 (1991)
7. Khasidashvili, Z.: Expression reduction systems. In: Proc. of I. Vekua Institute of Applied Mathematics, vol. 36, pp. 200–220 (1990)
8. Glauert, J., Kesner, D., Khasidashvili, Z.: Expression reduction systems and extensions: An overview. In: Middeldorp, A., van Oostrom, V., van Raamsdonk, F., de Vrijer, R. (eds.) Processes... (Klop Festschrift). LNCS, vol. 3838, pp. 496–553. Springer, Heidelberg (2005)
9. Mayr, R., Nipkow, T.: Higher-order rewrite systems and their confluence. Theor. Comp. Sci. 192(1), 3–29 (1998)
10. Hamana, M.: Semantic labelling for proving termination of combinatory reduction systems. In: Escobar, S. (ed.) WFLP 2009. LNCS, vol. 5979, pp. 62–78. Springer, Heidelberg (2010)
11. Nipkow, T., Prehofer, C.: Higher-order rewriting and equational reasoning. In: Bibel, W., Schmitt, P. (eds.) Automated Deduction — A Basis for Applications. Volume I: Foundations. J. of App. Log., vol. 8, pp. 399–430. Kluwer (1998)
12. Gabbay, M.J., Pitts, A.M.: A new approach to abstract syntax with variable binding. Formal Aspects of Computing 13(3-5), 341–363 (2002)
13. Pitts, A.M.: Nominal logic, a first order theory of names and binding. Information and Computation 186, 165–193 (2003)
14. Fernández, M., Gabbay, M.J.: Nominal rewriting. Inf. Comput. 205(6), 917–965 (2007)

15. Urban, C., Pitts, A.M., Gabbay, M.J.: Nominal unification. Theor. Comp. Sci. 323(13), 473–497 (2004)
16. Calvès, C., Fernández, M.: Matching and alpha-equivalence check for nominal terms. J. of Comp. and Syst. Sci. (2009); Special issue: Selected papers from WOL-LIC 2008
17. Cheney, J.: Relating nominal and higher-order pattern unification. In: Proc. of UNIF, pp. 104–119 (2005)
18. Fernández, M., Gabbay, M.J., Mackie, I.: Nominal rewriting systems. In: Proc. of 6th ACM SIGPLAN, PPDP 2004, pp. 108–119. ACM, New York (2004)
19. Levy, J., Villaret, M.: Nominal unification from a higher-order perspective. ACM Trans. Comput. Logic 13(2), 10:1–10:31 (2012)
20. Fernández, M., Rubio, A.: Nominal completion for rewrite systems with binders. In: Czumaj, A., Mehlhorn, K., Pitts, A., Wattenhofer, R. (eds.) ICALP 2012, Part II. LNCS, vol. 7392, pp. 201–213. Springer, Heidelberg (2012)
21. Domínguez, J.: A tool to apply nominal recursive path ordering to nominal rules (2014), http://www.inf.kcl.ac.uk/pg/domijesu/nrpo.gtz
22. Domínguez, J.: A tool to translate between closed nominal rewriting systems and combinatory reduction systems (2014), http://www.inf.kcl.ac.uk/pg/domijesu/NRS2CRS.tar.gz/
23. van Oostrom, V., van Raamsdonk, F.: Comparing combinatory reduction systems and higher-order rewrite systems. In: Heering, J., Meinke, K., Möller, B., Nipkow, T. (eds.) HOA 1993. LNCS, vol. 816, pp. 276–304. Springer, Heidelberg (1994)
24. Bertolissi, C., Cirstea, H., Kirchner, C.: Expressing combinatory reduction systems derivations in the rewriting calculus. Higher-Ord. and Symb. Comp. 19, 00110869 (2006)
25. Cirstea, H., Kirchner, C.: The rewriting calculus — Part I. J. of Pure and App. Logs. 9(3), 427–463 (2001)
26. Cirstea, H., Kirchner, C.: The rewriting calculus - Part II. IGPL 9(3), 377–410 (2001)
27. Bognar, M.: Contexts in Lambda Calculus. PhD thesis, Vrije Universiteit Amsterdam (2002)
28. van Raamsdonk, F.: Higher-order rewriting. In: Narendran, P., Rusinowitch, M. (eds.) RTA 1999. LNCS, vol. 1631, pp. 220–239. Springer, Heidelberg (1999)
29. Kop, C.: Simplifying algebraic functional systems. In: Winkler, F. (ed.) CAI 2011. LNCS, vol. 6742, pp. 201–215. Springer, Heidelberg (2011)
30. Jouannaud, J.-P.: Higher-order rewriting: Framework, confluence and termination. In: Middeldorp, A., van Oostrom, V., van Raamsdonk, F., de Vrijer, R. (eds.) Processes... (Klop Festschrift). LNCS, vol. 3838, pp. 224–250. Springer, Heidelberg (2005)

Expressivity and Succinctness of Order-Invariant Logics on Depth-Bounded Structures

Kord Eickmeyer[1], Michael Elberfeld[2], and Frederik Harwath[3]

[1] TU Darmstadt, Darmstadt, Germany
eickmeyer@mathematik.tu-darmstadt.de
[2] RWTH Aachen University, Aachen, Germany
elberfeld@informatik.rwth-aachen.de
[3] Goethe-Universität, Frankfurt am Main, Germany
harwath@cs.uni-frankfurt.de

Abstract. We study the expressive power and succinctness of order-invariant sentences of first-order (FO) and monadic second-order (MSO) logic on graphs of bounded tree-depth. Order-invariance is undecidable in general and, therefore, in finite model theory, one strives for logics with a decidable syntax that have the same expressive power as order-invariant sentences. We show that on graphs of bounded tree-depth, order-invariant FO has the same expressive power as FO, and order-invariant MSO has the same expressive power as the extension of FO with modulo-counting quantifiers. Our proof techniques allow for a fine-grained analysis of the succinctness of these translations. We show that for every order-invariant FO sentence there exists an FO sentence whose size is elementary in the size of the original sentence, and whose number of quantifier alternations is linear in the tree-depth. Our techniques can be adapted to obtain a similar quantitative variant of a known result that the expressive power of MSO and FO coincides on graphs of bounded tree-depth.

Keywords: Expressivity, succinctness, first-order logic, monadic second-order logic, order-invariance, tree-depth.

1 Introduction

Understanding the *expressivity* of logics on finite structures—the question of which properties are definable in a certain logic—plays an important role in database and complexity theory. In the former, logics are used to formulate queries; in the latter, they describe computational problems. Moreover, besides just studying a logic's expressivity, understanding its *succinctness*—the question how complex definitions of properties like queries and problems must be—is a requirement towards (theoretical) expressivity results of (potential) practical importance. The present work studies the succinctness of first-order logic (FO) as well as its succinctness compared to extensions allowing for the use of a linear

E. Csuhaj-Varjú et al. (Eds.): MFCS 2014, Part I, LNCS 8634, pp. 256–266, 2014.
© Springer-Verlag Berlin Heidelberg 2014

Table 1. Summary of our results: A formula φ of quantifier rank q is translated into a formula ψ that is equivalent to φ on graphs of tree-depth at most d

$\varphi \in$	$\leq$-inv-FO	MSO	$\leq$-inv-MSO
$\psi \in$	FO	FO	FO+MOD
$\|\psi\|$	$(2d+1)$-EXP(q)	$\mathcal{O}(d^2)$-EXP$(\mathrm{qr}(\varphi))$	non-elementary
$\mathrm{qad}(\psi)$	$\mathcal{O}(d)$	$\mathcal{O}(d)$	$\mathcal{O}(d)$

order and set quantifiers. This extends and refines recent studies on the expressivity of these logics [1,6] on restricted classes of structures. The structures we consider have bounded tree-depth, which is a graph invariant that measures how far a graph is from being a star in a similar way as tree-width measures how far a graph is from being a tree. Our results are summarised in Table 1.

In both database and complexity theory, one often assumes that structures come with a linear order and formulae are allowed to use this order as long as the properties defined by them do not depend on the concrete interpretation of the order in a structure. Such formulae are called *order-invariant*. Since testing order-invariance for given FO-formulae is undecidable in general, one tries to find logics that have the same expressive power as order-invariant formulae, but a decidable syntax. Several examples prove that order-invariant FO-formulae ($\leq$-inv-FO) are more expressive than FO-formulae without access to orders, cf. [11]. A common feature of these separating examples is that their Gaifman graphs contain large cliques, making them rather complicated from the point of view of graph structure theory. For tree structures, on the other hand, Benedikt and Segoufin [1] showed that the expressivity of FO and $\leq$-inv-FO coincide. We extend and refine this result by showing equal expressivity and succinctness results for FO and $\leq$-inv-FO on graphs of bounded tree-depth. The importance of the expressivity result is highlighted by the fact (proved in the full version of this paper) that order-invariance is undecidable even on graphs of tree-depth at most 2.

A logic that is commonly studied from the perspectives of algorithm design and language theory is monadic second-order logic (MSO), which extends FO-formulae by the ability to quantify over sets of elements instead of just single elements. While it has a rich expressivity that exceeds that of FO already on word structures, the expressive powers of FO and MSO coincide on any class of structures whose tree-depth is bounded [6]. We refine this by presenting a translation into succinct FO-formulae.

In [4], Courcelle raised the (still open) question whether $\leq$-inv-MSO has the same expressive power on graphs of bounded *tree-width* as the extension of MSO by first-order modulo-counting quantifiers (CMSO). We prove a stronger statement for graphs of bounded tree-depth: $\leq$-inv-MSO has the same expressive power as FO+MOD, i.e. the extension of FO by arbitrary first-order modulo-counting quantifiers.

Our results also have implications on FO itself. They imply that that the quantifier alternation hierarchy for FO of Chandra and Harel [3] collapses on graphs of bounded tree-depth, whereas by their result it is strict on trees of unbounded height. That means, for graphs of bounded tree-depth, we are able to turn any FO-formula into a formula whose size is bounded by the quantifier depth of the original formula and whose quantifier alternation depth is bounded by a linear function in the tree-depth.

Proof techniques and relation to prior works. Our proofs are based on fundamental techniques from finite model theory like interpretation arguments, logical types, and games. Compared to prior works like [6], we enrich the application of these techniques by a quantitative analysis, which results in succinct translations instead of just equal expressivity results.

The proofs of [6] use an involved constructive variant of the Feferman–Vaught composition theorem, which complicates a straightforward analysis of the formula size in the translation from MSO to FO. We also use composition arguments, but we get along with an easier non-constructive variant. There is another proof of the result of [6] in [7], but it relies on involved combinatorial insights that seem unsuited for both a tight analysis of succinctness as well as an adaptation to the ordered setting.

While our proofs are based on techniques from finite model theory, the results of [1] about the expressivity of $\leq$-inv-FO on trees use automata-theoretic and algebraic methods, which seem unsuited to obtain succinct translations. Due to the following reason, even our equal expressivity result for $\leq$-inv-FO and FO on graphs of bounded tree-depth is interesting: Benedikt and Segoufin [1] proved that on graphs of bounded tree-width every $\leq$-inv-FO sentence is equivalent to an MSO sentence. With the results of [6] this would imply our expressivity result (not the succinctness result). However, the result from [1] relies on an earlier proof of how to define tree decompositions of bounded width in MSO whose correctness has been doubted by Courcelle and Engelfriet [5]. A proof of our expressivity result for $\leq$-inv-FO along this lines is nevertheless possible. A direct FO-construction of a tree decomposition of bounded width and depth for graphs of bounded tree-depth will appear in the full version of this paper.

Organisation of this paper. The paper continues with a background section and, then, the results related to $\leq$-inv-FO, MSO, and $\leq$-inv-MSO are proved in Sections 3, 4, and 5, respectively. Due to space restrictions, some parts of the paper, such as proofs and definitions, are only sketched or omitted.

2 Background

General notation. The sets of natural numbers with and without 0 are denoted respectively by $\mathbb{N}$ and $\mathbb{N}^+$. Let $[i,j] := \{i,\ldots,j\}$ for all $i,j \in \mathbb{N}$ with $i \leq j$, and let $[j] := [1,j]$. For any $d \in \mathbb{N}$, the *class of functions that grow at most d-fold exponentially*, denoted by d-EXP(n), is made up by all functions $f \colon \mathbb{N} \to \mathbb{N}$ with $f(n) \leq c + d\text{-exp}(n^c)$ for some $c \in \mathbb{N}$; where d-exp$\colon \mathbb{N} \to \mathbb{N}$ is recursively defined via 0-exp$(n) := n$ and $(d+1)$-exp$(n) := 2^{d\text{-exp}(n)}$ for $d \in \mathbb{N}$. If we say that a relation is an *order*, we implicitly assume that it is *linear*.

Logic. For a reference on notation and standard methods in finite model theory, we refer to the book of Libkin [9]. Besides the standard logics FO and MSO, we also consider the logic FO+MOD that is obtained from FO by allowing the use of *modulo-counting quantifiers* $\exists^{i \,(\mathrm{mod}\ p)}$ for each $i \in \mathbb{N}, p \in \mathbb{N}^+$, where $\psi(\bar{y}) := \exists^{i \,(\mathrm{mod}\ p)} x \, \varphi(x, \bar{y})$ means that for a structure $\mathfrak{A}$ with universe A and a tuple of its elements $\bar{a}$, $\mathfrak{A} \models \psi(\bar{a})$ iff $|\{b \in A : \mathfrak{A} \models \varphi(b, \bar{a})\}| \equiv i \,(\mathrm{mod}\ p)$.

We write $\mathrm{qr}(\varphi)$ for the *quantifier rank* and $\|\varphi\|$ for the *size* (or *length*) of a formula φ. The *quantifier alternation depth* $\mathrm{qad}(\varphi)$ of a formula φ in *negation normal form* (NNF, i.e. all negations of φ occur directly in front of atomic formulae) is the maximum number of alternations between $\exists$- and $\forall$-quantifiers on all directed paths in the syntax tree of φ. If φ is not in NNF, we first find an equivalent formula φ' in NNF using a fixed conversion procedure and, then, define $\mathrm{qad}(\varphi) := \mathrm{qad}(\varphi')$.

For any logic $\mathsf{L} \in \{\mathsf{FO}, \mathsf{FO+MOD}, \mathsf{MSO}\}$, we write $\mathfrak{A} \equiv_q^{\mathsf{L}} \mathfrak{B}$ for $q \in \mathbb{N}$ to denote that structures $\mathfrak{A}$ and $\mathfrak{B}$ over the same signature σ satisfy the same $\mathsf{L}[\sigma]$-sentences of quantifier rank at most q. The $\equiv_q^{\mathsf{L}}$-equivalence class of $\mathfrak{A}$ is its (L, q)-*type* and denoted by $\mathrm{tp}_q^{\mathsf{L}}(\mathfrak{A})$. For $\mathsf{L} \in \{\mathsf{FO}, \mathsf{MSO}\}$, each (L, q)-type τ is definable by an L-sentence φ_τ with $\mathrm{qr}(\varphi_\tau) = q$, i.e. $\mathfrak{A} \models \varphi_\tau$ iff $\mathrm{tp}_q^{\mathsf{L}}(\mathfrak{A}) = \tau$; we identify each τ with one such sentence φ_τ. If the logic L has been fixed (as will be the case in most parts of this paper) or the concrete logic is not important for the discussion, we omit it in this and similar notation.

For every signature σ, we define the signature $\sigma^{\leq} := \sigma \cup \{\leq\}$, where $\leq$ is a binary relation symbol. A sentence $\varphi \in \mathsf{FO}[\sigma^{\leq}]$ is *order-invariant* exactly if the following holds for all finite σ-structures G and all linear orders $\preceq, \preceq'$ on the universe of G:
$$(G, \preceq) \models \varphi \quad \text{iff} \quad (G, \preceq') \models \varphi.$$
The set of all order-invariant $\varphi \in \mathsf{FO}[\sigma^{\leq}]$ is denoted by $\leq\text{-inv-}\mathsf{FO}[\sigma]$, and for such a φ and a σ-structure G we write $G \models_{\leq} \varphi$ if $(G, \preceq) \models \varphi$ for some (equivalently, for every) linear order $\preceq$ on G; $\leq\text{-inv-}\mathsf{MSO}$ is defined in the same way by using MSO instead of FO-formulae.

If ψ is a formula with a free variable z and φ is an arbitrary formula, then $\varphi|_\psi$ is the formula φ *relativised to* ψ. We construct $\varphi|_\psi$ by replacing subformulae $\exists z \, \varphi$ and $\forall z \, \varphi$ by $\exists z \, (\psi \wedge \varphi|_\psi)$ and $\forall z \, (\neg \psi \vee \varphi|_\psi)$, respectively.

Coloured and ordered graphs. The letter C will be used to denote a finite set of colours, and we define the signature $\sigma_C := \{E\} \cup \{P_c \mid c \in C\}$, where E is binary and every P_c is unary. A C-*coloured graph* is a σ_C-structure G with universe $V(G)$, symmetric and irreflexive edge relation $E(G)$, and such that the $P_c(G)$ form a partition of $V(G)$. We will simply speak of graphs when referring to C-coloured graphs, and write FO for $\mathsf{FO}[\sigma_C]$ etc. An order on a graph is an order on its vertex set. An *ordered graph* is a $\sigma_C^{\leq}$-structure $(G, \leq^G)$ where G is a graph and $\leq^G$ is an order on G.

The *restriction* of a binary relation R on a set M to a subset $N \subseteq M$ is the relation $R|_N := \{(x, y) \in R : x, y \in N\}$. For ease of notation we will sometimes drop the relativisation for orders on subgraphs and write $(H, \preceq)$ for $(H, \preceq|_H)$.

For two linear orders $\preceq$ and $\preceq'$ on disjoint sets M_1 and M_2, respectively, we define a linear order $\preceq \cdot \preceq'$ on $M_1 \cup M_2$, the *concatenation* of $\preceq$ and $\preceq'$, as $\preceq \cup \preceq' \cup (M_1 \times M_2)$.

Our formulae often speak about the distance between two vertices of a graph. To this end, we define existential FO-formulae $\varphi_{\mathrm{dist}\leq\ell}(x,y)$ by $\varphi_{\mathrm{dist}\leq 0}(x,y) := x = y$ and $\varphi_{\mathrm{dist}\leq\ell}(x,y) := \exists z\,(\varphi_{\mathrm{dist}\leq\ell-1}(x,z) \wedge (Ezy \vee z = y))$ for each $\ell \geq \mathbb{N}^+$.

Tree-depth. The following inductive definition is one of several equivalent ways to define the *tree-depth* $\mathrm{td}(G)$ of a graph (see [10] for a reference on tree-depth):

$$\mathrm{td}(G) := \begin{cases} 1 & \text{if } |V(G)| = 1 \\ 1 + \min_{r \in V(G)} \mathrm{td}(G \setminus r) & \text{if } G \text{ is connected and } |V(G)| > 1 \\ \max_{i \in [n]} \mathrm{td}(K_i) & \text{if } G \text{ has components } K_1, \ldots, K_n. \end{cases}$$

As an immediate consequence of this definition, each connected graph with $\mathrm{td}(G) > 1$ contains a vertex r with $\mathrm{td}(G \setminus r) = \mathrm{td}(G) - 1$. We denote the set of all such vertices by $\mathrm{roots}(G)$. Elements of $\mathrm{roots}(G)$ are called *tree-depth roots* of G. Furthermore, graphs of tree-depth 1 contain only isolated vertices. Another fact about tree-depth that we need is that there are only paths of length at most 2^d in graphs G with $\mathrm{td}(G) \leq d$. In particular, the diameter of such graphs is bounded by 2^d and hence the formula $\mathrm{reach}_d(x,y) := \varphi_{\mathrm{dist}\leq 2^d}(x,y)$ defines the relation containing all pairs $(u,v) \in V(G) \times V(G)$ such that u and v belong to the same (connected) component. Using this observation and the inductive definition of tree-depth, one can write down an FO-sentence $\varphi_{\mathrm{td}\leq d}$ with $\|\varphi_{\mathrm{td}\leq d}\| \in \mathcal{O}(d)$ that defines the class of graphs of tree-depth at most d on the class of all graphs, and an FO-formula $\varphi_{d\text{-roots}}(x)$ with $\|\varphi_{d\text{-roots}}\| \in \mathcal{O}(d)$ that defines the set $\mathrm{roots}(G)$ in a connected graph G with $1 < \mathrm{td}(G) \leq d$.

3 Order-Invariant First-Order Logic

We prove the following theorem in the present section.

Theorem 3.1. *For every $d \in \mathbb{N}^+$ and $\leq$-inv-FO-sentence φ with $\mathrm{qr}(\varphi) = q$, there is an FO-sentence ψ with $\|\psi\| \in (2d+1)\text{-}\mathrm{EXP}(q)$ and $\mathrm{qad}(\psi) \in \mathcal{O}(d)$ that is equivalent to φ on (coloured) graphs of tree-depth at most d.*

Several definitions and lemmas of this section are given in greater generality than needed here, because we will reuse them in later sections. In this section, whenever notation refer to a logic L and we omit it, assume that $\mathsf{L} = \mathsf{FO}$. The main ingredient for the proof of Theorem 3.1 is the following lemma which states that q-types of q-*ordered graphs* of tree-depth at most d, i.e. ordered graphs where the order is a q-*order*, which we define below, can be defined by FO-formulae without referring to a linear order. Let $\mathcal{T}_{C,q,d}$ denote the set of all q-types τ over the signature $\sigma_C^{\leq}$ such that there exists a q-ordered graph $(G, \preceq)$ with $\mathrm{td}(G) \leq d$ and $\mathrm{tp}_q(G, \preceq) = \tau$.

Lemma 3.2. *For all $q, d \in \mathbb{N}^+$ and $\tau \in \mathcal{T}_{C,q,d}$, there is an FO-sentence $\varphi_{\tau,d}$ with $\|\varphi_{\tau,d}\| \in (2d)\text{-}\mathrm{EXP}(q)$ and $\mathrm{qad}(\varphi_{\tau,d}) \in \mathcal{O}(d)$ that defines τ on graphs of tree-depth at most d.*

Here an FO-sentence φ_τ *defines τ on graphs of tree-depth at most d* if for each graph G with $\mathrm{td}(G) \le d$, we have $G \models \varphi_\tau$ iff there exists a q-order $\preceq$ such that $\mathrm{tp}_q(G, \preceq) = \tau$.

Before we discuss how to prove Lemma 3.2, let us first sketch how Theorem 3.1 can be proved with its help: For a given $\le$-inv-FO-sentence φ with $\mathrm{qr}(\varphi) = q$, we let ψ be the disjunction over all FO-sentences $\varphi_{\tau,d}$ for $\tau \in \mathcal{T}_{C,q,d}$ that are types of q-ordered graphs of tree-depth at most d satisfying φ. We have $\|\psi\| \in (2d+1)\text{-}\mathrm{EXP}(q)$ and $\mathrm{qad}(\psi) \in \mathcal{O}(d)$; since φ is order-invariant, ψ is equivalent to φ.

Encoding vertex information in extended colourings. During our proofs, we remove single vertices from a graph and encode information about them into colours of the remaining vertices. This allows us to recover the original graph using an FO-interpretation. Let $C' := C \times \{0, 1\}$. For a C-coloured graph G and $r \in V(G)$, define a C'-colouring of $G \setminus r$ by assigning to each vertex $v \in V(G \setminus r)$ of colour c in G the colour $(c, 1)$ if $\{r, v\} \in E(G)$, and $(c, 0)$, otherwise. The C'-coloured graph thus obtained is denoted by $G^{[r]}$. The following lemma is easy to prove following this definition.

Lemma 3.3. *Let $\mathsf{L} \in \{\mathsf{FO}, \mathsf{FO+MOD}\}$. For every $\mathsf{L}[\sigma_{C'}]$-sentence φ there is an $\mathsf{L}[\sigma_C]$-formula $\mathcal{I}(\varphi)(x)$ of the same quantifier rank and quantifier alternation depth such that*

$$G \models \mathcal{I}(\varphi)(r) \quad \text{iff} \quad G^{[r]} \models \varphi.$$

for all C-coloured graphs G and $r \in V(G)$.

Definition of q-orders. We fix orders $\preceq_{\mathsf{L},q}$, for any logic L, and $\preceq_C$ on, respectively, the set of (L, q)-types and any colour set C.

Definition 3.4 $((\mathsf{L}, q)\text{-order})$. *An order $\preceq$ of a graph G is an (L, q)-order if the following conditions are satisfied:*

1. *If G is a connected graph, then it contains either only one vertex, or it contains more than one vertex and the $\preceq$-least element r is an element of $\mathrm{roots}(G)$ whose colour is $\preceq_C$-minimal among the elements of $\mathrm{roots}(G)$, and $\mathrm{tp}_q^{\mathsf{L}}(G^{[r]}, \preceq) \preceq_{(\mathsf{L},q)} \mathrm{tp}_q^{\mathsf{L}}(G^{[r']}, \preceq)$ for all $r' \in \mathrm{roots}(G)$ of the same colour. Furthermore, $\preceq|_{V(G \setminus r)}$ is an (L, q)-order of $G \setminus r$.*

2. *Otherwise, if G has components $H_1, \ldots, H_\ell$, then, after suitably permuting the components, $\preceq = \preceq|_{H_1} \cdot \cdots \cdot \preceq|_{H_\ell}$, where each $\preceq|_{H_i}$ is an (L, q)-order of H_i, and $\mathrm{tp}_q^{\mathsf{L}}(H_i, \preceq) \preceq_{\mathsf{L},q} \mathrm{tp}_q^{\mathsf{L}}(H_j, \preceq)$ for $i \le j$.*

The least element of a q-order $\preceq$ is denoted by $r_\preceq$.

For each q-ordered C-coloured connected graph $(G, \preceq)$ with $\mathrm{td}(G) > 1$, we define an ordered C'-coloured graph

$$\tilde{G}_\preceq := (G^{[r_\preceq]}, \preceq).$$

Observe that $\mathrm{td}(G^{[r_{\preceq}]}) < \mathrm{td}(G)$ and that $\tilde{G}_{\preceq}$ is q-ordered. The following lemma, which states that $\tilde{G}_{\preceq}$ together with the colour of $r_{\preceq}$ determine the q-type of $(G, \preceq)$, can be proved using standard EF-game-based arguments.

Lemma 3.5. *Let* $\mathsf{L} \in \{\mathsf{FO}, \mathsf{MSO}\}$ *and* $q \in \mathbb{N}^+$. *Let* $(G, \preceq_G)$ *and* $(H, \preceq_H)$ *be* (L, q)-*ordered connected graphs such that* $\mathrm{td}(G), \mathrm{td}(H) > 1$ *and* $r_{\preceq_G}, r_{\preceq_H}$ *have the same colour. Then* $\tilde{G}_{\preceq_G} \equiv^{\mathsf{L}}_q \tilde{H}_{\preceq_H}$ *implies* $(G, \preceq_G) \equiv^{\mathsf{L}}_q (H, \preceq_H)$.

Using this, we can show that while there might be several q-orders of a given graph, up to $\equiv^{\mathsf{L}}_q$ they are all equivalent.

Lemma 3.6. *Let* $\mathsf{L} \in \{\mathsf{FO}, \mathsf{MSO}\}$, $q \in \mathbb{N}^+$. *For all* (L, q)-*orders* $\preceq, \preceq'$ *of a graph* G, *we have* $(G, \preceq) \equiv^{\mathsf{L}}_q (G, \preceq')$.

Threshold counting of components. We define an equivalence relation $\approx_{q,t}$ on ordered graphs that counts the number of components of different q-types up to a threshold value t. We show that there is a t depending on q, such that each $\equiv^{\mathsf{FO}}_q$-equivalence class of q-ordered graphs is a union of $\approx_{q,t}$-equivalence classes. Then we show, basically, that these equivalence classes are definable for graphs of bounded tree-depth. For every logic L and L-sentence φ, we let $n_\varphi(G)$ denote the number of components K of G such that $K \models \varphi$ and we let $n_{\varphi,t}(G) := \min\{n_\varphi(G), t\}$, for each $t \in \mathbb{N}$.

Definition 3.7 $(\approx_{\Phi,t}, \approx_{\mathsf{L},q,t})$. *Let* Φ *be a set of* L-*sentences and* $t \in \mathbb{N}$. *We say that two graphs* G *and* H *are* (Φ, t)-*similar (written* $G \approx_{\Phi,t} H$*) if*

$$n_{\varphi,t}(G) = n_{\varphi,t}(H)$$

for each $\varphi \in \Phi$. *In the special case that* Φ *is a set of* L-*sentences containing one sentence that defines* τ *for each* (L, q)-*type* τ, *we write* $\approx_{\mathsf{L},q,t}$ *instead of* $\approx_{\Phi,t}$; *whenever* L *is fixed, we write* $\approx_{q,t}$.

All these definitions are extended to ordered graphs by stipulating that a *component of an ordered graph* $(G, \preceq)$ is an ordered graph $(K, \preceq)$ where K is a component of G.

We show that in q-ordered graphs FO inherits its component counting capabilities from its capability to distinguish linear orders of different length. A proof of the lemma (based on different notation) is contained in the proof of [1, Thm. 5.5]; it requires only the fact that the components of a q-ordered graph are ordered according to their q-type.

Lemma 3.8. *Let* $q \in \mathbb{N}^+$ *and let* $t := 2^q + 1$. *If* $(G, \preceq_G)$ *and* $(H, \preceq_H)$ *are* q-*ordered graphs with* $(G, \preceq_G) \approx_{q,t} (H, \preceq_H)$, *then* $(G, \preceq_G) \equiv^{\mathsf{FO}}_q (H, \preceq_H)$.

The following lemma shows that $\approx_{\Phi,t}$-equivalence classes are definable for graphs of bounded tree-depth. It will be needed for the formula construction in the proof of Lemma 3.2 and in later sections.

Lemma 3.9. *Let* $\mathsf{L} \in \{\mathsf{FO}, \mathsf{FO+MOD}\}$. *For every* $d, t \in \mathbb{N}^+$, *set of* L-*sentences* $\Phi := \{\varphi_1, \ldots, \varphi_\ell\}$, *and* $\bar{n} := (n_1, \ldots, n_\ell) \in [0, t]^\ell$, *there is an* L-*sentence* $\psi_{\bar{n}, t}^\Phi$ *such that for each graph* G *with* $\mathrm{td}(G) \leq d$, *we have* $G \models \psi_{\bar{n}, t}^\Phi$ *iff* $n_{\varphi_i, t}(G) = n_i$ *for each* $i \in [\ell]$. *Moreover, the sentence has size* $\|\psi_{\bar{n}, t}^\Phi\| \in \ell \cdot \mathcal{O}(d \max_{i \in [\ell]} n_i^2 \|\varphi_i\|)$ *and* $\mathrm{qad}(\psi_{\bar{n}, t}^\Phi) \leq \max_{i \in [\ell]} \mathrm{qad}(\varphi_i) + 1$,

Finally, we can proof our main lemma.

Proof of Lemma 3.2. The proof proceeds by induction on the tree-depth d. Let $\mathcal{T}_{C, q, d}^{\mathrm{conn}}$ be defined analogously to $\mathcal{T}_{C, q, d}$ for q-ordered connected graphs.

Case 1: Connected graphs. As a first step, we prove the special case of the claim for connected graphs with a stronger upper bound on the formula size, i.e. we show that, on connected graphs of tree-depth at most d, each $\tau \in \mathcal{T}_{C, q, d}^{\mathrm{conn}}$ is defined by an FO-sentence $\varphi_{\tau, d}^{\mathrm{conn}}$ such that $\|\varphi_{\tau, d}^{\mathrm{conn}}\| \in (2(d-1)+1)\text{-EXP}(q)$ and $\mathrm{qad}(\varphi_{\tau, d}^{\mathrm{conn}}) \in \mathcal{O}(d)$. If $d = 1$, then any graph G of type τ consists of a single vertex of some colour $c \in C$; the FO-sentence $\varphi_{\tau, 1}^{\mathrm{conn}} := \exists x \, P_c(x) \wedge \forall y \, (x = y)$ defines τ since there is only one linear order on each such graph. Hence $\|\varphi_{\tau, 1}^{\mathrm{conn}}\|$ and $\mathrm{qad}(\varphi_{\tau, 1}^{\mathrm{conn}})$ are constant.

Now suppose that $d > 1$ and $\tau \in \mathcal{T}_{C, q, d}^{\mathrm{conn}}$. For each colour $\hat{c}$ we define a set $R_{\hat{c}} \subseteq \mathcal{T}_{C', q, d-1}$ that contains a q-type θ iff $\mathrm{tp}_q(H, \preceq) = \tau$ for a q-ordered C'-coloured connected graph $(H, \preceq)$ with $1 < \mathrm{td}(H) \leq d$ such that $\mathrm{tp}_q(\tilde{H}_{\prec}) = \theta$ and $r_{\preceq}$ has colour $\hat{c}$. We obtain an $\mathsf{FO}[\sigma_{C'}]$-sentence $\varphi_{\theta, d-1}$ by induction that defines θ on q-ordered C'-coloured graphs of tree-depth at most $d-1$, and that has size $\|\varphi_{\theta, d-1}\| \in (2(d-1))\text{-EXP}(q)$ and alternation-depth $\mathrm{qad}(\varphi_{\theta, d-1}) \in \mathcal{O}(d)$. Let $\varphi_{\tau, 1}$ be an FO-sentence with $\|\varphi_{\tau, 1}\| \in 2\text{-EXP}(q)$ and $\mathrm{qad}(\varphi_{\tau, 1}) \in \mathcal{O}(d)$, also given by induction, that defines τ on graphs of tree-depth 1. Now consider the following FO-sentence

$$\varphi_{\tau, d}^{\mathrm{conn}} := (\varphi_{\mathrm{td} \leq 1} \wedge \varphi_{\tau, 1}) \vee \bigvee_{\hat{c} \in C, \theta \in R_{\hat{c}}} \exists x \, \varphi_{d\text{-roots}}(x) \wedge P_{\hat{c}}(x) \wedge \mathcal{I}(\varphi_{\theta, d-1})(x),$$

where $\mathcal{I}$ is the operator defined in Lemma 3.3. Note that the size of $\varphi_{\tau, d}^{\mathrm{conn}}$ is $\|\varphi_{\tau, d}^{\mathrm{conn}}\| \in (2(d-1)+1)\text{-EXP}(q)$ (this is dominated by the maximal size of $|R_{\hat{c}}|$) and that $\mathrm{qad}(\varphi_{\tau, d}^{\mathrm{conn}}) \in \mathcal{O}(d)$. Using Lemma 3.3 and Lemma 3.5, it is not too hard to verify that $\varphi_{\tau, d}^{\mathrm{conn}}$ defines τ.

Case 2: Disconnected Graphs. Let $\mathcal{T}_{C, q, d}^{\mathrm{conn}} := \{\tau_1, \ldots, \tau_\ell\}$. Let Φ be a set that contains the formulae $\varphi_i := \varphi_{\tau_i, d}^{\mathrm{conn}}$ for each $i \in [\ell]$. For each graph G with $\mathrm{td}(G) \leq d$ and each component K of G, we have $K \models \varphi_i$ iff there is a q-order $\preceq$ of G such that $\mathrm{tp}_q(K, \preceq) = \tau_i$; due to Lemma 3.6, this holds iff $\mathrm{tp}_q(K, \preceq) = \tau_i$ for each q-order $\preceq$ of G. Thus $n_{\varphi_i}(G) = n_{\tau_i}(G, \preceq)$ for each q-order $\preceq$ of G. Let $t := 2^q + 1$ as in Lemma 3.8. For any ordered graph $(G, \preceq)$, let $\bar{n}(G, \preceq) := (n_{\tau_1, t}(G, \preceq), \ldots, n_{\tau_\ell, t}(G, \preceq))$, i.e. $\bar{n}(G, \preceq) \in [0, t]^\ell$.

Now consider a $\tau \in \mathcal{T}_{C, q, d}$. Let $R \subseteq [0, t]^\ell$ such that for each $\bar{n} \in [0, t]^\ell$, $\bar{n} \in R$ iff there exists a q-ordered graph $(G, \preceq)$ with $\mathrm{td}(G) \leq d$ and $\mathrm{tp}_q(G, \preceq) = \tau$ such that $\bar{n} = \bar{n}(G, \preceq)$. For each $\bar{n} \in R$, let $\psi_{\bar{n}, t}^\Phi(\bar{x})$ be the formula of Lemma 3.9.

Observe that $\|\psi_{\bar{n},t}^{\Phi}(\bar{x})\| \in \ell \cdot \mathcal{O}(dt^2 \max_{i \in [\ell]} \|\varphi_i\|)$. We hence have $\|\psi_{\bar{n},t}^{\Phi}(\bar{x})\| \in ((2d-1)+1)$-EXP$(q)$ and qad$(\psi_{\bar{n},t}^{\Phi}(\bar{x})) \in \mathcal{O}(d)$.

Define the FO-sentence $\varphi_{\tau,d} := \bigvee_{\bar{n} \in R} \psi_{\bar{n},t}^{\Phi}$. Since $|R| \in (2d)$-EXP(q), also $\|\varphi_{\tau,d}\| \leq |R| \cdot \max_{\bar{n} \in R} \|\psi_{\bar{n},t}^{\Phi}\| \in (2d)$-EXP$(q)$ and qad$(\varphi_{\tau,d}) \in \mathcal{O}(d)$.

We prove that $\varphi_{\tau,d}$ defines τ on graphs of tree-depth at most d. Let G be such a graph. Suppose first that there is a q-order $\preceq$ such that $(G, \preceq)$ has type τ. By the definition of R_τ there is a tuple $\bar{n} \in R_\tau$ such that $\bar{n} = \bar{n}(G, \preceq)$, so $G \models \psi_{\bar{n},t}^{\Phi}$ by Lemma 3.9.

Suppose now that $G \models \varphi_{\tau,d}$, i.e. say $G \models \psi_{\bar{n},t}^{\Phi}$ for some tuple $\bar{n} \in R_\tau$. By the definition of R, an ordered graph $(H, \preceq_H)$ with $(H, \preceq_H) \models \tau$, td$(H) \leq d$ and $\bar{n}(H, \preceq_H) = \bar{n}$ exists. By Lemma 3.9, we have $G \approx_{\Phi,t} H$. It follows from our choice of Φ that there is a q-order $\preceq_G$ on G such that $(H, \preceq_H) \approx_{q,t} (G, \preceq_G)$. Now $(H, \preceq_H) \equiv_q^{\mathsf{FO}} (G, \preceq_G)$ by Lemma 3.8. □

4   Monadic Second-Order Logic

The approach towards the results of the previous section can be adapted to obtain a quantitative variant of the result of [6] that MSO and FO have the same expressive power on the class of graphs of tree-depth at most d. Let $s(d) := \frac{d(d+1)}{2} + 2d$ for each $d \in \mathbb{N}$.

Theorem 4.1. *For each $d \in \mathbb{N}^+$ and MSO-sentence φ there is an FO-sentence ψ with $\|\psi\| \in (s(d)+1)$-EXP$(\mathrm{qr}(\varphi))$ and qad$(\psi) \in \mathcal{O}(d)$ that is equivalent to φ on graphs of tree-depth at most d.*

Much of the proof of Theorem 4.1 follows the proof of Theorem 3.1, but we are spared of the complications that arose in connection with the ordering of graphs. On the other hand, the proof of an analogue to Lemma 3.8 becomes more complicated. In Lemma 3.8, we did not use the fact that we consider only graphs of bounded tree-depth. Here naively ignoring the bounded tree-depth would lead to a non-elementary dependence of the counting threshold on q. We use the following lemma to avoid this.

Lemma 4.2. *For every $d, q \in \mathbb{N}^+$, there is a $t \in d$-EXP(q) such that, if G and H are graphs with td(G), td$(H) \leq d$ and $G \approx_{q,t} H$, then $G \equiv_q^{\mathsf{MSO}} H$.*

5   Order-Invariant Monadic Second-Order Logic

It is well-known that for each sentence in modulo-counting MSO (CMSO) there is an equivalent $\leq$-inv-MSO-sentence, and a conjecture of Courcelle implies that, on graphs of *bounded tree-width*, the converse of this statement is also true. In the special case where instead of bounded tree-width the graphs have bounded tree-depth, we show the following stronger result.

Theorem 5.1. *For every $d \in \mathbb{N}^+$ and $\leq$-inv-MSO-sentence φ there exists an FO+MOD-sentence ψ with qad$(\psi) \in \mathcal{O}(d)$ that is equivalent to φ on graphs of tree-depth at most d.*

Of course the analogue of this statement for more general classes of graphs is not true, e.g. graph connectivity is MSO-, but not FO+MOD-definable. In contrast to the previous sections, we do not analyse the formula size, because it is known from [8] that (plain) MSO can define the length of orders non-elementarily more succinct than FO. Again we need to understand $\leq$-inv-MSO's capabilities to count the components of a given q-type in q-ordered graphs. We say that ordered graphs $(G, \preceq_G)$ and $(G, \preceq_H)$ are (q,p)-*similar*, written $(G, \preceq_G) \approx_{q,p} (H, \preceq_H)$, if $n_\tau(G) \equiv n_\tau(H) \pmod{p}$, and $n_\tau(G) \geq p$ iff $n_\tau(H) \geq p$, for each q-type τ. The following lemma shows that MSO inherits its component counting capabilities on q-ordered graphs from its semilinear spectrum on linear orders.

Lemma 5.2. *For each $q \in \mathbb{N}^+$ a $p \in \mathbb{N}^+$ exists such that for all q-ordered graphs $(G, \preceq_G)$ and $(H, \preceq_H)$, if $(G, \preceq_G) \approx_{q,p} (H, \preceq_H)$ then $(G, \preceq_G) \equiv_q^{\mathsf{MSO}} (H, \preceq_H)$.*

The next lemma is a modulo-counting analogue of Lemma 3.9, and the two lemmas together can be used to define the $\approx_{q,p}$-equivalence class of a graph G from given sentences that define the q-types of the components.

Lemma 5.3. *For each $d, p \in \mathbb{N}^+$, each set $\Phi := \{\varphi_1, \ldots, \varphi_\ell\}$ of FO+MOD-sentences and each tuple of numbers $\bar{n} := (n_1, \ldots, n_\ell) \in [0, p-1]^\ell$ there is an FO+MOD-sentence $\chi_{\bar{n},p}^\Phi$ such that for each graph G with $\mathrm{td}(G) \leq d$, we have $G \models \chi_{\bar{n},p}^\Phi$ iff, for each $i \in [\ell]$, $n_{\varphi_i}(G) \equiv n_i \pmod{p}$. If $\mathrm{qad}(\varphi_i) = \mathcal{O}(d)$ for each $i \in [\ell]$, then $\mathrm{qad}(\chi_{\bar{n},p}^\Phi) \in \mathcal{O}(d)$.*

To prove the lemma, at first, it is not clear at all how modulo-counting quantifiers can be used to count the number of components satisfying a given FO+MOD-sentence. But it is shown in [2, Lem. 7] that the number of tree-depth roots of each component of a graph is bounded in terms of its tree-depth. For each component, we can use its roots as FO-definable representatives that allow us to perform the necessary counting. Using the previous lemmas, we can prove an analogue to Lemma 3.2, i.e. that each (MSO, q)-type of (MSO, q)-ordered graphs is FO+MOD-definable on bounded tree-depth graphs, in a very similar way to Lemma 3.2. This makes it possible to prove Theorem 5.1.

6   Final Remarks

We phrased our results for undirected (coloured) graphs to simplify notation, but their proofs generalise to structures with higher-arity relations (where the tree-depth of a structure is defined to be the tree-depth of its Gaifman graph). Furthermore, all our formula constructions imply algorithms to compute the formulas. It would be interesting to obtain corresponding lower bounds for our succinctness upper bounds.

Acknowledgements. We want to thank Isolde Adler for bringing the first two authors together with the third author, and Nicole Schweikardt for her helpful suggestions.

References

1. Benedikt, M.A., Segoufin, L.: Towards a characterization of order-invariant queries over tame graphs. Journal of Symbolic Logic 74(1), 168–186 (2009)
2. Bouland, A., Dawar, A., Kopczyński, E.: On tractable parameterizations of graph isomorphism. In: Thilikos, D.M., Woeginger, G.J. (eds.) IPEC 2012. LNCS, vol. 7535, pp. 218–230. Springer, Heidelberg (2012)
3. Chandra, A., Harel, D.: Structure and complexity of relational queries. JCSS 25(1), 99–128 (1982)
4. Courcelle, B.: The monadic second-order logic of graphs x: linear orderings. Theoretical Computer Science 160(1-2), 87–143 (1996)
5. Courcelle, B., Engelfriet, J.: Graph Structure and Monadic Second-Order Logic – A Language-Theoretic Approach. Cambridge University Press (2012)
6. Elberfeld, M., Grohe, M., Tantau, T.: Where first-order and monadic second-order logic coincide. In: Proc. LICS 2012, pp. 265–274. IEEE Computer Society (2012)
7. Gajarský, J., Hliněný, P.: Faster deciding MSO properties of trees of fixed height, and some consequences. In: Proc. FSTTCS 2012, pp. 112–123 (2012)
8. Grohe, M., Schweikardt, N.: The succinctness of first-order logic on linear orders. Logical Methods in Computer Science 1(1:6), 1–25 (2005)
9. Libkin, L.: Elements of Finite Model Theory. Springer (2004)
10. Nešetřil, J., Ossona de Mendez, P.: Sparsity: Graphs, Structures, and Algorithms. Springer, Heidelberg (2012)
11. Schweikardt, N.: A short tutorial on order-invariant first-order logic. In: Bulatov, A.A., Shur, A.M. (eds.) CSR 2013. LNCS, vol. 7913, pp. 112–126. Springer, Heidelberg (2013)

Two Recursively Inseparable Problems
for Probabilistic Automata[*]

Nathanaël Fijalkow[1,2], Hugo Gimbert[3], Florian Horn[1], and Youssouf Oualhadj[4]

[1] LIAFA, Université Paris 7, France
[2] University of Warsaw, Poland
[3] LaBRI, Université de Bordeaux, France
[4] Université de Mons, Belgium

Abstract. This paper introduces and investigates decision problems for *numberless* probabilistic automata, *i.e.* probabilistic automata where the *support* of each probabilistic transitions is specified, but the exact values of the probabilities are *not*. A numberless probabilistic automaton can be *instantiated* into a probabilistic automaton by specifying the exact values of the non-zero probabilistic transitions.
We show that the two following properties of numberless probabilistic automata are recursively inseparable:
- all instances of the numberless automaton have value 1,
- no instance of the numberless automaton has value 1.

1 Introduction

In 1963 Rabin [12] introduced the notion of probabilistic automata, which are finite automata able to randomise over transitions. A probabilistic automaton has a finite set of control states Q, and processes finite words; each transition consists in updating the control state according to a given probabilistic distribution determined by the current state and the input letter. This powerful model has been widely studied and has applications in many fields like software verification [3], image processing [5], computational biology [6] and speech processing [10].

Several algorithmic properties of probabilistic automata have been considered in the literature, sometimes leading to efficient algorithms. For instance, *functional equivalence* is decidable in polynomial time [13,14], and even faster with randomised algorithms, which led to applications in software verification [9].

However, many natural decision problems are undecidable, and part of the literature on probabilistic automata is about *intractability results*. For example the *emptiness*, the *threshold isolation* and the *value 1* problems are undecidable [11,2,8].

A striking result due to Condon and Lipton [4] states that, for every $\epsilon > 0$, the two following problems are recursively inseparable: given a probabilistic automaton $\mathcal{A}$,

---

[*] The research leading to these results has received funding from the European Union's Seventh Framework Programme (FP7/2007-2013) under grant agreement n° 259454 (GALE) and from the French Agence Nationale de la Recherche projects EQINOCS (ANR-11-BS02-004) and STOCH-MC (ANR-13-BS02-0011-01).

E. Csuhaj-Varjú et al. (Eds.): MFCS 2014, Part I, LNCS 8634, pp. 267–278, 2014.
© Springer-Verlag Berlin Heidelberg 2014

- does $\mathcal{A}$ accept some word with probability greater than $1 - \varepsilon$?
- does $\mathcal{A}$ accept every word with probability less than ε?

In the present paper we focus on *numberless* probabilistic automata, *i.e.* probabilistic automata whose non-zero probabilistic transitions are specified but the exact values of the probabilities are not. A numberless probabilistic automaton can be *instantiated* into a probabilistic automaton by specifying the exact values of the non-zero probabilistic transitions (see Section 2 for formal definitions).

The notion of numberless probabilistic automaton is motivated by the following example. Assume we are given a digital chip modelled as a finite state machine controlled by external inputs. The internal transition structure of the chip is known but the transitions themselves are not observable. We want to compute an initialisation input sequence that puts the chip in a particular initial state. In case some of the chip components have failure probabilities, this can be reformulated as a value 1 problem for the underlying probabilistic automaton: is there an input sequence whose acceptance probability is arbitrarily close to 1? Assume now that the failure probabilities are not fixed *a priori* but we can tune the quality of our components and choose the failure probabilities, for instance by investing in better components. Then we are dealing with a numberless probabilistic automaton and we would like to determine whether it can be instantiated into a probabilistic automaton with value 1, in other words we want to solve an *existential value 1 problem* for the numberless probabilistic automaton. If the failure probabilities are unknown then we are facing a second kind of problem called the *universal value 1 problem*: determine whether all instances of the automaton have value 1. We also consider variants where the freedom to choose transition probabilities is restricted to some intervals, that we call *noisy value 1 problems*.

One may think that relaxing the constraints on the exact transition probabilities makes things algorithmically much easier. However this is not the case, and we prove that the existential and universal value 1 problems are recursively inseparable: given a numberless probabilistic automaton $\mathcal{C}$,

- do all instances of $\mathcal{C}$ have value 1?
- does no instance of $\mathcal{C}$ have value 1?

This result is actually a corollary of a generic construction which constitutes the technical core of the paper and has the following properties. For every simple probabilistic automaton $\mathcal{A}$, we construct a numberless probabilistic automaton $\mathcal{C}$ such that the three following properties are equivalent:

(i) $\mathcal{A}$ has value 1,
(ii) one of the instances of $\mathcal{C}$ has value 1,
(iii) all instances of $\mathcal{C}$ have value 1.

The definitions are given in Section 2. The main technical result appears in Section 3, where we give the generic construction whose properties are described above.

In Section 4, we discuss the implications of our results, first for the noisy value 1 problems, and second for probabilistic Büchi automata [1].

2   Definitions

Let A be a finite alphabet. A (finite) word u is a (possibly empty) sequence of letters $u = a_0 a_1 \cdots a_{n-1}$; the set of finite words is denoted by A^*.

A probability distribution over a finite set Q is a function $\delta : Q \to \mathbb{Q}_{\geq 0}$ such that $\sum_{q \in Q} \delta(q) = 1$; we denote by $\frac{1}{3} \cdot q + \frac{2}{3} \cdot q'$ the distribution that picks q with probability $\frac{1}{3}$ and q' with probability $\frac{2}{3}$, and by q the trivial distribution picking q with probability 1. The support of a distribution δ is the set of states picked with positive probability, $i.e.$, $\mathrm{supp}(\delta) = \{q \in Q \mid \delta(q) > 0\}$. Finally, the set of probability distributions over Q is $\mathcal{D}(Q)$.

Definition 1 (Probabilistic automaton). *A probabilistic automaton (PA) is a tuple $\mathcal{A} = (Q, A, q_0, \Delta, F)$, where Q is a finite set of states, A is the finite input alphabet, $q_0 \in Q$ is the initial state, $\Delta : Q \times A \to D(Q)$ is the probabilistic transition function, and $F \subseteq Q$ is the set of accepting states.*

For convenience, we also use $\mathbb{P}_{\mathcal{A}}(s \xrightarrow{u} t)$ to denote the probability of going from the state s to the state t reading u, $\mathbb{P}_{\mathcal{A}}(s \xrightarrow{u} S)$ to denote the probability of going from the state s to a state in S reading u, and $\mathbb{P}_{\mathcal{A}}(u)$, the *acceptance probability* of a word $u \in A^*$, to denote $\mathbb{P}_{\mathcal{A}}(q_0 \xrightarrow{u} F)$.

We often consider the case of *simple probabilistic automata*, where the transition probabilities can only be 0, $1/2$, or 1.

Definition 2 (Value). *The value of a PA $\mathcal{A}$, denoted val($\mathcal{A}$), is the supremum acceptance probability over all input words*

$$val(\mathcal{A}) = \sup_{u \in A^*} \mathbb{P}_{\mathcal{A}}(u) \ .$$

The value 1 problem asks, given a (simple) PA $\mathcal{A}$ as input, whether val($\mathcal{A}$) = 1.

Theorem 1 ([8]). *The value 1 problem is undecidable for simple PA.*

Definition 3 (Numberless probabilistic automaton). *A numberless probabilistic automaton (NPA) is a tuple $\mathcal{A} = (Q, A, q_0, T, F)$, where Q is a finite set of states, A is the finite input alphabet, $q_0 \in Q$ is the initial state, $T \subseteq Q \times A \times Q$ is the numberless transition function, and $F \subseteq Q$ is the set of accepting states.*

The numberless transition function T is an abstraction of probabilistic transition functions. We say that Δ is consistent with T if for all letters a and states s and t, $\Delta(s, a, t) > 0$ if, and only if, $(s, a, t) \in T$.

A numberless probabilistic automaton is an equivalence class of probabilistic automata, which share the same set of states, input alphabet, initial and accepting states, and whose transition functions have the same support.

A NPA $\mathcal{A} = (Q, A, q_0, T, F)$ together with a probabilistic transition function Δ consistent with T defines a PA $\mathcal{A}[\Delta] = (Q, A, q_0, \Delta, F)$. Conversely, a PA $\mathcal{A} = (Q, A, q_0, \Delta, F)$ induces an underlying NPA $[\mathcal{A}] = (Q, A, q_0, T, F)$, where $T \subseteq Q \times A \times Q$ is defined by $(q, a, p) \in T$ if $\Delta(q, a)(p) > 0$.

We consider two decision problems for NPA:

- **The existential value 1 problem:** given a NPA $\mathcal{A}$, determine whether there exists Δ such that $\mathrm{val}(\mathcal{A}[\Delta]) = 1$.
- **The universal value 1 problem:** given a NPA $\mathcal{A}$, determine whether for all Δ, we have $\mathrm{val}(\mathcal{A}[\Delta]) = 1$.

Proposition 1. *There exists a NPA such that:*

- *there exists Δ such that $\mathrm{val}(\mathcal{A}[\Delta]) = 1$,*
- *there exists Δ' such that $\mathrm{val}(\mathcal{A}[\Delta']) < 1$.*

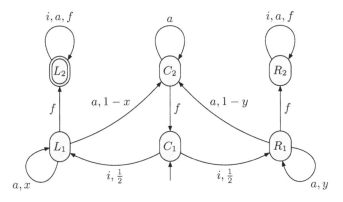

Fig. 1. This NPA has value 1 if and only if $x > y$

In this automaton, adapted from [8,7], the shortest word that can be accepted is $i \cdot f$, as i goes from C_1 to L_1, and f goes from L_1 to L_2. However, there are as much chances to go to R_2, so the value of $i \cdot f$ is $1/2$.

If x is strictly less than y, one can tip the scales to the left by adding a's between the i and f : each time, the run will have more chances to stay left than to stay right. After reading $i \cdot a^n \cdot f$, the probability of reaching L_2 is equal to x^n, while the probability of reaching R_2 is only y^n. There is also a very high chance that the run went back to C_1, but from there we can simply repeat our word an arbitrary number of times.

Let x, y, and ε be three real numbers such that $0 \le y < x \le 1$, and $0 < \varepsilon \le 1$. There is an integer n such that $x^n/(x^n + y^n)$ is greater than $1 - \varepsilon/2$, and an integer m such that $(1 - x^n - y^n)^m$ is less than $\varepsilon/2$. The word $(i \cdot a^n \cdot f)^m$ is accepted with probability greater than $1 - \varepsilon$.

On the other hand, if $x \le y$, there is no word with value higher than $1/2$.

3   Recursive Inseparability for Numberless Value 1 Problems

In this section, we prove the following theorem:

Theorem 2. *The two following problems for numberless probabilistic automata are recursively inseparable:*

- *all instances have value 1,*
- *no instance has value 1.*

Recall that two decision problems A and B are recursively inseparable if their languages L_A and L_B of accepted inputs are disjoint and there exists no recursive language L such that $L_A \subseteq L$ and $L \cap L_B = \emptyset$.

Note that it implies that both A and B are undecidable.

Equivalently, this means there exists no terminating algorithm which has the following behaviour on input x:

- if $x \in L_A$, then the algorithm answers "YES"
- if $x \in L_B$, then the algorithm answers "NO".

On an input that belongs neither to L_A nor to L_B, the algorithm's answer can be either "YES" or "NO".

3.1   Overall Construction

Lemma 1. *There exists an effective construction which takes as input a simple PA $\mathcal{A}$ and constructs a NPA $\mathcal{C}$ such that*

$$val(\mathcal{A}) = 1 \iff \forall \Delta, val(\mathcal{C}[\Delta]) = 1 \iff \exists \Delta, val(\mathcal{C}[\Delta]) = 1 .$$

We first explain how Lemma 1 implies Theorem 2. Assume towards contradiction that the problems "all instances have value 1" and "no instance has value 1" are recursively separable. Then there exists an algorithm A taking a NPA as input and such that: if all instances have value 1, then it answers "YES", and if no instance has value 1, then it answers "NO". We show using Lemma 1 that this would imply that the value 1 problem is decidable for simple PA, contradicting Theorem 1. Indeed, let $\mathcal{A}$ be a simple PA, applying the construction yields a NPA $\mathcal{C}$ such that

$$val(\mathcal{A}) = 1 \iff \forall \Delta, val(\mathcal{C}[\Delta]) = 1 \iff \exists \Delta, val(\mathcal{C}[\Delta]) = 1 .$$

In particular, either all instances of $\mathcal{C}$ have value 1, or no instance of $\mathcal{C}$ has value 1. Hence, if it answers "YES" then $val(\mathcal{A}) = 1$ and it if answers "NO" then $val(\mathcal{A}) < 1$, allowing to decide whether $\mathcal{A}$ has value 1 or not. This concludes the proof of Theorem 2 assuming Lemma 1.

The construction follows two steps.

The first step is to build from $\mathcal{A}$ a family of PA's $\mathcal{B}_\lambda$ whose transitions are all of the form $q \xrightarrow{a} (\lambda \cdot r, (1 - \lambda) \cdot s)$ in such a way that $\mathcal{B}_\lambda$ has the same value as $\mathcal{A}$ *for any value of* λ. Note that, while all the $\mathcal{B}_\lambda$'s belong to the same NPA, they are *not*, in general, a NPA: for example, if $\mathcal{A}$ were the simple version of the automaton of Figure 1, the $\mathcal{B}_\lambda$'s would be the instances where $x = y = \lambda$, while the underlying NPA would also include the cases where $x \neq y$.

The second step is to build from the $\mathcal{B}_\lambda$'s a NPA $\mathcal{C}$ such that, for each probabilistic transition function Δ, there is a λ such that $\mathcal{C}[\Delta]$ has value 1 if, and only if $\mathcal{B}_\lambda$ has value 1.

It follows that:

$$\exists \Delta, \mathrm{val}(\mathcal{C}[\Delta]) = 1 \implies \exists \lambda, \mathrm{val}(\mathcal{B}_\lambda) = 1$$
$$\implies \mathrm{val}(\mathcal{A}) = 1$$
$$\implies \forall \lambda, \mathrm{val}(\mathcal{B}_\lambda) = 1$$
$$\implies \forall \Delta, \mathrm{val}(\mathcal{C}[\Delta]) = 1 .$$

3.2   The Fair Coin Construction

Let $\mathcal{A} = (Q, A, q_0, \Delta, F)$ be a simple PA over the alphabet A. We construct a family of PAs $(\mathcal{B}_\lambda)_{\lambda \in]0,1[}$ over the alphabet $B = A \cup \{\sharp\}$, whose transitions have probabilities 0, λ, $1 - \lambda$ or 1, as follows.

The automaton $\mathcal{B}_\lambda$ is a copy of $\mathcal{A}$ where each transition of $\mathcal{A}$ is replaced by the gadget illustrated in Figure 2 (for simplicity, we assume that all the transitions of $\mathcal{A}$ are probabilistic). The initial and final states are the same in $\mathcal{A}$ and $\mathcal{B}_\lambda$.

The left hand side shows part of automaton $\mathcal{A}$: a probabilistic transition from q reading a, leading to r or s each with probability half. The right hand side shows how this behaviour is simulated by $\mathcal{B}_\lambda$: the letter a leads to an intermediate state q_a, from which we can read a new letter $\sharp$. Each time a pair of $\sharp$'s is read, the automaton $\mathcal{B}_\lambda$ goes to r with probability $\lambda \cdot (1-\lambda)$, goes to s with probability $(1 - \lambda) \cdot \lambda$, and stays in q_a with probability $\lambda^2 + (1 - \lambda)^2$. Reading a letter other than $\sharp$ while the automata is still in one of the new states leads to the new sink state $\perp$, which is not accepting. Thus, the probability of going to r is equal to the probability of going to s, and we can make the probability of a simulation failure as low as we want by increasing the number of $\sharp$'s between two successive "real" letters.

Let u be a word of A^*. We denote by $[u]^k$ the word of B^* where each letter $a \in A$ of u is replaced by $a \cdot \sharp^{2k}$. Conversely, if w is a word of B^*, we denote by $\tilde{w}$ the word obtained from w by removing all occurrences of the letter $\sharp$.

Intuitively, a run of $\mathcal{A}$ on the word u is simulated by a run of $\mathcal{B}_\lambda$ on the word $[u]^k$. Whenever there is a transition in $\mathcal{A}$, $\mathcal{B}_\lambda$ makes k attempts to simulate it through the gadget of Figure 2, and each attempt succeeds with probability $(1 - 2\lambda \cdot (1 - \lambda))$, so each transition fails with probability:

$$A_{\lambda,k} = 1 - (1 - 2\lambda \cdot (1 - \lambda))^k .$$

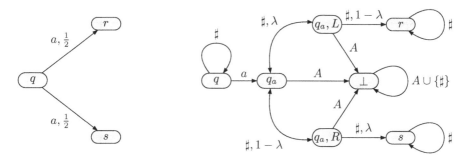

Fig. 2. The fair coin gadget

Proposition 2. *The probabilistic automaton $\mathcal{B}_\lambda$ satisfies:*

1. *For all $q, r \in Q$, $a \in A$, and $k \in \mathbb{N}$, $\mathbb{P}_{\mathcal{B}_\lambda}(q \xrightarrow{[a]^k} r) = A_{\lambda,k} \cdot \mathbb{P}_{\mathcal{A}}(q \xrightarrow{a} r)$,*
2. *For all $q, r \in Q$, $u \in A^*$, and $k \in \mathbb{N}$, $\mathbb{P}_{\mathcal{B}_\lambda}(q \xrightarrow{[u]^k} r) = A_{\lambda,k}^{|u|} \cdot \mathbb{P}_{\mathcal{A}}(q \xrightarrow{u} r)$.*
3. *For all $q, r \in Q$, $w \in (A \cup \{\sharp\})^*$, and $k \in \mathbb{N}$, $\mathbb{P}_{\mathcal{B}_\lambda}(q \xrightarrow{w} r) \leq \mathbb{P}_{\mathcal{A}}(q \xrightarrow{\tilde{w}} r)$.*

It follows from Proposition 2 that for any λ, the value of $\mathcal{B}_\lambda$ is equal to the value of $\mathcal{A}$.

3.3   The Simulation Construction

All the $\mathcal{B}_\lambda$'s induce the same NPA, that we denote by $\mathcal{B}$. The problem is that there are many other instances of $\mathcal{B}$, whose values may be higher than the value of $\mathcal{A}$ (recall the example of Figure 1, where the $\mathcal{B}_\lambda$ have value $1/2$, while there are instances of $\mathcal{B}$ with value 1). In this subsection, we construct a NPA $\mathcal{C}$ (over an extended alphabet C) whose instances simulate all the $\mathcal{B}_\lambda$'s, but only them.

The idea is that the new NPA should only have *one* probabilistic transition. An instance of this transition translates to a value for λ. Figure 3 describes a first attempt at this (notice that our convention is that an non-drawn transition means a loop, rather than a transition to a sink state).

In this automaton, that we call $\mathcal{B}'$, there are two copies of the set of states, and a single shared probabilistic transition $q_R \xrightarrow{\$} (\lambda \cdot s_0, (1 - \lambda) \cdot s_1)$ between them. In order to make all the probabilistic transitions of $\mathcal{B}$ happen in this center area, we use new letters to detect where the runs come from before the probabilistic transition, and where it should go afterwards.

For each pair of a letter b in B and a state q in Q, we introduce two new letters check(b, q) and apply(b, q). The letter check(b, q) loops over each state except the left copy of q, which goes to q_R. The letter apply(b, q) loops over each state except s_0, from where it goes to the λ-valued successor of (q, b) in $\mathcal{B}_\lambda$, and s_1, from where it goes to the $(1 - \lambda)$-valued successor of (q, b) in $\mathcal{B}_\lambda$. The new letter next_transition sends the run back to the left part once each possible state has been tested.

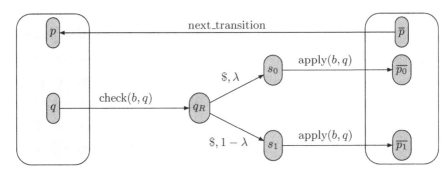

Fig. 3. Naive fusion of the probabilistic transitions

Thus, if we define the morphism $\widehat{}$ by its action on letters:

$$\widehat{b} = \mathrm{check}(b, q_0) \cdot \$ \mathrm{apply}(b, q_0) \cdots \mathrm{check}(b, q_{n-1}) \cdot \$ \cdot \mathrm{apply}(b, q_{n-1}) \cdot \mathrm{next\_transition} \,,$$

where the q_i's are the states of $\mathcal{B}_\lambda$, we get for any word u on the alphabet B, $\mathbb{P}_{\mathcal{B}_\lambda}(u) = \mathbb{P}_{\mathcal{B}'_\lambda}(\widehat{u})$.

The problem with this automaton is that one can "cheat", either by not testing an unwelcome state, or by changing state and letter between a check and the subsequent apply. In order to avoid this kind of behaviour, we change the automaton so that it is necessary to win an arbitrarily large number of successive times in order to approach the value 1, and we can test whether the word is fair after the first successful attempt. A side effect is that the simulation only works for the value 1: for other values, it might be better to cheat. The resulting automaton is described in Figure 4.

The structure of the automaton of Figure 3 is still there, but it has been augmented with an extra layer of scrutiny: each time we use the probabilistic transition, there is now a positive probability $(1 - \theta)$ to go to a new wait state. There is also a new letter next_word which has the following effect:

- if the run is in an accepting state, it goes to the initial state of the fairness checker $\mathcal{D}$;
- if the run is in a non-accepting state, it goes to the non-accepting sink state $\perp$ of $\mathcal{D}$;
- if the run is in the wait state, it goes back to the initial state.

The fairness checker $\mathcal{D}$ is a deterministic automaton which accepts the language $\{\widehat{u} \cdot \mathrm{next\_word} \mid u \in B^*\}^*$. Its final state is the only final state in all of $\mathcal{C}$.

Intuitively, a run can still cheat before the *first* next_word letter, but the benefits of doing so are limited: the probability that $\mathcal{C}[\lambda, \theta]$ accepts a word at that point is at most θ (except if the empty word is accepting, but that case is trivial). After that point, cheating is risky: if the run already reached $\mathcal{D}$, a false move will send it to the sink.

A more formal proof follows. A simple inspection of the construction of $\mathcal{C}$ yields Proposition 3:

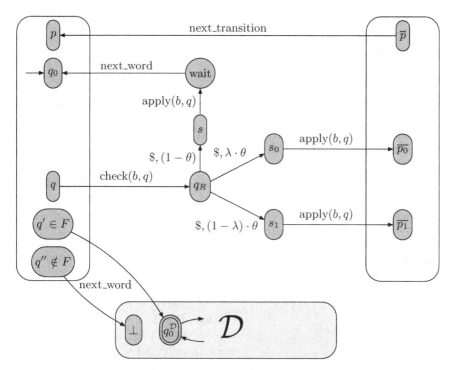

Fig. 4. The Numberless Probabilistic Automaton $\mathcal{C}$

Proposition 3. *Let u be a word of B^* of length k. We have:*

$$\mathbb{P}_{\mathcal{C}[\lambda,\theta]}(\widehat{u} \cdot next\_word) = \theta^k \cdot \mathbb{P}_{\mathcal{B}_\lambda}(u)$$
$$\mathbb{P}_{\mathcal{C}[\lambda,\theta]}((\widehat{u} \cdot next\_word)^\ell) = (1 - (1 - \theta^k)^\ell) \cdot \mathbb{P}_{\mathcal{B}_\lambda}(u) \ .$$

It follows from Proposition 3 that the value of $\mathcal{C}[\lambda, \theta]$ is at least the value of $\mathcal{B}_\lambda$.

Proposition 4. *Let u be a word of $(C \setminus \{next\_word\})^+$. We have:*

$$\mathbb{P}_{\mathcal{C}[\lambda,\theta]}(u) \le \theta \ .$$

Proposition 5 formalises the fact that there is no point in cheating after the first next_word letter:

Proposition 5. *Let $u_1, \dots, u_k$ be k words of $(C \setminus \{next\_word\})^*$ and w be the word $u_1 \cdot next\_word \cdots u_k \cdot next\_word$. Then, for any $1 \le i \le k$, if $u_i \notin \widehat{B^*}$, we have:*

$$\mathbb{P}_{\mathcal{C}[\lambda,\theta]}(w) \le \mathbb{P}_{\mathcal{C}[\lambda,\theta]}(u_i \cdot next\_word \cdots u_k \cdot next\_word) \ .$$

Proof. After reading $u_1 \cdots u_{i-1} \cdot next\_word$, a run must be in one of the following three states: q_0, $q_0^{\mathcal{D}}$, and $\perp$. As $u_i \notin \widehat{B^*}$, reading it from $q_0^{\mathcal{D}}$ will lead to $\perp$. Thus,

$$\mathbb{P}_{\mathcal{C}[\lambda,\theta]}(w) = \mathbb{P}_{\mathcal{C}[\lambda,\theta]}(q_0 \xrightarrow{u_1 \cdots u_{i-1} \cdot next\_word} q_0) \cdot \mathbb{P}_{\mathcal{C}[\lambda,\theta]}(u_i \cdots u_k \cdot next\_word) \ ,$$

and Proposition 5 follows.

Finally, Proposition 6 shows that $\mathcal{C}[\lambda, \theta]$ cannot have value 1 if $\mathcal{B}_\lambda$ does not.

Proposition 6. *For all words* $w \in C^*$ *such that* $\mathbb{P}_{\mathcal{C}[\lambda,\theta]}(w) > \theta$, *there exists a word* $v \in B^*$ *such that*

$$\mathbb{P}_{\mathcal{B}_\lambda}(u) \geq \frac{\mathbb{P}_{\mathcal{C}[\lambda,\theta]}(w) - \theta}{1 - \theta} .$$

Proof. Let us write $w = u_1 \cdot \text{next\_word} \cdots u_k \cdot \text{next\_word}$ with $u_1, \ldots, u_k \in (C \setminus \{\text{next\_word}\})^*$. By Proposition 4, $k > 1$, and by Proposition 5 we can assume that $u_2, \ldots, u_k$ belong to $\widehat{B}^*$. Let $v_2, \ldots, v_k$ be the words of B^* such that $u_i = \widehat{v_i}$. The $\mathcal{C}[\lambda, \theta]$-value of w can be seen as a weighted average of 1 (the initial cheat, with a weight of θ) and the $\mathcal{B}_\lambda$ values of the v_i's (the weight of $\mathbb{P}_{\mathcal{B}_\lambda}(v_i)$ is the probability that the run enters $\mathcal{D}$ while reading u_i). It follows that at least one of the v_i's has a $\mathcal{B}_\lambda$-value greater than the $\mathcal{C}[\lambda, \theta]$-value of w

Thus, for each λ and θ, the value of $\mathcal{C}[\lambda, \theta]$ is 1 if and only if the value of $\mathcal{B}_\lambda$ is 1. As all the $\mathcal{B}_\lambda$'s have the same value, which is equal to the value of $\mathcal{A}$, we get:

$$\exists \Delta, \text{val}(\mathcal{C}[\Delta]) = 1 \implies \exists \lambda, \text{val}(\mathcal{B}_\lambda) = 1$$
$$\implies \text{val}(\mathcal{A}) = 1$$
$$\implies \forall \lambda, \text{val}(\mathcal{B}_\lambda) = 1$$
$$\implies \forall \Delta, \text{val}(\mathcal{C}[\Delta]) = 1 .$$

Theorem 2 follows.

4   Consequences

In this section, we show several consequences of the recursive inseparability results from Theorem 2 and of the construction from Lemma 1. The first is a series of undecidability results for variants of the value 1 problem. The second is about probabilistic Büchi automata with probable semantics, as introduced in [1].

4.1   The Noisy Value 1 Problems

Observe that Theorem 2 implies the following corollary:

Corollary 1. *Both the universal and the existential value 1 problems are undecidable.*

We can go further. Note that the universal and existential value 1 problems quantify over all possible probabilistic transition functions. Here we define two more realistic problems for NPA, where the quantification is restricted to probabilistic transition functions that are ε-close to a given probabilistic transition function:

– **The noisy existential value 1 problem:** given a NPA $\mathcal{A}$, a probabilistic transition function Δ and $\varepsilon > 0$, determine whether there exists Δ' such that $|\Delta' - \Delta| \leq \varepsilon$ and $\text{val}(\mathcal{A}[\Delta']) = 1$.

– **The noisy universal value** 1 **problem:** given a NPA $\mathcal{A}$, a probabilistic transition function Δ and $\varepsilon > 0$, determine whether for all Δ' such that $|\Delta' - \Delta| \leq \varepsilon$, we have $\mathrm{val}(\mathcal{A}[\Delta']) = 1$.

It follows from Lemma 1 that both problems are undecidable:

Corollary 2. *Both the noisy universal and the noisy existential value* 1 *problems are undecidable.*

Indeed, we argue that the construction from Lemma 1 implies a reduction from either of these problems to the value 1 problem for simple PA, hence the undecidability. Let $\mathcal{A}$ be a simple PA, the construction yields a NPA $\mathcal{C}$ such that:

$$\mathrm{val}(\mathcal{A}) = 1 \iff \forall \Delta, \mathrm{val}(\mathcal{C}[\Delta]) = 1 \iff \exists \Delta, \mathrm{val}(\mathcal{C}[\Delta]) = 1 \ .$$

It follows that for any probabilistic transition function Δ and any $\varepsilon > 0$, we have:

$$\begin{aligned}\mathrm{val}(\mathcal{A}) = 1 &\iff (\forall \Delta', |\Delta' - \Delta| \leq \varepsilon \implies \mathrm{val}(\mathcal{C}[\Delta']) = 1) \\ &\iff (\exists \Delta', |\Delta' - \Delta| \leq \varepsilon \wedge \mathrm{val}(\mathcal{C}[\Delta']) = 1) \ .\end{aligned}$$

This completes the reduction.

4.2 Probabilistic Büchi Automata with Probable Semantics

We consider PA over infinite words, as introduced in [1]. A probabilistic Büchi automaton (PBA) $\mathcal{A}$ can be equipped with the so-called probable semantics, defining the language (over infinite words):

$$L^{>0}(\mathcal{A}) = \{w \in A^\omega \mid \mathbb{P}_{\mathcal{A}}(w) > 0\} \ .$$

It was observed in [1] that the value 1 problem for PA (over finite words) easily reduces to the emptiness problem for PBA with probable semantics (over infinite words).

Informally, from a PA $\mathcal{A}$, construct a PBA $\mathcal{A}'$ by adding a transition from every final state to the initial state labelled with a new letter $\sharp$. (From a non-final state, the letter $\sharp$ leads to a rejecting sink.) As explained in [1], this simple construction ensures that $\mathcal{A}$ has value 1 if and only if $\mathcal{A}'$ is non-empty, equipped with the probable semantics.

This simple reduction, together with Theorem 2, implies the following corollary:

Corollary 3. *The two following problems for numberless PBA with probable semantics are recursively inseparable:*

– *all instances have a non-empty language,*
– *no instance has a non-empty language.*

Acknowledgments. We would like to thank the referees for their helpful comments.

References

1. Baier, C., Bertrand, N., Größer, M.: Probabilistic ω-automata. Journal of the ACM 59(1), 1 (2012)
2. Bertoni, A., Mauri, G., Torelli, M.: Some recursive unsolvable problems relating to isolated cutpoints in probabilistic automata. In: Salomaa, A., Steinby, M. (eds.) ICALP 1977. LNCS, vol. 52, pp. 87–94. Springer, Heidelberg (1977)
3. Chatterjee, K., Doyen, L., Henzinger, T.A., Raskin, J.-F.: Algorithms for omega-regular games of incomplete information. Logical Methods in Computer Science, 3(3) (2007)
4. Condon, A., Lipton, R.J.: On the complexity of space bounded interactive proofs (extended abstract). In: Foundations of Computer Science, pp. 462–467 (1989)
5. Culik, K., Kari, J.: Digital images and formal languages, pp. 599–616. Springer-Verlag New York, Inc. (1997)
6. Durbin, R., Eddy, S.R., Krogh, A., Mitchison, G.: Biological Sequence Analysis: Probabilistic Models of Proteins and Nucleic Acids. Cambridge University Press (July 1999)
7. Fijalkow, N., Gimbert, H., Oualhadj, Y.: Deciding the value 1 problem for probabilistic leaktight automata. In: Logics in Computer Science, pp. 295–304 (2012)
8. Gimbert, H., Oualhadj, Y.: Probabilistic automata on finite words: Decidable and undecidable problems. In: Abramsky, S., Gavoille, C., Kirchner, C., Meyer auf der Heide, F., Spirakis, P.G. (eds.) ICALP 2010. LNCS, vol. 6199, pp. 527–538. Springer, Heidelberg (2010)
9. Kiefer, S., Murawski, A.S., Ouaknine, J., Wachter, B., Worrell, J.: Language equivalence for probabilistic automata. In: Gopalakrishnan, G., Qadeer, S. (eds.) CAV 2011. LNCS, vol. 6806, pp. 526–540. Springer, Heidelberg (2011)
10. Mohri, M.: Finite-state transducers in language and speech processing. Computational Linguistics 23, 269–311 (1997)
11. Paz, A.: Introduction to probabilistic automata. Academic Press (1971)
12. Rabin, M.O.: Probabilistic automata. Information and Control 6(3), 230–245 (1963)
13. Schützenberger, M.-P.: On the definition of a family of automata. Information and Control, 4 (1961)
14. Tzeng, W.-G.: A polynomial-time algorithm for the equivalence of probabilistic automata. SIAM Journal on Computing 21(2), 216–227 (1992)

Monadic Second-Order Logic
with Arbitrary Monadic Predicates*

Nathanaël Fijalkow[1,2] and Charles Paperman[1]

[1] LIAFA, Paris 7, France
[2] University of Warsaw, Poland

Abstract. We study Monadic Second-Order Logic (**MSO**) over finite words, extended with (non-uniform arbitrary) monadic predicates. We show that it defines a class of languages that has algebraic, automata-theoretic and machine-independent characterizations. We consider the *regularity question*: given a language in this class, when is it regular? To answer this, we show a *substitution property* and the existence of a *syntactical predicate*.

We give three applications. The first two are to give simple proofs of the Straubing and Crane Beach Conjectures for monadic predicates, and the third is to show that it is decidable whether a language defined by an **MSO** formula with morphic predicates is regular.

1 Introduction

The Monadic Second-Order Logic (**MSO**) over finite words equipped with the linear ordering on positions is a well-studied and understood logic. It provides a mathematical framework for applications in many areas such as program verification, database and linguistics. In 1962, Büchi [5] proved the decidability of the satisfiability problem for **MSO** formulae.

Uniform Monadic Predicates. In 1966, Elgot and Rabin [9] considered extensions of **MSO** with uniform monadic predicates. For instance, the following formula

$$\forall x, \ \mathbf{a}(x) \iff x \text{ is prime},$$

describes the set of finite words such that the letters a appear exactly in prime positions. The predicate "x is a prime number" is a uniform monadic predicate on the positions, it can be seen as a subset of $\mathbb{N}$.

Elgot and Rabin were interested in the following question: for a uniform monadic predicate $\mathbf{P} \subseteq \mathbb{N}$, is the satisfiability problem of $\mathbf{MSO}[\leq, \mathbf{P}]$ decidable? A series of papers gave tighter conditions on $\mathbf{P}$, culminating to two final answers: in 1984, Semenov [19] gave a characterization of the predicates $\mathbf{P}$

---

* The authors are supported by the project ANR 2010 BLAN 0202 02 (FREC). The first author is supported by the European Union's Seventh Framework Programme (FP7/2007-2013) under grant agreement 259454 (GALE) and 239850 (SOSNA). The second author is supported by Fondation CFM.

E. Csuhaj-Varjú et al. (Eds.): MFCS 2014, Part I, LNCS 8634, pp. 279–290, 2014.

such that $\mathbf{MSO}[\leq, \mathbf{P}]$ is decidable, and in 2006, Rabinovich and Thomas [15,17] proved that it is equivalent to the predicate $\mathbf{P}$ being effectively profinitely ultimately periodic.

Further questions on uniform monadic predicates have been investigated. For instance, Rabinovich [16] gave a solution to the Church synthesis problem for $\mathbf{MSO}[\leq, \mathbf{P}]$, for a large class of predicates $\mathbf{P}$.

In this paper, we consider non-uniform monadic predicates: such a predicate $\mathbf{P}$ is given, for each length $n \in \mathbb{N}$, by a predicate over the n first positions $\mathbf{P}_n \subseteq \{0, \ldots, n-1\}$. The set $\mathcal{M}$ of these predicates contains the set $\mathcal{M}^{\mathrm{unif}}$ of uniform monadic predicates.

Advice Regular Languages. We say that a language is *advice regular* if it is definable in $\mathbf{MSO}[\leq, \mathcal{M}]$. No computability assumptions are made on the monadic predicates, so this class contains undecidable languages.

Our first contribution is to give equivalent presentations of this class, which is a Boolean algebra extending the class of regular languages:

1. It has an equivalent automaton model: *automata with advice*.
2. It has an equivalent algebraic model: *one-scan programs*.
3. It has a machine-independent characterization, based on generalizations of Myhill-Nerode equivalence relations.

This extends the equivalence between automata with advice and Myhill-Nerode equivalence relations proved in [12] for the special case of uniform monadic predicates. We will rely on those characterizations to obtain several properties of the advice regular languages. Our main goal is the following regularity question: given an advice regular language L, when is L regular? To answer this question, we introduce two notions:

- The *substitution property*, which states that if a formula φ together with the predicate $\mathbf{P}$ defines a regular language $L_{\varphi, \mathbf{P}}$, then there exists a regular predicate $\mathbf{Q}$ such that $L_{\varphi, \mathbf{Q}} = L_{\varphi, \mathbf{P}}$.
- The *syntactical predicate* of a language L, which is the "simplest" predicate $\mathbf{P}_L$ such that $L \in \mathbf{MSO}[\leq, \mathbf{P}_L]$.

Our second contribution is to show that the class of advice regular languages has the substitution property, and that an advice regular language L is regular if and only if $\mathbf{P}_L$ is regular.

We apply these results to the case of morphic predicates [6], and obtain the following decidability result: given a language defined by an $\mathbf{MSO}$ formula with morphic predicates, one can decide whether it is regular.

Motivations from Circuit Complexity. Extending logics with predicates also appears in the context of circuit complexity. Indeed, a descriptive complexity theory initiated by Immermann [10] relates logics and circuits; it shows that a language is recognized by a Boolean circuit of constant depth and unlimited fan-in if and only if it can be described by a first-order formula with predicates (of any arity, so not only monadic ones), *i.e.* $\mathbf{AC^0} = \mathbf{FO}[\mathcal{N}]$.

This correspondence led to the study of two properties, which amount to characterize the regular languages (Straubing Conjecture) and the languages with a neutral letter (Crane Beach Conjecture) in several fragments of $\mathbf{FO}[\mathcal{N}]$. The Straubing Conjecture would, if true, give a deep understanding of many complexity classes inside $\mathbf{NC}^1$. Many cases of this conjecture are still open. On the other side, unfortunately the Crane Beach Conjecture does not hold in general, as shown by Barrington, Immermann, Lautemann, Schweikardt and Thérien [3]. On the positive side, both conjectures hold for uniform monadic predicates [3,20].

Our third contribution is to give simple proofs of the both the Straubing and the Crane Beach Conjectures for monadic predicates relying on our previous characterizations.

Outline. The Section 2 gives characterizations of advice regular languages, in automata-theoretic, algebraic and machine-independent terms. In Section 3, we study the regularity question, and give two different answers: one through the substitution property, and the other through the existence of a syntactical predicate. The last section, Section 4, provides applications of our results: easy proofs that the Straubing and the Crane Beach Conjectures hold for monadic predicates and decidability of the regularity problem for morphic regular languages.

2   Advice Regular Languages

In this section, we introduce the class of advice regular languages and give several characterizations.

Predicates. A monadic predicate $\mathbf{P}$ is given by $\mathbf{P} = (\mathbf{P}_n)_{n \in \mathbb{N}}$, where $\mathbf{P}_n \subseteq \{0, \ldots, n-1\}$. Since we mostly deal with monadic predicates, we often drop the word "monadic". In this definition the predicates are non-uniform: for each length n there is a predicate $\mathbf{P}_n$, and no assumption is made on the relation between $\mathbf{P}_n$ and $\mathbf{P}_{n'}$ for $n \neq n'$. A predicate $\mathbf{P}$ is uniform if there exists $\mathbf{Q} \subseteq \mathbb{N}$ such that for every n, $\mathbf{P}_n = \mathbf{Q} \cap \{0, \ldots, n-1\}$. We identify $\mathbf{P}$ and $\mathbf{Q}$, and see uniform predicates as subsets of $\mathbb{N}$.

For the sake of readability, we often define predicates as $\mathbf{P} = (\mathbf{P}_n)_{n \in \mathbb{N}}$ with $\mathbf{P}_n \subseteq \{0, 1\}^n$. In such case we can see $\mathbf{P}$ as a language over $\{0, 1\}$, which contains exactly one word for each length. Also, we often define predicates $\mathbf{P} = (\mathbf{P}_n)_{n \in \mathbb{N}}$ with $\mathbf{P}_n \in A^n$ for some finite alphabet A. This is not formally a predicate, but this amounts to define one predicate $\mathbf{P}^a$ for each letter a in A, with $\mathbf{P}_n^a(i) = 1$ if and only if $\mathbf{P}_n(i) = a$. This abuse of notations will prove very convenient. Similarly, any infinite word $w \in A^\omega$ can be seen as a uniform predicate.

Monadic Second-Order Logic. The formulae we consider are monadic second-order (**MSO**) formulae, obtained from the following grammar:

$$\varphi \;=\; \mathbf{a}(x) \mid x \leq y \mid P(x) \mid \varphi \wedge \varphi \mid \neg \varphi \mid \exists x, \; \varphi \mid \exists X, \; \varphi$$

Here $x, y, z, \ldots$ are first-order variables, which will be interpreted by positions in the word, and $X, Y, Z, \ldots$ are monadic second-order variables, which will interpreted by sets of positions in the word. We say that $\mathbf{a}$ is a letter symbol, $\leq$ the ordering symbol and $P, Q, \ldots$ are the numerical monadic predicate symbols, often refered to as predicate symbols.

The notation $\varphi(P^1, \ldots, P^\ell, x^1, \ldots, x^n, X^1, \ldots, X^p)$ means that in φ, the predicate symbols are among $P^1, \ldots, P^\ell$, the free first-order variables are among $x^1, \ldots, x^n$ and the free second-order variables are among $X^1, \ldots, X^p$. A formula without free variables is called a sentence.

We use the notation $\overline{P}$ to abbreviate $P^1, \ldots, P^\ell$, and similarly for all objects (variables, predicate symbols, predicates).

We now define the semantics. The letter symbols and the ordering symbol are always interpreted in the same way, as expected. For the predicate symbols, the predicate symbol P is interpreted by a predicate $\mathbf{P}$. Note that P is a syntactic object, while $\mathbf{P}$ is a predicate used as the interpretation of P.

Consider $\varphi(\overline{P}, \overline{x}, \overline{X})$ a formula, u a finite word of length n, $\overline{\mathbf{P}}$ predicates interpreting the predicate symbols from $\overline{P}$, $\overline{\mathbf{x}}$ valuation of the free first-order variables and $\overline{\mathbf{X}}$ valuation of the free second-order variables. We define $u, \overline{\mathbf{P}}, \overline{\mathbf{x}}, \overline{\mathbf{X}} \models \varphi$ by induction as usual, with

$$u, \overline{\mathbf{P}}, \overline{\mathbf{x}}, \overline{\mathbf{X}} \models P(y) \quad \text{if} \quad \mathbf{y} \in \mathbf{P}_n \; .$$

A sentence $\varphi(\overline{P})$ and a tuple of predicates $\overline{\mathbf{P}}$ interpreting the predicate symbols from $\overline{P}$ define a language

$$L_{\varphi, \overline{\mathbf{P}}} = \{u \in A^* \mid u, \overline{\mathbf{P}} \models \varphi\} \; .$$

Such a language is called advice regular, and the class of advice regular languages is denoted by $\mathbf{MSO}[\leq, \mathcal{M}]$.

Automata with Advice. We introduce automata with advice. Unlike classical automata, they have access to two more pieces of information about the word being read: its length and the current position. Both the transitions and the final states can depend on those two pieces of information. For this reason, they are (much) more expressive than classical automata, and recognize undecidable languages.

A non-deterministic automaton with advice is given by $\mathcal{A} = (Q, q_0, \delta, F)$ where Q is a finite set of states, $q_0 \in Q$ is the initial state, $\delta \subseteq \mathbb{N} \times \mathbb{N} \times Q \times A \times Q$ is the transition relation and $F \subseteq \mathbb{N} \times Q$ is the set of final states. In the deterministic case $\delta : \mathbb{N} \times \mathbb{N} \times Q \times A \to Q$.

A run over a finite word $u = u_0 \cdots u_{n-1} \in A^*$ is a finite word $\rho = q_0 \cdots q_n \in Q^*$ such that for all $i \in \{0, \ldots, n-1\}$, we have $(i, n, q_i, u_i, q_{i+1}) \in \delta$. It is accepting if $(n, q_n) \in F$.

One obtains a uniform model by removing one piece of information in the transition function: the length of the word. This automaton model is strictly weaker, and is (easily proved to be) equivalent to the one introduced in [12], where the automata read at the same time the input word and a fixed word called the advice. However, our definition will be better suited for some technical aspects: for instance, the number of Myhill-Nerode equivalence classes exactly correspond to the number of states in a minimal deterministic automaton.

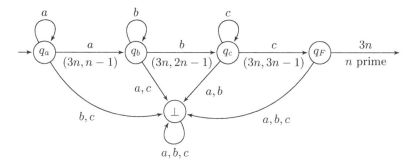

Fig. 1. The automaton for Example 2.1

Example 2.1. The language $\{a^n b^n c^n \mid n \text{ is a prime number}\}$ is recognized by a (deterministic) automaton with advice. The automaton is represented in figure 1. It has five states, q_a, q_b, q_c, q_F and $\bot$. The initial state is q_a. The transition function is defined as follows:

$$
\begin{aligned}
\delta(i, 3n, q_a, a) &= q_a & \text{if } i < n - 1 \\
\delta(n - 1, 3n, q_a, a) &= q_b \\
\delta(i, 3n, q_b, b) &= q_b & \text{if } n \le i < 2n - 1 \\
\delta(2n - 1, 3n, q_b, c) &= q_c \\
\delta(i, 3n, q_c, c) &= q_c & \text{if } 2n \le i < 3n - 1 \\
\delta(3n - 1, 3n, q_c, c) &= q_F
\end{aligned}
$$

All other transitions lead to $\bot$, the sink rejecting state. The set of final states is $F = \{(3n, q_F) \mid n \text{ is a prime number}\}$.

We mention another example, that appeared in the context of automatic structures [13]. They show that the structure $(\mathbb{Q}, +)$ is automatic with advice, which amounts to show that the language $\{\widehat{x} \natural \widehat{y} \natural \widehat{z} \mid z = x + y\}$, where $\widehat{x}$ denotes the factorial representation of the rational x, is advice regular.

One-scan Programs. Programs over monoids were introduced in the context of circuit complexity [1]: Barrington showed that any language in $\mathbf{NC^1}$ can be computed by a program of polynomial length over a non-solvable group. Here we present a simplification introduced in [20], adapted to the context of monadic predicates.

A one-scan program is given by $P = (M, (f_{i,n} : A \to M)_{i,n \in \mathbb{N}}, S)$ where M is a finite monoid and $S \subseteq M$. The function $f_{i,n}$ is used to compute the effect of the ith letter of an input word of length n. The program P accepts $u = u_0 \cdots u_{n-1}$ if $f_{0,n}(u_0) \cdots f_{n-1,n}(u_{n-1}) \in S$.

Note that this echoes the classical definition of recognition by monoids, where a morphism $f : A \to M$ into a finite monoid M recognizes the word $u = u_0 \cdots u_{n-1}$ if $f(u_0) \cdots f(u_{n-1}) \in S$. Here, a one-scan program uses different functions $f_{i,n}$, depending on the position i and the length of the word n.

Myhill-Nerode Equivalence Relations. Let $L \subseteq A^*$ and $p \in \mathbb{N}$, we define two equivalence relations:

- $u \sim_L v$ if for all $w \in A^*$, we have $uw \in L \iff vw \in L$,
- $u \sim_{L,p} v$ if for all $w \in A^p$, we have $uw \in L \iff vw \in L$.

The relation $\sim_L$ is called the (classical) Myhill-Nerode equivalence relation. Recall that $\sim_L$ contains finitely many equivalence classes if and only if L is regular, i.e. $L \in \mathbf{MSO}[\leq]$.

Theorem 2.2 (Advice Regular Languages). *Let L be a language of finite words, the following properties are equivalent:*

(1) $L \in \mathbf{MSO}[\leq, \mathcal{M}]$,
(2) L is recognized by a non-deterministic automaton with advice,
(3) L is recognized by a deterministic automaton with advice,
(4) There exists $K \in \mathbb{N}$ such that for all $i, p \in \mathbb{N}$, the restriction of $\sim_{L,p}$ to words of length i contains at most K equivalence classes.
(5) L is recognized by a one-scan program,

In such case, we say that L is advice regular.

This extends the Myhill-Nerode theorem proposed in [12], which proves the equivalence between (3) and (4) for uniform predicates.

3   The Regularity Question

In this section, we address the following question: given an advice regular language, when is it regular? We answer this question in two different ways: first by showing a substitution property, and second by proving the existence of a syntactical predicate.

Note that the regularity question is not a decision problem, as advice regular languages are not finitely presentable, so we can only provide (non-effective) characterizations of regular languages inside the advice regular languages.

In the next section, we will show how these two notions answer the regularity question: first by proving that the Straubing property holds in this case, and second by proving the decidability of the regularity problem for morphic regular languages.

3.1   A Substitution Property

In this subsection, we prove a substitution property for $\mathbf{MSO}[\leq, \mathcal{M}]$.

We say that a predicate $\mathbf{P} = (\mathbf{P}_n)_{n \in \mathbb{N}}$ is regular if the language $\mathbf{P} \subseteq \{0, 1\}^*$ is regular, defining the class $\mathcal{R}eg_1$ of regular *monadic* predicates (as defined in [14] and in [20]).

Theorem 3.1. *For all sentences* $\varphi(\overline{P})$ *in* $\mathbf{MSO}[\leq, \mathcal{M}]$ *and predicates* $\overline{\mathbf{P}} \in \mathcal{M}$ *such that* $L_{\varphi, \overline{\mathbf{P}}}$ *is regular, there exist* $\overline{\mathbf{Q}} \in \mathcal{R}eg_1$ *such that* $L_{\varphi, \overline{\mathbf{Q}}} = L_{\varphi, \overline{\mathbf{P}}}$.

The main idea of the proof is that among all predicates $\overline{\mathbf{Q}}$ such that $L_{\varphi, \overline{\mathbf{P}}} = L_{\varphi, \overline{\mathbf{Q}}}$, there is a minimal one with respect to a lexicographic ordering, which can be defined by an $\mathbf{MSO}$ formula. The key technical point is given by the following lemma, which can be understood as a regular choice function.

Lemma 3.2 (Regular Choice Lemma). *Let* M *be a regular language such that for all* $k \in \mathbb{N}$, *there exists a word* $w \in M$ *of length* k. *Then there exists* $M' \subseteq M$ *a regular language such that for all* $k \in \mathbb{N}$, *there exists exactly one word* $w \in M'$ *of length* k.

3.2   The Syntactical Predicate

In this subsection, we define the notion of syntactical predicate for an advice regular language. The word "syntactical" here should be understood in the following sense: the syntactical predicate $\mathbf{P}_L$ of L is the most regular predicate that describes the language L. In particular, we will prove that L is regular if and only if $\mathbf{P}_L$ is regular.

Let L be an advice regular language. We define the predicate $\mathbf{P}_L = (\mathbf{P}_{L,n})_{n \in \mathbb{N}}$. Thanks to Theorem 2.2, there exists $K \in \mathbb{N}$ such that for all $i, p \in \mathbb{N}$, the restriction of $\sim_{L,p}$ to words of length i contains at most K equivalence classes. Denote $Q = \{1, \ldots, K\}$ and $\Sigma = (Q \times A \to Q) \uplus Q$, where $Q \times A \to Q$ is the set of (partial) functions from $Q \times A$ to Q. We define $\mathbf{P}_{L,n} \in \Sigma^n$.

Let $i, n \in \mathbb{N}$. Among all words of length i, we denote by $u_1^{i,n}, u_2^{i,n}, \ldots$ the lexicographically minimal representants of the equivalence classes of $\sim_{L,n-i}$, enumerated in the lexicographic order:

$$u_1^{i,n} <_{\text{lex}} u_2^{i,n} <_{\text{lex}} u_3^{i,n} <_{\text{lex}} \cdots \tag{1}$$

In other words, $u_\ell^{i,n}$ is minimal with respect to the lexicographic order $<_{\text{lex}}$ among all words of length i in its equivalence class for $\sim_{L,n-i}$. Thanks to Theorem 2.2, there are at most K such words for each $i, n \in \mathbb{N}$.

We define $\mathbf{P}_{L,n}(i)$ (the ith letter of $\mathbf{P}_{L,n}$) by:

$$\mathbf{P}_{L,n}(i)(\ell, a) = k \quad \text{if } u_\ell^{i,n} \cdot a \sim_{L,n-i-1} u_k^{i+1,n}, \text{ for } i < n \tag{2}$$

$$\mathbf{P}_{L,n}(n-1)(\ell) \quad \text{if } u_\ell^{n,n} \in L . \tag{3}$$

Intuitively, the predicate $\mathbf{P}_L$ describes the transition function with respect to the equivalence relations $\sim_{L,p}$. We now give an example.

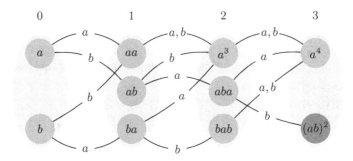

Fig. 2. The predicate $\mathbf{P}_L$ (here $\mathbf{P}_{L,4}$) for $L = (ab)^* + (ba)^*b$

Example 3.3. Consider the language $L = (ab)^* + (ba)^*b$. We represent $\mathbf{P}_{L,4}$ in figure 2. Each circle represents an equivalence class with respect to $\sim_{L,4}$, inside words of a given length. For instance, there are three equivalence classes for words of length 3: a^3, aba and bab. Note that these three words are the minimal representants of their equivalence classes with respect to the lexicographic order. For the last position (here 3), the equivalence class of $(ab)^2$ (which is actually reduced to $(ab)^2$ itself) is darker since it belongs to the language L.

Theorem 3.4. *Let L be an advice regular language. Then L is regular if and only if $\mathbf{P}_L$ is regular.*

The proof is split in two lemmas, giving each direction.

Lemma 3.5. *Let L be an advice regular language. Then $L \in \mathbf{MSO}[\leq, \mathbf{P}_L]$.*

Lemma 3.6. *Let L be an advice regular language defined with the predicates $\overline{\mathbf{P}}$. Then $\mathbf{P}_L \in \mathbf{MSO}[\leq, \overline{\mathbf{P}}]$.*

4 Applications

In this section we show several consequences of Theorem 2.2 (characterization of the advice regular languages), Theorem 3.1 (a substitution property for advice regular languages) and Theorem 3.4 (a syntactical predicate for advice regular predicates).

The first two applications are about two conjectures, the Straubing and the Crane Beach Conjectures, introduced in the context of circuit complexity. We first explain the motivations for these two conjectures, and show simple proofs of both in the special case of monadic predicates.

The third application shows that one can determine, given an **MSO** formula with morphic predicates, whether it defines a regular language.

4.1 The Straubing and Crane Beach Conjectures

We first quickly define some circuit complexity classes. The most important here is $\mathbf{AC^0}$, the class of languages defined by boolean circuits of bounded depth and polynomial size. From $\mathbf{AC^0}$, adding the modular gates gives rise to $\mathbf{ACC}$. Finally, the class of languages defined by boolean circuits of logarithmic depth, polynomial size and fan-in 2 is denoted by $\mathbf{NC^1}$. Separating $\mathbf{ACC}$ from $\mathbf{NC^1}$ remains a long-standing open problem.

One approach to better understand these classes is through descriptive complexity theory, giving a perfect correspondence between circuit complexity classes and logical formalisms. Unlike what we did so far, the logical formalisms involved here use predicates of any arity (we focused on predicates of arity one). A k-ary predicate $\mathbf{P}$ is given by $(\mathbf{P}_n)_{n \in \mathbb{N}}$, where $\mathbf{P}_n \subseteq \{0, \ldots, n-1\}^k$. We denote by $\mathcal{N}$ the class of all predicates, and by $\mathcal{R}eg$ the class of regular predicates as defined in [20].

Theorem 4.1 ([1,4,11]).

(1) $\mathbf{AC^0} = \mathbf{FO}[\mathcal{N}]$,
(2) $\mathbf{ACC} = (\mathbf{FO} + \mathbf{MOD})[\mathcal{N}]$.

Two conjectures have been formulated on the logical side, which aim at clarifying the relations between different circuit complexity classes. They have been stated and studied in special cases, we extrapolate them here to all fragments. Here the fragment $\mathbf{F}[\mathcal{P}]$ is described by a class $\mathbf{F}$ of formulae and a class $\mathcal{P}$ of predicates.

The first property, called the Straubing property, characterizes the regular languages (denoted by $\mathbf{REG}$) inside a larger fragment.

Definition 4.2 (Straubing Property). $\mathbf{F}[\mathcal{P}]$ *has the Straubing property if: all regular languages definable in* $\mathbf{F}[\mathcal{P}]$ *are also definable in* $\mathbf{F}[\mathcal{P} \cap \mathcal{R}eg]$. *In equation,* $\mathbf{F}[\mathcal{P}] \cap \mathbf{REG} = \mathbf{F}[\mathcal{P} \cap \mathcal{R}eg]$.

This statement appears for the first time in [2], where it is proved that $\mathbf{FO}[\mathcal{N}]$ has the Straubing property, relying on lower bounds for $\mathbf{AC^0}$ and an algebraic characterisation of $\mathbf{FO}[\mathcal{R}eg]$. Following this result, Straubing conjectures in [20] that $(\mathbf{FO} + \mathbf{MOD})[\mathcal{N}]$ and $\mathcal{B}\Sigma_k[\mathcal{N}]$ have the Straubing property for $k \geq 1$. It has been shown that several fragments have the Straubing property, as for instance, $\Sigma_1[\mathcal{N}]$, $\mathbf{FO}[\leq, \mathcal{M}^{\mathrm{unif}}]$ and $(\mathbf{FO} + \mathbf{MOD})[\leq, \mathcal{M}^{\mathrm{unif}}]$ (in [20]). We extend this result here, as a straightforward corollary of Theorem 3.1.

Theorem 4.3. *All fragments* $\mathbf{F}[\leq, \mathcal{M}]$ *have the Straubing property.*

In particular, for all $k \geq 1$, $\mathcal{B}\Sigma_k[\leq, \mathcal{M}]$ has the Straubing property. This result is, to the best of our knowledge, the first intermediary result towards a proof of the Straubing Conjecture for $\mathcal{B}\Sigma_k[\mathcal{N}]$.

The second property, called the Crane Beach property, characterizes the languages having a neutral letter, and has been proposed by Thérien for the special case of first-order logic. We say that a language L has a neutral letter e if for all words u, v, we have $uv \in L$ if and only if $uev \in L$.

Definition 4.4 (Crane Beach Property). $\mathbf{F}[\mathcal{P}]$ *has the Crane Beach property if: all languages having a neutral letter definable in* $\mathbf{F}[\mathcal{P}]$ *are definable in* $\mathbf{F}[\leq]$.

Unfortunately, the Crane Beach property does not hold in general.

Theorem 4.5 ([3,18]). *There exists a non-regular language having a neutral letter definable in* $\mathbf{FO}[\mathcal{N}]$.

A deeper understanding of the Crane Beach property specialized to first-order logic can be found in [3]. In particular, it has been shown that $\mathbf{FO}[\leq, \mathcal{M}^{\mathrm{unif}}]$ has the Crane Beach property. Here we obtain the following result as a simple corollary of Theorem 2.2.

Theorem 4.6. $\mathbf{MSO}[\leq, \mathcal{M}]$ *has the Crane Beach property.*

4.2 Morphic Regular Languages

In this subsection, we apply Theorem 3.4 to the case of morphic predicates, and obtain the following result: given an **MSO** formula with morphic predicates, it is decidable whether it defines a regular language.

The class of morphic predicates was first introduced by Thue in the context of combinatorics on words, giving rise to the HD0L systems. Formally, let A, B be two finite alphabets, $\sigma : A^* \to A^*$ a morphism, $a \in A$ a letter such that $\sigma(a) = a \cdot u$ for some $u \in A^+$ and $\varphi : A^* \to B^*$ a morphism. This defines the sequence of words $\varphi(a), \varphi(\sigma(a)), \varphi(\sigma^2(a)), \ldots$, which converges to a finite or infinite word. An infinite word obtained in this way is said morphic.

We see morphic words as predicates, and denote by HD0L the class of morphic predicates. The languages definable in $\mathbf{MSO}[\leq, \mathrm{HD0L}]$ are called morphic regular.

Theorem 4.7. *The following problem is decidable: given L a morphic regular language, is L regular?*

The proof of this theorem goes in two steps: first, we reduce the regularity problem for a morphic regular language L to deciding the ultimate periodicity of $\mathbf{P}_L$, and second, we show that $\mathbf{P}_L$ is morphic. Hence we rely on the following result: given a morphic word, it is decidable whether it is ultimately periodic. The decidability of this problem was conjectured 30 years ago and proved recently and simultaneously by Durand and Mitrofanov [8].

The first step is a direct application of Theorem 3.4. For the second step, observe that thanks to Lemma 3.6, we have $\mathbf{P}_L \in \mathbf{MSO}[\leq, \mathrm{HD0L}]$. We conclude with the following result from [7].

Lemma 4.8. HD0L *is closed under* **MSO***-interpretations, i.e. if* $\mathbf{P}$ *is an infinite word such that* $\mathbf{P} \in \mathbf{MSO}[\leq, \mathrm{HD0L}]$, *then* $\mathbf{P} \in$ HD0L.

Acknowledgments. This paper and its authors owe a lot to the numerous fruitful discussions we had with Olivier Carton, Thomas Colcombet, Sam van Gool and Jean-Éric Pin. The same goes for the anonymous referees, who improved the paper by their constructive comments.

References

1. Barrington, D.A.M.: Bounded-width polynomial-size branching programs recognize exactly those languages in NC^1. Journal of Computer and System Sciences 38(1), 150–164 (1989)
2. Barrington, D.A.M., Compton, K., Straubing, H., Thérien, D.: Regular languages in NC^1. Journal of Computer and System Sciences 44(3), 478–499 (1992)
3. Barrington, D.A.M., Immerman, N., Lautemann, C., Schweikardt, N., Thérien, D.: First-order expressibility of languages with neutral letters or: The Crane Beach conjecture. Journal of Computer and System Sciences 70(2), 101–127 (2005)
4. Barrington, D.A.M., Thérien, D.: Finite monoids and the fine structure of NC^1. Journal of the Association for Computing Machinery 35(4), 941–952 (1988)
5. Büchi, J.R.: On a decision method in restricted second-order arithmetic. In: Proceedings of the 1st International Congress of Logic, Methodology, and Philosophy of Science, CLMPS 1960, pp. 1–11. Stanford University Press (1962)
6. Carton, O., Thomas, W.: The monadic theory of morphic infinite words and generalizations. Information and Computation 176(1), 51–65 (2002)
7. Dekking, F.M.: Iteration of maps by an automaton. Discrete Mathematics 126(1-3), 81–86 (1994)
8. Durand, F.: Decidability of the HD0L ultimate periodicity problem. RAIRO Theor. Inform. Appl. 47(2), 201–214 (2013)
9. Elgot, C.C., Rabin, M.O.: Decidability and undecidability of extensions of second (first) order theory of (generalized) successor. Journal of Symbolic Logic 31(2), 169–181 (1966)
10. Immerman, N.: Languages that capture complexity classes. SIAM Journal of Computing 16(4), 760–778 (1987)
11. Koucký, M., Lautemann, C., Poloczek, S., Thérien, D.: Circuit Lower Bounds via Ehrenfeucht-Fraissé Games. In: IEEE Conference on Computational Complexity, pp. 190–201 (2006)
12. Kruckman, A., Rubin, S., Sheridan, J., Zax, B.: A myhill-nerode theorem for automata with advice. In: Faella, M., Murano, A. (eds.) GandALF. EPTCS, vol. 96, pp. 238–246 (2012)
13. Nies, A.: Describing groups. Bulletin of Symbolic Logic 13, 305–339, 9 (2007)
14. Péladeau, P.: Logically defined subsets of $\mathbb{N}^k$. Theoretical Computer Science 93(2), 169–183 (1992)
15. Rabinovich, A.: On decidability of monadic logic of order over the naturals extended by monadic predicates. Information and Computation 205(6), 870–889 (2007)
16. Rabinovich, A.: The Church problem for expansions of $(\mathbb{N}, <)$ by unary predicates. Information and Computation 218, 1–16 (2012)
17. Rabinovich, A., Thomas, W.: Decidable theories of the ordering of natural numbers with unary predicates. In: Ésik, Z. (ed.) CSL 2006. LNCS, vol. 4207, pp. 562–574. Springer, Heidelberg (2006)

18. Schweikardt, N.: On the Expressive Power of First-Order Logic with Built-In Predicates. PhD thesis, Gutenberg-Universtät in Mainz (2001)
19. Semenov, A.L.: Decidability of monadic theories. In: Chytil, M., Koubek, V. (eds.) MFCS 1984. LNCS, vol. 176, pp. 162–175. Springer, Heidelberg (1984)
20. Straubing, H.: Finite automata, formal logic, and circuit complexity. Birkhäuser Boston Inc., Boston (1994)

Transforming Two-Way Alternating Finite Automata to One-Way Nondeterministic Automata[*]

Viliam Geffert[1] and Alexander Okhotin[2]

[1] Department of Computer Science, Šafárik University, Košice, Slovakia
`viliam.geffert@upjs.sk`
[2] Department of Mathematics and Statistics, University of Turku,
Turku FI-20014, Finland
`alexander.okhotin@utu.fi`

Abstract. It is proved that a two-way alternating finite automaton (2AFA) with n states can be transformed to an equivalent one-way nondeterministic finite automaton (1NFA) with $f(n) = 2^{\Theta(n \log n)}$ states, and that this number of states is necessary in the worst case already for the transformation of a two-way automaton with universal nondeterminism ($2\Pi_1$FA) to a 1NFA. At the same time, an n-state 2AFA is transformed to a 1NFA with $(2^n - 1)^2 + 1$ states recognizing the complement of the original language, and this number of states is again necessary in the worst case. The difference between these two trade-offs is used to show that complementing a 2AFA requires at least $\Omega(n \log n)$ states.

1 Introduction

Two-way finite automata are one of the basic models of computation. The study of their descriptional complexity is important, in particular, because of its relationship to small-space complexity classes. The question of whether a two-way nondeterministic finite automaton (2NFA) can be transformed to an equivalent two-way deterministic automaton (2DFA) with polynomially many states is related to the L vs. NL problem in the complexity theory [7].

The basic fact about two-way alternating finite automata (2AFA) is that they can recognize only regular languages: Ladner, Lipton and Stockmeyer [9] showed how to transform an n-state 2AFA to an equivalent one-way deterministic automaton (1DFA) with 2^{n2^n} states. Later, Birget [2, Cor. 4.3(2)] developed another transformation of a *halting* n-state 2AFA to an equivalent one-way nondeterministic automaton (1NFA) with $2^{O(n)}$ states. The descriptional complexity of 2AFA has been a subject of recent research [3].

---

[*] This work is the result of the project implementation: Research and Education at UPJŠ—Heading towards Excellent European Universities, ITMS project code: 26110230056, supported by the Operational Program Education funded by the European Social Fund (ESF).

E. Csuhaj-Varjú et al. (Eds.): MFCS 2014, Part I, LNCS 8634, pp. 291–302, 2014.

This paper establishes two new simulations of 2AFA by 1NFA. The roots of the first simulation are in the transformation due to Birget [2], which is described in Section 3. A more sophisticated transformation, which does not require the 2AFA to halt on every input, is given in Section 4; it uses a new data structure—two stratified sets of states—to simulate an arbitrary n-state 2AFA by a 1NFA with around $2^{O(n \log n)}$ states. The precise number of states is given by a combinatorial expression, and the same number of states is shown to be necessary for this simulation in the worst case.

The other construction, presented in Section 5, transforms any given n-state 2AFA to a 1NFA with only $(2^n - 1)^2 + 1$ states, which recognizes the *complement* of the original language. This number of states is also shown to be precise. A similar method for transforming a 2NFA to a 1NFA recognizing the complement was discovered by Vardi [10]; the contribution of this paper is generalizing it to handle 2AFA and proving that it is optimal with respect to the number of states.

Finally, the two transformations given in this paper are used in Section 6 to prove that complementing an n-state 2AFA may require a 2AFA with at least $n(\log_2 n - 2)$ states.

2   Definitions

Two-way finite automata are equipped with an input tape delimited by end-markers, a reading head that scans one square at a time and moves by one square to the left or to the right at each step, and a finite-state control. The possible transitions of an automaton depend on its current state and the symbol currently scanned. On top of this, two-way alternating finite automata (2AFA) use *alternation*, that is, allow unrestricted existential and universal nondeterminism.

The usual way of defining an alternating model of computation is to separate its internal states into existential and universal ones. Multiple transitions from an existential state are then interpreted as a disjunction over the possible transitions ("at least one of them must lead to acceptance"), whereas a universal state defines a conjunction ("all of them must lead to acceptance"). This paper adopts a slightly relaxed definition, where any combination of existential and universal nondeterminism may be defined within a single transition, expressed as any monotone Boolean function (that is, not only a disjunction or a conjunction). The main reason for this extension is that all constructions in this paper naturally apply to all such automata.

Another minor detail of the definition is the condition for acceptance. Some authors define acceptance by entering one of several designated states while at the right end-marker, or while at either end-marker; other authors allow acceptance at any position on the tape. Though these details affect the size of the automaton by at most one state, this difference becomes essential in this paper, which is concerned with the exact size of automata. In this paper, the definition of a transition function is powerful enough to define acceptance in any state q while scanning any symbol a: it is sufficent to let the function $f = \delta(q, a)$ be constant 1.

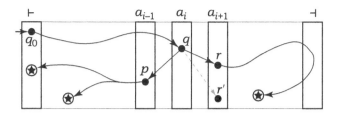

Fig. 1. A computation of a 2AFA, where $\delta(q, a_i)$ gives a function $(p, -1) \wedge ((r, +1) \vee (r', +1))$, which is true on $(p, -1) = 1$, $(r, +1) = 1$, $(r', +1) = 0$

Let $\mathbb{B} = \{0, 1\}$. A Boolean function of n variables $f : \mathbb{B}^n \to \mathbb{B}$ is said to be *monotone*, if, for any two Boolean n-tuples $(x_1, \ldots, x_n), (y_1, \ldots, y_n) \in \mathbb{B}^n$, if $x_i \leqslant y_i$ for all i, then $f(x_1, \ldots, x_n) \leqslant f(y_1, \ldots, y_n)$. For every finite set of variables X, denote by $\mathcal{MF}(X)$ the set of all monotone Boolean functions $f : \mathbb{B}^{|X|} \to \mathbb{B}$ of the variables in X.

Definition 1. *A* two-way alternating finite automaton *(2AFA) is a quadruple* $A = (\Sigma, Q, q_0, \delta)$, *in which*

- Σ *is a finite alphabet with* $\vdash, \dashv \notin \Sigma$, *where the left-end marker* $\vdash$ *and the right-end marker* $\dashv$ *are two special symbols;*
- Q *is a finite set of states;*
- $q_0 \in Q$ *is the initial state;*
- $\delta \colon Q \times (\Sigma \cup \{\vdash, \dashv\}) \to \mathcal{MF}(Q \times \{-1, +1\})$ *is a transition function.*

For an input string $w = a_1 \ldots a_\ell$, *let* $a_0 = \vdash$ *and* $a_{\ell+1} = \dashv$. *A configuration is a pair* (q, i) *of a state* $q \in Q$ *and a position in the input* $i \in \{0, 1, \ldots, \ell, \ell+1\}$. *An accepting computation is an acyclic graph of configurations, which satisfies the following conditions:*

- *the root is the initial configuration* $(q_0, 0)$;
- *for each node* (q, i), *let* $(q_1, i + d_1)$, ..., $(q_k, i + d_k)$ *be all its sons; then the Boolean function* $f = \delta(q, a_i)$ *evaluates to 1 on the substitution* $(q_j, d_j) = 1$ *for all* $j \in \{1, \ldots, k\}$ *and* $(q', d') = 0$ *for all other arguments.*

The language $L(A) \subseteq \Sigma^*$ *recognized by the automaton is the set of all strings, on which there is at least one accepting computation.*

An accepting computation of a 2AFA is illustrated in Figure 1. In this example, $\delta(q, a_i) = (p, -1) \wedge ((r, +1) \vee (r', +1))$. Thus, the transition in state q at a_i gives a function that evaluates to 1 for $(p, -1)$ and $(r, +1)$ true, and hence the rest of the computation follows these two branches, safely ignoring the branch using $(r', +1)$. In every leaf in the graph (that is, a node with no outgoing arcs, marked in the figure by a star), the function f must be constant 1; in other words, each branch ends with an accepting configuration. Note that an accepting computation is *acyclic*, that is, none of its branches may return to an earlier visited configuration.

If each transition $\delta(q, a)$ gives a Boolean function f that is a disjunction of some of its arguments, such an automaton is *nondeterministic* (2NFA, or $2\Sigma_1$FA), and its accepting computation is a line graph. If each function is a conjunction of some of its variables, this is a $2\Pi_1$FA. In a *deterministic* automaton (2DFA), each function f equals one of its arguments.

This paper is about transforming two-way automata to one-way nondeterministic finite automata. The latter can be defined by restricting 2NFA to move only to the right, by eliminating the end-markers, and by allowing acceptance only in the end of the input. This leads to the following simplified definition.

Definition 2. *A one-way nondeterministic finite automaton (1NFA) is a quintuple $B = (\Sigma, Q, Q_0, \delta, F)$, in which*

- *Σ is a finite alphabet;*
- *Q is a finite set of states;*
- *$Q_0 \subseteq Q$ is the set of initial states;*
- *$\delta\colon Q \times \Sigma \to 2^Q$ is a transition function;*
- *$F \subseteq Q$ is the set of accepting states.*

The automaton is said to reach a state $q' \in Q$ from a state $q \in Q$ by an input string $w = a_1 \ldots a_\ell$, with $\ell \geqslant 0$ and $a_i \in \Sigma$, if there exists a sequence of states $q^{(0)}, \ldots, q^{(\ell)}$, where $q^{(0)} = q$, $q^{(i+1)} \in \delta(q^{(i)}, a_i)$ for all i, and $q^{(\ell)} = q'$.

The language $L(A) \subseteq \Sigma^*$ recognized by the automaton is the set of all strings, by which one can reach some state in F from some state in Q_0.

3   Nondeterministic Simulation of Two-Way Automata

The general idea of simulating a two-way automaton A by a 1NFA, used in the constructions by Birget [2] and by Kapoutsis [6], is to guess A's accepting computation. This is done by calculating the crossing sequences of this computation at each symbol, while reading the input from left to right, guessing the continuation of the computation at each symbol, and linking different branches of the computation to each other.

Let $A = (\Sigma, Q, q_0, \delta, F)$ be a 2AFA. After reading a prefix $a_1 \ldots a_i$ of a string $a_1 \ldots a_i a_{i+1} \ldots a_\ell$, the simulating 1NFA shall remember two sets of states $P, R \subseteq Q$ visited by A in its guessed computation at positions i and $i + 1$, as illustrated in Figure 2(left). Here, the letter P stands for *provided* and R stands for *required*. First, the guessed computation of A on the unread suffix $a_{i+1} \ldots a_\ell \dashv$ is assumed to lead from each *required* state at position $i+1$ to acceptance or to any *provided* states at position i. At the same time, on the previously read prefix $\vdash a_1 \ldots a_i$, the computation from each *provided* state at position i should lead to acceptance or to any *required* states at position $i + 1$. Finally, from the initial configuration, each of the branches in the guessed computation must go to any *required* states at position $i + 1$, or halt and accept in the prefix $\vdash a_1 \ldots a_i$.

There is a serious complication with this data structure, illustrated in Figure 2(right). If the computation from a *provided* state $p \in P$ leads to a *required*

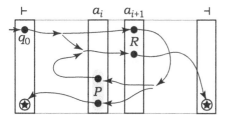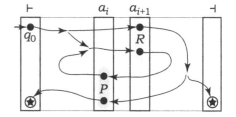

Fig. 2. Computations of a 2AFA and its crossing sections represented in the sets P and R

state $r \in R$, and the transition from the latter state goes back to the same *provided* state p, then an infinite loop has been formed, and there is no way to detect that. Then, some computations may be mislead into this infinite loop, under the wrong assumption that they accept there. In the special case of A being a 2NFA, where the computation being guessed is a single path rather than an acyclic graph, Kapoutsis [6] was able to rule out this problematic case by requiring that $|P| + 1 = |R|$, and by maintaining a bijection between these two sets; this, however, cannot be done for an alternating automaton. For a 2AFA, Birget [2, Thm. 2.3(4)] proved this construction to work under the assumption that the given 2AFA never loops, that is, cannot revisit the same configuration twice within a single computation path.

The first contribution of this paper is an extension of this construction to an arbitrary given 2AFA.

4   Simulation of 2AFA by 1NFA

The idea of the proposed simulation is to stratify the sets of states R and P, so that different states in R and P may be assigned to different strata, and then simulate only transitions from lower strata to higher strata. This will ensure that no cycles are ever formed in the guessed computation.

For that purpose, instead of the two sets R, P, the new construction shall use finite sequences of sets of states $(R_1, P_1, ..., R_k, P_k)$, where $R_1, \ldots, R_k \subseteq Q$ are pairwise disjoint sets that replace the set R, whereas the pairwise disjoint sets $P_1, \ldots, P_k \subseteq Q$ replace P. Then, for each required state $r \in R_j$ from a j-th stratum, the computation on the unread suffix beginning with r at position $i + 1$ is assumed lead to acceptance or to any provided states from the j-th or higher strata, at position i. For each *provided* state $p \in P_j$ from a j-th stratum, it is known that the computation beginning from state p at position i leads to acceptance or to any required states from strata $j + 1$ or higher, at position $i + 1$.

For example, the crossing section in Figure 2(left) can be represented with the first state from R as R_1, with the whole P as P_1, and with the second state from R as R_2. The last set P_2 is empty in this case. On the other hand, the sets P, R in Figure 2(right) admit no stratified representation.

Lemma 1. *For every 2AFA with a set of states Q, there exists a 1NFA with the set of states*

$$Q' = \big\{ (R_1, P_1, ..., R_k, P_k) \big| k \geqslant 0, R_i, P_i \subseteq Q, R_1, \ldots, R_k \neq \varnothing, P_1, \ldots, P_{k-1} \neq \varnothing,$$
$$P_1 \cup \ldots \cup P_k \neq Q, \ R_i \cap R_j = P_i \cap P_j = \varnothing \text{ for all } i \neq j \big\},$$

which recognizes the same language.

Proof. Let $A = (\Sigma, Q, q_0, \delta)$ be the given 2AFA, define a new 1NFA $B = (\Sigma, Q', Q_0', \delta', F')$. The goal of the construction is that B can reach a state $(R_1, P_1, \ldots R_k, P_k)$ after reading a prefix $a_1 \ldots a_m$ if and only if

- there is an acyclic computation graph beginning with the initial configuration $(q_0, 0)$ and restricted to the prefix $\vdash a_1 \ldots a_m$, where each branch either ends in an accepting configuration or in a configuration $(r, m+1)$, with $r \in R_1 \cup \ldots \cup R_k$ (the latter case means leaving the prefix from its last symbol a_m to the right), and
- for every i and for every state $p \in P_i$, there is an acyclic computation graph beginning with the configuration (p, m) and restricted to the prefix $\vdash a_1 \ldots a_m$, where each branch either ends in an accepting configuration or in a configuration $(r, m+1)$, with $r \in R_{i+1} \cup \ldots \cup R_k$.

The initial states Q_0', the transition function δ' and the accepting states F' of B are defined by a unified construction. Consider a state $(R_1, P_1, \ldots, R_k, P_k)$, which represents a guessed computation on a prefix $x \in \{\varepsilon\} \cup \vdash \Sigma^*$, and the next symbol $a \in \Sigma \cup \{\vdash, \dashv\}$. The 1NFA B should guess further fragments of this computation occurring between the rightmost symbol of x and the symbol a, and represent the resulting computation on xa as the next state $(R_1', P_1', \ldots, R_\ell', P_\ell')$.

For this to be possible, one should first fix a certain *alignment* between the sets $R_1, P_1, \ldots, R_k, P_k$ and $R_1', P_1', \ldots, R_\ell', P_\ell'$, which determines the order in which they may refer to each other. This alignment is represented as a sequence α of $2k + 2\ell$ sets, where all sets $R_1, P_1, \ldots, R_k, P_k$ occur in α in their original order, and so do all sets $R_1', P_1', \ldots, R_\ell', P_\ell'$. Let X be any of the sets $R_1, \ldots, R_k, P_1', \ldots, P_\ell'$ (these are the sets of states visited on the symbol a). Then, for every state $q \in X$, the Boolean function $f = \delta(q, a)$ must evaluate to 1 under the following substitution:

- the argument $(p, -1)$ is set to 1 if and only if there is a set P_i that occurs in α later than X and contains p.
- the argument $(r, +1)$ is set to 1 if and only if there is a set R_i that occurs in α later than X and contains r.

If these conditions hold, this shall be called a *transition from* $(R_1, P_1, \ldots, R_k, P_k)$ to $(R_1', P_1', \ldots, R_\ell', P_\ell')$ *by* a, and denoted by $(R_1, P_1, \ldots, R_k, P_k) \xrightarrow{a} (R_1', P_1', \ldots, R_\ell', P_\ell')$.

Then, initial states of the 1NFA B are defined using a transition by the left end-marker $\vdash$ from the pair $(\{q_0\}, \varnothing)$ representing the initial configuration of A.

$$Q_0' = \big\{ (R_1, P_1, \ldots, R_k, P_k) \mid (\{q_0\}, \varnothing) \xrightarrow{\vdash} (R_1, P_1, \ldots, R_k, P_k) \big\}$$

The transition function of B directly uses the above notion of a transition.

$$\delta'((R_1, P_1, \ldots, R_k, P_k), a) = \{ (R'_1, P'_1, \ldots, R'_\ell, P'_\ell) \mid$$
$$(R_1, P_1, \ldots, R_k, P_k) \xrightarrow{a} (R'_1, P'_1, \ldots, R'_\ell, P'_\ell)\}$$

Finally, accepting states are defined by a transition by the right end-marker $\dashv$ to an empty sequence of sets, that is, to $(R'_1, P'_1, \ldots, R'_\ell, P'_\ell)$ with $\ell = 0$. This sequence means that all A's transitions have been successfully resolved into accepting computations.

$$F' = \{ (R_1, P_1, \ldots, R_k, P_k) \mid (R_1, P_1, \ldots, R_k, P_k) \xrightarrow{\dashv} \text{empty} \}$$

The above correctness statement can be proved by an induction on the length of the prefix of the string.   □

Lemma 2. *For every $n \geqslant 1$, there exists an alphabet Σ_n and an n-state $2\Pi_1$ FA over Σ_n, for which every 1NFA recognizing the same language must have at least as many states as there are sequences $(R_1, P_1, ..., R_k, P_k)$, with $k \geqslant 0$, $R_i, P_i \subseteq \{1, \ldots, n\}$, $R_1, \ldots, R_k \neq \varnothing$, $P_1, \ldots, P_{k-1} \neq \varnothing$, $P_1 \cup \ldots \cup P_k \neq \{1, \ldots, n\}$ and $R_i \cap R_j = P_i \cap P_j = \varnothing$ for all $i \neq j$.*

Proof. Let $n \geqslant 2$ and let $Q = \{q_0, \ldots, q_{n-1}\}$. For every sequence $(R_1, P_1, ..., R_k, P_k)$, as in the statement of the lemma, the alphabet contains the symbols $a_{R_1, P_1, \ldots, R_k, P_k}$ and $b_{R_1, P_1, \ldots, R_k, P_k}$. In addition, there is a symbol c_q for each state $q \in Q$.

Assume that the states in Q are ordered as $q_0 < \ldots < q_{n-1}$. The automaton to be constructed is designed to accept all strings of the form $c_{q_{init}} a_{R_1, P_1, \ldots, R_k, P_k} b_{R_1, P_1, \ldots, R_k, P_k}$, where $q_{init} = \max Q \setminus (P_1 \cup \ldots \cup P_k)$ and the subscripts of the symbols a and b match each other. At the same time, the automaton should reject every string $c_{q_{init}} a_{R_1, P_1, \ldots, R_k, P_k} b_{R'_1, P'_1, \ldots, R'_\ell, P'_\ell}$ with mismatched subscripts.

Define a $2\Pi_1$FA with the set of states $Q = \{q_0, \ldots, q_{n-1}\}$, of which q_0 is the initial state. The automaton has the following transitions.

$$\delta(q_0, \vdash) = (q_0, +1)$$
$$\delta(q_0, c_q) = (q, +1)$$
$$\delta(q_{n-1}, c_q) = (q_{n-1}, -1)$$

$$\delta(q, a_{R_1, P_1, \ldots, R_k, P_k}) = \begin{cases} \bigwedge_{r \in R_1 \cup \ldots \cup R_k} (r, +1), & \text{if } q = \min Q \setminus (P_1 \cup \ldots \cup P_k) \\ \bigwedge_{r \in R_{i+1} \cup \ldots \cup R_k} (r, +1), & \text{if } q \in R_i \\ (q_f, -1), & \text{if } q \in P_k \end{cases}$$

$$\delta(r, b_{R_1, P_1, \ldots, R_k, P_k}) = \bigwedge_{p \in P_i \cup \ldots \cup P_k} (p, -1) \qquad (\text{unless } i = k \text{ and } P_k = \varnothing)$$

$$\delta(r, b_{R_1, P_1, \ldots, R_k, P_k}) = (q_{n-1}, +1) \qquad (\text{for all } r \in R_k, \text{ if } P_k = \varnothing)$$

It remains to prove that every 1NFA recognizing the language $L(A)$ must have the required number of states. The proof is carried out using the standard *fooling set* method, which consists of constructing such a set $S \subseteq \Sigma^* \times \Sigma^*$ of pairs of strings, that the concatenation uv of every pair $(u, v) \in S$ belongs to $L(A)$, but for every two distinct pairs $(u, v), (u', v') \in S$, at least one of the mismatched concatenations uv' and $u'v$ is not in $L(A)$. Once such a set is constructed, it is evident that every 1NFA recognizing this language must have at least as many states as there are pairs in $L(A)$: indeed, the states entered by the 1NFA after reading the first component of each pair must be pairwise distinct.

In this case, the fooling set consists of pairs corresponding to different sequences $(R_1, P_1, ..., R_k, P_k)$, as in the statement of the lemma. For every such sequence, the set S contains a pair of a string $u_{R_1,P_1,...,R_k,P_k} = c_{q_{init}} a_{R_1,P_1,...,R_k,P_k}$, where $q_{init} = \max Q \setminus (P_1 \cup ... \cup P_k)$, and another string $v_{R_1,P_1,...,R_k,P_k} = b_{R_1,P_1,...,R_k,P_k}$.

Consider any two sequences $(R_1, P_1, ..., R_k, P_k) \neq (R_1', P_1', ..., R_\ell', P_\ell')$ with $k, \ell \geqslant 1$, and the associated pairs of strings. Let i be the least number, for which

$$R_i \cup ... \cup R_k \neq R_i' \cup ... \cup R_\ell' \quad \text{or}$$
$$P_i \cup ... \cup P_k \neq P_i' \cup ... \cup P_\ell'.$$

Consider the former case. Then there exists a state q belonging to one of these unions and not to the other. Assume, without loss of generality, that $q \in R_i \cup ... \cup R_k$ and $q \notin R_i' \cup ... \cup R_\ell'$. Then consider the concatenation

$$u_{R_1,P_1,...,R_k,P_k} v_{R_1',P_1',...,R_\ell',P_\ell'} = c_{q_{init}} a_{R_1,P_1,...,R_k,P_k} b_{R_1',P_1',...,R_\ell',P_\ell'}.$$

Assume that $i \geqslant 2$. In this case, one of the branches of the computation tree on this string contains a transition from some state in P_{i-1} at the symbol $a_{R_1,P_1,...,R_k,P_k}$ to the state q at the symbol $b_{R_1',P_1',...,R_\ell',P_\ell'}$. If $q \in R_1' \cup ... \cup R_{i-1}'$, then this transition is followed by a transition to the same state from P_{i-1} at the symbol $a_{R_1,P_1,...,R_k,P_k}$, and this loop in the computation tree shows that the string is rejected. Otherwise, $q \notin R_1' \cup ... \cup R_\ell'$, and hence the transition from q at $b_{R_1',P_1',...,R_\ell',P_\ell'}$ is undefined, and the string is rejected as well.

If $i = 1$, then there is a transition from the state q_{init} at $a_{R_1,P_1,...,R_k,P_k}$ to the state q at the symbol $b_{R_1',P_1',...,R_\ell',P_\ell'}$. Since $q \notin R_1' \cup ... \cup R_\ell'$ by the assumption, the string is rejected.

The second case of a state q belonging to one of the sets $P_i \cup ... \cup P_k$ and $P_i' \cup ... \cup P_\ell'$, but not to the other is handled similarly. $\qquad \square$

Having proved that the 2AFA–1NFA tradeoff is exactly the number of sequences given in Lemma 2, it remains to assess the number of such sequences. For every number $n \geqslant 1$, denote by $F(n)$ the number of distinct sequences $(R_1, P_1, ..., R_k, P_k)$, with $k \geqslant 0$, $R_i, P_i \subseteq \{1, ..., n\}$, $R_1, ..., R_k \neq \varnothing$, $P_1, ..., P_{k-1} \neq \varnothing$, $P_1 \cup ... \cup P_k \neq \{1, ..., n\}$ and $R_i \cap R_j = P_i \cap P_j = \varnothing$ for all $i \neq j$. The following lemma gives rough bounds on $F(n)$, which are sufficient to estimate its growth rate.

Lemma 3. *For every $n \geqslant 1$, the number $f(n)$ is confined within the following bounds.*

$$(n!)^2 \leqslant F(n) \leqslant 4^n(2n)!$$

Accordingly, $F(n) = 2^{\Theta(n \log n)}$.

Proof. For the upper bound, first, any two subsets of Q are chosen: these are the sets $R = R_1 \cup \ldots \cup R_k$ and $P = P_1 \cup \ldots \cup P_k$, and there are 4^n possible choices. Thus, at most $2n$ elements belonging to R or P are chosen, and they can be listed in any order; there are $(2n)!$ possible permutations of these elements.

For the lower bound argument, consider any permutation $(j_1, \ldots, j_n)$ of the first n numbers (there are $n!$ of them) and any permutation $(i_1, \ldots, i_{n-1})$ of any $n-1$ of the first n numbers ($n!$ of them as well). These permutations give rise to a sequence of singletons $(R_1, P_1, \ldots R_n, P_n)$, where $R_t = \{j_t\}$ for all t, $P_t = \{i_t\}$ for all $t \neq n$, and $P_n = \varnothing$. This sequence satisfies the definition, and hence there are at least as many sequences satisfying the definition as there are such pairs of permutations.

The function is estimated as $2^{\Theta(n \log n)}$ according to Stirling's approximation for both the lower and the upper bounds. $\square$

Altogether, the succinctness tradeoff between 2AFA and 1NFA is characterized in the following theorem.

Theorem 1. *The number of states in a 1NFA that is sufficient and in the worst case necessary to represent any language recognized by an n-state 2AFA is exactly $F(n) = 2^{\Theta(n \log n)}$. This bound is precise already for a transformation from a $2\Pi_1 FA$ to a 1NFA.*

5   Simulating the Complement of a 2AFA by a 1NFA

In the previous section, simulating a two-way automaton by a 1NFA involved guessing its computation, that is, a certain data object that witnesses acceptance. A 1NFA recognizing *the complement* of a language accepted by a two-way automaton would have to guess a certain *witness for rejection*. Such a witness was first found for a 2NFA by Vardi [10], who presented a construction of a 1NFA with $4^n + 2^n$ states recognizing the complement of a given n-state 2NFA.

This paper develops a similar transformation of a 2AFA to a 1NFA recognizing the complement. The general idea is to reuse the transformation of Birget [1], as presented in Section 3, which works incorrectly for a looping 2AFA. Once the goal is to recognize the complement, this incorrect operation is precisely what is needed, because looping computations can now be handled as if they are accepting.

Before proceeding with the construction, it is convenient to define a complementary notion to the accepting computation of a 2AFA: a *witness for rejection*. The 1NFA to be constructed is going to guess this object.

Lemma 4. *Let $A = (\Sigma, Q, q_0, \delta,)$ be a 2AFA and let $w = a_1 \ldots a_\ell$ be an input string. A witness for rejection of w is a graph of configurations, which satisfies the following conditions:*

- *the root is the initial configuration $(q_0, 0)$;*
- *for each internal node (q, i), let $(q_1, i + d_1)$, $\ldots$, $(q_k, i + d_k)$ be all its sons; then the Boolean function $f = \delta(q, a_i)$ evaluates to 0 on the substitution $(q_j, d_j) = 0$ for all $j \in \{1, \ldots, k\}$ and $(q', d') = 1$ for all other arguments.*

*Then, w is **not accepted** by A if and only if there exists a witness for rejection of w.*

Note that every path in a witness for rejection either ends in a rejecting configuration, where f is constant 0, or loops by returning to some earlier node on this path.

Lemma 5. *For every 2AFA A with a set of states Q, there exists a 1NFA with the set of states*

$$Q' = \{ (P, R) \mid P, R \subseteq Q, \ P, R \neq \varnothing \} \cup \{q_{\text{acc.all}}\},$$

which recognizes the complement of $L(A)$.

Proof. Let $A = (\Sigma, Q, q_0, \delta)$ be the given 2AFA. One should define a new 1NFA $C = (\Sigma, Q', Q'_0, \delta', F')$ recognizing $\overline{L(A)}$.

The goal of the construction is that whenever a state (P, R) is reachable by C after reading a prefix $a_1 \ldots a_m$,

- there is a computation graph beginning with the initial configuration $(q_0, 0)$ and restricted to the prefix $\vdash a_1 \ldots a_m$, where each path either ends in a rejecting configuration, or in a configuration $(r, m + 1)$, with $r \in R$, or returns to an earlier node on this path, *and*
- for every state $p \in P$, there is a computation graph beginning with the configuration (p, m) and restricted to the prefix $\vdash a_1 \ldots a_m$, where each path either ends in a rejecting configuration or in a configuration $(r, m + 1)$, with $r \in R$, or returns to an earlier node.

The construction resembles that in Lemma 1, but is substantially simpler due to no stratification. For every two states (P, R) and (P', R'), and for every symbol $a \in \Sigma \cup \{\vdash, \dashv\}$, define $(P, R) \xrightarrow{a} (P', R')$ by requiring that for every state $q \in P' \cup R$, the Boolean function $f = \delta(q, a)$ evaluates to 0 under the following substitution: $(p, -1)$ is 0 if and only if $p \in P$, and $(r, +1)$ is 0 if and only if $r \in R'$, whereas all remaining variables are set to 1.

Then, define

$$Q'_0 = \{ (P, R) \mid (\varnothing, \{q_0\}) \xrightarrow{\vdash} (P, R) \},$$

$$\delta'((P, R), a) = \{ (P', R') \mid (P, R) \xrightarrow{a} (P', R') \} \cup \{ q_{\text{acc.all}} \mid \text{if } (P, R) \xrightarrow{a} (\varnothing, \varnothing) \},$$

$$\delta'(q_{\text{acc.all}}, a) = \{q_{\text{acc.all}}\},$$

$$F' = \{ (P, R) \mid (P, R) \xrightarrow{\dashv} (\varnothing, \varnothing) \} \cup \{q_{\text{acc.all}}\}.$$

The following matching lower bound holds for the transformation of a 2NFA to a 1NFA recognizing the complement.

Lemma 6. *For every $n \geqslant 1$ there exists an n-state 2NFA with acceptance only on the end-markers, for which every 1NFA recognizing the complementary language must have $(2^n - 1)^2 + 1$ states.*

Proof. Let $Q = \{q_0, \ldots, q_{n-1}\}$. For all subsets $P, R \subseteq Q$, with $P, R \neq \varnothing$, the alphabet contains the symbols $a_{P,R}$ and $b_{P,R}$. For each state $q \in Q$, there is a symbol c_q. The automaton has the following transitions.

$$\delta(q_0, \vdash) = (q_0, +1)$$
$$\delta(q_0, c_q) = (q, +1)$$
$$\delta(p, a_{P,R}) = \bigvee_{r \in R} (r, +1) \qquad \text{(for all } p \in P)$$
$$\delta(r, b_{P,R}) = \bigvee_{p \in P} (p, -1) \qquad \text{(for all } r \in R)$$

All remaining transitions specify immediate acceptance. In order to restrict acceptance to the end-markers, it is sufficient to define all remaining transitions as $\delta(q, c_{q'}) = (q_{n-1}, -1)$ $\delta(q, a_{P,R}) = (q_{n-1}, -1)$ and $\delta(q, b_{P,R}) = (q_{n-1}, +1)$, and let q_{n-1} be accepting on both end-markers.

The lower bound on the size of every 1NFA recognizing $\overline{L(A)}$ is obtained using the fooling set method, as in Lemma 2. Here the fooling set consists of the following pairs:

- for each $P, R \subseteq Q$ with $P, R \neq \varnothing$, a pair $u_{P,R} = c_{\min P} a_{P,R}$, $v_{P,R} = b_{P,R}$.
- one more pair $u_{q_{\text{acc.all}}} = c_{q_0} a_{\{q_0\},\{q_0\}} b_{\{q_0\},\{q_0\}}$, $v_{q_{\text{acc.all}}} = \varepsilon$.

On a correct concatenation $u_{P,R} v_{P,R}$, all computations of the 2NFA loop between $a_{P,R}$ and $b_{P,R}$. Similarly, on the concatenation $u_{q_{\text{acc.all}}} v_{q_{\text{acc.all}}}$, the 2NFA loops between the last two symbols. Thus, as required by the definition of the fooling set, the concatenation of each pair is in $\overline{L(A)}$.

Consider mismatched concatenations between any pairs $(u_{P,R}, v_{P,R})$ and $(u_{P',R'}, v_{P',R'})$. If $R \neq R'$, assume, without loss of generality, that there is a state $r \in R \setminus R'$; then A accepts the string $u_{P,R} v_{P',R'}$. If $R = R'$ and $P \neq P'$, let $p \in P \setminus P'$; then the string $u_{P',R} v_{P,R}$ is accepted.

Finally, for any pair $(u_{P,R}, v_{P,R})$, the mismatched concatenation $u_{P,R} v_{q_{\text{acc.all}}}$ is accepted by A. $\qquad\square$

Theorem 2. *The number of states in a 1NFA that is sufficient and in the worst case necessary to represent the complement of any language recognized by an n-state 2AFA is exactly $(2^n - 1)^2 + 1$. The bound is precise already for a transformation from a 2NFA to a 1NFA.*

6   Implications on Complementing 2AFA

Assume that a language is recognized by an n-state 2AFA. How many states are necessary in the worst case to represent the complement of such a language by another 2AFA?

Complementing a one-way deterministic automaton (1DFA) is trivial, as it can be done by inverting the set of states. For other types of automata, this problem is more interesting. For 2DFA, the known construction for the complementation uses $4n$ states, and proceeds by making the automaton halt on every input [5], or, to be precise, by constructing a reversible 2DFA [8]; no lower bound on complementing 2DFA is known up to date. Complementing a 1NFA in the worst case requires 2^n states [1], that is, requires determinizing it. The problem of complementing a 2NFA is related to the 2NFA–2DFA tradeoff, on which very little is known; however, for a unary alphabet, a 2NFA can be implemented using n^8 states using inductive counting [5], and there is a lower bound of $\Omega(n^2)$ states.

This paper contributes the first lower bound on complementing a 2AFA, obtained from the gap between the two precise trade-offs in Theorems 1 and 2.

Theorem 3. *For every $n \geqslant 1$, the number of states in a 2AFA needed to represent the complement of every n-state 2AFA is greater than $n(\log_2 n - 2)$.*

References

1. Birget, J.-C.: Partial orders on words, minimal elements of regular languages, and state complexity. Theoretical Computer Science 119, 267–291 (1993)
2. Birget, J.-C.: State-complexity of finite-state devices, state compressibility and incompressibility. Mathematical Systems Theory 26(3), 237–269 (1993)
3. Geffert, V.: An alternating hierarchy for finite automata. Theoretical Computer Science 445, 1–24 (2012)
4. Geffert, V., Mereghetti, C., Pighizzini, G.: Converting two-way nondeterministic unary automata into simpler automata. Theoretical Computer Science 295(1-3), 189–203 (2003)
5. Geffert, V., Mereghetti, C., Pighizzini, G.: Complementing two-way finite automata. Information and Computation 205(8), 1173–1187 (2007)
6. Kapoutsis, C.A.: Removing bidirectionality from nondeterministic finite automata. In: Jedrzejowicz, J., Szepietowski, A. (eds.) MFCS 2005. LNCS, vol. 3618, pp. 544–555. Springer, Heidelberg (2005)
7. Kapoutsis, C.A.: Two-way automata versus logarithmic space. In: Kulikov, A., Vereshchagin, N. (eds.) CSR 2011. LNCS, vol. 6651, pp. 359–372. Springer, Heidelberg (2011)
8. Kunc, M., Okhotin, A.: Reversibility of computations in graph-walking automata. In: Chatterjee, K., Sgall, J. (eds.) MFCS 2013. LNCS, vol. 8087, pp. 595–606. Springer, Heidelberg (2013)
9. Ladner, R., Lipton, R., Stockmeyer, L.: Alternating pushdown and stack automata. SIAM Journal on Computing 13(1), 135–155 (1984)
10. Vardi, M.: A note on the reduction of two-way automata to one-way automata. Information Processing Letters 30(5), 261–264 (1989)

Measure Properties of Game Tree Languages

Tomasz Gogacz[1], Henryk Michalewski[2],
Matteo Mio[3], and Michał Skrzypczak[2,*]

[1] Institute of Computer Science, University of Wrocław, Poland
[2] Faculty of Mathematics, Informatics and Mechanics, University of Warsaw, Poland
[3] Computer Laboratory, University of Cambridge, England (UK)

Abstract. We introduce a general method for proving measurability of topologically complex sets by establishing a correspondence between the notion of *game tree languages* from automata theory and the σ-algebra of $\mathcal{R}$-*sets*, introduced by A. Kolmogorov as a foundation for measure theory. We apply the method to answer positively to an open problem regarding the game interpretation of the probabilistic μ-calculus.

1 Introduction

Among logics for expressing properties of nondeterministic (including concurrent) processes, represented as transition systems, Kozen's modal μ-calculus [14] plays a fundamental rôle. This logic enjoys an intimate connection with parity games, which offers an intuitive reading of fixed-points, and underpins the existing technology for model-checking μ-calculus properties. An abstract setting for investigating parity games, using the tools of descriptive set theory, is given by so-called *game tree languages* (see, e.g. [2]). The language $\mathcal{W}_{i,k}$ is the set of parity games with priorities in $\{i \dots k\}$, played on an infinite binary tree structure, which are winning for Player $\exists$. The (i, k)-indexed sets $\mathcal{W}_{i,k}$ form a strict hierarchy of increasing topological complexity called the *index hierarchy* of game tree languages (see [5,1,2]). Precise definitions are presented in Section 2.

For many purposes in computer science, it is useful to add probability to the computational model, leading to the notion of probabilistic nondeterministic transition systems (PNTS's). In an attempt to identify a satisfactory analogue of Kozen's μ-calculus for expressing properties of PNTS's, the third author has recently introduced in [18,19] a quantitative fixed-point logic called *probabilistic μ-calculus with independent product* (pLμ). A central contribution of [19] is the definition of a game interpretation of pLμ, given in terms of a novel class of games

---

$\star$ All the authors were supported by the *Expressiveness of Modal Fixpoint Logics* project realized within the 5/2012 Homing Plus programme of the Foundation for Polish Science, work done during an intership of the first author in WCMCS. Additionally, the third author thanks Alex Simpson for numerous suggestions and comments, and acknowledges with gratitude the support of Advanced Grant "ECSYM" of the ERC.

E. Csuhaj-Varjú et al. (Eds.): MFCS 2014, Part I, LNCS 8634, pp. 303–314, 2014.
© Springer-Verlag Berlin Heidelberg 2014

generalizing ordinary two-player *stochastic* parity games. While in ordinary two-player (stochastic) parity games the outcomes are infinite sequences of game-states, in pLμ-games the outcomes are infinite trees, called *branching plays*, whose vertices are labelled with game-states. This is because in pLμ some of the game-states, called *branching states*, are interpreted as generating distinct game-threads, one for each successor state of the branching state, which continue their execution *concurrently* and *independently*. The winning set of a pLμ-game is therefore a collection of branching plays specified by a combinatorial condition associated with the structure of the game arena.

Unlike winning sets of ordinary two-player (stochastic) parity games, which are well-known to be Borel sets[1], the winning sets of pLμ-games generally belong to the $\mathbf{\Delta}_2^1$-class of sets in the projective hierarchy of Polish spaces [19, Theorem 4.20]. This high topological complexity is a serious concern because pLμ-games are *stochastic*, i.e. the final outcome (the branching play) is determined not only by the choices of the two players but also by the randomized choices made by a probabilistic agent. A pair of strategies for $\exists$ and $\forall$, representing a play up-to the choice of the probabilistic agent, only defines a probability measure on the space of outcomes. For this reason, one is interested in the *probability* of a play to satisfy the winning condition. Under the standard Kolmogorov's measure-theoretic approach to probability theory, a set has a well-defined probability only if it is a *measurable set*[2]. Due to a result of Kurt Gödel (see [10, § 25]), it is consistent with Zermelo-Fraenkel Set Theory with the Axiom of Choice (ZFC) that there exists a $\mathbf{\Delta}_2^1$ set which is not measurable. This means that it is not possible to prove (in ZFC) that all $\mathbf{\Delta}_2^1$-sets are measurable. However it may be possible to prove that a *particular* set (or family of sets) in the $\mathbf{\Delta}_2^1$-class is measurable. In [18] the author asks the following question[3]:

Question: are the winning sets of pLμ-games provably measurable?

This problem provided the original motivation of our work. We will answer positively to the question by developing a general method for proving measurability of topologically complicated sets.

This type of questions has been investigated since the first developments of measure theory, in late 19th century, as the existence of non-measurable sets (e.g. Vitali sets [10]) was already known. The measure-theoretic foundations of probability theory are based around the concept of a σ-algebra of measurable events on a space of potential outcomes. Typically, the σ-algebra is assumed to contain all open sets. Hence the minimal σ-algebra under consideration consists of all Borel sets whereas the maximal consists, by definition, of the collection of all measurable sets. The Borel σ-algebra, while simple to work with, lacks important classes of measurable sets (e.g. $\mathbf{\Pi}_1^1$-complete sets). On the other hand, the full σ-algebra of measurable sets may be difficult to work with since there

---

[1] See, e.g., Remark 10.57 in [3] for a discussion about measurability in this context.

[2] More precisely, *universally measurable*, see Section 2.

[3] Statement "is mG-UM(Γ_p) true?", see Definition 5.1.18 and discussion at the end of Section 4.5 in [18]. See also Section 8.1 in [19].

is no constructive methodology for establishing its membership relation, i.e. for proving that a given set belongs to this σ-algebra.

This picture led to a number of attempts to find the largest σ-algebra, extending the Borel σ-algebra and including as many measurable sets as possible and, at the same time, providing practical techniques for establishing the membership relation. A general methodology for constructing such σ-algebras is to identify a family $\mathcal{F}$ of *safe* operations on sets which, when applied to measurable sets are guaranteed to produce measurable sets. When the operations considered have countable arity (e.g. countable union), the σ-algebra generated by the open sets closing under the operations in $\mathcal{F}$ admits a transfinite decomposition into ω_1 levels, and this allows the membership relation to be established inductively. The simplest case is given by the σ-algebra of Borel sets, with $\mathcal{F}$ consisting of the operations of complementation and countable union. Other less familiar examples include $\mathcal{C}$-sets studied by E. Selivanovski [20], Borel programmable sets proposed by D. Blackwell [4] and $\mathcal{R}$-sets proposed by A. Kolmogorov [13].

The σ-algebra of $\mathcal{R}$-sets is, to our knowledge, the largest ever considered. Most measurable sets arising in ordinary mathematics are $\mathcal{R}$-sets belonging to the finite levels of the transfinite hierarchy of $\mathcal{R}$-sets. For example, all Borel sets, analytic sets, co-analytic sets and Selivanovski's $\mathcal{C}$-sets lie in the first two levels [8]. Thus, for most practical purposes, the following principle is valid:

Principle: " all practically useful measurable sets belong to the finite
   levels of the transfinite hierarchy of Kolomogorov's $\mathcal{R}$-sets."

Contributions. The definition of $\mathcal{R}$-sets in [13], formulated in terms of operations on sets and transformations on operations (Section 3), is purely settheoretical. As a main technical contribution of this work, we provide an alternative game-theoretical characterization of the finite levels of the hierarchy of $\mathcal{R}$-sets in terms of game tree languages $\mathcal{W}_{i,k}$.

Theorem 1. $\mathcal{W}_{k-1,2k-1}$ *is complete for the k-th level of the hierarchy of $\mathcal{R}$-sets.*

As a consequence one can establish the measurability of a given set $A \subseteq X$ by constructing a continuous reduction to $\mathcal{W}_{i,k}$. This can be thought as a *coding* f of elements in X in terms of parity games with priorities in $\{i, \dots, k\}$ such that $x \in A$ if and only if $f(x)$ is winning for Player $\exists$. Parity games are well-known and relatively simple to work with. Thus the proof method allows for easier applications. Since $\mathcal{R}$-sets exhaust the realm of reasonable measurable sets, and the sets $\mathcal{W}_{i,k}$ are complete among $\mathcal{R}$-sets, the method should cover most cases.

Additionally, in Section 6, we investigate the special $\aleph_1$-continuity property of measures on $\mathcal{W}_{i,k}$ with respect to the approximations $\mathcal{W}_{i,k}^{\alpha}$, crucially required in the proof of determinacy of pLμ-games of [19,18]. As observed in [18], the property follows from the set-theoretic Martin Axiom at $\aleph_1$ (MA$_{\aleph_1}$). The problem of whether the property (and, as a consequence, the validity of the determinacy proof) holds in ZFC alone is left open in [18]. We provide a partial positive answer to this question proving the continuity property for $\mathcal{W}_{0,1}$ in ZFC alone.

Furthermore, we show that for higher ranks the property follows from a set-theoretic assumption weaker than $\text{MA}_{\aleph_1}$ which, unlike $\text{MA}_{\aleph_1}$, does not depend on cardinality assumptions such as the negation of the Continuum Hypothesis.

Applications. As already observed in [18, §5.4], the winning sets of pLμ-games reduce to game tree languages. Thus Theorem 1 settles the question posed in [18] about the measurability of pLμ winning sets. More generally, our result can find applications in solving similar problems. For example, in models of probabilistic concurrent computation (e.g. probabilistic Petri nets [15], probabilistic event structures [9], stochastic distributed games [21]), executions are naturally modelled by configurations of event structures (i.e. special kinds of acyclic graphs) and not by sequences. Many natural predicates on executions (e.g. the collection of *well-founded* graphs) are of high topological complexity.

Related Work. Beside the original work of Kolmogorov [13], the measure theoretic properties of $\mathcal{R}$-sets are investigated with set-theoretic methods by Lyapunov in [16]. A game-theoretic approach to $\mathcal{R}$-sets, closely related to this work, is developed by Burgess in [8] where the following characterization is stated as a remark without a formal proof: (1) every set $A \subseteq X$ belongs to a finite level of the hierarchy of $\mathcal{R}$-sets if and only if it is of the form $A = \partial(K)$, for some set $K \subseteq \omega^\omega$ which is a Boolean combination of F_σ sets, and (2) the levels of the hierarchy of $\mathcal{R}$-sets are in correspondence with the levels of the *difference hierarchy* (see [12, §22.E]) of F_σ sets. The operation ∂ is the so-called *game quantifier* (see [12, §20.D] and [6,7,11,17]). Admittedly, our characterization of $\mathcal{R}$-sets in terms of game tree languages $\mathcal{W}_{i,k}$, can be considered as a modern variant of the result of Burgess.[4] Having concrete examples of complete sets, however, sheds light on the concept of $\mathcal{R}$-sets and, in analogy with the study of complexity classes in computational complexity theory, may simplify further investigations. Lastly, it is suggestive to think that the origins of the concept of parity games, developed since the 80's in Computer Science to investigate ω-regular properties of transition systems, could be backdated to the original work of A. Kolmogorov.

2   Basic Notions from Descriptive Set Theory

We assume the reader is familiar with the basic notions of descriptive set theory and measure theory. We refer to [12] as a standard reference on these subjects.

Given two sets X and Y, we denote with X^Y the set of functions from Y to X. We denote with 2 and ω the two element set and the set of all natural numbers, respectively. The powerset of X will be denoted by both 2^X and $\mathcal{P}(X)$, as more convenient to improve readability. A topological space is *Polish* if it is separable and the topology is induced by a complete metric. A set is *clopen* if it

---

[4] The fact that $\mathcal{W}_{i,k}$ are $\mathcal{R}$-sets follows from the above formulation of Burgess' theorem. Also, our Theorem 1 can be easily inferred for $k = 1$. The case of $k = 2$ follows from Burgess's theorem in conjunction with [17]. Our proof of Theorem 1 yields an independent and formal argument backing the statement of Burgess' theorem.

is both closed and open. A space is *zero-dimensional* if the clopen subsets form a basis of the topology. In this work we limit our attention to zero-dimensional Polish spaces. Let X, Y be two topological spaces and $A \subseteq X$, $B \subseteq Y$ be two sets. We say that A is *Wadge reducible* to B, written as $A \leq_W B$, if there exists a continuous function $f \colon X \to Y$ such that $A = f^{-1}(B)$. Two sets A and B are *Wadge equivalent* (denoted $A \sim_W B$) if $A \leq_W B$ and $B \leq_W A$ hold. Given a family $\mathcal{C}$ of subsets of X, we say a set $A \in \mathcal{C}$ is *Wadge complete* if $B \leq_W A$ holds for all $B \in \mathcal{C}$. Given a Polish space X, we denote with $\mathcal{M}_{=1}(X)$ the Polish space of all Borel probability measures μ on X (see e.g. [12, Theorem 17.22]). A set $N \subseteq X$ is μ-*null* if there exists a Borel set $B \supseteq N$ such that $\mu(B) = 0$. A set $A \subseteq X$ is μ-*measurable* if $A = B \cup N$, for a Borel set B and a μ-null set N. A set $A \subseteq X$ is *universally measurable* if it is μ-measurable for all $\mu \in \mathcal{M}_{=1}(X)$. In what follows we omit the "universally" adjective.

Given two natural numbers $i < k$, the set $\mathrm{Tr}_{i,k}$ of all complete (i.e. without leaves) binary trees whose vertices are labelled by elements of $\{\exists, \forall\} \times \{i, \ldots, k\}$ is endowed with the standard 0-dimensional Polish topology (see e.g. [2]). Each $t \in \mathrm{Tr}_{i,k}$ can be interpreted as a two-player parity game with priorities in $\{i, \ldots, k\}$, with players $\exists$ and $\forall$ controlling vertices labelled by $\exists$ and $\forall$, respectively.

Definition 1. *Given two natural numbers $i < k$, the* game tree language $\mathcal{W}_{i,k}$ *is the subset of* $\mathrm{Tr}_{i,k}$ *consisting of all parity games admitting a winning strategy for $\exists$. The pair (i, k) is called the (Rabin–Mostowski) index of $\mathcal{W}_{i,k}$.*

Clearly, there is a natural Wadge equivalence between the languages $\mathcal{W}_{i,k}$ and $\mathcal{W}_{i+2,k+2}$. Therefore, we identify indices (i, k) and $(i + 2j, k + 2j)$ for every $i \leq k$ and $j \in \omega$. Indexes can be partially ordered by defining $(i, k) \subseteq (i', k')$ if and only if $\{i, \ldots, k\} \subseteq \{i', \ldots, k'\}$.

3   Definition and Basic Properties of $\mathcal{R}$–sets

As outlined in the introduction, the σ-algebra of $\mathcal{R}$-sets is generated by a family $\mathcal{F}$ of *operations* on subsets having countable arity. Following Kolmogorov, we define $\mathcal{F}$ as the family generated by the operation $\bigcup \circ \bigcap$ and closing under a *transformation* co-$\mathcal{R}$. It will be convenient to assume that the countably many inputs of an operation Γ are indexed by a countable set (called the *arena*) denoted by $\mathbb{A}_\Gamma$. Thus an operation Γ has type $\Gamma \colon \mathcal{P}(X)^{\mathbb{A}_\Gamma} \to \mathcal{P}(X)$. The operations of countable union and intersections are denoted by $\bigcup$ and $\bigcap$, respectively, and their arena is defined as $\mathbb{A}_{\bigcup} = \mathbb{A}_{\bigcap} = \omega$.

Definition 2. *Given two operations Γ and Θ their composition $\Theta \circ \Gamma$ is the operation with arena $\mathbb{A}_\Gamma \times \mathbb{A}_\Theta$ defined as:* $\Theta \circ \Gamma(\{A_{s,s'} \mid s \in \mathbb{A}_\Gamma, s' \in \mathbb{A}_\Theta\}) = \Theta(\{ \Gamma(\{A_{s,s'} \mid s \in \mathbb{A}_\Gamma\}) \mid s' \in \mathbb{A}_\Theta\}).$

Definition 3. *A* basis *for an operation Γ is a set $N_\Gamma \subseteq 2^{\mathbb{A}_\Gamma}$ such that*

$$\Gamma(\{A_s : s \in \mathbb{A}_\Gamma\}) = \bigcup_{S \in N_\Gamma} \bigcap_{s \in S} A_s \tag{1}$$

Not all operations have a basis but a family $N \subseteq 2^{\mathbb{A}}$ uniquely determines an operation Γ with arena $\mathbb{A}$ and basis N. In what follows we only consider operations Γ with a basis. One can check that $N_{\bigcup} = \{\{n\} \mid n \in \omega\}$ and $N_{\bigcap} = \{\omega\}$.

Definition 4. *For a given operation Θ with arena $\mathbb{A}$ and basis N_Θ, we define a dual operation* co-Θ *with the same arena $\mathbb{A}$ and basis $N_{\text{co-}\Theta}$ defined as $N_{\text{co-}\Theta} \stackrel{\text{def}}{=} \{S \in 2^{\mathbb{A}} \mid \forall T \in N_\Theta \ T \cap S \neq \emptyset\}$. One can notice that equivalently we can define* co-$\Theta(\{A_s : s \in \mathbb{A}\}) = \bigcap_{S \in N_\Theta} \bigcup_{s \in S} A_s$.

As an illustration, the equalities co-$\bigcup = \bigcap$ and co-$\bigcap = \bigcup$ hold.

Definition 5. *The $\mathcal{R}$-transformation of an operation Θ with basis N_Θ is the operation $\mathcal{R}\Theta$, with arena $\mathbb{A}_{\mathcal{R}\Theta} = (\mathbb{A}_\Theta)^*$ (finite sequences of elements in $\mathbb{A}_\Theta$) uniquely determined by the basis:*

$$N_{\mathcal{R}\Theta} \stackrel{\text{def}}{=} \{S \subseteq (\mathbb{A}_\Theta)^* \mid \exists T \subseteq S. \ \epsilon \in T \wedge \forall t \in T \ \{v \in \mathbb{A}_\Theta : \ tv \in T\} \in N_\Theta\} \quad (2)$$

where ϵ denotes the empty sequence and tv the concatenation of $t \in (\mathbb{A}_\Theta)^*$ with $v \in \mathbb{A}_\Theta$. We denote with co-$\mathcal{R}$ *the composition* co-$(\mathcal{R}(\Theta))$ *and define the iteration*

$$\Theta_k \stackrel{\text{def}}{=} (\text{co-}\mathcal{R})^k \left(\bigcup \circ \bigcap \right).$$

Definition 6. *For a positive number $k \geq 1$, we say that a set $A \subseteq X$ is an $\mathcal{R}$-set of k-th level if and only if $A = \Theta_k(\{U_s : s \in \mathbb{A}_{\Theta_k}\})$ for some clopen sets $U_s \subseteq X$.*

In what follows by $\mathcal{R}$-sets we mean $\mathcal{R}$-sets of finite levels.

Lemma 1 ([8]). *The k–th level of $\mathcal{R}$-sets is closed under pre-images of continuous functions.*

We say that an operation Γ *preserves measurability* if for any family $\mathcal{E} = \{A_s\}_{s \in \mathbb{A}_\Gamma}$ of measurable sets, the set $\Gamma(\mathcal{E})$ is measurable. The following property motivates the notion of $\mathcal{R}$-sets:

Theorem 2 ([16, Theorem 4]). *If Γ and Θ preserve measurability then $\Gamma \circ \Theta$, $\mathcal{R}\Gamma$, and* co-Γ *preserve measurability.*

Corollary 1. *All $\mathcal{R}$-sets are measurable.*

4   Matryoshka Games

In this section we define *Matryoshka games*, a variant of parity games which make it easier to establish a connection with the operations Θ_k defined in Section 3.

A Matryoshka game $\mathcal{G}$ is the familiar structure of a two-player parity game played on an infinite countably branching graph, extended with a *labelling function* assigning to each finite play (i.e. every sequence of game-states ending in a terminal state) a *play label*. Formally, a Matryoshka game $\mathcal{G}$ is a structure:

$$\mathcal{G} = \{V^\mathcal{G} = V_\exists^\mathcal{G} \sqcup V_\forall^\mathcal{G}, F^\mathcal{G}, E^\mathcal{G}, v_I^\mathcal{G}, \Omega^\mathcal{G}, \mathbb{A}^\mathcal{G}, \text{label}^\mathcal{G}\},$$

such that $\{V^{\mathcal{G}} = V_{\exists}^{\mathcal{G}} \sqcup V_{\forall}^{\mathcal{G}}, F^{\mathcal{G}}, E^{\mathcal{G}}, v_I^{\mathcal{G}}, \Omega^{\mathcal{G}}\}$ is a standard parity game with initial state $v_I^{\mathcal{G}}$, terminal positions $F^{\mathcal{G}} \subseteq V^{\mathcal{G}}$, and priority assignment $\Omega^{\mathcal{G}}$. Additionally, $\mathbb{A}^{\mathcal{G}}$ is a set of *play labels*, and label$^{\mathcal{G}} : (V^{\mathcal{G}})^* F^{\mathcal{G}} \to \mathbb{A}^{\mathcal{G}}$ is a function assigning to finite plays their *play labels*. We assume that for every $v \in V^{\mathcal{G}}$ there is at least one $v' \in V^{\mathcal{G}} \cup F^{\mathcal{G}}$ such that $(v, v') \in E^{\mathcal{G}}$, so that the only terminal game-states are in $F^{\mathcal{G}}$. As for standard parity games, the pair (i, k) containing the minimal and maximal values of Ω is called the *index* of the game. By $P \in \{\exists, \forall\}$ we denote the *players* of the game. The opponent of P is denoted by $\bar{P}$.

A *play* is defined as usual as a maximal path in the arena, i.e., either as a finite sequence in $(V^{\mathcal{G}})^* F^{\mathcal{G}}$ or as an infinite sequence in $(V^{\mathcal{G}})^\omega$. Similarly, a strategy σ for Player P is a function $\sigma : (V^{\mathcal{G}})^* V_P^{\mathcal{G}} \to V^{\mathcal{G}} \cup F^{\mathcal{G}}$ defined as expected.

The novelty in Matryoshka games is given by the set of play labels $\mathbb{A}^{\mathcal{G}}$ and the associated labelling function label$^{\mathcal{G}}$. These are used to define *parametric* winning conditions in the Matryoshka game, as we now describe.

A set of play labels $X \subseteq \mathbb{A}^{\mathcal{G}}$ is called a *promise*. A finite play π is *winning for* $\exists$ *with promise* X if label$(\pi) \in X$. An infinite play π is winning for $\exists$ if $\left(\limsup_{n \to \infty} \Omega^{\mathcal{G}}(\pi(n))\right)$ is even, as usual. If a play is not winning for $\exists$ then it is winning for $\forall$. A strategy σ for Player P is *winning in the Matryoshka game* $\mathcal{G}$ *with promise* X if, for every counter-strategy τ of $\bar{P}$, the resulting play $\pi(\sigma, \tau)$ is winning for P with promise X, in the sense just described. The following proposition directly follows from the well-known determinacy of parity games.

Proposition 1. *If $\mathcal{G}$ is a Matryoshka game with play labels $\mathbb{A}^{\mathcal{G}}$ and $X \subseteq \mathbb{A}^{\mathcal{G}}$ then exactly one of the players has a winning strategy in $\mathcal{G}$ with promise X.*

The point of having parametrized winning conditions in Matryoshka games is the possibility of defining set-theoretic operations with a direct game interpretation. Given a Polish space X, the *operation* on sets (see Section 2) associated with a Matryoshka game $\mathcal{G}$ has arena $\mathbb{A}^{\mathcal{G}}$ and is defined as follows:

$$\mathcal{G}(\mathcal{E}) \overset{\text{def}}{=} \{x \in X : \exists \text{ has a w. s. in } \mathcal{G} \text{ with promise } \{s \in \mathbb{A}^{\mathcal{G}} : x \in E_s\}\} \quad (3)$$

where $\mathcal{E} = \{E_s : s \in \mathbb{A}^{\mathcal{G}}\}$ is a family of subsets of X.

We now sketch the definition of a Matryoshka game, called $\mathcal{G}_0$, whose associated operation is precisely the operation $(\bigcup \circ \bigcap)$ of Section 2. The structure of $\mathcal{G}_0$ is depicted in Figure 1. This is a simple two-steps game where $\exists$ chooses a number n and $\forall$ responds choosing a number m. Every play is finite and of the form $\langle \epsilon, n, n.m \rangle$. The set of play labels $\mathbb{A}^{\mathcal{G}_0}$ is defined as $\omega \times \omega$ and label$^{\mathcal{G}}(\langle \epsilon, n, n.m \rangle) = (n, m)$.

We now introduce *transformations* on games which directly match the corresponding transformations on operations defined in Section 2. Due to space limitations we only sketch the definitions.

For a Matryoshka game $\mathcal{G}$ of index (i, k), we define co-$\mathcal{G}$ as the game obtained from $\mathcal{G}$ by replacing the sets $V_\exists \leftrightarrow V_\forall$ and increasing all priorities in Ω by 1. Note that the index of co-$\mathcal{G}$ is $(i + 1, k + 1)$, and that the sets of plays in the two games are equal. We define $\mathbb{A}^{\text{co-}\mathcal{G}} \overset{\text{def}}{=} \mathbb{A}^{\mathcal{G}}$ and label$^{\text{co-}\mathcal{G}}(\pi) \overset{\text{def}}{=}$ label$^{\mathcal{G}}(\pi)$.

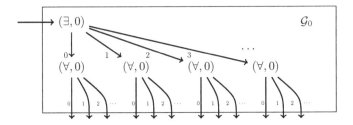

Fig. 1. The game $\mathcal{G}_0$ corresponding to the operation $\bigcup \circ \bigcap$

We now define the $\mathcal{R}$ transformation on games. Let us take a Matryoshka game $\mathcal{G}$ of index (i, k). Let $2j$ be the minimal even number such that $k \leq 2j$. The game $\mathcal{RG}$ is depicted on Figure 2.

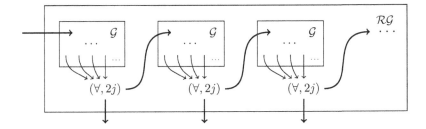

Fig. 2. The game $\mathcal{RG}$

A play in the game $\mathcal{RG}$ starts from a first copy of $\mathcal{G}$. In this *inner* game, the play π can either be infinite (in which case π is a valid play in $\mathcal{RG}$ and is winning for Player P iff it is winning for P in $\mathcal{G}$) or terminate in a terminal state of $\mathcal{G}$. In this latter case, Player $\forall$ can either conclude the game $\mathcal{RG}$, or start another session of the inner game $\mathcal{G}$. Observe that if $\forall$ always chooses to start a new session, they lose because the even priority $2j$ is maximal in $\mathcal{RG}$.

The set of play labels $\mathbb{A}^{\mathcal{RG}}$ is defined as $(\mathbb{A}^{\mathcal{G}})^*$, i.e., the set of finite sequences of play labels in $\mathcal{G}$. Let π be a play in $\mathcal{RG}$ that passes through n copies of $\mathcal{G}$ and then ends in a terminal position of $\mathcal{RG}$. In that case π can be decomposed into n plays $\pi_0, \ldots, \pi_{n-1}$ in $\mathcal{G}$. We then define the labeling function of $\mathcal{RG}$ as follows:

$$\mathrm{label}^{\mathcal{RG}}(\pi) \stackrel{\mathrm{def}}{=} \left(\mathrm{label}^{\mathcal{G}}(\pi_0), \ \mathrm{label}^{\mathcal{G}}(\pi_1), \ldots, \mathrm{label}^{\mathcal{G}}(\pi_{n-1})\right). \tag{4}$$

Given the basic Matryoshka game G_0 and the two transformations of games co- and $\mathcal{R}$, we can construct more and more complex "nested" games. This fact motivates the name of this class of games. We denote with $\mathcal{G}_k$ the game obtained from G_0 by iterating k-times the composed transformation co-$\mathcal{R}$.

By the definition, the game $\mathcal{G}_k$ for $k > 0$ consists of infinitely many copies of $\mathcal{G}_{k-1}$ and an additional set of new vertices as depicted on Figure 2. These new vertices are called the k-*layer* of the game. Therefore, by unfolding the definition, each vertex v of $\mathcal{G}_k$ is either a vertex of a copy of $\mathcal{G}_0$ or it belongs to a j-layer for some $1 \leq j \leq k$. Observe that if v is in a j-layer of $\mathcal{G}_k$ then

$$\Omega^{\mathcal{G}_k}(v) = k+j-1 \quad \text{and} \quad v \in V_\forall^{\mathcal{G}_k} \Leftrightarrow k+j-1 \equiv 0 \pmod 2. \tag{5}$$

We are now ready to state the expected correspondence between the operation Θ_k of Section 2 and the Matryoshka game $\mathcal{G}_k$.

Theorem 3. *For every $k \in \omega$ the basis N_{Θ_k} of the Θ_k operation equals the family* promise$(\mathcal{G}_k) \overset{\text{def}}{=} \{X \subseteq \mathbb{A}_k : \ \exists \text{ has a winning strategy in } \mathcal{G}_k \text{ with promise } X\}$.

Corollary 2. *For each k and $(E_s)_{s \in \mathbb{A}_k}$ we have $\Theta_k\big((E_s)_{s \in \mathbb{A}_k}\big) = \mathcal{G}_k\big((E_s)_{s \in \mathbb{A}_k}\big)$.*

5 Relation between $\mathcal{R}$–sets and the Index Hierarchy

In this section we prove the main result of this work, that is Theorem 1 stated in Section 1. As a preliminary step, it is convenient to define a variant of game tree languages defined on countable trees. This will simplify the connection with Matryoshka games which are played on countably branching structures. Let $\text{Tr}_{i,k}^\omega$ be the space of labelled ω-trees $t \colon \omega^* \to \{\exists, \forall\} \times \{i, \ldots, k, \top, \bot\}$. Each $t \in \text{Tr}_{i,k}^\omega$ is naturally interpreted as a parity game on the countable tree structure, with the possibility of terminating at leaves, labelled by $\top$ and $\bot$, which are winning for $\exists$ and $\forall$, respectively. We also require (1) that in the root there is a vertex (P, k) where $P = \exists$ if i is even and $P = \forall$ if i is odd and (2) that the tree is alternating, that is $\exists$ and $\forall$ make moves in turns.

Definition 7. $\mathcal{W}_{i,k}^\omega \subseteq \text{Tr}_{i,k}^\omega$ *is the set of ω-trees such that $\exists$ has a w.s.*

An easy argument shows that dropping conditions (1) and (2) gives a Wadge equivalent language. The following routine lemma establishes the connection between ω-branching game tree languages $\mathcal{W}_{i,k}^\omega$ and binary (as in Section 2) game tree languages $\mathcal{W}_{i,k}$.

Lemma 2. *For $i < k$ the language $\mathcal{W}_{i,k}$ is Wadge equivalent to $\mathcal{W}_{i+1,k}^\omega$. In particular $\mathcal{W}_{0,1} \sim_W \mathcal{W}_{1,1}^\omega$ and $\mathcal{W}_{1,3} \sim_W \mathcal{W}_{0,1}^\omega$.*

The fact that $\mathcal{W}_{i,k}$ corresponds to $\mathcal{W}_{i+1,k}^\omega$ reflects the cost of the translation of ω-branching games into binary games: an extra priority is required to mimic countably many choices by iterating binary choices. Thanks to this lemma, in Theorem 1 one can replace the languages $\mathcal{W}_{k-1,2k-1}$ with the languages $\mathcal{W}_{k,2k-1}^\omega$.

First, we show that every $\mathcal{W}_{k,2k-1}^\omega$ is indeed an $\mathcal{R}$-set. We will do so by explicitly constructing a family $\mathcal{E}_k = \{E_s \mid s \in \mathbb{A}_k\}$ of clopen sets in $\text{Tr}_{k,2k-1}^\omega$ such that $\Theta_k(\mathcal{E}_k) = \mathcal{W}_{k,2k-1}^\omega$, where $\mathbb{A}_k$ is the arena of the operation Θ_k. The construction requires some effort. First we recall, from Section 3 that the arena of the operation $\bigcup \circ \bigcap$ is $\mathbb{A}_0 = \{\langle n, m \rangle : n, m \in \omega\}$ (pairs of natural numbers) and from the definition of the transformation $\mathcal{R}$ we have $\mathbb{A}_k = (\mathbb{A}_{k-1})^*$. Thus, for all $k \in \omega$, $\mathbb{A}_k$

is a set of nested sequences of pairs of natural numbers. For a sequence $s \in \mathbb{A}_k$ we define the maps `flatten` and `prioritiesMap` such that $\texttt{flatten}(s) \in \mathbb{A}_0^*$ and $\texttt{prioritiesMap}(s) \in \omega^*$. The map `flatten` takes a nested sequence in $\mathbb{A}_k$ and returns the "flattened" sequence, that is all the braces are removed, for example $\texttt{flatten}(((((\langle 2,15\rangle)),(((\langle 7,5\rangle),(\langle 6,4\rangle))))) = (\langle 2,15\rangle,\langle 7,5\rangle,\langle 6,4\rangle)$. The function `prioritiesMap` computes the number of closing brackets after each pair of natural numbers: e.g., $\texttt{prioritiesMap}\big(((((\langle 2,15\rangle)),(((\langle 7,5\rangle),(\langle 6,4\rangle)))))\big) = (2,1,3)$.

We also define $\texttt{treeMap}(t,s)$ where $t \in \mathrm{Tr}^{\omega}_{k,2k-1}$ and $s \in \mathbb{A}_k$. Since we limited our attention to alternating trees, each vertex in the ω-branching tree t can be identified with a sequence of pairs of natural numbers. Then, if $s \in \mathbb{A}_k$, the function $\texttt{treeMap}(t,s)$ computes first $\texttt{flatten}(s)$ and returns the sequence of priorities assigned to the vertices along the path of t indicated by $\texttt{flatten}(s)$. On Figure 3 we have an example of a tree t where

$$\texttt{treeMap}\big(t,(((((\langle 2,15\rangle)),(((\langle 7,5\rangle),(\langle 6,4\rangle)))))\big) = (2,1,3).$$

Define $\mathcal{E}_k = \{E_s : s \in \mathbb{A}_k\}$ such that for $t \in \mathrm{Tr}^{\omega}_{k,2k-1}$ we have $t \in E_s$ iff for

- $v = \texttt{prioritiesMap}(s)$,
- $b = \texttt{treeMap}(t,s)$,
- $L = \min\{k \in \omega : v(k) \neq b(k)\}$

$v \neq b$ holds, and either $b(L)=\top$ or

$$\min(b(L),v(L)) \equiv 0 \pmod 2. \qquad (6)$$

It is simple to verify that the sets E_s are indeed clopen in the space $\mathrm{Tr}^{\omega}_{k,2k-1}$ (for a definition of the topology see, e.g. [2]).

Theorem 4. $\forall_{k\geq 1}\ \Theta_k(\mathcal{E}_k) = \mathcal{W}^{\omega}_{k,2k-1}$.

Proof. The proof is based on Matryoshka games. Consider a tree $t \in \mathrm{Tr}^{\omega}_{k,2k-1}$ and assume that Player $P \in \{\exists,\forall\}$ has a winning strategy σ on the tree t. We claim that P has a winning strategy in the Matryoshka game $\mathcal{G}_k$ with promise $\mathcal{E}_k$. From this fact the theorem will follow by an application of Corollary 2 and Proposition 1. We consider the case $P=\exists$. The opposite case is analogous.

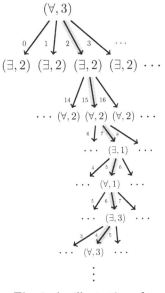

Fig. 3. An illustration of `treeMap`

We will simulate the game on t in the Matryoshka game $\mathcal{G}_k$. A play in $\mathcal{G}_k$ consists of playing pairs of numbers (corresponding to moves in t) in the copies of $\mathcal{G}_0$ and, additionally, deciding whether to *exit* an j-layer of the game or not. We say that a play in $\mathcal{G}_k$ is *fair* if whenever the players encounter a priority $k+j$ in t then they exit exactly j first layers of $\mathcal{G}_k$ (i.e. the layer $j+1$ is reached) and if they encounter a symbol $\bot$ or $\top$ then the players exit all the layers of $\mathcal{G}_k$.

Let $\exists$ use the original strategy σ in the copies of $\mathcal{G}_0$ and play "fairly" as long as $\forall$ does. If $\forall$ also plays "fairly" then the play is winning for $\exists$: either $\top$ is reached in t and $\exists$ wins since $t \in E_s$ or the play is infinite and $\exists$ wins by the parity condition — the priorities visited in $\mathcal{G}_k$ agree with those visited in t, see (5).

If $\forall$ does not play "fairly" (i.e. when a priority $k+j$ is reached in t they don't exit the l-layer of $\mathcal{G}_k$ with $l \leq j$ or they exit the $(j+1)$-layer of $\mathcal{G}_k$) then $\exists$ uses the following counter-strategy: whenever possible $\exists$ exits the current layer of $\mathcal{G}_k$. There are two possible developments of such a play. The first case is that $\forall$ allows to exit the whole game and then $\exists$ wins thanks to (6). Now assume that $\forall$ never allows the game to reach a terminal position. In that case, let j be maximal such that the j-layer of $\mathcal{G}_k$ is visited infinitely often. By (5) we know that the limit superior of the priorities visited in $\mathcal{G}_k$ is $k+j-1$ and, since $\forall$ is the owner of the vertices in the j-layer of $\mathcal{G}_k$, it holds that $k+j-1 \equiv 0 \pmod 2$. Therefore, $\exists$ wins the play by the parity condition. □

Theorem 5. *Let $L = \Theta_k(E_s)$ be a set obtained using the Θ_k operation applied to a family of clopen subsets $(E_s)_{s \in \mathbb{A}_k}$ with $E_s \subseteq Y$ in a Polish space Y. Then, there exists a continuous reduction $f \colon Y \to \mathrm{Tr}_{k,2k-1}^\omega$ such that $f^{-1}\left(\mathcal{W}_{k,2k-1}^\omega\right) = L$.*

Sketch. The operation Θ_k is presented as the corresponding Matryoshka game using Theorem 3 and Corollary 2. This is a parametrized family of parity games and thus continuously reducible to $\mathcal{W}_{k,2k-1}^\omega$. □

Theorems 4 and 5 imply that the language $\mathcal{W}_{k,2k-1}^\omega$ is complete for the k-th level of the hierarchy of $\mathcal{R}$-sets. Theorem 1 follows from Lemmas 1 and 2.

6   Continuity of measures on $\mathcal{W}_{i,k}$

For an odd $k \in \omega$ the language $\mathcal{W}_{i,k}$ admits a natural transfinite decomposition into simpler approximant sets $\mathcal{W}_{i,k}^\alpha$, for $\alpha < \omega_1$ (see [18, §6.2,3]). The proof of determinacy of pLμ games of [18] relies on the following special *continuity property*: $\sup_{\alpha < \omega_1} \mu\left(\mathcal{W}_{i,k}^\alpha\right) = \mu\left(\mathcal{W}_{i,k}\right)$. Since the increasing chain $\mathcal{W}_{i,k}^\alpha$ is uncountable, the property does not follow from the standard σ-continuity of measures. As observed in [18], the property follows from Martin Axiom at $\aleph_1$ (MA$_{\aleph_1}$). The problem of whether the property holds in ZFC alone is left open (see Item 2 of Section 8.2 in [18]). The following theorem gives a partial answer to this problem.

Theorem 6. *The continuity property holds in ZFC for $\mathcal{W}_{0,1}$. Let k be an odd number, $i < k$. For $\mathcal{W}_{i,k}$ the continuity property holds assuming the determinacy of Harrington's games[5] with arbitrary analytic winning sets.*

6.1   Improvement

After submitting the paper the authors have realised that it is possible to prove the above theorem without the additional assumption of determinacy. This proof will be included in the journal version of the paper.

---

[5] See, e.g., [10, Section 33.5] for details about this type of games.

7   Conclusion

The notion of $\mathcal{R}$-sets is a robust concept and admits natural variations. One can equivalently work in arbitrary (not zero-dimensional) Polish spaces and start from a basis of, e.g. Borel sets rather than clopens. The family of operations $\Theta_k = (\text{co-}\mathcal{R})^k(\bigcup \circ \bigcap)$ can be replaced by, e.g. either $(\text{co-}\mathcal{R})^k(\bigcup)$ or $(\text{co-}\mathcal{R})^k(\bigcap)$. Similarly, one can consider binary rather than countably branching, Matryoshka games. The notion of $\mathcal{R}$-sets remains unchanged in these alternative setups.

References

1. Arnold, A.: The μ-calculus alternation-depth hierarchy is strict on binary trees. ITA 33(4/5), 329–340 (1999)
2. Arnold, A., Niwiński, D.: Continuous separation of game languages. Fundamenta Informaticae 81(1-3), 19–28 (2007)
3. Baier, C., Katoen, J.-P.: Principles of Model Checking. The MIT Press (2008)
4. Blackwell, D.: Borel–programmable functions. Ann. of Probability 6, 321–324 (1978)
5. Bradfield, J.: The modal mu-calculus alternation hierarchy is strict. In: Sassone, V., Montanari, U. (eds.) CONCUR 1996. LNCS, vol. 1119, pp. 233–246. Springer, Heidelberg (1996)
6. Bradfield, J.C.: Fixpoints, games and the difference hierarchy. ITA 37(1), 1–15 (2003)
7. Bradfield, J.C., Duparc, J., Quickert, S.: Transfinite extension of the mu-calculus. In: Ong, L. (ed.) CSL 2005. LNCS, vol. 3634, pp. 384–396. Springer, Heidelberg (2005)
8. Burgess, J.P.: Classical hierarchies from a modern standpoint. II. R-sets. Fund. Math. 115(2), 97–105 (1983)
9. Varacca, D., Völzer, H., Winskel, G.: Probabilistic event structures and domains. In: Gardner, P., Yoshida, N. (eds.) CONCUR 2004. LNCS, vol. 3170, pp. 481–496. Springer, Heidelberg (2004)
10. Jech, T.: Set Theory. Springer Monographs in Mathematics. Springer (2002)
11. Kanoveĭ, V.G.: A. N. Kolmogorov's ideas in the theory of operations on sets. Uspekhi Mat. Nauk. 43(6(264)), 93–128 (1988)
12. Kechris, A.: Classical descriptive set theory. Springer, New York (1995)
13. Kolmogorov, A.: Operations sur des ensembles. Mat. Sb. 35, 415–422 (1928); (in Russian, summary in French)
14. Kozen, D.: Results on the propositional mu-calculus. In: Theoretical Computer Science, pp. 333–354 (1983)
15. Kudlek, M.: Probability in Petri Nets. Fundamenta Informaticae, 67(1) (2005)
16. Lyapunov, A.A.: $\mathcal{R}$-sets. Trudy Mat. Inst. Steklov. 40, 3–67 (1953)
17. Michalewski, H., Niwiński, D.: On topological completeness of regular tree languages. In: Constable, R.L., Silva, A. (eds.) Kozen Festschrift. LNCS, vol. 7230, pp. 165–179. Springer, Heidelberg (2012)
18. Mio, M.: Game Semantics for Probabilistic μ-Calculi. PhD thesis, School of Informatics, University of Edinburgh (2012)
19. Mio, M.: Probabilistic Modal μ-Calculus with Independent product. Logical Methods in Computer Science 8(4) (2012)
20. Selivanowski, E.: Sur une classe d'ensembles effectifs (ensembles C). Mat. Sb. 35, 379–413 (1928) (in Russian, summary in French)
21. Winskel, G.: Distributed probabilistic strategies. In: Proceedings of MFPS (2013)

On Upper and Lower Bounds on the Length of Alternating Towers

Štěpán Holub [1,*], Galina Jirásková [2,**], and Tomáš Masopust [3,***]

[1] Dept. of Algebra, Charles University, Sokolovská 83, 175 86 Praha, Czech Republic
holub@karlin.mff.cuni.cz
[2] Mathematical Institute, Slovak Academy of Sciences
Grešákova 6, 040 01 Košice, Slovak Republic
jiraskov@saske.sk
[3] Institute of Mathematics, ASCR, Žižkova 22, 616 62 Brno, Czech Republic,
and TU Dresden, Germany
masopust@math.cas.cz

Abstract. A tower between two regular languages is a sequence of strings such that all strings on odd positions belong to one of the languages, all strings on even positions belong to the other language, and each string can be embedded into the next string in the sequence. It is known that if there are towers of any length, then there also exists an infinite tower. We investigate upper and lower bounds on the length of finite towers between two regular languages with respect to the size of the automata representing the languages in the case there is no infinite tower. This problem is relevant to the separation problem of regular languages by piecewise testable languages.

1 Introduction

The separation problem appears in many disciplines of mathematics and computer science, such as algebra and logic [8,9], or databases and query answering [4]. Given two languages K and L and a family of languages $\mathcal{F}$, the problem asks whether there exists a language S in $\mathcal{F}$ such that S includes one of the languages K and L, and it is disjoint with the other. Recently, it has been independently shown in [4] and [8] that the separation problem for two regular languages given as NFAs and the family of piecewise testable languages is decidable in polynomial time with respect to both the number of states and the size of the alphabet. It should be noted that an algorithm polynomial with respect to the number of states and exponential with respect to the size of the alphabet has been known in the literature [1,3]. In [4], the separation problem has been shown to be equivalent to the non-existence of an infinite tower between the input languages. Namely, the languages have been shown separable by a

---

* Research supported by the Czech Science Foundation grant number 13-01832S.
** Research supported by grant APVV-0035-10.
*** Research supported by RVO 67985840 and by the DFG in grant KR 4381/1-1.

E. Csuhaj-Varjú et al. (Eds.): MFCS 2014, Part I, LNCS 8634, pp. 315–326, 2014.

piecewise testable language if and only if there does not exist an infinite tower. In [8], another technique has been used to prove the polynomial time bound for the decision procedure, and a doubly exponential upper bound on the index of the separating piecewise testable language has been given. This information can then be further used to construct a separating piecewise testable language.

However, there exists a simple (in the meaning of description, not complexity) method to decide the separation problem and to compute the separating piecewise testable language, whose running time depends on the length of the longest finite tower. The method is recalled in Section 3. This observation has motivated the study of this paper to investigate the upper bound on the length of finite towers in the presence of no infinite tower. So far, to the best of our knowledge, the only published result in this direction is a paper by Stern [12], who has given an exponential upper bound $2^{|\Sigma|^2 N}$ on the length of the tower between a piecewise testable language and its complement, where N is the number of states of the minimal deterministic automaton.

Our contribution in this paper are upper and lower bounds on the length of maximal finite towers between two regular languages in the case no infinite towers exist. These bounds depend on the size of the input (nondeterministic) automata. The upper bound is exponential with respect to the size of the input alphabet. More precisely, it is polynomial with respect to the number of states with the cardinality of the input alphabet in the exponent (Theorem 1). Concerning the lower bounds, we show that the bound is tight for binary languages up to a linear factor (Theorem 2), that a cubic tower with respect to the number of states exists (Theorem 3), and that an exponential lower bound with respect to the size of the input alphabet can be achieved (Theorem 4).

2   Preliminaries

We assume that the reader is familiar with automata and formal language theory. The cardinality of a set A is denoted by $|A|$ and the power set of A by 2^A. An alphabet Σ is a finite nonempty set. The free monoid generated by Σ is denoted by Σ^*. A string over Σ is any element of Σ^*; the empty string is denoted by ε. For a string $w \in \Sigma^*$, $\mathrm{alph}(w) \subseteq \Sigma$ denotes the set of all letters occurring in w.

We define *(alternating subsequence) towers* as a generalization of Stern's alternating towers [12]. For strings $v = a_1 a_2 \cdots a_n$ and $w \in \Sigma^* a_1 \Sigma^* \cdots \Sigma^* a_n \Sigma^*$, we say that v is a *subsequence* of w or that v can be *embedded* into w, denoted by $v \preccurlyeq w$. For languages K and L and the subsequence relation $\preccurlyeq$, we say that a sequence $(w_i)_{i=1}^k$ of strings is an *(alternating subsequence) tower between K and L* if $w_1 \in K \cup L$ and, for all $i = 1, \ldots, k-1$,

- $w_i \preccurlyeq w_{i+1}$,
- $w_i \in K$ implies $w_{i+1} \in L$, and
- $w_i \in L$ implies $w_{i+1} \in K$.

We say that k is the *length* of the tower. Similarly, we define an infinite sequence of strings to be an *infinite (alternating subsequence) tower between K*

and L. If the languages are clear from the context, we omit them. Notice that the languages are not required to be disjoint, however, if there exists a $w \in K \cap L$, then there exists an infinite tower, namely $w, w, w, \ldots$.

For two languages K and L, we say that the *language* K *can be embedded into the language* L, denoted $K \preccurlyeq L$, if for each string w in K, there exists a string w' in L such that $w \preccurlyeq w'$. We say that a *string* w *can be embedded into the language* L, denoted $w \preccurlyeq L$, if $\{w\} \preccurlyeq L$.

A *nondeterministic finite automaton* (NFA) is a 5-tuple $M = (Q, \Sigma, \delta, Q_0, F)$, where Q is the finite nonempty set of states, Σ is the input alphabet, $Q_0 \subseteq Q$ is the set of initial states, $F \subseteq Q$ is the set of accepting states, and $\delta : Q \times \Sigma \to 2^Q$ is the transition function that can be extended to the domain $2^Q \times \Sigma^*$. The language *accepted* by M is the set $L(M) = \{w \in \Sigma^* \mid \delta(Q_0, w) \cap F \neq \emptyset\}$. A *path* π is a sequence of states and input symbols $q_0, a_0, q_1, a_1, \ldots, q_{n-1}, a_{n-1}, q_n$, for some $n \geq 0$, such that $q_{i+1} \in \delta(q_i, a_i)$, for all $i = 0, 1, \ldots, n-1$. The path π is *accepting* if $q_0 \in Q_0$ and $q_n \in F$. We also use the notation $q_0 \xrightarrow{a_1 a_2 \cdots a_{n-1}} q_n$ to denote a path from q_0 to q_n under a string $a_1 a_2 \cdots a_{n-1}$.

The NFA M has a *cycle over an alphabet* $\Gamma \subseteq \Sigma$ if there exists a state q and a string w over Σ such that $\mathrm{alph}(w) = \Gamma$ and $q \xrightarrow{w} q$.

We assume that there are no useless states in the automata under consideration, that is, every state appears on an accepting path.

3   Computing a Piecewise Testable Separator [1]

We now motivate our study by recalling a "simple" method [5] solving the separation problem of regular languages by piecewise testable languages and computing a piecewise testable separator, if it exists. Our motivation to study the length of towers comes from the fact that the running time of this method depends on the maximal length of finite towers.

Let K and L be two languages. A language S *separates* K *from* L if S contains K and does not intersect L. Languages K and L are *separable by a family* $\mathcal{F}$ if there exists a language S in $\mathcal{F}$ that separates K from L or L from K.

A regular language is *piecewise testable* if it is a finite boolean combination of languages of the form $\Sigma^* a_1 \Sigma^* a_2 \Sigma^* \cdots \Sigma^* a_k \Sigma^*$, where $k \geq 0$ and $a_i \in \Sigma$, see [10,11] for more details.

Given two disjoint regular languages L_0 and R_0 represented as NFAs. We construct a decreasing sequence of languages $\ldots \preccurlyeq R_2 \preccurlyeq L_2 \preccurlyeq R_1 \preccurlyeq L_1 \preccurlyeq R_0$ as follows, show that a separator exists if and only if from some point on all the languages are empty, and use them to construct a piecewise testable separator.

For $k \geq 1$, let $L_k = \{w \in L_{k-1} \mid w \preccurlyeq R_{k-1}\}$ be the set of all strings of L_{k-1} that can be embedded into R_{k-1}, and let $R_k = \{w \in R_{k-1} \mid w \preccurlyeq L_k\}$, see Fig. 1. Let K be a language accepted by an NFA $A = (Q, \Sigma, \delta, Q_0, F)$, and let $\varepsilon(K)$ denote the language accepted by the NFA $A_\varepsilon = (Q, \Sigma, \delta_\varepsilon, Q_0, F)$,

---

[1] The method recalled here is not the original work of this paper and the credit for this should go to the authors of [5], namely to Wim Martens and Wojciech Czerwiński.

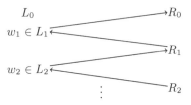

Fig. 1. The sequence of languages; an arrow stands for the embedding relation $\preccurlyeq$

where $\delta_\varepsilon(q, a) = \delta(q, a)$ and $\delta_\varepsilon(q, \varepsilon) = \bigcup_{a \in \Sigma} \delta(q, a)$. Then $L_k = L_{k-1} \cap \varepsilon(R_{k-1})$ (analogously for R_k), hence the languages are regular.

We now show that there exists a constant $B \geq 1$ such that $L_B = L_{B+1} = \ldots$, which also implies $R_B = R_{B+1} = \ldots$. Assume that no such constant exists. Then there are infinitely many strings $w_\ell \in L_\ell \setminus L_{\ell+1}$, for all $\ell \geq 1$, as depicted in Fig. 1. By Higman's lemma [6], there exist $i < j$ such that $w_i \preccurlyeq w_j$, hence $w_i \preccurlyeq R_{j-1}$, which is a contradiction because $w_i \not\preccurlyeq R_i$ and $R_{j-1} \subseteq R_i$.

By construction, languages L_B and R_B are mutually embeddable into each other, $L_B \preccurlyeq R_B \preccurlyeq L_B$, which describes a way how to construct an infinite tower. Thus, if there is no infinite tower, languages L_B and R_B must be empty.

The constant B depends on the length of the longest finite tower. Let $(w_i)_{i=1}^r$ be a maximal finite tower between L_0 and R_0 and assume that w_r belongs to L_0. In the first step, the method eliminates all strings that cannot be embedded into R_0, hence w_r does not belong to L_1, but $(w_i)_{i=1}^{r-1}$ is a tower between L_1 and R_0. Thus, in each step of the algorithm, all maximal strings of all finite towers (belonging to the language under consideration) are eliminated, while the rests of towers still form towers between the resulting languages. Therefore, as long as there is a maximal finite tower, the algorithm can make another step.

Assume that there is no infinite tower ($L_B = R_B = \emptyset$). We use the languages computed above to construct a piecewise testable separator. For a string $w = a_1 a_2 \cdots a_\ell$, we define $L_w = \Sigma^* a_1 \Sigma^* a_2 \Sigma^* \cdots \Sigma^* a_\ell \Sigma^*$, which is piecewise testable by definition. Let $up(L) = \bigcup_{w \in L} L_w$. The language $up(L)$ is regular and its NFA is constructed from an NFA for L by adding self-loops under all letters to all states, see [7] for more details. By Higman's lemma [6], $up(L)$ can be written as a finite union of languages of the form L_w, for some $w \in L$, hence it is piecewise testable. For $k = B, B - 1, \ldots, 1$, we define the piecewise testable languages $S_k = up(R_0 \setminus R_k) \setminus up(L_0 \setminus L_k)$ and show that $S = \bigcup_{k=1}^B S_k$ is a piecewise testable separator of L_0 and R_0.

To this end, we show that $L_0 \cap S_k = \emptyset$ and $R_0 \subseteq S$. To prove the former, let $w \in L_0$. If $w \in L_0 \setminus L_k$, then $w \in up(L_0 \setminus L_k)$, hence $w \notin S_k$. If $w \in L_k$ and $w \in up(R_0 \setminus R_k)$, then there is $v \in R_0 \setminus R_k$ such that $v \preccurlyeq w$. However, $R_k = \{u \in R_0 \mid u \preccurlyeq L_k\}$, hence $v \in R_k$, a contradiction. Thus $L_0 \cap S_k = \emptyset$. To prove the later, we show that $R_{k-1} \setminus R_k \subseteq S_k$. Then $R_0 = \bigcup_{k=1}^B (R_{k-1} \setminus R_k) \subseteq S$. To show this, we have $R_{k-1} \setminus R_k \subseteq R_0 \setminus R_k \subseteq up(R_0 \setminus R_k)$. If $w \in R_{k-1}$ and $w \in up(L_0 \setminus L_k)$, then there is $v \in L_0 \setminus L_k$ such that $v \preccurlyeq w$. However,

$L_k = \{u \in L_0 \mid u \preccurlyeq R_{k-1}\}$, hence $v \in L_k$, a contradiction. Thus, we have shown that $L_0 \cap S = \emptyset$ and $R_0 \subseteq S$. Moreover, S is piecewise testable because it is a finite boolean combination of piecewise testable languages.

4   The Length of Towers

Recall that it was shown in [4] that there is either an infinite tower or a constant bound on the length of any tower. We now establish an upper bound on the length of finite towers.

Theorem 1. *Let A_0 and A_1 be NFAs with at most n states over an alphabet Σ of cardinality m, and assume that there is no infinite tower between the languages $L(A_0)$ and $L(A_1)$. Let $(w_i)_{i=1}^r$ be a tower between $L(A_0)$ and $L(A_1)$ such that $w_i \in L(A_{i \bmod 2})$. Then $r \leq \frac{n^{m+1}-1}{n-1}$.*

Proof. First, we define some new concepts. We say that $w = v_1 v_2 \cdots v_k$ is a *cyclic factorization* of w with respect to a pair of states (q, q') in an automaton A, if there is a sequence of states $q_0, \ldots, q_{k-1}, q_k$ such that $q_0 = q$, $q_k = q'$, and $q_{i-1} \xrightarrow{v_i} q_i$, for each $i = 1, 2, \ldots k$, and either v_i is a letter, or the path $q_{i-1} \xrightarrow{v_i} q_i$ contains a cycle over $\mathrm{alph}(v_i)$. We call v_i a *letter factor* if it is a letter and $q_{i-1} \neq q_i$, and a *cycle factor* otherwise. The factorization is *trivial* if $k = 1$. Note that this factorization is closely related to the one given in [1], see also [2, Theorem 8.1.11].

We first show that if $q' \in \delta(q, w)$ in some automaton A with n states, then w has a cyclic factorization $v_1 v_2 \cdots v_k$ with respect to (q, q') that contains at most n cycle factors and at most $n - 1$ letter factors. Moreover, if w does not admit the trivial factorization with respect to (q, q'), then $\mathrm{alph}(v_i)$ is a strict subset of $\mathrm{alph}(w)$ for each cycle factor v_i, $i = 1, 2, \ldots, k$.

Consider a path π of the automaton A from q to q' labeled by a string w. Let $q_0 = q$. Define the factorization $w = v_1 v_2 \cdots v_k$ inductively by the following greedy strategy. Assume we have defined the factors $v_1, v_2 \ldots, v_{i-1}$ such that $w = v_1 \cdots v_{i-1} w'$ and $q_0 \xrightarrow{v_1 v_2 \cdots v_{i-1}} q_{i-1}$. The factor v_i is defined as the label of the longest possible initial segment π_i of the path $q_{i-1} \xrightarrow{w'} q'$ such that either π_i contains a cycle over $\mathrm{alph}(v_i)$ or $\pi_i = q_{i-1}, a, q_i$, where $v_i = a$, so v_i is a letter. Such a factorization is well defined, and it is a cyclic factorization of w.

Let p_i, $i = 1, \ldots, k$, be a state such that the path $q_{i-1} \xrightarrow{v_i} q_i$ contains a cycle $p_i \to p_i$ over $\mathrm{alph}(v_i)$ if v_i is a cycle factor, and $p_i = q_{i-1}$ if v_i is a letter factor. If $p_i = p_j$ with $i < j$ such that v_i and v_j are cycle factors, then we have a contradiction with the maximality of v_i since $q_{i-1} \xrightarrow{v_i v_{i+1} \cdots v_j} q_j$ contains a cycle $p_i \to p_i$ from p_i to p_i over the alphabet $\mathrm{alph}(v_i v_{i+1} \cdots v_j)$. Therefore the factorization contains at most n cycle factors.

Note that v_i is a letter factor only if the state p_i, which is equal to q_{i-1} in such a case, has no reappearance in the path $q_{i-1} \xrightarrow{v_i \cdots v_k} q'$. This implies that there are at most $n - 1$ letter factors. Finally, if $\mathrm{alph}(v_i) = \mathrm{alph}(w)$, then $v_i = v_1 = w$ follows from the maximality of v_1.

We now define inductively cyclic factorizations of w_i, such that the factorization of w_{i-1} is a refinement of the factorization of w_i. Let $w_r = v_{r,1} v_{r,2} \cdots v_{r,k_r}$ be a cyclic factorization of w_r defined, as described above, by some accepting path in the automaton $A_{r \bmod 2}$. Factorizations $w_{i-1} = v_{i-1,1} v_{i-1,2} \cdots v_{i-1,k_{i-1}}$ are defined as follows. Let

$$w_{i-1} = v'_{i,1} v'_{i,2} \cdots v'_{i,k_i},$$

where $v'_{i,j} \preccurlyeq v_{i,j}$, for each $j = 1, 2, \ldots, k_i$; note that such a factorization exists since $w_{i-1} \preccurlyeq w_i$. Then $v_{i-1,1} v_{i-1,2} \cdots v_{i-1,k_{i-1}}$ is defined as a concatenation of cyclic factorizations of $v'_{i,j}$, $j = 1, 2, \ldots, k_i$, corresponding to an accepting path of w_{i-1} in $A_{i-1 \bmod 2}$. The cyclic factorization of the empty string is defined as empty. Note also that a letter factor of w_i either disappears in w_{i-1}, or it is "factored" into a letter factor.

In order to measure the height of a tower, we introduce a weight function f of factors in a factorization $v_1 v_2 \cdots v_k$. First, let

$$g(x) = n \frac{n^x - 1}{n - 1}.$$

Note that g satisfies $g(x + 1) = ng(x) + (n - 1) + 1$. Now, let $f(v_i) = 1$ if v_i is a letter factor, and let $f(v_i) = g(|\operatorname{alph}(v_i)|)$ if v_i is a cycle factor. Note that, by definition, $f(\varepsilon) = 0$. The weight of the factorization $v_1 v_2 \cdots v_k$ is then defined by

$$W(v_1 v_2 \cdots v_k) = \sum_{i=1}^{k} f(v_i).$$

Let

$$W_i = W(v_{i,1} v_{i,2} \cdots v_{i,k_i}).$$

We claim that $W_{i-1} < W_i$ for each $i = 2, \ldots, r$. Let $v_1 v_2 \cdots v_k$ be the fragment of the cyclic factorization of w_{i-1} that emerged as the cyclic factorization of $v'_{i,j} \preccurlyeq v_{i,j}$. If the factorization is not trivial, then, by the above analysis,

$$W(v_1 v_2 \cdots v_k) \le n - 1 + n \cdot g(|\operatorname{alph}(v_{i,j})| - 1) < g(|\operatorname{alph}(v_{i,j})|) = f(v_{i,j}).$$

Similarly, we have $f(v'_{i,j}) < f(v_{i,j})$ if $|\operatorname{alph}(v'_{i,j})| < |\operatorname{alph}(v_{i,j})|$. Altogether, we have $W_{i-1} < W_i$ as claimed, unless

- $k_{i-1} = k_i$,
- the factor $v_{i-1,j}$ is a letter factor if and only if $v_{i,j}$ is a letter factor, and
- $\operatorname{alph}(v_{i-1,j}) = \operatorname{alph}(v_{i,j})$ for all $j = 1, 2, \ldots, k_i$.

Assume that such a situation takes place, and show that it leads to an infinite tower. Let L be the language of strings $z_1 z_2 \cdots z_{k_i}$ such that $z_j = v_{i,j}$ if $v_{i,j}$ is a letter factor, and $z_j \in (\operatorname{alph}(v_{i,j}))^*$ if $v_{i,j}$ is a cycle factor. Since $w_i \in L(A_{i \bmod 2})$ and $w_{i-1} \in L(A_{i-1 \bmod 2})$ holds, the definition of a cycle factor implies that, for each $z \in L$, there is some $z' \in L(A_0) \cap L$ such that $z \preccurlyeq z'$, and also $z'' \in L(A_1) \cap L$ such that $z \preccurlyeq z''$. The existence of an infinite tower follows. We have therefore proved $W_{i-1} < W_i$.

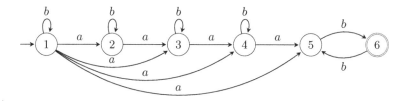

Fig. 2. Automaton A_0; $n - 1 = 6$

The proof is completed, since $W_r \leq f(w_r) \leq g(m)$, $W_1 \geq 0$, and the bound in the claim is equal to $g(m) + 1$. ⬜

For binary regular languages, we now show that there exists a tower of length at least $n^2 - O(n)$ between two binary regular languages having no infinite tower and represented by automata with at most n states.

Theorem 2. *The upper bound $\frac{n^3 - 1}{n - 1}$ on the length of a maximal tower is tight for binary languages up to a linear factor.*

Proof. Let n be an odd number and define the automata A_0 and A_1 with $n - 1$ and n states as depicted in Figs. 2 and 3, respectively.

The automaton $A_0 = (\{1, 2, \ldots, n - 1\}, \{a, b\}, \delta_0, 1, \{n - 1\})$ consists of an a-path from state 1 through states $2, 3, \ldots, n - 3$, respectively, to state $n - 2$, of a-transitions from state 1 to all states but itself and the final state, of self-loops under b in all but the states $n - 2$ and $n - 1$, and of a b-cycle from $n - 2$ to $n - 1$ and back to $n - 2$.

The automaton $A_1 = (\{1, 2, \ldots, n\}, \{a, b\}, \delta_1, 1, \{1, n\})$ consists of a b-path from state 1 through states $2, 3, \ldots, n - 1$, respectively, to state n, of an a-transition from state n to state 1, and of b-transitions going from state 1 to all even-labeled states.

Consider the string

$$(b^{n-1}a)^{n-3}(b^{n-1}b).$$

This string consists of $n - 2$ parts of length n and belongs to $L(A_0)$. Note that deleting the last letter b results in a string that belongs to $L(A_1)$. Deleting another letter b from the right results in a string belonging again to the language

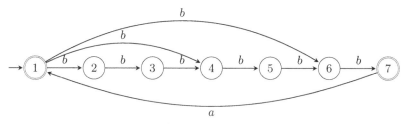

Fig. 3. Automaton A_1; $n = 7$

$L(A_0)$. We can continue in this way alternating between the languages until the letter a is the last letter, that is, until the string $(b^{n-1}a)^{n-3}$, which belongs to $L(A_1)$. Now, we delete the last two letters, namely the string ba, which results in a string from $L(A_0)$, and we can continue with deleting the last letters b again as described above. Moreover, we cannot accept the prefix b^{n-2} in A_0, hence the length of the tower is at least $n(n-2) - (n-3) - (n-2) = n^2 - 4n + 5$.

To show that there is no infinite tower between the languages $L(A_0)$ and $L(A_1)$, we can use the techniques described in [4,8], or to use the algorithm presented in Section 3. We can also notice that letter a can appear at most $n-3$ times in any string from $L(A_0)$ and that after at most $n-1$ occurrences of letter b, letter a must appear in a string from $L(A_1)$. As the languages are disjoint, any infinite tower would have to contain a string from $L(A_1)$ of length more than $n \cdot (n-3) + (n-1)$. But any such string in $L(A_1)$ must contain at least $n-2$ occurrences of letter a, hence it cannot be embedded into any string of $L(A_0)$. This means that there cannot be an infinite tower. □

In Theorem 2, we have shown that there exists a tower of a quadratic length between two binary languages having no infinite tower. Now we show that there exist two quaternary languages having a tower of length more than quadratic.

Theorem 3. *There exist two languages with no infinite tower having a finite tower of a cubic length.*

Proof. Let n be a number divisible by four and define the automata A_0 and A_1 with $n-1$ and n states as shown in Figs. 4 and 5, respectively.

The automaton $A_0 = (\{1, 2, \ldots, n-1\}, \{a, b, c, d\}, \delta_0, 1, \{n-1\})$ consists of an a-path through states $1, 2, \ldots, n-2$, respectively, of a-transitions from state 1 to all other states but itself and the final state, of self-loops under symbols b, c, d in all but the final state, and of a b-transition from all, but the final state, to the final state.

The automaton $A_1 = (\{1, 2, \ldots, n\}, \{a, b, c, d\}, \delta_1, 1, \{\frac{n}{2}, n\})$ consists of two parts. The first part is constituted by states $1, 2, \ldots, \frac{n}{2}$ with a d-path through states $1, 2, \ldots, \frac{n}{2}$, respectively, by self-loops under b, c in states $1, 2, \ldots, \frac{n}{2} - 1$, and by d-transitions from state 1 to all of states $2, 3, \ldots, \frac{n}{2}$. The second part is constituted by states $\frac{n}{2}, \ldots, n$ with a bc-path through states $\frac{n}{2}, \ldots, n-2$, respectively, by a-transitions from state $n-1$ to states 1 and n, by a c-transition from state $n-1$ to state n, and by b-transitions from state $\frac{n}{2}$ to all odd-numbered states between $\frac{n}{2}$ and $n-1$.

Note that the languages are disjoint since A_0 accepts strings ending with b, while A_1 accepts strings ending with a, c, or d.

Consider the string

$$\left[\left(bd(bc)^{\frac{n}{4}} \right)^{\frac{n}{2} - 2} bd(bc)^{\frac{n}{4} - 1} ba \right]^{n-3} \cdot \left(bd(bc)^{\frac{n}{4}} \right)^{\frac{n}{2} - 2} bd(bc)^{\frac{n}{4} - 1} bcb .$$

This string belongs to $L(A_0)$ and consists of $n-3$ parts each of length $\frac{n^2}{4} + \frac{n}{2} - 2$, plus one part of length $\frac{n^2}{4} + \frac{n}{2} - 1$. We can delete the last letters one by one,

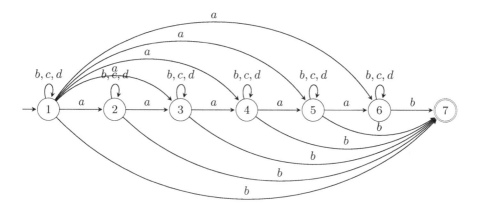

Fig. 4. Automaton A_0; $n - 1 = 7$

obtaining strings alternating between $L(A_1)$ and $L(A_0)$. Hence the length of this tower is $(n - 2) \cdot (\frac{n^2}{4} + \frac{n}{2} - 2) + 1$, which results in a tower of length $\Omega(n^3)$.

To show that there is no infinite tower between the languages, we can use the techniques described in [4,8], or the algorithm presented in Section 3.   □

As the last result of this paper, we prove an exponential lower bound with respect to the cardinality of the input alphabet.

Theorem 4. *There exist two languages with no infinite tower having a finite tower of an exponential length with respect to the size of the alphabet.*

Proof. For every non-negative integer m, we define a pair of nondeterministic automata A_m and B_m over the input alphabet $\Sigma_m = \{a_1, a_2, \ldots, a_m\} \cup \{b, c\}$ with a tower of length 2^{m+2} between $L(A_m)$ and $L(B_m)$, and such that there is no infinite tower between the two languages.

The two-state automaton $A_m = (\{1, 2\}, \Sigma_m, \delta_m, 1, \{2\})$ has self-loops under all symbols in state 1 and a b-transition from state 1 to state 2. The automaton is shown in Fig. 6 (left), and it accepts all strings over Σ_m ending with b.

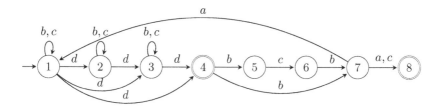

Fig. 5. Automaton A_1; $n = 8$ and $F = \{\frac{n}{2}, n\}$

Fig. 6. The two-state NFA A_m, for $m \geq 0$ (left), and the automaton B_0 (right)

The automata B_m are constructed inductively as follows. The automaton $B_0 = (\{p, q, r\}, \{b, c\}, \gamma_0, \{p\}, \{p, r\})$ accepts the finite language $\{\varepsilon, bc\}$, and it is shown in Fig. 6 (right).

Assume that we have constructed the nondeterministic finite automaton $B_m = (Q_m, \Sigma_m, \gamma_m, S_m, \{p, r\})$. We construct the nondeterministic automaton $B_{m+1} = (Q_m \cup \{m+1\}, \Sigma_m \cup \{a_{m+1}\}, \gamma_{m+1}, S_m \cup \{m+1\}, \{p, r\})$ by adding a new initial state $m + 1$ to Q_m, and transitions on a fresh input symbol a_{m+1}. The transition function γ_{m+1} extends γ_m so that it defines self-loops under all letters of Σ_m in the new state $m + 1$, and adds the transitions on input a_{m+1} from state $m + 1$ to all the states of S_m, that is, to all the initial states of B_m. The first two steps of the construction, that is, automata B_1 and B_2, are shown in Figs. 7 and 8, respectively. Note that $L(B_m) \subseteq L(B_{m+1})$ since all the initial states of B_m are initial in B_{m+1} as well, and the set of final states is $\{p, r\}$ in both automata.

By induction on m, we show that there exists a tower between the languages $L(A_m)$ and $L(B_m)$ of length 2^{m+2}. More specifically, we prove that there exists a sequence $(w_i)_{i=1}^{2^{m+2}}$ such that w_i is a prefix of w_{i+1} and $|w_{i+1}| = |w_i| + 1$ for all $i = 1, \ldots, 2^{m+2} - 1$, $w_1 = \varepsilon$, so $w_1 \in L(B_m)$, and $w_{2^{m+2}} \in L(A_m)$. Thus, the tower is fully characterized by its longest string $w_{2^{m+2}}$. Moreover, by definition, the letter b appears on all odd positions of $w_{2^{m+2}}$.

If $m = 0$, then such a tower is ε, b, bc, bcb, and it is of length 2^2. Assume that for some m, we have a sequence of prefixes of length 2^{m+2} as required above, and such that the length of its longest string wb is $2^{m+2} - 1$. Consider the automata A_{m+1} and B_{m+1} and the string

$$wba_{m+1}wb.$$

The length of this string is $2^{(m+1)+2} - 1$, which results in $2^{(m+1)+2}$ prefixes.

By the assumption, every odd position is occupied by letter b, hence every prefix of an odd length belongs to $L(A_{m+1})$. It remains to show that all even-length prefixes belong to $L(B_{m+1})$. Let x be such a prefix. If x does not contain a_{m+1}, then it is a prefix of wb and belongs to $L(B_m)$ by the induction hypothesis.

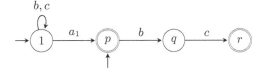

Fig. 7. Automaton B_1

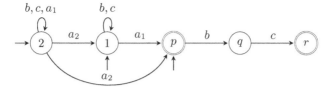

Fig. 8. Automaton B_2

If $x = wba_{m+1}y$, then B_{m+1} reads the string wb in state $m + 1$. Then, on input a_{m+1}, it goes to an initial state of B_m. From this initial state, the string y is accepted as a prefix of wb by the induction hypothesis. Thus x is in $L(B_{m+1})$.

To complete the proof, it remains to show that there is no infinite tower between the languages. We can either use the techniques described in [4,8], or the algorithm presented in Section 3. However, to give a brief idea why it is so, we can give an inductive argument. Since $L(B_0)$ is finite, there is no infinite tower between $L(A_0)$ and $L(B_0)$. Consider a tower between $L(A_{m+1})$ and $L(B_{m+1})$. If every string of the tower belonging to $L(B_{m+1})$ is accepted from an initial state different from $m + 1$, then it is a tower between $L(A_m)$ and $L(B_m)$, so it is finite. Thus, if there exists an infinite tower, there also exists an infinite tower where all strings belonging to $L(B_{m+1})$ are accepted only from state $m + 1$. However, every such string is of the form $(\{a_1, \ldots, a_m\} \cup \{b, c\})^* a_{m+1}y$, where the string y is accepted from an initial state different from $m+1$. Cutting off the prefixes from $(\{a_1, \ldots, a_m\} \cup \{b, c\})^* a_{m+1}$ results in an infinite tower between $L(A_m)$ and $L(B_m)$, which is a contradiction. $\qquad\square$

5   Conclusions

The definition of towers can be generalized from subsequences to basically any relation on strings, namely to prefixes, suffixes, etc. Notice that our lower-bound examples in Theorems 2, 3, and 4 are actually towers of prefixes, hence they give a lower bound on the length of towers of prefixes as well.

On the other hand, the upper-bound results cannot be directly used to prove the upper bounds for towers of prefixes. Although every tower of prefixes is also a tower of subsequences, the condition that there are no infinite towers is weaker for prefixes. The bound for subsequences therefore does not apply to languages that allow an infinite tower of subsequences but only finite towers of prefixes.

Finally, note that the lower-bound results are based on nondeterminism. We are aware of a tower of subsequences (prefixes) showing the quadratic lower bound for deterministic automata. However, it is an open question whether a longer tower can be found or the upper bound is significantly different for deterministic automata.

References

1. Almeida, J.: Implicit operations on finite J-trivial semigroups and a conjecture of I. Simon. Journal of Pure and Applied Algebra 69, 205–218 (1990)
2. Almeida, J.: Finite semigroups and universal algebra. Series in Algebra, vol. 3. World Scientific (1995)
3. Almeida, J., Zeitoun, M.: The pseudovariety J is hyperdecidable. RAIRO – Theoretical Informatics and Applications 31(5), 457–482 (1997)
4. Czerwiński, W., Martens, W., Masopust, T.: Efficient separability of regular languages by subsequences and suffixes. In: Fomin, F.V., Freivalds, R., Kwiatkowska, M., Peleg, D. (eds.) ICALP 2013, Part II. LNCS, vol. 7966, pp. 150–161. Springer, Heidelberg (2013), http://arxiv.org/abs/1303.0966
5. Czerwiński, W., Martens, W., Masopust, T.: Personal communication (2013)
6. Higman, G.: Ordering by divisibility in abstract algebras. Proceedings of the London Mathematical Society s3-2(1), 326–336 (1952)
7. Karandikar, P., Schnoebelen, P.: On the state complexity of closures and interiors of regular languages with subwords. In: Proc. of DCFS, http://arxiv.org/abs/1406.0690 (to appear, 2014)
8. Place, T., van Rooijen, L., Zeitoun, M.: Separating regular languages by piecewise testable and unambiguous languages. In: Chatterjee, K., Sgall, J. (eds.) MFCS 2013. LNCS, vol. 8087, pp. 729–740. Springer, Heidelberg (2013)
9. Place, T., Zeitoun, M.: Separating regular languages with first-order logic. In: Proc. of CSL-LICS, http://arxiv.org/abs/1402.3277 (accepted, 2014)
10. Simon, I.: Hierarchies of Events with Dot-Depth One. Ph.D. thesis, Dept. of Applied Analysis and Computer Science, University of Waterloo, Canada (1972)
11. Simon, I.: Piecewise testable events. In: Brakhage, H. (ed.) GI Conference on Automata Theory and Formal Languages. LNCS, vol. 33, pp. 214–222. Springer, Heidelberg (1975)
12. Stern, J.: Characterizations of some classes of regular events. Theoretical Computer Science 35, 17–42 (1985)

L^{ax}F: Side Conditions and External Evidence as Monads

Furio Honsell[1], Luigi Liquori[2], and Ivan Scagnetto[1]

[1] Università di Udine, Italy
{furio.honsell,ivan.scagnetto}@uniud.it
[2] INRIA, Sophia Antipolis Méditerranée, France
Luigi.Liquori@inria.fr

Abstract. We extend the constructive dependent type theory of the Logical Framework LF with a family of *monads* indexed by predicates over typed terms. These monads express the *effect* of factoring-out, postponing, or delegating to an *external oracle* the verification of a constraint or a side-condition. This new framework, called *Lax Logical Framework*, L^{ax}F, is a conservative extension of LF, and hence it is the appropriate metalanguage for dealing formally with side-conditions or external evidence in logical systems. L^{ax}F is the natural strengthening of LF_P (the extension of LF introduced by the authors together with Marina Lenisa and Petar Maksimovic), which arises once the *monadic* nature of the *lock* constructors of LF_P is fully exploited. The nature of these monads allows to utilize the *unlock* destructor instead of Moggi's monadic *let_T*, thus simplifying the equational theory. The rules for the unlock allow us, furthermore, to remove the monadic constructor once the constraint is satisfied. By way of example we discuss the encodings in L^{ax}F of call-by-value λ-calculus, Hoare's Logic, and Elementary Affine Logic.

1 Introduction

The system LF_P [18] is a conservative extension of LF. It was introduced to factor out neatly judgements whose justification can be delegated to an *external oracle*. This allows us to recover within a Logical Framework many different *proof cultures* that otherwise can be embedded only very deeply [14] or axiomatically [20]. In particular, recourse in formal proofs to external sources of justification and external evidence such as diagrams, physical analogies, explicit computations according to *Poincaré Principle* [5], and to external proof search tools can thus be explicitly *invoked* and *recorded* in a LF type-theoretic framework. Methodologically this is a simple, but quite significant move, since in dealing with logics one has to rely on external objects more often than one may think. Any *adequacy* result or even the very *execution* of the most obvious rule relies ultimately on some *external* unformalizable convention, as captured by *Münchausen trilemma* [1] or the story of *Achilles and the Tortoise* narrated by Lewis Carroll [7].

The idea behind LF_P is to express explicitly, by means of a new type constructor $\mathcal{L}^{\mathcal{P}}_{M,\sigma}[\tau]$, that in order to obtain a term of type τ it is necessary to

E. Csuhaj-Varjú et al. (Eds.): MFCS 2014, Part I, LNCS 8634, pp. 327–339, 2014.
© Springer-Verlag Berlin Heidelberg 2014

verify the constraint $\mathcal{P}(\Gamma \vdash_\Sigma M : \sigma)$. This idea grew out of a series of papers, [6,17,19,18,15], on extensions of LF published by the authors in recent years.

In this paper we introduce a new system, called *Lax Logical Framework*, $\mathsf{L}^{ax}\mathsf{F}$, which amounts to the natural generalization and strengthening of $\mathsf{LF}_\mathcal{P}$, once the *monadic* nature of the $\mathcal{L}^\mathcal{P}_{M,\sigma}[N]$ constructors is recognized and fully exploited. Hence $\mathsf{L}^{ax}\mathsf{F}$ is the extension of LF with a family of *monads* indexed by predicates over typed terms, which capture the *effect* of factoring out and postponing, or delegating to an external oracle the verification of the constraint or side-condition $\mathcal{P}(\Gamma \vdash_\Sigma M : \sigma)$.

Our basic idea is that any side condition $\mathcal{P}$ can be viewed as a monad $T_\mathcal{P}$, where the categorical natural transformation $\eta_{T_\mathcal{P}} : A \to T_\mathcal{P}(A)$ expresses the fact that a judgement can always be asserted *weakly*, *i.e.* subject to the satisfaction of a given constraint. While the other natural transformation characterizing a monad $\mu_{T_\mathcal{P}} : T^2_\mathcal{P}(A) \to T_\mathcal{P}(A)$, expresses the fact that it is useless to verify twice a given constraint.

The main extension with respect to the language of $\mathsf{LF}_\mathcal{P}$ is that, for $N : \tau$, the destructor $\mathcal{U}^\mathcal{P}_{M,\sigma}[N]$ of a particular lock-type, can be used freely provided it is *guarded*, *i.e.* it appears within a subterm whose type has the same lock-type constructor, *i.e.* $\mathcal{L}^\mathcal{P}_{M,\sigma}[\rho]$. Thereby, checking predicates in locks can be postponed and, most usefully, functions which output terms of lock-type can be "applied" also to locked-arguments. The nature of these monads allows us to utilize the $\mathcal{U}^\mathcal{P}_{M,\sigma}[N]$ destructor instead of the usual monadic let_T, thus greatly simplifying the equational theory. Moreover, as in the case of $\mathsf{LF}_\mathcal{P}$, but in addition to what happens with ordinary monads, the rules concerning $\mathcal{U}^\mathcal{P}_{M,\sigma}[N]$ allow us to drop the monadic constructor if the constraint is satisfied.

We give classical examples of encodings in $\mathsf{L}^{ax}\mathsf{F}$ of logical systems, thereby showing that $\mathsf{L}^{ax}\mathsf{F}$ is the appropriate metalanguage for dealing formally with side-conditions, and external evidence. Because of the extra expressive power given by guarded terms of the form $\mathcal{U}^\mathcal{P}_{M,\sigma}[N]$, signatures become much more flexible, thus achieving the full modularity that we have been looking for in recent years. We discuss briefly also the intriguing case of Elementary Affine Logic [3].

In conclusion, in this paper we extend:

• the well understood principles of the LF paradigm for explaining a logic, *i.e. judgments as types*, and *rules* or *hypothetical judgements as higher order types*, and *schemata as higher order functions*, and *quantified variables as bound metalanguage variables*, with the new clause: *side conditions* and *external evidence as monads*;

• the capacity of logical systems to combine and relate two software tools using a simple communication paradigm via "wrappers".

Related Work. This paper builds on earlier work of the authors [17,19,18,15] and was inspired by the very extensive work on Logical Frameworks by [24,27,8,23,25,26]. The term *"Lax"* is borrowed from [9,21], and indeed our system can be viewed as a generalization, to a family of lax operators, of the work carried out there, as well as Moggi's *partial* λ-calculus [22]. A correspondence between lax modalities and monads in functional programming was pointed out

$$\Sigma \in Signatures \qquad \Sigma ::= \emptyset \mid \Sigma, a{:}K \mid \Sigma, c{:}\sigma$$
$$K \in Kinds \qquad K ::= \mathsf{Type} \mid \Pi x{:}\sigma.K$$
$$\sigma, \tau, \rho \in Families\ (Types) \qquad \sigma ::= a \mid \Pi x{:}\sigma.\tau \mid \sigma N \mid \mathcal{L}^{\mathcal{P}}_{N,\sigma}[\rho]$$
$$M, N \in Objects \qquad M ::= c \mid x \mid \lambda x{:}\sigma.M \mid M\ N \mid \mathcal{L}^{\mathcal{P}}_{N,\sigma}[M] \mid \mathcal{U}^{\mathcal{P}}_{N,\sigma}[M]$$

Fig. 1. The pseudo-syntax of L^{ax}F

$$(\lambda x{:}\sigma.M)\,N \to_{\beta\mathcal{L}} M[N/x]\ (\beta{\cdot}O{\cdot}Main) \qquad \mathcal{U}^{\mathcal{P}}_{N,\sigma}[\mathcal{L}^{\mathcal{P}}_{N,\sigma}[M]] \to_{\beta\mathcal{L}} M\ (\mathcal{L}{\cdot}O{\cdot}Main)$$

Fig. 2. Main one-step-$\beta\mathcal{L}$-reduction rules in L^{ax}F

in [2,12]. In [23,11,10] the connection between constraints and monads in logic programming was considered, but to our knowledge this is the first paper which clearly establishes the correspondence between side conditions and monads in a *higher order dependent type theory* and logical frameworks.

Synopsis. In Section 2, we present the syntax and the typing system of L^{ax}F. In Section 3 we discuss the changes in the metatheory of the framework LF$_{\mathcal{P}}$, induced by the new typing rule, The conservativity of L^{ax}F both w.r.t. LF$_{\mathcal{P}}$ and to LF is discussed at the end of Section 3. Three case studies are presented in Section 4. Concluding remarks and directions for future work are in Section 5[1].

2   L^{ax}F

In this section, we introduce the syntax and the rules of L^{ax}F: in Figure 1, we give the syntactic categories of L^{ax}F, namely signatures, contexts, kinds, families (*i.e.*, types) and objects (*i.e.*, terms), while the main one-step $\beta\mathcal{L}$-reduction rules appear in Figure 2. The language of L^{ax}F is the same as that of LF$_{\mathcal{P}}$ [18]. In particular, w.r.t. classical LF, we add the *lock-types* constructor ($\mathcal{L}$) for building types of the shape $\mathcal{L}^{\mathcal{P}}_{N,\sigma}[\rho]$, where $\mathcal{P}$ is a predicate on typed judgements, and correspondingly at object level the constructor lock ($\mathcal{L}$) and destructor unlock ($\mathcal{U}$). The intended meaning of the $\mathcal{L}^{\mathcal{P}}_{N,\sigma}[\cdot]$ constructors is that of *logical filters*. Locks can be viewed also as a generalization of the Lax modality of [9,21]. One of the points of this paper is to show that they can be viewed also as *monads*.

Following the standard specification paradigm of Constructive Type Theory, we define lock-types using introduction, elimination, and equality rules. Namely, we introduce a lock-constructor for building objects $\mathcal{L}^{\mathcal{P}}_{N,\sigma}[M]$ of type $\mathcal{L}^{\mathcal{P}}_{N,\sigma}[\rho]$, via the introduction rule (*O·Lock*). Correspondingly, we introduce an unlock-destructor $\mathcal{U}^{\mathcal{P}}_{N,\sigma}[M]$ via the elimination rule (*O·Guarded·Unlock*). These rules give evidence to the understanding of *locks as monads*[2]. The introduction rule of lock-types immediately corresponds to the introduction rule of monads, but this

---

[1] A web appendix is available (for interested readers) at
http://www.dimi.uniud.it/scagnett/pubs/MonadixLFP-Appendix.pdf
[2] Given a predicate $\mathcal{P}$ and $\Gamma \vdash_\Sigma N : \sigma$, the intended monad $(T_{\mathcal{P}}, \eta, \mu)$ can be naturally defined on the term model of L^{ax}F viewed as a category. In particular $\eta_\rho \triangleq \lambda x{:}\rho.\mathcal{L}^{\mathcal{P}}_{N,\sigma}[x]$ and $\mu_\rho \triangleq \lambda x{:}\mathcal{L}^{\mathcal{P}}_{N,\sigma}[\mathcal{L}^{\mathcal{P}}_{N,\sigma}[\rho]].\ \mathcal{L}^{\mathcal{P}}_{N,\sigma}[\mathcal{U}^{\mathcal{P}}_{N,\sigma}[x]]$.

Signature rules

$$\frac{}{\emptyset \text{ sig}} \ (S{\cdot}Empty)$$

$$\frac{\Sigma \text{ sig} \quad \vdash_\Sigma K \quad a \notin \text{Dom}(\Sigma)}{\Sigma, a{:}K \text{ sig}} \ (S{\cdot}Kind)$$

$$\frac{\Sigma \text{ sig} \quad \vdash_\Sigma \sigma{:}\text{Type} \quad c \notin \text{Dom}(\Sigma)}{\Sigma, c{:}\sigma \text{ sig}} \ (S{\cdot}Type)$$

Context rules

$$\frac{\Sigma \text{ sig}}{\vdash_\Sigma \emptyset} \ (C{\cdot}Empty)$$

$$\frac{\vdash_\Sigma \Gamma \quad \Gamma \vdash_\Sigma \sigma{:}\text{Type} \quad x \notin \text{Dom}(\Gamma)}{\vdash_\Sigma \Gamma, x{:}\sigma} \ (C{\cdot}Type)$$

Kind rules

$$\frac{\vdash_\Sigma \Gamma}{\Gamma \vdash_\Sigma \text{Type}} \ (K{\cdot}Type)$$

$$\frac{\Gamma, x{:}\sigma \vdash_\Sigma K}{\Gamma \vdash_\Sigma \Pi x{:}\sigma.K} \ (K{\cdot}Pi)$$

Family rules

$$\frac{\vdash_\Sigma \Gamma \quad a{:}K \in \Sigma}{\Gamma \vdash_\Sigma a : K} \ (F{\cdot}Const)$$

$$\frac{\Gamma, x{:}\sigma \vdash_\Sigma \tau : \text{Type}}{\Gamma \vdash_\Sigma \Pi x{:}\sigma.\tau : \text{Type}} \ (F{\cdot}Pi)$$

$$\frac{\Gamma \vdash_\Sigma \sigma : \Pi x{:}\tau.K \quad \Gamma \vdash_\Sigma N : \tau}{\Gamma \vdash_\Sigma \sigma N : K[N/x]} \ (F{\cdot}App)$$

$$\frac{\Gamma \vdash_\Sigma \rho : \text{Type} \quad \Gamma \vdash_\Sigma N : \sigma}{\Gamma \vdash_\Sigma \mathcal{L}_{N,\sigma}^{\mathcal{P}}[\rho] : \text{Type}} \ (F{\cdot}Lock)$$

$$\frac{\Gamma \vdash_\Sigma \sigma : K \quad \Gamma \vdash_\Sigma K' \quad K =_{\beta\mathcal{L}} K'}{\Gamma \vdash_\Sigma \sigma : K'} \ (F{\cdot}Conv)$$

Object rules

$$\frac{\vdash_\Sigma \Gamma \quad c{:}\sigma \in \Sigma}{\Gamma \vdash_\Sigma c : \sigma} \ (O{\cdot}Const)$$

$$\frac{\vdash_\Sigma \Gamma \quad x{:}\sigma \in \Gamma}{\Gamma \vdash_\Sigma x : \sigma} \ (O{\cdot}Var)$$

$$\frac{\Gamma, x{:}\sigma \vdash_\Sigma M : \tau}{\Gamma \vdash_\Sigma \lambda x{:}\sigma.M : \Pi x{:}\sigma.\tau} \ (O{\cdot}Abs)$$

$$\frac{\Gamma \vdash_\Sigma M : \Pi x{:}\sigma.\tau \quad \Gamma \vdash_\Sigma N : \sigma}{\Gamma \vdash_\Sigma M N : \tau[N/x]} \ (O{\cdot}App)$$

$$\frac{\Gamma \vdash_\Sigma M : \sigma \quad \Gamma \vdash_\Sigma \tau : \text{Type} \quad \sigma =_{\beta\mathcal{L}} \tau}{\Gamma \vdash_\Sigma M : \tau} \ (O{\cdot}Conv)$$

$$\frac{\Gamma \vdash_\Sigma M : \rho \quad \Gamma \vdash_\Sigma N : \sigma}{\Gamma \vdash_\Sigma \mathcal{L}_{N,\sigma}^{\mathcal{P}}[M] : \mathcal{L}_{N,\sigma}^{\mathcal{P}}[\rho]} \ (O{\cdot}Lock)$$

$$\frac{\Gamma \vdash_\Sigma M : \mathcal{L}_{N,\sigma}^{\mathcal{P}}[\rho] \quad \mathcal{P}(\Gamma \vdash_\Sigma N : \sigma)}{\Gamma \vdash_\Sigma \mathcal{U}_{N,\sigma}^{\mathcal{P}}[M] : \rho} \ (O{\cdot}Top{\cdot}Unlock)$$

$$\frac{\Gamma, x{:}\tau \vdash_\Sigma M : \mathcal{L}_{S,\sigma}^{\mathcal{P}}[\rho] \quad \Gamma \vdash_\Sigma N : \mathcal{L}_{S,\sigma}^{\mathcal{P}}[\tau]}{\Gamma \vdash_\Sigma M[\mathcal{U}_{S,\sigma}^{\mathcal{P}}[N]/x] : \mathcal{L}_{S,\sigma}^{\mathcal{P}}[\rho[\mathcal{U}_{S,\sigma}^{\mathcal{P}}[N]/x]]} \ (O{\cdot}Guarded{\cdot}Unlock)$$

Fig. 3. The $\mathsf{L}^{\mathsf{ax}}\mathsf{F}$ Type System

is not so immediate for the elimination rule which is normally given for monads using a let_T-construct. The correspondence becomes clear once we realize that $let_{T_{\mathcal{P}(\Gamma \vdash S{:}\sigma)}} x = M$ in N can be safely replaced by $N[\mathcal{U}_{S,\sigma}^{\mathcal{P}}[M]/x]$ since the $\mathcal{L}_{S,\sigma}^{\mathcal{P}}[\cdot]$-monads satisfy the property $let_{T_{\mathcal{P}}} \ x = M$ in $N \to N$ if $x \notin FV(N)$, provided x occurs *guarded* in N, *i.e.* within subterms of the appropriate locked-type.

Finally, to recover the intended meaning of $\mathcal{L}_{N,\sigma}^{\mathcal{P}}[\cdot]$, we need to introduce in $\mathsf{L}^{\mathsf{ax}}\mathsf{F}$ also $(O{\cdot}Top{\cdot}Unlock)$, which allows for the elimination of the lock-type constructor if the predicate $\mathcal{P}$ is verified, possibly *externally*, on an appropriate and derivable judgement. Figure 3 shows the full typing system of $\mathsf{L}^{\mathsf{ax}}\mathsf{F}$. The type equality rule of $\mathsf{L}^{\mathsf{ax}}\mathsf{F}$ uses a notion of conversion which derives from $\beta\mathcal{L}$-reduction, a combination of standard β-reduction, $(\beta{\cdot}O{\cdot}Main)$, with another notion of reduction ($\mathcal{L} \cdot O \cdot Main$), called $\mathcal{L}$-reduction. The latter behaves as a lock-releasing mechanism, erasing the $\mathcal{U}$-$\mathcal{L}$ pair in a term of the form $\mathcal{U}_{N,\sigma}^{\mathcal{P}}[\mathcal{L}_{N,\sigma}^{\mathcal{P}}[M]]$.

Since external predicates affect reductions in L^{ax}F, they must be well-behaved in order to preserve subject reduction. And this property is needed for decidability, possibly *up to an oracle*, which is essential in LF's.

Definition 1 (Well-behaved predicates, [18]). *A finite set of predicates $\{\mathcal{P}_i\}_{i \in I}$ is well-behaved if each $\mathcal{P}$ in the set satisfies the following conditions:*

- *Closure under signature and context weakening and permutation:*
  1. *If Σ and Ω are valid signatures such that $\Sigma \subseteq \Omega$ and $\mathcal{P}(\Gamma \vdash_\Sigma \alpha)$, then $\mathcal{P}(\Gamma \vdash_\Omega \alpha)$.*
  2. *If Γ and Δ are valid contexts such that $\Gamma \subseteq \Delta$ and $\mathcal{P}(\Gamma \vdash_\Sigma \alpha)$, then $\mathcal{P}(\Delta \vdash_\Sigma \alpha)$.*
- *Closure under substitution: If $\mathcal{P}(\Gamma, x{:}\sigma', \Gamma' \vdash_\Sigma N : \sigma)$ and $\Gamma \vdash_\Sigma N' : \sigma'$, then $\mathcal{P}(\Gamma, \Gamma'[N'/x] \vdash_\Sigma N[N'/x] : \sigma[N'/x])$.*
- *Closure under reduction:*
  1. *If $\mathcal{P}(\Gamma \vdash_\Sigma N : \sigma)$ and $N \rightarrow_{\beta\mathcal{L}} N'$, then $\mathcal{P}(\Gamma \vdash_\Sigma N' : \sigma)$.*
  2. *If $\mathcal{P}(\Gamma \vdash_\Sigma N : \sigma)$ and $\sigma \rightarrow_{\beta\mathcal{L}} \sigma'$, then $\mathcal{P}(\Gamma \vdash_\Sigma N : \sigma')$.*

3   Metatheory of L^{ax}F

The proofs of the metatheoretic properties of L^{ax}F follow the pattern of [18].

Strong Normalisation and Confluence
The proof of strong normalization relies on that of LF [13]. First, we introduce the function $\cdot^{-\mathcal{U}\mathcal{L}}$: L^{ax}F $\rightarrow$ LF, mapping L^{ax}F terms into LF terms by deleting the $\mathcal{L}$ and $\mathcal{U}$ symbols[3]. The proof then proceeds by contradiction, assuming a term T with an infinite $\beta\mathcal{L}$-reduction sequence. Next, we prove that only a finite number of β-reductions can be performed within any given L^{ax}F term T. Whence, in order for T to have an infinite $\beta\mathcal{L}$-reduction sequence, it must have an infinite $\mathcal{L}$-sequence, which is impossible, obtaining the contradiction. We highlight only the crucial case of the rule ($O \cdot Guarded \cdot Unlock$). Its conclusion is translated to LF as follows:

$$\Gamma^{-\mathcal{U}\mathcal{L}} \vdash_{\Sigma^{-\mathcal{U}\mathcal{L}}} M^{-\mathcal{U}\mathcal{L}}[(\lambda x_f{:}\sigma^{-\mathcal{U}\mathcal{L}}.N^{-\mathcal{U}\mathcal{L}})\,S^{-\mathcal{U}\mathcal{L}}/x] : (\lambda y_f{:}\sigma^{-\mathcal{U}\mathcal{L}}.(\rho[\mathcal{U}_{S,\sigma}^{\mathcal{P}}[N]/x])^{-\mathcal{U}\mathcal{L}})\,S^{-\mathcal{U}\mathcal{L}}.$$

The latter judgment, through standard β-reduction yields:

$$\Gamma^{-\mathcal{U}\mathcal{L}} \vdash_{\Sigma^{-\mathcal{U}\mathcal{L}}} M^{-\mathcal{U}\mathcal{L}}[N^{-\mathcal{U}\mathcal{L}}/x] : \rho^{-\mathcal{U}\mathcal{L}}[(\lambda x_f{:}\sigma^{-\mathcal{U}\mathcal{L}}.N^{-\mathcal{U}\mathcal{L}})\,S^{-\mathcal{U}\mathcal{L}}/x],$$

---

[3] $\cdot^{-\mathcal{U}\mathcal{L}}$ is the identity over constants and variables and it is recursively applied to subterms of Π, λ and application constructors, preserving their structure. The only interesting cases are those involving the $\mathcal{L}$ and $\mathcal{U}$ constructors: $(\mathcal{L}_{N,\sigma}^{\mathcal{P}}[T])^{-\mathcal{U}\mathcal{L}} \triangleq (\lambda x_f{:}\sigma^{-\mathcal{U}\mathcal{L}}.T^{-\mathcal{U}\mathcal{L}})\,N^{-\mathcal{U}\mathcal{L}}$ and $(\mathcal{U}_{N,\sigma}^{\mathcal{P}}[T])^{-\mathcal{U}\mathcal{L}} \triangleq (\lambda x_f{:}\sigma^{-\mathcal{U}\mathcal{L}}.T^{-\mathcal{U}\mathcal{L}})\,N^{-\mathcal{U}\mathcal{L}}$, where x_f is a variable which *does not* occur free in T. Its purpose is to preserve the N and σ in the subscript of the $\mathcal{L}$ and $\mathcal{U}$ symbols, while being able to β-reduce to T in one step.

i.e. $\Gamma^{-\mathcal{UL}} \vdash_{\Sigma^{-\mathcal{UL}}} M^{-\mathcal{UL}}[N^{-\mathcal{UL}}/x] : \rho^{-\mathcal{UL}}[N^{-\mathcal{UL}}/x]$. Thus, only a finite number of β-reductions can be performed within the translation of any given $\mathsf{L}^{\mathsf{a\times}}\mathsf{F}$ term T and we can proceed by contradiction. Thus we have:

Theorem 1 (Strong Normalization of $\mathsf{L}^{\mathsf{a\times}}\mathsf{F}$).
1. If $\Gamma \vdash_{\Sigma} K$, then K is $\beta\mathcal{L}$-strongly normalizing.
2. if $\Gamma \vdash_{\Sigma} \sigma : K$, then σ is $\beta\mathcal{L}$-strongly normalizing.
3. if $\Gamma \vdash_{\Sigma} M : \sigma$, then M is $\beta\mathcal{L}$-strongly normalizing.

Confluence is proved as for $\mathsf{LF}_{\mathcal{P}}$, using *Newman's Lemma* ([4], Chapter 3), and showing that the reduction on "raw terms" is locally confluent. Hence, we have:

Theorem 2 (Confluence of $\mathsf{L}^{\mathsf{a\times}}\mathsf{F}$). *$\beta\mathcal{L}$-reduction is confluent, i.e.:*
1. If $K \twoheadrightarrow_{\beta\mathcal{L}} K'$ and $K \twoheadrightarrow_{\beta\mathcal{L}} K''$, then there exists a K''' such that $K' \twoheadrightarrow_{\beta\mathcal{L}} K'''$ and $K'' \twoheadrightarrow_{\beta\mathcal{L}} K'''$.
2. If $\sigma \twoheadrightarrow_{\beta\mathcal{L}} \sigma'$ and $\sigma \twoheadrightarrow_{\beta\mathcal{L}} \sigma''$, then there exists a σ''' such that $\sigma' \twoheadrightarrow_{\beta\mathcal{L}} \sigma'''$ and $\sigma'' \twoheadrightarrow_{\beta\mathcal{L}} \sigma'''$.
3. If $M \twoheadrightarrow_{\beta\mathcal{L}} M'$ and $M \twoheadrightarrow_{\beta\mathcal{L}} M''$, then there exists an M'''' such that $M' \twoheadrightarrow_{\beta\mathcal{L}} M''''$ and $M'' \twoheadrightarrow_{\beta\mathcal{L}} M''''$.

Subject Reduction

Inversion and subderivation properties play a key role in the proof of subject reduction (SR). However, in $\mathsf{L}^{\mathsf{a\times}}\mathsf{F}$, given a derivation of $\Gamma \vdash_{\Sigma} \alpha$ and a subterm N occurring in the subject of this judgement, we cannot always prove that there exists a derivation of the form $\Gamma \vdash_{\Sigma} N : \tau$ (for a suitable τ). Consider the following example. Clearly $\Gamma, x{:}\tau \vdash_{\Sigma} \mathcal{L}_{S,\sigma}^{\mathcal{P}}[x] : \mathcal{L}_{S,\sigma}^{\mathcal{P}}[\tau]$ holds, and assume that $\Gamma \vdash_{\Sigma} N : \mathcal{L}_{S,\sigma}^{\mathcal{P}}[\tau]$; we then have

$$\frac{\Gamma, x{:}\tau \vdash_{\Sigma} \mathcal{L}_{S,\sigma}^{\mathcal{P}}[x] : \mathcal{L}_{S,\sigma}^{\mathcal{P}}[\tau] \quad \Gamma \vdash_{\Sigma} N : \mathcal{L}_{S,\sigma}^{\mathcal{P}}[\tau]}{\Gamma \vdash_{\Sigma} \mathcal{L}_{S,\sigma}^{\mathcal{P}}[\mathcal{U}_{S,\sigma}^{\mathcal{P}}[N]] : \mathcal{L}_{S,\sigma}^{\mathcal{P}}[\tau]} \ (O{\cdot}Guarded{\cdot}Unlock)$$

but if $\mathcal{P}(\Gamma \vdash_{\Sigma} S : \sigma)$ does not hold, and τ is not a lock-type, then we cannot derive any judgement whose subject is $\mathcal{U}_{S,\sigma}^{\mathcal{P}}[N] : \tau$. Hence we have to restate point 6 of Proposition 3.11 (Subderivation, part 1) of [18] as follows:

Proposition 1 (Subderivation, part 1, point 6). *Given a derivation $\mathcal{D}$ of the judgement $\Gamma \vdash_{\Sigma} \alpha$, and a subterm N occurring in the subject of this judgement, we have that either there exists a derivation of a judgement having N as a subject, or there exists a derivation of a judgment having N' as a subject, where $N \equiv \mathcal{U}_{S,\sigma}^{\mathcal{P}}[N']$ (for suitable $\mathcal{P}$, S and σ).*

The proof is straightforward. And straightforward is also the extension to $\mathsf{L}^{\mathsf{a\times}}\mathsf{F}$ of the rest of the proof of SR for $\mathsf{LF}_{\mathcal{P}}$ in [18]. Thus we establish the fundamental:

Theorem 3 (Subject Reduction of $\mathsf{L}^{\mathsf{a\times}}\mathsf{F}$). *If predicates are well-behaved, then:*
1. If $\Gamma \vdash_{\Sigma} K$ and $K \rightarrow_{\beta\mathcal{L}} K'$, then $\Gamma \vdash_{\Sigma} K'$.
2. If $\Gamma \vdash_{\Sigma} \sigma : K$ and $\sigma \rightarrow_{\beta\mathcal{L}} \sigma'$, then $\Gamma \vdash_{\Sigma} \sigma' : K$.
3. If $\Gamma \vdash_{\Sigma} M : \sigma$ and $M \rightarrow_{\beta\mathcal{L}} M'$, then $\Gamma \vdash_{\Sigma} M' : \sigma$.

The issue of decidability for $\mathsf{L}^{\mathsf{a\times}}\mathsf{F}$ can be addressed as that for $\mathsf{LF}_{\mathcal{P}}$ in [18].

Conservativity

We recall that a system $\mathcal{S}'$ is a conservative extension of $\mathcal{S}$ if the language of $\mathcal{S}$ is included in that of $\mathcal{S}'$, and moreover for all judgements $\mathcal{J}$, in the language of $\mathcal{S}$, $\mathcal{J}$ is provable in $\mathcal{S}'$ if and only if $\mathcal{J}$ is provable in $\mathcal{S}$.

Theorem 4 (Conservativity of L^{ax}F). L^{ax}F *is a conservative extension of* LF.

Proof. (sketch) The *if* part is trivial. For the *only if* part, consider a derivation in L^{ax}F and drop all locks/unlocks (*i.e.* release the terms and types originally locked). This pruned derivation is a legal derivation in standard LF.

Notice that the above result holds independently of the particular nature or properties of the external oracles that we may *invoke* during the proof development (in L^{ax}F), *e.g.* decidability or recursive enumerability of $\mathcal{P}$.

Instead, L^{ax}F is *not* a conservative extension of LF$_\mathcal{P}$, since the new typing rule allows us to derive more judgements with unlocked-subject even if the predicate does not hold *e.g.*

$$\frac{\Gamma, x{:}\mathcal{L}^{\mathcal{P}}_{S,\sigma}[\tau] \vdash_\Sigma x : \mathcal{L}^{\mathcal{P}}_{S,\sigma}[\tau] \quad \Gamma \vdash_\Sigma N : \mathcal{L}^{\mathcal{P}}_{S,\sigma}[\mathcal{L}^{\mathcal{P}}_{S,\sigma}[\tau]]}{\Gamma \vdash_\Sigma x[\mathcal{U}^{\mathcal{P}}_{S,\sigma}[N]/x] : \mathcal{L}^{\mathcal{P}}_{S,\sigma}[\tau[\mathcal{U}^{\mathcal{P}}_{S,\sigma}[N]/x]]} \ (O\text{-}Guarded\text{-}Unlock)$$

Then, since x does not occur free in τ, $\mathcal{L}^{\mathcal{P}}_{S,\sigma}[\tau[\mathcal{U}^{\mathcal{P}}_{S,\sigma}[N]/x]] \equiv \mathcal{L}^{\mathcal{P}}_{S,\sigma}[\tau]$ and we get $\Gamma \vdash_\Sigma \mathcal{U}^{\mathcal{P}}_{S,\sigma}[N] : \mathcal{L}^{\mathcal{P}}_{S,\sigma}[\tau]$. We close this Section on L^{ax}F with a sort of "hygiene" theorem for the unguarded $\mathcal{U}$-destructor:

Theorem 5 (Soundness of unlock). *If* $\Gamma \vdash_\Sigma \mathcal{U}^{\mathcal{P}}_{N,\sigma}[M] : \tau$ *is derived in* L^{ax}F *and* Γ *does not contain variables ranging over lock-types (i.e.,* $x{:}\mathcal{L}^{\mathcal{P}}_{S,\sigma}[\rho] \notin \Gamma$), *then* $\mathcal{P}(\Gamma \vdash_\Sigma N : \sigma)$ *is true.*

Proof. The proof can be carried out by a straightforward induction on the derivation of $\Gamma \vdash_\Sigma \mathcal{U}^{\mathcal{P}}_{N,\sigma}[M] : \tau$.

4 Case Studies

In this section we discuss encodings of logics in L^{ax}F. Of course, all encodings given in [18] for LF$_\mathcal{P}$, carry over immediately to the setting of L^{ax}F, because the latter is a language extension of the former. So here, we do not present encodings for *modal* and *ordered linear logic*. However, the possibility of using guarded unlocks, *i.e.* the full power of the monad destructor, allows for significant simplifications in several of the encodings of logical systems given in LF$_\mathcal{P}$. We illustrate this point discussing call-by-value λ_v-calculus, which greatly benefits from the possibility of applying functions to locked-arguments, and Hoare's Logic, which combines various kinds of syntactical and semantical locks in its rules. We do not discuss adequacy of these encodings since it is a trivial variant of the one presented in [18]. Finally we discuss a very subtle natural deduction logic, *i.e.* Elementary Affine Logic. That encoding will illustrate how locks can be used to deal with *pattern matching*, and *terms rewriting*, and open up the road to embedding *logic programming* in type theory.

Call-by-value λ_v-calculus

We encode, using *Higher Order Abstract Syntax* (HOAS), the syntax of untyped λ-calculus: $M, N ::= x \mid M\ N \mid \lambda x.M$ as in [18], where natural numbers (through the constructor `free`) are used to represent free variables, while bound variables are rendered as metavariables of $\mathsf{L}^{\mathsf{a}\mathsf{x}}\mathsf{F}$ of type `term`:

Definition 2 ($\mathsf{L}^{\mathsf{a}\mathsf{x}}\mathsf{F}$ signature Σ_λ for untyped λ-calculus).

```
term :  Type                    nat  :  Type                0 : nat
   S :  nat -> nat              free :  nat -> term
 app :  term -> term -> term    lam  :  (term -> term) -> term
```

Definition 3 (Call-by-value reduction strategy). *The call-by-value (CBV) evaluation strategy is given by:*

$$\frac{}{\vdash_{CBV} M = M}\ (refl) \qquad\qquad \frac{\vdash_{CBV} N = M}{\vdash_{CBV} M = N}\ (symm)$$

$$\frac{\vdash_{CBV} M = N \quad \vdash_{CBV} N = P}{\vdash_{CBV} M = P}\ (trans) \qquad \frac{\vdash_{CBV} M = N \quad \vdash_{CBV} M' = N'}{\vdash_{CBV} M\ M' = N\ N'}\ (app)$$

$$\frac{v \text{ is a value}}{\vdash_{CBV} (\lambda x.M)\ v = M[v/x]}\ (\beta_v) \qquad \frac{\vdash_{CBV} M = N}{\vdash_{CBV} \lambda x.M = \lambda x.N}\ (\xi_v)$$

where values are either variables, constants, or abstractions.

The new typing rule ($O{\cdot}Guarded{\cdot}Unlock$) of $\mathsf{L}^{\mathsf{a}\mathsf{x}}\mathsf{F}$, allows to encode naturally the system as follows.

Definition 4 ($\mathsf{L}^{\mathsf{a}\mathsf{x}}\mathsf{F}$ signature Σ_{CBV} for λ-calculus CBV reduction). *We extend the signature of Definition 2 as follows:*

```
    eq : term->term->Type
  refl : ΠM:term.            (eq M M)
  symm : ΠM:term.ΠN:term.    (eq N M)->(eq M N)
 trans : ΠM,N,P:term.        (eq M N)->(eq N P) ->(eq M P)
eq_app : ΠM,N,M',N':term.    (eq M N)->(eq M'N')->(eq (app M M')(app N N'))
 betav : ΠM:  (term->term). ΠN:term.L^{Val}_{N,term}[(eq (app (lam M) N)(M N))]
  csiv : ΠM,N:(term->term).(Πx:term.L^{Val}_{x,term}[(eq (M x)(N x))])->(eq (lam M)(lam N))
```

where the predicate Val *is defined as follows:* Val$(\Gamma \vdash_\Sigma N : \text{term})$ *holds iff either* N *is an abstraction or a constant (i.e. a term of the shape* (free i)*).*

Notice the neat improvement w.r.t. to the encoding of $\mathsf{LF}_\mathcal{P}$, given in [18], as far as the rule `csiv`. The encoding of the rule ξ_v is problematic if bound variables are encoded using metavariables, because the predicate *Val* appearing in the lock cannot mention explicitly variables, for it to be well-behaved. In [18], since we could not apply the rules unless we had explicitly eliminated the *Val*-lock, in order to overcome the difficulty we had to make a detour using constants. In $\mathsf{L}^{\mathsf{a}\mathsf{x}}\mathsf{F}$, on the other hand, we can apply the rules "under *Val*", so to speak, and postpone the proof of the *Val*-checks till the very end, and then rather than checking *Val* we can get rid of the lock altogether, since the bound variable of the rule `csiv`, is allowed to be locked. Notice that this phrasing of the rule `csiv` amounts precisely to the fact that in λ_v variables range over values. As a concrete example of all this, we show how to derive the equation $\lambda x.z\,((\lambda y.y)\,x) = \lambda x.z\,x$. Using "pencil and paper" we would proceed as follows:

$$\dfrac{\dfrac{\dfrac{\overline{\quad-\quad}}{\vdash_{CBV} z = z}\,(refl)\qquad \dfrac{x\ \text{is a value}}{(\lambda y.y)\,x = y[x/y]}\,(\beta_v)}{\vdash_{CBV} z\,((\lambda y.y)\,x) = z\,x}\,(app)}{\vdash_{CBV} \lambda x.z\,((\lambda y.y)\,x) = \lambda x.z\,x}\,(\xi_v)$$

Similarly, in L^{ax}F, we can derive z:term $\vdash_\Sigma$ (refl z) : (eq z z) and

Γ, x:term $\vdash_\Sigma$ (betav (λy:term.y) x) : $\mathcal{L}_{\text{x,term}}^{Val}$[(eq (app (lam λy:term.y) x) ((λy:term.y) x))].

This far, in old LF$_\mathcal{P}$, we would be blocked if we could not prove that $Val\,(\Gamma$, x:term $\vdash_\Sigma$ x : term) holds, since eq_app cannot accept an argument with a locked-type. However, in L^{ax}F, we can apply the (O·Guarded·Unlock) rule obtaining the following proof term (from the typing environment Γ, x:term, z:term):

(eq_app z z (app (lam λy:term.y) x) x (refl z) $\mathcal{U}_{\text{x,term}}^{Val}$[(betav ($\lambda$y:term.y) x)]),

of type $\mathcal{L}_{\text{x,term}}^{Val}$[(eq (app z (app (lam λy:term.y) x)) (app z x))]. And abstracting x, a direct application of csiv yields the result.

Imp with Hoare Logic

An area of Logic which can greatly benefit from the new system L^{ax}F is *program logics*, because of the many syntactical checks which occur in these systems. For lack of space we can discuss only a few rules of Hoare's Logic for a very simple imperative language Imp, whose syntax is:

$p ::= skip \mid x := expr \mid p; p \mid$ null | assignment | sequence
       $if\ cond\ then\ p\ else\ p \mid while\ cond\ \{p\}$ cond | while

In [18] we presented an encoding of Hoare's logic for Imp in LF$_\mathcal{P}$ which delegated to external predicates the tedious and subtle checks that boolean expressions, in the *if* and *while* commands, are *quantifier free* (*QF* predicate) as well as the *non-interference* property in the assignment command. These syntactic constraints on the conditional and loop commands were rendered in LF$_\mathcal{P}$ as follows:

```
bool,prog : Type
Iif       : Πe:bool.prog -> prog ->𝓛QF  [prog]
                                    e,bool
Iwhile    : Πe:bool.prog -> 𝓛QF  [prog]
                             e,bool
```

where the predicate $QF(\Gamma \vdash_{\Sigma_{Imp}}$ e : bool) holds iff the formula e is *closed* and quantifier-free, *i.e.*, it does not contain the forall constructor. We can look at *QF* as a "good formation" predicate, filtering out *bad programs with invalid boolean expressions* by means of "stuck" terms. Thus, the encoding function $\epsilon_{\mathcal{X}}^{\text{prog}}$ mapping programs of the source language Imp, with free variables in $\mathcal{X}$, to the corresponding terms of LF$_\mathcal{P}$ could be defined very easily as follows[4]:

$\epsilon_{\mathcal{X}}^{\text{prog}}(if\ e\ then\ p\ else\ p') = \mathcal{U}_{\epsilon_{\mathcal{X}}^{\text{exp}}(e),\text{bool}}^{QF}[(\text{Iif }\epsilon_{\mathcal{X}}^{\text{exp}}(e)\ \epsilon_{\mathcal{X}}^{\text{prog}}(p)\ \epsilon_{\mathcal{X}}^{\text{prog}}(p'))]$ (*)

$\epsilon_{\mathcal{X}}^{\text{prog}}(while\ e\ \{p\})$ $= \mathcal{U}_{\epsilon_{\mathcal{X}}^{\text{exp}}(e),\text{bool}}^{QF}[(\text{Iwhile }\epsilon_{\mathcal{X}}^{\text{exp}}(e)\ \epsilon_{\mathcal{X}}^{\text{prog}}(p))]$ (*)

---

[4] For lack of space, we report only the cases of the conditional/loop commands.

(*) if e is a quantifier-free formula. However the terms on the right hand side cannot be directly expressed in $\mathsf{LF}_\mathcal{P}$ because if $QF(\Gamma \vdash_{\Sigma_{Imp}} \epsilon_\mathcal{X}^{exp}(e) : \texttt{bool})$ does not hold, we cannot use the unlock operator. Thus we could be left with two terms of type $\mathcal{L}^{QF}_{\epsilon_\mathcal{X}^{exp}(e),\texttt{bool}}[\texttt{prog}]$, instead of type $\texttt{prog}$. This is precisely the limit of the $\mathsf{LF}_\mathcal{P}$ encoding in [18]. Since a $\mathcal{U}$-term can only be introduced if the corresponding predicate holds, when we represent rules of Hoare Logic we are forced to consider only legal terms, and this ultimately amounts to restricting explicitly the object language in a way such that QF always returns true.

In $\mathsf{L}^a\mathsf{xF}$, instead, we can use naturally the following signature for representing Hoare's Logic, without assuming anything about the object language terms:

```
hoare : bool -> prog -> bool -> Type
```
$$\texttt{hoare\_Iif} : \Pi e,e',b:\texttt{bool}.\Pi p,p':\texttt{prog}.(\texttt{hoare (b and e) p e')} ->$$
$$(\texttt{hoare ((not b) and e) p' e')} ->$$
$$\mathcal{L}^{QF}_{b,\texttt{bool}}[(\texttt{hoare e }\mathcal{U}^{QF}_{b,\texttt{bool}}[(\texttt{Iif b p p')}]\ e')]$$
$$\texttt{hoare\_Iwhile} : \Pi e,b:\texttt{bool}.\Pi p:\texttt{prog}.(\texttt{hoare (e and b) p e)} ->$$
$$\mathcal{L}^{QF}_{b,\texttt{bool}}[(\texttt{hoare e }\mathcal{U}^{QF}_{b,\texttt{bool}}[(\texttt{Iwhile b p)}]\ ((\texttt{not b) and e)})]$$

Moreover, the $(O\cdot Guarded\cdot Unlock)$ rule allows also to "postpone" the verification that $QF(\Gamma \vdash_\Sigma e : \texttt{bool})$ holds (*i.e.*, that the formula e is quantifier-free).

Elementary Affine Logic

We provide a *shallow* encoding of *Elementary Affine Logic* as presented in [3]. This example will exemplify how locks can be used to deal with syntactical manipulations as in the promotion rule of Elementary Affine Logic, which clearly introduces a recursive processing of the context.

Definition 5 (Elementary Affine Logic). *Elementary Affine Logic can be specified by the following rules:*

$$\frac{}{A \vdash_{\mathrm{EAL}} A}\ (Var) \qquad \frac{\Gamma \vdash_{\mathrm{EAL}} B}{\Gamma, A \vdash_{\mathrm{EAL}} B}\ (Weak) \qquad \frac{\Gamma, A \vdash_{\mathrm{EAL}} B}{\Gamma \vdash_{\mathrm{EAL}} A \multimap B}\ (Abst)$$

$$\frac{\Gamma \vdash_{\mathrm{EAL}} A \quad \Delta \vdash_{\mathrm{EAL}} A \multimap B}{\Gamma, \Delta \vdash_{\mathrm{EAL}} B}\ (Appl) \qquad \frac{\Gamma \vdash_{\mathrm{EAL}} !A \quad \Delta, !A, \ldots, !A \vdash_{\mathrm{EAL}} B}{\Gamma, \Delta \vdash_{\mathrm{EAL}} B}\ (Contr)$$

$$\frac{A_1, \ldots, A_n \vdash_{\mathrm{EAL}} A \quad \Gamma_1 \vdash_{\mathrm{EAL}} !A_1 \quad \ldots \quad \Gamma_n \vdash_{\mathrm{EAL}} !A_n}{\Gamma_1 \ldots \Gamma_n \vdash_{\mathrm{EAL}} !A}\ (Prom)$$

Definition 6 ($\mathsf{L}^a\mathsf{xF}$ signature Σ_{EAL} for Elementary Affine Logic).

```
    o : Type                    T : o -> Type                    ! : o -> o
c_appl : ΠA,B :o. T(A) -> T(A ⊸ B) -> T(B)
c_abstr : ΠA,B :o. Πx:(T(A) -> T(B)). L^Light_{x,T(A)->T(B)}[T(A ⊸ B)]
c_prom : ΠA,A':o. Πx: T(A).L^Closed_{x,T(A)}[L^Prom_{<x,A,A'>,T(A)XoXo}[T(A')]]
```

where o *is the type of propositions,* $\multimap$ *and* ! *are the obvious syntactic constructors,* T *is the basic judgement, and* $< \mathsf{x}, \mathsf{y}, \mathsf{z} >$ *denotes any encoding of triples, whose type is denoted by* $\mu\mathsf{X}\sigma\mathsf{X}\tau$, *e.g.* $\lambda\mathsf{u}{:}\mu \to \sigma \to \tau \to \rho.\,\mathsf{u}\ \mathsf{x}\ \mathsf{y}\ \mathsf{z} : (\mu \to \sigma \to \tau \to \rho) \to \rho$. *The predicates involved in the locks are defined as follows:*

- *Light* $(\Gamma \vdash_{\Sigma_{\text{EAL}}} x : T(A) \to T(B))$ *holds iff if* A *is not of the shape* $!A$ *then the bound variable of* x *occurs at most once in the normal form of* x.
- *Closed* $(\Gamma \vdash_{\Sigma_{\text{EAL}}} x : T(A))$ *holds iff there are no free variables of type* $T(B)$, *for some* $B : o$ *in* x.
- *Prom* $(\Gamma \vdash_{\Sigma_{\text{EAL}}} < x, A, A' > : T(A)XoXo)$ *holds iff* $A \equiv (A_1 \multimap A_2 \multimap \ldots \multimap A_n)$ *and* $A' \equiv (!A_1 \multimap !A_2 \multimap \ldots \multimap !A_n)$ *and* $A_1, A_2, \ldots, A_n$ *are the arguments of the* *c_abstr-constructors in the derivation of* x.

A few remarks are mandatory. The promotion rule in [3] is in fact a family of natural deduction rules with an arbitrary number of assumptions. Our encoding achieves this via a number of application-rules. Adequacy for this signature can be achieved only in the general formulation of [18], namely:

Theorem 6 (Adequacy for Elementary Affine Logic). $A_1 \ldots A_n \vdash_{\text{EAL}} A$ *iff there exists* M *and* $A_1:o, \ldots, A_n:o, x_1:T(A_1), \ldots, x_n:T(A_n) \vdash_{\Sigma_{\text{EAL}}} M : T(A)$ *and all variables* x_i *(*$1 \le i \le n$*) occurring more than once in* M *have type of the shape* $T(!A_i)$.

The check on the context of the Adequacy Theorem is *external* to the system LᵃˣF, but this is in the nature of results which relate *internal* and *external* concepts. *E.g.* the very concept of LᵃˣF context, which appears in any Adequacy result, is external to LᵃˣF. This check is internalized if the term is closed.

5   Concluding Remarks

In this paper we have shown how to extend LF with a class of monads which capture the effect of delegating to an external oracle the task of providing part of the necessary evidence for establishing a judgment. Thus we have introduced an additional clause in the LF paradigm for encoding a logic, namely: *external evidence as monads*. This class of monads is very well-behaved and so we can simplify the equational theory of the system. But, in fact, all our metatheoretic results carry through also in the general case, where we deal with a generic monad using Moggi's let_T destructor, together with its equational theory. *I.e.* we have provided an extension of LF with monads.

In this paper we consider the verification of predicates in locks as purely *atomic actions*, *i.e.* each predicate *per se*. But of course predicates have a logical structure which can be reflected onto locks. *E.g.* we can consistently extend LᵃˣF by assuming that locks commute, combine, and entail, *i.e.* that the following types are inhabited: $\mathcal{L}^{\mathcal{P}}_{x,\sigma}[\tau] \to \mathcal{L}^{\mathcal{Q}}_{x,\sigma}[\tau]$ if $\mathcal{P}(\Gamma \vdash_\Sigma x : \sigma) \to \mathcal{Q}(\Gamma \vdash_\Sigma x : \sigma)$, $\mathcal{L}^{\mathcal{P}}_{x,\sigma}[\mathcal{L}^{\mathcal{Q}}_{x,\sigma}[M]] \to \mathcal{L}^{\mathcal{P}\&\mathcal{Q}}_{x,\sigma}[M]$, and $\mathcal{L}^{\mathcal{P}}_{x,\sigma}[\mathcal{L}^{\mathcal{Q}}_{y,\tau}[M]] \to \mathcal{L}^{\mathcal{Q}}_{y,\tau}[\mathcal{L}^{\mathcal{P}}_{x,\sigma}[M]]$.

We encoded call-by-value λ-calculus with Plotkin's classical notion of value. But the encoding remains the same, apart from what is delegated to the lock, if we consider other notions of value *e.g. closed normal forms* only for K-redexes [16]. This illustrates how monads handle side-conditions uniformly.

The way we dealt with the promotion rule in Elementary Affine Logic hints at the way to deal with *Pattern Matching* in LᵃˣF, and hence opens up the road to embedding *Logic Programming* and *Term Rewriting Systems* in type theory.

Acknowlegments. The authors would like to express sincere thanks to anonymous referees for their useful comments and useful remarks.

References

1. Albert, H.: Traktat über kritische Vernunft. J.C.B. Mohr (Paul Siebeckggu), Tübingen (1991)
2. Alechina, N., Mendler, M., de Paiva, V., Ritter, E.: Categorical and Kripke Semantics for Constructive S4 Modal Logic. In: Fribourg, L. (ed.) CSL 2001. LNCS, vol. 2142, pp. 292–307. Springer, Heidelberg (2001)
3. Baillot, P., Coppola, P., Lago, U.D.: Light logics and optimal reduction: Completeness and complexity. In: Proc. LICS, pp. 421–430. IEEE Computer Society (2007)
4. Barendregt, H.: Lambda Calculus: Its Syntax and Semantics. North Holland (1984)
5. Barendregt, H., Barendsen, E.: Autarkic computations in formal proofs. Journal of Automated Reasoning 28, 321–336 (2002)
6. Barthe, G., Cirstea, H., Kirchner, C., Liquori, L.: Pure Pattern Type Systems. In: Proc. POPL 2003, pp. 250–261. The ACM Press (2003)
7. Carroll, L.: What the Tortoise Said to Achilles. Mind 4, 278–280 (1895)
8. Cousineau, D., Dowek, G.: Embedding pure type systems in the lambda-Pi-calculus modulo. In: Della Rocca, S.R. (ed.) TLCA 2007. LNCS, vol. 4583, pp. 102–117. Springer, Heidelberg (2007)
9. Fairtlough, M., Mendler, M.: Propositional lax logic. Information and Computation 137(1), 1–33 (1997)
10. Fairtlough, M., Mendler, M., Cheng, X.: Abstraction and refinement in higher-order logic. In: Boulton, R.J., Jackson, P.B. (eds.) TPHOLs 2001. LNCS, vol. 2152, pp. 201–216. Springer, Heidelberg (2001)
11. Fairtlough, M., Mendler, M., Walton, M.: First-order Lax Logic as a Framework for Constraint Logic Programming. Tech. Rep., University of Passau (1997)
12. Garg, D., Tschantz, M.C.: From indexed lax logic to intuitionistic logic. Tech. Rep., DTIC Document (2008)
13. Harper, R., Honsell, F., Plotkin, G.: A framework for defining logics. Journal of the ACM 40(1), 143–184 (1993)
14. Hirschkoff, D.: A full formalisation of π-calculus theory in the calculus of constructions. In: Gunter, E.L., Felty, A.P. (eds.) TPHOLs 1997. LNCS, vol. 1275, pp. 153–169. Springer, Heidelberg (1997)
15. Honsell, F.: 25 years of formal proof cultures: Some problems, some philosophy, bright future. In: Proc. LFMTP 2013, pp. 37–42. ACM, New York (2013)
16. Honsell, F., Lenisa, M.: Semantical analysis of perpetual strategies in λ-calculus. Theoretical Computer Science 212(1), 183–209 (1999)
17. Honsell, F., Lenisa, M., Liquori, L.: A Framework for Defining Logical Frameworks. v. in Honor of G. Plotkin. ENTCS 172, 399–436 (2007)
18. Honsell, F., Lenisa, M., Liquori, L., Maksimovic, P., Scagnetto, I.: An Open Logical Framework. Journal of Logic and Computation (October 2013)
19. Honsell, F., Lenisa, M., Goel, G., Scagnetto, I.: A Conditional Logical Framework. In: Cervesato, I., Veith, H., Voronkov, A. (eds.) LPAR 2008. LNCS (LNAI), vol. 5330, pp. 143–157. Springer, Heidelberg (2008)
20. Honsell, F., Miculan, M., Scagnetto, I.: π-calculus in (Co)Inductive Type Theories. Theoretical Computer Science 253(2), 239–285 (2001)

21. Mendler, M.: Constrained proofs: A logic for dealing with behavioral constraints in formal hardware verification. In: Proc. Designing Correct Circuits, pp. 1–28. Springer (1991)

22. Moggi, E.: The partial lambda calculus. PhD thesis, University of Edinburgh. College of Science and Engineering. School of Informatics (1988)

23. Nanevski, A., Pfenning, F., Pientka, B.: Contextual Modal Type Theory. ACM Transactions on Computational Logic 9(3) (2008)

24. Pfenning, F., Schürmann, C.: System description: Twelf – a meta-logical framework for deductive systems. In: Ganzinger, H. (ed.) CADE 1999. LNCS (LNAI), vol. 1632, pp. 202–206. Springer, Heidelberg (1999)

25. Pientka, B., Dunfield, J.: Programming with proofs and explicit contexts. In: Proc. PPDP 2008, pp. 163–173. ACM (2008)

26. Pientka, B., Dunfield, J.: Beluga: A framework for programming and reasoning with deductive systems (system description). In: Giesl, J., Hähnle, R. (eds.) IJCAR 2010. LNCS, vol. 6173, pp. 15–21. Springer, Heidelberg (2010)

27. Watkins, K., Cervesato, I., Pfenning, F., Walker, D.: A Concurrent Logical Framework I: Judgments and Properties. Tech. Rep. CMU-CS-02-101, CMU (2002)

The Monoid of Queue Actions

Martin Huschenbett[1], Dietrich Kuske[1], and Georg Zetzsche[2]

[1] TU Ilmenau, Institut für Theoretische Informatik, Ilmenau, Germany
[2] TU Kaiserslautern, Fachbereich Informatik, Kaiserslautern, Germany

Abstract. We model the behavior of a fifo-queue as a monoid of transformations that are induced by sequences of writing and reading. We describe this monoid by means of a confluent and terminating semi-Thue system and study some of its basic algebraic properties such as conjugacy. Moreover, we show that while several properties concerning its rational subsets are undecidable, their uniform membership problem is NL-complete. Furthermore, we present an algebraic characterization of this monoid's recognizable subsets. Finally, we prove that it is not Thurston-automatic.

1 Introduction

Basic computing models differ in their storage mechanisms: there are finite memory mechanisms, counters, blind counters, partially blind counters, pushdowns, Turing tapes, queues and combinations of these mechanisms. Every storage mechanism naturally comes with a set of basic actions like reading a symbol from or writing a symbol to the pushdown. As a result, sequences of basic actions transform the storage. The set of transformations induced by sequences of basic actions then forms a monoid. As a consequence, fundamental properties of a storage mechanism are mirrored by algebraic properties of the induced monoid. For example, the monoid induced by a deterministic finite automaton is finite, a single blind counter induces the integers with addition, and pushdowns induce polycyclic monoids [10]. In this paper, we are interested in a queue as a storage mechanism. In particular, we investigate the monoid $\mathcal{Q}$ induced by a single queue.

The basic actions on a queue are writing the symbol a into the queue and reading the symbol a from the queue (for each symbol a from the alphabet of the queue). Since a can only be read from a queue if it is the first entry in the queue, these actions are partial. Hence, for every sequence of basic actions, there is a queue of shortest length that can be transformed by the sequence without error (i.e., without attempting to read a from a queue that does not start with a). Our first main result (Theorem 4.1) in section 4 provides us with a normal form for transformations induced by sequences of basic actions: The transformation induced by a sequence of basic actions is uniquely determined by the subsequence of write actions, the subsequence of read actions, and the length of the shortest queue that can be transformed by the sequence without error. The proof is based on a convergent finite semi-Thue system for the monoid $\mathcal{Q}$. In sections 3 and 5, we derive equations that hold in $\mathcal{Q}$. The main result in this direction is Theorem 5.3, which describes the normal form of the product of two sequences of basic actions in normal form, i.e., it describes the monoid operation in terms of normal forms.

E. Csuhaj-Varjú et al. (Eds.): MFCS 2014, Part I, LNCS 8634, pp. 340–351, 2014.

Sections 6 and 7 concentrate on the conjugacy problem in $\mathcal{Q}$. The fundamental notion of conjugacy in groups has been extended to monoids in two different ways: call x and y conjugate if the equation $xz = zy$ has a solution, and call them transposed if there are u and v such that $x = uv$ and $y = vu$. Then conjugacy $\approx$ is reflexive and transitive, but not necessarily symmetric, and transposition $\sim$ is reflexive and symmetric, but not necessarily transitive. These two relations have been considered, e.g., in [13,16,17,6,18,5]. We prove that conjugacy is the transitive closure of transposition and that two elements of $\mathcal{Q}$ are conjugate if and only if their subsequences of write and of read actions, respectively, are conjugate in the free monoid. This characterization allows in particular to decide conjugacy in polynomial time. In section 7, we prove that the set of solutions $z \in \mathcal{Q}$ of $xz = zy$ is effectively rational but not necessarily recognizable.

Section 8 investigates algorithmic properties of rational subsets of $\mathcal{Q}$. Algorithmic aspects of rational subsets have received increased attention in recent years; see [14] for a survey on the membership problem. Employing the fact that every element of $\mathcal{Q}$ has only polynomially many left factors, we can nondeterministically solve the rational subset membership problem in logarithmic space. Since the direct product of two free monoids embeds into $\mathcal{Q}$, all the negative results on rational transductions (cf. [1]) as, e.g., the undecidability of universality of a rational subset, translate into our setting (cf. Theorem 8.3). The subsequent section 9 characterizes the recognizable subsets of $\mathcal{Q}$. Recall that an element of $\mathcal{Q}$ is completely determined by its subsequences of write and read actions, respectively, and the length of the shortest queue that can be transformed without an error. Regular conditions on the subsequences of write and read actions, respectively, lead to recognizable sets in $\mathcal{Q}$. Regarding the shortest queue that can be transformed without error, the situation is more complicated: the set of elements of $\mathcal{Q}$ that operate error-free on the empty queue is not recognizable. Using an approximation of the length of the shortest queue, we obtain recognizable subsets $\Omega_k \subseteq \mathcal{Q}$. The announced characterization then states that a subset of $\mathcal{Q}$ is recognizable if and only if it is a Boolean combination of regular conditions on the subsequences of write and read actions, respectively, and sets Ω_k (cf. Theorem 9.4). In the final section 10, we prove that $\mathcal{Q}$ is not automatic in the sense of Thurston et al. [4] (it cannot be automatic in the sense of Khoussainov and Nerode [12] since the free monoid with two generators is interpretable in first order logic in $\mathcal{Q}$).

All missing proofs are contained in the complete version [9] of this paper.

2   Preliminaries

Let A be an alphabet. As usual, the set of finite words over A, i.e. the free monoid generated by A, is denoted A^*. Let $w = a_1 \ldots a_n \in A^*$ be some word. The *length* of w is $|w| = n$. The word obtained from w by reversing the order of its symbols is $w^R = a_n \ldots a_1$. A word $u \in A^*$ is a *prefix* of w if there is $v \in A^*$ such that $w = uv$. In this situation, the word v is unique and we refer to it by $u^{-1}w$. Similarly, u is a *suffix* of w if $w = vu$ for some $v \in A^*$ and we then put $wu^{-1} = v$. For $k \in \mathbb{N}$, we let $A^{\leq k} = \{ w \in A^* \mid |w| \leq k \}$ and define $A^{>k}$ similarly.

Let M be an arbitrary monoid. The *concatenation* of two subsets $X, Y \subseteq M$ is defined as $X \cdot Y = \{ xy \mid x \in X, y \in Y \}$. The *Kleene iteration* of X is the set

$X^* = \{ x_1 \cdots x_n \mid n \in \mathbb{N}, x_1, \ldots, x_n \in X \}$. In fact, X^* is a submonoid of M, namely the smallest submonoid entirely including X. Thus, X^* is also called *the submonoid generated by* X. The monoid M is *finitely generated*, if there is some finite subset $X \subseteq M$ such that $M = X^*$.

A subset $L \subseteq M$ is called *rational* if it can be constructed from the finite subsets of M using union, concatenation, and Kleene iteration only. The subset L is *recognizable* if there are a finite monoid F and a morphism $\phi \colon M \to F$ such that $\phi^{-1}(\phi(L)) = L$. The image of a rational set under a monoid morphism is again rational, whereas recognizability is retained under preimages of morphisms. It is well-known that every recognizable subset of a finitely generated monoid is rational. The converse implication is in general false. However, if M is the free monoid generated by some alphabet A, a subset $L \subseteq A^*$ is rational if and only if it is recognizable. In this situation, we call L *regular*.

3   Definition and Basic Equations

We want to model the behavior of a fifo-queue whose entries come from a finite set A with $|A| \geq 2$ (if A is a singleton, the queue degenerates into a partially blind counter). Consequently, the state of a valid queue is an element from A^*. In order to have a defined result even if a read action fails, we add the error state $\bot$. The basic actions are writing of the symbol $a \in A$ into the queue (denoted a) and reading the symbol $a \in A$ from the queue (denoted $\bar{a}$). Formally, $\overline{A}$ is a disjoint copy of A whose elements are denoted $\bar{a}$. Furthermore, we set $\Sigma = A \cup \overline{A}$. Hence, the free monoid Σ^* is the set of sequences of basic actions and it acts on the set $A^* \cup \{\bot\}$ by way of the function $. \colon (A^* \cup \{\bot\}) \times \Sigma^* \to A^* \cup \{\bot\}$, which is defined as follows:

$$q.\varepsilon = q \qquad q.au = qa.u \qquad q.\bar{a}u = \begin{cases} q'.u & \text{if } q = aq' \\ \bot & \text{otherwise} \end{cases} \qquad \bot.u = \bot$$

for $q \in A^*$, $a \in A$, and $u \in \Sigma^*$.

Example 3.1. Let the content of the queue be $q = ab$. Then $ab.\bar{a}c = b.c = bc.\varepsilon = bc$ and $ab.c\bar{a} = abc.\bar{a} = bc.\varepsilon = bc$, i.e., the sequences of basic actions $\bar{a}c$ and $c\bar{a}$ behave the same on the queue $q = ab$. In Lemma 3.5, we will see that this is the case for any queue $q \in A^* \cup \{\bot\}$. Differently, we have $\varepsilon.\bar{a}a = \bot \neq \varepsilon = \varepsilon.a\bar{a}$, i.e., the sequences of basic actions $a\bar{a}$ and $\bar{a}a$ behave differently on certain queues.

Definition 3.2. *Two words $u, v \in \Sigma^*$ are* equivalent *if $q.u = q.v$ for all queues $q \in A^*$. In that case, we write $u \equiv v$. The equivalence class wrt. $\equiv$ containing the word u is denoted $[u]$.*

Since $\equiv$ is a congruence on the free monoid Σ^*, we can define the quotient monoid $\mathcal{Q} = \Sigma^*/{\equiv}$ and the natural epimorphism $\eta \colon \Sigma^* \to \mathcal{Q} \colon w \mapsto [w]$. The monoid $\mathcal{Q}$ is called the monoid of queue actions.

Remark 3.3. Note that the concrete form of $\mathcal{Q}$ depends on the size of the alphabet A, so let $\mathcal{Q}_n$ denote the monoid of queue actions defined with $A = |n|$. As a consequence of Theorems 4.1 and 5.3 below, $\mathcal{Q}_n$ embeds into $\mathcal{Q}_2$ where the generators of $\mathcal{Q}_n$ are mapped to $[a^{n+i}ba^{n-i}b]$ and $[\bar{a}^{n+i}\bar{b}\bar{a}^{n-i}\bar{b}]$, respectively.

Intuitively, the basic actions a and $\overline{a}$ act "dually" on $A^* \cup \{\bot\}$. We formalize this intuition by means of the *duality map* $\delta\colon \Sigma^* \to \Sigma^*$, which is defined as follows: $\delta(\varepsilon) = \varepsilon$, $\delta(au) = \delta(u)\overline{a}$, and $\delta(\overline{a}u) = \delta(u)a$ for $a \in A$ and $u \in \Sigma^*$. Notice that $\delta(uv) = \delta(v)\delta(u)$ and $\delta(\delta(u)) = u$ (i.e., δ is an anti-isomorphism and an involution). In the following, we use the term "by duality" to refer to the proposition below.

Proposition 3.4. *For $u, v \in \Sigma^*$, we have $u \equiv v$ if and only if $\delta(u) \equiv \delta(v)$.*

Consequently, the duality map δ can be lifted to a map $\delta\colon \mathcal{Q} \to \mathcal{Q}\colon [u] \mapsto [\delta(u)]$. Also this lifted map is an anti-isomorphism of $\mathcal{Q}$ and an involution.

The second equivalence in the lemma below follows from the first one by duality.

Lemma 3.5. *Let $a, b \in A$. We have $ab\overline{b} \equiv a\overline{b}b$, $a\overline{a}\overline{b} \equiv \overline{a}a\overline{b}$, and if $a \neq b$ then $a\overline{b} \equiv \overline{b}a$.*

From the first and the last equivalence, we get $ab\overline{c} \equiv a\overline{c}b$ for any $a, b, c \in A$, even when $b = c$. Similarly, the second and the third equivalence imply $a\overline{b}\overline{c} \equiv \overline{b}a\overline{c}$.

Our computations in $\mathcal{Q}$ will frequently make use of alternating sequences of write- and read-operations on the queue. To simplify notation, we define the shuffle of two words over A and over $\overline{A}$ as follows: Let $a_1, a_2, \ldots, a_n, b_1, b_2, \ldots, b_n \in A$ with $v = a_1 a_2 \ldots a_n$ and $w = b_1 b_2 \ldots b_n$. We write $\overline{w}$ for $\overline{b_1}\, \overline{b_2}\, \ldots\, \overline{b_n}$ and set

$$\langle v, \overline{w} \rangle = a_1 \overline{b_1}\, a_2 \overline{b_2}\, \ldots\, a_n \overline{b_n}\,.$$

The following proposition describes the relation between the shuffle operation and the multiplication in $\mathcal{Q}$. Its proof works by induction on the lengths of x and y.

Proposition 3.6. *Let $u, v, x, y, x', y' \in A^*$.*

(1) if $xy = x'y'$ and $|x| = |y'| = |u|$, then $\langle u, \overline{x} \rangle\, \overline{y} \equiv \overline{x'}\, \langle u, \overline{y'} \rangle$.
(2) if $xy = x'y'$ and $|y| = |x'| = |v|$, then $x\, \langle y, \overline{v} \rangle = \langle x', \overline{v} \rangle\, y'$.
(3) If $|u| = |v|$ and $|x| = |y|$, then $x\, \langle u, \overline{v} \rangle\, \overline{y} \equiv \langle xu, \overline{vy} \rangle$.
(4) If $|x| = |y|$, then $\langle x, \overline{y} \rangle \equiv x\overline{y}$.

The first claim expresses that the sequence of write-operations u can be "moved along" the sequence of read-operations $\overline{xy} = \overline{x'y'}$, its dual (2) moves a sequence of read-operations $\overline{v}$ along a sequence of write-operations. The third claim expresses that write-operations from the left and read-operations from the right can be "swallowed" by a shuffle. The last one follows from (3) with $u = v = \varepsilon$.

Corollary 3.7. *Let $u, v, w \in A^*$. If $|w| = |v|$, then $\overline{u}v\overline{w} \equiv v\overline{u}\overline{w}$. If $|u| = |v|$, then $u\overline{v}w \equiv uw\overline{v}$.*

The first claim follows from $v\overline{w} \equiv \langle v, \overline{w} \rangle$ and the possibility to move v along the sequence of read-operations $\overline{uw}$, the second claim follows dually.

4   A Semi-thue System for $\mathcal{Q}$

We order the equations from Lemma 3.5 as follows (with $a \neq c$):

$$ab\overline{b} \to a\overline{b}b \qquad a\overline{a}\overline{b} \to \overline{a}a\overline{b} \qquad a\overline{c} \to \overline{c}a$$

Let R be the semi-Thue system with the above three types of rules. Note that a word over Σ is irreducible if and only if it has the form $\overline{u} \langle v, \overline{v} \rangle w$ for some $u, v, w \in A^*$. When doing our calculations, we found it convenient to think in terms of pictures as follows:

$\overline{u}$	$\overline{v}$	
	v	w

Here, the blocks represent the words $\overline{u}$, $\overline{v}$, v, and w, respectively, where we placed the read-blocks (i.e., words over $\overline{A}$) in the first line and write-blocks in the second. The shuffle $\langle v, \overline{v} \rangle$ is illustrated by placing the corresponding two blocks on top of each other.

Ordering the alphabet such that $\overline{a} < b$ for all $a, b \in A$, all rules are decreasing in the length-lexicographic order; hence R is terminating. It is confluent since the only overlap of left-hand sides have the form $ab\overline{b}\overline{c}$. Consequently, for any $u \in \Sigma^*$, there is a unique irreducible word $\mathsf{nf}(u)$ with $u \xrightarrow{*} \mathsf{nf}(u)$. We call $\mathsf{nf}(u)$ the *normal form* of u and denote the set of all normal forms by $\mathsf{NF} \subseteq \Sigma^*$, i.e.,

$$\mathsf{NF} = \{\, \mathsf{nf}(u) \mid u \in \Sigma^* \,\} = \overline{A}^* \{\, a\overline{a} \mid a \in A \,\}^* A^* .$$

By our construction of R from the equations in Lemma 3.5, $\mathsf{nf}(u) = \mathsf{nf}(v)$ implies $u \equiv v$ for any words $u, v \in \Sigma^*$. For the converse implication, let $u \equiv v$. Because of $u \equiv \mathsf{nf}(u)$, we can assume that u and v are in normal form, i.e., $u = \overline{u_1} \langle u_2, \overline{u_2} \rangle u_3$ and $v = \overline{v_1} \langle v_2, \overline{v_2} \rangle v_3$. Then one first shows $u_1 = v_1$ using $q.u = q.v$ for $q \in \{u_1, v_1\}$. The equation $u_2 = v_2$ follows from $u_1 = v_1$ and $q.u = q.v$ for $q \in \{u_1 u_2 a, v_1 v_2 a\}$ for each $a \in A$ (here we rely on the fact that $|A| \geq 2$). Finally, $u_3 = v_3$ follows from $u_1 = v_1$, $u_2 = v_2$, and $u_1.u = u_1.v$.

Consequently, $u \equiv v$ and $\mathsf{nf}(u) = \mathsf{nf}(v)$ are equivalent. Hence, the mapping $\mathsf{nf} \colon \Sigma^* \to \mathsf{NF}$ can be lifted to a mapping $\mathsf{nf} \colon \mathcal{Q} \to \mathsf{NF}$ by defining $\mathsf{nf}([u]) = \mathsf{nf}(u)$. In summary, we have the following theorem.

Theorem 4.1. *The natural epimorphism $\eta \colon \Sigma^* \to \mathcal{Q}$ maps the set NF bijectively onto $\mathcal{Q}$. The inverse of this bijection is the map $\mathsf{nf} \colon \mathcal{Q} \to \mathsf{NF}$.*

Let $\pi, \overline{\pi} \colon \Sigma^* \to A^*$ be the morphisms defined by $\pi(a) = \overline{\pi}(\overline{a}) = a$ and $\pi(\overline{a}) = \overline{\pi}(a) = \varepsilon$ for $a \in A$ (i.e., π is the projection of a word over Σ to its subword over A, and $\overline{\pi}$ is the projection to its subword over $\overline{A}$, with all the bars $^-$ deleted). By Theorem 4.1, these two morphisms can be lifted to morphisms $\pi, \overline{\pi} \colon \mathcal{Q} \to A^*$ by $\pi([u]) = \pi(u)$ and $\overline{\pi}([u]) = \overline{\pi}(u)$.

Definition 4.2. *Let $w \in \Sigma^*$ be a word and $\mathsf{nf}(w) = \overline{x} \langle y, \overline{y} \rangle z$ its normal form. The overlap width of w and of $[w]$ is the number $\mathsf{ow}(w) = \mathsf{ow}([w]) = |y|$.*

By Theorem 4.1, $q \in \mathcal{Q}$ is uniquely determined by $\pi(q)$, $\overline{\pi}(q)$, and $\mathsf{ow}(q)$. Let $\mathsf{nf}(q) = \overline{x} \langle y, \overline{y} \rangle z$. Then x is the shortest queue w with $w.q \neq \bot$. Furthermore, $\mathsf{ow}(q) = |\overline{\pi}(q)| - |x|$. Hence, q is also uniquely described by the two projections and the length of the shortest queue it transforms without error.

5 Multiplication

For two words u and v in normal form, we want to determine the normal form of uv. For this, the concept of *overlap* of two words will be important:

Definition 5.1. *For $u, v \in A^*$, let $\mathrm{ol}(v, u)$ denote the longest suffix of v that is also a prefix of u.*

For example, $\mathrm{ol}(ab, bc) = b$, $\mathrm{ol}(aba, aba) = aba$, and $\mathrm{ol}(ab, cba) = \varepsilon$. The following lemma describes the normal form of a word from $A^* \overline{A}^*$.

Lemma 5.2. *Let $u, v \in A^*$ and set $s = \mathrm{ol}(v, u)$, $r = vs^{-1}$ and $t = s^{-1}u$. Then $u\overline{v} \equiv \overline{r} \langle s, \overline{s} \rangle t$.*

The equation $u\overline{v} \equiv \overline{r} \langle s, \overline{s} \rangle t$ can be visualized as follows:

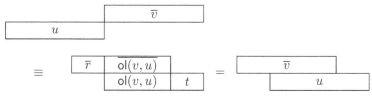

Our intuition is that the word $\overline{v}$ tries to slide along u to the left as far as possible. This movement is stopped as soon as we reach a word in normal form (which, for the first time, occurs when a suffix of v coincides with a prefix of u).

The proof of Lemma 5.2 first assumes $|u| = |v|$ and proceeds by induction on this length. The general case follows using Cor. 3.7. Applying Prop. 3.6(4), Cor. 3.7, and Lemma 5.2, one gets rather immediately the following description of the normal form of the product of two words in normal form.

Theorem 5.3. *Let $u_1, u_2, u_3, v_1, v_2, v_3 \in A^*$ and set $s = \mathrm{ol}(u_2 v_1 v_2, u_2 u_3 v_2)$, $r = u_2 v_1 v_2 s^{-1}$, and $t = s^{-1} u_2 u_3 v_2$. Then*

$$\overline{u_1} \langle u_2, \overline{u_2} \rangle u_3 \cdot \overline{v_1} \langle v_2, \overline{v_2} \rangle v_3 \equiv \overline{u_1 r} \langle s, \overline{s} \rangle t v_3 .$$

This theorem can be visualized as follows:

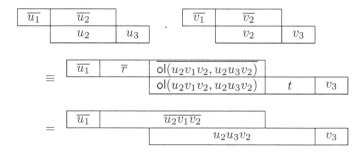

Here, first, v_2 moves to the left until it reaches the right border of u_3. Then $\overline{u_2}$ moves to the right until it reaches the left border of $\overline{v_1}$. Finally, the united block $\overline{u_2 v_1 v_2}$ tries to slide to the left along $u_2 u_3 v_2$ until a normal form is reached.

6 Conjugacy

The conjugacy relation in groups has two natural generalizations to monoids, which, when considered in $\mathcal{Q}$, we determine in this section. Let M be a monoid and $p, q \in M$. Then p and q are *conjugate*, which we denote by $p \approx q$, if there exists $x \in M$ such that $px = xq$. Furthermore, p and q are *transposed*, denoted by $p \sim q$, if there are $x, y \in M$ with $p = xy$ and $q = yx$. Moreover, $\overset{*}{\sim}$ is the transitive closure of $\sim$.

Observe that $\approx$ is reflexive and transitive whereas $\sim$ is reflexive and symmetric, and $\sim \; \subseteq \; \approx$. If M is actually a group, then both relations coincide and are equivalence relations, called conjugacy. The same is true for free monoids [15, Prop. 1.3.4] and even for special monoids [18], but there are monoids where none of this holds.

Example 6.1. Let $u, v, w \in A^*$. Then $\overline{u} \langle v, \overline{v} \rangle w \equiv \overline{u} v \overline{v} w \equiv v \overline{u} \overline{v} w$. Consequently, $\mathcal{Q} = \eta(A^* \overline{A}^* A^*)$ and dually $\mathcal{Q} = \eta(\overline{A}^* A^* \overline{A}^*)$. Furthermore, $v \overline{u} \overline{v} w \overset{*}{\sim} \overline{u} \overline{v} w v$. Hence, for every $q \in \mathcal{Q}$, there exists $q' \in \eta(\overline{A}^* A^*)$ with $q \overset{*}{\sim} q'$, i.e., $\mathcal{Q}$ is the closure of $\eta(\overline{A}^* A^*)$ under transposition.

Theorem 6.2. *For any* $p, q \in \mathcal{Q}$, *the following are equivalent:*

$$(1)\; p \overset{*}{\sim} q \qquad (2)\; p \approx q \qquad (3)\; q \approx p \qquad (4)\; (\pi(p) \sim \pi(q) \text{ and } \overline{\pi}(p) \sim \overline{\pi}(q))$$

The implication $(1) \Rightarrow (2)$ holds in every monoid since $\approx$ is transitive and since $\sim \; \subseteq \; \approx$. The implication $(2) \Rightarrow (4)$ holds since π and $\overline{\pi}$ are homomorphisms and since $\approx$, $\sim$, and $\overset{*}{\sim}$ coincide on the free monoid. To show $(4) \Rightarrow (1)$, we first invoke Example 6.1: we can assume $p = [\overline{\pi}(p)\pi(p)]$ and similarly for q. The crucial step is to show $[\overline{x}ay] \overset{*}{\sim} [\overline{x}ya]$ and $[\overline{a}\overline{x}y] \overset{*}{\sim} [\overline{x}\overline{a}y]$, i.e., that one can rotate a single letter in the read-part or in the write-part. This ensures the implication $(4) \Rightarrow (1)$. Since (2) and (4) are equivalent and since $\sim$ is symmetric on the free monoid, also the equivalence of (2) and (3) follows.

We obtain the following consequence of Theorem 6.2: Given two words $u, v \in \Sigma^*$, one can decide in quadratic time whether $\pi(u) \sim \pi(v)$ and $\overline{\pi}(u) \sim \overline{\pi}(v)$. Consequently, it is decidable in polynomial time whether $[u] \approx [v]$ holds. It remains an open question whether there is some number $k \in \mathbb{N}$ such that $p \overset{*}{\sim} q$ if and only if $p \overset{k}{\sim} q$.

7 Conjugators

Let M be a monoid and $x, y \in M$. An element $z \in M$ is a *conjugator of x and y* if $xz = zy$. The set of all conjugators of x and y is denoted $C(x, y) = \{ z \in M \mid xz = zy \}$.

Suppose that M is a free monoid A^* and consider $x, y \in A^*$. It is well-known that $C(x, y)$ is a finite union of sets of the form $u(vu)^*$ and hence regular. In contrast, the set of conjugators of $[\overline{a}]$ and $[\overline{a}]$ is not recognizable since $\eta^{-1}(C([\overline{a}], [\overline{a}])) \cap a^* \overline{a}^* = \{ a^k \overline{a}^\ell \mid k \le \ell \}$. In this section, we prove the following weaker result:

Theorem 7.1. *Let $x, y \in Q$. Then the set $C(x, y)$ is rational.*

The proof proceeds as follows: First note that, by Theorem 4.1, $xz = zy$ if and only if $\pi(xz) = \pi(zy)$, $\overline{\pi}(xz) = \overline{\pi}(zy)$, and $\mathrm{ow}(xz) = \mathrm{ow}(zy)$. The set $D(x, y)$ of all $z \in Q$ satisfying the first two conditions is recognizable (since x and y are fixed) and it remains to handle the third condition (under the assumption that the first two hold).

The crucial point in the proof of Theorem 7.1 is the regularity of the language

$$G_k = \{\, \mathrm{nf}(z) \mid z \in D(x, y), \mathrm{ow}(xz) - \mathrm{ow}(z) \geq k \,\} \ .$$

Having this, it follows that the languages

$$E_k = \{\, \mathrm{nf}(z) \mid z \in D(x, y), \mathrm{ow}(xz) - \mathrm{ow}(z) = k \,\} \text{ and}$$
$$F_k = \{\, \mathrm{nf}(z) \mid z \in D(x, y), \mathrm{ow}(zy) - \mathrm{ow}(z) = k \,\}$$

are regular. Consequently, the language

$$\bigcup_{0 \leq k \leq |\pi(x)|} E_k \cap F_k$$

is regular. Since one can also show that $0 \leq \mathrm{ow}(xz) - \mathrm{ow}(z) \leq |\pi(x)|$ for $z \in D(x, y)$, this language equals the language of all words $\mathrm{nf}(z)$ with $z \in C(x, y)$. Hence $C(x, y)$ is the image wrt. the natural epimorphism η of a regular language and therefore rational.

8   Rational Subsets

This section studies decision problems concerning rational subsets of Q.

Let $w \in \Sigma^*$. Then, by Theorem 5.3, the number of left-divisors of $[w]$ in Q is at most $|w|^3$. This allows to define a DFA with $|w|^3$ many states that recognizes the language $[w] = \{\, u \in \Sigma^* \mid u \equiv w \,\}$. Even more, this DFA can be constructed in logarithmic space. This fact allows to reduce the problem below in logarithmic space to the intersection problem of NFAs. Hence we get the following result, where completeness follows since A^* embeds into Q:

Theorem 8.1. *The following rational subset membership problem for Q is NL-complete:*

Input: A word $w \in \Sigma^*$ and an NFA $\mathcal{A}$ over Σ.
Question: Is there a word $v \in L(\mathcal{A})$ with $w \equiv v$?

We do not have a description of the submonoids of Q, but we get the following embedding of the direct product of two free monoids.

Proposition 8.2. *Let $\mathcal{R} \subseteq Q$ denote the submonoid generated by $\{[a], [ab], [\overline{b}], [\overline{abb}]\}$.*

(1) There exists an isomorphism α from $\{a, b\}^* \times \{c, d\}^*$ onto $\mathcal{R}$ with $\alpha((a, \varepsilon)) = [a]$, $\alpha((b, \varepsilon)) = [ab], \alpha((\varepsilon, c)) = [\overline{b}], \text{ and } \alpha((\varepsilon, d)) = [\overline{abb}]$.
(2) If $\mathcal{S} \subseteq \mathcal{R}$ is recognizable in $\mathcal{R}$, then it is recognizable in Q.

The proof makes heavy use of Theorem 4.1. This proposition implies in particular that rational transductions can be translated into rational subsets of $\mathcal{Q}$, resulting in the following undecidability results:

Theorem 8.3. *(1) The set of rational subsets of $\mathcal{Q}$ is not closed under intersection.*
(2) The emptiness of the intersection of two rational subsets of $\mathcal{Q}$ is undecidable.
(3) The universality of a rational subset of $\mathcal{Q}$ is undecidable.
   Consequently, inclusion and equality of rational subsets are undecidable.
(4) The recognizability of a rational subset of $\mathcal{Q}$ is undecidable.

We sketch the proof of statement (3), the other claims are proved along similar lines: Let $S \subseteq \{a, b\}^* \times \{c, d\}^*$ be rational. Then $\alpha(S)$ is rational. Due to Prop. 8.2 (2), the set $\mathcal{R}$ is recognizable in $\mathcal{Q}$. Thus, $\mathcal{Q} \setminus \mathcal{R}$ is recognizable and therefore, since $\mathcal{Q}$ is finitely generated, rational. Consequently, $\alpha(S) \cup \mathcal{Q} \setminus \mathcal{R}$ is rational as well. This rational set equals $\mathcal{Q}$ if and only if $\alpha(S) = \mathcal{R}$, i.e., $S = \{a, b\}^* \times \{c, d\}^*$. But this latter question is undecidable by [1, Theorem 8.4(iv)].

9   Recognizable Subsets

In this section, we aim to describe the recognizable subsets of $\mathcal{Q}$. Clearly, sets of the form $\pi^{-1}(L)$ or $\overline{\pi}^{-1}(L)$ for some regular $L \subseteq A^*$ as well as Boolean combinations thereof are recognizable. This does not suffice to produce all recognizable subsets: for instance, the singleton set $\{[\overline{a}a]\}$ is recognizable but any Boolean combination of inverse projections containing $[\overline{a}a]$ also includes $[a\overline{a}]$. However, we will see in the main result of this section, namely Theorem 9.4, that incorporating certain sets that can impose a simple restriction on relative positions of write and read symbols suffices to generate the recognizable sets as a Boolean algebra.

Recall that any $q \in \mathcal{Q}$ is completely determined by $\pi(q)$, $\overline{\pi}(q)$, and $\text{ow}(q)$. Consequently, it would seem natural to incorporate sets which restrict the overlap width. Unfortunately, this does not work since the set of all $q \in \mathcal{Q}$ with $\text{ow}(q) = k$ is not recognizable (for any $k \in \mathbb{N}$).

Nevertheless, the subsequent definition provides a slight variation of this idea which conduces to our purpose. To simplify notation, we say two elements $p, q \in \mathcal{Q}$ *have the same projections* and write $p \sim_\pi q$ if $\pi(p) = \pi(q)$ and $\overline{\pi}(p) = \overline{\pi}(q)$.

Definition 9.1. *For each $k \in \mathbb{N}$, the set $\Omega_k \subseteq \mathcal{Q}$ is given by*

$$\Omega_k = \{ q \in \mathcal{Q} \mid \forall p \in \mathcal{Q} \colon p \sim_\pi q \,\&\, \text{ow}(q) \le \text{ow}(p) \le k \implies p = q \} .$$

Observe that $\mathcal{Q} = \Omega_0 \supseteq \Omega_1 \supseteq \Omega_2 \supseteq \ldots$. Intuitively, for fixed projections $\pi(q)$ and $\overline{\pi}(q)$ the set Ω_k contains all q with $\text{ow}(q) \ge k$ as well as the unique q with maximal $\text{ow}(q) \le k$.

Example 9.2. (1) The queue action $q = [\overline{a}ba\overline{a}ba]$ satisfies $\text{ow}(q) = 1$ and hence $q \in \Omega_1$. The only $p \in \mathcal{Q}$ with $p \sim_\pi q$ and $\text{ow}(p) \ge \text{ow}(q)$ is $p = [a\overline{a}b\overline{b}a\overline{a}]$. Since $\text{ow}(p) = 3$, this implies $q \in \Omega_2$ but $q \notin \Omega_3$.
(2) For every $k \ge 1$, we have $[(\overline{a}a)^k] \in \Omega_{k-1} \setminus \Omega_k$.
(3) All queue actions of the form $q = [u\overline{v}]$ with $u, v \in A^*$ satisfy $q \in \Omega_k$ for every $k \in \mathbb{N}$.

Remark 9.3. We know that $q \in \mathcal{Q}$ is uniquely described by $\pi(q)$, $\overline{\pi}(q)$, and $\mathrm{ow}(q)$. Somewhat surprisingly, we still have a unique description of q if we replace $\mathrm{ow}(q)$ by the maximal $k \in \mathbb{N}$ with $q \in \Omega_k$ or the fact that there is no such maximum.

The aforementioned main result characterizing the recognizable subsets of $\mathcal{Q}$ is the following.

Theorem 9.4. *For every subset $L \subseteq \mathcal{Q}$, the following are equivalent:*

(1) L is recognizable,
(2) L is wrw-recognizable, i.e., the language $\eta^{-1}(L) \cap A^* \overline{A}^* A^*$ is regular,
(3) $\eta^{-1}(L) \cap \overline{A}^* A^* \overline{A}^*$ is regular,
(4) L is simple, i.e., a Boolean combination of sets of the form $\pi^{-1}(R)$ or $\overline{\pi}^{-1}(R)$ for some regular $R \subseteq A^*$ and the sets Ω_k for $k \in \mathbb{N}$.

The implication "(1)$\Rightarrow$(2)" is trivial. Regarding wrw-recognizability note that $L = \eta(\eta^{-1}(L) \cap A^* \overline{A}^* A^*)$ by Example 6.1, i.e., the language $\eta^{-1}(L) \cap A^* \overline{A}^* A^*$ describes the set L completely. This is not the case if we replace $A^* \overline{A}^* A^*$ by $A^* \overline{A}^*$: The set $L = \{ [\overline{a}^n a \overline{a} a^n] \mid n \geq 1 \}$ is not recognizable, since the set of its normal forms is not regular. However, $\eta^{-1}(L) \cap A^* \overline{A}^*$ is empty and hence regular.

Note that the implication"(4)$\Rightarrow$(1)" follows easily from the following result:

Proposition 9.5. *For each $k \in \mathbb{N}$, the set $\eta^{-1}(\Omega_k)$ is regular.*

The crucial point in its proof is the following characterization of the language $\eta(\Omega_k)$: $w \in \Sigma^*$ belongs $\eta^{-1}(\Omega_k)$ if and only if, for every $u \in A^{\leq k}$, one of the following holds:

1. u is no prefix of $\pi(w)$ or
2. u is no suffix of $\overline{\pi}(w)$ or
3. the ith write symbol in w appears before the ith of the last $|u|$ read symbols (for all $1 \leq i \leq |u|$).

As an illustration, $a\overline{a}b\overline{a}b\overline{a}a\overline{a}$ belongs to $\eta^{-1}(\Omega_3)$ and $a\overline{a}b\overline{a}b\overline{a}aa$ does not. For every $u \in A^{\leq k}$ the language of words w satisfying one of the above three conditions is regular. Hence $\eta^{-1}(\Omega_k)$ is the intersection of finitely many regular languages and therefore regular.

The implication "(2)$\Rightarrow$(4)" of Theorem 9.4 is the following:

Proposition 9.6. *If $L \subseteq \mathcal{Q}$ is wrw-recognizable, then it is simple.*

Proof idea. Let k be the number of elements of the syntactic monoid of $\eta^{-1}(L) \cap A^* \overline{A}^* A^*$. Consider the following partition of L:

$$L = \left(L \cap \pi^{-1}\left(A^{<k}\right) \cap \Omega_k \right) \cup \left(L \cap \pi^{-1}\left(A^{\geq k}\right) \cap \Omega_k \right) \cup \bigcup_{0 \leq \ell < k} \left(L \cap \Omega_\ell \setminus \Omega_{\ell+1} \right).$$

One can show that all the parts are simple; we indicate how this is done for the first part, i.e., the set $L \cap \pi^{-1}\left(A^{<k}\right) \cap \Omega_k$:

Let $K = \eta^{-1}(L) \cap A^* \overline{A}^* A^*$ and $\phi \colon \Sigma^* \to M$ be a morphism recognizing K. We further consider the morphisms $\mu, \overline{\mu} \colon A^* \to M$ defined by $\mu(w) = \phi(w)$ and $\overline{\mu}(w) = \phi(\overline{w})$. We show the claim by establishing the equation

$$L \cap \pi^{-1}\left(A^{<k}\right) \cap \Omega_k = \bigcup_{\substack{u \in A^{<k}, m \in M \\ \mu(u)m \in \phi(K)}} \pi^{-1}(u) \cap \overline{\pi}^{-1}\left(\overline{\mu}^{-1}(m)\right) \cap \Omega_k \,.$$

Let X and Y denote the left and right hand side of this equation, respectively. Clearly, $X, Y \subseteq \pi^{-1}\left(A^{<k}\right) \cap \Omega_k$. Consider some $q \in \pi^{-1}\left(A^{<k}\right) \cap \Omega_k$. It suffices to show that $q \in X$ precisely if $q \in Y$.

To this end, let $u = \pi(q)$. Then $|u| < k$. Using $u \in \Omega_k$, one can show that there is $p \in Q$ such that $q = [u]\,p$. Clearly, $\pi(p) = \varepsilon$, i.e., $p = [\overline{y}]$ for some $y \in A^*$. Notice that $q = [u\overline{y}]$. Altogether,

$$
\begin{aligned}
q \in X \quad &\Longleftrightarrow \quad q = [u\overline{y}] \in L \\
&\Longleftrightarrow \quad \phi(u\overline{y}) = \mu(u)\,\overline{\mu}\left(\overline{\pi}(q)\right) \in \phi(K) \quad \Longleftrightarrow \quad q \in Y \,.
\end{aligned}
$$

The simplicity of the other sets is shown using similar arguments.

Since the implication "(1)$\Rightarrow$(2)" in Theorem 9.4 is trivial, we have the equivalence of (1), (2), and (4). Claim (3) can be added using duality arguments.

10   Thurston-Automaticity

Many groups of interest in combinatorial group theory turned out to be Thurston-automatic [4]. The more general concept of a Thurston-automatic semigroup was introduced in [3]. In this chapter, we prove that the monoid of queue-actions Q does not fall into this class.

Let M be a monoid, Γ an alphabet, $\theta \colon \Gamma^+ \to M$ a semigroup morphism, and $L \subseteq \Gamma^+$. The triple (Γ, θ, L) is an *automatic structure* for the monoid M if θ maps L bijectively onto M, if the language L is regular and if the relations

$$L_a = \left\{ \, (u, v) \in L^2 \mid \theta(ua) = \theta(v) \, \right\} \subseteq L^2$$

are synchronously rational (i.e., accepted by a synchronous transducer, cf. [1,8]) for all $a \in \Gamma$.[1] A monoid is *Thurston-automatic* if it has some automatic structure.

Theorem 10.1. *The monoid of queue actions Q is not Thurston-automatic.*

Proof idea. Suppose Q is Thurston-automatic. By [7], there exists an automatic structure $(\Sigma \cup \{\iota\}, \theta, L)$ for Q with $\theta(a) = [a]$, $\theta(\overline{a}) = [\overline{a}]$ for $a \in A$, and $\theta(\iota) = [\varepsilon]$. For $m, n \in \mathbb{N}$, let $u_{m,n} \in L$ be the unique word with $\theta(u_{m,n}) = [\overline{a}^m a^n]$. By Theorem 5.3, there are precisely $\min(m, n) + 1$ many $q \in Q$ with $[\overline{a}^m a^n \overline{b}] = q[\overline{b}]$. It follows that this is the number of words $w \in L$ with $u_{m,n}\overline{b} \equiv w\overline{b}$. Since the set of pairs $(u_{m,n}, w)$ satisfying this equation (with $m, n \in \mathbb{N}$ and $w \in (\Sigma \cup \{\iota\})^*$) is synchronously rational [3], one can construct a nondeterministic finite automaton $\mathcal{A}$ with $\min(m, n) + 1$ many runs on any word of the form $\overline{a}^m a^n$. This then leads to a contradiction. $\qquad\square$

---

[1] This is not the original definition from [3], but it is equivalent by [3, Prop. 5.4].

Recently, the notion of an automatic group has been extended to that of Cayley graph automatic groups [11]. This notion can easily be extended to monoids. It is not clear whether the monoid Q is Cayley graph automatic. A way to disprove this would be to show that the elementary theory of its Cayley graph is undecidable.

Note that Q is not automatic in the sense of Khoussainov and Nerode [12]: This is due to the fact that $\eta(A^*)$ is isomorphic to A^* and an element of Q is in $\eta(A^*)$ if and only if it cannot be written as $r\bar{a}s$ for $r, s \in Q$ and $a \in A$. Hence, using the $\bar{a}$ for $a \in A$ as parameters, A^* is interpretable in first order logic in Q. Therefore, since A^* is not automatic in this sense [2], neither is Q [12].

References

1. Berstel, J.: Transductions and context-free languages. Teubner Studienbücher, Stuttgart (1979)
2. Blumensath, A., Grädel, E.: Finite presentations of infinite structures: Automata and interpretations. ACM Transactions on Computer Systems 37(6), 641–674 (2004)
3. Campbell, C.M., Robertson, E.F., Ruškuc, N., Thomas, R.M.: Automatic semigroups. Theoretical Computer Science 250(1-2), 365–391 (2001)
4. Cannon, J.W., Epstein, D.B.A., Holt, D.F., Levy, S.V.F., Paterson, M.S., Thurston, W.P.: Word processing in groups. Jones and Barlett Publ., Boston (1992)
5. Choffrut, C.: Conjugacy in free inverse monoids. Int. J. Algebra Comput. 3(2), 169–188 (1993)
6. Duboc, C.: On some equations in free partially commutative monoids. Theoretical Computer Science 46, 159–174 (1986)
7. Duncan, A.J., Robertson, E.F., Ruškuc, N.: Automatic monoids and change of generators. Mathematical Proceedings of the Cambridge Philosophical Society 127, 403–409, 11 (1999)
8. Frougny, C., Sakarovitch, J.: Synchronized rational relations of finite and infinite words. Theoretical Computer Science 108, 45–82 (1993)
9. Huschenbett, M., Kuske, D., Zetzsche, G.: The monoid of queue actions. preprint arXiv:1404.5479 (2014)
10. Kambites, M.: Formal languages and groups as memory. Communications in Algebra 37, 193–208 (2009)
11. Kharlampovich, O., Khoussainov, B., Miasnikov, A.: From automatic structures to automatic groups. preprint arXiv:1107.3645 (2011)
12. Khoussainov, B., Nerode, A.: Automatic presentations of structures. In: Leivant, D. (ed.) LCC 1994. LNCS, vol. 960, pp. 367–392. Springer, Heidelberg (1995)
13. Lentin, A., Schützenberger, M.-P.: A combinatorial problem in the theory of free monoids. In: Combin. Math. Appl., Proc. Conf. Univ. North Carolina 1967, pp. 128–144 (1969)
14. Lohrey, M.: The rational subset membership problem for groups: A survey (2013) (to appear)
15. Lothaire, M.: Combinatorics on Words. Encyclopedia of Mathematics and its Applications, vol. 17. Addison-Wesley (1983)
16. Osipova, V.A.: On the conjugacy problem in semigroups. Proc. Steklov Inst. Math. 133, 169–182 (1973)
17. Otto, F.: Conjugacy in monoids with a special Church-Rosser presentation is decidable. Semigroup Forum 29, 223–240 (1984)
18. Zhang, L.: Conjugacy in special monoids. J. of Algebra 143(2), 487–497 (1991)

Undecidable Properties of Self-affine Sets and Multi-tape Automata

Timo Jolivet[1,2] and Jarkko Kari[1]

[1] Department of Mathematics, University of Turku, Finland
[2] LIAFA, Université Paris Diderot, France

Abstract. We study the decidability of the topological properties of some objects coming from fractal geometry. We prove that having empty interior is undecidable for the sets defined by two-dimensional graph-directed iterated function systems. These results are obtained by studying a particular class of self-affine sets associated with multi-tape automata. We first establish the undecidability of some language-theoretical properties of such automata, which then translate into undecidability results about their associated self-affine sets.

1 Introduction

A classical way to define fractals is to use an ***iterated function system (IFS)***, specified by a finite collection of maps $f_1, \ldots, f_n : \mathbb{R}^d \to \mathbb{R}^d$ which are all ***contracting***: there exists $0 \leqslant c < 1$ such that $\|f_i(x) - f_i(y)\| \leqslant c\|x - y\|$ for all $x, y \in \mathbb{R}^d$. The associated fractal set, called the ***attractor*** of the IFS, is the unique nonempty compact set X such that $X = \bigcup_{i=1}^{n} f_i(X)$. Such a set X always exists and is unique thanks to a famous result of Hutchinson [10], based on an application of the Banach fixed-point theorem; see also [7] or [1]. For example, the classical Cantor set can be defined as the unique compact set $X \subseteq \mathbb{R}$ satisfying the set equation $X = \frac{1}{3}X \cup (\frac{1}{3}X + \frac{2}{3})$. Two other examples are given in Figure 1.

A question of interest is to determine when the fractal set X has nonempty interior. This question arises in several areas, including tiling theory, dynamical systems, number theory and Fourier analysis (see [15,12] and references therein). A well studied case is when the contracting maps are affine mappings of the form $f_i(x) = M^{-1}(x + v_i)$ where $v_i \in \mathbb{Z}^d$ and M is an integer expanding matrix which is common to all the f_i, like in the examples of Figure 1. In this case, having nonempty interior is equivalent with having nonzero Lebesgue measure, and there are efficient algorithms to decide this [9,3].

Much less is known in the more general case of ***self-affine attractors***, where the maps f_i are only restricted to be affine (of the form $f_i = M_i x + v_i$ where the M_i are real matrices and $v_i \in \mathbb{R}^d$). No algorithm is known to decide nonempty interior in this case, and specific results such as computation of Hausdorff dimension are known only for some very specific families of self-affine sets [2,13,8].

E. Csuhaj-Varjú et al. (Eds.): MFCS 2014, Part I, LNCS 8634, pp. 352–364, 2014.
© Springer-Verlag Berlin Heidelberg 2014

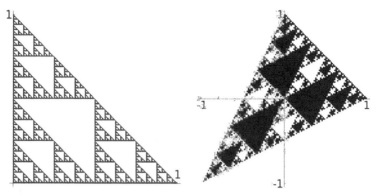

Fig. 1. Two self-affine sets defined by $X = \bigcup_{v \in \mathcal{D}} M^{-1}(X + v)$, where $M = \left(\begin{smallmatrix} 2 & 0 \\ 0 & 2 \end{smallmatrix}\right)$ and $\mathcal{D} = \{\left(\begin{smallmatrix} 0 \\ 0 \end{smallmatrix}\right), \left(\begin{smallmatrix} 1 \\ 0 \end{smallmatrix}\right), \left(\begin{smallmatrix} 0 \\ 1 \end{smallmatrix}\right)\}$ (left), and $\mathcal{D} = \{\left(\begin{smallmatrix} 0 \\ 0 \end{smallmatrix}\right), \left(\begin{smallmatrix} 1 \\ 0 \end{smallmatrix}\right), \left(\begin{smallmatrix} 0 \\ 1 \end{smallmatrix}\right), \left(\begin{smallmatrix} -1 \\ -1 \end{smallmatrix}\right)\}$ (right). The set on the left is the Sierpiński triangle and has empty interior. The set on the right is an example of a self-affine tile with nonempty interior (see [11]).

Our results. We are interested in the following question: to what extent can we decide if a self-affine set has nonemtpy interior?

We will answer this question by an undecidability result for a natural generalization of iterated function systems, which consist of a *finite system* of equations instead of just one, hence defining a finite number of attractors. This is formalized in the following definition: a d-dimensional **graph-directed iterated function system (GIFS)** is a directed graph in which each edge e is labelled by a contracting mapping $f_e : \mathbb{R}^d \to \mathbb{R}^d$. The **attractors** of the GIFS are the unique nonempty compact sets $\{X_q\}_{q \in \mathcal{Q}}$ such that $X_q = \bigcup_{r \in \mathcal{Q}} \bigcup_{e \in E_{q,r}} f_e(X_r)$, where $\mathcal{Q}$ is the set of vertices of the directed graph defining the GIFS, and $E_{q,r}$ denote the set of edges from vertex q to vertex r. Again, such a collection of compact sets $\{X_q\}_{q \in \mathcal{Q}}$ exists and is unique [6]. Fractals defined by GIFS are widely used to define various self-similar tilings of the plane, the study of which have applications in physics, dynamics and number theory. Note that the case of single-vertex graphs corresponds to classical iterated function systems.

Our main result, Theorem 4.2, states that it is undecidable if the attractors of a 2-dimensional, 3-state affine GIFS have empty interior. We follow an approach initiated by Dube [5] by associating self-affine sets with finite multi-tape automata. Then we relate some properties of the automaton with topological properties of its associated attractor, and we obtain the undecidability of the latter by proving the undecidability of the former.

In Section 2 we define multi-tape automata and we consider a variant of the Post correspondence problem in Section 2.1, which we use in Section 2.2 to prove undecidability results about multi-tape automata. We then relate some language-theoretical properties of an automaton with some topological properties of its attractor in Section 3. The main results are stated in Section 4.

2   Multi-tape Automata

A *d-**tape automaton*** $\mathcal{M}$ on alphabet $\mathcal{A} = A_1 \times \cdots \times A_d$ is defined by a finite set of ***states*** $\mathcal{Q}$, and a finite set of ***transitions*** $\mathcal{R} \subseteq \mathcal{Q} \times \mathcal{Q} \times (A_1^+ \times \cdots \times A_d^+)$. A d-tape automaton on state $\mathcal{Q}$ is conveniently represented by a directed graph with vertex set $\mathcal{Q}$ and an edge (q, r) labelled by $w_1 | \cdots | w_d$ for every transition $(q, r, (w_1, \ldots, w_d))$. This is illustrated in Example 2.1.

A ***configuration*** is an infinite sequence $c \in \mathcal{A}^{\mathbb{N}} = (A_1 \times \cdots \times A_d)^{\mathbb{N}}$. For $k \in \{1, \ldots, d\}$, the k***th tape*** of c refers to the infinite sequence $((c_n)_k)_{n \in \mathbb{N}}$, which is an infinite concatenation of words in $A_k^{\star}$. For convenience, configurations will be denoted by writing the tape components separated by the symbol "$|$". For example, $00 \cdots | 11 \cdots | 00 \cdots$ denotes the 3-tape configuration $(0, 1, 0), (0, 1, 0), \ldots \in (\{0, 1\} \times \{0, 1\} \times \{0, 1\})^{\mathbb{N}}$.

Let q be a state of $\mathcal{M}$. A configuration $c \in \mathcal{A}^{\mathbb{N}}$ is q-***accepted*** by $\mathcal{M}$ if there exists an infinite sequence of transitions $((q_n, r_n, (w_{n,1}, \ldots, w_{n,d})))_{n \geqslant 1}$ such that $q_1 = q$, $r_n = q_{n+1}$ for all $n \geqslant 1$, and for every $k \in \{1, \ldots, d\}$, the infinite word $w_{1,k} w_{2,k} \ldots$ is equal to the kth tape of c (that is, $w_{1,k} w_{2,k} \ldots = (c_1)_k (c_2)_k \ldots$). Such an infinite sequence of transitions will sometimes be referred to as a ***run of*** $\mathcal{M}$ ***starting at*** q. Note that we also forbid ε-transitions as the words $w_1, \ldots, w_d$ used in transitions are nonempty, so that each infinite run provides an infinite word on every tape.

Example 2.1. Consider the following 2-tape, 2-state automaton on alphabet $\mathcal{A} = \{0, 1\} \times \{0, 1, 2\}$, with state set $\mathcal{Q} = \{X, Y\}$ and transitions given by the following.

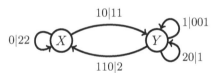

It is easy to check that the configuration $00 \cdots | 22 \cdots$ is not Y-accepted but is X-accepted by $\mathcal{M}$ (by repeatedly using the transition $(X, X, (0, 22))$). However, giving a precise description of the set of configurations which are accepted by $\mathcal{M}$ seems difficult.

Remark 2.2. Multi-tape automata are very powerful computational devices because of the fact that the words $w_1, \ldots, w_d$ in a transition are allowed to have different lengths. This is the fundamental feature that will allow us to establish several undecidability results about multi-tape automata later in this section. On the other hand, if the words $w_1, \ldots, w_d$ all have the same length in every transition, then it is easy to see that this model is not more powerful than a classical finite automaton on a product alphabet.

2.1   Post Correspondence Problems

The undecidability results of this article are all derived from the undecidability of the following decision problems. The ***Post correspondence problem (PCP)*** is: given n pairs of nonempty words $(u_1, v_1), \ldots, (u_n, v_n)$, decide if there exist $m \geqslant 1$ and a word $i_1 \cdots i_m$ such that $u_{i_1} \cdots u_{i_m} = v_{i_1} \cdots v_{i_m}$. This is a well-known undecidable problem [14].

We will need a slight variant of PCP, the **prefix Post correspondence problem (prefix-PCP)**: given n pairs of nonempty words $(u_1, v_1), \ldots, (u_n, v_n)$, decide if there exist $m, m' \geqslant 1$ and two words $i_1 \cdots i_m$ and $i_1 \cdots i_{m'}$ such that $u_{i_1} \cdots u_{i_m} = v_{i_1} \cdots v_{i_{m'}}$ and one of the two words $i_1 \cdots i_m$ and $i_1 \cdots i_{m'}$ is a prefix of the other.

A positive PCP always yields a positive prefix-PCP instance (by taking $m = m'$), but the converse is not always true. For example, the instance $(u_1, v_1) = (a, abb)$, $(u_2, v_2) = (bb, aa)$ admits the prefix-PCP solution given by $u_1 u_2 u_1 u_1 = v_1 v_2 = abbaa$, that is, $m = 4$, $m' = 2$ and the two words $i_1 i_2 i_3 i_4 = 1211$ and $i_1 i_2 = 12$. However, this instance cannot admit any PCP solution because no pair of words ends by the same symbol.

Lemma 2.3. *Prefix-PCP is undecidable.*

Proof. We reduce PCP to prefix-PCP. Let $(u_1, v_1), \ldots, (u_n, v_n)$ be an instance of PCP on alphabet $\mathcal{A}$. Let $\mathcal{B} = \mathcal{A} \cup \{\#, *\}$ be a new alphabet, where $\#$ and $*$ are two new symbols not contained in $\mathcal{A}$. We construct a prefix-PCP instance $(A_1, B_1), \ldots, (A_n, B_n), (U_1, V_1), \ldots, (U_n, V_n), (Y, Z)$ on the new alphabet $\mathcal{B}$, defined by

$$A_i = \#x_1 * x_2 * \cdots * x_k \qquad U_i = *x_1 * x_2 * \cdots * x_k \qquad Y = *\#$$
$$B_i = \#y_1 * y_2 * \cdots * y_\ell * \qquad V_i = \ y_1 * y_2 * \cdots * y_\ell * \qquad Z = \ \#$$

for all $i \in \{1, \ldots, n\}$, where $u_i = x_1 \cdots x_n$ and $v_i = y_1 \cdots y_\ell$ and the x_j, y_j are in $\mathcal{A}$. We now prove that the PCP instance has a solution if and only if the prefix-PCP instance has a solution. Suppose that there exists a solution $i_1 \cdots i_m$ to the PCP instance, that is $u_{i_1} \cdots u_{i_m} = v_{i_1} \cdots v_{i_m}$. Then clearly the prefix-PCP also has a solution, given by $A_{i_1} U_{i_2} \cdots U_{i_m} Y = B_{i_1} V_{i_2} \cdots V_{i_m} Z$.

Conversely, suppose that the prefix-PCP instance has a solution. By construction, because of $\#$ and $*$, there must exist a prefix-PCP solution of the form $A_{i_1} U_{i_2} \cdots U_{i_m} Y = B_{i_1} V_{i_2} \cdots V_{i_{m'}} Z$, where $i_1 \cdots i_m$ is a prefix of $i_1 \cdots i_{m'}$ or vice-versa. But the pairs (U_i, V_i) do not contain any $\#$, so the pair (Y, Z) is used exactly once, both after mth pair and the m'th pair, so $m = m'$ and the PCP instance has a solution. $\qquad \square$

2.2 Undecidable Properties of Multi-tape Automata

Let $\mathcal{M}$ be a d-tape automaton on alphabet $\mathcal{A}$, and let q be a state of $\mathcal{M}$. State q is **universal** if every sequence in $\mathcal{A}^{\mathbb{N}}$ is q-accepted by $\mathcal{M}$. A finite sequence $x \in \mathcal{A}^*$ is a **universal prefix** for state q if for every infinite sequence $y \in \mathcal{A}^{\mathbb{N}}$, the infinite sequence xy is q-accepted by $\mathcal{M}$.

Example 2.4. Let $\mathcal{M}$ be a 1-tape, 1-state automaton on alphabet $\{0, 1\}$ with three transitions labelled by 1, 10 and 00. The single state of $\mathcal{M}$ is not universal because every sequence starting with 01 is rejected, but the word 1 is a universal prefix: any sequence starting with 1 is accepted, because any finite segment $10^n 1$ is accepted by transitions 1, $00 \times k$, 1 if $n = 2k$ or 10, $00 \times k$, 1 if $n = 2k + 1$, and any infinite tail of 0's or 1's is obviously accepted. Hence there exist some multi-tape automata without universal states but that admit universal prefixes. The self-affine set associated with this automaton is discussed in Example 3.5.

Theorem 2.5. *It is undecidable whether a given state of a given d-tape automaton is universal. This problem remains undecidable if we restrict to 2-tape automata with 3 states.*

Proof. We reduce prefix-PCP, which is undecidable thanks to Lemma 2.3. Let (u_1, v_1), ..., (u_n, v_n) be an instance of prefix-PCP where the u_i, v_i are words over $\mathcal{B}$. We define a 2-tape automaton $\mathcal{M}$ on 3 states (denoted by X, U, V). The alphabet of $\mathcal{M}$ is $A_1 \times A_2$, with $A_1 = \{1, \ldots, n\}$ and $A_2 = \mathcal{B} \cup \{\#\}$, where n is the size of the prefix-PCP instance, $\mathcal{B}$ is the alphabet of words u_i, v_i and $\#$ is a new symbol not in $\mathcal{B}$. The transitions of $\mathcal{M}$ are

(1) $X \xrightarrow{i|u_i} U$ and $U \xrightarrow{i|u_i} U$ for every $i \in A_1$;

(2) $X \xrightarrow{i|v_i} V$ and $V \xrightarrow{i|v_i} V$ for every $i \in A_1$;

(3) $U \xrightarrow{i|u} X$ for every $i \in A_1$ and $u \in A_2^+$ such that
   (i) $|u| \leqslant |u_i|$,
   (ii) u is not a prefix of u_i,
   (iii) u does not begin with $\#$;

(4) $X \xrightarrow{i|u} X$ for every $i \in A_1$ and $u \in A_2^+$ such that (i) and (ii) above hold;

(5) $V \xrightarrow{i|v} X$ for every $i \in A_1$ and $v \in A_2^+$ such that
   (i) $|v| \leqslant |v_i|$,
   (ii) v is not a prefix of v_i,
   (iii) v does not begin with $\#$;

(6) $X \xrightarrow{i|v} X$ for every $i \in A_1$ and $v \in A_2^+$ such that (i) and (ii) above hold.

We now prove that the prefix-PCP instance $(u_1, v_1), \ldots, (u_n, v_n)$ has a solution if and only if state X is **not** universal in $\mathcal{M}$.

($\Rightarrow$) Suppose that the prefix-PCP instance admits a solution: there exist $m, m' \geqslant 1$ and two words $i_1 \cdots i_m$ and $i_1 \cdots i_{m'}$ such that $u_{i_1} \cdots u_{i_m} = v_{i_1} \cdots v_{i_{m'}}$, and one of the two words $i_1 \cdots i_m$ and $i_1 \cdots i_{m'}$ is a prefix of the other. Without loss of generality we can assume that $m \geqslant m'$ and $i_1 \cdots i_{m'}$ is a prefix of $i_1 \cdots i_m$. We prove that $\mathcal{M}$ cannot accept any infinite sequence in $(A_1 \times A_2)^{\mathbb{N}}$ beginning with

$$i_1 \cdots i_m \mid u_{i_1} \cdots u_{i_m} \#$$

when starting from state X, so $\mathcal{M}$ is not universal. Indeed, let us describe the evolution of $\mathcal{M}$ when reading such a sequence.

− We start from X, so $\mathcal{M}$ necessarily uses a transition defined in (1) and (2) and moves to state U or V after having read $i_1|u_{i_1}$ or $i_1|v_{i_1}$, respectively. (The other transitions (4) and (6) cannot be used because of the conditions (i) and (ii).) Note that both u_{i_1} and v_{i_1} are prefixes of the content of the second tape.

- Now if $\mathcal{M}$ is in state U, the remaining input starts with some i on the first tape and starts with u_i on the second tape. So $\mathcal{M}$ must use transition (1): stay in state U and read $i|u_i$. (Transition (3) cannot be used because of the conditions (i) and (ii).) The same holds if $\mathcal{M}$ is in state V.

It follows that when $\mathcal{M}$ reads $i_1, \ldots, i_{m'}$ on the first tape, then it is either in state U and has read $u_{i_1} \cdots u_{i_{m'}}$ on the second tape, or it is in state V and has read $v_{i_1} \cdots v_{i_{m'}} = u_{i_1} \cdots u_{i_m}$ on the second tape. In the second case, the next symbol on the second tape is #, so $\mathcal{M}$ is "blocked" on this input (there is no suitable transition for this sequence because of (iii)). In the first case, the computation must continue in the same way as before, so eventually $\mathcal{M}$ is still in state U and has read $i_1 \cdots i_m | u_{i_1} \cdots u_{i_m}$, and again, $\mathcal{M}$ is blocked because the next symbol on the second tape is #.

($\Leftarrow$) Suppose that no solution exists for the prefix-PCP instance. The following strategy shows that a move by the automaton can always be made, whatever its tape contents is. If $\mathcal{M}$ is in state U or V, make any available move. In state X, if no loop in X is possible, then in the current configuration $(i_1 i_2 \cdots | w)$, both u_{i_1} and v_{i_1} must be prefixes of w, otherwise (4) or (6) could have been used. Write $w = u_{i_1} w' = v_{i_1} w''$. Then:

(a) if $u_{i_1} \cdots u_{i_k}$# is a prefix of w for some k, do not go to U by reading $i_1 | u_{i_1}$;
(b) if $v_{i_1} \cdots v_{i_k}$# is a prefix of w for some k, do not go to V by reading $i_1 | v_{i_1}$;

The only possible ways to be stuck at this point are:

- $\mathcal{M}$ is in state U or V and the next symbol on the second tape is #;
- $\mathcal{M}$ is in state X and (a), (b) prevent from moving to U or V.

The second case cannot happen because it implies the existence of a prefix-PCP solution. If we are in the first case, we can assume by symmetry that we are in state U. In the last step where $\mathcal{M}$ went from X to U, the configuration must start with $i_1 i_2 \cdots | u_{i_1} u_{i_2} \cdots u_{i_k}$# $\cdots$ for some k, because this is the only way to get stuck in U some k steps later. However, this contradicts the choice made in (a) above, because $\mathcal{M}$ should have moved to V instead of state U.   □

Theorem 2.6. *It is undecidable whether a given state of a given d-tape automaton admits a universal prefix. This problem remains undecidable if we restrict to 2-tape automata with 3 states.*

Proof. We modify the prefix-PCP reduction made in the proof of Theorem 2.5. Let $(u_1, v_1), \ldots, (u_n, v_n)$ be an instance of prefix-PCP where the u_i, v_i are words over $\mathcal{B}^*$. First we modify the u_i, v_i by adding a new symbol $*$ not in $\mathcal{B}$ after each letter of each u_i and each v_i (a word $x_1 x_2 \cdots x_k$ becomes $x_1 * x_2 * \cdots x_k *$). This modified instance is clearly equivalent to the original one, so we denote it again by $(u_1, v_1), \ldots, (u_n, v_n)$.

We now define a 2-tape automaton $\mathcal{M}$ on 3 states X, U, V. We take the same alphabet $A_1 \times A_2$ as in the other reduction, with a new symbol & for both A_1 and A_2, and the symbol * for A_2. This gives $A_1 = \{1, \ldots, n\} \cup \{\&\}$ and $A_2 = \mathcal{B} \cup \{\#, \&, *\}$, where n is the size of the prefix-PCP instance, $\mathcal{B}$ is the alphabet of the words u_i, v_i and $\#, \&, *$ are new symbol not in $\mathcal{B}$. The transitions of $\mathcal{M}$ consist of

- (1) and (2) like in the proof of Theorem 2.5, without allowing any symbol & or *;
- (3), (4), (5), (6) like in the proof of Theorem 2.5, where symbols & or * are allowed, except in the first letter of u or v;

plus the following transitions:

(7) $X \xrightarrow{a|\&} X$, $U \xrightarrow{a|\&} X$ and $V \xrightarrow{a|\&} X$ for every $a \in A_1$;

(8) $X \xrightarrow{\&|a} X$, $U \xrightarrow{\&|a} X$ and $V \xrightarrow{\&|a} X$ for every $a \in A_2 \setminus \{*\}$;

(9) $X \xrightarrow{a|*b} X$, $U \xrightarrow{a|*b} X$ and $V \xrightarrow{a|*b} X$ for every $a \in A_1$ and $b \in A_2$.

We now prove that the prefix-PCP instance $(u_1, v_1), \ldots, (u_n, v_n)$ has a solution if and only if state X does **not** have any universal prefix.

($\Rightarrow$) Suppose that the prefix-PCP instance has a solution: there exist $m, m' \geqslant 1$ and two words $i_1 \cdots i_m$ and $i_1 \cdots i_{m'}$ such that $u_{i_1} \cdots u_{i_m} = v_{i_1} \cdots v_{i_{m'}}$ and one of the two words $i_1 \cdots i_m$ and $i_1 \cdots i_{m'}$ is a prefix of the other. Consider the following claim.

> **Claim.** Let $x \in A_1^*$ and $y \in A_2^*$ be such that $x\&\& \cdots |y\&\& \cdots$ is X-accepted by at most $k \geqslant 1$ different runs of $\mathcal{M}$. Then there exist $x' \in A_1^*$ and $y' \in A_2^*$ such that $xx'\&\& \cdots |yy'\&\& \cdots$ is X-accepted by at most $k-1$ different runs.

This claim implies that X does not have any universal prefix, *i.e.*, that for every finite words $x \in A_1^*$ and $y \in A_2^*$, there exists a configuration starting with $x|y$ that is not X-accepted. Indeed, for every such x, y, there can be only finitely many different accepting runs (say k) for $x\&\& \cdots |y\&\& \cdots$, because $\mathcal{M}$ eventually loops on state X with transition $\&|\&$. So it suffices to apply the claim k times to obtain a configuration starting with $x|y$ which is not X-accepted.

We now prove the claim, using the prefix-PCP solution. Let $x \in A_1^*$ and $y \in A_2^*$ be such that $x\&\& \cdots |y\&\& \cdots$ is X-accepted by k different runs. Denote by $R_1, \ldots, R_k$ the finite prefixes of the k runs, each cut when $\mathcal{M}$ reaches the $\&\& \cdots |\&\& \cdots$ part. Let $s = i_1 \cdots i_m \in A_1^*$ and let $t = u_1 \cdots u_{i_m}$, which can be written in the form $t = a_1 * a_2 * \cdots * a_{|t|-1} * \in A_2^*$, where each a_i is in $A_2 \setminus \{\#, \&, *\}$, thanks to the modification made to the instance.

Let ℓ be the distance between the two tapes heads when $\mathcal{M}$ has completed the finite run R_1. (Note that the first head is always behind the second one because it can only move by one cell at at time.) Without loss of generality we can assume that R_1 is the run for which such an ℓ is minimal. We now construct a configuration c which will "block" any run starting with R_1, without giving the other runs any possibilities for new nondeterministic branching. Let $L, L' \geqslant 0$ such that s (on the first tape) begins ℓ positions behind t (on the second tape) in the configuration

$$c = x \&^L s \&\& \cdots \mid y \&^{L'} t \# \&\& \cdots ,$$

so that during any run starting with R_1, $\mathcal{M}$ starts reading s and $t\#$ exactly at the same time. It follows that R_1 cannot be extended to an accepting run for c, because s, t corresponds to a prefix-PCP solution, similarly as in the proof of Theorem 2.5. The same is true for any other run R_i for which such an ℓ is the same as R_1.

Let us now consider another accepting run R_i. By minimality of ℓ, the distance between the two tapes heads when $\mathcal{M}$ first reaches $\&\& \cdots \mid \&\& \cdots$ during run R_i is strictly larger than ℓ. We now prove that R_i can be extended in a unique way to an accepting run for c. Indeed, any run of $\mathcal{M}$ starting with R_i must evolve in the following way:

- when t starts being read the second tape, s is not yet being read on the first tape, so at this time $\mathcal{M}$ is reading $\&$ on the first tape and a_1 on the second tape;
- the only possible transition is (8), so $\mathcal{M}$ moves one step on both tapes, and is now reading $*$ on the second tape;
- the only possible transition is (9), so $\mathcal{M}$ moves one step on the first tape and two steps on the second, and is again reading $*$ on the second tape;
- this continues until the whole $t = a_1 * a_2 * \cdots * a_{|t|-1} *$ has been read on the second tape, and $\mathcal{M}$ is deterministically looping on $\& \mid \&$.

From this analysis, it follows that R_i can be extended in a *unique* way to an accepting run for c. Hence c is a configuration starting with $x \mid y$ with at most $k - 1$ accepting runs, because every accepting run for c must start with an R_i, each of which can be extended in at most one way if $i \in \{2, \ldots, k\}$, or in no way at all if $i = 1$. Thus the claim is proved by taking $x' = \&^L s$ and $y' = \&^{L'} t\#$.

($\Leftarrow$) Suppose that no solution exists for the prefix-PCP instance. The strategy described in the "$\Leftarrow$" direction of the proof of Theorem 2.5 can be applied to prove that state X is universal, with the additional case that if the tape begins by $\&$ or $*$, then the transition (7), (8) or (9) can always be used. $\square$

Remark 2.7. In the reduction made in the above proof of Theorem 2.6, if state X has a universal prefix, then in fact X is universal. Also, in this case, it is easy to see that any finite word satisfying (i), (ii) and (iii) of transition (3) is a universal prefix for U (and V), so X, U (and V) have a common universal prefix Hence we have the following: given a 2-tape automaton $\mathcal{M}$ on 3 states and two states q, r of $\mathcal{M}$, it is undecidable if q and r have a common universal prefix.

3 Affine GIFS Associated with Multi-tape Automata

Let $\mathcal{M}$ be a d-tape automaton on alphabet $\mathcal{A} = A_1 \times \ldots \times A_d$. We want to give a "numerical interpretation" to a finite word $u \in \mathcal{A}^*$ or to an infinite configuration

$c \in \mathcal{A}^{\mathbb{N}}$. We must first specify, for each $k \in \{1, \ldots, n\}$, a numerical interpretation of the letters of A_k by choosing a bijection $\delta_k : A_k \to \{0, \ldots, |A_k| - 1\}$. We then define $\Delta_k : A_k^\star \to \mathbb{R}$ by

$$\Delta_k(u) = \sum_{1 \leqslant i \leqslant |u|} \delta_k(u_i) |A_k|^{-i}.$$

Equivalently, for $u = u_1 \cdots u_n \in A_k^n$, the number $\Delta_k(u)$ is represented by $0.\delta_k(u_1) \cdots \delta_k(u_n)$ in base $|A_k|$. Finally, let $\Delta : A_1^+ \times \ldots \times A_d^+ \to \mathbb{R}^d$ be defined by $\Delta(w_1, \ldots, w_d) = (\Delta_1(w_1), \ldots, \Delta_d(w_d))$. The domains of Δ_k and Δ can naturally be extended to $A_k^{\mathbb{N}}$ and $\mathcal{A}^{\mathbb{N}}$, respectively.

In the examples that will follow, if the alphabets A_k are all of the form $\{0, \ldots, |A_k| - 1\}$ and the maps $\delta_k : A_k \to \{0, \ldots, |A_k| - 1\}$ are not specified, we will assume for convenience that they are identity mappings.

Definition 3.1. Let $\mathcal{M}$ be a d-tape automaton on state set $\mathcal{Q}$ and alphabet $\mathcal{A} = A_1 \times \cdots \times A_n$. The **GIFS associated with** $\mathcal{M}$ is the GIFS defined by the graph G with vertex set $\mathcal{Q}$ and, for every transition $R = (q, r, (w_1, \ldots, w_d))$ of $\mathcal{M}$, an edge (q, r) labelled by the map $f_R : [0, 1]^d \to [0, 1]^d$ defined by

$$f_R(x) = \begin{pmatrix} |A_1|^{-|w_1|} & & 0 \\ & \ddots & \\ 0 & & |A_d|^{-|w_d|} \end{pmatrix} x + \Delta(w_1, \ldots, w_d).$$

Example 3.2. Let $\mathcal{M}$ be a 2-tape automaton on alphabet $\mathcal{A} = \{0, 1\} \times \{0, 1\}$, and let $c \in \mathcal{A}^{\mathbb{N}}$ be configuration. If $\mathcal{M}$ contains a transition $R = (q, r, (1011, 11))$, then applying the contracting map f_R on $\Delta(c) = (0.x_1 x_2 \ldots, 0.y_1 y_2 \ldots) \in [0, 1]^2$ has the following effect:

$$f_R(\Delta(c)) = \begin{pmatrix} 1/16 & 0 \\ 0 & 1/4 \end{pmatrix} \begin{pmatrix} 0.x_1 x_2 \ldots \\ 0.y_1 u_2 \ldots \end{pmatrix} + \Delta(1011, 11)$$

$$= \begin{pmatrix} 0.0000 x_1 x_2 \ldots \\ 0.00 y_1 u_2 \ldots \end{pmatrix} + \begin{pmatrix} 0.1011 \\ 0.11 \end{pmatrix} = \begin{pmatrix} 0.1011 x_1 x_2 \ldots \\ 0.11 y_1 u_2 \ldots \end{pmatrix}.$$

This suggests that applying a sequence of mappings $f_{R_1} \cdots f_{R_n}(\Delta(c))$ corresponds to concatenating the words associated with the transitions R_n in the numerical interpretation $\Delta(c)$ of a configuration c. This is the key idea to establish a correspondence between the GIFS of an automaton and its accepted configurations. This is formalized in the next proposition.

Proposition 3.3. *Let $\mathcal{M}$ be a 2-tape automaton and let q be a state of $\mathcal{M}$. The GIFS attractor of $\mathcal{M}$ associated with q is equal to the set $\{\Delta(c) \in \mathbb{R}^d : c \in \mathcal{A}^{\mathbb{N}}$ is q-accepted by $\mathcal{M}\}$.*

Proof. Let $x \in [0, 1]^d$. It follows from a standard fact in the theory of iterated function systems [7, Chapter 9] that $x \in X_q$ if and only if there is an infinite run $(R_n)_{n \geqslant 1}$ starting at q such that $x = \bigcap_{n \geqslant 1} f_{R_1} \cdots f_{R_n}([0, 1]^d)$, where f_{R_n} is the mapping of the GIFS of $\mathcal{M}$ associated with run R_n. Moreover, by definition of the GIFS of $\mathcal{M}$, for every such run $(R_n)_{n \geqslant 1}$, the configuration $c = w_{1,1} w_{2,1} \cdots \mid \cdots \mid w_{1,d} w_{2,d} \cdots$ is such that $x = \Delta(c)$, where the $w_{n,k}$ are given by the transitions $(q_n, r_n, (w_{n,1}, \ldots, w_{n,d}))$ for all $n \geqslant 1$, so the proposition is proved because c is a q-accepted configuration. $\qquad\square$

Example 3.4. Let $\mathcal{M}$ be the 1-state, 2-tape automaton on alphabet $\{0,1\}$ with transitions $0|0$, $0|1$, $1|0$. The iterated function system associated with $\mathcal{M}$ consists of the maps $x \mapsto \frac{x}{2}$, $x \mapsto \frac{x}{2} + (\frac{1}{2}, 0)$, $x \mapsto \frac{x}{2} + (0, \frac{1}{2})$ and it can easily be seen that the associated attractor the Sierpiński triangle (see Figure 1).

Example 3.5. The 1-tape, 1-state automaton $\mathcal{M}$ on alphabet $\{0,1\}$ with three transitions 1, 10 and 00 (given in Example 2.4) is an example of a non-universal automaton which admits universal prefixes. This reflects in the attractor associated with $\mathcal{M}$ in the following way: it is not equal to $[0,1]$ but it has nonempty interior. This can be proved either by Proposition 3.7, or by proving directly that a configuration $x \in \{0,1\}^{\mathbb{N}}$ is accepted by $\mathcal{M}$ if and only if it does not start with $0^{2k+1}1$ for some $k \geqslant 0$, which implies that the attractor is equal to $\bigcup_{k \geqslant 0} [2^{-2k-1}, 2^{-2k}]$.

Remark 3.6. Given a d-tape automaton and a point $x \in [0,1]^d$, if there exists two configurations c, c' such that $x = \Delta(c) = \Delta(c')$ and such that the tapes components $c_k \in A_k^{\mathbb{N}}$ and $c_k' \in A_k^{\mathbb{N}}$ differ for some $k \in \{1, \ldots, d\}$, then c_k and c_k' are both stationary, ending with 0^{ω} or $(|A_k| - 1)^{\omega}$. In particular, $\Delta : \mathcal{A}^{\mathbb{N}} \to \mathbb{R}^d$ is finite-to-one.

The next proposition establishes the desired correspondence between word-theoretical properties of multi-tape automata and topological properties of the associated self-affine attractors.

Proposition 3.7. *Let $\mathcal{M}$ be a d-tape automaton on alphabet $\mathcal{A}$, let q be a state of $\mathcal{M}$, and let X_q be the associated GIFS attractor. We have:*

(1) q is universal if and only if $X_q = [0,1]^d$,
(2) q has a universal prefix if and only if X_q has nonemtpy interior.

Proof. (1) If state q is universal the expansion of every element of $[0,1]^d$ is q-accepted so $X_q = [0,1]^d$ thanks to Proposition 3.3. Conversely, suppose that there exists an infinite sequence c that is not q-accepted. By a compactness argument, there must exist a prefix w of c such that wc' is not q-accepted for any infinite sequence c'. Thanks to Remark 3.6, by choosing c' with no tape components ending by 0^{ω} or $(|A_k| - 1)^{\omega}$, the sequence wc' is the *only* sequence such that $x = \Delta(wc')$, so $\Delta(wc') \notin X_q$ because otherwise wc' would be q-accepted. It follows that $X_q \neq [0,1]^d$.

(2) For a finite word $w \in \mathcal{A}^{\star}$, define the *cylinder* $[w]$ to be equal to the set of configurations that start with w. If q admits a universal prefix w, then $\Delta([w]) \subseteq X_q$ by Proposition 3.3, so X_q has nonempty interior. Conversely, suppose that there exists a nonempty open set $U \subseteq X_q$, and let $w \in \mathcal{A}^{\star}$ be a finite word such that $\Delta([w]) \subseteq U$. By a reasoning similar as in the proof of (1), we can prove that w is a universal prefix for q. $\qquad\square$

4   Undecidability Results for Self-affine Sets

Thanks to the undecidability results obtained for multi-tape automata in Theorem 2.5 and to the correspondence between word-theoretical and topological properties in Proposition 3.7, we obtain the following undecidability results about topological properties of self-affine attractors.

The first result below states that it is undecidable if an attractor "takes up the whole space", that is, equals $[0,1]^d$. It follows directly from Theorem 2.5 and Proposition 3.7, (1).

Theorem 4.1. *The following problem is undecidable. Instance: a d-dimensional affine GIFS $\mathcal{G}$ specified by maps with rational coefficients, and a state q of $\mathcal{G}$. Question: is $X_q = [0,1]^d$? This problem remains undecidable if we restrict to 2-dimensional GIFS with 3 states.*

The next result states the undecidability of a fundamental topological property for self-affine sets: having empty interior. It is a direct corollary of Theorem 2.6 and Proposition 3.7, (2).

Theorem 4.2. *The following problem is undecidable. Instance: a d-dimensional affine GIFS $\mathcal{G}$ specified by maps with rational coefficients, and a state q of $\mathcal{G}$. Question: does X_q have empty interior? This problem remains undecidable if we restrict to 2-dimensional GIFS with 3 states.*

Remark 4.3. All the undecidability results above have been obtained via a reduction using affine GIFS associated with a multi-tape automaton. Hence it follows that undecidability holds even if we restrict to affine GIFS in which the linear part of the contractions f_i are diagonal matrices whose entries are negative powers of integers. By adding dummy duplicate symbols, undecidability holds even if the entries are negative powers of two.

Remark 4.4. We can deduce from Remark 2.7 that the following problem is undecidable. Instance: a d-dimensional affine GIFS $\mathcal{G}$ specified by maps with rational coefficients, and two states q, r of $\mathcal{G}$. Question: does $X_q \cap X_r$ have empty interior? Indeed, it can be shown that q and r have a common universal prefix if and only if $X_q \cap X_r$ has nonemtpy interior, similarly as in Proposition 3.7.

5   Conclusion

We conclude this article by some questions and perspectives for further work. Is nonempty interior decidable for 1-state GIFS? (That is, for classical affine IFS.) What about the 1-dimensional case? Using multi-tape automata may lead to an undecidability result for the 1-state case, but for not for the 1-dimensional case. Indeed, 1-tape automata are not more powerful than classical finite automata, for which the properties we used in this article are all decidable. Note that for 1-state multi-tape automata, universality is trivially decidable, but the status of prefix-universality is not known in this case.

Also, let us note that having nonempty interior is equivalent to having nonzero Lebesgue measure in the case of integer self-affine tiles (as mentioned in the introduction), but not in the more general setting of self-affine (G)IFS (see for example [4]). How do these properties relate in the case of self-affine sets arising from multi-tape automata?

Another interesting aspect is the computability of fractal dimension (such as Hausdorff dimension). For example, can we decide if the Hausdorff dimension of a 2-dimensional self-affine set is equal to 2? And in the case of a self-affine set with nonempty interior, can we compute the Hausdorff dimension of its boundary? Very few results are known in this direction, apart from some very specific families such as Bedford-McMullen carpets [2,13,8]. A possible approach towards undecidability would be to adapt the reductions of this article in such a way that the Hausdorff dimension can be controlled in the reductions, or to relate the entropy of the automaton language with the Hausdorff dimension of its attractor and prove that entropy is uncomputable.

Acknowledgements. Research supported by the Academy of Finland Grant 131558 and by project Fractals and Numeration ANR-12-IS01-0002.

References

1. Barnsley, M.F.: Fractals everywhere, 2nd edn. Academic Press Professional, Boston (1993)
2. Bedford, T.: Crinkly curves, Markov partitions and box dimensions in self-similar sets. Ph.D. thesis, University of Warwick (1984)
3. Bondarenko, I.V., Kravchenko, R.V.: On Lebesgue measure of integral self-affine sets. Discrete Comput. Geom. 46(2), 389–393 (2011)
4. Csörnyei, M., Jordan, T., Pollicott, M., Preiss, D., Solomyak, B.: Positive-measure self-similar sets without interior. Ergodic Theory Dynam. Systems 26(3), 755–758 (2006)
5. Dube, S.: Undecidable problems in fractal geometry. Complex Systems 7(6), 423–444 (1993)
6. Falconer, K.: Techniques in fractal geometry. John Wiley & Sons Ltd., Chichester (1997)
7. Falconer, K.: Fractal geometry, 2nd edn. Mathematical Foundations and Applications. John Wiley & Sons Inc., Hoboken (2003)
8. Fraser, J.M.: Dimension theory and fractal constructions based on self-affine carpets. Ph.D. thesis, The University of St Andrews (2013)
9. Gabardo, J.P., Yu, X.: Natural tiling, lattice tiling and Lebesgue measure of integral self-affine tiles. J. London Math. Soc. (2) 74(1), 184–204 (2006)
10. Hutchinson, J.E.: Fractals and self-similarity. Indiana Univ. Math. J. 30(5), 713–747 (1981)
11. Lagarias, J.C., Wang, Y.: Self-affine tiles in $\mathbf{R}^n$. Adv. Math. 121(1), 21–49 (1996)
12. Lai, C.K., Lau, K.S., Rao, H.: Spectral structure of digit sets of self-similar tiles on $\mathbb{R}^1$. Trans. Amer. Math. Soc. 365(7), 3831–3850 (2013)

13. McMullen, C.: The Hausdorff dimension of general Sierpiński carpets. Nagoya Math. J. 96, 1–9 (1984)
14. Post, E.L.: A variant of a recursively unsolvable problem. Bull. Amer. Math. Soc. 52, 264–268 (1946)
15. Wang, Y.: Self-affine tiles. In: Advances in Wavelets (Hong Kong, 1997), pp. 261–282. Springer, Singapore (1999)

Complexity and Expressivity of Uniform One-Dimensional Fragment with Equality

Emanuel Kieroński* and Antti Kuusisto**

Institute of Computer Science, University of Wrocław, Poland

Abstract. Uniform one-dimensional fragment $UF_1^=$ is a formalism obtained from first-order logic by limiting quantification to applications of blocks of existential (universal) quantifiers such that at most one variable remains free in the quantified formula. The fragment is closed under Boolean operations, but additional restrictions (called uniformity conditions) apply to combinations of atomic formulas with two or more variables. $UF_1^=$ can be seen as a canonical generalization of two-variable logic, defined in order to be able to deal with relations of arbitrary arities. $UF_1^=$ was introduced recently, and it was shown that the satisfiability problem of the equality-free fragment UF_1 of $UF_1^=$ is decidable. In this article we establish that the satisfiability and finite satisfiability problems of $UF_1^=$ are NEXPTIME-complete. We also show that the corresponding problems for the extension of $UF_1^=$ with counting quantifiers are undecidable. In addition to decidability questions, we compare the expressivities of $UF_1^=$ and two-variable logic with counting quantifiers FOC^2. We show that while the logics are incomparable in general, $UF_1^=$ is strictly contained in FOC^2 when attention is restricted to vocabularies with the arity bound two.

Keywords: Two-variable logics, complexity, expressivity.

1 Introduction

Two-variable logic FO^2 was introduced by Henkin in [9] and proved decidable in [11] by Mortimer. The satisfiability and finite satisfiability problems of FO^2 were shown to be NEXPTIME-complete in [6]. The extension of two-variable logic with counting quantifiers, FOC^2, was proved decidable in [7], [12]. It was subsequently shown to be NEXPTIME-complete in [13]. Research on extensions and variants of two-variable logic is *currently very active*. Recent research efforts have mainly concerned decidability and complexity issues in restriction to particular classes of structures, and also questions related to different built-in features

---

* Supported by the Polish National Science Centre grant DEC-2013/09/B/ST6/01535.
** The author acknowledges that this work was carried out during a tenure of the ERCIM "Alain Bensoussan" Fellowship Programme. The research reported below has received funding from the European Union Seventh Framework Programme (FP7/2007-2013) under grant agreement number 246016.

E. Csuhaj-Varjú et al. (Eds.): MFCS 2014, Part I, LNCS 8634, pp. 365–376, 2014.
© Springer-Verlag Berlin Heidelberg 2014

and operators that increase the expressivity of the base language. Recent articles in the field include for example [3], [4], [10], [15], and several others.

Typical systems of modal logic are contained in two-variable logic, or some variant of it, and hence investigations on two-variable logics have direct implications on various fields of computer science, including verification of software and hardware, distributed systems, knowledge representation and artificial intelligence. However, two-variable logics do not cope well with relations of arities greater than two, and therefore the *scope of related research is significantly restricted*. In database theory contexts, for example, two-variable logics as such are usually not directly applicable due to the *severe arity-related limitations*.

The recent article [8] introduces the *uniform one-dimensional fragment*, $UF_1^=$, which is a natural generalization of FO^2 to contexts with relations of arbitrary arities. The logic $UF_1^=$ is a fragment of first-order logic obtained by restricting quantification to blocks of existential (universal) quantifiers that *leave at most one free variable* in the resulting formula. Additionally, a *uniformity condition* applies to the use of atomic formulas: if $n, k \geq 2$, then a Boolean combination of atoms $R(x_1, ..., x_k)$ and $S(y_1, ..., y_n)$ is allowed only if $\{x_1, ..., x_k\} = \{y_1, ..., y_n\}$. Boolean combinations of formulas with at most one free variable can be formed freely, and the use of equality is unrestricted.

It was established in [8] that already the equality-free fragment UF_1 of $UF_1^=$ can define properties not expressible in FOC^2 and also properties not expressible in the recently introduced *guarded negation fragment* [2], which significantly generalizes the *guarded fragment* [1]. The article [8] also shows, inter alia, that the equality-free logic UF_1 is decidable, and furthermore, that minor modifications to the syntax of UF_1 lead to undecidable formalisms. Namely, the non-uniform *general one-dimensional fragment* and the *strongly uniform two-dimensional fragment* were shown undecidable.

In this article we establish that the satisfiability and finite satisfiability problems of the uniform one-dimensional fragment with equality ($UF_1^=$) are NEXPTIME-complete. These results are obtained by appropriately generalizing and modifying the construction in [6] that provides small models for satisfiable FO^2-formulas in *Scott normal form*. The NEXPTIME-completeness of FOC^2 raises the natural question whether the extension $UFC_1^=$ of $UF_1^=$ with *counting quantifiers* remains decidable. We answer this question in the negative by showing that the satisfiability and finite satisifiability problems of $UFC_1^=$ are complete for Π_1^0 and Σ_1^0, respectively. These results are established by tiling arguments that make an appropriate use of a *ternary* relation, together with the usual unary relations commonly employed in similar undecidability proofs.

We also study the expressivity of $UF_1^=$. We establish that while $UF_1^=$ and FOC^2 are incomparable in expressivity in general, in restriction to vocabularies with the arity bound two, we have $UF_1^= < FOC^2$.

The uniform one-dimensional fragment $UF_1^=$ canonically extends FO^2, and in fact the equality-free fragments of $UF_1^=$ and FO^2 coincide when attention is limited to binary vocabularies. We believe that $UF_1^=$ is an interesting fragment that can be used in order to *extend the scope of research on two-variable logics to the realm involving relations of arbitrary arities.*

2   Preliminaries

Let m and $n \geq m$ be integers. We let $[m, n]$ denote the set of integers i such that $m \leq i \leq n$. If φ and ψ are first-order formulas, then $\varphi \equiv \psi$ indicates that the formulas are equivalent. If $\mathcal{L}$ and $\mathcal{L}'$ are fragments of first-order logic, we write $\mathcal{L} \leq \mathcal{L}'$ to indicate that for every sentence of $\mathcal{L}$, there exists an equivalent sentence of $\mathcal{L}'$. We let $\text{VAR} := \{ v_i \mid i \in \mathbb{N} \}$ denote the set of first-order variable symbols. We mostly use *metavariables* x, y, z, x_i, y_i, z_i, etc., in order to refer to symbols in VAR. Notice that for example x_1 and x_2 may denote the same variable in VAR, while v_1 and v_2 are necessarily different variables. The set of free variables of a formula ψ is denoted by $\mathit{free}(\psi)$.

Let $X = \{x_1, ..., x_n\}$ be a finite set of variable symbols. Let R be a k-ary relation symbol. An atomic formula $R(x_{i_1}, ..., x_{i_k})$ is called an X-*atom* if $\{x_{i_1}, ..., x_{i_k}\} = X$. A finite set of X-atoms is an X-*uniform set*. When X is irrelevant or known from the context, we may simply talk about a *uniform set*. For example, if x, y, z are distinct variables, then $\{T(x, y), S(y, x)\}$ and $\{R(x, x, y), R(y, y, x), S(y, x)\}$ are uniform sets, while $\{R(x, y, z), R(x, y, y)\}$ and $\{S(x, y), x = y\}$ are not (uniform sets are not allowed to contain equality atoms). The empty set is an X-uniform set for every finite subset of VAR, including $\emptyset$.

Let $\mathbb{Z}_+$ denote the set of positive integers. Let $\mathcal{V}$ denote a *complete relational vocabulary*, i.e., $\mathcal{V} := \bigcup_{k \in \mathbb{Z}_+} \tau_k$, where τ_k denotes a countably infinite set of k-ary relation symbols. Every vocabulary τ we consider below is assumed to be a subset of $\mathcal{V}$. A k-*ary* τ-*atom* is an atomic τ-formula ψ such that $|\mathit{free}(\psi)| = k$. For example, if $P \in \tau$ is a unary and $R \in \tau$ a binary symbol, then $P(x)$, $x = x$, $R(x, x)$ are unary τ-atoms, and $R(v_1, v_2)$, $v_1 = v_2$ are binary τ-atoms. If τ is known form the context or irrelevant, we may simply talk about k-ary atoms.

Let $\tau \subseteq \mathcal{V}$. The set $\text{UF}_1^=(\tau)$, or the set of τ-formulas of the uniform one-dimensional fragment, is the smallest set $\mathcal{F}$ satisfying the following conditions.

1. Every unary τ-atom is in $\mathcal{F}$. Also $\bot, \top \in \mathcal{F}$.
2. Every identity atom $x = y$ is in $\mathcal{F}$.
3. If $\varphi \in \mathcal{F}$, then $\neg\varphi \in \mathcal{F}$. If $\varphi_1, \varphi_2 \in \mathcal{F}$, then $(\varphi_1 \wedge \varphi_2) \in \mathcal{F}$.
4. Let $X = \{x_0, ..., x_k\} \subseteq \text{VAR}$. Let U be a finite set of formulas $\psi \in \mathcal{F}$ whose free variables are in X. Let $V \subseteq X$. Let F be a V-uniform set of τ-atoms. Let φ be any Boolean combination of formulas in $U \cup F$. Then $\exists x_1...\exists x_k \, \varphi \in \mathcal{F}$ and $\exists x_0...\exists x_k \, \varphi \in \mathcal{F}$.

Let $\text{UF}_1^=$ denote the set $\text{UF}_1^=(\mathcal{V})$.

Let $\bar{x}$ denote a tuple of variables, and let $\chi := \exists\bar{x} \, \varphi$ be a $\text{UF}_1^=$-formula formed by using the rule 4 above. Assume that φ is quantifier-free. Then we call φ a $\text{UF}_1^=$-*matrix*. If φ does not contain k-ary atoms for any $k \geq 2$, with the possible exception of equality atoms $x = y$, then we define $S_\varphi := \emptyset$. Otherwise we define S_φ to be the set V used in the construction of χ (see rule 4). The set S_φ is the set of *live variables* of φ.

Let $\psi(x_0, \ldots, x_k)$ be a $\mathrm{UF}_1^=$-matrix, where $(x_0, \ldots, x_k)$ enumerates the variables of ψ. Let $\mathfrak{A}$ be a structure. Let $a_0, \ldots, a_k \in A$, where A is the domain of $\mathfrak{A}$. We let $\mathrm{live}\big(\psi(x_0, \ldots, x_k)[a_0, \ldots, a_k]\big)$ denote the set $T \subseteq \{a_0, \ldots, a_k\}$ such that $a_i \in T$ iff x_i is a live variable of $\psi(x_0, \ldots, x_k)$. We may write $\mathrm{live}\big(\psi[a_0, \ldots, a_k]\big)$ instead of $\mathrm{live}\big(\psi(x_0, \ldots, x_k)[a_0, \ldots, a_k]\big)$ when no confusion can arise. Notice that since the elements a_i are not required to be distinct, it is possible that $|\mathrm{live}(\psi[a_0, \ldots, a_k])|$ is smaller than the number of live variables in ψ.

Let $\tau \subseteq \mathcal{V}$ be a finite vocabulary. A 1-*type* over the vocabulary τ is a maximal satisfiable set of literals (atoms and negated atoms) over τ with the variable v_1. The set of all 1-types over τ is denoted by $\boldsymbol{\alpha}[\tau]$, or just by $\boldsymbol{\alpha}$ when τ is clear from the context. We identify 1-types α and conjunctions $\bigwedge \alpha$. A k-*table* over τ is a maximal satisfiable set of $\{v_1, \ldots, v_k\}$-atoms and negated $\{v_1, \ldots, v_k\}$-atoms over τ. Recall that a $\{v_1, \ldots, v_k\}$-atom must contain exactly all the variables in $\{v_1, \ldots, v_k\}$, and note that a 2-table does not contain equality formulas or negated equality formulas. We identify k-tables β and conjunctions $\bigwedge \beta$.

Let $\mathfrak{A}$ be a τ-structure, and let $a \in A$. Let α be a 1-type over τ. We say that a *realizes* α if α is the unique 1-type such that $\mathfrak{A} \models \alpha[a]$. We let $\mathrm{tp}_{\mathfrak{A}}(a)$ denote the 1-type realized by a. Similarly, for *distinct* elements $a_1, \ldots, a_k \in A$, we let $\mathrm{tb}_{\mathfrak{A}}(a_1, \ldots, a_k)$ denote the unique k-table *realized* by the tuple $(a_1, \ldots, a_k)$, i.e., the k-table $\beta(v_1, \ldots, v_k)$ such that $\mathfrak{A} \models \beta[a_1, \ldots, a_k]$. Note that we have $\mathrm{tp}_{\mathfrak{A}}(a) \equiv \mathrm{tb}_{\mathfrak{A}}(a)$ for every $a \in A$.

Let m be the maximum arity of symbols in τ. We observe that to fully define a τ-structure $\mathfrak{A}$ over a known domain A, it is sufficient to consider each set $B \subseteq A$, $|B| \leq m$, and first choose an enumeration $(b_1, \ldots, b_{|B|})$ of the elements of B, and then specify $\mathrm{tb}_{\mathfrak{A}}(b_1, \ldots, b_{|B|})$.

Observation 1. *Let $\psi(x_1, \ldots, x_k)$ be a $\mathrm{UF}_1^=$-matrix, where $(x_1, \ldots, x_k)$ enumerates the variables in ψ. Let $\mathfrak{A}$ be a τ-structure, where τ is the set of relation symbols in ψ. Let $a_1, \ldots, a_k \in A$ be a sequence of (not necessarily distinct) elements. Whether or not $\mathfrak{A} \models \psi[a_1, \ldots, a_k]$ holds, depends only on (i) the 1-types of the elements a_i, (ii) the list of pairs (a_i, a_j) such that $a_i = a_j$, and (iii) the table $\mathrm{tb}_{\mathfrak{A}}(b_1, \ldots, b_l)$, where $(b_1, \ldots, b_l)$ is an arbitrary enumeration of $\mathrm{live}(\psi[a_1, \ldots, a_k])$.*

3   Complexity of $\mathrm{UF}_1^=$

We now introduce a normal form for $\mathrm{UF}_1^=$ inspired by the *Scott normal form* for FO^2 [14]. We say that a $\mathrm{UF}_1^=$-formula φ is in *generalized Scott normal form* if

$$\varphi = \bigwedge_{1 \leq i \leq m_\exists} \forall x \exists y_1 \ldots y_{k_i} \varphi_i^\exists(x, y_1, \ldots, y_{k_i}) \wedge \bigwedge_{1 \leq i \leq m_\forall} \forall x_1 \ldots x_{l_i} \varphi_i^\forall(x_1, \ldots, x_{l_i}), \quad (1)$$

where formulas $\varphi_i^\exists$ and $\varphi_i^\forall$ are quantifier-free $\mathrm{UF}_1^=$-matrices.

Proposition 1. *Each $\mathrm{UF}_1^=$-formula φ translates in polynomial time to a $\mathrm{UF}_1^=$-formula φ' in generalized Scott normal form (over a signature extended by some fresh unary symbols) such that φ and φ' are satisfiable over the same domains.*

Proof. A simple adaptation of a well-known translation given, e.g., in [5].   □

Let φ be the $\mathrm{UF}_1^=$-formula in generalized Scott-normal form given in Equation 1. Assume $\mathfrak{A} \models \varphi$. We will build a small τ-model $\mathfrak{A}'$ of φ, where τ is the set of relation symbols in φ. Our construction modifies and generalizes the construction of a small model for a satisfiable FO^2-formula in Scott normal form from [6]. Let $a \in A$ and $b_1, \ldots, b_{k_i} \in A$ be elements such that $\mathfrak{A} \models \varphi_i^\exists[a, b_1, \ldots, b_{k_i}]$. We say that the structure $\mathfrak{B} := \mathfrak{A} {\restriction} \{a, b_1, \ldots, b_{k_i}\}$ is a *witness structure* for a and $\varphi_i^\exists$. The substructure of $\mathfrak{B}$ restricted to the elements in live($\varphi_i^\exists[a, b_1, \ldots, b_{k_i}]$) is called the *live part* of $\mathfrak{B}$. If the live part of $\mathfrak{B}$ does not contain a, then the live part is called *free*. Note that $|B|$ may be smaller than $k_i + 1$ (this may be even imposed by the use of equalities). Also, a may be a member of the live part of $\mathfrak{B}$ even if the variable x is not a live variable of $\varphi_i^\exists$.

The court. Let n be the width of φ, i.e., $n = \max(\{k_i + 1\}_{1 \le i \le m_\exists} \cup \{l_i\}_{1 \le i \le m_\forall})$. We assume, w.l.o.g., that $n \ge 2$. A 1-type α realized in $\mathfrak{A}$ is *royal* if it is realized at most $n - 1$ times in $\mathfrak{A}$. The points in A that realize a royal 1-type are called *kings*. Let K be the set of all kings in $\mathfrak{A}$. Clearly $|K| \le (n-1)|\boldsymbol{\alpha}|$

We then define a set $D \subseteq A$. For each pair $(\alpha, \varphi_i^\exists)$, where α is a 1-type realized in $\mathfrak{A}$, if it is possible, select an element $a \in A$ that realizes the 1-type α such that there exists a witness structure $\mathfrak{B}_{\alpha,i}$ for a and $\varphi_i^\exists$ whose live part $\tilde{\mathfrak{B}}_{\alpha,i}$ is free. Add the elements of $\bar{B}_{\alpha,i}$ to D. Since we add at most $n - 1$ elements for each pair $(\alpha, \varphi_i^\exists)$, the total size of D is bounded by $(n-1)m_\exists|\boldsymbol{\alpha}|$.

For each $a \in K \cup D$ and each $\varphi_i^\exists$, select a witness structure in $\mathfrak{A}$ and let $C_{a,i}$ denote its universe. Define $C := K \cup D \cup \bigcup_{a \in K \cup D, 1 \le i \le m_\exists} C_{a,i}$. We call $\mathfrak{C} := \mathfrak{A} {\restriction} C$ the *court* of $\mathfrak{A}$. Note that $|C| \le n|K \cup D| \le n((n-1) + (n-1)m_\exists)|\boldsymbol{\alpha}|$. We have $|C| \le 2|\varphi|^3 2^{|\varphi|}$.

Universe. The court $\mathfrak{C}$ of $\mathfrak{A}$ will be a substructure of $\mathfrak{A}'$. The remaining part of the universe of $\mathfrak{A}'$ consists of three fresh disjoint sets E, F, G. Each of them contains $m_\exists + n$ elements of type α for each non-royal α realized in $\mathfrak{A}$. The i-th element of type α ($1 \le i \le m_\exists + n$) in E (resp. F, G) is denoted $e_{\alpha,i}$ (resp. $f_{\alpha,i}$, $g_{\alpha,i}$). The size of each set E, F, G is bounded by $(n + m_\exists)|\boldsymbol{\alpha}| \le 2|\varphi|2^{|\varphi|}$. Thus the total size of $|A'|$ is bounded by $8|\varphi|^3 2^{|\varphi|}$, which is exponential in $|\varphi|$.

Witnesses. Our next aim is to provide witness structures for each element of $a \in A' \setminus (K \cup D)$ and each $\varphi_j^\exists$. We will choose elements in A' which will form the universe (say, of size s) of the live part of a witness structure for a and $\varphi_j^\exists$ and define the s-table on these elements. The remaining elements of the witness structure (elements not in the live part) will then be very easily found in A'.

Let $a' \in A' \setminus (K \cup D)$. We find a *pattern element* $a \in A$ of a' as follows. If $a' \in C$, then the pattern element is a' itself. If $a' \in E \cup F \cup G$, then we let an arbitrary $a \in A$ such that $\mathrm{tp}_{\mathfrak{A}}(a) = \mathrm{tp}_{\mathfrak{A}'}(a')$ be the pattern element of a'. For each $1 \le j \le m_\exists$, we find a witness structure $\mathfrak{B}_{a,j}$ for a and $\varphi_j^\exists$, and let $\tilde{\mathfrak{B}}_{a,j}$ be its live part. If this live part is free, then there is nothing to do; an appropriate live part for the witness structure of a' and $\varphi_j^\exists$ already exists in $\mathfrak{D} := \mathfrak{C} {\restriction} D$. Otherwise, let $r_1, \ldots, r_k$ be the kings included in $\tilde{\mathfrak{B}}_{a,j}$ (possibly $k = 0$), and let

$a, b_1, \ldots, b_l$ be the non-royal elements of $\bar{\mathfrak{B}}_{a,j}$ (possibly $l = 0$). Let α_i be the 1-type of b_i $(1 \leq i \leq l)$. We consider the following cases.

Case 1. If $l = 0$ and $a' \in C$, then there is nothing to do; a' forms the live part of the desired witness structure together with some elements in K.

Case 2. If $l = 0$ and $a' \notin C$, then we set $\mathrm{tb}_{\mathfrak{A}'}(a', r_1, \ldots, r_k) := \mathrm{tb}_{\mathfrak{A}}(a, r_1, \ldots, r_k)$.

Case 3. If $l > 0$ and $a' \in E$, then we define $b_1' := f_{\alpha_1, j}$ and choose $b_2', \ldots, b_l'$ to be distinct elements of types $\alpha_2, \ldots, \alpha_l$ from $S := \{f_{\alpha, s} : m_\exists + 1 \leq s \leq m_\exists + n, \alpha \text{ non-royal}\}$. This is possible since $l < n$ and S contains n realizations of each non-royal 1-type. We set

$$\mathrm{tb}_{\mathfrak{A}'}(a', r_1, \ldots, r_k, b_1', \ldots, b_l') := \mathrm{tb}_{\mathfrak{A}}(a, r_1, \ldots, r_k, b_1, \ldots, b_l).$$

Case 4. If $l > 0$ and $a' \in F$ (resp. $a' \in G \cup (C \setminus (K \cup D))$), then we proceed as in the previous case, but we take the elements b_i' from G (resp. E).

The described procedure of providing live parts of witness structures can be executed without conflicts. It is probably worth commenting why we prepared free live parts of witness structures in $\mathfrak{D}$ instead of building them using elements of $E \cup F \cup G$ in a "regular" way. One of the problematic situations arises, e.g., when an element a' from, say, E builds the live part of its witness structure for some $\varphi_i^\exists$ using an element $b' \in F$ and some kings $r_1, \ldots, r_k$. In this case $\mathrm{tb}_{\mathfrak{A}'}(b', r_1, \ldots, r_k)$ is defined. However, it may happen that b' needs to form the live part of its witness structure for some $\varphi_j^\exists$ using precisely the elements $b', r_1, \ldots, r_k$, which can lead to a conflict.

Completion. Let $a_1', \ldots, a_k'$ $(a_i' \neq a_j'$ for $i \neq j$, $1 < k \leq n)$ be elements in A' such that the table $\mathrm{tb}_{\mathfrak{A}'}(a_1', \ldots, a_k')$ has not yet been defined. Select distinct elements $a_1, \ldots, a_k$ of $\mathfrak{A}$ such that $\mathrm{tp}_{\mathfrak{A}}(a_i) = \mathrm{tp}_{\mathfrak{A}'}(a_i')$ $(1 \leq i \leq k)$. This is always possible due to our strategy of not introducing extra kings. Set $\mathrm{tb}_{\mathfrak{A}'}(a_1', \ldots, a_k') := \mathrm{tb}_{\mathfrak{A}}(a_1, \ldots, a_k)$.

It is easy to show that $\mathfrak{A}' \models \varphi$. Thus we have proved:

Theorem 2. $\mathrm{UF}_1^=$ *has the finite model property. Moreover, every satisfiable* $\mathrm{UF}_1^=$ *formula φ has a model whose size is bounded exponentially in $|\varphi|$.*

It is now possible to prove the following theorem.

Theorem 3. *The satisfiability problem (=finite satisfiability problem) for* $\mathrm{UF}_1^=$ *is* NEXPTIME-*complete.*

4   Expressivity

In this section we compare the expressivity of $\mathrm{UF}_1^=$ with the expressivities of FO^2 and FOC^2. Clearly $\mathrm{UF}_1^=$ contains FO^2, and it is not hard to see that the inclusion is strict; equalities can be used freely in $\mathrm{UF}_1^=$, and for example the property that there are precisely two elements in a unary relation P is expressible in $\mathrm{UF}_1^=$ but not in FO^2. The expressivities of $\mathrm{UF}_1^=$ and FOC^2 are related as follows.

Theorem 4. $\mathrm{UF}_1^{=}$ and FOC^2 are incomparable in expressivity.

Proof. It is straightforward to establish that FOC^2 cannot express the $\mathrm{UF}_1^{=}$-sentence $\exists x \exists y \exists z R(x, y, z)$, and therefore $\mathrm{UF}_1^{=} \not\leq \mathrm{FOC}^2$. To show that $\mathrm{FOC}^2 \not\leq \mathrm{UF}_1^{=}$, let R be a binary relation symbol and consider models over the signature $\{R\}$. We claim that $\mathrm{UF}_1^{=}$ cannot express the FOC^2-definable condition that the in-degree (w.r.t. the relation R) at every node is at most one. Assume $\varphi(R)$ is a $\mathrm{UF}_1^{=}$-formula that defines the condition. Consider the conjunction $\varphi(R) \wedge \forall x \exists y R(x, y) \wedge \exists x \forall y \neg R(y, x)$. It is easy to see that this formula does not have a finite model, and thereby the assumption that $\mathrm{UF}_1^{=}$ can express $\varphi(R)$ is false. $\square$

The rest of this section is devoted to the scenario in which the signature contains only unary and binary relation symbols. We will show that in such a case the expressivity of $\mathrm{UF}_1^{=}$ lies strictly between FO^2 and FOC^2.

Let τ be a finite relational vocabulary. Let β be a 2-table over τ, and let x and y be *distinct* variables. Let S be the set of atoms obtained from β by replacing occurrences of the variables v_1 and v_2 in β by x and y, respectively. We call S a *binary τ-diagram* in the variables (x, y), and denote it by $\beta(x, y)$. We identify binary diagrams and conjunctions over them. A *binary τ-arrow* in the variables x, y is a formula $R(x, y)$ (or $R(y, x)$), where $R \in \tau$. Notice that neither equality atoms nor atoms of the form $R(x, x), R(y, y)$ are binary τ-arrows. It is easy to show that if φ is a Boolean combination of binary τ-arrows in the variables x, y, then φ is equivalent to the disjunction of τ-diagrams $\beta(x, y)$ that entail φ, i.e., $\beta(x, y) \models \varphi$. Note that $\bigvee \emptyset = \bot$ is a legitimate disjunction of diagrams.

Let $\{x_0, ..., x_k\}$ be a (possibly empty) set of distinct variables. An *identity literal* over $\{x_0, ..., x_k\}$ is a formula of the type $x_i = x_j$ or $\neg x_i = x_j$, where $i, j \in [0, k]$. An identity literal is *non-trivial* if the variables in it are different. An *identity profile* over $\{x_0, ..., x_k\}$, or a $\{x_0, ..., x_k\}$-*profile*, is a maximal satisfiable set of non-trivial identity literals over $\{x_0, ..., x_k\}$. We identify identity profiles and conjunctions over them. We let $diff(x_0, ..., x_k)$ denote the conjunction of inequalities $x_i \neq x_j$, where $i, j \in [0, k]$, $i \neq j$. An identity profile is a *discriminate profile* if it is the formula $diff(x_0, ..., x_k)$ for some set $\{x_0, ..., x_k\}$ of distinct variables. Let I be a set of identity literals over $\{x_0, ..., x_k\}$. Let φ be a $\{x_0, ..., x_k\}$-profile. We say that φ is *consistent with* I if $\varphi \models \bigwedge I$.

A $\mathrm{UF}_1^{=}$-formula φ is a *block formula* if φ is of the type $\exists \overline{x} \, \psi$ or $\neg \exists \overline{x} \, \psi$. Here $\exists \overline{x}$ denotes a vector of one or more existentially quantified variables. formulas $\exists \overline{x} \, \psi$ are called *positive* blocks, while formulas $\neg \exists \overline{x} \, \psi$ are *negative* blocks. A $\mathrm{UF}_1^{=}$-formula is *simple* if it is a literal or a block formula.

Let τ be a finite relational vocabulary with the arity bound two. Let $x_0, ..., x_k$ be distinct variable symbols. Let $\varphi := \exists x_1...x_k \, \psi$ be a $\mathrm{UF}_1^{=}$-formula over τ. Let $x, y \in \{x_0, ..., x_k\}$ be distinct variables. We call φ a τ-*diagram block* if the formula ψ is a conjunction

$$\beta(x, y) \wedge diff(x_0, ..., x_k) \wedge \psi_0(x_0) \wedge ... \wedge \psi_k(x_k),$$

where $\beta(x, y)$ is a binary τ-diagram in the variables (x, y), and each formula $\psi_i(x_i)$ is a conjunction of simple formulas ψ' such that $free(\psi') \subseteq \{x_i\}$. Furthermore, if $\chi(x)$ is a τ-diagram block with the free variable x, then also the formula

$\exists x\, \chi(x)$ is a τ-diagram block. A $\mathrm{UF}_1^=$-formula φ is in *diagram normal form* if for every positive block formula φ' that occurs as a subformula in φ, there is a τ such that φ' is a τ-diagram block. The following lemma is easy to prove.

Lemma 1. *Every positive block formula is equivalent to a disjunction of diagram blocks.*

Corollary 1. *Each $\mathrm{UF}_1^=$-formula is equivalent to a formula in diagram normal form.*

Proof. By induction on the structure of formulas, using Lemma 1.   □

Let τ be a finite relational vocabulary with the arity bound 2. Let $k \in \mathbb{Z}_+$, and let $x_0, ..., x_k$ be distinct variable symbols. Let $\varphi(x_0, ..., x_k) := \mathit{diff}(x_0, ..., x_k) \wedge \psi$, where ψ is conjunction of τ-literals such that the following conditions hold.

1. The variables of each conjunct of ψ are in $\{x_0, ..., x_k\}$.
2. If ψ has $R(x, y)$ or $\neg R(x, y)$ as a conjunct, where $R \in \tau$ is a binary relation symbol and x, y distinct variables, then $x_0 \in \{x, y\}$.

Then we call $\varphi(x_0, ..., x_k)$ a τ-*star formula* in the variables $(x_0, ..., x_k)$. The variable x_0 is called the *centre variable* of φ. Consider then a quantifier-free τ-formula $\psi(x_0, ..., x_k) := \mathit{diff}(x_0, ..., x_k) \wedge \beta \wedge \alpha$ such that the following conditions are satisfied.

1. The formula β is a conjunction $\beta_1(x_0, x_1) \wedge ... \wedge \beta_k(x_0, x_k)$, where each $\beta_i(x_0, x_i)$ is a binary τ-diagram in the variables (x_0, x_i).
2. The formula α is a conjunction $\alpha_0(x_0) \wedge ... \wedge \alpha_k(x_k)$, where each $\alpha_i(x_i)$ is a 1-type over τ and in the variable x_i. (The variable v_1 is replaced by x_i.)

The formula $\psi(x_0, ..., x_k)$ is called a τ-*star type* in the variables $(x_0, ..., x_k)$. The variable x_0 is the *centre variable* of the τ-star type. It is straightforward to show that every τ-star formula in the variables $(x_0, ..., x_k)$ is equivalent to a disjunction of τ-star types in the variables $(x_0, ..., x_k)$.

Let $\varphi(x_0, ..., x_k)$ be a τ-star formula in the variables $(x_0, ..., x_k)$. Then the formula $\exists x_1 ... \exists x_k \varphi(x_0, ..., x_k)$ is called a τ-*star centre formula* of the width k. Let $\psi(x_0, ..., x_k)$ be a τ-star type in the variables $(x_0, ..., x_k)$. Then the formula $\exists x_1 ... \exists x_k \psi(x_0, ..., x_k)$ is called a τ-*star centre type* of the width k. The following lemma follows immediately from the fact that every τ-star formula is equivalent to a disjunction of τ-star types.

Lemma 2. *Every τ-star centre formula of the width k is equivalent to a disjunction of τ-star centre types of the width k.*

A 2-*type* over τ is a maximal satisfiable set of τ-literals in the variables v_1 and v_2 (equalities and negated equalities are considered to be τ-literals). If $\mathcal{T}$ is a 2-type over τ and x, y distinct variables, we let $\mathcal{T}(x, y)$ denote the set obtained from $\mathcal{T}$ by replacing all occurrences of v_1 and v_2 in $\mathcal{T}$ by x and y, respectively. Below we identify sets $\mathcal{T}(x, y)$ and conjunctions over them.

Let ψ be a τ-star type in the variables $(x_0, ..., x_k)$. Each pair (x_0, x_i), where $i \in [1, k]$, is called a *ray* of ψ. Let $\mathcal{T}$ be a 2-type over τ. We say that the ray (x_0, x_i) of ψ *realizes* $\mathcal{T}$ if $\psi \models \mathcal{T}(x_0, x_i)$.

Theorem 5. *Let τ be a relational vocabulary with the arity bound 2. Then $\mathrm{UF}_1^=(\tau) \leq \mathrm{FOC}^2(\tau)$. The inclusion is strict if τ contains a binary symbol.*

Proof. We have above shown that $\mathrm{FOC}^2(\tau) \not\leq \mathrm{UF}_1^=(\tau)$ if τ contains a binary relation symbol. Therefore it suffices to show that $\mathrm{UF}_1^=(\tau) \leq \mathrm{FOC}^2(\tau)$. The claim is established by induction on the structure of $\mathrm{UF}_1^=$-formulas in diagram normal form. We discuss the case involving quantifiers.

Let $\sigma \subseteq \tau$ and consider a σ-diagram block $\varphi(x_0) := \exists x_1 ... \exists x_k\, \psi$, where $\psi := \beta(x_0, x_1) \wedge \mathit{diff}(x_0, ..., x_k) \wedge \psi_0(x_0) \wedge ... \wedge \psi_k(x_k)$. Note that we assume that the free variable x_0 of $\varphi(x_0)$ occurs in the σ-diagram $\beta(x_0, x_1)$—unless σ does not contain binary relation symbols and thus $\beta(x_0, x_1) = \top$. The case where σ contains binary relation symbols and x_0 does not occur in the binary σ-diagram of ψ, is discussed later. Write each formula $\psi_i(x_i)$ in a form $\delta_i(x_i) \wedge \delta_i'(x_i)$, where $\delta_i(x_i)$ is a conjunction of the literals that occur as conjuncts in $\psi_i(x_i)$ and $\delta_i'(x_i)$ is the conjunction of the block formulas of $\psi_i(x_i)$. We have

$$\psi \equiv \beta(x_0, x_1) \wedge \mathit{diff}(x_0, ..., x_k) \wedge \big(\delta_0(x_0) \wedge \delta_0'(x_0)\big) \wedge ... \wedge \big(\delta_k(x_k) \wedge \delta_k'(x_k)\big).$$

Let $P_0, ..., P_k$ be fresh unary relation symbols. Consider the formula

$$\psi' := \beta(x_0, x_1) \wedge \mathit{diff}(x_0, ..., x_k) \wedge \big(\delta_0(x_0) \wedge P_0(x_0)\big) \wedge ... \wedge \big(\delta_k(x_k) \wedge P_k(x_k)\big).$$

Let σ' be the set of relation symbols in ψ'. Let us consider the formula $\chi(x_0) := \exists x_1 ... \exists x_k\, \psi'$. By Lemma 2, we have $\chi(x_0) \equiv \chi'(x_0) := \theta_0(x_0) \vee ... \vee \theta_m(x_0)$, where each $\theta_i(x_0)$ is a σ'-star centre type. We shall next show that each σ'-star centre type can be expressed in FOC^2. This will conclude the argument concerning the formula $\varphi(x_0)$, as the disjuncts $\theta_i(x_0)$ of $\chi'(x_0)$ can first be replaced by equivalent FOC^2-formulas, and after that, each subformula $P_i(z)$ $(0 \leq i \leq k)$ in the resulting formula can be replaced by an FOC^2-formula $\delta_i''(z) \equiv \delta_i'(z)$ obtained by the induction hypothesis. Here z is either of the variables in the two-variable formula we are constructing. If necessary, variables in δ_i' can be circulated to avoid variable capture. This way we obtain an FOC^2-formula equivalent to $\varphi(x_0)$.

The notion of a star centre type was of course designed to be expressible in FOC^2. Consider the σ'-star centre type $\exists x_1 ... \exists x_k\, \gamma$, where γ is the σ'-star type

$$\gamma := \mathit{diff}(x_0, ..., x_k) \bigwedge_{i \in \{1, ..., k\}} \beta_i(x_0, x_k) \wedge \bigwedge_{i \in \{0, ..., k\}} \alpha_i(x_i).$$

For each 2-type $\mathcal{T}$ over σ', let $\#\mathcal{T}$ denote the number of rays of γ that realize $\mathcal{T}$. Let T denote the set of all 2-types over σ'. Define

$$\gamma'(x_0) := \bigwedge_{\mathcal{T} \in T} \exists^{\geq \#\mathcal{T}} y\ \mathcal{T}(x_0, y).$$

It is easy to see that the FOC^2-formula $\gamma'(x_0)$ is equivalent to the σ'-star centre type $\exists x_1 ... \exists x_k\, \gamma$.

Let us then consider a σ-diagram block formula $\theta(x_0) := \exists x_1 ... \exists x_k\, \eta$, where $\eta := \beta(x_i, x_j) \wedge \mathit{diff}(x_0, ..., x_k) \wedge \psi_0(x_0) \wedge ... \wedge \psi_k(x_k)$, and $x_0 \notin \{x_i, x_j\}$. Let $\bar{x}$ denote a tuple containing exactly the variables in $\{x_0, ..., x_k\} \setminus \{x_i\}$. Consider the block formula $\theta'(x_i) := \exists \bar{x}\, \eta$. In this formula, the free variable x_i occurs in the part $\beta(x_i, x_j)$ of η. Thus, by our argument above, $\theta'(x_i)$ is equivalent to a formula $\theta''(x_i)$ of FOC^2.

By the induction hypothesis, there are FOC^2-formulas $\psi_0'(x_0) \equiv \psi_0(x_0)$ and $\psi_j'(x_j) \equiv \psi_j(x_j)$. Let $\psi_j'(x_0)$ denote the FOC^2-formula obtained from $\psi_j'(x_j)$ by changing the free variable x_j to x_0, and circulating variables, if necessary. Let the variables used in $\psi_0'(x_0)$ and $\psi_j'(x_0)$ be x_0 and x_i. Define $\mathcal{B}(x_i, x_0) := \beta(x_i, x_0) \wedge \psi_j'(x_0)$. The original formula $\theta(x_0)$ is equivalent to the FOC^2-formula

$$\psi_0'(x_0) \wedge \exists x_i \Big(x_i \neq x_0 \wedge \theta''(x_i) \wedge \big(\neg \mathcal{B}(x_i, x_0) \vee \exists^{\geq 2} x_0 \big(x_i \neq x_0 \wedge \mathcal{B}(x_i, x_0) \big) \big) \Big).$$

To conclude the proof, we need to discuss the case involving a block formula $\chi := \exists x_0 ... \exists x_k \chi'$ that does not contain a free variable. Assume, w.l.o.g., that $k \geq 1$. Convert the block formula $\exists x_1 ... \exists x_k \chi'$ to an FOC^2-formula $\pi(x_0)$. Thus the original formula χ is equivalent to the FOC^2-formula $\exists x_0 \pi(x_0)$.   □

5 Undecidability of UFC$_1^=$

Since FOC^2 and UF$_1^=$ are decidable, it is natural to ask whether the extension of UF$_1^=$ by counting quantifiers, UFC$_1^=$, remains decidable. Formally, UFC$_1^=$ is obtained from UF$_1^=$ by allowing the free substitution of quantifiers $\exists$ by quantifiers $\exists^{\geq k}, \exists^{\leq k}, \exists^{=k}$. We next show that both the general and the finite satisfiability problems of UFC$_1^=$ are undecidable.

For the proofs, we use the standard tiling and periodic tiling arguments. A *tile* is a mapping $t : \{R, L, T, B\} \to C$, where C is a countably infinite set of colours. We use the subscript notation $t_X := t(X)$ for $X \in \{R, L, T, B\}$. Intuitively, t_R, t_L, t_T and t_B are the colours of the right edge, left edge, top edge and bottom edge of the tile t, respectively.

Let $\mathfrak{S} := (S, H, V)$ be a structure with domain S and binary relations H and V. Let $\mathbb{T}$ be a finite nonempty set of tiles. A $\mathbb{T}$-*tiling* of $\mathfrak{S}$ is a function $f : S \to \mathbb{T}$ that satisfies the following conditions.

(T_H) For all $a, b \in S$, if $f(a) = t$, $f(b) = t'$ and $(a, b) \in H$, then $t_R = t_L'$.
(T_V) For all $a, b \in S$, if $f(a) = t$, $f(b) = t'$ and $(a, b) \in V$, then $t_T = t_B'$.

The *tiling problem* for $\mathfrak{S}$ asks, given a finite nonempty set $\mathbb{T}$ of tiles, whether there exists a $\mathbb{T}$-tiling of $\mathfrak{S}$.

The standard *grid* is the structure $\mathfrak{G} := (\mathbb{N} \times \mathbb{N}, H, V)$, where $H = \{ ((i, j), (i + 1, j)) \mid i, j \in \mathbb{N} \}$ and $V = \{ ((i, j), (i, j + 1)) \mid i, j \in \mathbb{N} \}$ are binary relations. It is well known that the tiling problem for $\mathfrak{G}$ is Π_1^0-complete. Let n be a positive integer. Let $T := [0, n-1] \times [0, n-1]$. An $(n \times n)$-*torus* is the structure (T, H, V) such that $H = \{ ((i, j), (i+1, j)) \mid (i, j) \in T \}$ and $V = \{ ((i, j), (i, j+1)) \mid (i, j) \in T \}$, where the sum is taken modulo n. The *periodic tiling problem* asks, given a finite nonempty set $\mathbb{T}$ of tiles, whether there exist an $n \in \mathbb{Z}_+$ such the $(n \times n)$-torus is $\mathbb{T}$-tilable. It is well known that the periodic tiling problem is Σ_1^0-complete.

We shall below define a UFC$_1^=$-formula η which axiomatizes a sufficiently rich class of grid-like structures. In order to encode grids *with* UFC$_1^=$-*formulas*, we employ a *ternary* predicate R and a unary predicate E. Intuitively, E labels elements that represent the nodes of the even rows of a grid, and R contains triples (a, b, c) such that b is the horizontal successor of a and c is the vertical successor of b, or b is the vertical successor of a and c is the horizontal successor of b. The following figure depicts an initial portion of an infinite structure $\mathfrak{A}_\mathfrak{G}$ (over the signature $\{R, E\}$) which is our intended encoding of the standard infinite grid $\mathfrak{G}$. An arrow from a node a via b to c means that $R(a, b, c)$ holds.

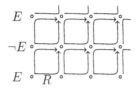

Define the formulas $\varphi_H(x, y) := \exists z\big((R(x, y, z) \lor R(z, x, y)) \land (E(x) \leftrightarrow E(y))\big)$ and $\varphi_V(x, y) := \exists z\big((R(x, y, z) \lor R(z, x, y)) \land (E(x) \leftrightarrow \neg E(y))\big)$. Note that these are not $\mathrm{UF}_1^=$-formulas. Let $\mathfrak{A}$ be a structure over the vocabulary $\{R, E\}$. We let $\mathfrak{A}^*$ be the structure over the vocabulary $\{H, V\}$ such that the $\mathfrak{A}^*$ has the same domain A as $\mathfrak{A}$, and the relation $H^{\mathfrak{A}^*}$ $(V^{\mathfrak{A}^*})$ is the set of pairs $(a, a') \in A$ such that $\mathfrak{A} \models \varphi_H[a, a']$ $(\mathfrak{A} \models \varphi_V[a, a'])$. Note that $\mathfrak{A}_\mathfrak{G}^*$ is the standard grid $\mathfrak{G}$.

We next define a $\mathrm{UFC}_1^=$-formula η that captures some essential properties of $\mathfrak{A}_\mathfrak{G}$. Let η be the conjunction of the formulas (2) – (8) below. Note that the syntactic restrictions of $\mathrm{UFC}_1^=$ are indeed met.

$$\exists x E(x) \tag{2}$$
$$\forall x \exists^{=1} y \exists z (R(x, y, z) \land (E(x) \leftrightarrow E(y))) \tag{3}$$
$$\forall x \exists^{=1} y \exists z (R(x, y, z) \land (E(x) \leftrightarrow \neg E(y))) \tag{4}$$
$$\forall x \exists^{=1} z \exists y R(x, y, z) \tag{5}$$
$$\forall x \forall y \forall z (R(x, y, z) \rightarrow (E(x) \leftrightarrow \neg E(z))) \tag{6}$$
$$\forall x \exists^{=1} y \exists z (((E(x) \leftrightarrow E(y)) \land (R(z, x, y) \lor R(x, y, z)))) \tag{7}$$
$$\forall x \exists^{=1} y \exists z (((E(x) \leftrightarrow \neg E(y)) \land (R(z, x, y) \lor R(x, y, z)))) \tag{8}$$

We claim that η has the following properties.

(i) There exists a model $\mathfrak{A} \models \eta$ such that $\mathfrak{A}^* = \mathfrak{G}$.
(ii) For every model $\mathfrak{A} \models \eta$, there is a homomorphism from $\mathfrak{G}$ to $\mathfrak{A}^*$.

Assume we can show that η indeed has the above properties. Let $\mathbb{T}$ be an arbitrary input to the tiling problem, and let $\mathcal{P}_\mathbb{T} := \{P_t \mid t \in \mathbb{T}\}$ be a set of fresh unary predicate symbols. Construct a $\mathrm{UFC}_1^=$-formula $\varphi_\mathbb{T} := \psi_0 \land \psi_H \land \psi_V$ over the vocabulary $\{R, E\} \cup \mathcal{P}_\mathbb{T}$ as follows.

1. ψ_0 states that each point of the model is in the interpretation of exactly one predicate symbol P_t, $t \in \mathbb{T}$.
2. $\psi_H \equiv \forall x \forall y \bigwedge_{t, t' \in \mathbb{T}, t_R \neq t'_L} \neg(\varphi_H(x, y) \land P_t(x) \land P_{t'}(y))$. Note that the right hand side here is not a $\mathrm{UFC}_1^=$-formula, but it can easily be modified so that the resulting formula is.
3. $\psi_V \equiv \forall x \forall y \bigwedge_{t, t' \in \mathbb{T}, t_T \neq t'_B} \neg(\varphi_V(x, y) \land P_t(x) \land P_{t'}(y))$.

It is easy to see that $\eta \wedge \varphi_{\mathbb{T}}$ has a model iff there exists a $\mathbb{T}$-tiling of $\mathfrak{G}$.

It is not difficult to show that η indeed has the properties (i) and (ii) listed above Thus the satisfiability problem of $\mathrm{UFC}_1^=$ is Π_1^0-hard. Since $\mathrm{UFC}_1^=$ is a fragment of first-order logic, the following theorem holds.

Theorem 6. *The satisfiability problem of* $\mathrm{UFC}_1^=$ *is* Π_1^0-*complete.*

The above argument leading to Theorem 6 can be used with minor modifications in order to show Σ_1^0-completeness of the finite satisfiability problem of $\mathrm{UFC}_1^=$. The argument uses the periodic tiling problem. Thus the following theorem holds.

Theorem 7. *The finite satisfiability problem of* $\mathrm{UFC}_1^=$ *is* Σ_1^0-*complete.*

References

1. Andréka, H., van Benthem, J., Németi, I.: Modal languages and bounded fragments of predicate logic. Journal of Philosophical Logic 27(3), 217–274 (1998)
2. Bárány, V., ten Cate, B., Segoufin, L.: Guarded negation. In: Aceto, L., Henzinger, M., Sgall, J. (eds.) ICALP 2011, Part II. LNCS, vol. 6756, pp. 356–367. Springer, Heidelberg (2011)
3. Benaim, S., Benedikt, M., Charatonik, W., Kieroński, E., Lenhardt, R., Mazowiecki, F., Worrell, J.: Complexity of two-variable logic on finite trees. In: Fomin, F.V., Freivalds, R., Kwiatkowska, M., Peleg, D. (eds.) ICALP 2013, Part II. LNCS, vol. 7966, pp. 74–88. Springer, Heidelberg (2013)
4. Charatonik, W., Witkowski, P.: Two-variable logic with counting and trees. In: LICS, pp. 73–82 (2013)
5. Ebbinghaus, H.-D., Flum, J.: Finite model theory. Perspectives in Mathematical Logic. Springer (1995)
6. Grädel, E., Kolaitis, P., Vardi, M.: On the decision problem for two-variable first-order logic. Bulletin of Symbolic Logic 3(1), 53–69 (1997)
7. Grädel, E., Otto, M., Rosen, E.: Two-variable logic with counting is decidable. In: LICS, pp. 306–317 (1997)
8. Hella, L., Kuusisto, A.: One-dimensional fragment of first-order logic. arXiv:1404.4004 (2014)
9. Henkin, L.: Logical systems containing only a finite number of symbols. Presses De l'Université De Montréal (1967)
10. Kieroński, E., Michaliszyn, J., Pratt-Hartmann, I., Tendera, L.: Two-variable first-order logic with equivalence closure. SIAM Journal of Computing 43(3) (2014)
11. Mortimer, M.: On languages with two variables. Mathematical Logic Quarterly 21(1), 135–140 (1975)
12. Pacholski, L., Szwast, W., Tendera, L.: Complexity of two-variable logic with counting. In: LICS, pp. 318–327. IEEE (1997)
13. Pratt-Hartmann, I.: Complexity of the two-variable fragment with counting quantifiers. Journal of Logic, Language and Information 14(3), 369–395 (2005)
14. Scott, D.: A decision method for validity of sentences in two variables. Journal Symbolic Logic 27, 477 (1962)
15. Szwast, W., Tendera, L.: FO^2 with one transitive relation is decidable. In: STACS, pp. 317–328 (2013)

A Unifying Approach for Multistack Pushdown Automata*

Salvatore La Torre[1], Margherita Napoli[1], and Gennaro Parlato[2]

[1] Dipartimento di Informatica, Università degli Studi di Salerno, Italy
[2] School of Electronics and Computer Science, University of Southampton, UK

Abstract. We give a general approach to show the closure under complement and decide the emptiness for many classes of multistack visibly pushdown automata (MVPA). A central notion in our approach is the *visibly path-tree*, i.e., a stack tree with the encoding of a path that denotes a linear ordering of the nodes. We show that the set of all such trees with a bounded size labeling is regular, and path-trees allow us to design simple conversions between tree automata and MVPA's. As corollaries of our results we get the closure under complement of ordered MVPA that was an open problem, and a better upper bound on the algorithm to check the emptiness of bounded-phase MVPA's.

1 Introduction

Pushdown automata working with multiple stacks (*multistack pushdown automata*, MPA for short) are a natural model of the control flow of shared-memory multithreaded programs. They are as much expressive as Turing machines already when only two stacks are used (the two stacks can act as the two halves of the tape portion in use). Therefore, the research on MPA's related to the development of formal methods for the analysis of multithreaded programs has mainly focused on decidable restricted versions of these models (as a sample of recent research see [2–6, 8, 9, 12, 14–18, 20]).

Formal language theories are a valuable source of tools for applications in other domains. Robust definitions, i.e., classes with decidable decision problems and closed under the main language operations (among all the Boolean operations), are particularly appealing. For instance, in the automata-theoretic approach to model-checking linear-time properties, the verification problem can be rephrased as a language inclusion or checking the emptiness of a language intersection, pattern-matching problems are often rephrased as membership queries. In a recent paper [10], the authors define a notion of *perfect* class of languages as a class that is closed under the Boolean operations and with a decidable emptiness problem, and investigate perfect classes modulo *bounded languages*.

Robust theories of MPA's introduced in the literature rely on both a restriction on the admitted behaviours [12–14] and the *visibility* [1] of stack operations (each symbol of the input alphabet explicitly identifies if a push onto stack i, or a pop from stack i, or no stack operation must happen on reading it).

---

* Partially supported by the FARB grants 2011-2013, Università degli Studi di Salerno.

E. Csuhaj-Varjú et al. (Eds.): MFCS 2014, Part I, LNCS 8634, pp. 377–389, 2014.

The restriction is imposed to gain the decidability of the emptiness problem. Visibility instead gains the closure under intersection, which does not hold also for a single stack pushdown automaton (with visibility, stack operations on a same stack synchronize). It is not a severe restriction for applications, the sequence of locations visited in the executions of programs being indeed visible.

In the literature, the results on MPA's are shown with different techniques for the different restrictions. Here, we introduce a unifying approach to show the two main technical challenges in proving robustness: emptiness decidability and closure under complement. We introduce the notion of *visibly path-tree*, that incidentally also allows us to define a robust class of *multistack visibly pushdown automata* (MVPA) that subsumes the main classes indentified in the literature.

A visibly path-tree is essentially a tree that encodes a visibly multi-stack word such that: (1) the left child of a node is its linear successor and the right-child relation captures the relations among matching *calls* (each causing a push transition on a specified stack) and *returns* (each causing a pop transition on a specified stack) and (2) it has an additional labeling that encodes a traversal of the tree that reconstructs the corresponding word. This labeling is formed by an ordered sequence of pairs, each composed of a direction (pointing to a neighbor in the tree) and an index (denoting the position of the pair to follow in the pointed neighbor). We define the class TMVPA by restricting MVPA to languages that for a given $k > 0$, contain only words that can be encoded into a visibly k-path-tree, i.e., a path-tree with at most k pairs in the labeling of each node.

Our first result is to construct, for an MVPA A over an n-stack alphabet, two tree automata P_k and $\mathcal{A}_k$. P_k accepts the set PT_k of all visibly k-path-trees and has size $2^{O(nk)}$. If the input is a k-path-tree, $\mathcal{A}_k$ accepts it iff it encodes a word accepted by A. $\mathcal{A}_k$ has size $O(|A|^k)$. Thus, we reduce the emptiness problem for TMVPA to checking the emptiness of the intersection of P_k and $\mathcal{A}_k$, that yields a $2^{O(k(n+\log|A|))}$ time solution. To show the closure under complement we first take the tree automaton for the intersection of P_k and the complement of $\mathcal{A}_k$, and then, from this, we construct $\bar{A}$ that accepts the complement of the language accepted by A. The size of $\bar{A}$ is exponential in the size of A and doubly exponential in k. From the effectiveness of these two proofs, we also get the decidability of containment, equivalence and universality problems.

Our approach is general, in the sense that it indeed works for each class of MVPA's that is defined by a restriction R that *refines* the bounded path-tree restriction used to define TMVPA, i.e., such that there is a bound k that suffices to encode in k-path-trees all the words satisfying R. This is sufficient for the complement since the actual restriction is captured by the resulting MVPA in the end. Instead for the emptiness, we also need to construct an additional tree automaton that exactly captures the restriction R on the k-path-trees.

As corollaries of our results, we show the closure under complement for ordered MVPA that was open[1] and an algorithm in time $2^{(n+\log|A|)2^{O(d)}}$ to check the

---

[1] A proof via determinization was given in [8], but that is wrong since the language of all the words $(ab)^i c^j d^{i-j} x^j y^{i-j}$ is both accepted by a 2-stack OMVPA and inherently nondeterministic for MVPA's [13].

emptiness for bounded-phase MVPA, improving the $2^{|A|2^{O(d)}}$ bound shown in [11, 12] and matching the bound that can be derived from the results of [9].

Our path-tree representation has been inspired by the tree decomposition of bounded-phase and ordered words given in [19]. Concerning the closure under complement, our construction is structured as that from [12] but differs from it for the tree encoding of runs (path-tree) and thus in the tree automata constructions. Further, the approach from [12] does not apply directly to the ordered restriction and a different encoding would be needed. We give (i) a unifying approach for bounded-phase and ordered restrictions by a less restrictive limitation, and (ii) sufficient conditions for applying our constructions to new restrictions that can be captured by bounded path-trees.

2 Preliminaries

For $i, j \in \mathbb{N}$, we let $[i, j] = \{d \in \mathbb{N} \mid i \leq d \leq j\}$ and $[j] = [1, j]$.

Words over call-return alphabets. Given a finite alphabet Σ and an integer $n > 0$, an n-stack call-return labeling is a $lab_{\Sigma,n} : \Sigma \to (\{ret, call\} \times [n]) \cup \{int\}$, and an n-stack call-return alphabet is $\widetilde{\Sigma}_n = (\Sigma, lab_{\Sigma,n})$. We fix the n-stack call-return alphabet $\widetilde{\Sigma}_n = (\Sigma, lab_{\Sigma,n})$ for the rest of the paper.

For $h \in [n]$, we denote $\Sigma_r^h = \{a \in \Sigma \mid lab_{\Sigma,n}(a) = (ret, h)\}$ (*set of returns*), $\Sigma_c^h = \{a \in \Sigma \mid lab_{\Sigma,n}(a) = (call, h)\}$ (*set of calls*), and $\Sigma_{int} = \{a \in \Sigma \mid lab_{\Sigma,n}(a) = int\}$ (*set of internals*). Moreover, $\Sigma_c = \bigcup_{h=1}^n \Sigma_c^h$, $\Sigma_r = \bigcup_{h=1}^n \Sigma_r^h$ and $\Sigma^h = \Sigma_c^h \cup \Sigma_r^h \cup \Sigma_{int}$.

A *stack-h context* is a word in $(\Sigma^h)^*$. For a word $w = a_1 \ldots a_m$ over $\widetilde{\Sigma}_n$, denoting $C_h = \{i \in [m] \mid a_i \in \Sigma_c^h\}$ and $R_h = \{i \in [m] \mid a_i \in \Sigma_r^h\}$, the *matching relation* $\sim_h$ defined by w is such that (1) $\sim_h \subseteq C_h \times R_h$, (2) if $i \sim_h j$ then $i < j$, (3) for each $i \in C_h$ and $j \in R_h$ s.t. $i < j$, there is an $i' \in [i, j]$ s.t. either $i' \sim_h j$ or $i \sim_h i'$, and (4) for each $i \in C_h$ (resp. $i \in R_h$) there is at most one $j \in [m]$ s.t. $i \sim_h j$ (resp. $j \sim_h i$). When $i \sim_h j$, we say that positions i and j *match* in w. If $i \in C_h$ and $i \not\sim_h j$ for any $j \in R_h$, then i is an *unmatched call*. Analogously, if $i \in R_h$ and $j \not\sim_h i$ for any $j \in C_h$, then i is an *unmatched return*.

Multi-stack visibly pushdown languages. A multi-stack visibly pushdown automaton over an n-stack call-return alphabet pushes a symbol on stack h when it reads a call of the stack h, and pops a symbol from stack h when it reads a return of the stack h. Moreover, it just changes its state, without modifying any stack, when reading an internal symbol. A special bottom-of-stack symbol $\perp$ is used: it is never pushed or popped, and is in each stack when computation starts. A *multi-stack visibly pushdown automaton* (MVPA) A over $\widetilde{\Sigma}_n$ is $(Q, Q_I, \Gamma, \delta, Q_F)$ where Q is a finite set of states, $Q_I \subseteq Q$ is the set of initial states, Γ is a finite stack alphabet containing $\perp$, $\delta \subseteq (Q \times \Sigma_c \times Q \times (\Gamma \setminus \{\perp\})) \cup (Q \times \Sigma_r \times \Gamma \times Q) \cup (Q \times \Sigma_{int} \times Q)$ is the transition function, and $Q_F \subseteq Q$ is the set of final states. Moreover, A is *deterministic* if $|Q_I| = 1$, and $|\{(q, a, q') \in \delta\} \cup \{(q, a, q', \gamma') \in \delta\} \cup \{(q, a, \gamma, q') \in \delta\}| \leq 1$, for each $q \in Q$, $a \in \Sigma$ and $\gamma \in \Gamma$.

A *configuration* of an MVPA A over $\widetilde{\Sigma}_n$ is a tuple $\alpha = \langle q, \sigma_1, \ldots, \sigma_n \rangle$, where $q \in Q$ and each $\sigma_h \in (\Gamma \setminus \{\perp\})^*.\{\perp\}$ is a *stack content*. Moreover, α is *initial* if $q \in Q_I$ and $\sigma_h = \bot$ for every $h \in [n]$, and *accepting* if $q \in Q_F$. A *transition* $\langle q, \sigma_1, \ldots, \sigma_n \rangle \xrightarrow{a}_A \langle q', \sigma_1', \ldots, \sigma_n' \rangle$ is such that one of the following holds:

[Push] $a \in \Sigma_c^h$, $\exists \gamma \in \Gamma \setminus \{\bot\}$ such that $(q, a, q', \gamma) \in \delta$, $\sigma_h' = \gamma \cdot \sigma_h$, and $\sigma_i' = \sigma_i$ for every $i \in ([n] \setminus \{h\})$.

[Pop] $a \in \Sigma_r^h$, $\exists \gamma \in \Gamma$ such that $(q, a, \gamma, q') \in \delta$, $\sigma_i' = \sigma_i$ for every $i \in ([n] \setminus \{h\})$, and either $\gamma \neq \bot$ and $\sigma_h = \gamma \cdot \sigma_h'$, or $\gamma = \sigma_h = \sigma_h' = \bot$.

[Internal] $a \in \Sigma_{int}$, $(q, a, q') \in \delta$, and $\sigma_h' = \sigma_h$ for every $h \in [n]$.

For a word $w = a_1 \ldots a_m$ in Σ^*, a *run* of A on w from α_0 to α_m, denoted $\alpha_0 \xrightarrow{w}_A \alpha_m$, is a sequence of transitions $\alpha_{i-1} \xrightarrow{a_i}_A \alpha_i$, for $i \in [m]$. A word w is accepted by A if there exist an initial configuration α and an accepting configuration α' such that $\alpha \xrightarrow{w}_A \alpha'$. The language accepted by A is denoted with $L(A)$. A language $L \subseteq \Sigma^*$ is a *multi-stack visibly pushdown language* (MVPL) if there exist an MVPA A over $\widetilde{\Sigma}_n = (\Sigma, lab_{\Sigma,n})$ such that $L = L(A)$.

3   Visibly Path-Trees

Trees. A (binary) *tree* T is any finite prefix-closed subset of $\{\swarrow, \searrow\}^*$. A *node* is any $x \in T$, the *root* is ε and the edge-relation is implicit: *edges* are pairs of the form $(v, v.d)$ with $v, v.d \in T$ and $d \in \{\swarrow, \searrow\}$; for a node v, $v.\swarrow$ is its *left-child* and $v.\searrow$ is its *right-child*. We also denote with $v.\uparrow$ the *parent* of v, and with $D = \{\uparrow, \swarrow, \searrow\}$ the set of directions. For a finite alphabet Υ, a Υ-*labeled tree* is a pair (T, λ) where T is a tree, and $\lambda : T \to \Upsilon$ is a labeling map.

Path-trees. For a tree T, a T-*path* π is any sequence $\pi = v_1, v_2, \ldots, v_\ell$ of T nodes s.t. (1) v_1 is the root of T, (2) for $i \in [\ell - 1]$, v_{i+1} is $v_i.d_i$ for some $d_i \in D$ (π corresponds to a traversal of T), (3) for $i \in [\ell - 1]$, $v_\ell \neq v_i$ (the last node occurs once in π), (4) π contains at least one occurrence of each node in T, and (5) for $i \in [1, \ell - 1]$, if v_i is the first occurrence of a node $v \in T$ that has a left child, i.e., $v.\swarrow \in T$, then v_{i+1} is the first occurrence of $v.\swarrow$ in π (in the T traversal, we first visit the left child of any newly discovered node).

For the tree T_1 in Fig. 1, $\pi_1 = \varepsilon, 1, 3, 1, 4, 1, \varepsilon, 2, 5, 2, \varepsilon, 1, 3, 6$ is a T_1-path. By deleting exactly one occurrence of any node in π_1 or concatenating more occurrences, the resulting sequence would not satisfy one of the above properties.

We introduce the notion of *path-tree*, that is, a labeled tree (T, λ) that encodes a T-path

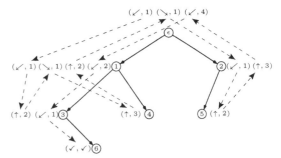

Fig. 1. A sample path-tree T_1

in its labels as follows. Denote $dir_{\swarrow}^+ = dir^+ \cup \{(\checkmark, \checkmark)\}$ where $dir = D \times \mathbb{N}$ and $\checkmark \notin D \cup \mathbb{N}$. Except for one node that is labeled with $(\checkmark, \checkmark)$, each other node has a label in dir^+. The labeling is such that by starting from the first pair of the root, we can build a chain of pairs ending at $(\checkmark, \checkmark)$. In such a chain, a pair (d, i) labeling a node u is followed by the i-th pair labeling $u.d$ (i.e., a child or the parent of u, depending on d). For example, a pair $(\swarrow, 2)$ at a node u denotes that the next pair in the chain is the second pair labeling its left child. The sequence of nodes visited by following such a chain is the path defined by λ in T. To ensure that the defined path is a T-path, we require some additional properties on λ which are detailed in the formal below. In Fig. 1, we give a path-tree T_1 and emphasize the chain defined by the labels of T_1 by linking the pairs with dashed arrows. The path defined by the labeling of T_1 is the path π_1 above, which is a T_1-path.

In the following, for a sequence $\rho = (d_1, i_1) \ldots (d_h, i_h) \in dir_{\checkmark}^+$, we let $|\rho| = h$ and denote with $\rho[j] = (d_j, i_j)$, for $j \in [h]$.

Definition 1. *A $dir_{\checkmark}^+$-labeled tree (T, λ) is a* path-tree *if there is exactly one node labeled with $(\checkmark, \checkmark)$ and, for every node v of T with $\lambda(v) = (d_1, i_1) \ldots (d_h, i_h)$, and for every $j \in [h]$, the following holds:*

1. *if $i_j \neq \checkmark$ then $v.d_j$ is a node of T and $i_j \leq |\lambda(v.d_j)|$ (pointed pair exists);*
2. *if $v \neq \varepsilon$ or $j > 1$, then there are exactly one node u and one index $i \leq |\lambda(u)|$ s.t. $\lambda(u)[i] = (d, j)$ and $u.d = v$ (except for the first pair labeling the root, every pair is pointed exactly from one adjacent node);*
3. *if $v = u.d$, for a node u of T and $d \in \{\swarrow, \searrow\}$, then there exists $i \leq |\lambda(u)|$ s.t. $\lambda(u)[i] = (d, 1)$ (except for the root the first pair in a label is always pointed from the parent);*
4. *if $v.\swarrow \in T$ then $\lambda(v)[1] = (\swarrow, 1)$ (the first pair in a label always points to the first pair of the left child, if any);*
5. *if $j < h$ there is a $i > i_j$ s.t. $\lambda(v.d_j)[i]$ is $(\uparrow, j + 1)$, if $d_j \in \{\swarrow, \searrow\}$, and $\lambda(v.d_j)[i]$ is $(\swarrow, j+1)$ (resp. $(\searrow, j+1)$), if $d_j = \uparrow$ and v is a left (resp. right) child (if a pair of v points to a pair β of an adjacent node $u = v.d_j$, the next pair of v is pointed from a pair β' of u that follows β in the u labeling); moreover, for all $\ell \in [i_j + 1, i - 1]$, $\lambda(v.d_j)[\ell]$ does not point to a pair of v.*

Path-trees define T-paths. We define a function tp that maps each path-tree (T, λ) into a corresponding sequence of T nodes, and show that indeed $tp(T, \lambda)$ is a T-path. Let $\pi = v_1, \ldots, v_\ell$, and $d_1, \ldots, d_\ell$, and $i_1, \ldots, i_\ell$ be the maximal sequences such that (1) v_1 is the root and $\lambda(v_1)[1] = (d_1, i_1)$, and (2) for $j \in [2, \ell]$, $v_j = v_{j-1}.d_{j-1}$ and $\lambda(v_j)[i_{j-1}] = (d_j, i_j)$. We define $tp(T, \lambda)$ as the sequence π. Also, we say that, in π, the occurrence v_1 *corresponds* to the first pair of the root and the occurrence v_{j+1} of a node $v \in T$ *corresponds* to the i_j-th pair of v, for $j \in [\ell - 1]$. The following lemma holds:

Lemma 1. *For each path-tree $\mathcal{T} = (T, \lambda)$, node u and $i \leq |\lambda(u)|$, the i-th occurrence of u in $tp(\mathcal{T})$ corresponds to the i-th pair of u, and $tp(\mathcal{T})$ is a T-path.*

From T-paths to path-trees. For a T-path π, we define $pt(\pi)$ as the tree whose labeling defines exactly π. We iteratively construct a sequence of labeling maps by concatenating a pair at each iteration.

Denote $dir_{\checkmark}^* = dir^* \cup \{(\checkmark, \checkmark)\}$. For a T-path $\pi = v_1, \ldots, v_\ell$ and $i \in [\ell]$, let $\lambda_i^\pi : T \to dir_{\checkmark}^*$ be the mapping defined as follows:

- $\lambda_1^\pi(v_1) = (d_1, 1)$, $v_2 = v_1.d_1$ and $\lambda_1^\pi(v) = \varepsilon$ for every $v \in T \setminus \{v_1\}$;
- for $i \in [2, \ell - 1]$, $\lambda_i^\pi(v_i) = \lambda_{i-1}^\pi(v_i).(d_i, j+1)$ where $j = |\lambda_{i-1}^\pi(v_{i+1})|$, v_{i+1} is $v_i.d_i$ and for every $v \in T \setminus \{v_i\}$, $\lambda_i^\pi(v) = \lambda_{i-1}^\pi(v)$;
- $\lambda_\ell^\pi(v_\ell) = (\checkmark, \checkmark)$, and $\lambda_\ell^\pi(v) = \lambda_{\ell-1}^\pi(v)$ for every $v \in T \setminus \{v_\ell\}$.

Define $pt(\pi)$ as (T, λ_ℓ^π). From the definitions we get:

Lemma 2. *For any T-path π and path-tree Z, $tp(pt(\pi)) = \pi$ and $pt(tp(Z)) = Z$.*

Visibly path-trees. Let $\mathcal{T} = (T, (\lambda_{dir}, \lambda_\Sigma))$ be such that (T, λ_{dir}) is a path-tree and λ_Σ maps each node of T to a symbol from $\widetilde{\Sigma}_n$. With $word_\mathcal{T}$ we denote the word $\lambda_\Sigma(v_1)\ldots\lambda_\Sigma(v_h)$ where $v_1 \ldots v_h$ is obtained from $tp(T, \lambda_{dir})$ by retaining only the first occurrences of each T node. Also, for a node z of T, we set $pos_\mathcal{T}(z) = i$ if $z = v_i$, that is, $pos_\mathcal{T}$ denotes the position corresponding to z within $word_\mathcal{T}$.

Intuitively, a visibly path-tree is a path-tree with an additional labeling such that the right child relation captures exactly the matching relations defined by the word corresponding to the encoded T-path. Formally, a *visibly path-tree* $\mathcal{T}$ over $\widetilde{\Sigma}_n$ is a labeled tree $(T, (\lambda_{dir}, \lambda_\Sigma))$ such that (1) (T, λ_{dir}) is a path-tree and (2) v is the right child of u if and only if $pos_\mathcal{T}(u) \sim_h pos_\mathcal{T}(v)$ in $word_\mathcal{T}$, for some $h \in [n]$ (*right-child relation corresponds to the matching relations of $word_\mathcal{T}$*).

For $k > 0$, a *visibly k-path-tree* is a visibly path-tree $\mathcal{T} = (T, (\lambda_{dir}, \lambda_\Sigma))$ such that each $\lambda_{dir}(v)$ contains at most k pairs.

Tree encoding of words. We can map each word $w = a_1 \ldots a_\ell$ over $\widetilde{\Sigma}_n$ to a visibly path-tree $wt(w) = (T, (\lambda_{dir}, \lambda_\Sigma))$ as follows. The labeled tree (T, λ_Σ) is such that $|T| = \ell$, a_1 labels the root of T and for $i \in [2, \ell]$: a_i labels the right child of the node labeled with a_j, $j < i$, if $j \sim_h i$ for some $h \in [n]$, and labels the left child of the node labeled with a_{i-1}, otherwise.

Define a path $\pi_w = v_1 \pi_2 \ldots \pi_\ell$ of T such that v_1 is the root of T and for $i \in [2, \ell]$, π_i is the ordered sequence of nodes that are visited on the shortest path in T from the node corresponding to a_{i-1} to that corresponding to a_i (first node excluded). It is simple to verify that indeed π_w is a T-path. Thus, we define λ_{dir} to encode π_w, i.e., such that $tp(T, \lambda_{dir}) = \pi_w$.

A *k-path-tree word* w over $\widetilde{\Sigma}_n$ is s.t. $wt(w)$ is a visibly k-path-tree over $\widetilde{\Sigma}_n$. Fig. 2 gives an example of

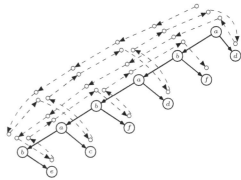

Fig. 2. The visibly path-tree $wt(w)$ for $w = (ab)^3\,cd^2\,ef^2$

the visibly 5-path-tree corresponding to the word $(ab)^3\,cd^2\,ef^2$ with call-return alphabet where a is a call and c, d are returns of stack 1, and b is a call and e, f are returns of stack 2.

In the following, we denote with $T_k(\widetilde{\Sigma}_n)$ the set of all k-path-tree words and with $PT_k(\widetilde{\Sigma}_n)$ the set of all the visibly k-path trees, over $\widetilde{\Sigma}_n$.

4   Two Base Constructions Used in Our Approach

We assume that the reader is familiar with the standard notion of nondeterministic tree automata (see [21]).

Regularity of $PT_k(\widetilde{\Sigma}_n)$. We construct a tree automaton P_k accepting $PT_k(\widetilde{\Sigma}_n)$ as the intersection of two automata P and R, where, for an input tree $\mathcal{T} = (T, (\lambda_{dir}, \lambda_\Sigma))$, P checks that (T, λ_{dir}) is indeed a path-tree and R checks that the right-child relation of $\mathcal{T}$ corresponds to the matching relations of $word_{\mathcal{T}}$.

Note that each property stated in Def. 1 is local to each node and its children. Thus, P can check them by storing in its states the label of the parent of the current node. Assuming a bound k on the number of pairs labeling each node, the size of P is thus exponential in k.

To construct R, we first construct an automaton for the negation of property (2) of the definition of visibly path-tree and then complement it.

Fix $\mathcal{T} = (T, (\lambda_{dir}, \lambda_\Sigma))$ and for any two nodes u, v of T, define $<$ s.t. $u < v$ holds iff the first occurrence of u precedes the first occurrence of v in $tp(T, \lambda_{dir})$.

We recall property (2): "a node v is the right child of u in T if and only if $pos_{\mathcal{T}}(u) \sim_h pos_{\mathcal{T}}(v)$ in $word_{\mathcal{T}}$, for some $h \in [n]$". By the definition of $\sim_h$, $h \in [n]$, the negation of property (2) holds iff either:

1. there are $u, v \in T$ s.t. v is the right child of u, $\lambda_\Sigma(u) \in \Sigma_c^h$ (call of stack h) and $\lambda_\Sigma(v) \notin \Sigma_r^h$ (not a return of stack h); or
2. there are $u, v \in T$ s.t. $u < v$, and (i) $\lambda_\Sigma(u) \in \Sigma_c^h$ and u has no right child, and (ii) $\lambda_\Sigma(v) \in \Sigma_r^h$ and v is not a right child (i.e., by the right-child relation, there are a call and a return of stack h that are both unmatched); or
3. there are $u, v \in T$ s.t. v is the right child of u, $\lambda_\Sigma(u) \in \Sigma_c^h$, and either:
   i. there is a $w \in T$ s.t. $u < w < v$ and either (a) $\lambda_\Sigma(w) \in \Sigma_c^h$ and w has no right child, or (b) $\lambda_\Sigma(w) \in \Sigma_r^h$ and w is not a right child (i.e., the right-child relation leaves unmatched either a call or a return occurring between a matched pair of the same stack h); or
   ii. there are $w, z \in T$ s.t. z is the right child of w, $\lambda_\Sigma(w) \in \Sigma_c^h$, and either $w < u < z < v$ or $u < w < v < z$ (i.e., the right-child relation restricted to stack h is not nested).

For $h \in [n]$ and assuming (T, λ_{dir}) is a path tree s.t. $|\lambda_{dir}(u)| \leq k$ for each $u \in T$, we construct an automaton B_h as the union of four automata, one for each of the above violations 1, 2, 3.*i* and 3.*ii*. Thus, B_h accepts $\mathcal{T}$ iff the right-child relation of $\mathcal{T}$ does not capture properly the matching relation $\sim_h$ of $word_{\mathcal{T}}$ (i.e., property (2) does not hold w.r.t. the matching relation $\sim_h$).

The first automaton nondeterministically guesses a node u and then accepts iff u has a right child, say v, and the labels of u and v witness violation 1. The number of states of this automaton is constant w.r.t. k and n.

In the other violations, the $<$ relation is used. We now describe an efficient construction to capture this relation by a tree automaton on k-path-trees and then conclude the discussion on the remaining violations.

Checking $u < v$. We first assume that the input tree has two marked nodes u and v. Observe that $u < v$ holds iff either (a) v is in the subtree rooted at u, or (b) there are a node w with two children and $i \leq |\lambda_{dir}(w)|$ s.t. u and v are in two different subtrees rooted at the children of w, and in $tp(\mathcal{T})$ the i-th occurrence of w occurs in between the first occurrence of u and the first occurrence of v.

Property (a) can be easily checked by a top-down tree automaton with a constant number of states. For property (b), we construct a tree automaton S that nondeterministically guesses the node w, its child w_u whose subtree contains u and its child w_v whose subtree contains v. Then, denoting $\lambda_{dir}(w) = (d_1, i_1) \ldots (d_\ell, i_\ell)$, it guesses two pairs $(d_r, i_r), (d_s, i_s)$ such that $r < s$, $w.d_r = w_u$ and $w.d_s = w_v$, with the meaning that: the first occurrence of u must be in between the r-th and the $(r+1)$-th occurrence of w, and the first occurrence of v must be in between the s-th and the $(s+1)$-th occurrence of w (if any). By Lemma 1, this is ensured by checking that u is visited on its first pair by starting from the i_r-th pair of w and before reaching the i_{r+1}-th pair of w, and similarly v w.r.t. the i_s-th and i_{s+1}-th pairs of w. The guessed i_r and the request of searching for the first occurrence of u are passed onto w_u, analogously i_s and v are passed onto w_v. Each such request is then passed along a nondeterministically guessed path in the respective subtrees, updating the indices according to the given intuition. S rejects the tree if it can visit the requested node but not on its first pair, or it reaches a leaf, and either (i) it has not guessed the node w yet or (ii) is on a selected path and the requested node was not found. In all the other cases it accepts.

Overall, we can construct S with an initial state (that is used also to store that w has not being guessed yet), an acceptance state, a rejection state and states of the form (i, x) where $i \in [1, k]$ and $x \in \{u, v\}$ (storing the request after w is guessed). Thus, in total, it has $3 + 2k$ states.

$<$ relations over many nodes. To check Boolean combinations of constraints of the form $u < v$, we can of course use the standard construction with unions and intersections of copies of S, that will yield an automaton with polynomially many states. A more efficient construction that is linear in k can be obtained by generalizing the approach used for S to capture the wished relation directly.

Handling the remaining violations. By using an automaton as above to check a proper ordering of the guessed nodes, we can design the tree automata for the rest of the violations quite easily. For example, consider the violation 3.*ii*. Denote with S' the automaton that checks $w < u < z < v$ and with S'' the automaton that checks $u < w < v < z$. Assuming that u, v, w, z are marked in the input tree, the properties v is the right child of u, z is the right child of w, and u, w are labeled with calls of stack h are local to the nodes u, w and their right children, thus can be easily checked by a tree automaton M with a constant number of states. Thus, we construct a tree automaton, that captures the intersection of M with the union of S' and S''. This automaton, by a direct construction of the automaton equivalent to the union of of S' and S'' as observed above, can be built with a number of states linear in k. From it, the automaton V_{3ii} checking for violation 3.*ii* can be obtained by removing the assumption on the marking of u, v, w, z as in the usual projection construction. Thus it can be constructed with a number of states that is linear in k.

Similarly for the other violations we get corresponding tree automata with $O(k)$ states, and thus also B_h also has $O(k)$ states.

For each tree $\mathcal{T} \in L(P)$ that is not accepted by B_h, we get that its right-child relation does not violate the $\sim_h$ relation. Thus, denoting with $\bar{B}_h$ the automaton obtained by complementing B_h, if we take the intersection of all $\bar{B}_h$ for $h \in [n]$, we get an automaton checking property (2) of the definition of visibly k-path-tree provided that the input tree $\mathcal{T} \in L(P)$, i.e., (T, λ_{dir}) is a path-tree. Since complementation causes an exponential blow-up in the number of states, the size of each $\bar{B}_h$ is $2^{O(k)}$, and the automaton resulting from their intersection has size $2^{O(n\,k)}$. Therefore, we get:

Theorem 1. *For $k \in \mathbb{N}$, there is an effectively constructible tree automaton accepting $PT_k(\widetilde{\Sigma}_n)$ of size exponential in n and k.*

Tree automaton for an MVPA. By assuming $\mathcal{T} \in PT_k(\widetilde{\Sigma}_n)$, we can construct a tree automaton $\mathcal{A}_k$ that captures the runs of A over $word_{\mathcal{T}}$.

Assume that our tree automaton can read the input tree $\mathcal{T}$ by moving along the path $tp(\mathcal{T})$ (not just top-down but as a tree-walking automaton that moves by following the encoded path). Also assume that each node labeled with a call is also labeled with a stack symbol (that is used to match push and pop transitions). We can then simulate any run of A by mimicking its moves at each node u when it is visited for the first time (from Lemma 1 this happens when a node is visited on its first pair). To construct a corresponding top-down tree automaton we use the fact that on each run of the above automaton we cross a node at most k times and according to the directions annotated in its labels. Thus we can use as states ordered tuples of at most k states of A, and design the transitions as in the standard construction from two-way to one way finite automata, moving top-down in the tree and matching the state of a node with the states of its children according to the directions in the labels. When such a matching is not possible, the automaton halts rejecting the input. On the only node labeled with $(\checkmark, \checkmark)$, it accepts iff A accepts. The tree automaton $\mathcal{A}_k$ is then obtained from this automaton by projecting out the stack symbols from the input, and therefore, its size is $O(|A|^k)$.

Lemma 3. *For an* MVPA *A and a k-path-tree $\mathcal{T}$, $\mathcal{A}_k$ accepts $\mathcal{T}$ iff $word_{\mathcal{T}} \in L(A)$. The size of $\mathcal{A}_k$ is $O(|A|^k)$.*

5   A General Approach for Complement and Emptiness

We first introduce two properties for classes of MVPL languages by using the notion of k-path-trees. Given an MVPL class $\mathcal{L}$, the first property requires that there is a k s.t. each word in a language of $\mathcal{L}$ can be encoded as a visibly k-path-tree. We show that each class that has such a property is closed under complement. The second property requires in addition that the language of all

k-path-trees corresponding to words in the languages of $\mathcal{L}$ is regular. We show that for each such class the emptiness problem is decidable.

MVPL classes and properties. We consider here a generic notation $\mathcal{X}_{\bar{p}}(\widetilde{\Sigma}_n)$ for sets of words over a call-return alphabet $\widetilde{\Sigma}_n$, each set being characterized by a particular restriction, captured by the symbol $\mathcal{X}$, and parameterized over a possibly empty tuple of integer-valued parameters $\bar{p}$. For example, we have already defined the set $T_k(\widetilde{\Sigma}_n)$ of the k-path-tree words. Another example is $C_b(\widetilde{\Sigma}_n)$ that denotes the set of all words $w \in \Sigma^*$ that can be split into $w_1 \ldots w_b$, where w_i contains calls and returns of at most one stack, for $i \in [b]$ (*bounded context-switching* [20]).

We denote with $\mathcal{X}$MVPL the class of all the languages $L \subseteq \Sigma^*$ such that there exist an n-stack call-return labeling $lab_{\Sigma,n}$, a valuation of the parameters $\bar{p}$, and an MVPA A over $\widetilde{\Sigma}_n = (\Sigma, lab_{\Sigma,n})$ for which $L = L(A) \cap \mathcal{X}_{\bar{p}}(\widetilde{\Sigma}_n)$. For example, TMVPL denotes such a class for $\mathcal{X}_{\bar{p}}(\widetilde{\Sigma}_n) = T_k(\widetilde{\Sigma}_n)$.

A class $\mathcal{X}$MVPL is *PT-covered* if for each $n > 0$ and $\bar{p}$, there exists a $k > 0$ such that $\mathcal{X}_{\bar{p}}(\widetilde{\Sigma}_n) \subseteq T_k(\widetilde{\Sigma}_n)$. We refer to such k as the *PT-parameter*. A class $\mathcal{X}$MVPL is *PT-definable* if it is PT-covered and there is an automaton $\mathcal{A}_{\mathcal{X}_{\bar{p}}(\widetilde{\Sigma}_n)}$ that accepts the set of all trees $wt(w)$ s.t. $w \in \mathcal{X}_{\bar{p}}(\widetilde{\Sigma}_n)$. Clearly, TMVPL is PT-definable.

Deterministic MVPA's do not capture all TMVPL. Let L_1 be the language $\{(ab)^i c^{i-j} d^j e^{i-j} f^j \mid i \geq j > 0\}$ and assume the call-return alphabet as in the example of Fig. 2. An MVPA A accepting L_1 just needs to guess the value of j (by nondeterministically switching to a different symbol to push onto both stacks) after reading a prefix $(ab)^j$ and then check exact matching with the returns. Also, notice that for each $w \in L_1$, w is also 5-path-tree (see Fig. 2), and since L_1 is inherently nondeterministic for MVPAs [13], we get:

Lemma 4. *The class of MVPA's that captures TMVPL is not determinizable.*

Complement. Consider a $\mathcal{X}$MVPL language L over a call-return alphabet $\widetilde{\Sigma}_n$. The *complement* of L in $\mathcal{X}$MVPL is $\bar{L} = \mathcal{X}_{\bar{p}}(\widetilde{\Sigma}_n) \setminus L$.

Despite of Lemma 4, we show that TMVPL, and in general any PT-covered MVPL class, are all closed under complement. For this, consider an MVPA A, and denote with P_k the tree automaton accepting $PT_k(\widetilde{\Sigma}_n)$ and $\mathcal{A}_k$, as in Section 4. We can complement $\mathcal{A}_k$ and then take the intersection with P_k, thus capturing all the trees $\mathcal{T} \in PT_k(\widetilde{\Sigma}_n)$ s.t. $word_{\mathcal{T}}$ is not accepted by A. The size of the resulting tree automaton $\bar{\mathcal{B}}_k$ is exponential in $|A|$ and doubly exponential in k.

The following lemma concludes the proof.

Lemma 5. *For any tree automaton $\mathcal{H}$ with $\mathcal{L}(\mathcal{H}) \subseteq PT_k(\widetilde{\Sigma}_n)$, there is an effectively constructible MVPA H over $\widetilde{\Sigma}_n$ such that $L(H)$ is the set of all the k-path-tree words $word_{\mathcal{T}}$ such that $\mathcal{T} \in \mathcal{L}(\mathcal{H})$. Moreover, the size of H is polynomial in the size of $\mathcal{H}$ and exponential in n.*

Intuitively, from $\mathcal{H}$, we can construct an MVPA H that mimics $\mathcal{H}$ transitions as follows: on internal symbols, H moves exactly as $\mathcal{H}$ (there is no right child);

on call symbols, H enters the state that $\mathcal{H}$ would enter on the left child and pushes onto a stack the one that $\mathcal{H}$ would enter on the right child; on return symbols, H acts as if the current state is the one popped from the stack. The correctness of this construction relies on the fact that for each tree $\mathcal{T} \in PT_k(\widetilde{\Sigma}_n)$, the successor position in $word_\mathcal{T}$ corresponds to the left child in $\mathcal{T}$, if any, or else, to a uniquely determined node (a right child) labeled with a return matching the most recent still unmatched call of the stack.

From the above lemma we can construct an MVPA $\bar{A}$ that accepts all words w such that $wt(w) \in L(\bar{\mathcal{B}}_k)$. Being the size of $\bar{A}$ polynomial in $|\bar{\mathcal{B}}_k|$ and exponential in n, we get (notice that since $\mathcal{X}_{\bar{p}}(\widetilde{\Sigma}_n) \subseteq T_k(\widetilde{\Sigma}_n)$, the words in $L(\bar{A})$ that are not in $\mathcal{X}_{\bar{p}}(\widetilde{\Sigma}_n)$ are ruled out by the intersection with this set):

Theorem 2. *Any PT-covered class $\mathcal{X}$MVPL is closed under complement. Also, for an MVPA A, there is an effectively constructible MVPA $\bar{A}$ s.t. $L(\bar{A}) \cap \mathcal{X}_{\bar{p}}(\widetilde{\Sigma}_n) = \mathcal{X}_{\bar{p}}(\widetilde{\Sigma}_n) \setminus L(A)$, and its size is exponential in $|A|$ and doubly exponential in the PT-parameter.*

As corollaries of the above theorem, we get the closure under complement of two well-known classes of MVPL: PMVPL defined by the sets $P_d(\widetilde{\Sigma}_n)$ of all words with a number of *phases* bounded by d [12], and OMVPL defined by the sets $O(\widetilde{\Sigma}_n)$ of all words for which when a pop transition happens it is always from the least indexed non-empty stack [7]. In fact, by the tree-decompositions given in [19] we get that OMVPL is PT-covered with $k = (n+1)2^{n-1} + 1$, where n is the number of stacks, and PMVPL is PT-covered for $k = 2^d + 2^{d-1} + 1$, where d is the bound on the number of phases.

Corollary 1. OMVPL *(resp. PMVPL) is closed under complement. Moreover, for an MVPA A, there is an effectively constructible MVPA $\bar{A}$ s.t. $L(\bar{A}) \cap O(\widetilde{\Sigma}_n) = O(\widetilde{\Sigma}_n) \setminus L(A)$ (resp. $L(\bar{A}) \cap P_d(\widetilde{\Sigma}_n) = P_d(\widetilde{\Sigma}_n) \setminus L(A)$), and its size is exponential in the size of A and triply exponential in n (resp. d, where d is the bound on the number of phases).*

Emtpiness. For PT-definable classes $\mathcal{X}$MVPL, we reduce the emptiness problem to the emptiness for tree automata by constructing a tree automaton given as intersection of $\mathcal{A}_{\mathcal{X}_{\bar{p}}(\widetilde{\Sigma}_n)}$, P_k and $\mathcal{A}_k$, where k is the PT-parameter.

Theorem 3. *The emptiness problem for any PT-definable class $\mathcal{X}$MVPL is decidable in $|\mathcal{A}_{\mathcal{X}_{\bar{p}}(\widetilde{\Sigma}_n)}| \, |A|^k \, 2^{O(nk)}$ time, where A is the starting MVPA and n is the number of stacks.*

The size of $\mathcal{A}_{\mathcal{X}_{\bar{p}}(\widetilde{\Sigma}_n)}$ is bounded by the size of P_k in both OMVPL and PMVPL (we can construct such automata from simple MSO formulas capturing the restrictions and using the linear successor relation, see [12, 19]), thus we get:

Corollary 2. *The emptiness problem for PMVPL (resp. OMVPL) is decidable in $2^{(n+\log|A|)2^{O(d)}}$ (resp. $|A|^{2^{O(n \log n)}}$) time, where A is the starting MVPA, n is the number of stacks and d is the bound on the number of phases.*

References

1. Alur, R., Madhusudan, P.: Visibly pushdown languages. In: STOC, pp. 202–211. ACM (2004)
2. Atig, M.F.: Model-checking of ordered multi-pushdown automata. Logical Methods in Computer Science 8(3) (2012)
3. Atig, M.F., Bollig, B., Habermehl, P.: Emptiness of multi-pushdown automata is 2Etime-complete. In: Ito, M., Toyama, M. (eds.) DLT 2008. LNCS, vol. 5257, pp. 121–133. Springer, Heidelberg (2008)
4. Atig, M.F., Bouajjani, A., Narayan Kumar, K., Saivasan, P.: Linear-time model-checking for multithreaded programs under scope-bounding. In: Chakraborty, S., Mukund, M. (eds.) ATVA 2012. LNCS, vol. 7561, pp. 152–166. Springer, Heidelberg (2012)
5. Bansal, K., Demri, S.: Model-checking bounded multi-pushdown systems. In: Bulatov, A.A., Shur, A.M. (eds.) CSR 2013. LNCS, vol. 7913, pp. 405–417. Springer, Heidelberg (2013)
6. Bollig, B., Kuske, D., Mennicke, R.: The complexity of model checking multi-stack systems. In: LICS, pp. 163–172. IEEE Computer Society (2013)
7. Breveglieri, L., Cherubini, A., Citrini, C., Crespi-Reghizzi, S.: Multi-push-down languages and grammars. Int. J. Found. Comput. Sci. 7(3), 253–292 (1996)
8. Carotenuto, D., Murano, A., Peron, A.: 2-visibly pushdown automata. In: Harju, T., Karhumäki, J., Lepistö, A. (eds.) DLT 2007. LNCS, vol. 4588, pp. 132–144. Springer, Heidelberg (2007)
9. Cyriac, A., Gastin, P., Kumar, K.N.: MSO decidability of multi-pushdown systems via split-width. In: Koutny, M., Ulidowski, I. (eds.) CONCUR 2012. LNCS, vol. 7454, pp. 547–561. Springer, Heidelberg (2012)
10. Esparza, J., Ganty, P., Majumdar, R.: A perfect model for bounded verification. In: LICS, pp. 285–294. IEEE (2012)
11. La Torre, S., Madhusudan, P., Parlato, G.: An infinite automaton characterization of double exponential time. In: Kaminski, M., Martini, S. (eds.) CSL 2008. LNCS, vol. 5213, pp. 33–48. Springer, Heidelberg (2008)
12. La Torre, S., Madhusudan, P., Parlato, G.: A robust class of context-sensitive languages. In: LICS, pp. 161–170. IEEE Computer Society (2007)
13. La Torre, S., Madhusudan, P., Parlato, G.: The language theory of bounded context-switching. In: López-Ortiz, A. (ed.) LATIN 2010. LNCS, vol. 6034, pp. 96–107. Springer, Heidelberg (2010)
14. La Torre, S., Napoli, M., Parlato, G.: Scope-Bounded Pushdown Languages. In: DLT. LNCS. Springer (2014)
15. La Torre, S., Napoli, M.: Reachability of multistack pushdown systems with scope-bounded matching relations. In: Katoen, J.-P., König, B. (eds.) CONCUR 2011. LNCS, vol. 6901, pp. 203–218. Springer, Heidelberg (2011)
16. La Torre, S., Napoli, M.: A temporal logic for multi-threaded programs. In: Baeten, J.C.M., Ball, T., de Boer, F.S. (eds.) TCS 2012. LNCS, vol. 7604, pp. 225–239. Springer, Heidelberg (2012)
17. La Torre, S., Parlato, G.: Scope-bounded multistack pushdown systems: Fixed-point, sequentialization, and tree-width. In: FSTTCS. LIPIcs, vol. 18, pp. 173–184. Schloss Dagstuhl - Leibniz-Zentrum fuer Informatik (2012)
18. Lal, A., Touili, T., Kidd, N., Reps, T.: Interprocedural analysis of concurrent programs under a context bound. In: Ramakrishnan, C.R., Rehof, J. (eds.) TACAS 2008. LNCS, vol. 4963, pp. 282–298. Springer, Heidelberg (2008)

19. Madhusudan, P., Parlato, G.: The tree width of auxiliary storage. In: POPL, pp. 283–294. ACM (2011)
20. Qadeer, S., Rehof, J.: Context-bounded model checking of concurrent software. In: Halbwachs, N., Zuck, L.D. (eds.) TACAS 2005. LNCS, vol. 3440, pp. 93–107. Springer, Heidelberg (2005)
21. Thomas, W.: Languages, automata, and logic. In: Handbook of Formal Languages, vol. 3, pp. 389–455. Springer (1997)

Definability and Transformations for Cost Logics and Automatic Structures

Martin Lang[1,*], Christof Löding[1], and Amaldev Manuel[2,**]

[1] RWTH Aachen University, Lehrstuhl für Informatik 7, D-52056 Aachen, Germany
[2] LIAFA, CNRS & Université Paris Diderot – Paris 7, France

Abstract. We provide new characterizations of the class of regular cost functions (Colcombet 2009) in terms of first-order logic. This extends a classical result stating that each regular language can be defined by a first-order formula over the infinite tree of finite words with a predicate testing words for equal length. Furthermore, we study interpretations for cost logics and use them to provide different characterizations of the class of resource automatic structures, a quantitative version of automatic structures. In particular, we identify a complete resource automatic structure for first-order interpretations.

1 Introduction

The theory of regular cost functions [1] has emerged in recent years as a general theory for extensions of automata and logics that have been studied in the context of boundedness problems. In these problems, the exact values of the functions are not of specific interest but rather whether the function is bounded on specific subsets of the domain. For this reason, two cost functions are considered to be equivalent if they are bounded on the same subsets of the domain. It turns out that this coarser view renders decision problems for some classes of automaton-definable cost functions decidable. The central automaton model in this setting is the one of B-automata, which associate a value to the input words using counters that can be incremented or reset by the transitions (the execution of the transitions, however, does not depend on the counter values).

Together with the development of regular cost functions, several logical formalisms appeared. The logics introduced in this area extend normal first- or monadic second-order logics by special quantitative operators. In this paper, we are concerned with two such logics, namely cost MSO (CMSO) and cost FO (CFO). In CMSO (cf. [1]) atomic formulas $|X| \leq N$ for a set variable X and a free variable N can be used. The value of a formula is the least value for N such that the formula becomes true (and infinity otherwise). In CFO (cf. [2]) a quantifier of the form $\forall^{\leq N} x \varphi(x)$ can be used, which states that $\varphi(x)$ is true

* Supported by DFG research grant *Automatentheoretische Verifikationsprobleme mit Ressourcenschranken.*
** Supported by funding from European Union's Seventh Framework Programme (FP7/2007-2013) under grant agreement №259454.

E. Csuhaj-Varjú et al. (Eds.): MFCS 2014, Part I, LNCS 8634, pp. 390–401, 2014.
© Springer-Verlag Berlin Heidelberg 2014

for almost all elements with at most N exceptions. As for CMSO, the value of the formula is the least value for N such that the formula is true. In order to ensure monotonicity, the newly introduced operators are only allowed to appear positively in a formula.

In the classical setting of languages, it is known that the regular languages are precisely those that are definable in MSO over word structures. This correspondence extends to B-automata and CMSO (cf. [1]). The FO definable languages correspond to the strict subclass of counter-free or aperiodic regular languages. There is also an analogue of this theorem in the cost setting, which is formulated using the temporal logic CLTL instead of CFO (cf. [3]) (the connection between CLTL and CFO was made in [2]).

However, there is a different way of looking at FO-definability of languages, as taken in [4]. Instead of considering the words as structures (with the word positions as elements), one can consider the set of all words as the domain of a structure. On such a structure one can define languages (subsets of the domain) by using formulas with one free element variable. If one equips the structure with successor relations for appending a letter to a word, the prefix relation, and a predicate for testing whether two words have the same length, then it turns out that the FO-definable languages are precisely the regular ones. Over a binary alphabet we refer to this structure as $\mathcal{T}_2^{el}$. More generally, one can show that $\mathcal{T}_2^{el}$ is complete for the class of automatic structures in the sense that all automatic structures can be defined (or interpreted) in $\mathcal{T}_2^{el}$ by FO formulas. An automatic structure is a structure whose domain and relations can be defined by finite automata (for accepting relations the automata read all the input words synchronously in parallel). See [5,6] for a more detailed introduction.

In this paper we mainly study this notion of definability and the class of automatic structures in the quantitative setting of regular cost functions. A notion of automatic structures with costs has already been introduced in [7]. There, the cost is not coming from specific operators in the logic, but is part of the structure: a tuple of elements is not simply in relation or not, but a value is associated to the tuple, which could be interpreted as the cost of being in relation (where the value infinity means that the tuple is not in the relation at all). This is achieved by using B-automata instead of classical automata in the definition of automatic structures. In [7] these costs have been considered as a model for the consumption of resources, and therefore these structures are called resource automatic structures and FO is referred to as FO with resource relations (FO+RR). As first main contribution, we define a complete resource automatic structure $c\mathcal{T}_2^{el}$ as extension of $\mathcal{T}_2^{el}$ and show that basically FO over $c\mathcal{T}_2^{el}$ has the same expressive power as CFO over $\mathcal{T}_2^{el}$.

Another way of obtaining a complete automatic structure is to consider finite sets of natural numbers as the elements of the structure (represented by finite words using the characteristic vector of the set) with the standard order of natural numbers on singleton sets, and the subset relation between sets. This structure can easily be defined in weak MSO over the structure $(\mathbb{N}, \mathrm{Succ})$ of the natural numbers with successor relation (in weak monadic second-order logic, set quantification only ranges over finite sets). In this transformation, we proceed from the structure $(\mathbb{N}, \mathsf{Succ})$ to its weak powerset structure (restricted to finite sets). By definition, WMSO on the original structure corresponds to FO on the powerset structure. This connection has already been observed in [8] and has been studied in more detail in [9]. Our second main contribution is a corresponding result for CWMSO and CFO. Furthermore, we extend the weak powerset structure by a size predicate for sets, show that this yields a complete resource automatic structure and establish the correspondence between CWMSO formulas and FO+RR formulas on this extended weak powerset structure.

The remainder of this paper is structured as follows. In Section 2 we give basic definitions and results. In Section 3 we characterize the class of regular cost functions in terms of CFO and FO+RR, thereby also relating these two logics over the class of resource automatic structures. In Section 4 we show how to obtain the class of resource automatic structures using CWMSO and establish a connection between CWMSO and the first-order logics CFO and FO+RR on the powerset structure. Furthermore, we would like to thank the anonymous reviewers for their constructive comments that helped us to improve this work.

2   Preliminaries

We start with providing a formal basis for the concepts mentioned before. A *cost function* is a function of the form $f : A \to \mathbb{N} \cup \{\infty\}$ that maps elements of its domain to natural numbers or infinity. In order to define the equivalence relation between cost functions, we first introduce the notion of a *correction function*. A correction function $\alpha : \mathbb{N} \cup \{\infty\} \to \mathbb{N} \cup \{\infty\}$ is a monotone mapping that maps ∞ and only ∞ to ∞. Let $f, g : A \to \mathbb{N} \cup \{\infty\}$ be two cost functions. We say f is α-dominated by g and write $f \preceq_\alpha g$ if for all $a \in A : f(a) \le \alpha(g(a))$. We call f and g α-equivalent and write $f \approx_\alpha g$ if $f \preceq_\alpha g$ and $g \preceq_\alpha f$. Additionally, we may also drop the annotation α to indicate that there exists an α such that the relation holds. Then, we have $f \approx g$ iff they are bounded on the same subsets of the domain. A proof of this fact can be found in [1]. In this work, we often have the case $\alpha(n) = 2^n$. We use $\approx_{\exp}$ as a shorthand notation for $\approx_\alpha$ with this α in order to provide explicit bounds without too much notational overhead.

The basic model we consider to define *regular* cost functions are B-automata. They can be seen as NFAs extended by a finite set of non-negative integer counters. These counters support three kinds of operations[1]. First, the counter can be incremented ($\mathtt{i}$). Second, the counter can be reset to zero ($\mathtt{r}$) and lastly the counter can be left unchanged (ε). Each transition assigns one of these operations to every counter. So, formally, we have:

Definition 1. *A B-automaton is a tuple* $\mathfrak{A} = (Q, \Sigma, \mathsf{In}, \Delta, \mathsf{Fin}, \Gamma)$. *Where the components* $Q, \Sigma, \mathsf{In}, \mathsf{Fin}$ *are defined as for normal NFAs. The component* Γ *is a finite set of counters and* Δ *is a subset of* $Q \times \Sigma \times Q \times \{\mathtt{i}, \mathtt{r}, \varepsilon\}^\Gamma$.

---

[1] Please note that we do not consider the *check* operation originally introduced for B-automata in order to simplify the notation. This does not change their expressive power.

A run of a B-automaton is defined as usual as a sequence of (connected) transitions. The counter operations are executed along a run. This way, we associate the value of a run with the maximal occurring counter value. In total, the B-automaton induces a function that maps every word w to the inf of the values of all accepting runs on w. We write $[\![\mathfrak{A}]\!](w)$ to refer to this value. We also consider B-automata that read tuples of words synchronously, and thus defining a cost function over tuples of words. We call these automata (synchronous) B-transducers. Formally, they can be seen as standard B-automata whose alphabet consists of tuples of letters, using a padding symbol to extend all words to the same length (see, e.g., [7] for a detailed definition).

The logics C(W)MSO and CFO extend usual (W)MSO and FO logic by special quantitative operators (with the abbreviations as used in the introduction). These quantitative operators are only allowed to appear positively in formulas (within an even number of negations). A C(W)MSO formula is built by the normal MSO quantifiers and connectives. It additionally may have the atomic operation $|X| \leq N$ for set variables X (and a special symbol N later interpreted as some natural number). For a C(W)MSO formula φ we write $\mathfrak{S}, n \models \varphi$ if the structure $\mathfrak{S}$ satisfies φ as normal (W)MSO formula when we replace all the $|X| \leq N$ by a formula which checks that X is at most of size n. Moreover, we write $[\![\varphi]\!]^{\mathfrak{S}}$ to indicate the infimum over $n \in \mathbb{N}$ such that $\mathfrak{S}, n \models \varphi$. For example, the formula $\exists X(|X| \leq N \wedge \forall x(x \in X))$ counts the number of elements in a finite structure and evaluates to infinity on infinite structures. A CFO formula is built by normal FO quantifiers and connectives. It additionally may have the new quantifier $\forall^{\leq N} x \psi$. Similar to C(W)MSO, we write $\mathfrak{S}, n \models \varphi$ if $\mathfrak{S}$ satisfies φ as normal FO formula when we replace all occurrences of $\forall^{\leq N} x \psi$ by "ψ is true for all x with at most n exceptions". The notation $[\![\varphi]\!]^{\mathfrak{S}}$ is used accordingly. For example, the formula $\forall^{\leq N} x(x \neq x)$ counts the number of elements in a structure as the above CMSO formula. We remark that the restriction to appear positively ensures the monotonicity of the model relation and thereby that the quantitative semantics is well-defined. As a convention, we use upper case letters for set variables and lower case variables for element variables.

The two structures that are of main interest to us are the infinite binary tree with equal level predicate ($\mathcal{T}_2^{\mathsf{el}}$) and the natural numbers with successor predicate ($\mathbb{N}^{+1}$). We formally define $\mathcal{T}_2^{\mathsf{el}} = (\{0,1\}^*, \preceq, S_0, S_1, \mathsf{el})$. This follows the idea to identify a node with the word that describes the path leading from the root to it. The letter 0 indicates that the path in the tree branches to the left and the letter 1 indicates a branch to the right, respectively. With this view on the universe, the relation $\preceq$ is the prefix relation on words, the relations S_0 and S_1 are appending one letter (0 or 1, respectively) and el is the equal length predicate for a pair of words. The structure $\mathbb{N}^{+1}$ is defined by $\mathbb{N}^{+1} = (\mathbb{N}, \mathsf{Succ})$ where $\mathsf{Succ}(x, y)$ holds iff $y = x + 1$.

In addition to the cost logics, we also consider the logic First-Order+Resource Relations (for short FO+RR), which is evaluated over quantitative structures. A resource- or cost-structure is similar to normal relational structures but the relations have a quantitative valuation.

Definition 2 ([7]). *A resource structure* $\mathfrak{S} = (S, R_1, \ldots, R_n)$ *is a tuple consisting of a universe* S *and relation symbols* R_1 *up to* R_n. *The relation symbols are valuated by functions* $R_i^{\mathfrak{S}} : S^k \to \mathbb{N} \cup \{\infty\}$ *where* k *is the arity of the relation* R_i.

The syntax of the logic FO+RR is normal first-order logic without negation. The semantics of the relations is given by the resource structure, the semantics of complete formulas is given inductively by:

$$[\![R_i x_1 \ldots x_{k_i}]\!]^{\mathfrak{S}} := R_i^{\mathfrak{S}}(x_1, \ldots, x_{k_i})$$

$$[\![x = y]\!]^{\mathfrak{S}} := \begin{cases} 0 & \text{if } x = y \\ \infty & \text{otherwise} \end{cases} \qquad [\![x \neq y]\!]^{\mathfrak{S}} := \begin{cases} \infty & \text{if } x = y \\ 0 & \text{otherwise} \end{cases}$$

$$[\![\varphi \wedge \psi]\!]^{\mathfrak{S}} := \max([\![\varphi]\!]^{\mathfrak{S}}, [\![\psi]\!]^{\mathfrak{S}}) \qquad [\![\varphi \vee \psi]\!]^{\mathfrak{S}} := \min([\![\varphi]\!]^{\mathfrak{S}}, [\![\psi]\!]^{\mathfrak{S}})$$

$$[\![\exists x \varphi(x)]\!]^{\mathfrak{S}} := \inf_{s \in S} [\![\varphi(s)]\!]^{\mathfrak{S}} \qquad [\![\forall x \varphi(x)]\!]^{\mathfrak{S}} := \sup_{s \in S} [\![\varphi(s)]\!]^{\mathfrak{S}}$$

Intuitively, the value of a formula is the amount of resources needed to make the formula true. Interpreting ∞ as false and 0 as true, one obtains the standard semantics of FO. Therefore, we can interpret classical relations also as resource relations using the characteristic function operator χ that transforms a relation R into a function that maps tuples in relation to 0 and all other tuples to ∞. Moreover, we remark that although negation is not allowed in FO+RR, we may use negation on relations that are defined as characteristic function of a normal relation because we can always add the characteristic function of the complement.

Logical interpretations are a classical tool to represent one relational structure in another (cf. [10]). The key idea is to use logical formulas with free variables to define the universe as well as the relations of some structure $\mathfrak{B}$ in another structure $\mathfrak{A}$. This provides a systematic way to transfer decidability results for the logic on $\mathfrak{A}$ to $\mathfrak{B}$. One can effectively rewrite questions about formulas on $\mathfrak{B}$ to questions about formulas on $\mathfrak{A}$ by just replacing the relations with their defining formulas and relativizing quantification to the definition of the new universe. We extend this basic concept to define resource structures in two ways: We can either start with a normal relational structure and use a quantitative logic such as CMSO to define the quantitative relations of a resource structure or we may start with a resource structure and use FO+RR to define another resource structure. Moreover, we have to provide an additional formula that describes the negation of the universe in order to be able to relativize universal formulas.

Definition 3. *Let* $\mathcal{L}$ *be a logic with quantitative semantics. A quantitative interpretation* $\mathcal{I}$ *is a tuple* $(\delta, \bar{\delta}, \varphi_1, \ldots, \varphi_k)$ *of* $\mathcal{L}$ *formulas. For an interpretation* $\mathcal{I}$ *and an appropriate structure* $\mathfrak{A} = (A, R_1, \ldots, R_n)$, *we define the resource structure* $\mathcal{I}(\mathfrak{A}) = (B, R_1, \ldots, R_k)$ *where* $B = \{a \in A \mid [\![\delta(a)]\!]^{\mathfrak{A}} = 0\}$ *and* $R_i^{\mathcal{I}(\mathfrak{A})}(\bar{x}) = [\![\varphi_i(\bar{x})]\!]^{\mathfrak{A}}$. *The formulas* δ *and* $\bar{\delta}$ *are only allowed to assume the values* 0 *and* ∞ *and have to be inverse in the sense that* $[\![\delta(x)]\!]^{\mathfrak{A}} = \infty \Leftrightarrow [\![\bar{\delta}(x)]\!]^{\mathfrak{A}} = 0$. *Moreover, we say a resource structure* $\mathfrak{S}$ *is* $\mathcal{L}$*-interpretable in* $\mathfrak{A}$ *if there is an interpretation* $\mathcal{I}$ *such that* $\mathfrak{S} \cong \mathcal{I}(\mathfrak{A})$.

In the same way as in the classical case, an algorithm to compute the value of a formula can be transferred with quantitative $\mathcal{L}$-interpretations. However, the semantics of the logical connectives $\wedge$, $\vee$ and the quantifications has to coincide in FO+RR and $\mathcal{L}$. Since this is the case for FO+RR and cost logics, we obtain:

Proposition 4. *Let $\mathfrak{A}$ be a relational or resource structure, $\mathcal{L}$ be a logic of* CMSO, CWMSO, CFO, FO+RR, *and $\mathcal{I}$ be an $\mathcal{L}$-interpretation for $\mathfrak{A}$. Then each* FO+RR *formula over $\mathcal{I}(\mathfrak{A})$ can be transformed into an equivalent $\mathcal{L}$-formula over $\mathfrak{A}$.*

Proof. As usual, we replace the occurrences of the $\mathcal{I}(\mathfrak{A})$ relations by their defining formulas given in the interpretation and relativize existential quantification by using δ and universal quantification by using $\bar{\delta}$. The resulting transformed $\mathcal{L}$ formula is then a formula over $\mathfrak{A}$. The equivalence of the semantics follows from a straight forward induction over the structure of φ (using an inductive definition for the semantics of the logic $\mathcal{L}$). Note that the domain of $\mathcal{I}(\mathfrak{A})$ is, in general, a subset of the domain of $\mathfrak{A}$. For the equivalence statement we view a cost function over the domain of $\mathcal{I}(\mathfrak{A})$ as a cost function of the domain of $\mathfrak{A}$ that maps all elements in the difference to ∞. $\square$

We remark that although the formula $\bar{\delta}$ is needed in an interpretation for technical reasons, we can omit it in most of the cases relevant to us. If $\mathcal{L}$ is one of the cost logics and the formula δ makes no use of the special quantitative operators, $\bar{\delta}$ can be obtained by taking the normal negation. This also applies to the case that $\mathcal{L}$ is FO+RR and the formula δ only uses relations that are defined using the χ operator. For the sake of simplicity, we do not define $\bar{\delta}$ explicitly in such situations.

3   Quantitative First-Order Logics

In the classical setting, it is known that FO formulas with one free variable over $\mathcal{T}_2^{\mathsf{el}}$ characterize the regular languages:

Theorem 5 ([4]). *For every regular language $L \subseteq \{0,1\}^*$ there is an FO formula $\varphi(x)$ with one free variable such that $w \in L$ iff $\mathcal{T}_2^{\mathsf{el}}, w \models \varphi$.*

We aim to show that this result extends in a very natural way to the setting of CFO and regular cost functions. Moreover, we relate CFO and FO+RR by introducing $c\mathcal{T}_2^{\mathsf{el}}$ as a variant of $\mathcal{T}_2^{\mathsf{el}}$ in form of a resource structure and show that the expressive power of CFO on $\mathcal{T}_2^{\mathsf{el}}$ equals FO+RR on $c\mathcal{T}_2^{\mathsf{el}}$ despite their apparent differences. Hence, both formalisms yield a new characterization of regular cost functions in terms of first-order like logics. We consider this to be an example for the robustness of the notion of regular cost functions.

   With the quantitative semantics of CFO in mind, each formula with one free variable defines a function from the universe of the structure to $\mathbb{N} \cup \{\infty\}$. We claim that the definable functions are exactly the regular cost functions:

Theorem 6. *For every regular cost function $f : \{0,1\}^* \to \mathbb{N} \cup \{\infty\}$, there is a CFO formula $\varphi(x)$ such that $f \approx_{\exp} \llbracket \varphi \rrbracket^{\mathcal{T}_2^{\mathsf{el}}}$. Moreover, every function $\llbracket \varphi \rrbracket^{\mathcal{T}_2^{\mathsf{el}}}$ defined by a CFO formula $\varphi(x)$ is a regular cost function.*

We do not give a direct proof of this fact here but focus on establishing the above mentioned connection to resource structures and FO+RR. Theorem 6 follows from the translation between CFO and FO+RR (Propositions 8 and 9) and the characterization of regular cost functions with FO+RR (Theorem 12). We start with formally defining $c\mathcal{T}_2^{\mathsf{el}}$. It consists of all the relations present in $\mathcal{T}_2^{\mathsf{el}}$ but now valuated with their characteristic function and one new (truly) quantitative relation $|\cdot|_1$ that counts the number of ones in a word. Formally, we define $c\mathcal{T}_2^{\mathsf{el}}$ by:

Definition 7. *Let $c\mathcal{T}_2^{\mathsf{el}} = (\{0,1\}^*, \preceq, S_0, S_1, \mathsf{el}, |\cdot|_1)$ with $|w|_1^{c\mathcal{T}_2^{\mathsf{el}}}$ counting the number of letters 1 in w and all the other relations valuated by the characteristic functions of their valuations in $\mathcal{T}_2^{\mathsf{el}}$.*

We observe that one can easily define $c\mathcal{T}_2^{\mathsf{el}}$ with CFO formulas in $\mathcal{T}_2^{\mathsf{el}}$, which means by Proposition 4 that FO+RR over $c\mathcal{T}_2^{\mathsf{el}}$ can be translated into CFO over $\mathcal{T}_2^{\mathsf{el}}$. In combination with Theorem 12 this yields one direction of Theorem 6.

Proposition 8. $c\mathcal{T}_2^{\mathsf{el}}$ *is CFO-interpretable in* $\mathcal{T}_2^{\mathsf{el}}$.

Proof. The universe of $c\mathcal{T}_2^{\mathsf{el}}$ is identical to $\mathcal{T}_2^{\mathsf{el}}$. Consequently, we can set $\delta(x) := x = x$. The relations $\preceq$, S_0, S_1, el directly define their quantitative counterparts because their quantitative semantics is just the characteristic function. So, it remains to define $|\cdot|_1$ as a CFO formula over $\mathcal{T}_2^{\mathsf{el}}$:

$$|w|_1 := \forall^{\leq N} x (\exists y (S_0 y x \vee S_1 y x) \wedge x \preceq w) \to \exists y S_0 y x$$

The idea behind this formula is that all elements x that are a predecessor of w (except the empty word) have to be 0-successors with at most N exceptions that are exactly the 1-positions in w. □

In order to provide the other direction of Theorem 6, we show how to translate CFO formulas over $\mathcal{T}_2^{\mathsf{el}}$ into equivalent FO+RR formulas over $c\mathcal{T}_2^{\mathsf{el}}$.

Proposition 9. *For every CFO formula φ, there is an FO+RR formula $\tilde{\varphi}$ such that $\llbracket \varphi \rrbracket^{\mathcal{T}_2^{\mathsf{el}}} \approx_{\exp} \llbracket \tilde{\varphi} \rrbracket^{c\mathcal{T}_2^{\mathsf{el}}}$.*

Proof (sketch). We provide an inductive translation from CFO to FO+RR. The key difficulty here arises from replacing the $\forall^{\leq N}$ quantifier with an FO+RR-expressible equivalent. We do so by approximating the set with the exception elements in form of one tree element. The path from the root to this tree element branches to the right in each level such that there are two exception elements with longest common ancestor in this level. With this idea, we can approximate the number of exceptions up to one exponential. □

Now that we established the connection between CFO and FO+RR, we show that FO+RR formulas with one free variable on $c\mathcal{T}_2^{\mathsf{el}}$ capture regular cost functions. The translation from FO+RR into B-automata can easily be done with the concept of resource automatic structures in mind. We recall the definition from [7] and show that it applies to $c\mathcal{T}_2^{\mathsf{el}}$:

Definition 10. *A resource structure* $\mathfrak{S} = (S, R_1, \ldots, R_n)$ *is called* resource automatic *if* $S \subseteq \{0,1\}^*$ *is a regular language and and there are synchronous B-transducers* $\mathfrak{T}_1, \ldots, \mathfrak{T}_n$ *such that* $R_i^{\mathfrak{S}}(\bar{x}) := [\![\mathfrak{T}_i]\!](\bar{x})$.

Proposition 11. $c\mathcal{T}_2^{\mathsf{el}}$ *is a resource automatic structure.*

Proof. It is known that $\mathcal{T}_2^{\mathsf{el}}$ is an automatic structure (cf. [6]). By interpreting the NFAs defining the classical relations as B-automata in which the counters are not used, we directly obtain automata for the characteristic functions. The relation $|\cdot|_1$ can be defined by an automaton that just counts the letters 1. □

With this result in mind we can now characterize regular cost functions in terms of FO+RR on $c\mathcal{T}_2^{\mathsf{el}}$:

Theorem 12. *For every regular cost function* $f : \{0,1\}^* \to \mathbb{N} \cup \{\infty\}$, *there is an* FO+RR *formula* $\varphi(x)$ *such that* $f = [\![\varphi]\!]^{c\mathcal{T}_2^{\mathsf{el}}}$. *Moreover, the function* $[\![\varphi]\!]^{c\mathcal{T}_2^{\mathsf{el}}}$ *defined by an* FO+RR *formula with one free variable is a regular cost function.*

Proof. We first show the second part. Let $\varphi(x)$ be a FO+RR formula with one free variable. Since $c\mathcal{T}_2^{\mathsf{el}}$ is resource automatic, the inductive automaton translation from [7] provides us with a B-automaton $\mathfrak{A}$ such that $[\![\varphi]\!]^{c\mathcal{T}_2^{\mathsf{el}}} \approx_\alpha [\![\mathfrak{A}]\!]$.

For the converse, we encode the run of a B-automaton in $c\mathcal{T}_2^{\mathsf{el}}$. Let $\mathfrak{A} = (Q, \Sigma, \mathsf{In}, \Delta, \mathsf{Fin}, \Gamma)$ be a B-automaton with $|Q| = n$ and $|\Gamma| = m$. The basic construction follows the classical approach. We simulate the behavior of $\mathfrak{A}$ on a word w with a formula $\varphi(w)$ by existentially guessing a run, verifying that it is accepting and now additionally computing its value. The state sequence is encoded along the levels in the tree in the following way: For each of the n states, we guess a position p_i on the same level as w. The path to p_i branches to the right in all levels in which the run is currently in the i-th state. Now we only have to verify that in every level up to $|w|$ exactly one of the paths branches right and that there are transitions that enable the respective state change given the letter of w in the corresponding level. We extend this with $2m$ additional positions on the level of w to describe the behavior of the counters. For each counter, there is a position c_i^{i} that branches right in every level where the transition increments the respective counter and a position c_i^{r} that branches right in every level where the transition resets the counter. We can then calculate the value of a run by selecting maximal segments of the path given by c_i^{i} that are not interrupted by a right-branch of c_i^{r}. We use the $|\cdot|_1$ relation on these segments to count the increments. □

A Complete Resource Automatic Structure

The study of *complete* structures for certain classes of logical structures provides insight into the whole class of structures by looking at a single structure.

Hence, we are interested in finding a complete structure for resource automatic structures. This not only provides a characterization of the expressive power of the formalism but also enables us to better understand the type of quantitative extension realized by resource automatic structures. First, we formally fix the notion of completeness:

Definition 13. *Let $\mathcal{C}$ be a class of resource structures. We call a structure $\mathfrak{S}$ complete for $\mathcal{C}$ if $\mathfrak{S} \in \mathcal{C}$ and for all structures $\mathfrak{A} \in \mathcal{C}$, there is an FO+RR-interpretation $\mathcal{I}$ such that $\mathfrak{A} \cong \mathcal{I}(\mathfrak{S})$.*

By Proposition 11, we already know that $c\mathcal{T}_2^{\mathsf{el}}$ is a resource automatic structure. In order to show that it is complete, we have to extend the ideas of Theorem 12 to synchronous B-transducers. Consider an arbitrary resource automatic structure $\mathfrak{S}$. The universe is represented by a regular language. As a consequence, the classical Theorem 5 provides us with a formula δ to encode the domain. It remains to find formulas that define the relations in $c\mathcal{T}_2^{\mathsf{el}}$. For this we take the synchronous B-transducer that defines the relation. Since this is essentially only a B-automaton working over a vector of the original alphabet, the same approach as in Theorem 12 can be used to obtain the formula φ that defines the relation in $c\mathcal{T}_2^{\mathsf{el}}$. Although this involves some technical difficulties such as necessary padding when working with synchronous transducers, no new ideas are required in principle. Altogether, we obtain:

Corollary 14. *The structure $c\mathcal{T}_2^{\mathsf{el}}$ is complete for resource automatic structures.*

This result concisely illustrates the quantitative extension that is provided by resource automatic structures compared to the standard model. The idea of characteristic functions provides an embedding that shows that resource automatic structures extend the classical concept. The quantitative aspect boils down to a relation that counts the number of letters 1 in the word presentation of the elements.

4   (Finite)Set Transformations

In the classical setting, it is a well-known fact that (W)MSO formulas over a structure are equivalent to FO formulas over the (weak) powerset structure (cf. [11]). We aim at providing a generalization of this fact to the area of cost logics and resource structures. First, we fix the notation used in this section. Let $\mathfrak{A} = (A, R_1, \ldots, R_k)$ be a relational structure. For a j-ary relation R, let the *set extension* of R be given by $\mathcal{P}(R) := \{(\{x_1\}, \ldots, \{x_j\}) \mid (x_1, \ldots, x_j) \in R\}$. The powerset structure of $\mathfrak{A}$ is $\mathcal{P}(\mathfrak{A}) = (\mathcal{P}(A), \mathrm{Sing}, \subseteq, \mathcal{P}(R_1), \ldots, \mathcal{P}(R_k))$ where $\subseteq$ is the normal subset relation and Sing is a unary predicate that indicates singleton sets. Additionally, we also show that there is a correspondence to a canonical resource extension of the powerset structure. The resource powerset structure of $\mathfrak{A}$ is $c\mathcal{P}(\mathfrak{A}) = (\mathcal{P}(A), \mathrm{Sing}, \mathrm{size}, \subseteq, \mathcal{P}(R_1), \ldots, \mathcal{P}(R_k))$ where size is a unary resource relation mapping a set to its size, the other relations are valuated with the characteristic functions of their valuations in the classical powerset structure. Analogously, we also consider the weak variant with only finite subsets of A in the universe. We denote the weak variants by an index w.

Proposition 15. *The following correspondence holds for all relational structures* $\mathfrak{A}$. *For every* CMSO *formula* φ, *there is a* CFO *formula* φ_1 *(respectively a* FO+RR *formula* φ_2*) and vice versa such that for all* $X_1, \ldots, X_k \subseteq A$ *and all* $x_1, \ldots, x_\ell \in A$ *it is the case that*

$$[\![\varphi(X_1, \ldots, X_k, x_1, \ldots, x_\ell)]\!]^{\mathfrak{A}} \approx_{\exp} [\![\varphi_1(X_1, \ldots, X_k, \{x_1\}, \ldots, \{x_\ell\})]\!]^{\mathcal{P}(\mathfrak{A})}$$

$$[\![\varphi(X_1, \ldots, X_k, x_1, \ldots, x_\ell)]\!]^{\mathfrak{A}} = [\![\varphi_2(X_1, \ldots, X_k, \{x_1\}, \ldots, \{x_\ell\})]\!]^{c\mathcal{P}(\mathfrak{A})}.$$

The same holds for CWMSO *and the respective weak powerset structures.*

We obtain the previous result by inductively transforming logical formulas in a way that preserves the semantics up to α. Most of the translations are relatively straightforward encodings of the missing operators. However, transforming a CFO formula over the powerset structure back into a CMSO formula over the original structure involves counting the number of "exceptions" in $\forall^{\leq N} x \varphi(x)$ formulas. Since the x are sets of the original structure, we cannot simply existentially quantify the set of exceptions and bound its size. We solve this by instead bounding the size of sets that contain only elements with a distinct membership profile w.r.t. the exception sets, i.e., a pair of element z, z' can be member in this set iff there is an exception that contains exactly one of z, z'. For these sets, we recognize that their size approximates the number of exception sets up to an exponentiation. This can be seen as a refinement of the approximation idea used in Proposition 2.1 in [12].

The translation from Proposition 15 has the following immediate consequence for the definable cost functions:

Corollary 16. *Let* f *be a cost function. The following are equivalent:*

1. f *is definable in (weak)* CMSO *over a structure* $\mathfrak{A}$.
2. f *is definable in* CFO *over (weak)* $\mathcal{P}(\mathfrak{A})$.
3. f *is definable in* FO+RR *over (weak)* $c\mathcal{P}(\mathfrak{A})$.

In order to close our study of logical formalisms for regular cost functions, we connect our results in the area of first-order logics with CWMSO on $\mathbb{N}^{+1}$. For this we recognize that $c\mathcal{T}_2^{\text{el}}$ is almost $c\mathcal{P}_w(\mathbb{N}^{+1})$. A word from $\{0, 1\}$ can be seen as a set over the natural numbers that contains the positions in the word with letter 1. Nevertheless, we additionally need WMSO-definable coding in order to distinguish tree elements with trailing zeros. Hence, we obtain that $c\mathcal{T}_2^{\text{el}}$ and $c\mathcal{P}_w(\mathbb{N}^{+1})$ are FO+RR-interpretable in each other. The connection between the two first-order logics stated in Theorem 17 below was already established in Proposition 9 and Proposition 8.

Theorem 17. *Logical formulas can be transformed in a semantics preserving way among* CFO *on* $\mathcal{T}_2^{\text{el}}$, FO+RR *on* $c\mathcal{T}_2^{\text{el}}$ *and* CWMSO *on* $\mathbb{N}^{+1}$.

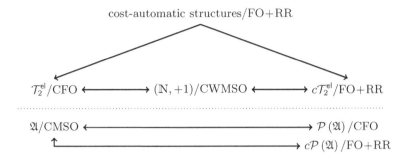

Fig. 1. Cost Logics Correspondences Overview

The interpretations to transform cT_2^{el} in $c\mathcal{P}_w\left(\mathbb{N}^{+1}\right)$ and vice versa in the proof of the previous theorem also yield the following result as immediate consequence.

Corollary 18. $c\mathcal{P}_w\left(\mathbb{N}^{+1}\right)$ *is a complete resource automatic structure.*

As an example application of Theorem 17, we settle the open question of the last section in [7]. Based on the results known at that time, it was unclear whether the full *bounded reachability problem* for pushdown systems with B-counters can be encoded in form of a CWMSO formula on T_2. The problem asks, given a pushdown system with counter operations as in B-automata, and two regular sets A, B of pushdown configurations, whether there is a uniform upper bound $k \in \mathbb{N}$ such that it is possible to reach from all elements in A some element in B with a configuration sequence whose counter values are bounded by k. In [7] it was already established that the problem can be encoded in FO+RR over the configuration graph, which is a resource automatic structure. With Corollary 14 and Theorem 17 we obtain that the problem is expressible in CWMSO over $\mathbb{N}^{+1}$. Hence, it is also expressible in CWMSO over T_2 since $\mathbb{N}^{+1}$ is CWMSO definable in T_2. However, the argument here uses general properties of resource automatic structures and does not provide the insight of the direct formulation in CWMSO over T_2 given in [7] for the special case of a single counter.

5 Conclusion

In this work we established connections among the different logics that arose around regular cost functions. Figure 1 provides an overview of the obtained results. We showed that the two quantitative first-order logics CFO and FO+RR are essentially equally expressive on the infinite binary tree with equal level predicate despite their rather different mechanisms for defining costs. Their expressive power on the tree could be summarized on an intuitive level by first-order queries plus the ability to count elements satisfying first-order properties. Furthermore, both formalisms provide another characterization for regular cost functions. The extension of this result to cost functions over tuples of words provided the insight that $c\mathcal{T}_2^{el}$ is a complete structure for the class of resource automatic structures. These results nicely extend the classical results for regular languages and automatic structures, and can be seen as another sign that the notion of regular cost functions is a good quantitative generalization of regular languages.

In the second part, we showed extensions of the classical results that allow to exchange between MSO logic and FO logic over the power set structure. These results enabled particular transformations among CFO on $\mathcal{T}_2^{el}$, FO+RR on $c\mathcal{T}_2^{el}$ and CWMSO on $\mathbb{N}^{+1}$.

We are currently working on continuing this research in two directions. First, we want to extend the ideas of resource automatic structures towards resource tree automatic structures. This could lead to new decision methods for quantitative WMSO logics over the infinite tree. Second, we try to connect our results to the world of (W)MSO with the unbounding quantifier (for short (W)MSO+U) as introduced by Mikołaj Bojańczyk in [13].

References

1. Colcombet, T.: Regular cost functions over words. Online Manuscript (2009)
2. Kuperberg, D.: Study of classes of regular cost functions. PhD thesis, LIAFA Paris (December 2012)
3. Kuperberg, D.: Linear temporal logic for regular cost functions. In: STACS. LIPIcs, vol. 9, pp. 627–636. Schloss Dagstuhl - Leibniz-Zentrum fuer Informatik (2011)
4. Eilenberg, S., Elgot, C.C., Shepherdson, J.C.: Sets recognized by n-tape automata. J. Algebra 13, 447–464 (1969)
5. Khoussainov, B., Nerode, A.: Automatic presentations of structures. In: Leivant, D. (ed.) LCC 1994. LNCS, vol. 960, pp. 367–392. Springer, Heidelberg (1995)
6. Blumensath, A., Grädel, E.: Automatic structures. In: Proceedings of the 15th Annual IEEE Symposium on Logic in Computer Science, pp. 51–62. IEEE (2000)
7. Lang, M., Löding, C.: Modeling and verification of infinite systems with resources. Logical Methods in Computer Science 9(4) (2013)
8. Elgot, C.C., Rabin, M.O.: Decidability and undecidability of extensions of second (first) order theory of (generalized) successor. Journal of Symbolic Logic 31(2), 169–181 (1966)
9. Colcombet, T., Löding, C.: Transforming structures by set interpretations. Logical Methods in Computer Science 3(2) (2007)
10. Grädel, E.: Finite Model Theory and Descriptive Complexity. In: Finite Model Theory and its Applications, pp. 125–230. Springer (2007)
11. Blumensath, A., Colcombet, T., Löding, C.: Logical theories and compatible operations. In: Logic and Automata, pp. 73–106 (2008)
12. Bárány, V., Kaiser, L., Rabinovich, A.: Expressing Cardinality Quantifiers in Monadic Second-Order Logic over Trees. Fundamenta Informaticae 100, 1–18 (2010)
13. Bojańczyk, M.: A bounding quantifier. In: Marcinkowski, J., Tarlecki, A. (eds.) CSL 2004. LNCS, vol. 3210, pp. 41–55. Springer, Heidelberg (2004)

Generalised Lyndon-Schützenberger Equations[*]

Florin Manea[1], Mike Müller[1], Dirk Nowotka[1], and Shinnosuke Seki[2,3]

[1] Christian-Albrechts-Universität zu Kiel, Institut für Informatik, Kiel, Germany
{flm,mimu,dn}@informatik.uni-kiel.de
[2] Helsinki Institute for Information Technology (HIIT), P.O. Box 15600, 00076,
Aalto, Finland
[3] Department of Information and Computer Science, Aalto University
P. O. Box 15400, FI-00076, Aalto, Finland shinnosuke.seki@aalto.fi

Abstract. We fully characterise the solutions of the generalised Lyndon-Schützenberger word equations $u_1 \cdots u_\ell = v_1 \cdots v_m w_1 \cdots w_n$, where $u_i \in \{u, \theta(u)\}$ for all $1 \leq i \leq \ell$, $v_j \in \{v, \theta(v)\}$ for all $1 \leq j \leq m$, $w_k \in \{w, \theta(w)\}$ for all $1 \leq k \leq n$, and θ is an antimorphic involution. More precisely, we show for which ℓ, m, and n such an equation has only θ-periodic solutions, i.e., $u, v, w \in \{t, \theta(t)\}^*$ for some word t, closing an open problem by Czeizler et al. (2009).

1 Introduction

A word is a *power* (or *repetition*) if it can be written as a repeated concatenation of one of its prefixes. A word w is a *pseudo-power* (more precisely, f-power) if it can be written as a repeated concatenation of one of its prefixes t and its image $f(t)$ under a morphic or antimorphic function f, thus $w \in t\{t, f(t)\}^+$. Introduced in [4], the latter notion is a natural generalisation of the former: when f is the identity morphism, pseudo-powers are, indeed, classical powers. More interestingly, when f is the reversal (identity antimorphism), pseudo-powers are repeated concatenations of a word and a reversed version of it, so, in a sense, generalised palindromic structures. To this end, the study of combinatorial properties of pseudo-powers supports the development of a generalised periodicity theory.

The initial motivation of studying pseudo-repetitions (according to [4]) came from two important biological concepts: tandem repeat, i.e., the consecutive repetition of the same sequence of nucleotides in a DNA strand, and the inverted repeat, i.e., a sequence of nucleotides whose reversed Watson-Crick complement occurred before in the strand, both occurrences encoding, essentially, the same genetic information. Noting that the Watson-Crick complement can be abstracted as an antimorphic involution on the DNA-alphabet, pseudo-powers formalise generalised tandem repeats, in which one sequence is followed by several consecutive occurrences of either its copy or its reversed complement.

The study of combinatorial properties of pseudo-powers was mostly concerned with the case when f is an involution; this case seems the most motivated, modelling both the original repetitions and palindromic structures, and in particular

[*] Supported by DFG grants 596676 (Manea), 582014 (Müller), and 590179 (Nowotka) and Academy of Finland Postdoctoral Researcher Grant 13266670/T30606 (Seki).

E. Csuhaj-Varjú et al. (Eds.): MFCS 2014, Part I, LNCS 8634, pp. 402–413, 2014.
© Springer-Verlag Berlin Heidelberg 2014

the Watson-Crick complement. In this context, generalisations of the Fine and Wilf theorem, avoidability results, and methods to solve word equations were derived ([4], [1], [2], respectively). Also, efficient methods of testing whether a word is or contains a pseudo-power were recently developed (see, e.g., [5,11]).

Naturally, the study of pseudo-powers was focused on translating classical periodicity results into this new and more general setting. Accordingly, Czeizler et al. [3] investigated a generalisation of Lyndon and Schützenberger's equations. Lyndon and Schützenberger [9] showed that in all solutions of an equation $u^l = v^m w^n$, with $l, m, n \geq 2$, in a free group, u, v, and w are necessarily powers of a common element. Their result also holds, if u, v and w are elements of a free semigroup [8]. Its extension of the form $x^k = z_1^{k_1} z_2^{k_2} \cdots z_n^{k_n}$ was studied in [6]. The generalisation by Czeizler et al. [3] is $u_1 \cdots u_\ell = v_1 \cdots v_m w_1 \cdots w_n$, where $u_i \in \{u, \theta(u)\}$ for all $1 < i \leq \ell$, $v_j \in \{v, \theta(v)\}$ for all $1 \leq j \leq m$, $w_k \in \{w, \theta(w)\}$ for all $1 \leq k \leq n$, and θ is an antimorphic involution. Following the classical case, they studied under which conditions $u, v, w \in \{t, \theta(t)\}^+$ for some word t. In other words, they studied which equations have only solutions where u, v, w are pseudo-powers, or more precisely, θ-powers of the same word; we call such solutions θ-periodic.

The results obtained on these generalised equations in [3,7,10] are summarised in Table 1. One can note from this table that the more problematic equations are those with $\ell = 3$. More precisely, equations which allow non-θ-periodic solutions and equations having only θ-periodic solutions were identified; however, no precise characterisation was obtained. Hence, this case seems to be especially intricate and interesting, as the separating border between the cases when the equation has only θ-periodic solutions and the cases when it may also have non-θ-periodic solutions is drawn here. Our work closes the gap providing a full characterisation of the generalised Lyndon-Schützenberger equations having only θ-periodic solutions. This seems to us a relevant step towards the aforementioned development of a generalised periodicity theory.

Table 1. Results on the equation $u_1 \cdots u_\ell = v_1 \cdots v_m w_1 \cdots w_n$

ℓ	m	n	$u, v, w \in \{t, \theta(t)\}^+$?	
			Answer	This paper
≥ 4	≥ 3	≥ 3	Yes [3,7]	
3	≥ 12	≥ 12	Yes [10]	
3	$5 \leq \min\{m,n\}$, m or n odd		Yes [10]	
3	$5 \leq \min\{m,n\} < 12$, m and n even		Open (A)	Yes (Thm. 1)
3	4	≥ 5 and odd	Open (B)	Yes (Thm. 2)
3	4	≥ 4 and even	No [7]	
3	3	≥ 3	No [7]	
	one of $\{\ell, m, n\}$ equals 2		No [3,7]	

Our Results. As mentioned above, we are interested in solutions of the equation

$$u_1 u_2 u_3 = v_1 \cdots v_m w_1 \cdots w_n, \tag{1}$$

where $5 \leq \min\{m, n\} < 12$ and m and n are even, or $m = 4$ and $n \geq 5$ is odd, $u_1, u_2, u_3 \in \{u, \theta(u)\}$, $v_j \in \{v, \theta(v)\}$ for all $1 \leq j \leq m$ and $w_k \in \{w, \theta(w)\}$ for all $1 \leq k \leq n$, and θ an antimorphic involution.

In [10] it is already shown that if $m, n \geq 5$ and $u_1u_2u_3 \neq uuu$ or $m|v| \geq 2|u|$ then (1) has only θ-periodic solutions. Thus, the open case addressed by our Theorem 1 regards the θ-periodicity of the solutions of (1) for $u_1 = u_2 = u_3$ and $m|v| < 2|u|$. Due to space restrictions, we can not present the complete proofs for this case in this extended abstract. In the light of the results from [10], the positive answer we give in this case was somehow expected. We give a brief overview of our approach in Sect. 3. The following theorem is obtained.

Theorem 1. *If $\ell = 3$, $5 \leq \min\{m, n\} < 12$, and both m and n even, then (1) implies that $u, v, w \in \{t, \theta(t)\}^*$ for some word t.*

For the open case addressed by our Theorem 2, where $\ell = 3, m = 4, n \geq 5$ and odd, we give here the main proofs. This case seems very interesting to us for two reasons. Firstly, the fact that (1) has only θ-periodic solutions under these restrictions seems surprising, as it shows a different behaviour from the case when the same bounds apply to ℓ, m, and n, but n is even; this shows exactly where the equations having only θ-periodic solutions are separated from the ones that may also have other solutions. Secondly, due to the small number of factors v or $\theta(v)$, it seems that a different (at least partly) approach is needed in this case. Indeed, the common approach in the proofs of the results of [3,7,10] or in those supporting Theorem 1 was to find a long enough factor of $u_1u_2u_3$ that reflects an alignment between some of the factors $v_1, \ldots, v_m$ and some of the factors $w_1, \ldots, w_n$, and then to apply periodicity results in the lines of Theorems 4 or 5 to get that u, v, and w are all θ-powers of the same word. Employing such a strategy seems more difficult when we only have few occurrences of v or $\theta(v)$ (or, alternatively, of w or $\theta(w)$). Our proofs show how this can be done. While the basic techniques we rely on are the usual ones of combinatorics on words, including detailed case analyses, we also make use of novel arguments regarding θ-periodic words and exhibit more general ways of applying known θ-periodicity arguments. We show the following.

Theorem 2. *If $\ell = 3$, $m = 4$ and $n \geq 5$ odd, then (1) implies that $u, v, w \in \{t, \theta(t)\}^*$ for some word t.*

Theorems 1 and 2 and Table 1 fully characterise the θ-periodicity of the solutions of (1).

Theorem 3. *Equation (1) implies that $u, v, w \in \{t, \theta(t)\}^*$ for some word t if and only if (1) $\ell \geq 4$ and $m, n \geq 3$, or (2) $\ell \geq 3$ and $m, n \geq 5$, or (3) $\ell = 3, m = 4$, and n is an odd number at least 5.*

2   Preliminaries

For a finite alphabet Σ, we denote by Σ^* and Σ^+ the set of all words and the set of all non-empty words over Σ, respectively. The empty word is denoted by ε and the length of a word w is denoted by $|w|$. For a word $w = uvz$ we say that u is a *prefix* of w, v is a *factor* of w, and z is a *suffix* of w. We denote that by $u \leq_p w$, $v \leq_f w$, and $z \leq_s w$, respectively. If $vz \neq \varepsilon$ we call u a *proper prefix*, and we denote that by $u <_p w$, and symmetrically for suffixes. Similarly, v is called a *proper factor* of w, denoted by $v <_f w$, if $u \neq \varepsilon, z \neq \varepsilon$. A word w is called *primitive*, if $w = u^k$ implies $k = 1$, so $u = w$; otherwise, w is called *power* or *repetition*. For a word w, we define the word w^ω as the infinite word whose prefix of length $n|w|$ is w^n, for all $n \in \mathbb{N}$.

Primitive words are characterised as follows:

Proposition 1. *If w is primitive and $ww = xwy$, then either $x = \varepsilon$ or $y = \varepsilon$.*

A function $f : \Sigma^* \to \Sigma^*$ is an *antimorphism*, if $f(uv) = f(v)f(u)$ for all $u, v \in \Sigma^*$. Furthermore, f is an *involution*, if $f(f(a)) = a$ for all $a \in \Sigma$. A function that is both and antimorphism and an involution will be called an *antimorphic involution*. Let θ be such an antimorphic involution. A word w is called θ-*primitive* if $w = u_1 \cdots u_k$ with $u_i \in \{u, \theta(u)\}$ for all $1 \leq i \leq k$ implies $k = 1$ and hence $u \in \{w, \theta(w)\}$; otherwise, w is a θ-*power*. A θ-primitive word is primitive, but the converse does not hold: for instance, the word $w = abba$ is primitive but $w = ab\theta(ab)$, for θ being the reversal. Any nonempty word w admits a *unique* θ-primitive word t such that $w \in t\{t, \theta(t)\}^*$, and the t is called the θ-*primitive root* of w. A word w is a θ-*palindrome* if $w = \theta(w)$.

Kari et al. [7] characterised θ-primitive words similarly to Proposition 1:

Lemma 1. *For a θ-primitive word $x \in \Sigma^+$, neither $x\theta(x)$ nor $\theta(x)x$ can be a proper factor of a word in $\{x, \theta(x)\}^3$.*

Furthermore, Czeizler et al. [4] showed the following:

Lemma 2. *Let $x \in \Sigma^+$ be a θ-primitive word, and $x_1, x_2, x_3, x_4 \in \{x, \theta(x)\}$. If $x_1 x_2 y = z x_3 x_4$ for some words $y, z \in \Sigma^+$ with $|y|, |z| < |x|$, then $x_2 \neq x_3$.*

Two words x, y are *conjugates*, denoted by $x \sim y$, if $xz = zy$ holds for some word $z \in \Sigma^*$. Then there exist $p, q \in \Sigma^*$ such that $x = pq$, $y = qp$, and $z = (pq)^i p$ for some $i \geq 0$. For this or the next theorem, see, e.g., [8].

Theorem 4. *If $\alpha \in u\{u, v\}^*$ and $\beta \in v\{u, v\}^*$ have a common prefix of length at least $|u| + |v| - \gcd(|u|, |v|)$, where $\gcd$ denotes the greatest common divisor, then $u, v \in \{t\}^+$ for a word t.*

Czeizler et al. [4] proved the following generalisation of Theorem 4:

Theorem 5. *Let $u, v \in \Sigma^+$ with $|u| \geq |v|$. If $\alpha \in \{u, \theta(u)\}^+$ and $\beta \in \{v, \theta(v)\}^+$ have a common prefix of length at least $\min\{2|u| + |v| - \gcd(|u|, |v|), \mathrm{lcm}(|u|, |v|)\}$, where lcm denotes the least common multiple, then $u, v \in t\{t, \theta(t)\}^+$ for some θ-primitive word $t \in \Sigma^+$.*

This theorem immediately solves (1) when $v_1 \cdots v_m$ or $w_1 \cdots w_n$ is very long.

Corollary 1. *If (1) holds with $|u_1 u_2| < |v_1 \cdots v_{m-1}|$ or $|u_2 u_3| < |w_2 \cdots w_n|$, then $u, v, w \in \{t, \theta(t)\}^+$ for some word t.*

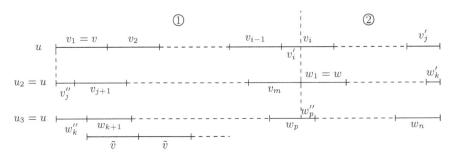

Fig. 1. Visualisation of (2), where $j = m - i + 1$

We will also make frequent use of the following results from [3]:

Proposition 2 (Prop. 20 and 21 in [3]). *Let* $u, v \in \Sigma^+$ *so that* v *is* θ-*primitive,* $u_1, u_2, u_3 \in \{u, \theta(u)\}$ *and* $v_1, \ldots, v_m \in \{v, \theta(v)\}$ *for some* $m \geq 3$. *Assume that* $v_1 \cdots v_m <_p u_1 u_2 u_3$ *and* $2|u| < m|v| < 2|u| + |v|$. *Then:*

- *If* m *is odd, then* $u_2 \neq u_1$ *and* $v_1 = \ldots = v_m = z\theta(z)p$, *where* $p = \theta(p)$.
- *If* m *is even, then one of the following holds:*
    1. $u_1 \neq u_2$ *and* $v_1 = \ldots = v_m = xz\theta(z)$, *where* $x = \theta(x)$, *or*
    2. $u_1 = u_2$, $v_1 = \ldots = v_{\frac{m}{2}}$ *and* $v_{\frac{m}{2}+1} = \ldots = v_m = \theta(v_1)$ *such that*
       $v_1^{m/2}\theta(v_1)^{m/2} = u^2 x^2$ *for some* θ-*palindrome* x.

A symmetrical result (i.e., $v_1 \ldots v_m <_s u_1 u_2 u_3$ in the hypothesis and u_1 and u_2 replaced by u_2 and u_3, respectively, in the conclusions) can be easily derived.

3   Basic Proof Strategy for Theorems 1 and 2

Partial results of Theorem 1 were given in [10] as: (1) with $m, n \geq 5$ admits only θ-periodic solutions if u_2 or u_3 is $\theta(u)$ (u_1 is fixed to u throughout the paper without loss of generality) or $|u_1 u_2| < |v_1 \cdots v_m|$ (or symmetrically, $|w_1 \cdots w_n| > |u_2 u_3|$). So, it seems more difficult to handle the case $u_1 = u_2 = u_3 = u$, that is, to solve equations of the form:

$$u^3 = v_1 \cdots v_m w_1 \cdots w_n \tag{2}$$

under the condition $|u| < |v_1 \cdots v_m| < 2|u|$. This hardness intuitively stems from the fact that (2) provides no information about $\theta(u)$ and neither $v_1 \cdots v_m$ nor $w_1 \cdots w_n$ is long enough to apply Theorem 5. In this section, we present a strategy to tackle this hard case appearing in our proofs for Theorems 1 and 2.

Note first that u is assumed to be θ-primitive throughout the paper. If not, the given equation is immediately reduced to the equivalent equation $u_1' \cdots u_\ell' = v_1 \cdots v_m w_1 \cdots w_n$ for some $\ell \geq 6$ and words $u_1', \ldots, u_\ell'$ each of which is either the θ-primitive root of u or its θ-image. The reduced equation is known to admit only θ-periodic solutions (see Table 1). Furthermore, we will always assume that $v_1 = v$ and $w_1 = w$ in this article: As θ is an involution, the other cases can be reduced to this one by a simple renaming of v to $\theta(v)$ or of w to $\theta(w)$. Finally, as already remarked in [3], if we show that two of the words u, v, and w are

θ-powers of a word t, then so is the third. Hence we will conclude whenever we established that two of u, v, w are θ-powers of a common word.

Back to the case above, due to the length condition, the border between v_m and w_1 is inside u_2 and splits u_2 into the prefix, denoted by ①, and the remaining suffix, denoted by ②, as shown in Fig. 1. There exists an index $1 \leq i \leq m$ such that ① $= v_1 \cdots v_{i-1} v_i' = v_{m-i+1}'' v_{m-i+2} \cdots v_m$ for some proper prefix v_i' of v_i and proper suffix v_{m-i+1}'' of v_{m-i+1} (for clarity, let $j = m-i+1$ below). With v_i' empty, ① $= v_1 \cdots v_{i-1}$ and ② $= v_i \cdots v_j$ so that $u = $ ①② would not be θ-primitive. Thus, v_i' must be nonempty. Likewise, ② $= w_1 \cdots w_{k-1} w_k' = w_p'' w_{p+1} \cdots w_n$ for some $1 \leq k \leq n$ and nonempty proper prefix w_k' of w_k, where $p = n-k+1$.

Lemma 3. *Let* ① *and* ② *be as above and* $|u| < m|v| < 2|u|$. *If* $m, n \geq 3$, *then both* ① *and* ② *are* θ-palindromes, *and* $v_1 = \cdots = v_{i-1}$, $v_{j+1} = \cdots = v_m = \theta(v_1)$, $v_j = \theta(v_i)$, $w_1 = \cdots = w_{k-1}$, $w_{p+1} = \cdots = w_n = \theta(w_1)$, *and* $w_p = \theta(w_k)$.

Proof. The symmetric roles of $v_1 \cdots v_m$ and $w_1 \cdots w_n$ allow us to assume that $|v_1 \cdots v_m| \geq |w_1 \cdots w_n|$, that is, |①| $\geq$ |②|. This means that $v_1 <_p$ ① since $m \geq 3$ and $v_i' \neq \varepsilon$. In order for u to be θ-primitive, $v_i' \notin \{\rho_\theta(v), \theta(\rho_\theta(v))\}^+$, where $\rho_\theta(v)$ is the θ-primitive root of v. Regarding $v_1 \cdots v_i$ and $v_j \cdots v_m$ rather as repeated concatenations of $\rho_\theta(v)$ and $\theta(\rho_\theta(v))$ and taking $v_1 = v$, Lemma 2 implies that $v_1 \cdots v_{i-1}$ is a power of v while $v_{j+1} \cdots v_m$ is a power of $\theta(v)$. In this way, $v_1 = \cdots = v_{i-1} = v$ and $v_{j+1} = \cdots = v_m = \theta(v)$ were obtained. Hence, v_i' is a suffix of $\theta(v)$, that is, $\theta(v_i')$ is a prefix of v of the same length as v_j''. Thus, $\theta(v_i') = v_j''$. Since $i \geq 2$ and $|v_i'| < |v|$, now we have that ① is a θ-palindrome.

As for ②, the above argument works as long as w_1 is its prefix or $w_n = \theta(w_1)$. If w_1 is longer than ② and $w_n = w_1$, then we can let $w_1 = $ ②w_1'' and $w_n = w_n'$② for some w_1'', w_n' and in fact $w_n' = \theta(w_1'')$ because they are a prefix and a suffix of the θ-palindrome ①. Substituting these into $w_n = w_1$ yields a conjugacy ②$w_1'' = \theta(w_1'')$②, and as mentioned in Sect. 2, it is solved as ② $= (\beta\alpha)^s\beta$, $w_1'' = \alpha\beta$, and $\theta(w_1'') = \beta\alpha$ for some $s \geq 0$ and words α, β. Then $w_1'' = \alpha\beta = \theta(\alpha)\theta(\beta)$, and hence, both α and β are θ-palindromes. Therefore, ② is a θ-palindrome.

As ① and ② are θ-palindromes, it follows that $v_i = \theta(v_j'')\theta(v_j') = \theta(v_j)$. □

See Fig. 1 and observe that Lemma 3 aligned $w_{k+1} \cdots w_p$ with the power of a conjugate $\tilde{v}$ of v. Intuitively speaking, for n large enough, we can apply Theorem 5 to this alignment and obtain $\tilde{v}, w \in \{t, \theta(t)\}^+$ for some t. This is quite close to the actual goal $v, w \in \{t, \theta(t)\}^+$ (tools for the purpose are obtained in [10]), but the final steps are case-specific and omitted due to the page limit. Only the cases with small m and n remain to be examined. Combinatorial arguments are required for them, but fortunately, the number of these cases is small. Theorem 1 follows in this manner, without any dependency on the proof of Theorem 2. The proof of the latter is given next.

4   Proof of Theorem 2

The case (B) from Table 1, left open in [7,10], was that of the equations

$$u_1 u_2 u_3 = v_1 v_2 v_3 v_4 w_1 \cdots w_n, \qquad\qquad (3)$$

where $u_i \in \{u, \theta(u)\}$ for all $1 \le i \le 3$, $v_j \in \{v, \theta(v)\}$ for all $1 \le j \le 4$ and $w_k \in \{w, \theta(w)\}$ for all $1 \le k \le n$ and $n \ge 5$ is odd. Note that there are instances of (3) with $n = 3$ that also have non-θ-periodic solutions, by symmetry to the case $\ell = 3, m = 3, n = 4$ (see Example 50 in [7]).

We first remark that we can assume v to be θ-primitive. Otherwise, replacing v with its θ-primitive root, we end up with an equation involving a larger number of the factors v or $\theta(v)$, and these have only θ-periodic solutions, by Theorem 1.

To solve (3), we analyse all possible values of u_2 and u_3, and for all these we look at the different relations between $|v_1 \cdots v_4|$ and $|u|$ separately.

▶ **The case** $u_1 u_2 u_3 = u\theta(u)u$. We actually show a more general result:

Lemma 4. *For any* $n, m \ge 3$, *if* $u\theta(u)u = v_1 \cdots v_m w_1 \cdots w_n$, *then* $u, v, w \in \{t, \theta(t)\}^+$ *for some word* t.

Proof. Applying θ to the equation gives $\theta(u)u\theta(u) = \theta(w_n) \cdots \theta(w_1)\theta(v_m) \cdots \theta(v_1)$, and we catenate this to the original equation to yield:

$$(u\theta(u))^3 = v_1 \cdots v_m w_1 \cdots w_n \theta(w_n) \cdots \theta(w_1)\theta(v_m) \cdots \theta(v_1).$$

Cyclic shift converts this into $x^3 = \theta(v_m) \cdots \theta(v_1)v_1 \cdots v_m w_1 \cdots w_n \theta(w_n) \cdots \theta(w_1)$, where x is a conjugate of $u\theta(u)$. This is an extended Lyndon-Schützenberger equation of the form $x^3 = v_1 \cdots v_{m'} w_1 \cdots w_{n'}$, with $m' = 2m \ge 6$ and $n' = 2n \ge 6$. The known results displayed in Table 1 (if $\min\{m', n'\} \ge 12$) and Theorem 1 (if $6 \le \min\{m', n'\} < 12$) provide a characterisation of the solutions of such equations, showing that $v, w \in \{t, \theta(t)\}^+$. □

▶ **The case** $u_1 u_2 u_3 = uuu$. The analysis of this case is split in three subcases depending on whether $4|v| < |u|$, $|u| < 4|v| < 2|u|$, or $4|v| > 2|u|$ holds.

Let us begin with the case when $4|v| < |u|$. Then there is $k \ge 1$ such that the first u ends on w_k. Let $u^2 = w_k'' w_{k+1} \cdots w_n$. If w is θ-primitive, then Proposition 2 implies that $n - k + 1$ is even, and hence, k is also even as n is assumed to be odd. Thus, $w_1 \cdots w_n$ is a suffix of u^3 and its length is at least $2|u| + |w|$. Theorem 5 concludes $u, w \in \{t, \theta(t)\}^+$ for some t. If w is not θ-primitive, then replacing $w_1, \ldots, w_n$ with the representations as the θ-power of the θ-primitive root of w reduces the given equation into the one handled in the next lemma.

Lemma 5. *If* $u_1 u_2 u_3 = uuu$, $4|v| < |u|$, $n \ge 10$, *and* (3) *holds, then* $u, v, w \in \{t, \theta(t)\}^+$ *for some* t.

Proof. If the first u ends to the right of w_1, then $w_1 \cdots w_n$ is a long enough suffix of u^3 to apply Theorem 5 as just explained above. As $2|u| < n|w|$, what remains to be examined is the case when it ends on w_1. By Proposition 2, n is even as w is assumed to be θ-primitive, and moreover $w_1 = \cdots = w_{n/2} = w$ and $w_{n/2+1} = \cdots = w_n = \theta(w)$, and

$$w_1 \cdots w_n = w^{n/2}\theta(w)^{n/2} = x^2 u^2 \tag{4}$$

for some θ-palindrome x. Since the first u ends on w_1, we can let $w_1 = w = x^2 z$ for some z. Substituting this to (4) gives $u = (zx^2)^{n/2-1} zx$. Now we have

$$u = (zx^2)^{n/2-1} zx = v_1 \cdots v_4 x^2. \tag{5}$$

Since $n \geq 10$, this means that $(zx^2)^3 \leq_p v_1 \cdots v_4 \leq_p (zx^2)^{n/2-1} zx$. Thus, $v_1 \cdots v_4$ is long enough for Theorem 5 to imply $zx^2 \in \{v, \theta(v)\}^+$. That is, $|zx^2|$ is a multiple of $|v|$, but a simple length argument implies $|zx^2| = |v|$ in order for (5) to hold. However, then (5) would give $z = x$ so that $w = x^2 z$ would not be θ-primitive. $\quad\square$

Lemma 6. *If $u_1 u_2 u_3 = uuu$, $|u| < 4|v| < 2|u|$, and (3) holds, then $u, v, w \in \{t, \theta(t)\}^+$ for some word t.*

Proof. We either have $u_2 = v_3'' v_4 w_1 \cdots w_k'$ or $u_2 = v_4'' w_1 \cdots w_k'$ for some k. By Lemma 3, we get that ① and ② are θ-palindromes. So, $w_1 \cdots w_n = $②①② is also θ-palindrome. As n is odd, we get that $w_{\frac{n+1}{2}} = \theta(w_{\frac{n+1}{2}})$ and therefore $w = \theta(w)$. Thus, we have $u^3 = v_1 \cdots v_4 w^n$, with $n \geq 5$.

If $|w^n| \geq |u| + |w|$, then Theorem 4 implies $u, w \in \{t\}^+$ for some t and we are done. Hence, we assume $|w^n| < |u| + |w|$, that is, $|u| > (n-1)|w|$. Then $|①| = 2|u| - n|w| > |u| - |w| > \frac{n-2}{n-1}|u| > \frac{1}{2}|u|$. Hence, it suffices to consider the case $u = v_1 v_2 v_3'$. Then Lemma 2 implies $v_4 = \theta(v)$. Also, if $v_2 = v$, we can apply Theorem 4 to v^2 and $\tilde{w}^\omega$, where $\tilde{w}$ is a conjugate of w, to get that v is not primitive, a contradiction. So we also assume $v_2 = \theta(v)$. Now if $v_3 = v$, then the given equation turns into a classical Lyndon-Schützenberger equation $u^3 = (v\theta(v))^2 w^5$, which is solved as $u, v\theta(v), w \in \{t\}^+$ for some t and we are done. Otherwise, since $v_1 v_2 v_3 v_4 = $①②① is a θ-palindrome, we get $v = \theta(v)$, and once again we obtain a classical Lyndon-Schützenberger equation $u^3 = v^4 w^5$ and we are done. $\quad\square$

The next lemma follows easily by Proposition 2, its proof is therefore omitted.

Lemma 7. *If $u_1 u_2 u_3 = uuu$, $4|v| > 2|u|$, and (3) holds, then $u, v, w \in \{t, \theta(t)\}^+$ for some word t.*

▶ **The case $u_1 u_2 u_3 = uu\theta(u)$.** Here, the case when $2|u| < 4|v|$ is already covered in [7] in a more general form as follows.

Lemma 8 (Proposition 51 in [7]). *For any $m, n \geq 3$, if $u_1 u_2 u_3 = uu\theta(u)$, (1) holds, and $m|v| > 2|u|$, then $u, v, w \in \{t, \theta(t)\}^+$ for some word t.*

Lemma 9. *If $u_1 u_2 u_3 = uu\theta(u)$, $4|v| < |u|$, and (3) holds, then $u, v, w \in \{t, \theta(t)\}^+$ for some t.*

Proof. By Corollary 1, we can assume that $4|v| < |u| < 4|v| + |w|$, that is, the first u ends inside w_1 (i.e., $|w_1 \cdots w_n| > |u\theta(u)| > |w_2 \cdots w_n|$). Hence, $2|u| > (n-1)|w|$, and this means $w_{n-1} w_n <_s \theta(u)$. Thus, $\theta(w_n)\theta(w_{n-1}) <_p u$.

If $n \geq 7$, we have $|w| < \frac{1}{3}|u|$, and hence, $|v_1 \cdots v_4| > \frac{2}{3}|u|$. Note that the argument here does not rely on the parity of n.

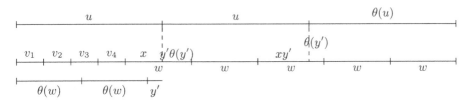

Fig. 2. The equation $uu\theta(u) = v_1v_2v_3v_4w_1\cdots w_n$ with $n = 5$ and $4|v| < |u|$

Assume first that $2|w| \leq |v_1v_2v_3|$. Since $\theta(u)$ is a suffix of $w_1\cdots w_n$, the prefix $v_1\cdots v_4$ of u is a prefix of $\theta(w_n)\cdots\theta(w_1)$, which is long enough for Theorem 5 to give $v, w \in \{t, \theta(t)\}^+$ for some t.

If $2|w| > |v_1v_2v_3|$, then we have $|v_1v_2v_3| < 2|w| < |v_1\cdots v_4|$. Proposition 2 gives $v_1 = v_2 = v$ while $v_3 = v_4 = \theta(v)$. Thus, $v_1\cdots v_4$ is a θ-palindrome. It has a prefix $\theta(w_n)$ and hence a suffix w_n. Now we have that $w_nw_1w_2\cdots w_n$ is a suffix of $uu\theta(u)$ and its length is at least $2|u| + |w|$, enough for Theorem 5.

The case $n = 5$ (see Fig. 2) remains. We can assume that w is θ-primitive; otherwise, the examination is reduced to the case when $n \geq 7$. Proposition 2 is hence applicable to $w_1\cdots w_5$ and gives $w_1 = \cdots = w_5 = w$. As shown in Fig. 2, now we have $w = xy$ for some θ-palindromes x, y such that $y = y'\theta(y')$. If $|y'| \geq |x|$, then we get $\theta(w)^2 \leq_p v_1\cdots v_4$, and the same reasoning as in the previous paragraph applies. Thus, we assume $|y'| < |x|$. See $v_1\cdots v_4x = \theta(w)^2y'$ in Fig. 2. Since $\theta(w) = yx$, this gives $x \leq_s xy'$, and this implies that xy' is a θ-palindrome because $|y'| < |x|$.

The centre of the prefix $\theta(w)^2$ of u is either on v_3 or on v_4 (it cannot be to the right of v_4 since $|w| < \frac{1}{2}|u|$). If it is inside v_3 as depicted in Fig. 2, we can let $\theta(w) = v_1v_2v_3'$ and $v_3''v_4 \leq_p \theta(w)$ for some v_3', v_3'' whose catenation is v_3. Lemma 2 implies that $v_3''v_4$ is a θ-palindrome by fixing $v_4 = \theta(v_1)$. Thus, $v_3''v_4$ is a suffix of w, and this extends the power $w_1\cdots w_5$ of w to the left sufficiently for Theorem 5 to imply $u, w \in \{t, \theta(t)\}^+$ for some t.

To handle the last case when the centre is inside v_4, from $v_1\cdots v_4x = \theta(w)^2y'$, we first derive the equation $v_1\cdots v_4 = y'\theta(y')xy'\theta(y')\theta(y')$. In order for the centre to be there, we need $|v_4| > |y'\theta(y')\theta(y')|$, and hence, $|y| < |v_1|$. This has two consequences. One is that v_2v_3 is a factor of x, and hence, trivially, that of the θ-palindrome xy'. In addition, the equation implies that the centre of v_2v_3 is at the centre of xy'. Thus, $v_3 = \theta(v_2)$. The other consequence is that $v_1v_2v_3 <_p yxy <_p v_1v_2v_3v_4$. Hence, $\theta(v_3)\theta(v_2)\theta(v_1)$ is a proper factor of $v_1\cdots v_4$ so that Lemma 1 implies $v_2 = v_3$. Consequently, $v = \theta(v)$. Now $v_1\cdots v_4$ is a θ-palindrome, and it has $\theta(w)$ as a prefix, and hence, w as its suffix. Theorem 5 is now applied to $ww_1\cdots w_5$ to obtain $u, w \in \{t, \theta(t)\}^+$. $\square$

The next result follows by combining arguments used in the corresponding subcase for $u_1u_2u_3 = uuu$ and in the proof of the previous Lemma.

Lemma 10. *If $u_1u_2u_3 = uu\theta(u)$, $|u| < 4|v| < 2|u|$, and (3) holds, then $u, v, w \in \{t, \theta(t)\}^+$ for some t.*

▶ **The case** $u_1u_2u_3 = u\theta(u)\theta(u)$. The subcase when $4|v| < |u|$ is simple. We just have to apply θ to both sides of the equations and then use Lemma 8.

Lemma 11. *If* $u_1u_2u_3 = u\theta(u)\theta(u)$, $4|v| < |u|$, *and* (3) *holds, then* $u, v, w \in \{t, \theta(t)\}^+$ *for some* t.

Lemma 12. *If* $u_1u_2u_3 = u\theta(u)\theta(u)$, $|u| < 4|v| < 2|u|$, *and* (3) *holds, then* $u, v, w \in \{t, \theta(t)\}^+$ *for some* t.

Proof. We consider two separate main cases: The first is when $v_3''v_4 \leq_p u_2$, where v_3'' is a suffix of $v_3 = v_3'v_3''$. As v is assumed to be θ-primitive, $|v_3'| \neq |v_3''|$. In this case $|v| > \frac{1}{3}|u|$ and so $4|v| > \frac{4}{3}|u|$. It follows that $n|w| < \frac{5}{3}|u|$ and since $n \geq 5$, we get that $|w| < \frac{1}{3}|u| < |v|$. Further, we can assume that $|u| < 2|v| + |w|$, otherwise we get the claimed result by Theorem 5. This means that $|w| > |v_3'|$.

We will now prove that $v_1 = v_3 = v_4 = v$ holds, as follows. Suppose, towards a contradiction, that $v_4 = \theta(v)$. Since $\theta(w_n) \leq_p v$, this assumption implies that v_4 ends with w_n. Then $w_n w_1 \ldots w_k' \leq_s \theta(u)$, where $w_k = w_k'w_k''$ for some k, and, thus, $w_n w_1 \ldots w_k' \leq_s w_1 \cdots w_n$. By Lemma 1, we get that $w_n = w_1 = \ldots = w_{k-1}$ and $w_{n-1} = w_n$. So, by Proposition 1, $w_n \neq w_n$ must hold, a contradiction. Thus, $v_4 = v$. On the θ-palindrome $u\theta(u)$, $\theta(v_4)\theta(v_3)$ lies inside $v_2v_3v_4$ in such a way that Lemma 1 implies $v_3 = v_4$ (recall that $|v_3'| \neq |v_3''|$). Hence, $v_1 = v_3 = v_4 = v$.

We split the discussion further, according to the relation between $|v_3'|$ and $|v_3''|$:

If $|v_3'| > |v_3''|$, then we have also $|w| > |v_3''|$, as we already established $|w| > |v_3'|$. Now, $|v| = |v_3'| + |v_3''|$ and $|u| = 3|v_3'| + 2|v_3''|$. From this we get that $n|w| = 3|u| - 4|v| = 5|v_3'| + 2|v_3''|$. However, $|w| > |v_3'|$ and $|w| > |v_3''|$, so n must be 5.

Assume $3|w| \leq 2|v|$, that is $6|w| \leq 4|v|$. Now, as $|v_3'| > |v_3''|$, we have that $\frac{5}{2}|v| < |u|$, and therefore $4|v| < \frac{8}{5}|u|$. Since $6|w| \leq 4|v|$, we get $6|w| < \frac{8}{5}|u|$, hence $5|w| < \frac{8}{6}|u|$. It follows that $4|v| + 5|w| < \frac{8}{5}|u| + \frac{8}{6}|u| = \frac{88}{30}|u| < 3|u|$, a contradiction. Thus $3|w| > 2|v|$ must hold, and as $\theta(w_5)\theta(w_4)\theta(w_3) \leq_p vv_2v$, we can apply Proposition 2 to get $v_2 = \theta(v)$ and $w_3 = w_4 = w_5$. The fact that w_1w_2' is a suffix of $w_3w_4w_5$ leads to $w_1 = w$ and $w_3 = w_4 = w_5 = \theta(w)$. Now we have that $w_5 \leq_p v_1$, so $w \leq_p v$ and we have two occurrences of w inside $u_2 = \theta(u)$: One (the prefix of v_4) is after the prefix of length $3|v| - |u|$ and the other one (w_1) is after the prefix of length $4|v| - |u|$. Both these occurrences fall inside $w_3w_4w_5 = \theta(w)^3$ inside $u_3 = \theta(u)$. In order for w to be primitive (as assumed), the length of the factor between those two occurrences must be a multiple of $|w|$. This factor is of length $|v|$ though, and since $|v| > |w|$, but $2|w| > |v|$ (as $|w| > |v_3'|$ and $|w| > |v_3''|$), this is impossible.

We reached the case when $|v_3'| < |v_3''|$; by Lemma 1 we get that $v_2 = v_3 = v_4$. We first establish that $v_1 = v_2$ holds as well. Assume towards a contradiction, that $v_1 = \theta(v_2)$. Then, as $\theta(w_n) \leq_p v_1$, we have $w_n \leq_s v_4 = v_2 = \theta(v_1)$. Now, if w_1 is a factor of u_2, the word $w_n w_1 \cdots w_k'$ for some $k \geq 2$ is a suffix of $\theta(u)$. However, also $\theta(u) = w_k'' \cdots w_n$. By Lemma 1, we get that $w_n = w_1 = \ldots = w_{k-1} = w$, and also that $w_{n-1} = w_n$. But then $w_{k-1} = w_n$ appears as a proper factor inside $w_{n-1}w_n = w_nw_n$, contradicting the primitivity of w. If w_1 is not a factor of u_2, that is, $w_1' \leq_s u_2$, where $w_1 = w_1'w_1''$, then $w_n w_1'$ is a suffix of $w_{n-1}w_n$, so $xw_nw_1 = w_{n-1}w_ny$ where $y = w_1''$ and x is the prefix of length $|w_1''|$ of w_{n-1}. By Lemma 2 we get that $w_n \neq w_n$, a contradiction. Thus, $v_1 = \ldots = v_4 = v$.

If the prefix of length $|v| + |w|$ of $\theta(w_1 \cdots w_n)$ is a prefix of a word in $\{w\}^+$ or $\{\theta(w)\}^+$, then we can apply Theorem 4 to get the claimed result. Thus we assume that $w\theta(w)$ or $\theta(w)w$ appears as a factor inside vv after a prefix that is strictly shorter than $|v|$. Without loss of generality we assume that it is $w\theta(w)$ that occurs there, and we focus on the first occurrence of this factor inside vv.

We first rule out the possibility that $w\theta(w)$ is a prefix of vv: We observe that $|u| \geq 2|w| + |v|$. To see this, assume that $|u| < 2|w| + |v|$. This means that $3|u| < 6|w| + 3|v| < 5|w| + 4|v|$, which is a contradiction. Therefore, if $w\theta(w)$ is a prefix of vv, it has another appearance inside u after a prefix of length $|v|$. However, as $|v|$ is not divisible by $|w|$ (otherwise $v \in \{w, \theta(w)\}^+$ and $|v| > |w|$, so v would not be θ-primitive), this other occurrence of $w\theta(w)$ is a proper factor of some word $w_i w_{i+1} w_{i+2}$ inside $u_3 = \theta(u)$. By Lemma 1 this is impossible if w is θ-primitive. Hence, we assume that $xw\theta(w)$ is a prefix of vv, where $x \in \{w\}^+$. As $|w| > |v_3'|$, we have an occurrence of $w\theta(w)$ in $u_2 = \theta(u)$ after a prefix of length $|x| - |v_3'|$. So in u, we have an occurrence of $w\theta(w)$ after the prefix of length $|x|$, and after the prefix of length $|u| - (|x| - |v_3'|) - 2|w|$. As $x \in \{w\}^+$, we must have $|u| - (|x| - |v_3'|) - 2|w| \geq |x|$. Now as u is prefix of $\theta(w_1 \cdots w_n)$, in order to avoid a contradiction with Lemma 1, the difference between the lengths of those two prefixes must be divisible by $|w|$. However, this difference is $|u| - (|x| - |v_3'|) - 2|w| - |x| = |u| - 2|x| + |v_3'| - 2|w|$ and as $x \in \{w\}^+$, the term $|u| + |v_3'|$ must be divisible by $|w|$. Now $|u| = 2|v| + |v_3'|$, thus $2|u| = 4|v| + 2|v_3'|$, and as $|w_1 \cdots w_n| = 3|u| - 4|v|$, we have $|w_1 \cdots w_n| = |u| + 2|v_3'|$. If now $|u| + |v_3'|$ is divisible by $|w|$, then also $|v_3'|$ must be divisible by $|w|$. This however is a contradiction, as we assumed that $|v_3'| < |w|$.

The other main case is when $u_1 = u$ ends on v_4. The proof below does not rely on n being odd so that we can assume that w is θ-primitive. Then, $|v_1 v_2 v_3| < |u|$, that is, $|v| < \frac{1}{3}|u|$. We also have $n|w| < 2|u|$, that is, $|w| < \frac{2}{n}|u| \leq \frac{2}{5}|u|$. Let $u\theta(u) = v_1 \cdots v_4 w_1 \cdots w_{k-1} w_k'$ for some k and a prefix w_k' of w_k. Since $|v_1 \cdots v_4| < \frac{4}{3}|u|$ and $|w| \leq \frac{2}{5}|u|$, k must be at least 2. Note that $w_1 \cdots w_{k-1} w_k'$ is a suffix of $\theta(u)$, and $\theta(u)$ is a suffix of $w_1 \cdots w_n$. That is, $w_1 \cdots w_{k-1} w_k' <_s w_1 \cdots w_n$. Lemmas 1 and 2 imply $w_1 = \cdots = w_{k-1} = w$, $w_k' <_p w$, and $\alpha = w_1 \cdots w_{k-1} w_k'$ is a θ-palindrome. Then $u\theta(u) = v_1 \cdots v_4 \alpha$. Thus, α is also a prefix of this θ-palindrome $u\theta(u)$, and moreover, $\alpha <_p v_1 \cdots v_4$. We now have that $u\theta(u) <_p (v_1 \cdots v_4)^2$. Since u is assumed to be θ-primitive, Proposition 2 gives $v_1 = v_2 = v_3 = v_4 = v$. Now $\alpha <_p v^4$ and $\alpha <_p w^k$. For the sake of Theorem 4, it suffices to prove $|\alpha| \geq |v| + |w|$. In fact, from $n|w| = 3|u| - 4|v|$ and $|\alpha| = 2|u| - 4|v|$, we get $|\alpha| - (|v| + |w|) = 2|u| - 4|v| - \frac{3}{n}|u| + \frac{4}{n}|v| - |v| = \left(\frac{1}{3} - \frac{5}{3n}\right)|u|$, which is at least 0 since $n \geq 5$.   □

Lemma 13. *If* $u_1 u_2 u_3 = u\theta(u)\theta(u)$, $2|u| < 4|v| < 2|u| + |v|$, *and* (3) *holds, then* $u, v, w \in \{t, \theta(t)\}^+$ *for some* t.

Proof. By Proposition 2, we have $v_1 = \ldots = v_4 = v$ and $v = xz\theta(z)$, for some words x and z, with $x = \theta(x)$. Furthermore, $u = vxz = xz\theta(z)xz$. Therefore,

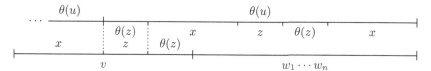

Fig. 3. The word $\theta(u)$

the situation in $u_3 = \theta(u)$ looks as illustrated in Fig. 3. We see that $z = \theta(z)$ in this case. Now $\theta(u) = zxzzx = zzw_1 \cdots w_n$, and so $xz^2x = zw_1 \cdots w_n$.

Since both xz^2x and z are θ-palindromes and $|z| \leq |w_1 \cdots w_n|$, the equation implies $z \leq_s w_1 \cdots w_n$. Hence, there exists an integer k such that $z = w_k'' w_{k+1} \cdots w_n$ where $w_k = w_k' w_k''$ for some w_k', w_k''. If w_k'' was empty, we would get $z = w_{k+1} \cdots w_n$, but substituting this into $\theta(u) = zzw_1 \cdots w_n$ would make $\theta(u)$ not θ-primitive. Thus, $w_k'' \neq \varepsilon$. Then $xz^2x = w_k'' w_{k+1} \cdots w_n w_1 \cdots w_n$ implies $w_1 = \ldots = w_n = w$. Now we have $xz^2x = zw^n$ and concatenating z^2 to the left on both sides results in $(z^2x)^2 = z^3w^n$. This is a conventional Lyndon-Schützenberger equation, and it implies $z^2x, z \in \{w\}^+$. Substituting this into the equation, however, implies $z^2x \in \{w\}^+$, and $|z^2x| \geq 3|w|$. Hence, z^2x is not primitive and neither is its conjugate v, a contradiction.   □

▶ **Conclusion.** By the results of Lemmas 4–13 we get that Theorem 2 holds.

References

1. Bischoff, B., Currie, J.D., Nowotka, D.: Unary patterns with involution. Int. J. Found. Comput. Sci. 23(8), 1641–1652 (2012)
2. Blondin Massé, A., Gaboury, S., Hallé, S., Larouche, M.: Solving equations on words with morphisms and antimorphisms. In: Dediu, A.-H., Martín-Vide, C., Sierra-Rodríguez, J.-L., Truthe, B. (eds.) LATA 2014. LNCS, vol. 8370, pp. 186–197. Springer, Heidelberg (2014)
3. Czeizler, E., Czeizler, E., Kari, L., Seki, S.: An extension of the Lyndon-Schützenberger result to pseudoperiodic words. DLT 2009 209, 717–730 (2011); Its Conference version is in: Diekert, V., Nowotka, D. (eds.) DLT 2009. LNCS, vol. 5583, pp. 183–194. Springer, Heidelberg (2009)
4. Czeizler, E., Kari, L., Seki, S.: On a special class of primitive words. Theor. Comput. Sci. 411(3), 617–630 (2010)
5. Gawrychowski, P., Manea, F., Nowotka, D.: Testing generalised freeness of words. In: Proc. STACS. LIPIcs, vol. 25, pp. 337–349 (2014)
6. Harju, T., Nowotka, D.: On the equation $x^k = z_1^{k_1} z_2^{k_2} \cdots z_n^{k_n}$ in a free semigroup. Theor. Comput. Sci. 330(1), 117–121 (2005)
7. Kari, L., Masson, B., Seki, S.: Properties of pseudo-primitive words and their applications. Int. J. Found. Comput. Sci. 22(2), 447–471 (2011)
8. Lothaire, M.: Combinatorics on words. Cambridge University Press (1997)
9. Lyndon, R.C., Schützenberger, M.P.: The equation $a^m = b^n c^p$ in a free group. Michigan Math. J. 9(4), 289–298 (1962)
10. Manea, F., Müller, M., Nowotka, D.: On the pseudoperiodic extension of $u^l = v^m w^n$. In: Proc. FSTTCS. LIPIcs, vol. 24, pp. 475–486 (2013)
11. Xu, Z.: A minimal periods algorithm with applications. In: Amir, A., Parida, L. (eds.) CPM 2010. LNCS, vol. 6129, pp. 51–62. Springer, Heidelberg (2010)

Complexity of Equivalence and Learning for Multiplicity Tree Automata

Ines Marusic and James Worrell

Department of Computer Science, University of Oxford, UK
{ines.marusic,jbw}@cs.ox.ac.uk

Abstract. We consider the complexity of equivalence and learning for multiplicity tree automata, i.e., weighted tree automata with weights in a field. We first show that the equivalence problem for multiplicity tree automata is logspace equivalent to polynomial identity testing. Secondly, we consider the problem of learning multiplicity tree automata in Angluin's exact learning model. Here we give lower bounds on the number of queries, both for the case of an arbitrary and a fixed underlying field. We also present a learning algorithm in which trees are represented succinctly as DAGs. Assuming a Teacher that represents counterexamples as succinctly as possible, our algorithm uses exponentially fewer queries than the best previously known procedure, leaving only a polynomial gap with the above-mentioned lower bound. Moreover, fixing the alphabet rank, the query complexity of our algorithm matches the lower bound up to a constant factor.

1 Introduction

Trees are a natural model of structured data, including syntactic structures in natural language processing and XML data on the web. Many of those applications require representing functions from trees into the real numbers. A broad class of such functions can be defined by *multiplicity tree automata*, which generalise classical finite tree automata by having transitions with weights in a field.

A fundamental problem concerning multiplicity tree automata is *equivalence*: given two automata, do they define the same function. Seidl [13] proved that equivalence for multiplicity tree automata is decidable in randomised polynomial time. No finer analysis of the complexity of this problem exists to date.

Our first contribution (Section 3) is to show that the equivalence problem for multiplicity tree automata is logspace equivalent to polynomial identity testing, i.e., the problem of deciding whether a polynomial given as an arithmetic circuit is zero. The latter is known to be solvable in randomised polynomial time whereas solving it in deterministic polynomial time is a well-studied and longstanding open problem [3].

Equivalence is closely connected to the problem of *learning* in Angluin's exact learning model [2]. In this model, a Learner actively collects information about the target function from a Teacher through *membership queries*, which ask for the

E. Csuhaj-Varjú et al. (Eds.): MFCS 2014, Part I, LNCS 8634, pp. 414–425, 2014.
© Springer-Verlag Berlin Heidelberg 2014

value of the function on a specific input, and *equivalence queries*, which suggest a hypothesis to which the Teacher provides a counterexample if one exists. Exact learnability of multiplicity word automata has been extensively studied. Beimel *et al.* [4] show that multiplicity word automata can be learned efficiently and apply this to learn various classes of DNF formulae and polynomials. These results were generalised by Klivans and Shpilka [11] to show learnability of restricted algebraic branching programs and noncommutative set-multilinear arithmetic formulae. Bisht, Bshouty, and Mazzawi [5] give an almost tight (up to a *log* factor) lower bound for the number of queries made by any exact learning algorithm for the class of multiplicity word automata.

In Section 5 we give lower bounds on the number of queries needed to learn multiplicity tree automata in the exact learning model, both for the case of an arbitrary and a fixed underlying field. The bound in the latter case is proportional to the automaton size for trees of a fixed maximal branching degree.

Habrard and Oncina [9] present an exact learning algorithm for multiplicity tree automata, using a number of equivalence queries that is bounded by the number of states in a minimal representation of the automaton being learned, and a number of membership queries that is proportional to the size of the minimal representation and the length of the longest counterexample returned by the Teacher. However, the smallest counterexample to an equivalence query may be exponential in the size of the target automaton, if given explicitly in a standard representation as a tree. Thus the algorithm of [9] is not polynomial in the size of the target automaton, even for a Teacher that returns counterexamples of minimal size. Moreover, there is an exponential gap between the query complexity of this algorithm and the above-mentioned lower bound.

Given two inequivalent multiplicity tree automata with n states in total, Seidl's algorithm [13] produces a subtree-closed set of trees of cardinality at most n that contains a tree on which the automata differ. It follows that the counterexample contained in the set has at most n subtrees, and hence can be represented as a DAG with at most n nodes. Thus it is natural to consider a Teacher that can return succinctly represented counterexamples, i.e., trees represented as DAGs. Tree automata that run on DAG representations of finite trees were introduced in [7].

In Section 4 we present a new exact learning algorithm for multiplicity tree automata that achieves the same bound in the number of equivalence queries as the algorithm of [9], while using polynomially many membership queries in the counterexample size, even when counterexamples are given succinctly. Assuming that the Teacher provides minimal succinct representations of counterexamples, our algorithm therefore makes polynomially many queries in the target size. This is exponentially fewer queries than the best previously known algorithm [9] and leaves only a polynomial gap with the above-mentioned lower bound. Moreover, for a fixed maximal branching degree the number of queries matches the lower bound to within a multiplicative constant.

Like the algorithm of [9], our algorithm constructs a Hankel matrix of the target automaton. However on receiving a counterexample tree z, the former algorithm adds a new column to the Hankel matrix for every suffix of z, while our algorithm adds (at most) one new row for each subtree of z. But the number of suffixes may be exponential in the size of the DAG representation of z, whereas the number of subtrees is linear.

The full version of this paper is available as [12].

2   Preliminaries

For every $n \in \mathbb{N}$, we write $[n]$ for the set $\{1, 2, ..., n\}$ and I_n for the identity matrix of size n. For every $i \in [n]$, we write e_i for the i-th n-dimensional coordinate vector. For any matrix A, we write A_i for its ith row, A^j for its jth column, and $A_{i,j}$ for its $(i, j)^{\text{th}}$ entry. Given nonempty subsets I and J of the rows and columns of A, respectively, we write $A_{I,J}$ for the submatrix $(A_{i,j})_{i \in I, j \in J}$ of A. For singletons, we denote simply $A_{i,J} := A_{\{i\},J}$ and $A_{I,j} := A_{I,\{j\}}$.

Let $n_1, ..., n_k \in \mathbb{N}$ and $(i_1, ..., i_k) \in [n_1] \times \cdots \times [n_k]$. If a matrix A has $n_1 \cdot ... \cdot n_k$ rows, then we write $A_{(i_1,...,i_k)}$ for its $(\sum_{l=1}^{k-1}(i_l - 1) \cdot (\prod_{p=l+1}^{k} n_p) + i_k)^{\text{th}}$ row.

Kronecker Product. Let A be a matrix of dimension $m_1 \times n_1$ and B a matrix of dimension $m_2 \times n_2$. The *Kronecker product* of A by B, written as $A \otimes B$, is a matrix of dimension $(m_1 m_2) \times (n_1 n_2)$ where $(A \otimes B)_{(i_1,i_2),(j_1,j_2)} = A_{i_1,j_1} \cdot B_{i_2,j_2}$ for every $i_1 \in [m_1]$, $i_2 \in [m_2]$, $j_1 \in [n_1]$, and $j_2 \in [n_2]$.

The Kronecker product is bilinear, associative, and has the following *mixed-product property*: Given matrices A, B, C, and D where products $A \cdot C$ and $B \cdot D$ are defined, it holds that $(A \otimes B)(C \otimes D) = (A \cdot C) \otimes (B \cdot D)$.

For every $k \in \mathbb{N}_0$ we define the k-*fold Kronecker product* of a matrix A by itself, written $A^{\otimes k}$, as $A^{\otimes 0} = I_1$ and $A^{\otimes k} = A^{\otimes(k-1)} \otimes A$ for $k \geqslant 1$.

Tree Series. A *ranked alphabet* is a tuple (Σ, rk) where Σ is a nonempty finite set of symbols and $rk : \Sigma \to \mathbb{N}_0$. For every $k \in \mathbb{N}_0$, we define the set of all k-*ary* symbols $\Sigma_k := rk^{-1}(\{k\})$. The set of Σ-*trees* (*trees* for short), written T_Σ, is the smallest set T such that (i) $\Sigma_0 \subseteq T$ and (ii) if $k \geqslant 1$, $\sigma \in \Sigma_k$, $t_1, ..., t_k \in T$ then $\sigma(t_1, ..., t_k) \in T$. A *subtree* of a tree t is a tree consisting of a node in t and all of its descendants in t.

Let Σ be a ranked alphabet and $\mathbb{F}$ a field. A *tree series over Σ with coefficients in $\mathbb{F}$* is a mapping $f : T_\Sigma \to \mathbb{F}$. For $t \in T_\Sigma$, $f(t)$ is called the *coefficient* of t in f. The set of all tree series over Σ with coefficients in $\mathbb{F}$ is denoted by $\mathbb{F}\langle\langle T_\Sigma \rangle\rangle$.

We define tree series *height* $\in \mathbb{Q}\langle\langle T_\Sigma \rangle\rangle$, *size* $\in \mathbb{Q}\langle\langle T_\Sigma \rangle\rangle$, and $\#_\sigma \in \mathbb{Q}\langle\langle T_\Sigma \rangle\rangle$ where $\sigma \in \Sigma$, by (i) if $t \in \Sigma_0$ then $height(t) = 0$, $size(t) = 1$, and $\#_\sigma(t) = \mathbb{1}_{\{t=\sigma\}}$, and (ii) if $t = a(t_1, ..., t_k)$ where $k \geqslant 1$ then $height(t) = 1 + max_{i \in [k]} height(t_i)$, $size(t) = 1 + \sum_{i \in [k]} size(t_i)$, and $\#_\sigma(t) = \mathbb{1}_{\{a=\sigma\}} + \sum_{i \in [k]} \#_\sigma(t_i)$ respectively.

Let $\square$ be a nullary symbol not contained in Σ. The set C_Σ of Σ-*contexts* (*contexts* for short) is the set of $(\{\square\} \cup \Sigma)$-trees in which $\square$ occurs exactly once. Given $c \in C_\Sigma$ and $t \in T_\Sigma \cup C_\Sigma$, we write $c[t]$ for the tree obtained by substituting t for $\square$ in c. A *suffix* of a Σ-tree t is a Σ-context c such that $t = c[t']$ for some Σ-tree t'. We define the *Hankel matrix* of a tree series $f \in \mathbb{F}\langle\langle T_\Sigma \rangle\rangle$ to be the matrix $H : T_\Sigma \times C_\Sigma \to \mathbb{F}$ such that $H_{t,c} = f(c[t])$ for every $t \in T_\Sigma$ and $c \in C_\Sigma$.

Multiplicity Tree Automata. Let $\mathbb{F}$ be a field. An $\mathbb{F}$-*multiplicity tree automaton* ($\mathbb{F}$-*MTA*) is a quadruple (n, Σ, μ, γ) which consists of the *dimension* $n \in \mathbb{N}$ representing the number of states, a ranked alphabet Σ, for every $\sigma \in \Sigma$ a transition matrix $\mu(\sigma) \in \mathbb{F}^{n^{rk(\sigma)} \times n}$, and the *final weight vector* $\gamma \in \mathbb{F}^{1 \times n}$.

Let $A = (n, \Sigma, \mu, \gamma)$ be an $\mathbb{F}$-MTA. We extend μ to T_Σ by $\mu(\sigma(t_1, ..., t_k)) := (\mu(t_1) \otimes \cdots \otimes \mu(t_k)) \cdot \mu(\sigma)$ for every $\sigma \in \Sigma_k$ and $t_1, ..., t_k \in T_\Sigma$. The tree series $\|A\| \in \mathbb{F}\langle\langle T_\Sigma \rangle\rangle$ *recognised* by A is defined by $\|A\|(t) = \mu(t) \cdot \gamma^\top$ for every $t \in T_\Sigma$. The *size* of A, written as $|A|$, is the total number of entries in all transition matrices and the final weight vector of A, i.e., $|A| := \sum_{\sigma \in \Sigma} n^{rk(\sigma)+1} + n$.[1]

We extend μ to C_Σ by treating $\Box$ as a unary symbol and defining $\mu(\Box) := I_n$. This allows to define $\mu(c) \in \mathbb{F}^{n \times n}$ for every $c \in C_\Sigma$ inductively as $\mu(c) := (\mu(t_1) \otimes \cdots \otimes \mu(t_k)) \cdot \mu(\sigma)$ for every $c = \sigma(t_1, ..., t_k) \in C_\Sigma$. It is easy to see that for every $t \in T_\Sigma$ and $c \in C_\Sigma$, $\mu(c[t]) = \mu(t) \cdot \mu(c)$.

Let $A_1 = (n_1, \Sigma, \mu_1, \gamma_1)$ and $A_2 = (n_2, \Sigma, \mu_2, \gamma_2)$ be $\mathbb{F}$-MTAs. A_1 and A_2 are said to be *equivalent* if $\|A_1\| \equiv \|A_2\|$. The *Cartesian product* of A_1 and A_2, written as $A_1 \times A_2$, is the $\mathbb{F}$-MTA $(n_1 \cdot n_2, \Sigma, \{\mu_1(\sigma) \otimes \mu_2(\sigma) : \sigma \in \Sigma\}, \gamma_1 \otimes \gamma_2)$.

Proposition 1. *[8, Chapter 9] Let A_1 and A_2 be $\mathbb{Q}$-multiplicity tree automata over a ranked alphabet Σ. Then $\|A_1 \times A_2\|(t) = \|A_1\|(t) \cdot \|A_2\|(t)$ for all $t \in T_\Sigma$. Furthermore, automaton $A_1 \times A_2$ can be computed in logarithmic space.*

Recognisable Tree Series. A tree series f is called *recognisable* if it is recognised by some MTA; such an automaton is called an *MTA-representation* of f. A *minimal* MTA-representation of f is one of minimal dimension. The set of all recognisable tree series in $\mathbb{F}\langle\langle T_\Sigma \rangle\rangle$ is denoted by $Rec(\Sigma, \mathbb{F})$.

Theorem 1. *[9] Let $f \in \mathbb{F}\langle\langle T_\Sigma \rangle\rangle \setminus \{0\}$ and H be the Hankel matrix of f. Then, $f \in Rec(\Sigma, \mathbb{F})$ if and only if H has finite rank over $\mathbb{F}$. In case $f \in Rec(\Sigma, \mathbb{F})$, the dimension of a minimal MTA-representation of f is $rank(H)$ (over $\mathbb{F}$).*

The following result of [6, Proposition 4] states that for any recognisable tree series, its minimal MTA-representation is unique up to change of basis.

Theorem 2. *Let $f \in Rec(\Sigma, \mathbb{F}) \setminus \{0\}$ have a Hankel matrix of rank r. Suppose $\mathbb{F}$-MTA $A_1 = (r, \Sigma, \mu_1, \gamma_1)$ recognises f. Then for any $\mathbb{F}$-MTA $A_2 = (r, \Sigma, \mu_2, \gamma_2)$, A_2 recognises f if and only if there exists an invertible matrix $U \in \mathbb{F}^{r \times r}$ such that $\gamma_2 = \gamma_1 \cdot U^\top$ and $\mu_2(\sigma) = U^{\otimes rk(\sigma)} \cdot \mu_1(\sigma) \cdot U^{-1}$ for every $\sigma \in \Sigma$.*

DAG Representations of Trees. A *DAG representation* of a Σ-tree (Σ-*DAG* or *DAG* for short) is a directed acyclic ordered multigraph whose nodes are labelled with symbols from Σ such that any node labelled with a k-ary symbol has outdegree k. Note that Σ-trees are a subclass of Σ-DAGs.

A *sub-DAG* of a DAG G is a DAG consisting of a node in G and all of its descendants in G. The *size* of a DAG G is the number of nodes in G.

---

[1] We measure size assuming explicit rather than sparse representations of the transition matrices and final weight vector because minimal automata are only unique up to change of basis; cf. Theorem 2.

The computation of an MTA on DAGs is as follows. A *run* of an $\mathbb{F}$-MTA $A = (n, \Sigma, \mu, \gamma)$ on a Σ-DAG $G = (V, E)$ is a mapping $\rho : V \to \mathbb{F}^n$ such that for every $v \in V$, if v is labelled σ and has successors $v_1, ..., v_k \in V$ with multiplicities $i_1, ..., i_k$ respectively, then $\rho(v) = (\rho(v_1)^{\otimes i_1} \otimes \cdots \otimes \rho(v_k)^{\otimes i_k}) \cdot \mu(\sigma)$. We define $\|A\|(G) := \rho(v_0) \cdot \gamma^{\top}$ where v_0 is the root of G. It is straightforward to show that the value of an MTA on a DAG is equal to its value on the tree unfolding.

Because each context has exactly 1 occurrence of symbol $\square$, any *DAG representation of a Σ-context* has a unique path from the root to the $\square$-labelled node and every edge along that path has multiplicity 1.

Arithmetic Circuits. An *arithmetic circuit* is a finite directed acyclic vertex-labelled multigraph whose vertices, called *gates*, have indegree 0 or 2. Vertices of indegree 0 are called *input gates* and are labelled with a constant 0 or 1, or a variable from the set $\{x_i : i \in \mathbb{N}\}$. Vertices of indegree 2 are called *internal gates* and are labelled with one of the arithmetic operations $+$, $\times$, or $-$. We assume that there is a unique gate with outdegree 0 called the *output gate*. An arithmetic circuit is called *variable-free* if all input gates are labelled 0 or 1.

The *size* of an arithmetic circuit C is the number of gates in C. The depth of a gate v in C, denoted $depth(v)$, is the length of a longest directed path from v to the output gate of C. The *depth* of C is the maximal depth of a gate in C. The *output* of C, written f_C, is the polynomial computed by the output gate of C. The *Arithmetic Circuit Identity Testing (ACIT)* problem asks whether the output of a given arithmetic circuit is equal to the zero polynomial.

Exact Learning Model. Let f be a *target function*. A *Learner* (*learning algorithm*) may, in each step, propose a hypothesis function h by making an *equivalence query* (EQ) to a *Teacher*. If h is equivalent to f, then the Teacher returns YES and the Learner succeeds and halts. Otherwise, the Teacher returns NO with a *counterexample*, which is an assignment z such that $h(z) \neq f(z)$. Moreover, the Learner may query the Teacher for the value of the function f on a particular assignment z by making a *membership query* (MQ) on z. The Teacher returns the value $f(z)$. We say that the Learner *learns* a class of functions $\mathscr{C}$, if for every function $f \in \mathscr{C}$, the Learner outputs a hypothesis h that is equivalent to f and does so in time polynomial in the *size* of a shortest representation of f and the length of a longest counterexample returned by the Teacher.

3   MTA Equivalence Is Interreducible with ACIT

In this section we show that the equivalence problem for $\mathbb{Q}$-multiplicity tree automata is logspace interreducible with **ACIT**. An analogous result, characterising equivalence of probabilistic visibly pushdown automata on words in terms of polynomial identity testing, has been shown in [10]. On several occasions in this section, we will implicitly make use of the fact that a composition of two logspace reductions is again a logspace reduction [3, Lemma 4.17].

First, we give a logspace reduction from the equivalence problem for $\mathbb{Q}$-MTAs to **ACIT**. We define $T_{\Sigma}^{\leq n} := \{t \in T_{\Sigma} : height(t) < n\}$ for any ranked alphabet Σ

and integer $n \in \mathbb{N}$. Seidl [13, Theorem 4.2] gave the following decision procedure for the equivalence of two $\mathbb{F}$-MTAs with $\mathbb{F}$ being an arbitrary field.

Proposition 2. *Suppose A and B are multiplicity tree automata of dimension n_1 and n_2, respectively, and over a ranked alphabet Σ. Then, A and B are equivalent if and only if $\|A\|(t) = \|B\|(t)$ for every $t \in T_\Sigma^{<n_1+n_2}$.*

Lemma 1. *Given a $\mathbb{Q}$-multiplicity tree automaton A over a ranked alphabet Σ and $n \in \mathbb{N}$ one can compute, in logarithmic space in $|A|$ and n, a variable-free arithmetic circuit that has output $\sum_{t \in T_\Sigma^{<n}} \|A\|(t)$.*

Proof. Let $A = (r, \Sigma, \mu, \gamma)$ and $\Sigma = \Sigma_0 \cup \cdots \cup \Sigma_m$. For every $i \in \mathbb{N}$, we have $T_\Sigma^{<i+1} = \{\sigma(t_1, \ldots, t_k) : k \in \{0, \ldots, m\}, \sigma \in \Sigma_k, t_1, \ldots, t_k \in T_\Sigma^{<i}\}$ and thus by bilinearity of Kronecker product

$$
\sum_{t \in T_\Sigma^{<i+1}} \mu(t) = \sum_{k=0}^{m} \sum_{\sigma \in \Sigma_k} \sum_{t_1 \in T_\Sigma^{<i}} \cdots \sum_{t_k \in T_\Sigma^{<i}} (\mu(t_1) \otimes \cdots \otimes \mu(t_k)) \cdot \mu(\sigma)
$$

$$
= \sum_{k=0}^{m} \sum_{\sigma \in \Sigma_k} \left(\left(\sum_{t_1 \in T_\Sigma^{<i}} \mu(t_1) \right) \otimes \cdots \otimes \left(\sum_{t_k \in T_\Sigma^{<i}} \mu(t_k) \right) \right) \cdot \mu(\sigma)
$$

$$
= \sum_{k=0}^{m} \left(\sum_{t \in T_\Sigma^{<i}} \mu(t) \right)^{\otimes k} \sum_{\sigma \in \Sigma_k} \mu(\sigma). \tag{1}
$$

Since $\sum_{t \in T_\Sigma^{<1}} \mu(t) = \sum_{\sigma \in \Sigma_0} \mu(\sigma)$, it follows from (1) that one can compute, in logarithmic space in $|A|$ and n, a variable-free arithmetic circuit with output $\sum_{t \in T_\Sigma^{<n}} \mu(t)$. More details of this reduction are given in the full version [12].

The result follows by observing that $\sum_{t \in T_\Sigma^{<n}} \|A\|(t) = (\sum_{t \in T_\Sigma^{<n}} \mu(t)) \cdot \gamma^\top$. ∎

Proposition 3. *The equivalence problem for $\mathbb{Q}$-multiplicity tree automata is logspace reducible to **ACIT**.*

Proof. Let A and B be $\mathbb{Q}$-multiplicity tree automata over a ranked alphabet Σ and n be the sum of their dimensions. Proposition 1 implies that

$$
\sum_{t \in T_\Sigma^{<n}} (\|A\|(t) - \|B\|(t))^2 = \sum_{t \in T_\Sigma^{<n}} \|A\|(t)^2 + \|B\|(t)^2 - 2\|A\|(t)\|B\|(t)
$$

$$
= \sum_{t \in T_\Sigma^{<n}} \|A \times A\|(t) + \|B \times B\|(t) - 2\|A \times B\|(t).
$$

Thus by Proposition 2, automata A and B are equivalent if and only if

$$
\sum_{t \in T_\Sigma^{<n}} \|A \times A\|(t) + \sum_{t \in T_\Sigma^{<n}} \|B \times B\|(t) - 2 \sum_{t \in T_\Sigma^{<n}} \|A \times B\|(t) = 0. \tag{2}
$$

We know from Proposition 1 that automata $A \times A$, $B \times B$, and $A \times B$ can be computed in logarithmic space. Thus by Lemma 1, one can compute in logarithmic space in $|A|$ and $|B|$ variable-free arithmetic circuits that have outputs $\sum_{t \in T_\Sigma^{\leq n}} \|A \times A\|(t)$, $\sum_{t \in T_\Sigma^{\leq n}} \|B \times B\|(t)$, and $\sum_{t \in T_\Sigma^{\leq n}} \|A \times B\|(t)$ respectively. Using (2) we can now easily construct a variable-free arithmetic circuit that has output 0 if and only if automata A and B are equivalent.   ∎

In the remainder of this section we present a converse reduction: from **ACIT** to equivalence testing for $\mathbb{Q}$-multiplicity tree automata.

Allender *et al.* [1, Proposition 2.2] give a logspace reduction of the general **ACIT** problem to the special case of **ACIT** for variable-free circuits. The latter can, by representing arbitrary integers as differences of two nonnegative integers, be reformulated as the problem of deciding whether two variable-free arithmetic circuits with only + and ×-internal gates compute the same number.

Proposition 4. *ACIT is logspace reducible to the equivalence problem for $\mathbb{Q}$-multiplicity tree automata.*

Proof. Let C_1 and C_2 be two variable-free arithmetic circuits whose internal gates are labelled with + or ×. By padding with extra gates, without loss of generality we can assume that in each circuit the children of a depth-i gate both have depth $i + 1$, +-gates have even depth, ×-gates have odd depth, and input gates all have the same even depth d.

In the following we reduce the problem of deciding whether C_1 and C_2 have the same output to equivalence testing for $\mathbb{Q}$-MTAs A_1 and A_2. Automata A_1 and A_2 are both defined over a ranked alphabet $\Sigma = \{\sigma_0, \sigma_1, \sigma_2\}$ where σ_0 is a nullary, σ_1 a unary, and σ_2 a binary symbol. Intuitively, A_1 and A_2 both recognise the common 'tree-unfolding' of C_1 and C_2.

We now derive A_1 from C_1; A_2 is analogously derived from C_2. Let $\{v_1, ..., v_r\}$ be the set of gates of C_1 where v_r is the output gate. Automaton A_1 has a state q_i for every gate v_i of C_1. Formally, $A_1 = (r, \Sigma, \mu, e_r)$ where for every $i \in [r]$, $\mu(\sigma_0)_i = 1$ if v_i is an input gate with label 1, otherwise $\mu(\sigma_0)_i = 0$. If v_i is a +-gate with children v_{j_1} and v_{j_2} then $\mu(\sigma_1)_{j_1,i} = \mu(\sigma_1)_{j_2,i} = 1$ if $j_1 \neq j_2$, $\mu(\sigma_1)_{j_1,i} = 2$ if $j_1 = j_2$, and $\mu(\sigma_1)_{l,i} = 0$ for every $l \notin \{j_1, j_2\}$. If v_i is an input or a ×-gate then $\mu(\sigma_1)^i = 0$. If v_i is a ×-gate with children v_{j_1} and v_{j_2} then $\mu(\sigma_2)_{(j_1,j_2),i} = 1$ and $\mu(\sigma_2)_{(l_1,l_2),i} = 0$ for every $(l_1, l_2) \neq (j_1, j_2)$. If v_i is an input or a +-gate then $\mu(\sigma_2)^i = 0$.

Define a sequence $(t_n)_{n \in \mathbb{N}_0} \subseteq T_\Sigma$ by $t_0 = \sigma_0$, $t_{n+1} = \sigma_1(t_n)$ for n odd and $t_{n+1} = \sigma_2(t_n, t_n)$ for n even. It can easily be seen that $\|A_1\|(t_d) = f_{C_1}$. (See the full version [12] for more details.) It is moreover clear by construction that for every $t \in T_\Sigma \setminus \{t_d\}$, $\|A_1\|(t) = 0$. Analogously we have that $\|A_2\|(t_d) = f_{C_2}$ and for every $t \in T_\Sigma \setminus \{t_d\}$, $\|A_2\|(t) = 0$. Therefore, automata A_1 and A_2 are equivalent if and only if C_1 and C_2 have the same output.   ∎

4   The Learning Algorithm

In this section we give an exact learning algorithm for multiplicity tree automata. Over an arbitrary field $\mathbb{F}$, the algorithm can be seen as running on a BSS machine that can store field elements and perform arithmetic operations and equality tests at unit cost [3]. Over $\mathbb{Q}$, the algorithm can be implemented in randomised polynomial time by representing rationals as arithmetic circuits [1].

Let $f \in \text{Rec}(\Sigma, \mathbb{F})$ be the target function. The algorithm learns an MTA-representation of f using its Hankel matrix H. Note that, by Theorem 1, H has finite rank over $\mathbb{F}$. At each stage the algorithm maintains the following data: an integer $n \in \mathbb{N}$, a set of n 'rows' $X = \{t_1, ..., t_n\} \subseteq T_\Sigma$, a finite set of 'columns' $Y \subseteq C_\Sigma$, and a submatrix $H_{X,Y}$ of H that has full row rank. These data determine a *hypothesis automaton* A whose states correspond to the rows of $H_{X,Y}$, with the i-th row being the state reached after reading t_i. Formally, algorithm LMTA is given in the following table. Here for any k-ary symbol $\sigma \in \Sigma$ we define $\sigma(X, ..., X) := \{\sigma(t_{i_1}, ..., t_{i_k}) : (i_1, ..., i_k) \in [n]^k\}$.

Algorithm LMTA

Target: $f \in \text{Rec}(\Sigma, \mathbb{F})$, where $\Sigma = \Sigma_0 \cup \cdots \cup \Sigma_m$ and $\mathbb{F}$ is a field

1. Make an equivalence query EQ(0).
   If the answer is YES then halt with **output** 0.
   Otherwise the answer is NO and z is a counterexample. Initialise:
   $n \leftarrow 1$, $t_n \leftarrow z$, $X \leftarrow \{t_n\}$, $Y \leftarrow \{\square\}$.
2. 2.1. For every $k \in \{0, ..., m\}$, $\sigma \in \Sigma_k$, and $(i_1, ..., i_k) \in [n]^k$:
   If $H_{\sigma(t_{i_1}, ..., t_{i_k}), Y}$ is not a linear combination of $H_{t_1, Y}, ..., H_{t_n, Y}$ then
   $n \leftarrow n + 1$, $t_n \leftarrow \sigma(t_{i_1}, ..., t_{i_k})$, $X \leftarrow X \cup \{t_n\}$.
   2.2. Define an $\mathbb{F}$-MTA $A = (n, \Sigma, \mu, \gamma)$ as follows.
   2.2.1. $\gamma^\top = H_{X,\square}$.
   2.2.2. For every $k \in \{0, ..., m\}$ and $\sigma \in \Sigma_k$:
   Define the matrix $\mu(\sigma) \in \mathbb{F}^{n^k \times n}$ by the equation

   $$\mu(\sigma) \cdot H_{X,Y} = H_{\sigma(X, ..., X), Y}.$$

3. 3.1. Make an equivalence query EQ($\|A\|$).
   If the answer is YES then halt with **output** $\|A\|$.
   Otherwise the answer is NO and z is a counterexample. Find a tree
   $\sigma(\tau_1, ..., \tau_k)$ which is a subtree of z and satisfies:
   (i) For every $j \in [k]$, $H_{\tau_j, Y} = \mu(\tau_j) \cdot H_{X,Y}$, and
   (ii) For some $c \in Y$, $H_{\sigma(\tau_1, ..., \tau_k), c} \neq \mu(\sigma(\tau_1, ..., \tau_k)) \cdot H_{X,c}$.
   3.2. For every $j \in [k]$ and $(i_1, ..., i_{j-1}) \in [n]^{j-1}$:
   $Y \leftarrow Y \cup \{c[\sigma(t_{i_1}, ..., t_{i_{j-1}}, \square, \tau_{j+1}, ..., \tau_k)]\}$.
   3.3. For every $j \in [k]$:
   If $H_{\tau_j, Y}$ is not a linear combination of $H_{t_1, Y}, ..., H_{t_n, Y}$ then
   $n \leftarrow n + 1$, $t_n \leftarrow \tau_j$, $X \leftarrow X \cup \{t_n\}$.
   3.4. Go to 2.

Correctness. On target $f \in \text{Rec}(\Sigma, \mathbb{F})$ algorithm LMTA halts with output f after at most $\text{rank}(H)$ iterations of the main loop. The proof is given in [12].

Succinct Representations. We assume that the counterexamples are represented as DAGs. Accordingly, we represent the trees in X as Σ-DAGs and the contexts in Y as $(\{\Box\} \cup \Sigma)$-DAGs.

As shown in Section 2, multiplicity tree automata can run directly on DAGs. Crucially, as explained in the proof of Theorem 3, Step 3.1 can be run directly on the DAG representation of the counterexample without unfolding. Therefore, the correctness proof holds for DAG representations as well.

Complexity Analysis. For every $n \in \mathbb{N}$, we denote by $M(n)$ the complexity of multiplying two matrices of dimension $n \times n$.

Theorem 3. *Let $f \in Rec(\Sigma, \mathbb{F})$ where Σ has rank m. Let H be the Hankel matrix of f and $r := rank(H)$ (over $\mathbb{F}$). Let A be a minimal MTA-representation of f. Then, f is learnable by algorithm* LMTA *making $r + 1$ equivalence queries, $|A|^2 + |A| \cdot s$ membership queries and $O(|A|^2 + r \cdot M(r) + |A| \cdot r \cdot s)$ arithmetic operations, where s denotes the size of a largest counterexample, represented as a DAG, that is returned during the execution of the algorithm.*

Proof. As noted above, LMTA halts with output f after at most r iterations of the loop consisting of Step 2 and Step 3, thus making at most $r + 1$ EQs.

Matrix $H_{X,Y}$ has full row rank (see [12] for details). This implies $|X| \leqslant r$. Note that in each iteration of Step 3.2 the cardinality of Y increases by at most $(n^m - 1)/(n - 1)$. Since the number of iterations of Step 3.2 is at most $r - 1$, on completion of the algorithm we have $|Y| \leqslant r^m$.

The number of membership queries made in Step 2 over the whole algorithm is $(\sum_{\sigma \in \Sigma} |\sigma(X, ..., X)| + |X|) \cdot |Y|$ since we need to ask for the values of the entries of $H_{X,Y}$ and $H_{\sigma(X,...,X),Y}$ for every $\sigma \in \Sigma$.

In Step 3.1 the procedure for finding a required sub-DAG of z is as follows: Check whether $H_{\tau,Y} = \mu(\tau) \cdot H_{X,Y}$ for all sub-DAGs τ of z in a nondecreasing order of height; stop when a sub-DAG τ is found such that $H_{\tau,Y} \neq \mu(\tau) \cdot H_{X,Y}$. Thus, the number of MQs made in one iteration of Step 3 is $size(z) \cdot |Y| \leqslant s \cdot |Y|$ since for every sub-DAG τ of z we need to ask for the values of the entries of $H_{\tau,Y}$. From here it follows by a simple calculation that the total number of MQs is at most $|A|^2 + |A| \cdot s$.

The bound on the number of arithmetic operations is proved in [12].   ■

The complexity of LMTA should be compared to the complexity of the algorithm of [9], which makes $r + 1$ EQs, $|A| \cdot s$ MQs and a number of arithmetic operations polynomial in $|A|$ and s, where s is the size of the largest counterexample given as a tree. Note that the algorithm of [9] cannot be straightforwardly adapted to work directly with DAG representations of trees since when given a counterexample z, every suffix of z is added to the set of columns. However, the tree unfolding of a DAG can have exponentially many different suffixes.

5   Lower Bounds on Query Complexity of Learning MTA

In this section, we study lower bounds on the query complexity of learning multiplicity tree automata in the exact learning model. Our results generalise the corresponding lower bounds for learning multiplicity word automata [5], and make no assumption about the computational model of the learning algorithm.

First, we give a lower bound on the query complexity of learning multiplicity tree automata over an arbitrary field.

Theorem 4. *Any exact learning algorithm that learns the class of multiplicity tree automata of dimension at most r, over a ranked alphabet (Σ, rk) and any field, must make at least $\sum_{\sigma \in \Sigma} r^{rk(\sigma)+1} - r^2$ queries.*

Proof. Take an arbitrary exact learning algorithm L that learns the class of multiplicity tree automata of dimension at most r, over a ranked alphabet (Σ, rk) and over any field. Let $\mathbb{F}$ be any field.

Define an extension field $\mathbb{K} := \mathbb{F}(\{z_{i,j}^{\sigma} : \sigma \in \Sigma, i \in [r^{rk(\sigma)}], j \in [r]\})$ of $\mathbb{F}$, where the set $\{z_{i,j}^{\sigma} : \sigma \in \Sigma, i \in [r^{rk(\sigma)}], j \in [r]\}$ is algebraically independent over $\mathbb{F}$. Define a $\mathbb{K}$-multiplicity tree automaton $A := (r, \Sigma, \mu, \gamma)$ where $\gamma = e_1 \in \mathbb{F}^{1 \times r}$ and $\mu(\sigma) = [z_{i,j}^{\sigma}]_{i,j} \in \mathbb{K}^{r^{rk(\sigma)} \times r}$ for every $\sigma \in \Sigma$. Define a tree series $f := \|A\|$. It is clear from the structure of A that the Hankel matrix of f has rank r.

We run algorithm L on the target function f. By assumption, the output of L is an MTA $A' = (r, \Sigma, \mu', \gamma')$ such that $\|A'\| \equiv f$. Let n be the number of queries made by L on target f. Let $t_1, ..., t_n \in T_{\Sigma}$ be the trees which L either made an MQ on, or received as counterexample to an EQ. Then for every $l \in [n]$, there exists a multivariate polynomial $p_l \in \mathbb{F}[(z_{i,j}^{\sigma})_{i,j,\sigma}]$ such that $f(t_l) = p_l$.

Note that A and A' are two minimal MTA-representations of f. Thus by Theorem 2, there exists an invertible matrix $U \in \mathbb{K}^{r \times r}$ such that $\gamma^{\top} = U \cdot (\gamma')^{\top}$ and $\mu(\sigma) = U^{\otimes rk(\sigma)} \cdot \mu'(\sigma) \cdot U^{-1}$ for every $\sigma \in \Sigma$. This implies that the entries of matrices $\mu(\sigma)$, $\sigma \in \Sigma$, are generated by the entries of U and $\{p_l : l \in [n]\}$. Since the entries of matrices $\mu(\sigma)$, $\sigma \in \Sigma$, form an algebraically independent set over $\mathbb{F}$, their number is at most $r^2 + n$. ∎

Secondly, we give a lower bound on the query complexity of learning multiplicity tree automata over a specific field $\mathbb{F}$.

Theorem 5. *Let $\mathbb{F}$ be an arbitrary field. Any exact learning algorithm that learns the class of $\mathbb{F}$-multiplicity tree automata of dimension at most r, over a ranked alphabet (Σ, rk) with at least one unary symbol and maximal rank m, must make number of queries at least*

$$\frac{1}{2^{m+1}} \cdot \left(\sum_{\sigma \in \Sigma} r^{rk(\sigma)+1} - r^2 - r \right). \tag{3}$$

Proof. Without loss of generality we assume that r is even, and write $n := r/2$. Let L be an exact learning algorithm for the class of $\mathbb{F}$-multiplicity tree automata of dimension at most r, over a ranked alphabet (Σ, rk) with $rk^{-1}(\{1\}) \neq \emptyset$. We will identify a family of functions $\mathcal{C}$ such that L has to make at least $\sum_{\sigma \in \Sigma} n^{rk(\sigma)+1} - n^2 - n$ queries to distinguish between the members of $\mathcal{C}$.

Let $\sigma_0, \sigma_1 \in \Sigma$ be nullary and unary symbols respectively. Let $P \in \mathbb{F}^{n \times n}$ be the permutation matrix corresponding to the cycle $(1, 2, ..., n)$. Define $\mathcal{A}$ to be the set of all $\mathbb{F}$-multiplicity tree automata $(2n, \Sigma, \mu, \gamma)$ where

- $\gamma = \begin{bmatrix} 1 & 0 \end{bmatrix} \otimes e_1$, $\mu(\sigma_0) = \begin{bmatrix} 1 & 0 \end{bmatrix} \otimes e_1$, and $\mu(\sigma_1) = I_2 \otimes P$;
- For each k-ary symbol $\sigma \in \Sigma \setminus \{\sigma_0, \sigma_1\}$, there exists $B(\sigma) \in \mathbb{F}^{n^k \times n}$ such that

$$\mu(\sigma) = \begin{bmatrix} 1 & 1 \end{bmatrix} \otimes \left(\begin{bmatrix} I_n \\ -I_n \end{bmatrix}^{\otimes k} \cdot B(\sigma) \right).$$

We define a set of recognisable tree series $\mathcal{C} := \{\|A\| : A \in \mathcal{A}\}$.

In Lemma 2, the proof of which is given in the full version [12], we state some properties of the functions in $\mathcal{C}$. More precisely, we show that the coefficient of a tree $t \in T_\Sigma$ in any $f \in \mathcal{C}$ fundamentally depends on whether t has 0, 1, or at least 2 nodes whose label is not σ_0 or σ_1. Here for every $i \in \mathbb{N}_0$ and $t \in T_\Sigma$, we use $\sigma_1^i(t)$ to denote the tree $\underbrace{\sigma_1(\sigma_1(...\sigma_1(t)...))}_{i}$.

Lemma 2. *The following properties hold for every $f \in \mathcal{C}$ and $t \in T_\Sigma$.*

(i) If $t = \sigma_1^j(\sigma_0)$ where $j \in \{0, 1, ..., n-1\}$, then $f(t) = \mathbb{1}_{\{j=0\}}$;

(ii) If $t = \sigma_1^j(\sigma(\sigma_1^{i_1}(\sigma_0), ..., \sigma_1^{i_k}(\sigma_0)))$ where $j, i_1, ..., i_k \in \{0, 1, ..., n-1\}$ and $\sigma \in \Sigma_k \setminus \{\sigma_0, \sigma_1\}$, then $f(t) = B(\sigma)_{(1+i_1, ..., 1+i_k), (1+n-j) \bmod n}$;

(iii) If $\sum_{\sigma \in \Sigma \setminus \{\sigma_0, \sigma_1\}} \#_\sigma(t) \geq 2$, then $f(t) = 0$.

Remark 1. As $P^n = I_n$, we have $\mu(\sigma_1)^n = I_{2n}$. Thus for every $f \in \mathcal{C}$, $\sigma \in \Sigma_k \setminus \{\sigma_0, \sigma_1\}$, and $j, i_1, ..., i_k \in \mathbb{N}_0$, it holds that $f(\sigma_1^j(\sigma_0)) = f(\sigma_1^{j \bmod n}(\sigma_0))$ and $f(\sigma_1^j(\sigma(\sigma_1^{i_1}(\sigma_0), ..., \sigma_1^{i_k}(\sigma_0)))) = f(\sigma_1^{j \bmod n}(\sigma(\sigma_1^{i_1 \bmod n}(\sigma_0), ..., \sigma_1^{i_k \bmod n}(\sigma_0))))$.

Run L on a target $f \in \mathcal{C}$. Lemma 2 and Remark 1 imply that when L makes an MQ on $t \in T_\Sigma$ where $\sum_{\sigma \in \Sigma \setminus \{\sigma_0, \sigma_1\}} \#_\sigma(t) \geq 2$ then the Teacher returns 0, while when L makes an MQ on $t = \sigma_1^j(\sigma_0)$ the Teacher returns 1 if $j \bmod n = 0$ and returns 0 otherwise. In these cases, L does not gain any new information about f since every function in $\mathcal{C}$ satisfies the values returned by the Teacher.

When L makes an MQ on $t = \sigma_1^j(\sigma(\sigma_1^{i_1}(\sigma_0), ..., \sigma_1^{i_k}(\sigma_0)))$ where $\sigma \in \Sigma_k \setminus \{\sigma_0, \sigma_1\}$, the Teacher returns an arbitrary number in $\mathbb{F}$ if $f(t)$ is not already known from an earlier query. Lemma 2 and Remark 1 imply that L thereby learns the entry $B(\sigma)_{(1+(i_1 \bmod n), ..., 1+(i_k \bmod n)), (1+n-j) \bmod n}$.

When L makes an EQ on a hypothesis function $h \in \mathcal{C}$, the Teacher finds some entry $B(\sigma)_{(i_1, ..., i_k), j}$ that L does not know from previous queries and returns tree $\sigma_1^{1+n-j}(\sigma(\sigma_1^{i_1-1}(\sigma_0), ..., \sigma_1^{i_k-1}(\sigma_0)))$ as counterexample.

With each query, L learns at most one entry of $B(\sigma)$ for some $\sigma \in \Sigma \setminus \{\sigma_0, \sigma_1\}$. The number of queries made by L on target f is therefore at least the total number of entries of $B(\sigma)$ for all $\sigma \in \Sigma \setminus \{\sigma_0, \sigma_1\}$, which is bounded below by the expression (3). ∎

Note that both lower bounds are exponential in the alphabet rank and polynomial in the automaton size. When the maximal rank is fixed, the lower bound in Theorem 5 is the same up to a constant factor as for learning over an arbitrary field. Moreover in this case, and assuming a Teacher that represents counterexamples as succinctly as possible, both lower bounds match the query complexity of algorithm LMTA to within a constant factor.

6 Future Work

Beimel *et al.* [4] apply their learning algorithm for multiplicity word automata to show exact learnability of various classes of DNF formulae and polynomials. Given that tree automata are more expressive than word automata, a natural direction for future work is to seek to apply our learning algorithm to derive new results for exact learning of DNF formulae and polynomials in the commutative and noncommutative cases.

Acknowledgements. The authors would like to thank Michael Benedikt for stimulating discussions and helpful advice.

References

1. Allender, E., Bürgisser, P., Kjeldgaard-Pedersen, J., Bro Miltersen, P.: On the complexity of numerical analysis. SIAM J. Comput. 38(5), 1987–2006 (2009)
2. Angluin, D.: Queries and concept learning. Machine Learning 2(4), 319–342 (1988)
3. Arora, S., Barak, B.: Computational Complexity: A Modern Approach. Cambridge University Press (2009)
4. Beimel, A., Bergadano, F., Bshouty, N.H., Kushilevitz, E., Varricchio, S.: Learning functions represented as multiplicity automata. J. ACM 47 (2000)
5. Bisht, L., Bshouty, N.H., Mazzawi, H.: On optimal learning algorithms for multiplicity automata. In: Learning Theory, pp. 184–198 (2006)
6. Bozapalidis, S., Alexandrakis, A.: Représentations matricielles des séries d'arbre reconnaissables. RAIRO-Theoretical Informatics and Applications-Informatique Théorique et Applications 23(4), 449–459 (1989)
7. Charatonik, W.: Automata on DAG representations of finite trees. Research Report MPI-I-1999-2-001, Max-Planck-Institut für Informatik, Saarbrücken (1999)
8. Droste, M., Kuich, W., Vogler, H.: Handbook of weighted automata. Springer (2009)
9. Habrard, A., Oncina, J.: Learning multiplicity tree automata. In: Sakakibara, Y., Kobayashi, S., Sato, K., Nishino, T., Tomita, E. (eds.) ICGI 2006. LNCS (LNAI), vol. 4201, pp. 268–280. Springer, Heidelberg (2006)
10. Kiefer, S., Murawski, A., Ouaknine, J., Wachter, B., Worrell, J.: On the complexity of equivalence and minimisation for $\mathbb{Q}$-weighted automata. Logical Methods in Computer Science 9(1) (2013)
11. Klivans, A.R., Shpilka, A.: Learning restricted models of arithmetic circuits. Theory of Computing 2(1), 185–206 (2006)
12. Marusic, I., Worrell, J.: Complexity of equivalence and learning for multiplicity tree automata. CoRR, abs/1405.0514 (2014)
13. Seidl, H.: Deciding equivalence of finite tree automata. SIAM J. Comput. 19(3), 424–437 (1990)

Monadic Datalog and Regular Tree Pattern Queries

Filip Mazowiecki, Filip Murlak, and Adam Witkowski

University of Warsaw, Poland

Abstract. Containment of monadic datalog programs over trees is decidable. The situation is more complex when tree nodes carry labels from an infinite alphabet that can be tested for equality. Then, it matters whether descendant relation is allowed or not: descendant relation can be eliminated easily from monadic programs only when label equalities are not used. With descendant, even containment of linear monadic programs in unions of conjunctive queries is undecidable and positive results are known only for bounded-depth trees.

We show that without descendant containment of connected monadic programs is decidable over ranked trees, but over unranked trees it is so only for linear programs. With descendant it becomes decidable over unranked trees under restriction to downward programs: each rule only moves down from the node in the head. This restriction is motivated by regular tree pattern queries, a recent formalism in the area of ActiveXML, which we show to be equivalent to linear downward programs.

1 Introduction

Among logics with fixpoint capabilities, one of the most prominent is datalog, obtained by adding fixpoint operator to unions of conjunctive queries (positive existential first order formulae). Datalog originated as a declarative programming language, but later found many applications in databases as a query language. Unfortunately, with increased expressive power comes undecidability of basic properties of queries. Classical static analysis problems of containment and equivalence are undecidable [18]. It is also undecidable if a given datalog program is equivalent to some non-recursive datalog program [13] (equivalence to a given non-recursive program is decidable [10]). Since these problems are subtasks in important data management tasks, like query minimization, data integration, or data exchange, the negative results for full datalog fuel interest in restrictions [5,7,8]. Important restrictions include *monadic* programs, using only unary intensional predicates; *linear* programs, with at most one use of an intensional predicate per rule; and *connected* programs, where within each rule all mentioned nodes are connected with each other.

When the class of considered structures is restricted to words or trees, even satisfiability is undecidable (see e.g. [1]), but for monadic datalog programs we regain decidability of containment [14]. This decidability result does not carry over to the case when tree nodes carry labels from an infinite alphabet that

E. Csuhaj-Varjú et al. (Eds.): MFCS 2014, Part I, LNCS 8634, pp. 426–437, 2014.

can be tested for equality. In this setting it matters whether descendant relation is allowed or not: descendant relation can be easily eliminated from monadic programs only if they do not use label equality. In a recent paper by Abiteboul et al. [1], it is shown that with descendant even containment of linear monadic programs in unions of conjunctive queries is undecidable. Restriction to bounded-depth trees restores decidability, even for containment of arbitrary programs in monadic programs, but this is not very surprising: without sibling order datalog programs cannot make much of unbounded branching.

In this paper we show that on trees over an infinite alphabet

1. containment is undecidable for child-only monadic programs, but it becomes decidable (in 3-ExpTime) under restriction to linear programs;
2. with descendant it becomes decidable (in 2-ExpTime) under restriction to *downward* programs: each rule only moves down from the node in the head.

To provide a broader background for these results, we first consider ranked trees, where the number of children is fixed. There, like for words, containment of monadic programs is decidable without descendant, but with descendant it is undecidable even for downward linear programs.

While forbidding descendant is a natural restriction (pointed out as an open problem in [1]), downward programs may seem exotic. But in fact, they were our initial point of interest. Our original motivation comes from the area of ActiveXML, where testing equivalence of ActiveXML systems amounts to solving the containment problem for regular tree pattern queries (RTPQs): essentially, regular expressions over the set of tree patterns using child and descendant axes and data equalities. In [3] it is shown that the containment problem is in ExpTime for RTPQs with restricted data comparisons, mimicking data comparisons allowed in XPath (an XML query language used widely in practice and extensively studied [6,12,16,17]). This result essentially relies on Figueira's ExpTime-completeness of satisfiability for RegXPath($\downarrow$, =), i.e., XPath with child axis and data equality, extended with Kleene star [12]. We show that RTPQs are equivalent to linear downward programs (a special case of a more general correspondence between monadic programs and a natural extension of RTPQs). Thus, our result on downward datalog can be seen as an extension of Figueira's result, immediately giving decidability of the general ActiveXML problem.

Organization. In Section 2 we introduce datalog and RTPQs, and explore the connections between them. Then, after a brief glance at containment over ranked trees in Section 3, we move on to unranked trees to discuss our main results in Section 4. We conclude in Section 5 with possible directions for future research. All the missing proofs can be found in the appendix, available on-line.

2   Datalog and RTPQs

Both formalisms work over finite unranked trees labeled with letters from an infinite alphabet Σ. We write $nodes_t$ for the set of nodes of tree t and lab_t :

$nodes_t \to \Sigma$ for the function assigning labels to nodes. We use standard notation for axes: $\downarrow, \downarrow_+, \uparrow, \uparrow^+$ stand for child, descendant, parent and ancestor relations. Binary relation $\sim$ holds between nodes with identical labels. We also have a unary predicate a for each $a \in \Sigma$, holding for the nodes labeled with a.

Many papers on tree-structured databases work with a slightly different data model, called "data trees", where each node has a label from a finite alphabet and a data value from an infinite data domain. In that model labels can be used explicitly in the formulas, but cannot be directly tested for equality, and data values can be tested for equality, but cannot be used explicitly (as constants). These two models are very similar, but not directly comparable for query languages with limited negation. However, we can quite easily incorporate additional finite alphabet to our setting, obtaining a generalization of the two settings, and the complexity results do not change. Forbidding the use of constants from the infinite alphabet may affect some of our lower bounds but it does not affect decidability. For the undecidable fragments of datalog we give an additional sketch of the proof for "data trees" in the appendix.

We begin with a brief description of the syntax and semantics of datalog; for more details see [2] or [9]. A **datalog program** $\mathcal{P}$ over a relational signature S is a set of rules of the form *head* $\leftarrow$ *body*, where *head* is an atom over S and *body* is a (possibly empty) conjunction of atoms over S written as a comma-separated list. There is a designated rule called the **goal** of the program. All variables in the body that are not used in the head, are implicitly quantified existentially. The relational symbols, or predicates, in S fall into two categories. **Extensional** predicates are the ones explicitly stored in the database; they are never used in the heads of rules. In our setting they come from $\{\downarrow, \downarrow_+, \sim\} \cup \Sigma$. Alphabet Σ is infinite, but program $\mathcal{P}$ uses only its finite subset $\Sigma_{\mathcal{P}}$. **Intensional** predicates, used in the heads, are defined by the rules. The program is evaluated by generating all the atoms (over intensional predicates) that can be inferred from the underlying structure (tree) by applying the rules repeatedly, to the point of saturation, and then taking atoms matching the head of the goal rule. Each inferred atom can be witnessed by a **proof tree**: an atom inferred by rule r from intensional atoms $A_1, A_2, \ldots, A_n$ is witnessed by a proof tree whose root has label r, and its children are the roots of proof trees for atoms A_i (if r has no intensional predicates in its body, the root has no children).

Example 1. The program below computes nodes from which one can reach label a along a path such that each node on the path has a child with identical label and a descendant with label b (or has label b itself).

$$P(X) \leftarrow X{\downarrow}Y, P(Y), X{\downarrow}Y', X \sim Y', Q(X) \quad (p_1)$$
$$P(X) \leftarrow a(X) \quad (p_2)$$
$$Q(X) \leftarrow X{\downarrow}Y, Q(Y) \quad (q_1)$$
$$Q(X) \leftarrow b(X) \quad (q_2)$$

The intensional predicates are P and Q, and the goal is P. The proof tree shown in the center witnesses that P holds in the root of the tree on the right.

In this paper we consider only **monadic** programs, i.e., programs whose intensional predicates are at most unary. Moreover, throughout the paper we assume that the programs do not use 0-ary intensional predicates. For general programs this is merely for the sake of simplicity: one can always turn 0-ary predicate Q to unary predicate $Q(X)$ by introducing a dummy variable X. For connected and downward programs (described below) this restriction matters.

A datalog program is **linear**, if the right-hand side of each rule contains at most one atom with an intensional predicate (proof trees for such programs are single branches). For a datalog rule r, let G_r be the graph whose vertices are the variables used in r and edge is placed between X and Y if the body of r contains an atomic formula $X{\downarrow}Y$ or $X{\downarrow_+}Y$. A program $\mathcal{P}$ is **connected** if for each rule $r \in \mathcal{P}$, G_r is connected. We say that $\mathcal{P}$ is **downward** if for each rule $r \in \mathcal{P}$, G_r is a directed tree whose root is the variable used in the head of r. The program from Example 1 is connected and downward, but not linear. In fact, each downward program is connected.

Previous work on datalog on arbitrary structures often considered the case of connected programs [11,13]. A practical reason is that real-life programs tend to be connected and linear (cf. [4]). Also, rules that are not connected combine pieces of unrelated data, corresponding to the *cross product*, an unnatural operation in the database context. It seems even more natural to assume connectedness, as we work with tree-structured databases. We shall do so.

We write $\mathsf{Datalog}(\downarrow, \downarrow_+)$ for the class of *connected monadic* datalog programs, and $\mathsf{Datalog}(\downarrow)$ for *connected monadic* programs that do not use the relation $\downarrow_+$. For linear or downward programs we shall use combinations of letters L and D, e.g., $\mathsf{LD\text{-}Datalog}(\downarrow)$ means linear downward programs from $\mathsf{Datalog}(\downarrow)$.

Recall that **conjunctive queries** (CQs) are existential first order formulas of the form $\exists x_1 \ldots x_k \varphi$, where φ is a conjunction of atoms. Sometimes we will also speak of **unions of conjunctive queries** (UCQs), corresponding to programs with a single intensional predicate (goal), which is never used in the bodies of rules. Like for datalog we use notation $\mathsf{CQ}(\downarrow, \downarrow_+)$, $\mathsf{CQ}(\downarrow)$, $\mathsf{UCQ}(\downarrow, \downarrow_+)$, $\mathsf{UCQ}(\downarrow)$.

We now move to the second formalism. Let $\Delta = \{x_1, x_2, \ldots\}$ be an infinite set of variables and let $\tau \subseteq \{\downarrow, \downarrow_+, \uparrow, \uparrow^+\}$ be a non-empty set of axes. A **graph pattern** over τ is a directed multigraph with vertices labeled with elements of $\Sigma \cup \Delta$ and edges labeled with elements of τ. We write $vertices_\pi$ for the set of vertices of pattern π and $lab_\pi : vertices_\pi \to \Sigma \cup \Delta$ for the labeling function. Additionally, each pattern π has two distinguished nodes: $in(\pi)$ and $out(\pi)$.

Definition 1. *A **homomorphism** $h : \pi \to t$ from a pattern π over τ to a tree t is a function $h : vertices_\pi \to nodes_t$ such that for all vertices v, w in π*

- *if $lab_\pi(v) \in \Sigma$, then $lab_t(h(v)) = lab_\pi(v)$;*
- *if $lab_\pi(v) = lab_\pi(w) \in \Delta$, then $h(v) \sim h(w)$; and*
- *h preserves binary relations from τ.*

We write $t, v, w \models \pi$ if there is a homomorphism from π to t that maps $in(\pi)$ to v and $out(\pi)$ to w. If this is the case we also write $t, v \models \pi$ and $t \models \pi$.

A **tree pattern query** (TPQ) over τ is a graph pattern π over τ that is a directed tree, whose *in* node is the root. The set of all TPQs over τ is denoted by $\mathsf{TPQ}(\tau)$. Let $\Pi = \pi_1 \cdot \ldots \cdot \pi_n$ be a word over the alphabet $\mathsf{TPQ}(\tau)$, with $\pi_i \in \mathsf{TPQ}(\tau)$. A homomorphism $h : \Pi \to t$ is a sequence of homomorphisms $h_i : \pi_i \to t$ for $i = 1, \ldots, n$ such that $h_i(out(\pi_i)) = h_{i+1}(in(\pi_{i+1}))$ for all $i < n$. We write $t \models \Pi$ if there is a homomorphism $h : \Pi \to t$; $t, v \models \Pi$ if h maps $in(\pi_1)$ to v; and $t, v, w \models \Pi$ if additionally h maps $out(\pi_n)$ to w.

Definition 2. *A **regular tree pattern query** (RTPQ) φ over τ is a regular expression over the alphabet $\mathsf{TPQ}(\tau)$. $L(\varphi)$ denotes the language generated by φ. We write $t, v, w \models \varphi$ iff there is $\Pi \in L(\varphi)$ such that $t, v, w \models \Pi$, and similarly for $t, v \models \varphi$ and $t \models \varphi$. The set of all RTPQs over τ is denoted by $\mathsf{RTPQ}(\tau)$.*

Example 2. To illustrate the power of RTPQs, we show how to simulate the n-bit binary counter, enumerating values from 0 to $2^n - 1$, with an RTPQ of size $\mathcal{O}(n)$. To increase the counter, we need to find the least significant 0, change it to 1 and change all less significant bits to 0; all more significant bits remain unchanged. Skipping $\downarrow$ to ease the notation, we can express this with patterns

$$inc_i = (X_1)^{in} X_2 \cdots X_{n-i} 01 \cdots 1 (X_1)_{out} X_2 \cdots X_{n-i} 10 \cdots 0 \quad \text{for } i < n,$$
$$inc_n = (0)^{in} 1 \cdots 1 (1)_{out} 0 \cdots 0.$$

Then, we combine patterns inc_i into a counter going from 0 to $2^n - 1$:

$$c_n = val_0 \cdot \big(inc_1 \cup inc_2 \cup \cdots \cup inc_n \big)^* \cdot val_{2^n-1}$$

where val_k is number k stored in n bits in binary, e.g., $val_0 = (0)^{in}_{out} 0 \cdots 0$. This expression works if we run it on a tree consisting of a single branch: if such tree satisfies c_n, then it must contain all values from 0 to $2^n - 1$. Note that by locating the *out* node in inc_i on the $(n+1)$-th position, we pass many values between two consecutive TPQs, because the following nodes must overlap with the initial n nodes of the next pattern. On arbitrary trees this is no longer the case, but we can make the expression work by running it up the tree, i.e., replacing $\downarrow$ with $\uparrow$.

The usual correspondence between patterns and CQs holds also for trees over infinite alphabet, except that relation $\sim$ in patterns is not represented explicitly, but by repeated labels from Δ. Hence, translation from patterns to CQs involves quadratic blowup. As observed in [15], each satisfiable graph pattern can be expressed as a union of TPQs reflecting different ways of mapping the pattern to a tree. While the size of the TPQs can be bounded by the size of the pattern, their number is exponential. Given this, the following fact is almost immediate.

Proposition 1. *The following classes of queries have the same expressive power:*
- *$\mathsf{RTPQ}(\downarrow, \downarrow_+, \uparrow, \uparrow^+)$ and $\mathsf{L\text{-}Datalog}(\downarrow, \downarrow_+)$;*   *$\mathsf{RTPQ}(\downarrow, \uparrow)$ and $\mathsf{L\text{-}Datalog}(\downarrow)$;*
- *$\mathsf{RTPQ}(\downarrow, \downarrow_+)$ and $\mathsf{DL\text{-}Datalog}(\downarrow, \downarrow_+)$;*   *$\mathsf{RTPQ}(\downarrow)$ and $\mathsf{DL\text{-}Datalog}(\downarrow)$.*

Translations to datalog are polynomial; translations to RTPQs are exponential.

When rules are translated into patterns, and *vice versa*, variables correspond to nodes; the variable in the head corresponds to the *in* node, and the variable used in the intensional atom (unique in linear programs) corresponds to the *out* node. Thus, we may speak of *in* or *out* variables, and head or intensional nodes.

The connection between datalog and patterns goes beyond linear programs. Consider patterns with an additional relation ϵ; we indicate the use of ϵ-edges by including ϵ in the signature, e.g., $\mathsf{TPQ}(\downarrow, \downarrow_+, \epsilon)$. A homomorphism from such a pattern π is a family of homomorphisms from all maximal ϵ-free subpatterns of π, such that nodes connected by ϵ-edges are mapped to the same tree node.

Since proof trees of linear programs are single branches, they can be interpreted as words of graph patterns (corresponding to rule bodies). Each such word has a natural representation as a pattern with ϵ-edges: simply add an ϵ-edge from the *out* node of each pattern to the *in* node of the following pattern in the word. This gives a pattern representation for proof trees of linear programs. To generalize it to non-linear programs, let π_r be the graph pattern corresponding to the conjunction of extensional atoms in the body of rule r, and let $\mathcal{G}_\mathcal{P}$ be the set of all π_r's. Then, to each proof tree one can associate a graph pattern by replacing each r-labeled node by the pattern π_r together with appropriate ϵ-edges between π_r's intensional nodes and its children's head nodes. Patterns obtained this way are called **witnessing patterns**. Note that in witnessing patterns, each maximal connected ϵ-free subpattern corresponds to a rule of $\mathcal{P}$.

3 Ranked Trees

Throughout the paper we work with the Boolean variant of containment, i.e., satisfiability of $\mathcal{P} \wedge \neg \mathcal{Q}$, where $\mathcal{P}$ and $\mathcal{Q}$ are treated as Boolean queries, with all variables quantified out existentially. Using well known methods one can reduce the unary variant to the Boolean one (not by simply rewriting queries, though).

To warm up, we first look at the case of words, where many ideas can be illustrated without much of the technical difficulty of trees. Over words, the relations $\downarrow$ and $\downarrow_+$ are interpreted as "next position" and "following position".

Over data words and data trees, containment of monadic datalog programs with descendant is undecidable – even containment of linear monadic programs in unions of conjunctive queries [1, Proposition 3.3]. As discussed in Section 2, despite obvious similarities between our setting and the setting of data trees, neither lower bounds nor upper bounds carry over immediately. The reduction in [1] relies on the presence of finite alphabet and cannot be directly adapted to our setting, but with a little effort the use of finite alphabet can be eliminated (see Appendix). Once descendant is disallowed, we immediately regain decidability.

Proposition 2. *Containment over words is*
1. *undecidable for* $\mathsf{LD\text{-}Datalog}(\downarrow, \downarrow_+)$ *(even containment in UCQs);*
2. *in* EXPSPACE *for* $\mathsf{Datalog}(\downarrow)$ *and* PSPACE-*complete for* $\mathsf{L\text{-}Datalog}(\downarrow)$.

Proof. An easy reduction based on the binary counter encoding from Sect. 2 shows that even satisfiability for $\mathsf{L\text{-}Datalog}(\downarrow)$ programs is PSPACE-hard (see Appendix). Here we show that containment for $\mathsf{L\text{-}Datalog}(\downarrow)$ is in PSPACE.

Let $\mathcal{P}, \mathcal{Q} \in$ L-Datalog($\downarrow$). Recall that $\mathcal{G}_{\mathcal{P}}$ is the set of patterns corresponding to the bodies of $\mathcal{P}$'s rules and for each $\pi \in \mathcal{G}_{\mathcal{P}}$, $in(\pi)$ and $out(\pi)$ are the nodes corresponding to the head variable and the variable in the intensional atom in the rule giving rise to π. As we work over words and disallow $\downarrow_+$ (following position), w.l.o.g. we can assume that patterns in $\mathcal{G}_{\mathcal{P}}$ are words (with the in and out nodes placed in arbitrary positions). Since the programs are linear, the witnessing patterns are essentially words as well: concatenations of words from $\mathcal{G}_{\mathcal{P}}$, decorated with ϵ-edges connecting the out node of the preceding pattern to the in node of the subsequent pattern. We denote the set of words corresponding to witnessing patterns by $L(\mathcal{P})$. We write $L_{pref}(\mathcal{P}), L_{inf}(\mathcal{P}), L_{suf}(\mathcal{P})$ for the languages of prefixes, infixes, and suffixes of words from $L(\mathcal{P})$.

Satisfiability of $\varphi = \mathcal{P} \wedge \neg \mathcal{Q}$ can be verified over finite alphabet $\Sigma_0 = \Sigma_\varphi \,\dot{\cup}\, \{a_1, a_2, \ldots, a_n\}$, where Σ_φ is the set of labels used explicitly by rules in $\mathcal{P}, \mathcal{Q}$; and n is the maximal number of nodes in a pattern in $\mathcal{G}_{\mathcal{P}} \cup \mathcal{G}_{\mathcal{Q}}$. To relabel a word satisfying φ so that it uses only labels from Σ_0, simply process it from left to right, replacing each letter not in Σ_φ with the *least* recently used letter a_i: when an occurrence of a letter b is to be replaced by a_i, first change all later occurrences of a_i to a fresh letter, then replace all later occurrences of b with a_i. This policy ensures that equality of labels within distance n is not affected.

For program $\mathcal{P}$ we construct a deterministic automaton $\mathcal{A}_{\mathcal{P}}$ recognizing words over alphabet Σ_0 satisfying $\mathcal{P}$. Let w be the whole word read by $\mathcal{A}_{\mathcal{P}}$ and let w_s be the recently read n-letter suffix. States of $\mathcal{A}_{\mathcal{P}}$ have two components. The first component is a word from $\bigcup_{k \leq n}(\Sigma_0)^k$, corresponding to the suffix w_s (initially, there are less then n letters to remember). The second component is a subset of

$$\{1, 2, \ldots, n+1\} \times \mathcal{G}_{\mathcal{P}} \times \{1, 2, \ldots, n+1\} \times \mathcal{G}_{\mathcal{P}}.$$

For $p, p' \leq n$, each element (p, π, p', π') of this set corresponds to a subpattern $\Pi \in L_{inf}(\mathcal{P})$, which starts from π, ends with π', and a homomorphism $h : \Pi \to w$ that maps $in(\pi)$ to $w[|w| - |w_s| + p]$ and $out(\pi')$ to $w[|w| - |w_s| + p']$. The intended meaning is that the state q stores the information about the first and last subpattern of Π from $\mathcal{G}_{\mathcal{P}}$ for every homomorphism from patterns in $L_{inf}(\mathcal{P})$ to w such that the in node from the first subpattern and the out node from the last subpattern are mapped to w_s. This information is sufficient to build information about larger patterns by induction.

The additional value $n+1$ in the first coordinate indicates that the represented homomorphism is from $\Pi \in L_{pref}(\mathcal{P})$ to w, and we only remember the information about the last part of Π. Similarly, $n + 1$ in the third component indicates that $\Pi \in L_{suf}(\mathcal{P})$. Thus, tuples with $n + 1$ at the first and third coordinate represent $\Pi \in L(\mathcal{P})$; states containing such tuples are accepting.

The automaton starts in state $(\varepsilon, \emptyset)$. When it reads letter a in state (u, Φ), it moves to state (v, Ψ), where v is the suffix of ua of length at most n and Ψ is defined as follows. Set Ψ contains all elements of Φ, but the first and third coordinates that are at most n are decreased by 1 (unless $|v| < n$). If either drops below 1, the element is discarded. Moreover, for all $\pi \in \mathcal{G}_{\mathcal{P}}$, if there is a homomorphism η from π to w_s and the image contains the new letter, then

Ψ contains all elements $(\eta(in(\pi)), \pi, \eta(out(\pi)), \pi)$. If π corresponds to a non-recursive rule, we also add tuple $(\eta(in(\pi)), \pi, n+1, \pi)$. If π corresponds to the goal rule, we add tuple $(n+1, \pi, \eta(out(\pi)), \pi)$. If both of these hold, we add tuple $(n+1, \pi, n+1, \pi)$. Finally, we close Ψ under legal concatenations: if Ψ contains (p_1, π_1, p_2, π_2) and (p_3, π_3, p_4, π_4) such that $p_2 = p_3$ and $\pi_2 \cdot \pi_3 \in L_{inf}(\varphi_i)$, then Ψ should also contain (p_1, π_1, p_4, π_4).

We construct an analogous automaton for $\mathcal{Q}$. Since the automata are deterministic, we easily get a product automaton equivalent to $\mathcal{P} \wedge \neg \mathcal{Q}$. The size of $\mathcal{A}_{\mathcal{P}}$ is exponential in size of $\mathcal{P}$, but its states and transitions can be generated on the fly in polynomial space. To check its emptiness we make a simple reachability test, which is in NLOGSPACE. Altogether, this gives a PSPACE algorithm.   □

The results for words can be lifted to ranked trees: complexities are higher, but the general picture remains the same.

Theorem 1. *Over ranked trees containment is*
1. *undecidable for* LD-Datalog$(\downarrow, \downarrow_+)$ *(even containment in UCQs);*
2. *in* 3-EXPTIME *for* Datalog$(\downarrow)$ *and* 2-EXPTIME-*complete for* L-Datalog$(\downarrow)$.

Note that undecidability for LD-Datalog$(\downarrow, \downarrow_+)$ over words immediately gives undecidability for L-Datalog$(\downarrow, \downarrow_+)$ over trees (ranked or unranked): we simply run the programs up the tree (when we go up, a tree looks like a word); showing undecidability for LD-Datalog$(\downarrow, \downarrow_+)$ over ranked trees requires additional effort.

4   Unranked Trees

We have seen in Section 3 that over ranked trees containment is undecidable for downward programs, but decidable for child-only programs. Over unranked trees exactly the opposite happens. Containment for general programs, even for L-Datalog$(\downarrow, \downarrow_+)$, remains undecidable as explained in Section 3, but the reduction for LD-Datalog$(\downarrow, \downarrow_+)$ in Theorem 1 relies heavily on the fixed number of children and does not go through for unranked trees. In Section 4.1 we obtain decidability even for D-Datalog$(\downarrow, \downarrow_+)$. More surprisingly, we lose decidability for Datalog$(\downarrow)$: using non-linearity and recursion we can simulate $\downarrow_+$ to a point sufficient to repeat (upwards) the reduction for words.

Proposition 3. *Over unranked trees containment of* Datalog$(\downarrow)$ *programs in* UCQ$(\downarrow)$ *queries is undecidable.*

In Section 4.2 we show how to get decidability for *linear* Datalog$(\downarrow)$ programs.

4.1   Downward Programs

The aim of this section is to sketch out the proof of the following theorem.

Theorem 2. *Over unranked trees containment is* 2-EXPTIME-*complete for* D-Datalog$(\downarrow, \downarrow_+)$, *and* EXPSPACE-*complete for* LD-Datalog$(\downarrow, \downarrow_+)$.

In fact, in the non-linear case we do not even need $\downarrow_+$ to show hardness. This is not so surprising: already in Proposition 3 we simulate $\downarrow_+$ using non-linearity.

The proof of the positive claims of Theorem 2 bears some similarity to the argument showing decidability of containment of datalog programs in UCQs over arbitrary structures [10]. We show that satisfiability of $\mathcal{P} \wedge \neg \mathcal{Q}$ can be tested over special trees, so called canonical models, and the set of suitable encodings of canonical models satisfying $\mathcal{P} \wedge \neg \mathcal{Q}$ is recognized by a tree automaton.

So far our decidability results relied on bounding the number of labels in the models of $\mathcal{P} \wedge \neg \mathcal{Q}$; this requires a bound on the number of children. Over unranked trees we do the opposite: we use the fact that witnessing patterns of downward programs admit injective (up to ϵ-edges) homomorphisms into unranked trees, and we make the labels as different from each other as permitted by $\mathcal{P}$.

Definition 3. *A tree is a* **canonical model** *for a satisfiable pattern π from* $\mathsf{TPQ}(\downarrow, \downarrow_+, \epsilon)$*, if it can be obtained in the course of the following procedure:*

1. *rename variables so that maximal ϵ-free subpatterns use disjoint sets of variables (do not change equalities within maximal ϵ-free subpatterns);*
2. *unify labels of nodes connected with ϵ-edges (since π is satisfiable, whenever end-points of an ϵ-edge have different labels, at least one is labeled with a variable: replace all occurrences of this variable with the other label);*
3. *merge all nodes connected with ϵ-edges;*
4. *substitute each variable with a fresh label;*
5. *replace each $\downarrow_+$ edge with a sequence of nodes labeled with fresh variables, connected with $\downarrow$ edges.*

Note that a pattern from $\mathsf{TPQ}(\downarrow, \downarrow_+, \epsilon)$ represents a family of canonical models: fresh labels and, more importantly, the lengths of $\downarrow_+$ paths are chosen arbitrarily.

Lemma 1. *$\mathcal{P} \wedge \neg \mathcal{Q}$ is satisfiable over unranked trees iff it is satisfiable in a canonical model of a witnessing pattern of $\mathcal{P}$.*

Let N be the maximal number of variables in a rule in $\mathcal{P}$. Let $\mathsf{TPQ}^{\mathcal{P}}(\downarrow, \downarrow_+, \epsilon)$ be the class of patterns from $\mathsf{TPQ}(\downarrow, \downarrow_+, \epsilon)$ with branching bounded by N and labels coming from $\Sigma_{\mathcal{P}} \subseteq \Sigma$ (labels used explicitly in $\mathcal{P}$) or a fixed set $\Delta_0 \subseteq \Delta$ of size N. As $\mathcal{P}$ is downward, its witnessing patterns are elements of $\mathsf{TPQ}^{\mathcal{P}}(\downarrow, \downarrow_+, \epsilon)$.

Patterns in $\mathsf{TPQ}^{\mathcal{P}}(\downarrow, \downarrow_+, \epsilon)$ can be viewed as trees over the alphabet $(\Sigma_{\mathcal{P}} \cup \Delta_0) \times \{\downarrow, \downarrow_+, \epsilon\}$, where the first component is the label of the node in the pattern, and the second component determines the kind of edge between the node and its parent (in the root the second component plays no role, we may assume that is ϵ). Conversely, each tree over this alphabet corresponds to a pattern from $\mathsf{TPQ}^{\mathcal{P}}(\downarrow, \downarrow_+, \epsilon)$ (up to the second component of the root label). The technical core of Theorem 2 is the following lemma.

Lemma 2. *The set of patterns $\pi \in \mathsf{TPQ}^{\mathcal{P}}(\downarrow, \downarrow_+, \epsilon)$ such that $\mathcal{P} \wedge \neg \mathcal{Q}$ is satisfied in a canonical model of π is recognized by a double exponential tree automaton, whose states and transitions can be enumerated in exponential working memory.*

To test satisfiability of $\mathcal{P} \wedge \neg \mathcal{Q}$, we generate the automaton and test its emptiness. Since emptiness can be tested in PTIME, this is a 2-EXPTIME algorithm.

If $\mathcal{P}$ is linear ($\mathcal{Q}$ may be non-linear), we can do better. As proof trees of linear programs are words, their witnessing patterns have a very particular shape: one main branch (corresponding to the recursive calls of intensional predicates) and small subtrees off this branch with at most N branches (fragments of patterns corresponding $\mathcal{P}$'s rules). Over such trees emptiness of a given automaton $\mathcal{B}$ can be tested by a non-deterministic algorithm in space $\mathcal{O}(N \log |\mathcal{B}|)$. Indeed, the claim is well known (and easy to prove) for trees with at most N branches (see e.g., [17]). In our case, the algorithm can keep guessing the main path bottom-up, together with the states in the nodes just off the path, and use the previous result as a subprocedure to test if $\mathcal{B}$ can accept from these states.

Generating the automaton from Lemma 2 on the fly in exponential working memory, we can test its emptiness on trees of special shape in EXPSPACE. A natural generalization of the claim above gives EXPSPACE algorithm for *nested linear programs*, where any predicate defined by a linear subprogram can be used freely outside of this subprogram. Nested linear programs are a more robust class then linear programs, e.g., they are closed under conjunction.

4.2   Linear Child-Only Programs

For child-only programs, we get decidability if we assume linearity.

Theorem 3. *Over unranked trees containment for* L-Datalog($\downarrow$) *is in 3-*EXPTIME*; containment of* L-Datalog($\downarrow$) *programs in* UCQ($\downarrow$) *queries is 2-*EXPTIME*-complete.*

The general strategy of the proof is as for downward programs, but this time we face a new difficulty: witnessing patterns are not trees any more and they need not admit injective homomorphisms into trees. Indeed, rules of L-Datalog($\downarrow$) programs can be turned into tree patterns by merging nodes sharing a child, and witnessing patterns are sequences of tree patterns connected with ϵ-edges between *out* nodes and *in* nodes—but this time *in* nodes need not be roots.

The first step to fix this is to adjust canonical models: in item 3 of Definition 3 we merge not only nodes connected with ϵ-edges, but also nodes with a common child; item 5 becomes void. After this modification we can reprove Lemma 1 for Datalog($\downarrow$) programs (linearity plays no role here).

Unlike for downward programs, each witnessing pattern has a unique canonical model (up to the choice of fresh labels). To test satisfiability of $\mathcal{P} \wedge \neg \mathcal{Q}$ we need to find a witnessing pattern π for $\mathcal{P}$ whose canonical model does not satisfy $\mathcal{Q}$. This is also more involved than for downward programs: nodes arbitrarily far apart in π may represent the same node of the canonical model. An automaton cannot compute this correspondence; we need to make it explicit.

One can think of an L-Datalog($\downarrow$) program as a tree-walking automaton that moves from node to node checking some local conditions, until it reaches an accepting state: a satisfied non-recursive rule. Each time the node to move to is determined by the **spine** of the current rule: the shortest path between the *in* and

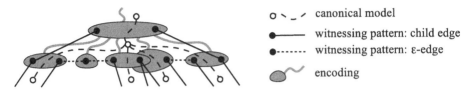

Fig. 1. Encoding of witnessing patterns

out nodes of the corresponding tree pattern. Let $Spine(\mathcal{P})$ be the disjoint union of the nodes in the spines of $\mathcal{P}$'s rules. In our encoding of witnessing patterns (Figure 1), an unranked tree over the alphabet $\left(Spine(\mathcal{P}) \cup \{\top\}\right)^2$ represents the way the sequence of spines is mapped to the corresponding canonical model (the remaining nodes can be computed on the fly by the automaton); $\top$ is used in nodes traversed once, only going down or only going up. The proof of Theorem 3 amounts to showing the following lemma.

Lemma 3. *The set of encodings of witnessing patterns of $\mathcal{P}$ whose canonical models satisfy $\mathcal{P} \wedge \neg \mathcal{Q}$ is regular. The recognizing (unranked) tree automaton can be computed in* 3-EXPTIME *in general, and in* 2-EXPTIME *if* $\mathcal{Q} \in \mathsf{UCQ}(\downarrow)$.

5 Conclusions

The containment problem for connected monadic datalog on trees over infinite alphabet is undecidable. We considered two restrictions: downward programs, $\mathsf{D\text{-}Datalog}(\downarrow, \downarrow_+)$, and child-only programs, $\mathsf{Datalog}(\downarrow)$. Table 1 summarizes our results. It is not difficult to extend the algorithm for linear $\mathsf{Datalog}(\downarrow)$ over unranked trees to cover non-connected programs; for the remaining algorithms this remains an open question.

Table 1. Complexity of containment for datalog fragments

	Unranked trees		Ranked trees	
	linear	non-linear	linear	non-linear
$\mathsf{D\text{-}Datalog}(\downarrow, \downarrow_+)$	EXPSPACE	2-EXPTIME	UNDEC.	UNDEC.
$\mathsf{Datalog}(\downarrow)$	in 3-EXPTIME	UNDEC.	2-EXPTIME	in 3-EXPTIME

We also investigated connections between monadic datalog and extensions of regular tree pattern queries, discovering natural translations between the two formalisms; translations to RTPQs involve exponential blow-up. Thus, results on datalog give direct corollaries about corresponding classes of RTPQs. In particular, the result on linear $\mathsf{D\text{-}Datalog}(\downarrow, \downarrow_+)$ gives an EXPSPACE upper bound for containment of $\mathsf{RTPQ}(\downarrow, \downarrow_+)$, which solves an open problem from [3].

For child-only queries we provide few tight complexity bounds. Some gaps stand out in Table 1, but we would like to point out one that is less apparent, but more intriguing. Like all our lower bounds, the ones for $\mathsf{Datalog}(\downarrow)$ work for containment in UCQs, but these bounds do not carry over to unary queries. For example, containment over unranked trees of linear Datalog($\downarrow$) programs in UCQ($\downarrow$) is 2-ExpTime-complete, but containment of unary linear Datalog($\downarrow$) programs in unary UCQ($\downarrow$) queries could have much lower complexity. Also, for linear D-Datalog($\downarrow$) we have only PSpace-hardness.

Acknowledgments. First and foremost we thank Pierre Bourhis for inspiring discussions and for encouraging us to generalize our results to non-linear programs. We also thank the anonymous referees for their helpful comments. We acknowledge the support of grant DEC-2013/09/N/ST6/01170 from Poland's National Science Center.

References

1. Abiteboul, S., Bourhis, P., Muscholl, A., Wu, Z.: Recursive queries on trees and data trees. In: ICDT 2013, pp. 93–104 (2013)
2. Abiteboul, S., Hull, R., Vianu, V.: Foundations of Databases. Addison Wesley (1995)
3. Abiteboul, S., ten Cate, B., Katsis, Y.: On the equivalence of distributed systems with queries and communication. In: ICDT 2011, pp. 126–137 (2011)
4. Bancilhon, F., Ramakrishnan, R.: An amateur's introduction to recursive query processing strategies. ACM SIGMOD 1986, 16–52 (1986)
5. Benedikt, M., Bourhis, P., Senellart, P.: Monadic datalog containment. In: Czumaj, A., Mehlhorn, K., Pitts, A., Wattenhofer, R. (eds.) ICALP 2012, Part II. LNCS, vol. 7392, pp. 79–91. Springer, Heidelberg (2012)
6. Benedikt, M., Fan, W., Geerts, F.: XPath satisfiability in the presence of DTDs. J. ACM 55(2) (2008)
7. Bonatti, P.: On the decidability of containment of recursive datalog queries - preliminary report. In: PODS 2004, pp. 297–306 (2004)
8. Calvanese, D., De Giacomo, G., Vardi, M.: Decidable containment of recursive queries. Theor. Comput. Sci. 336(1), 33–56 (2005)
9. Ceri, S., Gottlob, G., Tanca, L.: Logic programming and databases. Springer (1990)
10. Chaudhuri, S., Vardi, M.: On the equivalence of recursive and nonrecursive datalog programs. In: PODS 1992, pp. 55–66 (1992)
11. Cosmadakis, S., Gaifman, H., Kanellakis, P., Vardi, M.: Decidable optimization problems for database logic programs (preliminary report). In: STOC 1988, pp. 477–490 (1988)
12. Figueira, D.: Satisfiability of downward XPath with data equality tests. In: PODS 2009, pp. 197–206 (2009)
13. Gaifman, H., Mairson, H., Sagiv, Y., Vardi, M.: Undecidable optimization problems for database logic programs. J. ACM 40(3), 683–713 (1993)
14. Gottlob, G., Koch, C.: Monadic datalog and the expressive power of languages for web information extraction. J. ACM 51(1), 74–113 (2004)
15. Gottlob, G., Koch, C., Schulz, K.: Conjunctive queries over trees. J. ACM 53(2), 238–272 (2006)
16. Miklau, G., Suciu, D.: Containment and equivalence for a fragment of XPath. J. ACM 51(1), 2–45 (2004)
17. Neven, F., Schwentick, T.: On the complexity of XPath containment in the presence of disjunction, DTDs, and variables. Log. Meth. Comput. Sci. 2(3) (2006)
18. Shmueli, O.: Equivalence of datalog queries is undecidable. J. Log. Program. 15(3), 231–241 (1993)

Model Checking Concurrent Recursive Programs Using Temporal Logics

Roy Mennicke

Technische Universität Ilmenau, Germany

Abstract. We consider two bounded versions of the model checking problem of a fixed temporal logic TL whose modalities are MSO-definable and which is specifying properties of multiply nested words, i.e., of runs of pushdown automata with multiple stacks. One of the problems asks, given a multi-stack system $\mathcal{A}$, a temporal formula F from TL, and a bound k, whether all nested words ν which are accepted by $\mathcal{A}$ and whose split-width is bounded by k satisfy F. We describe the connection between the complexity of this problem and the maximal number n of monadic quantifier alternations in the definitions of the modalities of TL. In fact, we present almost tight upper and lower bounds for every natural number n. Regarding the other model checking problem considered, we require ν to be a k-scope nested word. In this case, we can infer the same lower and upper bounds.

1 Introduction

Boolean concurrent programs with recursive procedure calls can be modeled by pushdown automata with multiple stacks. The idea is that there is one stack for every process resp. thread of the program. A procedure call is then simulated by a push operation carried out on the appropriate stack, whereas the return of the same procedure corresponds to the matching pop operation. An individual execution of such a pushdown automaton can be considered as a word with multiple nesting relations. There is an edge between two positions of the word if they represent matching push and pop operations on the corresponding stack.

Since the general model checking problem for multi-stack systems and simple specification languages is undecidable, several restrictions considering only a subset of the runs of the system were introduced. For instance, one investigated nested words (NWs) which can be divided into a bounded number of contexts [14,3]. The latter is an interval in which all operations refer to the same stack. Later, the notion of context-bounded NWs was generalized to NWs consisting of a bounded number of phases [11,4,3]. In contrast to a context, a phase only requires pops to happen on the same stack. Another restriction is scope-boundedness [12,1,13]. Here, every nesting edge may only span a bounded number of contexts. Even more (orthogonal) approaches exist [5,2,7,13]. Recently, the notion of split NWs resp. split-width was introduced [7,6]. Split NWs are allowed to contain gaps in their linear order. Each interval between two gaps can be seen as a component. Adjacent components can be merged by closing the

E. Csuhaj-Varjú et al. (Eds.): MFCS 2014, Part I, LNCS 8634, pp. 438–450, 2014.

gap between them. A new split NW can be obtained by shuffling the components of two source split NWs. By k-BS, one denotes the set of all split NWs which can be built from atomic split NWs by repeatedly applying the shuffle and merge operations such that all intermediate split NWs consist of at most k components. In [7], the authors showed that the model checking problem for multi-stack systems restricted to NWs from k-BS and monadic second-order logic can be solved in non-elementary time. This was done by representing NWs as so-called proof trees exhibiting an alphabet of exponential size. They asked whether temporal logics could be used instead in order to gain a reasonable complexity.

This question marks the beginning of our investigations of the model checking problem of bounded split-width multi-stack systems and temporal logics whose modalities are MSO-definable. MSO-definable temporal logics go back to Gabbay et al. [8] and subsume virtually all temporal logics considered so far (see [4] for examples of MSO-definable modalities). In fact, we resume research by Gastin and Kuske [9,10] who considered Mazurkiewicz traces and investigations by Bollig, Kuske, and Mennicke [4] regarding phase-bounded NWs. For this, we extend and enhance the technique used in [4]. In particular, we drastically reduce the influence of the size of the temporal formula on the complexity.

If TL is a fixed MSO-definable temporal logic, then the bounded split-width model checking problem of TL (SW-MC(TL) for short) asks, given a multi-stack automaton $\mathcal{A}$, a temporal formula F from TL, and a split-width bound k, whether all NWs from k-BS which are accepted by $\mathcal{A}$ satisfy F. In order to solve SW-MC(TL), we introduce an alternative definition of k-BS. In contrast to the proof trees encoding from [7], this allows us to represent split-width bounded NWs as trees over an alphabet of constant size. This, in turn, enables us to transform atomic formulas specifying properties of NWs into small tree automata. Finally, SW-MC(TL) can then be reduced to the emptiness problem of tree automata. Using the above technique, we are able to show that, if TL is $M\Sigma_n$-definable, i.e., if its modalities can be defined using formulas from the n-th level of the monadic quantifier alternation hierarchy, then we can solve SW-MC(TL) in $(n+1)$-fold exponential time. Conversely, we prove that, for every $n \geq 1$, there exists an $M\Sigma_n$-definable temporal logic TL such that SW-MC(TL) is hard for n-fold exponential space.

By exploiting a connection between scope-boundedness and bounded split-width stated in [7], we can infer the same lower and upper bounds for SCOPE-MC. The definition of SCOPE-MC can be taken from SW-MC verbatim with the exception that "split-width" is replaced by "scope". The consideration of ordered NWs [5] is left for future work. However, we believe that the corresponding bounded model checking problem can be solved in at most $(n+2)$-fold exponential time. Unfortunately, in terms of the lower bound, our proof idea cannot be easily transferred to the setting of ordered NWs.

The missing proofs can be found in a technical report with the same title.

2   Preliminaries

If $n \in \mathbb{N}$, then $[n]$ denotes the set $\{1, 2, \ldots, n\}$. Furthermore, if R is a binary relation over the set M, then $\mathrm{supp}(R)$ is the set of all elements $x \in M$ for which there exists $y \in M$ such that $(x, y) \in R \cup R^{-1}$. The function tower $: \mathbb{N}^2 \to \mathbb{N}$ is inductively defined by $\mathrm{tower}(0, m) = m$ and $\mathrm{tower}(\ell + 1, m) = 2^{\mathrm{tower}(\ell, m)}$ for all $\ell, m \in \mathbb{N}$. We let $\mathrm{poly}(n)$ denote the set of polynomial functions in one argument.

A *word over an alphabet* Γ is a structure $w = (P, \to, \lambda)$ such that P is the set of *positions*, $(P, \to^*)$ is a finite, linearly ordered set, $\to$ is the direct successor relation wrt. $\to^*$, and $\lambda : P \to \Gamma$ is a mapping. Note that in order to represent the empty word ε, we allow the empty structure $(\emptyset, \emptyset, \emptyset)$. If the alphabet Γ is clear from the context, we call w a word instead of a word over Γ.

2.1   Split Nested Words

Split nested words were recently introduced in [7]. Intuitively, a k-split nested word is a nested word split into k components. We first define split words without nesting relations.

If $k \geq 0$, then a *k-split word over* Γ is a structure $\overline{w} = (P, \to, \dashrightarrow, \lambda)$ such that $(P, \to \cup \dashrightarrow, \lambda)$ is a word over Γ, $\to \cap \dashrightarrow = \emptyset$, and $|\dashrightarrow| = k - 1$ if $k \geq 1$, and $P = \to = \dashrightarrow = \lambda = \emptyset$ otherwise. If k is clear from the context or if we do not care about the exact k, then we sometimes omit k and call $\overline{w}$ a *split word*. The minimal (maximal) element of w with respect to $(\to \cup \dashrightarrow)^*$ is denoted by $\min(\overline{w})$ ($\max(\overline{w})$, resp.). If $w = (P, \to, \lambda)$ is a word, then we identify w with the 1-split word $(P, \to, \emptyset, \lambda)$. Further, we identify isomorphic split words.

Now, we define two different concatenation operations carried out on two split words $\overline{w}_0$ and $\overline{w}_1$. Both result in a split word consisting of the components of $\overline{w}_0$ followed by the components of $\overline{w}_1$. Intuitively, the operation denoted by "$\odot$" puts $(\max(\overline{w}_0), \min(\overline{w}_1))$ into $\to$, whereas "$\circledcirc$" inserts $(\max(\overline{w}_0), \min(\overline{w}_1))$ into $\dashrightarrow$. More precisely: Let $\overline{w}_i = (P_i, \to_i, \dashrightarrow_i, \lambda_i)$ be a split word for $i \in \{0, 1\}$. If $P_i \neq \emptyset$ for all $i \in \{0, 1\}$, then we define:

$$\overline{w}_0 \odot \overline{w}_1 = (P_0 \uplus P_1, \to_0 \cup \to_1 \cup \{(\max(\overline{w}_0), \min(\overline{w}_1))\}, \dashrightarrow_0 \cup \dashrightarrow_1, \lambda_0 \cup \lambda_1)$$

$$\overline{w}_0 \circledcirc \overline{w}_1 = (P_0 \uplus P_1, \to_0 \cup \to_1, \dashrightarrow_0 \cup \dashrightarrow_1 \cup \{(\max(\overline{w}_0), \min(\overline{w}_1))\}, \lambda_0 \cup \lambda_1)$$

Otherwise, if $\overline{w}_i = \varepsilon$ for some $i \in \{0, 1\}$, then we set $\overline{w}_0 \odot \overline{w}_1 = \overline{w}_0 \circledcirc \overline{w}_1 = \overline{w}_{1-i}$. Let $\overline{w} = (P, \to, \dashrightarrow, \lambda)$ be a split word and $\leq = (\to \cup \dashrightarrow)^*$. A *nesting relation* $\curvearrowright$ over $\overline{w}$ is a binary relation such that, for all $i, j, i', j' \in P$, the following holds: (1) if $i \curvearrowright j$, then $i < j$ and (2) if $i \curvearrowright j$, $i' \curvearrowright j'$, and $i \leq i'$, then $i < i' < j' < j$ or $i < j < i' < j'$ or $(i = i'$ and $j = j')$. If $i \curvearrowright j$, then we say that i is a *call* with *matching return* j. The idea is that the split word $\overline{w}$ describes the execution of some recursive program. Then $i \curvearrowright j$ shall mean that, at time i, the execution calls some procedure and, at time j, the control is returned to the calling program. Having this in mind, condition (1) expresses that every return occurs after its matching call. Condition (2) ensures that no position is both, a call and a return, every call has exactly one matching return and vice versa,

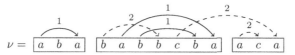

Fig. 1. The 3-split 2-NW $\nu = (\overline{w}, \curvearrowright_1, \curvearrowright_2)$ from Example 2.1. Note that the relation $\curvearrowright_1$ ($\curvearrowright_2$, resp.) is represented by the solid (dashed) edges labeled by 1 (2). Furthermore, the individual components of ν are depicted by rectangles.

and calls and matching returns are well nested. In the following, we will consider words with not only one, but with $\sigma \in \mathbb{N}$ many nesting relations.

For $\sigma \in \mathbb{N}$ and $k \geq 0$, a *k-split σ-nested word over Γ (k-split σ-NW)* is a tuple $\nu = (\overline{w}, \curvearrowright_1, \curvearrowright_2, \ldots, \curvearrowright_\sigma)$ such that (1) $\overline{w}$ is a k-split word over Γ, (2) $\curvearrowright_s$ is a nesting relation over $\overline{w}$ for all $s \in [\sigma]$, and (3) $\text{supp}(\curvearrowright_s) \cap \text{supp}(\curvearrowright_{s'}) = \emptyset$ for all $1 \leq s < s' \leq \sigma$. If ν is a 1-split σ-NW, then we also call it a σ-NW. Note that the above condition (3) expresses that a position can only be related to at most one stack. Let us assume that, at every call position of ν, the λ-label of the very same position is pushed onto the corresponding stack. Then, at every position i, the current content of stack s is implicitly given by the labels of the call positions of stack s which are preceding i and whose matching return positions are strict successors of i (w.r.t. the linear ordering of ν).

In the following, we fix the number of stacks $\sigma \in \mathbb{N}$ and the alphabet Γ and often speak of NWs instead of σ-NWs over Γ (i.e., we omit the number of stacks σ and the alphabet Γ).

Example 2.1. Figure 1 shows the 3-split 2-NW $\nu = ([13], \rightarrow, \dashrightarrow, \lambda, \curvearrowright_1, \curvearrowright_2)$ where $\lambda = \{(1, a), (2, b), \ldots\}$, $\rightarrow = \{(1, 2), (2, 3), \ldots\}$, $\dashrightarrow = \{(3, 4), (10, 11)\}$, $\curvearrowright_1 = \{(1, 3), (5, 10), \ldots\}$, and $\curvearrowright_2 = \{(4, 7), \ldots\}$.

We now define two operations called *merge* and *shuffle* which are carried out on split NWs. Note that the definitions of these operations in [7] are phrased differently; however, it can be easily checked that they coincide.

The Shuffle Operation. The shuffle operation works by mixing the components of two split NWs without changing the relative order of the components belonging to the same split nested word and by taking care of the restrictions regarding nesting edges (i.e., we may not introduce intersecting nesting edges). If $\nu_i = (w_{i,1} \odot w_{i,2} \odot \ldots \odot w_{i,k_i}, \curvearrowright_{i,1}, \ldots, \curvearrowright_{i,\sigma})$ is a k_i-split nested word and $w_{i,j}$ is a word for all $i \in [2]$ and $j \in [k_i]$, then $\text{shf}(\nu_1, \nu_2)$ is the set of all $(k_1 + k_2)$-split nested words $\nu = (v_1 \odot v_2 \odot \ldots \odot v_{k_1+k_2}, \curvearrowright_1, \ldots, \curvearrowright_\sigma)$ where $\curvearrowright_s = \curvearrowright_{1,s} \cup \curvearrowright_{2,s}$ for all $s \in [\sigma]$ and there exists a bijection $f \colon ((\{1\} \times [k_1]) \cup (\{2\} \times [k_2])) \rightarrow [k_1 + k_2]$ such that $w_{i,j} = v_{f(i,j)}$ and $f(i, j) < f(i, \ell)$ for all $i \in [2]$ and $1 \leq j < \ell \leq k_i$.

The Merge Operation. Intuitively, the merge operation allows to merge two or more components of a given split NW. More precisely, if $\nu = (w_1 \odot \ldots \odot w_k, \curvearrowright_1, \ldots, \curvearrowright_\sigma)$ is a k-split NW where $w_1, \ldots, w_k$ are words, then $\text{mrg}(\nu)$ denotes the set of all split NWs $\nu' = (w', \curvearrowright_1, \ldots, \curvearrowright_\sigma)$ such that there exists a total function $g \colon \{2, 3, \ldots, k\} \rightarrow \{\odot, \circledcirc\}$ fulfilling the following conditions: (i) there exists $i \in \{2, 3, \ldots, k\}$ with $g(i) = \odot$ and (ii) $w' = w_1\, g(2)\, w_2\, g(3)\, w_3\, g(4) \ldots g(k)\, w_k$.

Because of (i), at least one pair of adjacent components of ν has been turned into a single component using the operation $\odot$.

A NW ν is *atomic* if it is of the form $\boxed{a}$, $\boxed{\overset{\overset{s}{\frown}}{a \quad b}}$, or $\boxed{a}\,\boxed{\overset{\overset{s}{\frown}}{b}}$ where $a, b \in \Gamma$ and $s \in [\sigma]$. If $k \geq 2$, then k-BS is the least set fulfilling the following conditions: (i) all atomic NWs are contained in k-BS, (ii) if $\nu_i \in k$-BS is a k_i-split NW for $i \in [2]$ and $k_1 + k_2 \leq k$, then $\mathrm{shf}(\nu_1, \nu_2) \subseteq k$-BS, and (iii) if $\nu \in k$-BS, then $\mathrm{mrg}(\nu) \subseteq k$-BS. In contrast to [7], we explicitly add the atomic NW $\boxed{\overset{\overset{s}{\frown}}{a \quad b}}$ to k-BS in order to keep the following definitions and proofs simple.

2.2 Scope-Bounded NWs

The notion of scope-boundedness goes back to La Torre and Napoli [12]. Here, we recall a slightly adapted definition used in [7]. Intuitively, a NW is scope-bounded for some $\gamma \geq 1$ if all nesting edges span at most γ contexts. The latter is an interval within a NW in which at most one stack is involved. More precisely, let $\nu = (P, \to, \lambda, \curvearrowright_1, \ldots, \curvearrowright_\sigma)$ be a NW. If $\gamma \geq 1$, then ν is γ-*scope bounded* if for all $s \in [\sigma]$, $(i, j) \in \curvearrowright_s$, and $x_1, x_2, \ldots, x_{\gamma+1} \in \bigcup_{s \in [\sigma]} \mathrm{supp}(\curvearrowright_s)$ with $i < x_1 < x_2 < \ldots < x_{\gamma+1} < j$ the following holds: there exists $k \in [\gamma]$ and $s' \in [\sigma]$ such that $x_k, x_{k+1} \in \mathrm{supp}(\curvearrowright_{s'})$.

Example 2.2. Let ν be the split NW from Fig. 1. Consider the edge connecting the positions 5 and 10. The positions 6, 7, and 9 witness the fact that ν is not a 2-scope bounded NW. However, it can be shown that it is 3-scope bounded.

Theorem 2.3 ([7]). *If ν is a γ-scope (1-split) NW, then $\nu \in (\gamma + 2)$-BS.*

The above connection between bounded scope and bounded split-width was established in [7]. It should be investigated whether the more permissive definition of bounded scope given in [13] admits a similar theorem.

2.3 Monadic Second-Order Logic and Temporal Logics

We fix the set $\{x, y, z, \ldots\}$ of individual and the set $\{X, Y, Z, \ldots\}$ of set variables. By MSO, we denote the *set of all monadic second-order formulas* which can be built from the atomic formulas $\lambda(x) = a$, $x \to y$, $x \curvearrowright_s y$, $\mathrm{call}_s(x)$, $\mathrm{ret}_s(x)$, $\min(x)$, $\max(x)$, $x = y$, and $x \in X$ where $a \in \Gamma$ and $s \in [\sigma]$. The *set of all first-order formulas* FO $\subseteq$ MSO contains all formulas without second-order quantification $\exists X$. We use the usual abbreviations such as $\forall x\, \varphi$ for $\neg \exists x\, \neg \varphi$.

Let $\nu = (P, \to, \lambda, \curvearrowright_1, \ldots, \curvearrowright_\sigma)$ be a NW and $\varphi(X_1, \ldots, X_\ell, x_1, \ldots, x_k)$ be a formula with free variables from $\{X_1, \ldots, X_\ell, x_1, \ldots, x_k\}$. Furthermore, let $I_1, \ldots, I_\ell \subseteq P$ and $i_1, \ldots, i_k \in P$. Then we write $\nu, I_1, \ldots, I_\ell, i_1, \ldots, i_k \models \varphi$ if φ evaluates to true when interpreting the variables by $I_1, \ldots, I_\ell \subseteq P$ and $i_1, \ldots, i_k \in P$, resp. For instance, we have $\nu, i \models \mathrm{call}_s(x)$ if there exists $j \in P$ with $i \curvearrowright_s j$. The semantics of the remaining formulas are as expected. An MSO-formula is an *m-ary modality definition* if it has at most m free set variables $X_1, \ldots, X_m$ and one free individual variable x. An *MSO-definable temporal logic* is a tuple $\text{TL} = (B, \text{arity}, [\![-]\!])$ where B is a finite set of modalities, the mapping arity: $B \to \mathbb{N}$ specifies the arity of every modality from B, and $[\![-]\!] : B \to \text{MSO}$ is a mapping such that $[\![M]\!]$ is an m-ary modality definition whenever $\text{arity}(M) = m$ for $M \in B$. The set of all formulas from TL is the least set such that the following conditions hold: (i) if $M \in B$ with $\text{arity}(M) = 0$, then M is a formula from TL and (ii) if $M \in B$, $\text{arity}(M) > 0$, and $F_1, F_2, \ldots, F_{\text{arity}(M)}$ are formulas from TL, then $M(F_1, \ldots, F_{\text{arity}(M)})$ is a formula from TL. The size $|F|$ of a temporal formula F is its number of subformulas.

Let $\nu = (P, \to, \lambda, \curvearrowright_1, \ldots, \curvearrowright_\sigma)$ be a NW and F be a formula from TL. The semantics F^ν of F in ν is the set of positions from P where F holds. More formally: If $F = M(F_1, \ldots, F_m)$ where $M \in B$ is of arity $m \geq 0$, then $F^\nu = \{i \in P \mid \nu, i, F_1^\nu, \ldots, F_m^\nu \models [\![M]\!]\}$. We write $\nu, i \models F$ for $i \in F^\nu$ and $\nu \models F$ for $\nu, \min(P, \to, \lambda) \models F$. Examples of MSO-definable modalities can be found in [4].

Now, let TL be some MSO-definable temporal logic. We define the following *model checking problems* for TL where we kindly refer the reader to [4] for a formal definition of σ-stack automata.

- SW-MC(TL) is the set of triples $(\mathcal{A}, F, k)$ where $\mathcal{A}$ is a σ-stack automaton, F is a temporal formula from TL, and $k \geq 2$ such that every NW $\nu \in k\text{-BS}$ accepted by $\mathcal{A}$ satisfies F (where k is encoded in unary).
- SCOPE-MC(TL) is the set of triples $(\mathcal{A}, F, \gamma)$ where $\mathcal{A}$ is a σ-stack automaton, F is a temporal formula from TL, and $\gamma \geq 1$ such that every γ-scope NW accepted by $\mathcal{A}$ satisfies F (where k is encoded in unary).

Finally, we recall the definition of the *monadic quantifier alternation hierarchy*. An MSO-formula φ belongs to $M\Sigma_n$ if it is of the form $\exists \overline{X}_1 \forall \overline{X}_2 \ldots \exists/\forall \overline{X}_n \psi$ where $\psi \in \text{FO}$ and the $\overline{X}_i$'s are tuples of individual and set variables. In contrast, it belongs to $M\Pi_n$ if it is of the form $\forall \overline{X}_1 \exists \overline{X}_2 \ldots \forall/\exists \overline{X}_n \psi$. Let $\mathcal{L}$ be some fragment of MSO such as FO or $M\Sigma_n$ etc. An MSO-definable temporal logic $\text{TL} = (B, \text{arity}, [\![-]\!])$ is *$\mathcal{L}$-definable* if $[\![M]\!] \in \mathcal{L}$ for all modalities $M \in B$.

3   The Model Checking Problem

This section mainly deals with the proof of an upper bound for the model checking problems SW-MC(TL) and SCOPE-MC(TL). A complementing lower bound can be found at the end of this section.

3.1   An Equivalent Definition of Bounded Split-Width

In order to solve the model checking problems of an MSO-definable temporal logic TL, we adapt and enhance the technique used in [4]. For this, we represent split nested words as trees. Our encoding will be in the spirit of the *proof trees* defined in [7]. However, proof trees exhibit an alphabet of size exponential in the split-width bound k. This leads to the undesirable effect that a tree automaton with polynomially many states still exhibits exponentially many transitions;

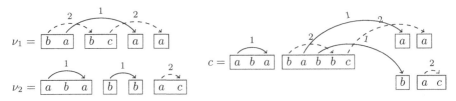

Fig. 2. The split NWs ν_1, ν_2, and the nw-container $c = (\overline{w}_0, \overline{w}_1, \overline{w}_2, \curvearrowright_1, \curvearrowright_2)$ from Example 3.2 where $\overline{w}_0$ is depicted on the left and $\overline{w}_1$ ($\overline{w}_2$) on the upper (lower) right

hence, it cannot be constructed in space $O(\log k)$. Therefore, we develop an encoding using trees over an alphabet of size independent from k.

For this, we simulate the merge and shuffle operations defined for split NWs stepwise. More precisely, given two split nested words ν_1 and ν_2, a new split NW ν is obtained as follows: We begin with an empty split NW. In each step one of the remaining components of ν_1 and ν_2 is chosen and added to ν. The latter is done using either the operation $\odot$ or $\circledcirc$. During the whole process, it is ensured that the relative order of the components of the input split NWs is maintained and that no intersecting nesting edges referring to the same stack are introduced. A so-called nw-container represents an intermediate configuration of the above described iterative process.

Definition 3.1. *A k-split nw-container is a structure* $(\overline{w}_0, \overline{w}_1, \overline{w}_2, \curvearrowright_1, \ldots, \curvearrowright_\sigma)$ *where*

(1) $\overline{w}_i = (P_i, \to_i, \dashrightarrow_i, \lambda_i)$ *is a k_i-split word for all $i \in \{0, 1, 2\}$,*
(2) $k \geq 1$, $k_0 + k_1 + k_2 = k$, *and* $P_i \cap P_j = \emptyset$ *for all $i, j \in \{0, 1, 2\}$ with $i \neq j$,*
(3) $\curvearrowright_s \subseteq P_0^2 \cup ((P_0 \cup P_1) \times P_1) \cup ((P_0 \cup P_2) \times P_2)$ *for all $s \in [\sigma]$,*
(4) *and, for all $i \in [2]$, $(\overline{w}_0 \odot \overline{w}_i, \curvearrowright_1', \ldots, \curvearrowright_\sigma')$ is a $(k_0 + k_i)$-split NW where, for all $s \in [\sigma]$, $\curvearrowright_s' = \curvearrowright_s \cap (P_0^2 \cup ((P_0 \cup P_i) \times P_i))$.*

If $\nu = (\overline{w}, \curvearrowright_1, \ldots, \curvearrowright_\sigma)$ is a split NW, then, for convenience, we denote by $ct(\nu)$ the nw-container $(\overline{w}, \varepsilon, \varepsilon, \curvearrowright_1, \ldots, \curvearrowright_\sigma)$.

Example 3.2. Let $\nu_i = (w_{i,1} \circledcirc w_{i,2} \circledcirc w_{i,3} \circledcirc w_{i,4}, \curvearrowright_{i,1}, \curvearrowright_{i,2})$ be the 4-split 2-NW from Fig. 2 for $i \in [2]$. Then $c = (\overline{w}_0, \overline{w}_1, \overline{w}_2, \curvearrowright_1, \curvearrowright_2)$ is the 6-split 2-nw-container shown in Fig. 2 where $\overline{w}_0 = w_{2,1} \circledcirc w_{1,1} \odot w_{2,2} \odot w_{1,2}$, $\overline{w}_1 = w_{1,3} \circledcirc w_{1,4}$, $\overline{w}_2 = w_{2,3} \circledcirc w_{2,4}$, $\curvearrowright_1 = \curvearrowright_{1,1} \cup \curvearrowright_{1,2}$, and $\curvearrowright_2 = \curvearrowright_{2,1} \cup \curvearrowright_{2,2}$.

We now define three different operations carried out on nw-containers. The first one is a binary operation called union operation. It can only be applied to containers c for which there exists a split NW ν with $c = ct(\nu)$. Basically, the union operation puts two split NWs (resp., their nw-containers) into one nw-container such that, using the not yet defined operations add and madd, their components can be shuffled and merged.

The union Operation. If $c_i = (\overline{w}_i, \varepsilon, \varepsilon, \curvearrowright_{i,1}, \ldots, \curvearrowright_{i,\sigma})$ is a k_i-split nw-container, $\overline{w}_i = (P_i, \to_i, \dashrightarrow_i, \lambda_i)$ for all $i \in [2]$, and $P_1 \cap P_2 = \emptyset$, then $union(c_1, c_2)$ denotes the (k_1+k_2)-split nw-container $(\varepsilon, \overline{w}_1, \overline{w}_2, \curvearrowright_1, \ldots, \curvearrowright_\sigma)$ where $\curvearrowright_s = \curvearrowright_{1,s} \cup \curvearrowright_{2,s}$ for all $s \in [\sigma]$.

Now, let $k \geq 1$, $c = (\overline{w}_0, \overline{w}_1, \overline{w}_2, \curvearrowright_1, \ldots, \curvearrowright_\sigma)$ be a k-split nw-container, $m \in [2]$, and $\overline{w}_m = v_1 \odot v_2 \odot \ldots \odot v_n$ where $v_1, \ldots, v_n$ are words and $n \geq 1$.

The add Operation. Intuitively, this operation appends either the first component of $\overline{w}_1$ or the first component of $\overline{w}_2$ to the split word $\overline{w}_0$ using the $\odot$ operation. More formally, consider the structure $c' = (\overline{w}_0', \overline{w}_1', \overline{w}_2', \curvearrowright_1, \ldots, \curvearrowright_\sigma)$ where (1) $\overline{w}_0' = \overline{w}_0 \odot v_1$, (2) $\overline{w}_m' = v_2 \odot v_3 \odot \ldots \odot v_n$, (3) $\overline{w}_j' = \overline{w}_j$, and $j = 3 - m$ is the other index. If c' is an nw-container (i.e., if we did not introduce intersecting nesting edges referring to the same stack), then we set $\text{add}_m(c) = c'$. Otherwise, $\text{add}_m(c)$ is undefined.

The madd Operation. This operations appends either the first component of $\overline{w}_1$ or the first component of $\overline{w}_2$ to the split word $\overline{w}_0$ by merging it with the last component of $\overline{w}_0$ (i.e., the $\odot$ operation is used). The formal definition can be taken from the add operation verbatim with the exception that $\overline{w}_0' = \overline{w}_0 \odot v_1$ needs to be replaced by $\overline{w}_0' = \overline{w}_0 \odot v_1$ and $\text{add}_m(c) = c'$ by $\text{madd}_m(c) = c'$.

Example 3.3. Let $\nu_i = (w_{i,1} \odot w_{i,2} \odot w_{i,3} \odot w_{i,4}, \curvearrowright_{i,1}, \curvearrowright_{i,2})$ be the split NW from Fig. 2 for $i \in [2]$. Then $(\text{madd}_1 \circ \text{madd}_2 \circ \text{add}_1 \circ \text{add}_2 \circ \text{union})(ct(\nu_1), ct(\nu_2))$ is the nw-container c from Fig. 2. Note that $\text{madd}_1(c)$ is not defined, because condition (4) of Def. 3.1 would not be fulfilled. Intuitively, the nesting edges of stack 1 from $w_{1,1}$ to $w_{1,3}$ and from $w_{2,2}$ to $w_{2,3}$ would intersect each other. Of course, the same holds for $\text{add}_1(c)$.

By C_k, we denote the set of all those nw-containers which can be built from the atomic split NWs (resp., their nw-containers) by applying the above three operations and not exceeding the split-width bound k during this process.

Proposition 3.4. *For all $k \geq 2$, k-BS $= \{\nu \mid \nu$ is a split NW, $ct(\nu) \in C_k\}$.*

3.2   The Encoding of NWs as Trees

Let the alphabet $\Lambda = \Gamma \cup \{\text{union}, \text{add}_1, \text{add}_2, \text{madd}_1, \text{madd}_2\} \cup \{\curvearrowright_s, \overset{m}{\curvearrowright}_s \mid s \in [\sigma]\}$ be fixed. We now inductively define the set E_k of trees over Λ representing nw-containers from C_k and the mapping val specifying the abstract nw-container represented by a tree from E_k. Intuitively, E_k can be considered the set of parse trees of the terms over Λ defining split NWs from k-BS.

Definition 3.5. *If $k \geq 2$, then we define E_k to be the least set such that:*

- *For all $a \in \Gamma$, we have $a \in E_k$ and set $val(a) = ct(\boxed{a})$.*
- *For all $a, b \in \Gamma$, $s \in [\sigma]$, we have $t = \curvearrowright_s(a, b) \in E_k$ and $val(t) = ct(\overset{\curvearrowright^s}{\boxed{a}\ \boxed{b}})$.*
- *For all $a, b \in \Gamma$, $s \in [\sigma]$, we have $t = \overset{m}{\curvearrowright}_s(a, b) \in E_k$ and $val(t) = ct(\overset{\curvearrowright^s}{\boxed{a\ \ b}})$.*
- *If $t' \in E_k$, $h \in \{\text{add}_1, \text{add}_2, \text{madd}_1, \text{madd}_2\}$, and $h(val(t'))$ is defined, then $t = h(t') \in E_k$. We set $val(t) = h(val(t'))$.*

   – If $t_1, t_2 \in E_k$, $val(t_i) = ct(\nu_i)$, ν_i is a k_i-split NW for $i \in [2]$, and $k_1 + k_2 \leq k$, then $t = union(t_1, t_2) \in E_k$. We set $val(t) = union(val(t_1), val(t_2))$.

Clearly, the leaves of $t \in E_k$ represent the positions of $val(t)$. The inner nodes of t specify the relations between these positions in such a way that they can be recovered by small tree automata. In order to make this correspondence explicit, we give the following definitions and the subsequent theorem. The latter constructs, for instance, a tree automaton $\mathcal{B}_{\rightarrow}$ that runs on trees $t \in E_k$ with two marked nodes x and y. It accepts such a marked tree if and only if x and y are leaves (and therefore positions in $val(t)$) that satisfy $x \rightarrow y$.

In the following, we also consider trees from E_k to be labeled graphs. If $t \in E_k$, $X_1, \ldots, X_m$ are sets of nodes, and $x_1, \ldots, x_n$ are nodes of t, then we define $t' = (t, X_1, \ldots, X_m, x_1, \ldots, x_n)$ to be the tree over the alphabet $\Lambda \times \{0,1\}^{m+n}$ which is obtained by extending the labels of the nodes v of t as follows: the label of v is enriched with $m + n$ bits $b_1, \ldots, b_m, c_1, \ldots, c_n$ such that (1) $b_i = 1$ if and only if $v \in X_i$ for $i \in [m]$ and (2) $c_i = 1$ if and only if $v = x_i$ for $i \in [n]$. Note that a tree over the alphabet $\Lambda \times \{0,1\}^{m+n}$ is of the above form if and only if, for all $i \in [n]$, there is a unique node whose bit c_i is set to 1. Hence, the set of these trees forms a regular tree language that can be accepted by a tree automaton with 2^n states.

Definition 3.6. *If $\mathcal{B}$ is a tree automaton over the alphabet $\Lambda \times \{0,1\}^n$, then we define, for every tree t, $R(\mathcal{B}, t) = \{(x_1, \ldots, x_n) \mid (t, x_1, \ldots, x_n) \in L(\mathcal{B})\}$.*

Now, several tree automata are constructed. Intuitively, if $t \in E_k$ represents a container $((P_i, \rightarrow_i, \dashrightarrow_i, \lambda_i)_{i \in \{0,1,2\}}, \curvearrowright_1, \ldots, \curvearrowright_\sigma)$, then, for all $i \in \{0,1,2\}$, $\mathcal{B}_{\overline{w}_i}$ recovers P_i, $\mathcal{B}_{\rightarrow}$ reconstructs $\rightarrow_i$, $\mathcal{B}_{\dashrightarrow}$ restores $\dashrightarrow_i$, and $\mathcal{B}_a$ with $a \in \Gamma$ retrieves all positions labeled by a. Similarly, $\mathcal{B}_s$ is used to obtain $\curvearrowright_s$ for all $s \in [\sigma]$.

Theorem 3.7. *Given $k \geq 2$, one can construct tree automata $\mathcal{B}_{\rightarrow}$, $\mathcal{B}_{\dashrightarrow}$, $\mathcal{B}_{\overline{w}_i}$ for every $i \in \{0,1,2\}$, $\mathcal{B}_s$ for every $s \in [\sigma]$, and $\mathcal{B}_a$ for every $a \in \Gamma$ in space $O(\log k)$ such that, for all $t \in E_k$, $val(t)$ is isomorphic to the nw-container*

$$expval_k(t) = \big((P_i, \rightarrow_i, \dashrightarrow_i, \lambda_i)_{i \in \{0,1,2\}}, \curvearrowright_1, \ldots, \curvearrowright_\sigma\big)$$

where $P_i = R(\mathcal{B}_{\overline{w}_i}, t)$, $\rightarrow_i = R(\mathcal{B}_{\rightarrow}, t) \cap P_i^2$, $\dashrightarrow_i = R(\mathcal{B}_{\dashrightarrow}, t) \cap P_i^2$, $\lambda_i = \{(v, a) \mid a \in \Gamma, v \in R(\mathcal{B}_a, t) \cap P_i\}$ for all $i \in \{0,1,2\}$, and $\curvearrowright_s = R(\mathcal{B}_s, t)$ for all $s \in [\sigma]$.

Proof. Concerning the automaton $\mathcal{B}_s$ with $s \in [\sigma]$, we have $i \curvearrowright_s j$ if and only if (1) i and j share the same parent node labeled by a symbol from $\{\curvearrowright_s, \overset{m}{\curvearrowright}_s\}$ and (2) i is a left child and j is a right child. The above two conditions can be easily checked by a tree automaton of size independent from k.

    As another example, consider the automaton $\mathcal{B}_{\rightarrow}$ which can be constructed in space $O(\log k)$. The idea is that the automaton keeps track of the positions i and j for which $i \rightarrow j$ shall be checked. More precisely, the automaton is in the state $(\ell, m_1, \ell_1, m_2, \ell_2)$ where $\ell \in [k]$, $\ell_1, \ell_2 \in [k] \cup \{\bot\}$, and $m_1, m_2 \in \{0, 1, 2, \bot\}$ if and only if the current node represents an nw-container $c = (\overline{w}_0, \overline{w}_1, \overline{w}_2, \curvearrowright_1, \ldots, \curvearrowright_\sigma)$ such that

- $\overline{w}_0$ is an ℓ-split word,
- either (a) $m_1 \in \{0, 1, 2\}$, $\ell_1 \in [k]$ and position i is the right-most position of the ℓ_1-th component of $\overline{w}_{m_1}$ or (b) $m_1 = \ell_1 = \bot$ and there do not exist $i \in \{0, 1, 2\}$ and $\ell \in [k]$ such that i is the right-most position of the ℓ-th component of $\overline{w}_i$, and
- either (a) $m_2 \in \{0, 1, 2\}$, $\ell_2 \in [k]$ and position j is the left-most position of the ℓ_2-th component of $\overline{w}_{m_2}$ or (b) $m_2 = \ell_2 = \bot$ and there do not exist $i \in \{0, 1, 2\}$ and $\ell \in [k]$ such that j is the left-most position of the ℓ-th component of $\overline{w}_i$.

The automaton $\mathcal{B}_\rightarrow$ is in the state $\top$ if and only if the current node represents an nw-container exhibiting a component which contains the positions i and j such that $i \rightarrow j$. The sole final state of $\mathcal{B}_\rightarrow$ is $\top$. It is straightforward to specify a transition relation implementing the above described behavior.   □

Definition 3.8. *For all NWs ν, we set $tree_k(\nu) = \{t \in E_k \mid expval_k(t) = ct(\nu)\}$ where $expval_k$ is the mapping from Theorem 3.7.*

3.3   From Temporal Formulas to Tree Automata

Now, our intermediate goals is to directly transform an arbitrary MSO-formula $\varphi(X_1, \ldots, X_m, x_1, \ldots, x_n)$ into a tree automaton $\mathcal{B}_\varphi$ such that, for all NWs $\nu \in k\text{-BS}$ and $t \in tree_k(\nu)$, we have $\nu, X_1, \ldots, X_m, x_1, \ldots, x_n \models \varphi$ if and only if $(t, X_1, \ldots, X_m, x_1, \ldots, x_n) \in L(\mathcal{B}_\varphi)$.

Proposition 3.9. *Given $k \geq 2$ and a fixed (possibly negated) atomic MSO-formula φ, one can construct in space $O(\log k)$ a tree automaton $\mathcal{B}$ with the following property: If $\nu = (P, \rightarrow, \lambda, \curvearrowright_1, \ldots, \curvearrowright_\sigma)$ is a NW from k-BS, $I \subseteq P$, $i, j \in P$, and $t \in tree_k(\nu)$, then $(t, I, i, j) \in L(\mathcal{B}) \iff \nu, I, i, j \models \varphi$.*

Proof. We have $\nu, i \models (\lambda(x) = a)$ if and only if $i \in R(\mathcal{B}_a, t)$ where $\mathcal{B}_a$ is the automaton from Theorem 3.7. From the label of i in t one can immediately tell whether $\nu, i, I \models x \in X$ or $\nu, i, j \models (x = y)$ holds. We have $\nu, i \models call_s(x)$ if and only if i is the left child of a node labeled by a symbol from $\{\curvearrowright_s, \overset{m}{\curvearrowright}_s\}$. Analogously, one can construct a tree automaton for $ret_s(x)$. Moreover, $\nu, i, j \models (x \curvearrowright_s y)$ and $\nu, i, j \models (x \rightarrow y)$ can be decided using the automata $\mathcal{B}_s$ and $\mathcal{B}_\rightarrow$ from Theorem 3.7.

Concerning the formula $min(x)$, we construct a tree automaton $\mathcal{B}$ with the set of states $\{\bot, 0, 1, 2\}$. It is in state $i \in \{0, 1, 2\}$ if the current node represents a container $c = (\overline{w}_0, \overline{w}_1, \overline{w}_2, \curvearrowright_1, \ldots, \curvearrowright_\sigma)$ such that x is the minimal position of $\overline{w}_i$. Otherwise, it is in state $\bot$. The state 0 is the sole accepting state of $\mathcal{B}$. Analogously, one can construct an automaton for the formula $max(x)$.

The negations of the above atomic formulas can be dealt with similarly.   □

Note that the sizes of all but the tree automata constructed for the formulas $x \rightarrow y$ and $\neg(x \rightarrow y)$ are independent from the split-with bound k. Now, we give a definition of a second measure of the complexity of formulas from MSO, namely the *full quantifier alternation hierarchy*. In contrast to the monadic quantifier alternation hierarchy, we do not only consider monadic quantifiers but also quantification over individual variables.

Definition 3.10. *By Σ_n, we denote all formulas $\varphi \in MSO$ which are of the form $\exists \overline{X}_1 \forall \overline{X}_2 \ldots \exists/\forall \overline{X}_n \psi$ where $\overline{X}_i$ are tuples of individual and set variables and ψ is quantifier-free. In contrast, Π_n denotes all formulas of the form $\forall \overline{X}_1 \exists \overline{X}_2 \ldots \forall/\exists \overline{X}_n \psi$.*

By exploiting the existence of the atomic formulas $\min(x)$, $\max(x)$, $\mathrm{call}_s(x)$, and $\mathrm{ret}_s(x)$ and by using Hanf's theorem, one can transform every $M\Sigma_n$-formula into an equivalent Σ_{n+1}-formula. The following proposition shows that every formula of an arbitrary but fixed $M\Sigma_n$-definable temporal logic can be transformed into an equivalent MSO-sentence of a certain kind in polynomial time:

Proposition 3.11. *Let $TL = (B, arity, \llbracket - \rrbracket)$ be some $M\Sigma_n$-definable temporal logic. From a temporal formula F of TL, one can compute in time $poly(|F|)$ an MSO-sentence $\psi = \exists \overline{X} \left[\bigwedge_{i \in [[F]]} \left(\psi_{1,i}(\overline{X}) \wedge \forall y \, \psi_{2,i}(y, \overline{X}) \right) \right]$ (where $\overline{X}$ is an $|F|$-tuple of set variables) such that, for all $i \in [[F]]$, $\psi_{1,i} \in \Pi_{n+1}$, $\psi_{2,i} \in \Sigma_{n+1}$, and, for all NWs ν, we have $\nu \models F$ if and only if $\nu \models \psi$.*

Note that the above proposition differs from [4, Proposition 1] where formulas ψ_1 and ψ_2 are constructed whose sizes are linear in $|F|$. In contrast, the sizes of our formulas $\psi_{1,i}$ and $\psi_{2,i}$ are independent from $|F|$. Later, this allows us to construct a tree automaton for F in space $|F| \cdot \mathrm{tower}_n(\mathrm{poly}(k))$ instead of $\mathrm{tower}_{n+1}(\mathrm{poly}(|F| \cdot \log k))$, i.e., the influence of the size of F on the space complexity is vastly reduced.

Using Propositions 3.9 and 3.11, standard automata constructions for Boolean combinations and projection, and a special handling of formulas of the form $\forall y \, \varphi$ (like they occur in the formula ψ from Prop. 3.11), one can show the following:

Theorem 3.12. *Let TL be a fixed $M\Sigma_n$-definable temporal logic. From a formula F from TL and $k \geq 2$, one can construct in space $|F| \cdot \mathrm{tower}_n(\mathrm{poly}(k))$ a tree automaton $\mathcal{B}_F$ with the following property: For all NWs ν from k-BS and $t \in \mathrm{tree}_k(\nu)$, we have $\nu \models F \iff t \in L(\mathcal{B}_F)$.*

3.4   The Decision Procedure

Proposition 3.13. *If $k \geq 2$ and $\mathcal{A}$ is a σ-stack automaton, then we can construct in space $O(k \cdot \log |\mathcal{A}|)$ a tree automaton $\mathcal{B}_{\mathcal{A},k}$ with $L(\mathcal{B}_{\mathcal{A},k}) = \{t \mid t \in \mathrm{tree}_k(\nu), \nu \in k\text{-}BS$ is accepted by $\mathcal{A}\}$.*

Let TL be some fixed $M\Sigma_n$-definable temporal logic over the fixed alphabet Γ. Given a σ-stack automaton $\mathcal{A}$, a formula F from TL, and $k \geq 2$, one can construct the automata $\mathcal{B}_{\neg F}$ and $\mathcal{B}_{\mathcal{A},k}$ from Theorem 3.12 and Prop. 3.13, respectively. Note that the negation $\neg$ can be easily defined using the FO-modality $\llbracket \neg \rrbracket(X_1, x) = \neg(x \in X_1)$. We have $(\mathcal{A}, F, k) \in$ SW-MC(TL) if and only if $L(\mathcal{B}_{\mathcal{A},k}) \cap L(\mathcal{B}_{\neg F}) = \emptyset$. The automaton $\mathcal{B}_{\mathcal{A},k}$ can be constructed in space $O(k \cdot \log |\mathcal{A}|)$ and $\mathcal{B}_{\neg F}$ can be obtained in space $|\neg F| \cdot \mathrm{tower}_n(\mathrm{poly}(k))$. Since the emptiness problem for tree automata can be solved in polynomial time, the model checking problem SW-MC(TL) can be decided in time $|\mathcal{A}|^{|F| \cdot \mathrm{tower}_n(\mathrm{poly}(k))}$. Similar arguments hold for SCOPE-MC(TL). Hence, the following holds:

Theorem 3.14. *Let $n \geq 0$ and TL be some fixed $M\Sigma_n$-definable temporal logic. Then SW-MC(TL) and SCOPE-MC(TL) are both solvable in $(n+1)$-EXPTIME. More precisely, given a σ-stack automaton $\mathcal{A}$, a formula F from TL, and a bound k, the decision procedure runs in time $|\mathcal{A}|^{|F| \cdot tower_n(poly(k))}$.*

By adapting the technique from [4], we polynomially reduce the satisfiability problem of temporal logics over labeled grids to the satisfiability problems of temporal logics over NWs. The following lower bound can then be inferred:

Theorem 3.15. *For all $n \geq 1$, alphabets Γ with $|\Gamma| \geq 3$, and number of stacks $\sigma \geq 2$, there is an $M\Sigma_n$-definable temporal logic TL such that SW-MC(TL) resp. SCOPE-MC(TL) is n-EXPSPACE-hard.*

Acknowledgements. The author likes to express his sincere thanks to his doctoral adviser Dietrich Kuske for his guidance and valuable advice. Furthermore, he is grateful to the anonymous referees for their detailed reviews and helpful remarks.

References

1. Atig, M.F., Bouajjani, A., Narayan Kumar, K., Saivasan, P.: Linear-time model-checking for multithreaded programs under scope-bounding. In: Chakraborty, S., Mukund, M. (eds.) ATVA 2012. LNCS, vol. 7561, pp. 152–166. Springer, Heidelberg (2012)
2. Atig, M.F., Narayan Kumar, K., Saivasan, P.: Adjacent ordered multi-pushdown systems. In: Béal, M.-P., Carton, O. (eds.) DLT 2013. LNCS, vol. 7907, pp. 58–69. Springer, Heidelberg (2013)
3. Bansal, K., Demri, S.: Model-checking bounded multi-pushdown systems. In: Bulatov, A.A., Shur, A.M. (eds.) CSR 2013. LNCS, vol. 7913, pp. 405–417. Springer, Heidelberg (2013)
4. Bollig, B., Kuske, D., Mennicke, R.: The complexity of model checking multi-stack systems. In: LICS, pp. 163–172. IEEE Computer Society (2013)
5. Breveglieri, L., Cherubini, A., Citrini, C., Crespi-Reghizzi, S.: Multi-push-down languages and grammars. Int. J. Found. Comput. Sci. 7(3), 253–292 (1996)
6. Cyriac, A.: Verification of Communicating Recursive Programs via Split-width. PhD thesis, ENS Cachan (2014)
7. Cyriac, A., Gastin, P., Kumar, K.N.: MSO decidability of multi-pushdown systems via split-width. In: Koutny, M., Ulidowski, I. (eds.) CONCUR 2012. LNCS, vol. 7454, pp. 547–561. Springer, Heidelberg (2012)
8. Gabbay, D., Hodkinson, I., Reynolds, M.: Temporal Logic: Mathematical Foundations and Computational Aspects, vol. 1. Oxford University Press (1994)
9. Gastin, P., Kuske, D.: Uniform satisfiability in PSPACE for local temporal logics over Mazurkiewicz traces. Fundam. Inform. 80(1-3), 169–197 (2007)
10. Gastin, P., Kuske, D.: Uniform satisfiability problem for local temporal logics over Mazurkiewicz traces. Inf. Comput. 208(7), 797–816 (2010)
11. La Torre, S., Madhusudan, P., Parlato, G.: A robust class of context-sensitive languages. In: LICS, pp. 161–170. IEEE Computer Society (2007)
12. La Torre, S., Napoli, M.: Reachability of multistack pushdown systems with scope-bounded matching relations. In: Katoen, J.-P., König, B. (eds.) CONCUR 2011. LNCS, vol. 6901, pp. 203–218. Springer, Heidelberg (2011)

13. La Torre, S., Napoli, M.: A temporal logic for multi-threaded programs. In: Baeten, J.C.M., Ball, T., de Boer, F.S. (eds.) TCS 2012. LNCS, vol. 7604, pp. 225–239. Springer, Heidelberg (2012)
14. Qadeer, S., Rehof, J.: Context-bounded model checking of concurrent software. In: Halbwachs, N., Zuck, L.D. (eds.) TACAS 2005. LNCS, vol. 3440, pp. 93–107. Springer, Heidelberg (2005)

Decidability of the Interval Temporal Logic AĀBB̄ over the Rationals

Angelo Montanari[1], Gabriele Puppis[2], and Pietro Sala[3]

[1] University of Udine, Italy
angelo.montanari@uniud.it
[2] LaBRI / CNRS, University of Bordeaux, France
gabriele.puppis@labri.fr
[3] University of Verona, Italy
pietro.sala@univr.it

Abstract. The classification of the fragments of Halpern and Shoham's logic with respect to decidability/undecidability of the satisfiability problem is now very close to the end. We settle one of the few remaining questions concerning the fragment AĀBB̄, which comprises Allen's interval relations "meets" and "begins" and their symmetric versions. We already proved that AĀBB̄ is decidable over the class of all finite linear orders and undecidable over ordered domains isomorphic to $\mathbb{N}$. In this paper, we first show that AĀBB̄ is undecidable over $\mathbb{R}$ and over the class of all Dedekind-complete linear orders. We then prove that the logic is decidable over $\mathbb{Q}$ and over the class of all linear orders.

1 Introduction

Even though it has been authoritatively and repeatedly claimed that interval-based formalisms are the most appropriate ones for a variety of application domains, e.g., [6], until very recently interval temporal logics were a largely unexplored land. There are at least two explanations for such a situation: computational complexity and technical difficulty. On the one hand, the seminal work by Halpern and Shoham on the interval logic of Allen's interval relations (HS for short) showed that such a logic is highly undecidable over all meaningful classes of linear orders [5], and ten years later Lodaya proved that a restricted fragment of it, denoted BE, featuring only two modalities (those for Allen's relations *begins* and *ends*), suffices for undecidability [7]. On the other hand, formulas of interval temporal logics express properties of pairs of time points rather than of single time points, and are evaluated as sets of such pairs, that is, binary relations. As a consequence, there is no reduction of the satisfiability/validity in interval logics to monadic second-order logic, and thus Rabin's theorem (the standard proof machinery) is not applicable here.

In the last decade, a systematic investigation of HS fragments has been carried out. Their classification with respect to the decidability/undecidability of their satisfiability problem is now very close to the end. The outcome of the analysis is that undecidability rules over HS fragments [1, 8], but some meaningful exceptions exist [2, 3, 4, 10, 11]. While setting the status of most and least expressive

E. Csuhaj-Varjú et al. (Eds.): MFCS 2014, Part I, LNCS 8634, pp. 451–463, 2014.

interval logics is relatively straightforward, e.g., undecidability of full HS can be shown by a reduction from the non-halting problem for Turing machines, decidability of the logic of Allen's relations *begins* and *begun by* $B\bar{B}$ can be proved by a reduction to the (point-based) linear temporal logic of future and past, dealing with those fragments that lie on the marginal land between decidability and undecidability is much more difficult. (Un)decidability of HS fragments depends on two factors: their set of interval modalities and the class of linear orders over which they are interpreted. While the first one is fairly obvious, the second one is definitively less immediate. Some HS fragments behave the same over all classes of linear orders. This is the case with the logic of temporal neighbourhood $A\bar{A}$, which is NEXPTIME-complete over all relevant classes of linear orders [3]. A real character is, on the contrary, the temporal logic of sub-intervals D: its satisfiability problem is PSPACE-complete over the class of dense linear orders [2] and undecidable over the classes of finite and discrete linear orders [8] (it is still unknown over the class of all linear orders).

In this paper, we focus our attention on the satisfiability problem for the logic $A\bar{A}B\bar{B}$, which pairs the decidable fragments $A\bar{A}$ and $B\bar{B}$. In [11], we proved that the problem is decidable, but not primitive recursive, over finite linear orders, and undecidable over the natural numbers. Here, we first show that undecidability can be lifted to the temporal domain $\mathbb{R}$, as well as to the class of all Dedekind-complete linear orders. Then, we consider the order $\mathbb{Q}$. We devise two semi-decision procedures: the first one terminates if and only if the input formula is unsatisfiable over $\mathbb{Q}$, while the second one terminates if and only if the input formula is satisfiable over $\mathbb{Q}$. Running the two procedures in parallel gives a decision algorithm for $A\bar{A}B\bar{B}$ over $\mathbb{Q}$. We conclude the paper by showing that decidability over the class of all linear orders follows from that over $\mathbb{Q}$.

2   The Logic

We begin by introducing the logic $A\bar{A}B\bar{B}$. Let Σ be a set of proposition letters. The logic $A\bar{A}B\bar{B}$ consists of formulas built up from letters in Σ using the Boolean connectives $\neg$ and $\vee$ and the unary modalities $\langle A \rangle$, $\langle \bar{A} \rangle$, $\langle B \rangle$, and $\langle \bar{B} \rangle$. We will often make use of shorthands like $\varphi_1 \wedge \varphi_2 = \neg(\neg\varphi_1 \vee \neg\varphi_2)$, $[A]\varphi = \neg\langle A \rangle\neg\varphi$, $[B]\varphi = \neg\langle B \rangle\neg\varphi$, true $= a \vee \neg a$, and false $= a \wedge \neg a$, for $a \in \Sigma$.

To define the semantics of $A\bar{A}B\bar{B}$ formulas, we consider a linear order $\mathbb{D} = (D, <)$, called *temporal domain*, and we denote by $\mathbb{I}_\mathbb{D}$ the set of all closed intervals $[x, y]$ over $\mathbb{D}$, with $x \leq y$. We call *interval structure* any Kripke structure of the form $\mathcal{I} = (\mathbb{I}_\mathbb{D}, \sigma, A, \bar{A}, B, \bar{B})$, where $\sigma : \mathbb{I}_\mathbb{D} \to \mathscr{P}(\Sigma)$ is a function mapping intervals to sets of proposition letters and A, $\bar{A}$, B, and $\bar{B}$ are the Allen's relations "meet", "met by", "begun by", and "begins", which are defined as follows: $[x, y]\ A\ [x', y']$ iff $y = x'$, $[x, y]\ \bar{A}\ [x', y']$ iff $x = y'$, $[x, y]\ B\ [x', y']$ iff $x = x' \wedge y' < y$, and $[x, y]\ \bar{B}\ [x', y']$ iff $x = x' \wedge y < y'$. Formulas are interpreted over a given interval structure $\mathcal{I} = (\mathbb{I}_\mathbb{D}, \sigma, A, \bar{A}, B, \bar{B})$ and a given initial interval $I \in \mathbb{I}_\mathbb{D}$ in the natural way, as follows: $\mathcal{I}, I \vDash a$ iff $a \in \sigma(I)$, $\mathcal{I}, I \vDash \neg\varphi$ iff $\mathcal{I}, I \nvDash \varphi$, $\mathcal{I}, I \vDash \varphi_1 \vee \varphi_2$ iff $\mathcal{I}, I \vDash \varphi_1$ or $\mathcal{I}, I \vDash \varphi_2$, and, most importantly, for all relations $R \in \{A, \bar{A}, B, \bar{B}\}$,

$$\mathcal{I}, I \models \langle R \rangle \varphi \qquad \text{iff} \qquad \text{there is } J \in \mathbb{I}_\mathbb{D} \text{ such that } I \, R \, J \text{ and } \mathcal{I}, J \models \varphi.$$

We say that a formula φ is *satisfiable* over a class $\mathscr{C}$ of interval structures if $\mathcal{I}, I \models \varphi$ for some $\mathcal{I} = (\mathbb{I}_\mathbb{D}, \sigma, A, \bar{A}, B, \bar{B})$ in $\mathscr{C}$ and some interval $I \in \mathbb{I}_\mathbb{D}$.

For example, the formula [B]false (hereafter abbreviated π) hold over all and only the *singleton* intervals $[x, x]$. Similarly, the formula [B][B]false (abbreviated unit) holds over the *unit-length* intervals of a discrete order, e.g. over the intervals of $\mathbb{Z}$ of the form $[x, x+1]$. The formula $[\bar{A}][\bar{A}][A][A]\varphi$ ($[G]\varphi$ for short) forces φ to hold universally, that is, over all intervals. The formula $[G] (\neg\pi \rightarrow \langle B \rangle \neg\pi)$ (φ_{dense} for short) holds over all and only the interval structures with a dense domain, e.g., the order $\mathbb{Q}$ of the rationals.

Logical Types. We now introduce basic terminology and notation that are common in the temporal logic setting. The *closure* of a formula φ is defined as the set closure(φ) of all sub-formulas of φ and all their negations (we identify $\neg\neg\psi$ with ψ, $\neg\langle A \rangle\psi$ with $[A]\neg\psi$, etc.). For a technical reason that will be clear soon, we also introduce the *extended closure* of φ, denoted closure$^+(\varphi)$, that extends closure(φ) by adding all formulas of the form $\langle R \rangle\psi$ and $[R]\psi$, with $R \in \{A, \bar{A}, B, \bar{B}\}$ and $\psi \in$ closure(φ).

Let $\mathcal{I} = (\mathbb{I}_\mathbb{D}, \sigma, A, \bar{A}, B, \bar{B})$ be an interval structure. We associate with each interval $I \in \mathbb{I}_\mathbb{D}$ its φ-*type* type$_{\mathcal{I}}^{\varphi}(I)$, defined as the set of all formulas $\psi \in$ closure$^+(\varphi)$ such that $\mathcal{I}, I \models \psi$ (when no confusion arises, we omit the parameters $\mathcal{I}$ and φ). A particular role will be played by those types F that contain the subformula [B]false, which are necessarily associated with singleton intervals. When no interval structure is given, we can still try to capture the concept of type by means of a maximal "locally consistent" subset of closure$^+(\varphi)$. Formally, we call φ-*atom* any set $F \subseteq$ closure$^+(\varphi)$ such that (i) $\psi \in F$ iff $\neg\psi \notin F$, for all $\psi \in$ closure$^+(\varphi)$, (ii) $\psi \in F$ iff $\psi_1 \in F$ or $\psi_2 \in F$, for all $\psi = \psi_1 \vee \psi_2 \in$ closure$^+(\varphi)$, (iii) if [B]false $\in F$ and $\psi \in F$, then $\langle A \rangle\psi \in F$ and $\langle\bar{A}\rangle\psi \in F$, for all $\psi \in$ closure(φ), (iv) if [B]false $\in F$ and $\langle A \rangle\psi \in F$, then $\psi \in F$ or $\langle\bar{B}\rangle\psi \in F$, for all $\psi \in$ closure(φ). We call π-*atoms* those atoms that contain the formula [B]false, which are thus candidate types of singleton intervals. We denote by atoms(φ) the set of all φ-atoms.

Given an atom F and a relation $R \in \{A, \bar{A}, B, \bar{B}\}$, we let req$_R(F)$ be the set of *requests of F along direction R*, namely, the formulas $\psi \in$ closure(φ) such that $\langle R \rangle\psi \in F$. Similarly, we let obs($F$) be the set of *observables of F*, namely, the formulas $\psi \in F \cap$ closure(φ) – intuitively, the observables of F are those formulas $\psi \in F$ that fulfil requests of the form $\langle R \rangle\psi$ from other atoms. Note that, for all π-atoms F, we have req$_A(F) =$ obs(F) $\cup$ req$_{\bar{B}}(F)$ and req$_{\bar{A}}(F) \supseteq$ obs(F).

Compass Structures. Formulas of interval temporal logics can be equivalently interpreted over the so-called compass structures [14]. These structures can be seen as two-dimensional spaces in which points are labelled with complete logical types (atoms). Such an alternative interpretation exploits the existence of a natural bijection between the intervals $I = [x, y]$ over a temporal domain $\mathbb{D}$ and the points $p = (x, y)$ in the $\mathbb{D} \times \mathbb{D}$ grid such that $x \leq y$. It is convenient to introduce a *dummy atom* $\varnothing$, distinct from all other atoms, and assume that it labels all and only the points (x, y) such that $x > y$, which do not correspond to intervals. We fix the convention that $\mathsf{obs}(\varnothing) = \varnothing$ and $\mathsf{req}_R(\varnothing) = \varnothing$ for all $R \in \{A, \bar{A}, B, \bar{B}\}$.

Formally, a *compass φ-structure over a linear order* $\mathbb{D}$ is a labelled grid $\mathcal{G} = (\mathbb{D} \times \mathbb{D}, \tau)$, where the function $\tau : \mathbb{D} \times \mathbb{D} \to \mathsf{atoms}(\varphi) \uplus \{\varnothing\}$ maps any point (x, y) to either a φ-atom (if $x \leq y$) or the dummy atom $\varnothing$ (if $x > y$).

We observe that Allen's relations over intervals have analogue relations over points. Figure 1 gives a geometric interpretation of relations $A, \bar{A}, B, \bar{B}$ (by a slight abuse of notation, we use the same letters for the corresponding relations over the points of a compass structure). Thanks to such an interpretation, any interval structure $\mathcal{I}$ can be converted to a compass one $\mathcal{G} = (\mathbb{D} \times \mathbb{D}, \tau)$ by simply letting $\tau(x, y) = \mathsf{type}([x, y])$ for all $x \leq y \in \mathbb{D}$. The converse, however, is not true in general, as the atoms associated with points in a compass structure may be inconsistent with respect to the underlying geometrical interpretation of Allen's relations. To ease a correspondence between interval and compass structures, we enforce suitable *consistency conditions* on compass structures. For this, we introduce two relations over atoms F, G:

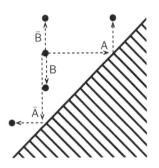

Fig. 1. Geometric interpretation of Allen's relations

$F \uparrow G$ iff
$$\begin{cases} \mathsf{req}_{\bar{B}}(F) \supseteq \mathsf{obs}(G) \cup \mathsf{req}_{\bar{B}}(G) \\ \mathsf{req}_B(G) \supseteq \mathsf{obs}(F) \cup \mathsf{req}_B(F) \\ \mathsf{req}_{\bar{A}}(F) = \mathsf{req}_{\bar{A}}(G) \end{cases}$$

$F \rightarrowtail G$ iff
$$\begin{cases} \mathsf{req}_A(F) = \mathsf{obs}(G) \cup \mathsf{req}_B(G) \cup \mathsf{req}_{\bar{B}}(G) \\ \mathsf{req}_{\bar{A}}(G) \supseteq \mathsf{obs}(F). \end{cases}$$

Note that the relation $\uparrow$ is transitive, while $\rightarrowtail$ only satisfies $\rightarrowtail \circ \uparrow \subseteq \rightarrowtail$. Observe also that, for all interval structures $\mathcal{I}$ and all intervals I, J in it, if $I \bar{B} J$ (resp., $I A J$), then $\mathsf{type}_{\mathcal{I}}(I) \uparrow \mathsf{type}_{\mathcal{I}}(J)$ (resp., $\mathsf{type}_{\mathcal{I}}(I) \rightarrowtail \mathsf{type}_{\mathcal{I}}(J)$). Hereafter, we tacitly assume that every compass structure $\mathcal{G} = (\mathbb{D} \times \mathbb{D}, \tau)$ satisfies analogous consistency properties with respect to its atoms, namely, for all points $p = (x, y)$ and $q = (x', y')$ in $\mathbb{D} \times \mathbb{D}$, with $x \leq y$ and $x' \leq y'$, if $p \bar{B} q$ (resp., $p A q$), then $\tau(p) \uparrow \tau(q)$ (resp., $\tau(p) \rightarrowtail \tau(q)$). In addition, we say that a request $\psi \in \mathsf{req}_R(\tau(p))$ of a point p in a compass structure $\mathcal{G} = (\mathbb{D} \times \mathbb{D}, \tau)$ is *fulfilled* if there is another point q such that $p R q$ and $\psi \in \mathsf{obs}(\tau(q))$ – in this case, we say that q is a *witness of fulfilment of ψ from p*. The compass structure $\mathcal{G}$ is said to be *globally fulfilling* if all requests of all its points are fulfilled.

We can now recall the standard correspondence between interval and compass structures (the proof is based on a simple induction on sub-formulas):

Proposition 1 ([11]). *Let φ be an $A\bar{A}B\bar{B}$ formula. For every globally fulfilling compass structure $\mathcal{G} = (\mathbb{D} \times \mathbb{D}, \tau)$, there is an interval structure $\mathcal{I} = (\mathbb{I}_{\mathbb{D}}, \sigma, A, \bar{A}, B, \bar{B})$ such that, for all $x \leq y \in \mathbb{D}$ and all $\psi \in \mathsf{closure}^+(\varphi)$, $\mathcal{I}, [x, y] \models \psi$ iff $\psi \in \tau(x, y)$.*

In view of Proposition 1, the satisfiability problem for a given $A\bar{A}B\bar{B}$ formula φ reduces to the problem of deciding the existence of a globally fulfilling compass

$\tilde{\varphi}$-structure $\mathcal{G} = (\mathbb{D} \times \mathbb{D}, \tau)$, with $\tilde{\varphi} = \langle G \rangle \varphi$ ($\langle G \rangle \varphi$ is a shorthand for $\neg [G] \neg \varphi$), that features the observable $\tilde{\varphi}$ in every point, that is, $\tilde{\varphi} \in \text{obs}(\tau(x, y))$ for all $x \leq y \in \mathbb{D}$.

3 Satisfiability over Finite and Dedekind-Complete Orders

The satisfiability problem for A̅A̅B̅B̅ was originally addressed in [11]. We first proved that A̅A̅B̅B̅ is decidable if interpreted over finite linear orders, but not primitive recursive. The decidability result rests on a contraction method that, given a formula φ and a finite compass structure satisfying φ, shows that, under suitable conditions, the compass structure can be reduced in size while preserving consistency and fulfilment properties. This leads to a non-deterministic procedure that decides whether φ is satisfiable by exhaustively searching all contraction-free compass structures. The proof of termination relies on Dickson's lemma, while non-primitive recursiveness is proved via a reduction from the reachability problem for lossy counter machines [13]. Then, we showed that the problem becomes undecidable if we interpret A̅A̅B̅B̅ over a temporal domain isomorphic to $\mathbb{N}$ (in fact, this is already the case with the proper fragment A̅A̅B). The proof is based on a reduction from an undecidable variant of the reachability problem for lossy counter machines, called structural termination [9], which consists of deciding whether a given lossy counter machine admits a halting computation starting from a given location and some arbitrary initial assignment for the counters. Due to an oversight, in [11] we claimed that such an undecidability result can be transferred to any class of linear orders in which $\mathbb{N}$ can be embedded. As a matter of fact, Dedekind completeness is a necessary condition. The following theorem properly states undecidability results for A̅A̅B.

Theorem 1. *The satisfiability problem for A̅A̅B interpreted over $\mathbb{N}$, $\mathbb{R}$, and the class of all Dedekind-complete linear orders is undecidable.*

In view of the above theorem and the decidability results in [11], the satisfiability problem for A̅A̅B̅B̅ over $\mathbb{Q}$, as well as over the class of all interval structures, remains open. In the next section, we will show that, quite surprisingly, both problems are decidable with non-primitive recursive complexity.

4 Satisfiability over the Rationals and All Linear Orders

We begin by describing a fairly simple semi-decision procedure for the *unsatisfiability* of A̅A̅B̅B̅ formulas over interval structures with a dense temporal domain. The crucial observation is that, whenever a formula φ is unsatisfiable over $\mathbb{Q}$, this can be witnessed by a *finite* set of intervals with inconsistent requests. Based on this observation, one can enumerate all finite compass structures that witness φ and are distinct up to isomorphism, following the partial order induced by the embedding relation (this relation is defined as an isomorphism between the smaller structure and the restriction of the larger structure to a suitable subset of its temporal domain). The only way the enumeration

procedure can terminate is when
no refinement is applicable: in this
case, one proves that the input
formula φ is not satisfiable. Conversely, if the enumeration procedure does not terminate, then
the formula φ is satisfied by some

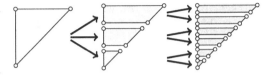

Fig. 2. Decomposition of a compass structure

compass structure that is obtained from the limit of an infinite series of refinements (suitable fairness conditions for the generated refinements guarantee that
the temporal domain of the limit compass structure is isomorphic to $\mathbb{Q}$). We
refer the reader to [12] for the details of the above enumeration procedure of
unsatisfiable formulas over $\mathbb{Q}$.

The rest of the section is devoted to finding a semi-decision procedure that
receives an input formula φ and terminates (successfully) iff φ is satisfiable
over an interval structure with a dense temporal domain. Differently from the
previous procedure, this one is based on enumerating suitable *finite abstractions*
of compass structures, which is far from being an easy task.

A first step consists of simplifying the consistency and fulfilment conditions.
More precisely, we show how to turn them into more "local" constraints, so as
to ease, later, the abstraction task. To this end, recall that the rational line is
isomorphic to any countable dense ordering with neither a minimal element nor
a maximal one. This means that, for the purpose of studying satisfiability over
$\mathbb{Q}$, it does not matter if we consider interval structures over $\mathbb{Q}$ or over any subset
of it that is dense and contains no extremal elements. Similarly, the complexity
of the satisfiability problem does not change if we add minimal and maximal
elements to the underlying temporal domain – for the sake of brevity, we call the
resulting order a *dense order with endpoints*. Now, to turn the consistency and
fulfilment conditions into local constraints, we decompose any dense order with
endpoints $\mathbb{D}$ into some infinite, finitely-branching tree T whose nodes represent
pairs of elements of $\mathbb{D}$ of the form $s = (y_1, y_2)$, with $y_1 < y_2$, and whose edges
connect nodes $(y_1, y_2) \in T$ to tuples of nodes $s_1 = (z_1, z_2), ..., s_n = (z_n, z_{n+1})$,
with $n \geq 2$ and $y_1 = z_1 < z_2 < ... < z_n < z_{n+1} = y_2$ (see Figure 2). Note that the
domain $\mathbb{D}$ is not necessarily entirely covered by the time points that appear in
the nodes of a decomposition T. Moreover, since all dense orders with endpoints
are isomorphic, we will not be concerned with the coordinates of the nodes of T
and we will often overlook them in the constructions that follow.

Using decompositions of temporal domains we can extract "horizontal slices"
of a compass structure. More precisely, given a compass structure $\mathcal{G} = (\mathbb{D} \times \mathbb{D}, \tau)$
and a node $s = (y_1, y_2)$ of a decomposition T of $\mathbb{D}$, we define the *slice of $\mathcal{G}$ in*
s as the induced sub-structure $\mathcal{G}_s = (\mathbb{D} \times \{y_1, y_2\}, \tau)$. Intuitively, the slice $\mathcal{G}_s$
is obtained from $\mathcal{G}$ by selecting the rows with coordinates y_1 and y_2 and by
restricting the labelling function τ to them (to reduce the notational overload,
we denote such a restriction of the labelling function by τ).

Below, we introduce suitable abstractions, called profiles, for the labels that
can appear in a slice of a compass structure. Intuitively, for each slice $\mathcal{G}_s =$

$(\mathbb{D} \times \{y_1, y_2\}, \tau)$ and each pair of atoms (F, G), where possibly $F = \varnothing$ or both $F = \varnothing$ and $G = \varnothing$ (dummy atoms), we keep track of the *number* of coordinates $x \in \mathbb{D}$ such that $\tau(x, y_1) = F$ and $\tau(x, y_2) = G$. In particular, in these abstractions, we forget the occurrence order of the pairs of atoms along the x-axis. To this end, we make extensive use of *multisets*. Given a multiset M and an element e in M, we denote by $M(e)$ the *multiplicity*, that is, the number of occurrences, of e in M, and we write $M(e) = \infty$ when M contains infinitely many occurrences of e. We freely use set-theoretic notations with multisets. For example, we denote membership by $e \in M$, containment by $M \subseteq N$, etc. Moreover, given a multiset M of n-tuples and a set $I \subseteq \{1, \ldots, n\}$ of component indices, we denote by $M|_I$ the *projection* of M onto I, that is, the multiset that contains exactly $\sum_{e|_I = f} M(e)$ occurrences of each I-tuple f (note that the sum ranges over all n-tuples e that coincide with f on the components indexed in I). Note that, differently from set projections, projections of multisets are injective, as they send distinct occurrences of tuples to distinct occurrences of tuples. In particular, $|_I$ defines a *bijection* from multiset M to multiset $M|_I$. Finally, we denote by $\mathsf{set}(M)$ the *support* of a multiset M, that is, the set of all elements e such that $M(e) \geq 1$.

We associate with each slice $\mathcal{G}_s = (\mathbb{D} \times \{y_1, y_2\}, \tau)$ of a globally fulfilling compass structure $\mathcal{G}$, the multiset M defined by $M(F, G) = |\{x \in \mathbb{D} : \tau(x, y_1) = F, \tau(x, y_2) = G\}|$ for all $(F, G) \in (\mathsf{atoms}(\varphi) \uplus \{\varnothing\})^2$. We call this multiset *the profile of the slice $\mathcal{G}_s$* and we denote it by $\mathsf{profile}(\mathcal{G}_s)$. Note that the projection $\mathsf{profile}(\mathcal{G}_s)|_1$ (resp., $\mathsf{profile}(\mathcal{G}_s)|_2$) onto the first (resp., second) component is a multiset that represents the number of occurrences of each atom along the lower (resp., upper) row of the slice $\mathcal{G}_s$. Definition 1 below captures a more general notion of profile that does not refer to a particular compass structure. We will then introduce trees labelled with profiles as abstractions of compass structures.

Definition 1. A *profile* is a multiset M of pairs of (possibly dummy) atoms $(F, G) \in (\mathsf{atoms}(\varphi) \uplus \{\varnothing\})^2$ such that: (i) for all $(F, G) \in M$, if $F \neq \varnothing$, then $G \neq \varnothing$ and $F \uparrow G$; (ii) for all $(F, G), (F', G') \in M$, $\mathsf{req}_A(F) = \mathsf{req}_A(F')$ and $\mathsf{req}_A(G) = \mathsf{req}_A(G')$; (iii) M contains infinitely many occurrences of pairs $(\varnothing, G)$ with $G \neq \varnothing$; (iv) M contains *exactly one* occurrence of a pair (F, G) with F π-atom and *exactly one* occurrence of a pair $(\varnothing, H)$ with H π-atom (for short, we denote the two pairs (F, G) and $(\varnothing, H)$ by M_π and M^π, respectively); (v) if $M_\pi = (F, G)$, then $\mathsf{req}_{\bar{A}}(F) = \bigcup_{(F', G') \in M} \mathsf{obs}(F')$; similarly, if $M^\pi = (\varnothing, H)$, then $\mathsf{req}_{\bar{A}}(H) = \bigcup_{(F', G') \in M} \mathsf{obs}(G')$.

Definition 2. A *profile tree* is an infinite finitely-branching tree $\mathcal{T} = (T, N, E)$, where T is a decomposition of some dense order with endpoints, N is a function mapping nodes of T to profiles, and E is a function mapping nodes of T to multisets of tuples of atoms, such that:

- **(profile-match)** every node $s \in T$ has at least two children, say, $s_1, \ldots, s_n$, with $n \geq 2$, and $E(s)$ is a multiset of $(n + 1)$-tuples such that $E(s)|_{1, n+1} = N(s)$ and $E(s)|_{i, i+1} = N(s_i)$ for all $1 \leq i \leq n$;
- **(profile-finite-req)** for every node $s \in T$ and pair $(F_1, F_{n+1}) \in N(s)$, with $F_1 \neq \varnothing$, if $N(s)(F_1, F_{n+1}) < \infty$, then $E(s)$ contains *exactly* $N(s)(F_1, F_{n+1})$

occurrences of tuples $(F_1, \ldots, F_{n+1})$ such that $\text{req}_{\bar{B}}(F_1) = \bigcup_{2 \leq i \leq n+1} \text{obs}(F_i) \cup \text{req}_{\bar{B}}(F_{n+1})$ and $\text{req}_B(F_{n+1}) = \bigcup_{1 \leq i \leq n} \text{obs}(F_i) \cup \text{req}_B(F_1)$;

- **(profile-infinite-req)** for every node $s \in T$ and pair $(F_1, F_{n+1}) \in N(s)$, with $F_1 \neq \varnothing$, if $N(s)(F_1, F_{n+1}) = \infty$, then $E(s)$ contains *at least one* occurrence of a tuple $(F_1, \ldots, F_{n+1})$ such that $\text{req}_{\bar{B}}(F_1) = \bigcup_{2 \leq i \leq n+1} \text{obs}(F_i) \cup \text{req}_{\bar{B}}(F_{n+1})$ and $\text{req}_B(F_{n+1}) = \bigcup_{1 \leq i \leq n} \text{obs}(F_i) \cup \text{req}_B(F_1)$;

- **(profile-dummy)** for every node $s \in T$ and pair $(\varnothing, G) \in N(s)$, with $G \neq \varnothing$, $E(s)$ contains at least one occurrence of a π-*tuple*, i.e., a tuple with a π-atom, of the form $(F_1, \ldots, F_{n+1})$, with $F_1 = \varnothing$ and $F_{n+1} = G$.

In addition, if the profile at the root s_0 of $\mathcal{T}$ contains only the pair $N^\pi(s_0) = (F, G)$, with F π-atom and $\text{req}_{\bar{B}}(G) = \varnothing$, and some pairs of the form $(\varnothing, H)$, with $H \neq \varnothing$ and $\text{req}_{\bar{B}}(H) = \varnothing$, then $\mathcal{T}$ is said to be a *full* profile tree.

The first item of Definition 2 enforces the matching conditions between the pairs in the profile of a node and the pairs in the profiles of its children. The second item requires that all requests that appear in a pair (F, G) of the profile of a node s are either "locally fulfilled" by the observables of corresponding pairs in the profiles of the children or transferred to other nodes of the profile tree at the same level as s. This condition, however, concerns only those pairs (F, G) that have finite multiplicity in the profile; for the remaining pairs, we enforce a similar, but weaker condition (third item of the definition). Finally, the fourth item requires that for each atom G, if the profile $N(s)$ contains the pair $(\varnothing, G)$, then at least one occurrence of this pair is "refined" in the multiset $E(s)$ by an occurrence of a tuple of the form $(\varnothing, \ldots, \varnothing, F, \ldots, G)$ that contains a π-atom F (such a tuple is called for short π-*tuple*) and that ends with the atom G (possibly $F = G$). We will see later that this condition is necessary for the fulfilment of the requests along the direction $\bar{A}$.

Below, we show that full profile trees are correct (though not yet finite) abstractions of globally fulfilling compass structures. We present this result with two statements showing, respectively, completeness and a weak form of soundness of profile trees. Note that the two-way correspondence is sufficient for witnessing satisfiability of $A\bar{A}B\bar{B}$ formulas by means of profile trees.

Proposition 2. For every globally fulfilling compass structure $\mathcal{G} = (\mathbb{D} \times \mathbb{D}, \tau)$ over a dense order with endpoints $\mathbb{D}$, there is a full profile tree $\mathcal{T} = (T, N, E)$ such that T is a decomposition of $\mathbb{D}$ and, for all nodes $s \in T$, $N(s) = \text{profile}(\mathcal{G}_s)$. Conversely, for every full profile tree $\mathcal{T} = (T, N, E)$, with T decomposition of some dense order with endpoints $\mathbb{D}$, there is a globally fulfilling compass structure $\mathcal{G} = (\mathbb{D}' \times \mathbb{D}', \tau)$, with $\mathbb{D}' \subseteq \mathbb{D}$ dense order with endpoints, such that $\text{set}(\text{profile}(\mathcal{G}_s)) = \text{set}(N(s))$ for all $s \in T$.

Below, we show how to further restrict ourselves to a complete subset of full profile trees and derive *finite* representations of them. The general idea is to normalise profile trees so as to obtain structures that are sufficiently "regular" to be represented by finite trees. To this end, we introduce a finite variant of the notion of profile tree, called *finite profile tree*, that is obtained by enforcing the conditions of Definition 2 to internal nodes only (accordingly, since the multisets $E(s)$ that are associated with the leaves s in a finite profile tree are not anymore relevant, one can assume that the function E is undefined on the leaves). We also introduce a strengthening of the containment relation on multisets, denoted by $\sqsubseteq$ and defined as follows: $M \sqsubseteq N$ iff $\mathsf{set}(M) = \mathsf{set}(N)$ and $M(\bar{F}) \leq N(\bar{F})$ (resp., $M(\bar{F}) = N(\bar{F})$) for all tuples $\bar{F}$ (resp., π-tuples $\bar{F}$). The following definition captures precisely the set of profile trees we are interested in.

Definition 3. Let $\mathcal{T}$ be a finite or infinite profile tree. We say that $\mathcal{T}$ is *pseudo-regular* iff for all paths π, there are $s, s' \in \pi$, with s proper ancestor of s', such that $N(s) \sqsubseteq N(s')$ and $N(s)(\varnothing, G) = N(s')(\varnothing, G)$ for all atoms G.

In the following, we mainly work with profiles that appear at the roots of infinite profile trees (*feasible* profiles for short). We observe that the restriction of the partial order $\sqsubseteq$ to feasible profiles is a *well* partial order: indeed, the definition of profile tree implies that every π-tuple has multiplicity either 0 or 1 in any feasible profile, which in turn means that $\sqsubseteq$ is the conjunction of the well partial order $\subseteq$ and an equivalence of finite index. Hence, by a combination of Dickson's and König's lemmas, every infinite pseudo-regular tree has a finite prefix that is also pseudo-regular (a prefix of a tree is any restriction of it to an upward-closed set of nodes). A converse result also holds:

Proposition 3. For every finite pseudo-regular profile tree $\mathcal{T}$, there is an infinite profile tree $\mathcal{T}'$ that has the same profile as $\mathcal{T}$ at the root.

The crux of our semi-decision procedure for testing the satisfiability of AĀBB̄ formulas is to enumerate all atoms that appear in feasible profiles. Proposition 3 allows us to use finite pseudo-regular profile trees as witnesses of existence of some of these atoms. Unfortunately, this is not yet the end of the story, because not all profile trees are pseudo-regular and hence, a priori, there might exist atoms that appear only in infinite profile trees that are not pseudo-regular. The last piece of the puzzle amounts at showing that this is not the case and that we can indeed safely restrict ourselves to atoms appearing in pseudo-regular profile trees. We will prove this result by normalizing infinite profile trees via a series of operations that "inflate" the profiles as much as possible.

An important aspect that must be taken into account while inflating the profiles in a tree is that there are matching constraints to satisfy. As a matter of fact, these constraints induce dependencies between the multiplicities of pairs $(\varnothing, F)$ in the profile associated with a node s and the multiplicities of corresponding pairs (F, G) in the profile associated with the right sibling of s. As a consequence, there will be differences in the treatment of pairs of the form $(\varnothing, F)$ and pairs of the form (F, G), with $F \neq \varnothing$. We take a brief interlude to give an example of the type of dependencies that can be enforced.

Example 1. Consider a formula φ that contains, among other conjuncts, the subformula $[\mathsf{G}](a \rightarrow [\mathsf{B}]\neg a \wedge [\bar{\mathsf{B}}]\neg a)$. Figure 3 describes a slice of a compass structure that may satisfy φ, with some distinguished points annotated with observables and requests. The formula requires that all a-labelled points lie on distinct vertical axes; on the other hand, it allows arbitrarily many a-labelled points to be horizontally aligned. This is a representative example because, in general, forbidding multiple occurrences of an observable along the same horizontal line can be only done using the modal operator $[\mathsf{E}]$, which is not available in the logic. As concerns the multiplicities of the example profile, we observe that by inserting multiple a-labelled points along a single horizontal line and by accordingly modifying the upper part of the compass structure, one can get as many pairs

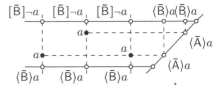

Fig. 3. Dependencies between multiplicities

of atoms (F, G), where $\langle \bar{B} \rangle a \in F$ and $[\bar{B}] \neg a \in G$. On the other hand, to increase the number of pairs $(\varnothing, F)$, where $\langle \bar{B} \rangle a \in F$, one has to introduce new horizontal lines ending with π-atoms H such that $\langle \bar{A} \rangle a \in H$: this is not always possible as other conjuncts of φ may enforce bounds to the number of π-atoms H.

As shown by the above example, the simplest way one can inflate a profile, while preserving its feasibility, is by increasing the multiplicites associated with the pairs (F, G), where $F \neq \varnothing$. We formalise this in the next lemma.

Lemma 1. If N is a feasible profile and N' is a profile such that $N \sqsubseteq N'$ and $N(\varnothing, G) = N'(\varnothing, G)$ for all atoms G, then N' is feasible too. Moreover, a profile tree with root profile N' can be obtained from a profile tree with root profile N without modifying the underlying decomposition tree.

We describe a second inflation method, which depends on the previous one and can be used to further simplify the reasoning on the matching conditions of a profile tree $\mathcal{T} = (T, N, E)$. In particular, it shows that w.l.o.g. one can assume that the *finiteness* of the multiplicity of any tuple $(F_1, \ldots, F_{n+1})$ in a multiset $E(s)$ depends only on the multiplicity of the first component F_1 in $E(s)|_1$. This property is formalized below by the definition of "pointwise fair" profile tree, followed by a corresponding lemma that shows how to enforce the property.

Definition 4. A multiset E of $(n + 1)$-tuples is *fair* if for all $(n + 1)$-tuples $(F_1, \ldots, F_{n+1}) \in E$, with $F_1 \neq \varnothing$, $E|_1(F_1) = \infty$ implies $E(F_1, \ldots, F_{n+1}) = \infty$. A profile tree $\mathcal{T} = (T, N, E)$ is *pointwise fair* if all multisets $E(s)$ are fair.

Lemma 2. For every feasible profile N, there is an infinite pointwise fair profile tree that has root profile $N' \sqsupseteq N$. Moreover, one can assume that, for all pairs of atoms (F, G), if $N|_1(F) < \infty$, then $N(F, G) = N'(F, G)$.

A third inflation method makes use of the fact that the partial order $\sqsubseteq$ restricted to the set of feasible profiles is ω-complete.

Lemma 3. Every sequence of feasible profiles $N_0 \sqsubseteq N_1 \sqsubseteq \ldots$ has a supremum $\sup_i N_i$, defined by $(\sup_i N_i)(F, G) = \sup_{i \in \mathbb{N}} \big(N_i(F, G) \big)$ for all atoms F, G, that is a feasible profile.

We have described three ways of increasing the multiplicities of profiles at the roots of profile trees. In general, these techniques are not applicable to nodes that are strictly below the root. This is why we introduce a new partial order $\trianglelefteq$, incomparable with $\sqsubseteq$, that is defined only over feasible profiles N, N' as follows:

$$N \trianglelefteq N' \quad \text{iff} \quad \begin{cases} N \sqsubseteq N' \\ \operatorname{set}(N|_2) = \operatorname{set}(N'|_2) \\ N(F, G) = N'(F, G) \quad \text{for all atoms } F, G \neq \varnothing. \end{cases}$$

We observe that from any infinite ⊴-chain of feasible profiles, one can extract an infinite sub-sequence that is also a ⊑-chain. Thus, an immediate consequence of Lemma 3 is that every ⊴-chain has an *upper bound*. In its turn, the existence of upper bounds on ⊴-chains implies the existence of feasible profiles that are *maximal* with respect to ⊴ (this can be seen as a consequence of Zorn's Lemma):

Corollary 1. For all feasible profiles N, there is a ⊴-maximal profile $N' \trianglerighteq N$.

Based on existence of ⊴-maximal profiles, we say that a profile tree is *pointwise ⊴-maximal* if all its profiles are ⊴-maximal. Below, we show that all atoms of feasible profiles appear at the roots of some pseudo-regular profile trees.

Proposition 4. For every infinite pointwise fair profile tree with root profile N, there is an infinite pointwise fair and pointwise ⊴-maximal profile tree with root profile $N' \trianglerighteq N$.

Proposition 5. Every infinite pointwise fair and pointwise ⊴-maximal profile tree is pseudo-regular.

Wrapping up, we can devise a semi-decision procedure that tests the satisfiability of a formula φ over $\mathbb{Q}$. The procedure works as follows. It first transforms φ into an equi-satisfiable formula $\varphi_{][}$ interpreted over a dense order with endpoints $\mathbb{D}$. Then, the procedure enumerates all finite full pseudo-regular trees, until a tree is found that contains the formula $\langle G \rangle \varphi_{][}$ as an observable of one of its atoms. The above semi-decision procedure is correct, namely, it terminates successfully iff the input formula φ is satisfiable over $\mathbb{Q}$. Indeed, if φ is satisfiable over $\mathbb{Q}$, then $\varphi_{][}$ is satisfiable over a dense order with endpoints $\mathbb{D}$, and hence there is a globally fulfilling compass structure $\mathcal{G}$ that contains $\langle G \rangle \varphi_{][}$ as an observable of all its atoms. By Propositions 4 and 5, there is also an infinite, pseudo-regular full profile tree $\mathcal{T}$ that witnesses $\langle G \rangle \varphi_{][}$ at the root profile. By the remarks that follow Definition 3, there is also a prefix of $\mathcal{T}$ that is a finite pseudo-regular full profile tree, and eventually this tree must be discovered by the procedure. Conversely, if the procedure terminates with a finite pseudo-regular full profile tree witnessing $\langle G \rangle \varphi_{][}$, then by Proposition 3 there is an infinite full profile tree $\mathcal{T}$, and hence a compass structure $\mathcal{G}$, that witness the satisfiability of $\langle G \rangle \varphi_{][}$ over $\mathbb{D}$. One can then conclude that φ is satisfiable over $\mathbb{Q}$.

A full decision procedure that solves the satisfiability problem for AĀBB̄ over $\mathbb{Q}$ can simply run in parallel the two semi-decision procedures that we described for unsatisfiability and satisfiability of AĀBB̄ formulas.

As for the satisfiability problem over the class of all interval structures, one can simply observe the following. The logic AĀBB̄, as any other HS fragment, can be viewed as a fragment of first-order logic that uses binary relations to express properties of pairs of elements of the underlying temporal domain. The relation $<$ of the temporal domain can be easily constrained by a first-order formula so as to define a linear order, and Allen's relations can be expressed in first-order logic in term of $<$. From Löwenheim-Skolem theorem, it follows that every interval structure can be assumed to contain only *countably* many intervals. Moreover, since every countable linear order can be embedded inside $\mathbb{Q}$, satisfiability of formulas of a given HS fragment over the class of all linear orders can be reduced to their satisfiability over $\mathbb{Q}$, provided that the fragment is powerful enough to express such an embedding. This is the case with $A\bar{A}B\bar{B}$: it suffices to introduce a distinguished proposition letter $\#$, to constrain all $\#$-labelled intervals to be singletons ($[G](\# \to \pi)$), and to relativize all modalities to intervals with endpoints labelled by $\#$ (intervals that satisfy $\langle B\rangle\# \wedge \langle A\rangle\#$). We conclude by establishing the precise complexity of the satisfiability problem.

Theorem 2. The satisfiability problem for $A\bar{A}B\bar{B}$ interpreted over $\mathbb{Q}$, as well as over the class of all linear orders, is decidable, but not primitive recursive.

5   Conclusions

In this paper we close the open questions concerning the satisfiability problem for the interval temporal logic $A\bar{A}B\bar{B}$. First, we generalized the undecidability result from [11] to $\mathbb{R}$ and to the class of all Dedekind-complete linear orders, and then we proved that it is decidable in two interesting cases: $\mathbb{Q}$ and the class of all interval structures. To decide satisfiability of $A\bar{A}B\bar{B}$ formulas over $\mathbb{Q}$ we used a combination of techniques from [4] (tree-shaped decomposition of models) and [11] (encoding of models by systems with counters), plus new key ingredients (separation into two semi-decision procedures, Konig's lemma). As concerns the second result, the decidability of $A\bar{A}B\bar{B}$ over the class of all interval structures follows from the decidability over $\mathbb{Q}$ and from Löwenheim-Skolem theorem, which allows us to assume, without loss of generality, that the interval structures are countable and hence embeddable inside $\mathbb{Q}$. The fact that $A\bar{A}B\bar{B}$ is powerful enough to express the embedding of a countable order inside $\mathbb{Q}$ completes the reduction. It is worth pointing out that the same technique cannot be applied to all HS fragments; for instance, the satisfiability problem for the temporal logic of sub-/super-intervals $D\bar{D}$ is known to be decidable over $\mathbb{Q}$ [2, 10], but it is open for the class of all interval structures.

References

[1] Bresolin, D., Della Monica, D., Goranko, V., Montanari, A., Sciavicco, G.: The dark side of Interval Temporal Logic: sharpening the undecidability border. In: Proc. of the 18th TIME, pp. 131–138. IEEE (2011)

[2] Bresolin, D., Goranko, V., Montanari, A., Sala, P.: Tableaux for logics of subinterval structures over dense orderings. Journal of Logic and Computation 20(1), 133–166 (2010)

[3] Bresolin, D., Goranko, V., Montanari, A., Sciavicco, G.: Propositional interval neighborhood logics: Expressiveness, decidability, and undecidable extensions. Annals of Pure and Applied Logic 161(3), 289–304 (2009)

[4] Bresolin, D., Montanari, A., Sala, P., Sciavicco, G.: What's decidable about Halpern and Shoham's interval logic? The maximal fragment $AB\bar{B}\bar{L}$. In: Proc. of the 26th LICS, pp. 387–396. IEEE (2011)

[5] Halpern, J., Shoham, Y.: A propositional modal logic of time intervals. Journal of the ACM 38(4), 935–962 (1991)

[6] Kamp, H., Reyle, U.: From Discourse to Logic: Introduction to Model-theoretic Semantics of Natural Language, Formal Logic and Discourse Representation Theory. Studies in Linguistics and Philosophy, vol. 42. Springer (1993)

[7] Lodaya, K.: Sharpening the undecidability of interval temporal logic. In: Kleinberg, R.D., Sato, M. (eds.) ASIAN 2000. LNCS, vol. 1961, pp. 290–298. Springer, Heidelberg (2000)

[8] Marcinkowski, J., Michaliszyn, J.: The ultimate undecidability result for the Halpern-Shoham logic. In: Proc. of the 26th LICS, pp. 377–386. IEEE (2011)

[9] Mayr, R.: Undecidable problems in unreliable computations. Theoretical Computer Science 297(1-3), 337–354 (2003)

[10] Montanari, A., Puppis, G., Sala, P.: A decidable spatial logic with cone-shaped cardinal directions. In: Grädel, E., Kahle, R. (eds.) CSL 2009. LNCS, vol. 5771, pp. 394–408. Springer, Heidelberg (2009)

[11] Montanari, A., Puppis, G., Sala, P.: Maximal decidable fragments of halpern and shoham's modal logic of intervals. In: Abramsky, S., Gavoille, C., Kirchner, C., Meyer auf der Heide, F., Spirakis, P.G. (eds.) ICALP 2010, Part II. LNCS, vol. 6199, pp. 345–356. Springer, Heidelberg (2010)

[12] Montanari, A., Puppis, G., Sala, P.: Decidability of the interval temporal logic $A\bar{A}B\bar{B}$ over the rationals. Tech. Rep. RR01/2014, Univ. of Udine (2014)

[13] Schnoebelen, P.: Lossy counter machines decidability cheat sheet. In: Kučera, A., Potapov, I. (eds.) RP 2010. LNCS, vol. 6227, pp. 51–75. Springer, Heidelberg (2010)

[14] Venema, Y.: Two-dimensional modal logics for relation algebras and temporal logic of intervals. ITLI prepublication series LP-89-03, Univ. of Amsterdam (1989)

Reachability in Pushdown Register Automata*

Andrzej S. Murawski[1], Steven J. Ramsay[1], and Nikos Tzevelekos[2]

[1] University of Warwick
[2] Queen Mary University of London

Abstract. We investigate reachability in pushdown automata over infinite alphabets: machines with finite control, a finite collection of registers and pushdown stack. First we show that, despite the stack's unbounded storage capacity, in terms of reachability/emptiness these machines can be faithfully represented by using only $3r$ elements of the infinite alphabet, where r is the number of registers. Moreover, this bound is tight. Next we settle the complexity of the associated reachability/emptiness problems. In contrast to register automata, where differences in register storage policies gave rise to differing complexity bounds, the emptiness problem for pushdown register automata is EXPTIME-complete in all cases. We also provide a solution to the global reachability problem, based on representing pushdown configurations with a special register automaton. Finally, we examine extensions of pushdown storage to higher orders and show that reachability is undecidable already at order 2, unlike in the finite alphabet case.

1 Introduction

Recent years have seen lively interest in automata over infinite alphabets, driven by applications in quite diverse areas where abstraction by a finite domain was deemed unsatisfactory. A case in point are markup languages [18,4], most notably XML, which permit the use of potentially unbounded data values in documents and allow queries to perform comparison tests on such data. A similar scenario occurs in reference-based programming languages, such as object-oriented [6,2,12] or ML-like languages [16,17], where memory is managed with the help of abstract addresses (reference names) that can be created afresh, compared for equality but little else. Other examples include array-accessing programs [1] as well as programs with restricted integer parameters [7].

Such applications call for a robust theory of automata over infinite alphabets, which will match our understanding of the finite-alphabet setting. Thus the limits will be exposed and a complexity-theoretic guide established for applications. A lot of the groundwork, surveyed in [20,3], was already dedicated to uncovering a notion of "regularity" in the infinite-alphabet case. One way to extend the concept of finite memory to such a setting consists of introducing a fixed number of *registers* for storing elements of the alphabet [13]. Another strand of work aimed to identify the infinite-alphabet "context-free" languages. Cheng and Kaminski [8] introduced context-free grammars over infinite alphabets and defined a corresponding notion of pushdown automata. Segoufin presents a similar definition in [20], albeit couched in a way suitable to process data words.

---

* Research supported by the Engineering and Physical Sciences Research Council (EP/J019577/1) and the Royal Academy of Engineering (RF: Tzevelekos).

E. Csuhaj-Varjú et al. (Eds.): MFCS 2014, Part I, LNCS 8634, pp. 464–473, 2014.

Our paper is devoted to studying exactly such computational scenarios through a study of *pushdown register systems* (PDRS), devices in which registers are integrated with a pushdown store. Although of foundational nature, the work is largely motivated by the pertinence of such machines to software model checking [6,2], and in particular their application to game-semantics-based verification [17]. We present several new results on the complexity of reachability testing. Altogether they fill a gap in the theory of "context-free" languages over infinite alphabets. More specifically, we make the following contributions.

Alphabet Distinguishability. A finite-memory automaton [13] with r registers can store r elements of the infinite alphabet at any instant. In fact, such automata are only capable of remembering r elements of the infinite alphabet over the course of a run — for any accepting run one can construct another one involving only r elements of the alphabet. Even though pushdown register systems have no bound on the number of elements of the alphabet that can be stored at any instant, we show that, over the course of a run, they can nevertheless remember at most $3r$ of them. More precisely, we show that for any run of a PDRS with r registers there exists an equivalent run involving only $3r$ elements. Moreover, no smaller number is enough: we exhibit a family of PDRS whose runs require remembering at least $3r$ elements.

Reachability Testing. The above-mentioned result yields an obvious methodology for reductions to the finite-alphabet setting, which immediately implies decidability of associated reachability problems, and language emptiness. While the decidability of emptiness has already been proved in [8] using context-free grammars, we provide exact complexity bounds for the problem, namely, EXPTIME-completeness.

In the pushdown-free setting, language nonemptiness was known to be NL-, NP- and PSPACE-complete, depending on the register discipline. In contrast, in the pushdown case, such distinctions do not affect the complexity: even if identical elements can be kept in different registers, the problem can still be solved in EXPTIME, while it is EXPTIME-hard already in the case where only distinct elements are allowed. In the last case, the hardness proof is technically involved since sequences of distinct names do not provide a supportive framework for representing memory content (as needed in reduction arguments using computation histories).

We show how to conduct *global* reachability analysis, which asks for a representation of all configurations from which a specified set of configurations can be reached. In the finite-alphabet case, it is well known that, if the target set is regular, the set of configurations that reach it can be captured by a finite automaton [5]. We prove an analogous result in the infinite-alphabet setting using a variant of register automata.

Higher-Order. Higher-order pushdown automata [15] take the idea of pushdown storage further by allowing for nesting. Standard pushdown store is considered to be order 1, while the elements stored in an order-k ($k > 1$) pushdown store are $(k-1)$-pushdown stores. In the finite alphabet setting this leads to an infinite hierarchy of decidable models of computation with a $(k-1)$-EXPTIME-complete problem at order k. We examine how the model behaves in the infinite alphabet setting, after the addition of a fixed number of registers for storing elements of the infinite alphabet.

We first observe that one can no longer establish a uniform bound on the number of symbols of the infinite alphabet that suffice to represent arbitrary runs. The existence of such a bound would imply decidability of the associated reachability problems, but the lack of a bound is not sufficient for establishing undecidability: indeed, the *decidable* class of data automata from [4] contains an automaton that can recognize all words consisting of distinct letters. Still, we show that the reachability problem for higher-order register pushdown automata is undecidable, already at order 2 and with one register.

2   Basic Definitions

Let us assume a countably infinite alphabet $\mathcal{D}$ of **data values** or **names**. We introduce a simple formalism for computations based on a finite number of $\mathcal{D}$-valued registers and a pushdown store. Writing $[r]$ for $\{1, \cdots, r\}$, by an **r-register assignment** we mean an injective map from $[r]$ to $\mathcal{D}$. We write Reg_r for the set of all such assignments.

Definition 1. *A* **pushdown r-register system** *(r-PDRS) is a tuple* $\mathcal{S} = \langle Q, q_I, \tau_I, \delta \rangle$, *where:*

- *Q is a finite set of* states, *with $q_I \in Q$ being initial,*
- *$\tau_I \in Reg_r$ is the initial r-register assignment,*
- *and $\delta \subseteq Q \times Op_r \times Q$ is the transition relation,*

with $Op_r = \{ i^\bullet, push(i), pop(i) \mid 1 \leq i \leq r \} \cup \{ pop^\bullet \}$.[1]

The operations executed in each transition have the following meaning: – the $i^\bullet$ operation *refreshes* the content of the ith register; – $push(i)$ pushes the symbol currently in the ith register on the stack; – $pop(i)$ pops the stack if the top symbol is the same as that stored in the ith register; – $pop^\bullet$ pops the stack if the top of the stack is currently not present in any of the registers. This semantics is given formally below.

Definition 2. *A* **configuration** *of an r-PDRS $\mathcal{S}$ is a triple $(q, \tau, s) \in Q \times Reg_r \times \mathcal{D}^*$. We say that (q_2, τ_2, s_2) is a successor of (q_1, τ_1, s_1), written $(q_1, \tau_1, s_1) \vdash (q_2, \tau_2, s_2)$, if $(q_1, op, q_2) \in \delta$ for some $op \in Op_r$ and one of the following conditions holds.*

- *$op = i^\bullet, \forall j. \tau_2(i) \neq \tau_1(j), \forall j \neq i. \tau_2(j) = \tau_1(j)$ and $s_2 = s_1$.*
- *$op = push(i), \tau_2 = \tau_1$ and $s_2 = \tau_1(i)s_1$.*
- *$op = pop(i), \tau_2 = \tau_1$ and $\tau_1(i)s_2 = s_1$.*
- *$op = pop^\bullet, \tau_2 = \tau_1$ and, for some $d \in \mathcal{D}, \forall j. \tau_1(j) \neq d$ and $ds_2 = s_1$.*

A **transition sequence** *of $\mathcal{S}$ is a sequence $\rho = \kappa_0, \cdots, \kappa_k$ of configurations with $\kappa_j \vdash \kappa_{j+1}$, for all $0 \leq j < k$. We say that ρ ends in a state q if $q_k = q$, where q_k is the state in κ_k. We call ρ a* **run** *if $\kappa_0 = (q_I, \tau_I, \epsilon)$.*

Remark 3. r-PDRS is meant to be a minimalistic model allowing us to study reachability in the infinite-alphabet setting with registers and pushdown storage. Existing related models [8], [20] feature transitions of a more compound shape, which can be readily translated into sequences of PDRS transitions.

---

[1] For technical reasons, it is convenient to have ε-transitions. However, to keep the definition minimal, we observe that they can be simulated with $push(1)$ followed by $pop(1)$.

For instance, a transition of an infinite-alphabet pushdown automaton [8] typically involves a refreshment $(i^\bullet)$ followed by pop $(pop(j))$ and a sequence of pushes $(push(j))$. This decomposition leads to a linear blow-up in size for translations of reachability questions into the r-PDRS setting. For register pushdown automata [20], an additional complication is their use of *non-injective* register assignments. Observe, though, that transitions in the non-injective framework can be easily mimicked using injective register assignments provided we keep track of the partitions determined by duplicated values in the original automaton. The book-keeping can be implemented inside the control state, which leads to an exponential blow-up in the size of the system, because the number of all possible partitions is exponential. Note that the number of registers does not change during such a simulation. Another difference is that register pushdown automata [20] are tailored towards data languages, i.e. a stack symbol is an element of $\mathcal{D}$ paired up with a tag drawn from a finite set. From this perspective, r-PDRSs use a singleton set of tags. Still, richer tag sets could be encoded via sequences of elements of $\mathcal{D}$ (for example, to simulate the ith out of k tags, we could push sequences of the form $d_1^i d_2$ for $d_1, d_2 \in \mathcal{D}$ with $d_1 \neq d_2$). This reduction is achievable in polynomial time.

Following [13,8,18], we mostly use *injective* register assignments. This is done to allow us to explore whether the restriction still leads to asymptotically more efficient reachability testing, as in the pushdown-free case. On a foundational note, injectivity gives a more essential treatment of freshness with respect to a set of registers: non-injective assignments can easily be used to encode PSPACE computations that have little to do with the interaction between finite control (and pushdown) and freshness.

Name permutations. There is a natural action of the group of permutations of $\mathcal{D}$ on stacks, assignments, runs, etc. For instance, given permutation $\pi : \mathcal{D} \to \mathcal{D}$ and an assignment τ, the result of applying π to τ is the register assignment $\pi \cdot \tau$ given by $\{(i, \pi(d)) \mid (i, d) \in \tau\}$. Similarly, $\pi \cdot s = \pi(d_n) \cdots \pi(d_1)$ for any stack $s = d_n \cdots d_1$ while, on the other hand, $\pi \cdot q = q$ for all states q. Hence, $\pi \cdot (q, \tau, s) = (q, \pi \cdot \tau, \pi \cdot s)$ and, for $\rho = \kappa_0 \vdash \cdots \vdash \kappa_n$ a transition sequence, $\pi \cdot \rho$ is the sequence $\pi \cdot \kappa_0, \cdots, \pi \cdot \kappa_n$.

Note that, as long as our constructions involve finitely many names, they will always have a finite support: we say that a set $S \subseteq \mathcal{D}$ *supports* some (nominal) element x if, for all permutations π, if $\pi(n) = n$ for all $n \in S$ then $\pi \cdot x = x$. Accordingly, **the support** $\nu(x)$ of x is the smallest set S supporting x. For example, $\nu(\tau) = \{\tau(i) \mid i \in [r]\}$, for all assignments τ. The support of a run $\rho = \kappa_0 \vdash \cdots \vdash \kappa_n$ is $\nu(\rho) = \bigcup_{j=0}^{n} \nu(\kappa_j)$, i.e. it consists of all elements of $\mathcal{D}$ that occur in it. The finite-support setting can be formally described by means of *nominal sets* [11] and closure results such as the following hold.

Fact 4 (Closure Under Permutations). *Fix an r-PDRS and let ρ be a transition sequence and $\pi : \mathcal{D} \to \mathcal{D}$ a permutation. Then $\pi \cdot \rho$ is also a transition sequence.*

3 Distinguishability

Devices with r registers but without pushdown storage, such as finite-memory automata [13], can take advantage of the registers to distinguish r elements of $\mathcal{D}$ from the rest. Consequently, any run can be replaced with a run that ends in the same state, yet is supported by merely r elements of the infinite alphabet [13, Proposition 4].

With extra pushdown storage, an r-PDRS is capable of storing unboundedly many elements of $\mathcal{D}$. Nevertheless, the restricted nature of the stack makes it possible to place a finite bound on the size of the support needed for a run to a given state, which is again a function of the number of registers.

Lemma 5 (Limited Distinguishability). *Fix an r-PDRS. For every transition sequence* $\rho = (q_0, \tau_0, \epsilon) \vdash^n (q_n, \tau_n, \epsilon)$, *there is a transition sequence* $\rho' = (q_0, \tau_0', \epsilon) \vdash^n (q_n, \tau_n', \epsilon)$ *with* $\tau_0' = \tau_0$, $\tau_n' = \tau_n$ *and* $|\nu(\rho')| \leq 3r$.

Proof. The proof is by induction on n. For $n \leq 1$ the result is trivial. Otherwise, the difficult case arises when the transition sequence is of the form: $(q_0, \tau_0, \epsilon) \vdash^k (q_k, \tau_k, \epsilon) \vdash^{n-k} (q_n, \tau_n, \epsilon)$ with $0 < k < n$. It follows from the induction hypothesis that there are sequences: $\rho_1 = (q_0, \tau_0', \epsilon) \vdash^k (q_k, \tau_k', \epsilon)$ and $\rho_2 = (q_k, \tau_k', \epsilon) \vdash^{n-k} (q_n, \tau_n', \epsilon)$ with $\tau_0' = \tau_0$, $\tau_n' = \tau_n$, $\tau_k' = \tau_k$ and which each, individually, use no more than $3r$ names. Let $N \supseteq \nu(\tau_0) \cup \nu(\tau_k) \cup \nu(\tau_n)$ be a set of names of size $3r$. We aim to map $\nu(\rho_1)$ and $\nu(\rho_2)$ into N by injections i and j respectively. For i we set $i(a) = a$ for any $a \in \nu(\tau_0) \cup \nu(\tau_k)$ and otherwise choose some *distinct* $b \in N \setminus (\nu(\tau_0) \cup \nu(\tau_k))$. Similarly, for j we set $j(a) = a$ for any $a \in (\nu(\tau_k) \cup \nu(\tau_n))$ and otherwise choose some distinct $b \in N \setminus (\nu(\tau_k) \cup \nu(\tau_n))$. Note that these choices are always possible because $|\nu(\rho_1)| \leq |N| \geq |\nu(\rho_2)|$. Finally, we extend i and j to permutations π_i and π_j on $\mathcal{D}$. Since transition sequences are closed under permutations (Fact 4): $(q_0, \pi_i \cdot \tau_0, \epsilon) \vdash^k (q_k, \pi_i \cdot \tau_k = \pi_j \cdot \tau_k, \epsilon) \vdash^{n-k} (q_n, \pi_j \cdot \tau_n, \epsilon)$ is a valid transition sequence with $\pi_i \cdot \tau_0 = \tau_0$, $\pi_j \cdot \tau_n = \tau_n$ and which is supported by a subset of N. $\square$

Corollary 6. *Fix an r-PDRS $\mathcal{S}$ and a state q of $\mathcal{S}$. If there is a run of $\mathcal{S}$ ending in q then there is a run of $\mathcal{S}$ ending in q that is supported by at most $3r$ distinct names.*

The $3r$ bound given above is optimal in the sense that there exists an r-PDRS such that all runs to a certain state will have to rely on $3r$ elements of $\mathcal{D}$.

Lemma 7 (Most Discriminating r-PDRS). *There exists an r-PDRS $\langle Q, q_I, \tau_I, \epsilon \rangle$ and $q \in Q$ such that $|\nu(\rho)| = 3r$ for any run ρ ending in q.*

Proof. Consider the following high-level description of an r-PDRS. The machine proceeds as follows:

1. Push registers in numerical order, twice, to obtain stack $\tau_I(r) \cdot \tau_I(1)\tau_I(r) \cdot \tau_I(1)$.
2. Refresh registers by performing $i^\bullet$ for all $1 \leq i \leq r$. Let the new assignment be τ_1.
3. Perform $pop^\bullet$ r-times, thus ensuring that, for each $1 \leq i, j \leq r$, $\tau_I(i) \neq \tau_1(j)$.
4. Push all registers in numerical order, to obtain stack $\tau_1(r) \cdots \tau_1(1)\tau_I(r) \cdots \tau_I(1)$.
5. Refresh all registers. Let the new assignment be τ_2.
6. Perform $pop^\bullet$ $2r$-times, thus, for each $i, , j$, $\tau_2(i) \neq \tau_1(j)$ and $\tau_2(i) \neq \tau_I(j)$.
7. Silently transition to state q.

Now observe that the conditions in steps 3 and 6 and the fact that register assignments are injective ensure that $|\nu(\tau_I) \cup \nu(\tau_1) \cup \nu(\tau_2)| = 3r$. Hence, any run reaching q is supported by exactly $3r$ distinct names. $\square$

Remark 8. The $3r$ bound given above can be adapted to the automata presentations of [8,20] yielding bounds $3r + \Theta(1)$. An adaptation of Lemma 7 improves upon Example 6 of [8], where a language requiring $2r\ 1$ different symbols was presented.

Being able to bound the number of registers is useful for obtaining reachability algorithms as it allows us to remove the complications of the infinite alphabet and reduce problems to the well-studied finite alphabet setting (e.g. Theorem 9).

4   Reachability is EXPTIME-complete

We consider the following decision problem, call it r-PDRS REACH:

Given an r-PDRS S and $q \in Q$, is there a run of S ending in q?

We shall show that the problem (and its counterparts for all the other closely related machine models) is EXPTIME-complete. Note that reachability is equivalent to language non-emptiness in the automata case.

Theorem 9. r-PDRS REACH *and language emptiness for infinite-alphabet pushdown automata [8] and register pushdown automata [20] are solvable in exponential time.*

Proof. Lemma 5 yields an exponential-time reduction of r-PDRS REACH to the classic reachability problem for pushdown systems over finite alphabets [5]: one can replace the r D-valued registers with r $[3r]$-valued registers, and then incorporate them into the finite control (for a singly-exponential blow-up of the state space). Since the latter problem is solvable in polynomial time, it follows that r-PDRS REACH is in EXPTIME.

By Remark 3, the emptiness problem for infinite-alphabet pushdown automata [8] can be reduced to r-PDRS REACH in polynomial time, immediately yielding the EXPTIME upper bound[2]. For register pushdown automata [20] we have an exponential-time reduction to r-PDRS REACH, which does not yield the required bound. However, recall that the translation into r-PDRS preserves the number of registers, so Lemma 5 still implies a linear upper bound for the number of D-values needed for finding an accepting run. Consequently, we can reduce language emptiness of register pushdown automata to a reachability problem for pushdown systems at an exponential cost. Since the latter is in P, the former is in EXPTIME.   □

The bound given above is tight: we simulate a polynomial-space Turing machine with a stack (aka polynomial-space auxiliary pushdown automaton [9]), which has an EXPTIME-complete halting problem[3].

Theorem 10. r-PDRS REACH *is EXPTIME-hard.*

Proof (sketch). For simplicity, let us assume a binary tape alphabet. The main challenge in the proof is the modelling of n tape cells using $p(n)$ registers, for a polynomial p. Recall that register assigments are injective, so it is not clear which registers represent 0's and which represent 1's. Thus, to encode n bits $b_1, \cdots, b_n$, we shall use a special encoding scheme based on $2n$ names $r_1, \cdots, r_{2n} \in D$ stored in registers and an auxiliary

---

[2] Through a careful reading of the argument for emptiness in [8] one can infer an exponential upper bound, but here Lemma 5 gives a direct argument.

[3] A reduction from the more familiar alternating polynomial-space Turing machines would also be possible, but Cook's model is closer to r-PDRS, which allows us to concentrate on the main issue of encoding binary memory content without the need to model alternation.

"mask" of names $m_1, \cdots, m_{2n} \in \mathcal{D}$ stored on the stack. The registers and masks will be related by $\{r_{2j-1}, r_{2j}\} = \{m_{2j-1}, m_{2j}\}$ and $b_j = 0$ will be represented by the case $r_{2j-1} = m_{2j-1}, r_{2j} = m_{2j}$. Note that, due to injectivity, both r_j's and m_j's cannot be present in registers at the same time and hence the latter will be pushed on the stack. However, the stack is also needed for pushing and popping ordinary stack symbols by the Turing machine, so masks will not always be at the top of stack at the time when they are needed for decoding[4]. We overcome this obstacle by employing 3 different masks for encoding memory: one is used whilst simulating push-transitions (*push-mode*), one for pops (*pop-mode*) and an auxiliary one to ensure continuity between the different instances of masks. Let us call these masks M_1, M_2 and M_3 respectively.

In push-mode, instead of popping M_1 from the stack in order to compare it with the registers and hence decode the memory, we will be *guessing* it and pushing the guess onto the stack, on the understanding that the correctness of each guess (call it $\hat{M}_1$) is to be verified later in the corresponding pop steps. Moreover, in push-mode we will also be pushing the mask M_2 so that it is readily available for pop-mode. When it is time to switch to pop-mode, the tape content so far encoded with mask M_1 will be re-encoded with M_2 so that the forthcoming pop-move can be simulated with M_2. During pop-transitions, in addition to stack symbols and the mask M_2 used for decoding, we will also pop the accompanying guessed mask $\hat{M}_1$ and verify its correctness by comparing it with the last unverified M_1, which is stored in registers apart from the simulated memory. Because at the bottom of the stack we have the actual mask M_1, such equality comparisons will eventually assert that $\hat{M}_1 = M_1$ for all guesses $\hat{M}_1$.

A final complication arises when we want to switch from pop-mode to push-mode. We said that, when popping, we verify the guesses $\hat{M}_1$. Thus, if a push follows a pop, the mask $\hat{M}_1$ that resides in the registers needs to be pushed back on the stack so that it can be verified later once we return to pop-mode. At the same time, we need to store in our registers some content X, so that X and $\hat{M}_1$ encode the current tape content. However, the formation of X destroys $\hat{M}_1$ in the registers. To prevent the information from being lost, we make another guess $\hat{\hat{M}}_1$ and use the third mask M_3 to check that the guess was correct (more precisely, on the stack we store M_3 and some $\hat{M}_3$ such that $\hat{M}_1 = \hat{\hat{M}}_1$ iff $\hat{M}_3 = M_3$). Whether $\hat{M}_3 = M_3$ holds is verified in a later pop step. $\quad\square$

The EXPTIME-hardness carries over to the language emptiness problem associated with infinite-alphabet pushdown automata [8] and register pushdown automata [20]. Since the latter allows for storage of identical values in different registers, their hardness can also be established more directly by encoding relative to two fixed data values for 0 and 1. These different policies for register management are known to lead to different complexity bounds for emptiness testing in the absence of pushdown store: NP-completeness [19][5] (injective assignment) vs PSPACE-completeness (non-injective assignment) [10]. Perhaps surprisingly, we have shown the presence of pushdown store cushions the differences and there is no gap analogous to that between [8] and [20].

---

[4] For example, after simulating a push-transition, the mask used for realising the transition will be hidden by the pushed symbol and thus unavailable to support the next transition.

[5] This result is affected by registers initially containing a special *undefined* value, without which the emptiness problem is reducible to that for finite automata and, consequently, NL-complete.

5 Global Reachability

We now move on to investigate global reachability for r-PDRS. We show that, given an r-PDRS $\mathcal{S}$ and a representation $\mathcal{C}$ of a set of configurations of $\mathcal{S}$, one can construct, in exponential time, a representation of the set of configurations $\mathsf{Pre}^*_{\mathcal{P}}(\mathcal{C})$ from which $\mathcal{S}$ can reach a configuration in $\mathcal{C}$. To that end we extend the methodology of Bouajjani, Esparza and Maler [5] to the infinite alphabet setting.

The developments in this section rely on an auxiliary variant of (stack-free) register automata which feature symbolic transitions representing multiple rearrangements of registers. In order to describe them, let us introduce r-*register manipulations*, which are partial functions $R \in [r] \times [r] \hookrightarrow \{0, 1\}$ such that $R^{-1}\{1\}$ is a partial injection. We denote the set of all such partial functions by RegMan_r and use R^b to refer to $R^{-1}\{b\}$, for $b \in \{0, 1\}$. Given $R, S \in \mathsf{RegMan}_r$, we define $R \; ; \; S$ as follows.

$$(R \; ; \; S)(i, \, j) = \begin{cases} 1 & (S^1 \circ R^1)(i) = j \\ 0 & \exists k \in [r]. \, (R^1(i) = k \wedge S^0(k) = j) \vee (R^0(i) = k \wedge S^1(k) = j) \end{cases}$$

Moreover, given $i \in [r]$, we shall write $R_{i\bullet}$ for the partial function defined by, for all $j \in [r]$, $R_{i\bullet}(j, i) = 0$ and, for all $j \neq i$, $R_{i\bullet}(j, j) = 1$.

Register manipulations can be seen as abstract predicates on register assignments. In particular, given two register assignments τ, τ', we write $\tau \; R \; \tau'$ just if, for all $(i, j) \in \mathbf{dom}\, R$, $R(i, \, j) = 0$ implies $\tau(i) \neq \tau'(j)$ and $R(i, \, j) = 1$ implies $\tau(i) = \tau'(j)$.

Definition 11. *A* **register-manipulating r-register automaton** *(r-RMRA) is a tuple* $\langle Q, \, F, \, \Delta \rangle$ *with Q a finite set of states, $F \subseteq Q$ a subset of final states and $\Delta \subseteq Q \times \mathsf{OP}_r \times Q$ the transition relation, with $\mathsf{OP}_r = [r] \cup \{\bullet\} \cup \mathsf{RegMan}_r$.*

The operations of RMRAs generalise the stack-free operations of PDRSs: $i \in [r]$ specifies reading a name already present in the ith register, $\bullet$ reads a locally fresh name and $R \in \mathsf{RegMan}_r$ is an internal action such that if $q \xrightarrow{R} q'$ then any configuration $(q, \, \tau)$ may transition to any configuration $(q, \, \tau')$ satisfying $\tau \; R \; \tau'$. In what follows, we will start RMRAs from various initial configurations, so we do not include an initial state or register assignment in their specifications.

Definition 12. *Given an r-RMRA $\mathcal{A} = \langle Q, F, \Delta \rangle$, a state $q \in Q$ and an r-register assignment τ, we set:* $\mathcal{L}(\mathcal{A})(q, \, \tau) = \{w \in \mathcal{D}^* \mid w \text{ is accepted by } \mathcal{A} \text{ from } (q, \, \tau)\}$. *Moreover, given an r-PDRS $\mathcal{S} = \langle P, \, q_I, \tau_I, \, \delta \rangle$ such that $P \subseteq Q$, we say that $\mathcal{A}$* **represents** *the $\mathcal{S}$-configuration $(p, \, \tau, \, s)$ whenever $s \in \mathcal{L}(\mathcal{A})(p, \, \tau)$. We write $\mathcal{C}(\mathcal{A})$ for the set of $\mathcal{S}$-configurations represented by $\mathcal{A}$.*

Given an r-RMRA characterising a set of configurations of an r-PDRS $\mathcal{S}$, our aim is to construct another RMRA that represents exactly those configurations of $\mathcal{S}$ that can reach configurations in $\mathcal{C}(\mathcal{A})$, i.e. we aim to construct a representation of $\mathsf{Pre}^*_{\mathcal{P}}(\mathcal{C}(\mathcal{A}))$.

We shall do this in the "saturation" style of the classical construction of [5] but we need more notation in order to deal with the infinite alphabet. Given $R \in \mathsf{RegMan}_r$, we say that R is *consistent* with the statement $i = j$ (respectively $i^\bullet$) just if $R(i, j) \neq 0$ and $[i \in \mathbf{dom}\, R^1 \vee j \in \mathbf{ran}\, R^1]$ implies $R^1(i) = j$ (resp. $i \notin \mathbf{dom}\, R^1$) and in that case we write $R \; || \; i = j$ (resp. $R \; || \; i^\bullet$). So, the meaning of $R \; || \; i^\bullet$, is that i

in the situation before R may be locally fresh with respect to the situation after R. If $R \parallel i = j$ (resp. $R \parallel i^\bullet$) then we write $R[i = j]$ (resp. $R[i^\bullet]$) for $R \cup \{(i, j) \mapsto 1\}$ (resp. $R \cup \{(i, j) \mapsto 0 \mid j \in [r]\}$). Note the difference between $R_{i\bullet}$ and $R[i^\bullet]$. We write $q \xrightarrow{R}^* q'$ just if there is some finite, possibly empty, sequence $\langle q_i \rangle_{i \in [n]}$ such that $q_1 = q$ and $q_n = q'$ and, for all $i \in [n-1]$, $q_i \xrightarrow{R_i} q_{i+1}$ and $R_1 ; \cdots ; R_{n-1} = R$.

Definition 13. *Given an r-PDRS S over states P and an r-RMRA $\mathcal{A}$ over states Q and transitions Δ and such that $P \subseteq Q$ and Δ contains no transitions to states in P, we construct another r-RMRA $\mathsf{SAT}(\mathcal{A})$ by induction (note that op ranges over OP_r):*

$$\frac{p \xrightarrow[\mathcal{A}]{op} p'}{p \xrightarrow[\mathsf{SAT}(\mathcal{A})]{op} p'} \ (N) \qquad \frac{p \xrightarrow[\mathcal{S}]{i^\bullet} p'}{p \xrightarrow[\mathsf{SAT}(\mathcal{A})]{R_{i\bullet}} p'} \ (i) \qquad \frac{p \xrightarrow[\mathcal{S}]{push(i)} p' \quad p' \xrightarrow[\mathsf{SAT}(\mathcal{A})]{R}^* q \quad q \xrightarrow[\mathsf{SAT}(\mathcal{A})]{j} q'}{p \xrightarrow[\mathsf{SAT}(\mathcal{A})]{R[i=j]} q'} \ (ii)$$

$$\frac{p \xrightarrow[\mathcal{S}]{push(i)} p' \quad p' \xrightarrow[\mathsf{SAT}(\mathcal{A})]{R}^* q \quad q \xrightarrow[\mathsf{SAT}(\mathcal{A})]{\bullet} q'}{p \xrightarrow[\mathsf{SAT}(\mathcal{A})]{R[i^\bullet]} q'} \ (iii) \qquad \frac{p \xrightarrow[\mathcal{S}]{pop(i)} p'}{p \xrightarrow[\mathsf{SAT}(\mathcal{A})]{i} p'} \ (iv) \qquad \frac{p \xrightarrow[\mathcal{S}]{pop^\bullet} p'}{p \xrightarrow[\mathsf{SAT}(\mathcal{A})]{\bullet} p'} \ (v)$$

where we additionally require $R \parallel i = j$ in rule (ii), and $R \parallel i^\bullet$ in rule (iii).

The above construction can be carried out in exponential time: consider that there are at most $|Q \times \mathsf{OP}_r \times Q|$ many transitions added, which is at most exponential in the size of the input. For each transition, computation is either trivial or, in (ii) and (iii), involves computing exponentially many graph reachability queries.

Theorem 14. *Given r-PDRS S and r-RMRA $\mathcal{A}$ as above, $\mathcal{C}(\mathsf{SAT}(\mathcal{A})) = \mathsf{Pre}_{\mathcal{P}}^*(\mathcal{C}(\mathcal{A}))$.*

We can thus verify whether one can reach a configuration represented by $\mathcal{A}$ from a given configuration: construct the corresponding $\mathsf{SAT}(\mathcal{A})$ and check membership. To implement the latter in nondeterministic space, given a source configuration (q, τ, w), we need $O(\log |Q_{\mathsf{SAT}(\mathcal{A})}| + p(r) + \log |w|)$ bits to track the state, register assignment and position in w respectively. This is polynomial space in $S, \mathcal{A}, w$ which, along with the construction of $\mathsf{SAT}(\mathcal{A})$, yields an exponential-time reachability testing routine.

Finally, let us remark that RMRAs are no more expressive than register automata with nondeterministic reassignment [14]. An r-RMRA $\mathcal{A} = \langle Q, F, \Delta \rangle$ can be seen as an *r-register automaton with nondeterministic reassignment* (r-RA$_{nr}$) if $\Delta \subseteq Q \times \mathsf{OP}_r^- \times Q$, with $\mathsf{OP}_r^- = [r] + \{ R_{i\bullet} \mid i \in [r] \}$.

Lemma 15. *For any r-RMRA $\mathcal{A}$, one can construct a $(2r+1)$-RA$_{nr}$ $\hat{\mathcal{A}}$ such that, for each $\mathcal{A}$-configuration κ there exists a $\hat{\mathcal{A}}$-configuration $\hat{\kappa}$ satisfying $\mathcal{L}(\mathcal{A})(\kappa) = \mathcal{L}(\hat{\mathcal{A}})(\hat{\kappa})$.*

References

1. Alur, R., Cerný, P., Weinstein, S.: Algorithmic analysis of array-accessing programs. ACM Trans. Comput. Log. 13(3) (2012)
2. Atig, M.F., Bouajjani, A., Qadeer, S.: Context-Bounded Analysis for Concurrent Programs with Dynamic Creation of Threads. Log. Meth. Comput. Sci. 7(4) (2011)

3. Björklund, H., Schwentick, T.: On notions of regularity for data languages. Theor. Comput. Sci. 411(4-5) (2010)
4. Bojańczyk, M., David, C., Muscholl, A., Schwentick, T., Segoufin, L.: Two-variable logic on data words. ACM Trans. Comput. Log. 12(4) (2011)
5. Bouajjani, A., Esparza, J., Maler, O.: Reachability analysis of pushdown automata: Application to model-checking. In: Mazurkiewicz, A., Winkowski, J. (eds.) CONCUR 1997. LNCS, vol. 1243, pp. 135–150. Springer, Heidelberg (1997)
6. Bouajjani, A., Fratani, S., Qadeer, S.: Context-bounded analysis of multithreaded programs with dynamic linked structures. In: Damm, W., Hermanns, H. (eds.) CAV 2007. LNCS, vol. 4590, pp. 207–220. Springer, Heidelberg (2007)
7. Bouajjani, A., Habermehl, P., Mayr, R.: Automatic verification of recursive procedures with one integer parameter. Theor. Comput. Sci. 295, 85–106 (2003)
8. Cheng, E.Y.C., Kaminski, M.: Context-free languages over infinite alphabets. Acta Inf. 35(3), 245–267 (1998)
9. Cook, S.A.: Characterizations of pushdown machines in terms of time-bounded computers. J. ACM 18(1), 4–18 (1971)
10. Demri, S., Lazić, R.: LTL with the freeze quantifier and register automata. ACM Trans. Comput. Log. 10(3) (2009)
11. Gabbay, M.J., Pitts, A.M.: A new approach to abstract syntax with variable binding. Formal Aspects of Computing 13 (2002)
12. Grigore, R., Distefano, D., Petersen, R.L., Tzevelekos, N.: Runtime verification based on register automata. In: Piterman, N., Smolka, S.A. (eds.) TACAS 2013. LNCS, vol. 7795, pp. 260–276. Springer, Heidelberg (2013)
13. Kaminski, M., Francez, N.: Finite-memory automata. Theor. Comput. Sci. 134(2) (1994)
14. Kaminski, M., Zeitlin, D.: Finite-memory automata with non-deterministic reassignment. Int. J. Found. Comput. Sci. 21(5) (2010)
15. Maslov, A.N.: Multilevel stack automata. Probl. of Inf. Transm. 12 (1976)
16. Murawski, A.S., Tzevelekos, N.: Algorithmic nominal game semantics. In: Barthe, G. (ed.) ESOP 2011. LNCS, vol. 6602, pp. 419–438. Springer, Heidelberg (2011)
17. Murawski, A.S., Tzevelekos, N.: Algorithmic games for full ground references. In: Czumaj, A., Mehlhorn, K., Pitts, A., Wattenhofer, R. (eds.) ICALP 2012, Part II. LNCS, vol. 7392, pp. 312–324. Springer, Heidelberg (2012)
18. Neven, F., Schwentick, T., Vianu, V.: Finite state machines for strings over infinite alphabets. ACM Trans. Comput. Log. 5(3) (2004)
19. Sakamoto, H., Ikeda, D.: Intractability of decision problems for finite-memory automata. Theor. Comput. Sci. 231(2) (2000)
20. Segoufin, L.: Automata and logics for words and trees over an infinite alphabet. In: Ésik, Z. (ed.) CSL 2006. LNCS, vol. 4207, pp. 41–57. Springer, Heidelberg (2006)
21. Tan, T.: On pebble automata for data languages with decidable emptiness problem. J. Comput. Syst. Sci. 76(8) (2010)

A Generalization of the Łoś-Tarski Preservation Theorem over Classes of Finite Structures

Abhisekh Sankaran, Bharat Adsul, and Supratik Chakraborty

Dept. of Computer Science and Engg., IIT Bombay, India
{abhisekh,adsul,supratik}@cse.iitb.ac.in

Abstract. We present a logic-based combinatorial property of classes of finite structures that allows an effective generalization of the Łoś-Tarski preservation theorem to hold over classes satisfying the property. The well-studied classes of words and trees, and structures of bounded tree-depth are shown to satisfy the property. We also show that starting with classes satisfying this property, the classes generated by applying composition operations like disjoint union, cartesian and tensor products, inherit the property. We finally show that all classes of structures that are well-quasi-ordered under the embedding relation satisfy a natural generalization of our property.

1 Introduction

Preservation theorems in first-order logic (henceforth called FO) have been extensively studied in classical model theory [3]. An FO preservation theorem asserts that the collection of FO definable classes closed under a model-theoretic operation corresponds to the collection of classes definable by a syntactic fragment of FO. A classical preservation theorem is the Łoś-Tarski theorem, which states that over the class of all structures, the class defined by an FO sentence is preserved under substructures if, and only if, the sentence is equivalent to a universal sentence [3]. It was conjectured in [11], and subsequently proved in [10], that this theorem can be generalized using a simple yet delicate semantic notion of *preservation under substructures modulo k-sized "cruxes"* (in [11,10], 'cruxes' are called 'cores'). This notion reduces to the usual notion of preservation under substructures when k equals 0. The generalized Łoś-Tarski theorem, proved in [10], states that over the class of all structures and for all natural numbers k, the class defined by an FO sentence is preserved under substructures modulo k-sized cruxes if, and only if, the sentence is equivalent to an $\exists^k \forall^*$ sentence (i.e., a prenex sentence having quantifier prefix of the form $\exists^k \forall^*$). Since finite structures are important from a computational perspective, it is interesting to study preservation theorems over classes of finite structures. Unfortunately, most preservation theorems, including the Łoś-Tarski theorem, fail over the class of all finite structures. Earlier work [1,5,6,2] has therefore studied preservation theorems over special classes of finite structures. In this paper, we undertake a similar study for the generalized Łoś-Tarski theorem. Specfically, we identify a logic-based combinatorial property that allows the generalized Łoś-Tarski theorem to hold over any class of finite structures satisfying the property. We show that several well-studied

E. Csuhaj-Varjú et al. (Eds.): MFCS 2014, Part I, LNCS 8634, pp. 474–485, 2014.
© Springer-Verlag Berlin Heidelberg 2014

classes satisfy this property. Furthermore, the property permits an *effective* translation of an FO sentence defining a class that is preserved under substructures modulo k-sized cruxes, to an equivalent $\exists^k \forall^*$ sentence.

In [1], Atserias, Dawar and Grohe considered classes of finite structures that are acyclic, of bounded degree (more generally, "wide") or of bounded tree-width. They showed that under suitable closure assumptions, each of these classes admits the Łoś-Tarski theorem. Subsequently, Harwath, Heimberg and Schweikardt [6] studied the bounds for an effective version of the Łoś-Tarski theorem over bounded degree structures. In [5], Duris showed that the Łoś-Tarski theorem holds for structures that are acyclic in a more general sense. Unfortunately, as discussed in Section 2, none of the above classes, in general, admits the generalized Łoś-Tarski theorem. This motivates us to ask: *Can we identify properties that allow an effective version of the generalized Łoś-Tarski theorem to hold, and are also satisfied by interesting classes of finite structures?* This paper answers this question affirmatively. Interestingly, the classes of structures studied here are incomparable to those studied in [1,6,5].

The primary results of this paper can be summarized as follows.

1. In Section 3, we present a parameterized logic-based combinatorial property of classes of finite structures, and show that this property entails an effective version of the generalized Łoś-Tarski theorem. Intuitively, if a class $\mathcal{S}$ satisfies this property for parameter k, denoted $\mathcal{P}_{logic}(\mathcal{S}, k)$, then for every natural number m, given any structure in $\mathcal{S}$ and k elements of it, there always exists an m-equivalent bounded substructure, containing these elements, that is in $\mathcal{S}$. Further, the bound is a computable function of m.

2. In Sections 4 and 5, we respectively show that the following interesting classes of structures satisfy $\mathcal{P}_{logic}(\cdot, k)$ for all k: (i) the classes of all words and trees over a finite alphabet, and (ii) any substructure-closed class of relational structures whose Gaifman graphs have bounded tree-depth.

3. In Section 6, we show that for all k, the property $\mathcal{P}_{logic}(\cdot, k)$ is preserved under natural composition operators on structures, like disjoint union, cartesian and tensor products. This allows us to construct additional classes of structures that satisfy the generalized Łoś-Tarski theorem, from known classes. Interesting examples of such constructed classes are grids of bounded dimension and various classes of co-graphs like all co-graphs, complete graphs, complete n-partite graphs for each n, threshold graphs etc. It is important to note that the classes considered in Sections 4, 5 and 6 lie beyond those studied in [1,6,5], and yet satisfy the Łoś-Tarski theorem.

4. In Section 7, we briefly discuss two other parameterized properties, denoted $\mathcal{P}_{wqo}(\cdot, k)$ and $\mathcal{P}_{logic}^{gen}(\cdot, k)$, each of which entails the generalized Łoś-Tarski theorem "(not necessarily an effective version though)". The property $\mathcal{P}_{wqo}(\cdot, k)$ is based on well-quasi-ordering of the embedding relation on structures, while $\mathcal{P}_{logic}^{gen}(\cdot, k)$ is a generalization of $\mathcal{P}_{logic}(\cdot, k)$. An interesting result is that $\mathcal{P}_{wqo}(\cdot, k)$ is subsumed by $\mathcal{P}_{logic}^{gen}(\cdot, k)$, yielding a logic-based tool to show that certain classes are not well-quasi-ordered under the embedding relation.

To prove the above results, we use a combination of techniques. For example, to show that trees and words satisfy $\mathcal{P}_{logic}(\cdot, k)$, we present a composition lemma for trees in Section 4, and use it to show that certain "prunings" of trees preserve m-equivalence. In Section 5, to prove that the class of structures with Gaifman graphs of tree-depth at most n satisfies $\mathcal{P}_{logic}(\cdot, k)$, we introduce the notion of a *twin of a structure with respect to a given element* and use inductive reasoning over n. In Section 6, the proof of closure of $\mathcal{P}_{logic}(\cdot, k)$ under natural composition operators uses a *tree representation* of structures generated by applying the operators, and uses results for trees proved earlier in Section 4. For lack of space, we defer the full proofs of our results to the journal version of the paper.

2   Notation and Preliminaries

Let $\mathbb{N}$ denote the natural numbers *including zero*. We assume that the reader is familiar with standard notation and terminology of first-order logic. We consider only finite vocabularies, represented by τ, that contain only predicate symbols of *positive* arity (and no constants or functions), unless explicitly stated otherwise. We denote by $FO(\tau)$ the set of all FO formulae over τ. A sequence $(x_1, \ldots, x_k)$ of variables is written as $\bar{x}$. We abbreviate a block of quantifiers $Qx_1 \ldots Qx_k$ by $Q^k \bar{x}$, where $Q \in \{\forall, \exists\}$. Given $k, p \in \mathbb{N}$, let $\exists^k \forall^p$ denote the set of all $FO(\tau)$ sentences in prenex normal form whose quantifier prefix has k existential quantifiers followed by p universal quantifiers. We use $\exists^k \forall^*$ to denote $\bigcup_{p \in \mathbb{N}} \exists^k \forall^p$.

Standard notions of τ-structures, substructures and extensions (see [3]) are used throughout. As in [3], by substructures, we mean *induced* substructures. Given a τ-structure $\mathfrak{A}$, we use $U_{\mathfrak{A}}$ to denote the universe of $\mathfrak{A}$ and $|\mathfrak{A}|$ to denote its cardinality. Given τ-structures $\mathfrak{A}$ and $\mathfrak{B}$, we use $\mathfrak{A} \subseteq \mathfrak{B}$ to denote that $\mathfrak{A}$ is a substructure of $\mathfrak{B}$. Given a τ-structure $\mathfrak{A}$ and an $FO(\tau)$ sentence φ, if $\mathfrak{A} \models \varphi$, we say that $\mathfrak{A}$ is a *model* of φ. We focus only on recursive (or decidable) classes of finite τ-structures in this paper. All classes of τ-structures, and subclasses thereof, are also assumed to be closed under isomorphisms.

The following notion is central to our work.

Definition 1. *Let $\mathcal{S}$ be a class of τ-structures and $k \in \mathbb{N}$. A subclass $\mathcal{C}$ of $\mathcal{S}$ is said to be* preserved under substructures modulo k-sized cruxes over $\mathcal{S}$ *if every τ-structure $\mathfrak{A} \in \mathcal{C}$ has a subset Crux of $U_{\mathfrak{A}}$ such that (i) $|\mathsf{Crux}| \leq k$, and (ii) for every $\mathfrak{B} \in \mathcal{S}$, if $\mathfrak{B} \subseteq \mathfrak{A}$ and $\mathsf{Crux} \subseteq U_{\mathfrak{B}}$, then $\mathfrak{B} \in \mathcal{C}$. The set Crux is called a k-crux of $\mathfrak{A}$ with respect to $\mathcal{C}$ over $\mathcal{S}$.*

As an example, if $\mathcal{S}$ is the class of all graphs, then the subclass $\mathcal{C}$ of $\mathcal{S}$ comprising graphs containing a k-length cycle as a subgraph is preserved under substructures modulo k-sized cruxes over $\mathcal{S}$. Like Definition 1, most other definitions, discussions and results in this paper are stated with respect to an underlying class $\mathcal{S}$ of structures. When $\mathcal{S}$ is clear from the context, we omit the mention of $\mathcal{S}$. Note that Definition 1 is an adapted version of Definition 2 of [11]; the notion of 'core' in the latter is exactly the notion of 'crux' in the former when the underlying class $\mathcal{S}$ is the class of all structures. We avoid using the word 'core' for a crux to prevent confusion with existing notions of cores in the literature [2].

Given a class $\mathcal{S}$ of structures, let $PSC(k)$ denote the collection of the sub-classes of $\mathcal{S}$ that are preserved under substructures modulo k-sized cruxes over $\mathcal{S}$, and that are definable over $\mathcal{S}$ by FO sentences. We interchangeably talk of $PSC(k)$ as a collection of classes and as a set of the defining FO sentences. Similarly, we interchangeably use $\exists^k\forall^p$ (and $\exists^k\forall^*$) to denote a set of FO sentences and the corresponding subclasses of $\mathcal{S}$ defined by these sentences. The generalized Łoś-Tarski theorem, proved in [10], can now be stated as follows.

Theorem 1. *Over the class of all structures, for all $k \in \mathbb{N}$, $PSC(k) = \exists^k\forall^*$.*

Note that the notion of cruxes, central to Theorem 1, differs from that of ex-istential witnesses. If φ is an $\exists^k\forall^*$ sentence and $\mathfrak{A} \models \varphi$, then every witness of the existential variables of φ forms a k-crux of $\mathfrak{A}$. The converse, however, need not be true [11]. Specifically, let $\tau = \{E\}$, where E is a binary predicate. Con-sider the $FO(\tau)$ sentence $\varphi \equiv \exists x\,\forall y\, E(x,y)$, and the τ-structure $\mathfrak{A}$ defined by $U_{\mathfrak{A}} = \{0,1\}$ and $E^{\mathfrak{A}} = \{(0,0),(0,1),(1,1)\}$. Clearly, $\mathfrak{A} \models \varphi$ and $\mathfrak{A}$ has only one witness for variable x of φ, viz. 0. Yet, both $\{0\}$ and $\{1\}$ are 1-cruxes of $\mathfrak{A}$!

Significantly, Theorem 1 fails, in general, over the classes studied in [1,6,5]. To see why this is so, let $\mathcal{S}$ be the class of graphs that are disjoint unions of undirected paths. Observe that $\mathcal{S}$ is closed under substructures and disjoint unions, is acyclic and has degree bounded by 2. Consider the subclass $\mathcal{C}$ of $\mathcal{S}$ comprising graphs containing at least 2 connected components. The subclass $\mathcal{C}$ is definable over $\mathcal{S}$ by an FO sentence ψ asserting that any model either has at least 3 end points or has at least 2 isolated vertices. Further, for any graph G in $\mathcal{C}$, any two vertices belonging to distinct components of G form a 2-crux of G; hence $\mathcal{C}$ is in $PSC(2)$. However, as shown in [11], there exists no $\exists^2\forall^*$ sentence that defines $\mathcal{C}$ over $\mathcal{S}$. Likewise, one can show that the class of all directed graphs of tree-width 1 fails to satisfy $PSC(2) = \exists^2\forall^*$. This motivates our quest for alternative properties of classes of finite structures over which Theorem 1 holds.

3   A Logic Based Combinatorial Property

We begin by recalling from standard FO terminology [7] that if m is a natural number, two τ-structures $\mathfrak{A}$ and $\mathfrak{B}$ are said to be *m-equivalent*, denoted $\mathfrak{A} \equiv_m \mathfrak{B}$, iff $\mathfrak{A}$ and $\mathfrak{B}$ agree on the truth of every $FO(\tau)$ sentence of quantifier rank at most m. We can now define a parameterized logic-based combinatorial property of classes of finite structures as follows.

Definition 2. *Let k be a natural number and $\mathcal{S}$ be a class of finite structures. We say that $\mathcal{P}_{logic}(\mathcal{S}, k)$ holds if there exists a computable function $\theta_k : \mathbb{N} \to \mathbb{N}$ such that for each $m \in \mathbb{N}$, for each structure $\mathfrak{A}$ of $\mathcal{S}$ and for each subset W of $U_{\mathfrak{A}}$ of size at most k, there exists $\mathfrak{B} \subseteq \mathfrak{A}$ such that (i) $\mathfrak{B} \in \mathcal{S}$, (ii) $W \subseteq U_{\mathfrak{B}}$, (iii) $|\mathfrak{B}| \leq \theta_k(m)$, and (iv) $\mathfrak{B} \equiv_m \mathfrak{A}$. We call θ_k a witness function of $\mathcal{P}_{logic}(\mathcal{S}, k)$.*

Remark: If $\mathcal{P}_{logic}(\mathcal{S}, k)$ holds and if $\mathcal{S}'$ is a subclass of $\mathcal{S}$ that is closed under substructures over $\mathcal{S}$, then it is easy to see that $\mathcal{P}_{logic}(\mathcal{S}', k)$ also holds.

We list below two simple examples of classes satisfying $\mathcal{P}_{logic}(\cdot, k)$ for every $k \in \mathbb{N}$. Various non-trivial examples are presented in Sections 4, 5 and 6.

1. Let $\mathcal{S}$ be a finite class of finite structures. Clearly, $\mathcal{P}_{logic}(\mathcal{S}, k)$ holds for all $k \in \mathbb{N}$, with $\theta_k(m)$ giving the size of the largest structure in $\mathcal{S}$.
2. Let $\mathcal{S}$ be the class of all finite linear orders. Then $\mathcal{P}_{logic}(\mathcal{S}, k)$ holds for all $k \in \mathbb{N}$, with $\theta_k(m) = \max(2^m, k)$.

The next theorem is one of the main results of this paper. Before stating the theorem, we make two observations. First, given a recursive class $\mathcal{S}$ of finite structures and a natural number n, the subclass of all structures in $\mathcal{S}$ of size at most n is definable by an effectively computable FO sentence in $\exists^n \forall^*$. We call this sentence $\xi_{\mathcal{S},n}$. Second, given a sentence ψ of $FO(\tau)$ and any sequence $\bar{x}$ of variables, one can effectively compute a quantifier-free $FO(\tau)$ formula with free variables $\bar{x}$ such that this formula evaluates to true in a τ-structure $\mathfrak{A}$ with $\bar{x}$ interpreted as $\bar{a}$ iff ψ holds in the substructure of $\mathfrak{A}$ induced by $\bar{a}$. Following notation in [11], we denote this formula as $\psi|_{\bar{x}}$, read as ψ *relativized to $\bar{x}$*.

Theorem 2. *Let $\mathcal{S}$ be a recursive class of finite structures and $k \in \mathbb{N}$ be such that $\mathcal{P}_{logic}(\mathcal{S}, k)$ holds. Then $PSC(k) = \exists^k \forall^*$ over $\mathcal{S}$, and the translation from $PSC(k)$ to $\exists^k \forall^*$ is effective. Specifically, if a witness function for $\mathcal{P}_{logic}(\mathcal{S}, k)$ is θ_k, then an FO sentence χ of quantifier rank m in $PSC(k)$ is equivalent (over $\mathcal{S}$) to the sentence $\exists^k \bar{x} \forall^p \bar{y} \; \psi|_{\bar{x}\bar{y}}$, where $p = \theta_k(m)$ and $\psi \equiv (\xi_{\mathcal{S},p} \to \chi)$.*

Proof: It is obvious that $\exists^k \forall^* \subseteq PSC(k)$ over $\mathcal{S}$. Towards the converse, consider a sentence χ, of quantifier rank m, in $PSC(k)$ over $\mathcal{S}$. Consider the sentence $\varphi \equiv \exists^k \bar{x} \forall^p \bar{y} \; \psi|_{\bar{x}\bar{y}}$, where p and ψ are as stated above. Since χ is in $PSC(k)$ over $\mathcal{S}$, every model $\mathfrak{A}$ of χ in $\mathcal{S}$ also satisfies φ. This is because the elements of any k-crux of $\mathfrak{A}$ can serve as witnesses of the existential quantifiers in φ. To show $\varphi \to \chi$ over $\mathcal{S}$, suppose $\mathfrak{A}$ is a model of φ in $\mathcal{S}$. Let W be a set of witnesses in $\mathfrak{A}$ for the k existential variables in φ. Clearly, $|W| \leq k$. Since $\mathcal{P}_{logic}(\mathcal{S}, k)$ holds, there exists $\mathfrak{B} \subseteq \mathfrak{A}$ such that (i) $\mathfrak{B} \in \mathcal{S}$, (ii) $W \subseteq \mathsf{U}_{\mathfrak{B}}$, (iii) $|\mathfrak{B}| \leq \theta_k(m) = p$, and (iii) $\mathfrak{B} \equiv_m \mathfrak{A}$. Since $\mathfrak{A} \models \varphi$, by instantiating the universal variables in φ with the elements of $\mathsf{U}_{\mathfrak{B}}$, we have $\mathfrak{B} \models \chi$. Since the quantifier rank of χ is m and $\mathfrak{B} \equiv_m \mathfrak{A}$, it follows that $\mathfrak{A} \models \chi$. Therefore, χ is equivalent to φ over $\mathcal{S}$. Finally, since m is effectively computable from χ, so are p, $\xi_{\mathcal{S},p}$ and φ. ∎

4   Words and Trees over a Finite Alphabet

Given an alphabet Σ, let $Words(\Sigma)$ and $Trees(\Sigma)$ denote the set of all finite words and trees, respectively, over Σ. The key result of this section is as follows.

Theorem 3. *For every finite alphabet Σ, both $\mathcal{P}_{logic}(Words(\Sigma), k)$ and $\mathcal{P}_{logic}(Trees(\Sigma), k)$ hold for every natural number k.*

For purposes of our discussion, we use a poset-theoretic representation of trees. A tree is a finite poset $P = (A, \leq)$ with a unique minimal element (called "root"), and for every $a, b, c \in A$, $\big((a \leq c) \wedge (b \leq c)\big) \to \big(a \leq b \vee b \leq a\big)$. Informally, the Hasse diagram of P is an (inverted) tree with every parent p connected to its child c. A tree over Σ, henceforth called a Σ-tree, is a pair (P, λ) where

$P = (A, \leq)$ is a tree and $\lambda : A \to \Sigma$ is a labeling function. The elements of A are also called nodes (or elements) of the Σ-tree. In the special case where the underlying poset is a linear order, a Σ-tree is called a Σ-word. We denote trees by either $\mathfrak{s}$ or $\mathfrak{t}$. A Σ-forest $\mathfrak{f}$ is a (finite) disjoint union of Σ-trees.

Let τ be the vocabulary $\{\leq\} \cup \{Q_a \mid a \in \Sigma\}$, where $\leq$ is a binary predicate and each Q_a is a unary predicate. A Σ-tree $\mathfrak{s} = ((A_\mathfrak{s}, \leq_\mathfrak{s}), \lambda_\mathfrak{s})$ has a natural representation as a structure $\mathfrak{A}_\mathfrak{s}$ over τ. To represent a Σ-forest $\mathfrak{f}$ as a τ-structure $\mathfrak{A}_\mathfrak{f}$, we use the disjoint union (denoted $\sqcup$) of the τ-structures representing the Σ-trees in $\mathfrak{f}$. For clarity of exposition, we use $\mathfrak{s}$ (resp. $\mathfrak{f}$) to denote both a Σ-tree $\mathfrak{s}$ (resp. Σ-forest $\mathfrak{f}$) and its corresponding τ-structure $\mathfrak{A}_\mathfrak{s}$ (resp. $\mathfrak{A}_\mathfrak{f}$).

We use the standard notions of height, degree and subtree of a given tree. Given two Σ-trees $\mathfrak{s} = ((A_\mathfrak{s}, \leq_\mathfrak{s}), \lambda_\mathfrak{s})$ and $\mathfrak{t} = ((A_\mathfrak{t}, \leq_\mathfrak{t}), \lambda_\mathfrak{t})$ with disjoint sets of nodes, and an element e of $\mathfrak{s}$, the *join of* $\mathfrak{t}$ *to* $\mathfrak{s}$ *at* e, denoted $\mathfrak{s} \cdot_e \mathfrak{t}$, is the Σ-tree obtained from $\mathfrak{s}$ by adding $\mathfrak{t}$ as a new "child subtree" of the element e of $\mathfrak{s}$. Given a Σ-tree $\mathfrak{s}$, a Σ-forest $\mathfrak{f} = \bigsqcup_{i=1}^{n} \mathfrak{t}_i$ and an element e of $\mathfrak{s}$, the *join of* $\mathfrak{f}$ *to* $\mathfrak{s}$ *at* e, denoted $\mathfrak{s} \cdot_e \mathfrak{f}$, is the Σ-tree $((((\mathfrak{s} \cdot_e \mathfrak{t}_1) \cdot_e \mathfrak{t}_2) \cdots) \cdot_e \mathfrak{t}_n)$.

The proof of Theorem 3 uses two key auxiliary lemmas. The first is a *composition* lemma for trees. This lemma intuitively states that if $\mathfrak{t}$ is a tree, a is a node of $\mathfrak{t}$, and $\mathfrak{f}$ is a forest, then the $\equiv_m$ class of $\mathfrak{t} \cdot_a \mathfrak{f}$ is completely determined by the $\equiv_m$ classes of $(\mathfrak{t}, a)$ and $\mathfrak{f}$. Composition results of this kind were first studied by Feferman and Vaught, and subsequently by others (see [8] for a survey).

Lemma 1. *Let $\mathfrak{t}_i$ be a non-empty Σ-tree containing element a_i, and $\mathfrak{f}_i$ be a non-empty Σ-forest containing element b_i, for $i \in \{1,2\}$. Let $\mathfrak{s}_i = \mathfrak{t}_i \cdot_{a_i} \mathfrak{f}_i$ for $i \in \{1,2\}$. Suppose $(\mathfrak{t}_1, a_1) \equiv_m (\mathfrak{t}_2, a_2)$. Then the following hold.*

1. *If $(\mathfrak{f}_1, b_1) \equiv_m (\mathfrak{f}_2, b_2)$, then $(\mathfrak{s}_1, a_1, b_1) \equiv_m (\mathfrak{s}_2, a_2, b_2)$.*
2. *If $\mathfrak{f}_1 \equiv_m \mathfrak{f}_2$, then $(\mathfrak{s}_1, a_1) \equiv_m (\mathfrak{s}_2, a_2)$.*

The proof of Lemma 1 uses the Ehrenfeucht-Fraïssé theorem [7] and is similar to the proof of the composition lemma for words. Before stating the next auxiliary lemma, we introduce some notation. Given an alphabet Σ and a natural number m, let $\Delta(m, \Sigma)$ denote the set of all equivalence classes of the $\equiv_m$ relation over $Trees(\Sigma)$. Let Alph denote the set of all finite alphabets, and $g : \mathbb{N} \times \text{Alph} \to \mathbb{N}$ be a computable function such that $g(m, \Sigma) \geq |\Delta(m, \Sigma)|$. It is known that g exists (see proof of Lemma 3.13 in [7]). We now state our next auxiliary lemma.

Lemma 2. *Let $\mathfrak{s}$ be a Σ-tree. For every $m \in \mathbb{N}$, each of the following exist.*

(a) A subtree $\mathfrak{t}_1$ of $\mathfrak{s}$ such that $\mathfrak{t}_1$ has degree $\leq m \cdot g(m, \Sigma)$ and $\mathfrak{t}_1 \equiv_m \mathfrak{s}$.
(b) A subtree $\mathfrak{t}_2$ of $\mathfrak{s}$ such that $\mathfrak{t}_2$ has height $\leq g(m, \Sigma)$ and $\mathfrak{t}_2 \equiv_m \mathfrak{s}$.

Proof Sketch: (a) Let d denote $m \cdot g(m, \Sigma)$. If each node of $\mathfrak{s}$ has at most d children, then taking $\mathfrak{t}_1$ to be $\mathfrak{s}$, we are done. Else, let a be a node of $\mathfrak{s}$, having $> d$ children. Let $\Gamma(a)$ denote the set of all subtrees of $\mathfrak{s}$ rooted at the children of a in $\mathfrak{s}$, and let $\mathfrak{f}$ be the forest whose trees are exactly the members of $\Gamma(a)$. Let $\mathfrak{t}$ be the tree such that $\mathfrak{s} = \mathfrak{t} \cdot_a \mathfrak{f}$. For every $\delta \in \Delta(m, \Sigma)$, let $\Gamma(a, \delta)$ be the set consisting of the members of $\Gamma(a)$ whose $\equiv_m$ class is δ. Construct the forest $\mathfrak{f}_1$

by performing the following operation on $\mathfrak{f}$ for each $\delta \in \Delta(m, \Sigma)$: retain $\Gamma(a, \delta)$ entirely in $\mathfrak{f}$ if $|\Gamma(a, \delta)| < m$, else retain exactly (any) m members of $\Gamma(a, \delta)$ in $\mathfrak{f}$ and remove the rest. It is easy to see that $\mathfrak{f} \equiv_m \mathfrak{f}_1$. Let $\mathfrak{s}_1 = \mathfrak{t} \cdot_a \mathfrak{f}_1$. Then using Lemma 1, we get that $\mathfrak{s}_1 \equiv_m \mathfrak{s}$. Observe that $\mathfrak{s}_1$ has strictly fewer nodes that have $> d$ children, compared to $\mathfrak{s}$. Recursing on $\mathfrak{s}_1$, we are eventually done.

(b) Let A be the underlying set of $\mathfrak{s}$. Define the function $h : A \to \Delta(m, \Sigma)$ such that $h(a)$ is the $\equiv_m$ class of the subtree of $\mathfrak{s}$ rooted at a, for every $a \in A$. If for each path in $\mathfrak{s}$, no two distinct elements on the path have the same h value, then the height of $\mathfrak{s}$ is at most $g(m, \Sigma)$. Then the desired subtree $\mathfrak{t}_2$ can be chosen to be $\mathfrak{s}$ itself. Otherwise, there exist distinct $a, b \in A$ such that (i) $\mathfrak{s} \models (a \leq b)$ and (ii) $h(a) = h(b)$. Let $\mathfrak{s}_1$ be the subtree of $\mathfrak{s}$ obtained by 'replacing' the subtree rooted at a with the subtree rooted at b. By Lemma 1, we get $\mathfrak{s}_1 \equiv_m \mathfrak{s}$. Also, $\mathfrak{s}_1$ has strictly fewer nodes than $\mathfrak{s}$. Recursing on $\mathfrak{s}_1$, we are eventually done. ∎

The proof of Theorem 3 for $Trees(\Sigma)$ is now completed as follows. Given $m \in \mathbb{N}$, a Σ-tree $\mathfrak{s} = (P, \lambda)$ and a set W of at most k elements of $\mathfrak{s}$, let $\mathfrak{s}' = (P, \lambda')$ be the tree over $\Sigma' = \Sigma \times \{0, 1\}$ such that for every $a \in P$, $\lambda'(a) = (\lambda(a), 1)$ if $a \in W$, and $\lambda'(a) = (\lambda(a), 0)$ otherwise. Let $n = \max(m, k)$. By Lemma 2, there exists a subtree $\mathfrak{t}'$ of $\mathfrak{s}'$ with degree at most $n \cdot g(n, \Sigma')$ and height at most $g(n, \Sigma')$, that is n-equivalent to $\mathfrak{s}'$. It is easy to check that by dropping the second component of the labels of all elements of $\mathfrak{t}'$, we get a subtree $\mathfrak{t}$ of $\mathfrak{s}$ containing W. In addition, $\mathfrak{t} \equiv_m \mathfrak{s}$ and $|\mathfrak{t}| \leq \theta_k(m)$, where $\theta_k(m) = (n \cdot g(n, \Sigma') + 1)^{g(n, \Sigma') + 1}$ and $n = \max(m, k)$. Then $\mathcal{P}_{logic}(Trees(\Sigma), k)$ holds with θ_k as the witness function. Whence, $\mathcal{P}_{logic}(Words(\Sigma), k)$ holds by the remark following Definition 2. ∎

5   Structures of Bounded Tree-Depth

Nešetřil and de Mendez introduced the notion of *tree-depth* of an undirected graph in [9]. Intuitively, the tree-depth of a graph G, denoted $td(G)$, is a measure of how far G is from being a star. The following is an inductive definition of tree-depth, given by Lemma 2.2 of [9]. In this definition, $G = (V, E)$ denotes a graph, $\mathsf{Comp}(G)$ denote the set of all connected components of G, and $G \setminus v$ denotes the graph obtained by removing the vertex v from G.

$$td(G) = \begin{cases} 1 & \text{if } G \text{ has a single vertex} \\ 1 + \min_{v \in V} td(G \setminus v) & \text{if } G \text{ is connected and has multiple vertices} \\ \max_{G' \in \mathsf{Comp}(G)} td(G') & \text{if } G \text{ is disconnected} \end{cases}$$

The *Gaifman graph* $\mathcal{G}(\mathfrak{A})$ of a relational structure $\mathfrak{A}$ is an undirected graph whose nodes are the elements of $\mathfrak{A}$, and in which two nodes are adjacent if, and only if, they appear together in some tuple of some relation of $\mathfrak{A}$ [7]. We say that a structure $\mathfrak{A}$ is *connected* if $\mathcal{G}(\mathfrak{A})$ is connected, else we say $\mathfrak{A}$ is *disconnected*. A substructure $\mathfrak{B}$ of $\mathfrak{A}$ is said to be a *connected component* of $\mathfrak{A}$ if $\mathcal{G}(\mathfrak{B})$ is a connected component of $\mathcal{G}(\mathfrak{A})$. We say that $\mathfrak{A}$ has tree-depth n if $\mathcal{G}(\mathfrak{A})$ has tree-depth n. We say that a class $\mathcal{S}$ of structures has *bounded tree-depth* if there exists a natural number n such that all structures in $\mathcal{S}$ have tree-depth at most n. The main result of this section can now be stated as follows.

Theorem 4. *Let $\mathcal{S}$ be a substructure-closed class of finite structures, of bounded tree-depth. Then $\mathcal{P}_{logic}(\mathcal{S}, k)$ holds for every natural number k.*

In this section, we allow the vocabulary τ to contain 0-ary predicate symbols. To prove Theorem 4, we introduce the notion of *twin-structures*. Given a vocabulary τ and a predicate R in τ, let $\#R$ denote the arity of R. If $\#R > 0$, then for each subset T of $\{1, \ldots \#R\}$, let R_T denote a predicate of arity $\#R - |T|$. Define $\hat{\tau}$ to be the vocabulary $\{R \mid R \in \tau, \#R = 0\} \cup \{R_T \mid R \in \tau, \#R > 0, T \subseteq \{1, \ldots, \#R\}\}$. Given a τ-structure $\mathfrak{A}$ and an element a of it, let $\mathfrak{A} \setminus a$ denote the substructure of $\mathfrak{A}$ induced by $U_{\mathfrak{A}} \setminus \{a\}$. Given a predicate R in τ, a subset T of $\{1, \ldots, \#R\}$ and a $(\#R - |T|)$-tuple $\bar{b}$ from $\mathfrak{A} \setminus a$, let $ex_a(\bar{b}, T)$ denote the expansion of $\bar{b}$ with a at the positions in T. Formally, $ex_a(\bar{b}, T)$ is the $\#R$-tuple whose i^{th} component is a for each $i \in T$, and whose sub-tuple obtained by dropping all the a's, is exactly $\bar{b}$. Then the *twin-structure of $\mathfrak{A}$ with respect to a*, denoted $\mathsf{twin}(\mathfrak{A}, a)$, is a $\hat{\tau}$-structure defined as: (i) The universe of $\mathsf{twin}(\mathfrak{A}, a)$ is $U_{\mathfrak{A}} \setminus \{a\}$ (ii) For every 0-ary predicate R in τ, we have $\mathsf{twin}(\mathfrak{A}, a) \models R$ iff $\mathfrak{A} \models R$ (iii) For every predicate R_T in $\hat{\tau}$ and for every $(\#R - |T|)$-tuple $\bar{b}$ of elements of $\mathsf{twin}(\mathfrak{A}, a)$, we have $\mathsf{twin}(\mathfrak{A}, a) \models R_T(\bar{b})$ iff $\mathfrak{A} \models R(ex_a(\bar{b}, T))$. Observe that $\mathfrak{A}$ and $\mathsf{twin}(\mathfrak{A}, a)$ uniquely identify each other, upto isomorphism. The following lemma is easy.

Lemma 3. *Let $\mathfrak{A}$ and $\mathfrak{B}$ be given structures and let a and b be elements of $\mathfrak{A}$ and $\mathfrak{B}$ respectively. Then given $m \in \mathbb{N}$, if $\mathsf{twin}(\mathfrak{A}, a) \equiv_m \mathsf{twin}(\mathfrak{B}, b)$, then $\mathfrak{A} \equiv_m \mathfrak{B}$.*

Let $\Delta(m, \tau)$ be the set of equivalence classes of the $\equiv_m$ relation over the class of all τ-structures. Let Vocab be the set of all finite vocabularies. Consider a computable function $g_1 : \mathbb{N} \times \mathsf{Vocab} \to \mathbb{N}$ such that $g_1(m, \tau) \geq |\Delta(m, \tau)|$. It is known that g_1 exists (see proof of Lemma 3.13 in [7]). Define the function $f_1 : \mathbb{N} \times \mathbb{N} \times \mathsf{Vocab} \to \mathbb{N}$ as $f_1(n, m, \tau) = 1 + m \cdot g_1(m, \hat{\tau}) \cdot f_1(n - 1, m, \hat{\tau})$, where $f_1(1, m, \tau) = 1$. We now have the following lemma.

Lemma 4. *Given $m \in \mathbb{N}$ and a τ-structure $\mathfrak{A}$ that is connected and has tree-depth n, there exists $\mathfrak{B} \subseteq \mathfrak{A}$ such that (i) $\mathfrak{B} \equiv_m \mathfrak{A}$ and (ii) $|\mathfrak{B}| \leq f_1(n, m, \tau)$.*

Proof Sketch: The proof is by induction on n. The base case is trivial. Assume that the result holds for all tree-depths from 1 to $n - 1$ and for all vocabularies τ. Let $\mathfrak{A}$ be as given. Since $\mathfrak{A}$ has tree-depth n, by definition, there exists $a \in \mathfrak{A}$ such that $\mathcal{G}(\mathfrak{A}) \setminus a$ has tree-depth at most $n - 1$. Consider $\mathsf{twin}(\mathfrak{A}, a)$. It is easy to show that for any $\hat{\tau}$-structure $\hat{\mathfrak{D}}$, if $\hat{\mathfrak{D}} \subseteq \mathsf{twin}(\mathfrak{A}, a)$, then $\hat{\mathfrak{D}} = \mathsf{twin}(\mathfrak{C}, a)$ for some $\mathfrak{C} \subseteq \mathfrak{A}$, containing a. Then by a "degree reduction" argument similar to the one in the proof of Lemma 2(a), there exists $\mathfrak{A}' \subseteq \mathfrak{A}$, containing a such that (i) each connected component of $\mathsf{twin}(\mathfrak{A}', a)$ is a connected component of $\mathsf{twin}(\mathfrak{A}, a)$ (ii) the set Y of the connected components of $\mathsf{twin}(\mathfrak{A}', a)$ has size $\leq m \cdot g_1(m, \hat{\tau})$, and (iii) $\mathsf{twin}(\mathfrak{A}', a) \equiv_m \mathsf{twin}(\mathfrak{A}, a)$. From Lemma 3, it follows that $\mathfrak{A}' \equiv_m \mathfrak{A}$.

Now for $\hat{\mathfrak{D}} \in Y$, let $\mathfrak{C} \subseteq \mathfrak{A}'$ be such that $\mathfrak{C}$ contains a and $\hat{\mathfrak{D}} = \mathsf{twin}(\mathfrak{C}, a)$. Observing that $\mathcal{G}(\mathfrak{C}) \setminus a$ and $\mathcal{G}(\mathsf{twin}(\mathfrak{C}, a))$ are the same graph, it follows that the tree-depth of $\hat{\mathfrak{D}}$ is at most $n - 1$. Applying the induction hypothesis on $\hat{\mathfrak{D}}$, there exists $\hat{\mathfrak{D}}_1 \subseteq \hat{\mathfrak{D}}$ such that (i) $\hat{\mathfrak{D}}_1 \equiv_m \hat{\mathfrak{D}}$ and (ii) $|\hat{\mathfrak{D}}_1| \leq f_1(n - 1, m, \hat{\tau})$.

If $\widehat{\mathfrak{D}}_2 = \bigsqcup_{\widehat{\mathfrak{D}} \in Y} \widehat{\mathfrak{D}}_1$, then $\widehat{\mathfrak{D}}_2 \equiv_m \mathsf{twin}(\mathfrak{A}', a)$. Since $\widehat{\mathfrak{D}}_2 \subseteq \mathsf{twin}(\mathfrak{A}', a)$, there exists $\mathfrak{A}'' \subseteq \mathfrak{A}'$ containing a such that $\widehat{\mathfrak{D}}_2 = \mathsf{twin}(\mathfrak{A}'', a)$. Invoking Lemma 3 on $\mathsf{twin}(\mathfrak{A}'', a)$ and $\mathsf{twin}(\mathfrak{A}', a)$, it follows that $\mathfrak{A}'' \equiv_m \mathfrak{A}'$. Then $\mathfrak{A}'' \subseteq \mathfrak{A}$ and $\mathfrak{A}'' \equiv_m \mathfrak{A}$. It is easy to see that $|\mathfrak{A}''| \le f_1(n, m, \tau)$. Taking $\mathfrak{B}$ as $\mathfrak{A}''$, the proof is completed. ∎

Proof of Theorem 4 : Let $\mathcal{S}$ be a substructure-closed class of τ-structures of tree-depth at most n. For notational convenience, let $f_2(n, m, \tau)$ denote $m \cdot g_1(m, \tau) \cdot f_1(n, m, \tau)$, for all $m, n \in \mathbb{N}$ and $\tau \in \mathsf{Vocab}$. We show below that $\mathcal{P}_{logic}(\mathcal{S}, k)$ holds for each $k \in \mathbb{N}$, with the witness function given by $\theta_k(m) = f_2(n, m, \tau)$ if $k = 0$, and by $\theta_k(m) = f_2(n, r, \nu)$ otherwise, where $r = \max(m, k)$ and $\nu = \tau \cup \{P\}$ for a unary predicate P not in τ. The result for $k = 0$ follows from Lemma 4 and from the fact that upto m-equivalence, any τ-structure has at most $m \cdot g_1(m, \tau)$ connected components. Suppose we are given $\mathfrak{A} \in \mathcal{S}$ and a subset W of at most k elements of $\mathfrak{A}$, for $k > 0$. Then consider the ν-structure $\mathfrak{A}'$ whose τ-reduct is $\mathfrak{A}$ and in which P is interpreted to be exactly W. By the result for $k = 0$, there exists $\mathfrak{B}' \subseteq \mathfrak{A}'$ such that (i) $\mathfrak{A}' \equiv_r \mathfrak{B}'$ and (ii) $|\mathfrak{B}'| \le f_2(n, r, \nu)$. It is clear that the τ-reduct of $\mathfrak{B}'$, say $\mathfrak{B}$, is such that (i) $\mathfrak{B} \subseteq \mathfrak{A}$ (hence $\mathfrak{B} \in \mathcal{S}$) (ii) $W \subseteq \mathsf{U}_{\mathfrak{B}}$ (iii) $\mathfrak{A} \equiv_m \mathfrak{B}$ and (iv) $|\mathfrak{B}| \le f_2(n, r, \nu) = \theta_k(m)$. Finally, since $g_1(\cdot, \cdot)$, $f_1(\cdot, \cdot, \cdot)$ and $f_2(\cdot, \cdot, \cdot)$ are easily seen to be computable, we are done. ∎

Remark: The classes of bounded tree-depth considered in this section were not studied earlier in [1]. While these classes in general are not acyclic, nor of bounded degree (more generally, not wide too), they are certainly of bounded tree-width [9]. However, [1] talks only about the class of *all* structures of tree-width n for each $n \in \mathbb{N}$, and not about any subclasses of it.

6   Generating New Classes of Structures

We consider some natural ways of generating new classes of structures from a base class $\mathcal{S}$ of structures. The primary result of this section is that classes generated by these techniques inherit the $\mathcal{P}_{logic}(\cdot, k)$ property of the base class.

We focus on disjoint union ($\sqcup$), complement (!), cartesian product ($\times$) and tensor product ($\otimes$) of τ-structures coming from a base class $\mathcal{S}$. The definition of $\sqcup$ is standard. The definitions of !, $\times$ and $\otimes$ below are inspired by their definitions for graphs. Let $\mathfrak{A}$ and $\mathfrak{B}$ be τ-structures. The *complement* of $\mathfrak{A}$, denoted $!\mathfrak{A}$, is the τ-structure such that (i) $\mathsf{U}_{!\mathfrak{A}} = \mathsf{U}_{\mathfrak{A}}$, and (ii) for every n-ary predicate R in τ, for every n-tuple $(a_1, \dots a_n) \in \mathsf{U}_{\mathfrak{A}}^n$, $!\mathfrak{A} \models R(a_1, \dots a_n)$ iff $\mathfrak{A} \not\models R(a_1, \dots a_n)$. The *cartesian product* of $\mathfrak{A}$ and $\mathfrak{B}$, denoted $\mathfrak{A} \times \mathfrak{B}$, is the structure $\mathfrak{C}$ defined as follows: (i) $\mathsf{U}_{\mathfrak{C}} = \mathsf{U}_{\mathfrak{A}} \times \mathsf{U}_{\mathfrak{B}}$, and (ii) for each n-ary predicate R in τ, for each n-tuple $((a_1, b_1), \dots, (a_n, b_n))$ of $\mathsf{U}_{\mathfrak{C}}$, we have $\mathfrak{C} \models R((a_1, b_1), \dots, (a_n, b_n))$ iff $((a_1 = \cdots = a_n \wedge \mathfrak{B} \models R(b_1, \dots, b_n)) \vee (\mathfrak{A} \models R(a_1, \dots, a_n) \wedge b_1 = \cdots = b_n))$. The *tensor product* of $\mathfrak{A}$ and $\mathfrak{B}$, denoted $\mathfrak{A} \otimes \mathfrak{B}$, is defined similar to the cartesian product, except that $\mathfrak{A} \otimes \mathfrak{B} \models R((a_1, b_1), \dots, (a_n, b_n))$ iff $\mathfrak{A} \models R(a_1, \dots, a_n)$ and $\mathfrak{B} \models R(b_1, \dots, b_n)$.

Let Op be the set $\{\sqcup, !, \times, \otimes\}$. The following properties of operations in Op are important for our purposes. Let $\circledast$ be a binary operation in Op and $m \in \mathbb{N}$.

P1) If $\mathfrak{A}_1 \subseteq \mathfrak{B}_1$ and $\mathfrak{A}_2 \subseteq \mathfrak{B}_2$, then (i) $!\mathfrak{A}_1 \subseteq !\mathfrak{B}_1$ and (ii) $(\mathfrak{A}_1 \circledast \mathfrak{A}_2) \subseteq (\mathfrak{B}_1 \circledast \mathfrak{B}_2)$.

P2) If $\mathfrak{A}_1 \equiv_m \mathfrak{B}_1$ and $\mathfrak{A}_2 \equiv_m \mathfrak{B}_2$, then (i) $!\mathfrak{A}_1 \equiv_m !\mathfrak{B}_1$ and (ii) $(\mathfrak{A}_1 \circledast \mathfrak{A}_2) \equiv_m (\mathfrak{B}_1 \circledast \mathfrak{B}_2)$.

Given a class $\mathcal{S}$, let $!\mathcal{S}$ denote the class $\{!\mathfrak{A} \mid \mathfrak{A} \in \mathcal{S}\}$. Given classes $\mathcal{S}_1$ and $\mathcal{S}_2$ and a binary operation $\circledast \in \mathsf{Op}$, let $\mathcal{S}_1 \circledast \mathcal{S}_2$ denote the class $\{\mathfrak{A} \circledast \mathfrak{B} \mid \mathfrak{A} \in \mathcal{S}_1, \mathfrak{B} \in \mathcal{S}_2\}$. We now have the following important lemma.

Lemma 5. *Let $\mathcal{S}_1, \mathcal{S}_2$ be classes of structures. Let $\circledast$ be a binary operation in Op and $k \in \mathbb{N}$. If $\mathcal{P}_{logic}(\mathcal{S}_i, k)$ holds for $i \in \{1, 2\}$, then each of $\mathcal{P}_{logic}(!\mathcal{S}_1, k)$, $\mathcal{P}_{logic}(!\mathcal{S}_2, k)$ and $\mathcal{P}_{logic}(\mathcal{S}_1 \circledast \mathcal{S}_2, k)$ holds. In addition, $\mathcal{P}_{logic}(\mathcal{S}_1 \cup \mathcal{S}_2, k)$ holds.*

Given a class $\mathcal{S}$ of structures satisfying $\mathcal{P}_{logic}(\cdot, k)$, it follows that any class $\mathcal{S}'$ of structures obtained by finitely many applications of the operations in Op and by taking finite unions of the classes obtained thereof, also satisfies $\mathcal{P}_{logic}(\cdot, k)$. However, there are interesting classes that can be generated only by allowing *infinite* unions of these derived classes. For example, the class of all co-graphs is generated from the class of single vertex graphs by finitely many applications of $\sqcup$ and $!$, and then taking the infinite union of all the classes of graphs thus obtained. The rest of this section is motivated by such infinite unions.

Given a class $\mathcal{S}$ of structures and $O \subseteq \mathsf{Op}$, an *expression tree over* $(\mathcal{S}, O)$ is a tree over O whose leaf nodes are labelled with specific structures from $\mathcal{S}$ and internal nodes are labelled with operations from O (i.e. elements of O). [1] If $\mathfrak{s}$ is an expression tree over $(\mathcal{S}, O)$, let $\mathfrak{C}_\mathfrak{s}$ denote the structure naturally represented by $\mathfrak{s}$ upto isomorphism. We denote by $Z_{\mathcal{S},O}$ the class of all structures defined by all possible expression trees over $(\mathcal{S}, O)$.

Theorem 5. *Let $\mathcal{S}$ be a given class of structures and let $O = \{\sqcup, !\}$. For each $k \in \mathbb{N}$, if $\mathcal{P}_{logic}(\mathcal{S}, k)$ holds, then so does $\mathcal{P}_{logic}(Z_{\mathcal{S},O}, k)$.*

Proof Sketch: Consider $\mathfrak{A} \in Z_{\mathcal{S},O}$ and $m \in \mathbb{N}$. Let $W \subseteq U_\mathfrak{A}$ be a set of size at most k. Let $\mathfrak{s}$ be an expression tree of $\mathfrak{A}$, i.e. $\mathfrak{C}_\mathfrak{s} = \mathfrak{A}$. The proof is in two parts:

(I) We first construct a bounded sized sub-expression-tree $\mathfrak{t}$ of $\mathfrak{s}$ such that (i) W is contained in the leaves of $\mathfrak{t}$, (ii) $\mathfrak{C}_\mathfrak{t} \subseteq \mathfrak{A}$ and (ii) $\mathfrak{C}_\mathfrak{t} \equiv_m \mathfrak{A}$. To do this, we label each node a of $\mathfrak{s}$ by the pair (δ, i), where δ is the $\equiv_m$ class of $\mathfrak{C}_{\mathfrak{s}_a}$, $\mathfrak{s}_a$ is the subtree of $\mathfrak{s}$ rooted at a, and i is the number of leaves of $\mathfrak{s}_a$ that contain any element of W. We then do a "height reduction" as in the proof of Lemma 2(b) to get $\mathfrak{t}$.

(II) We create a tree $\mathfrak{t}_1$ from $\mathfrak{t}$ by replacing the leaves of $\mathfrak{t}$ with m-equivalent bounded substructures ensuring that $\mathfrak{C}_{\mathfrak{t}_1}$ contains W (using the $\mathcal{P}_{logic}(\mathcal{S}, k)$ assumption). By a hierarchical compositional reasoning (using properties P1 and P2), we show that $\mathfrak{C}_{\mathfrak{t}_1}$ is the 'right' substructure of $\mathfrak{C}_\mathfrak{s}$.   ∎

We list below examples of classes of structures satisfying $\mathcal{P}_{logic}(\cdot, k)$ that can be generated by applying the above results to simple classes of structures. In all these examples, we assume a finite set of colours.

---

[1] We think of a tree in the poset-theoretic sense, as in Section 4. The number of children of any internal node is equal to the arity of the operation labeling the node.

1. The class of all coloured co-graphs, obtained using expression trees with $\sqcup$ and ! as operations at the internal nodes and coloured isolated nodes as leaves. By the remark following Definition 2, any class of coloured co-graphs closed under induced subgraphs is also an example. Special cases include the classes of coloured complete graphs, coloured complete n-partite graphs for any $n \in \mathbb{N}$ and coloured threshold graphs [4].
2. The class of r-dimensional grid posets for every $r \in \mathbb{N}$, where an r-dimensional grid poset is a tensor product of r linear orders.

Using ideas similar to those in the proof of Theorem 5, it can be shown that for $k = 0$ or 1, if $\mathcal{P}_{logic}(\mathcal{S}, k)$ holds, then so does $\mathcal{P}_{logic}(Z_{\mathcal{S}, \mathsf{Op}}, k)$, where Op is as defined earlier. An interesting corollary of this result is that the class of all finite dimensional grid posets satisfies $\mathcal{P}_{logic}(\cdot, k)$ when $k = 0$ or 1.

7   Related Properties: $\mathcal{P}_{wqo}(\mathcal{S}, k)$ and $\mathcal{P}_{logic}^{gen}(\mathcal{S}, k)$

In this section, we investigate some other natural properties which are also sufficient to guarantee a generalization of the Łoś-Tarski theorem. Towards this, we first introduce the property $P_{logic}^{gen}(\mathcal{S}, k)$.

Definition 3. *Let $P_{logic}^{gen}(\mathcal{S}, k)$ be the property obtained by dropping the computability restriction of the witness function θ_k in the definition of $P_{logic}(\mathcal{S}, k)$.*

Clearly, $P_{logic}(\mathcal{S}, k)$ implies $P_{logic}^{gen}(\mathcal{S}, k)$. It is also clear from the proof of Theorem 2 that $P_{logic}^{gen}(\mathcal{S}, k)$ entails $PSC(k) = \exists^k \forall^*$, though the former need not entail an effective form of the latter. Also, it turns out that the converse of this entailment is not true in general; the class $\mathcal{S}$ of all undirected cycles satisfies $PSC(k) = \exists^k \forall^*$ for all k, but fails to satisfy $\mathcal{P}_{logic}^{gen}(\mathcal{S}, k)$ for any k.

We now turn our attention to another seemingly unrelated property. We begin with some notation. Given a vocabulary τ and $k \in \mathbb{N}$, let τ_k denote the vocabulary obtained by adding k new constant symbols to τ. Let $\mathcal{S}$ be a class of structures. We use $\mathcal{S}_k$ to denote the class of all τ_k-structures whose τ-reducts are structures in $\mathcal{S}$. Given $\mathfrak{A}, \mathfrak{B} \in \mathcal{S}_k$, we say that $\mathfrak{A}$ *embeds* in $\mathfrak{B}$ if $\mathfrak{A}$ is isomorphic to a substructure of $\mathfrak{B}$. Notationally, we represent this as $\mathfrak{A} \hookrightarrow \mathfrak{B}$. Observe that $(\mathcal{S}_k, \hookrightarrow)$ is a pre-order. We now define the property $\mathcal{P}_{wqo}(\mathcal{S}, k)$ via the notion of a pre-order being a *well-quasi-order* (w.q.o.) [4].

Definition 4. *We say that $\mathcal{P}_{wqo}(\mathcal{S}, k)$ holds if $(\mathcal{S}_k, \hookrightarrow)$ is a well-quasi-order.*

Basic examples of classes satisfying $\mathcal{P}_{wqo}(\mathcal{S}, k)$ are (i) a finite class, and (ii) the class of all finite linear orders. Let Σ be a finite alphabet. In our notation, the celebrated results such as Higman's lemma and Kruskal's tree theorem [4] simply say that $\mathcal{P}_{wqo}(Words(\Sigma), 0)$ and $\mathcal{P}_{wqo}(Trees(\Sigma), 0)$ respectively hold. *A priori*, there is no reason to expect any relation between the w.q.o.-based and the logic-based properties defined above. Surprisingly, we have the following result.

Theorem 6. *For any class $\mathcal{S}$ and any $k \in \mathbb{N}$, $\mathcal{P}_{wqo}(\mathcal{S}, k)$ implies $\mathcal{P}_{logic}^{gen}(\mathcal{S}, k)$.*

It turns out however that $\mathcal{P}_{logic}(\mathcal{S}, k)$ and $\mathcal{P}_{wqo}(\mathcal{S}, k)$ are mutually incompatible. We will present the proofs of these results in the journal version of the paper. An important consequence of the above discussion is that $\mathcal{P}_{wqo}(\mathcal{S}, k)$ entails $PSC(k) = \exists^k \forall^*$. Note that this entailment also need not necessarily give us an effective generalization of the Łoś-Tarski theorem. This highlights the importance of our central notion, namely $P_{logic}(\mathcal{S}, k)$.

8   Conclusion

The study of preservation theorems over special classes of finite structures has recently seen a revival of interest. This paper contributes to this line of work by studying a logic-based combinatorial property that permits an effective version of the generalized Łoś-Tarski theorem to hold over several well-studied classes of finite structures. As future work, we wish to understand better the boundaries of when the generalized Łoś-Tarski theorem, and more importantly, an effective version of it, holds over classes of finite structures. The notion of well-quasi-ordering has turned out to be of central importance in several areas of computer science. In this context, Theorem 6 provides a new logic-based tool for proving that certain classes are not well-quasi-ordered under the embedding relation on structures. It also suggests that our formulations of the logic-based properties might have applications even outside the realm of preservation theorems.

Acknowledgment. The authors thank Ajit A. Diwan for insightful discussions. The authors also thank the anonymous referees for their comments.

References

1. Atserias, A., Dawar, A., Grohe, M.: Preservation under extensions on well-behaved finite structures. SIAM J. Comput. 38(4), 1364–1381 (2008)
2. Atserias, A., Dawar, A., Kolaitis, P.G.: On preservation under homomorphisms and unions of conjunctive queries. J. ACM 53(2), 208–237 (2006)
3. Chang, C.C., Keisler, H.J.: Model Theory, 3rd edn. Elsevier Science Publishers (1990)
4. Diestel, R.: Graph Theory. Springer (2010)
5. Duris, D.: Extension preservation theorems on classes of acyclic finite structures. SIAM J. Comput. 39(8), 3670–3681 (2010)
6. Harwath, F., Heimberg, L., Schweikardt, N.: Preservation and decomposition theorems for bounded degree structures. In: Proc. of CSL-LICS (2014)
7. Libkin, L.: Elements of Finite Model Theory. Springer (2004)
8. Makowsky, J.A.: Algorithmic uses of the Feferman-Vaught theorem. Ann. Pure Appl. Logic 126(1-3), 159–213 (2004)
9. Nesetril, J., de Mendez, P.O.: Tree-depth, subgraph coloring and homomorphism bounds. Eur. J. Comb. 27(6), 1022–1041 (2006)
10. Sankaran, A., Adsul, B., Chakraborty, S.: Generalizations of the Łoś-Tarski preservation theorem. CoRR, abs/1302.4350 (2013)
11. Sankaran, A., Adsul, B., Madan, V., Kamath, P., Chakraborty, S.: Preservation under substructures modulo bounded cores. In: Ong, L., de Queiroz, R. (eds.) WoLLIC 2012. LNCS, vol. 7456, pp. 291–305. Springer, Heidelberg (2012)

Determinising Parity Automata[*]

Sven Schewe and Thomas Varghese

University of Liverpool

Abstract. Parity word automata and their determinisation play an important role in automata and game theory. We discuss a determinisation procedure for nondeterministic parity automata through deterministic Rabin to deterministic parity automata. We prove that the intermediate determinisation to Rabin automata is optimal. We show that the resulting determinisation to parity automata is optimal up to a small constant. Moreover, the lower bound refers to the more liberal Streett acceptance. We thus show that determinisation to Streett would not lead to better bounds than determinisation to parity. As a side-result, this optimality extends to the determinisation of Büchi automata.

1 Introduction

The quest for optimal complementation [19,21,14] and determinisation [11,12,9,15,3] of nondeterministic automata has been long and fruitful. The quest for optimal Büchi complementation techniques seems to have been settled with matching upper [14] and lower [21] bounds. A similar observation might, on first glance, be made for Büchi determinisation, as matching upper [15] and lower [3] bounds were established shortly after those for complementation. However, while these bounds are tight to the state, they refer to deterministic Rabin automata only, with exponentially many Rabin pairs in the states of the initial automaton.

Choosing Rabin automata as targets is not the only natural choice. The dual Streett acceptance condition is a similarly natural goal, and determinising to parity automata seems to be an even more attractive target, as emptiness games for parity automata have a lower computational complexity compared to emptiness games for Streett or Rabin automata. For parity and Streett automata, however, no similarly tight result is known. Indeed, the best known algorithm [9] provides an $O(n!^2)$ bound on the states [15] (for state-based acceptance; the bound can be improved to $O(n!(n-1)!)$ when transition based acceptance is used) of a deterministic parity automaton obtained from a nondeterministic Büchi automaton with n states, as compared to the approximately $(1.65n)^n$ states of deterministic Rabin automaton [15,3].

Another argument for using parity or Streett conditions is that determinisation constructions are often nested. E.g., in distributed synthesis [10,7,5], several co-determinisation (determinisation of the complement language) steps are used. Using Rabin automata as a target in one step, one has to use a determinisation

---

[*] A full version with proofs is available at `http://arxiv.org/abs/1401.5394`.

E. Csuhaj-Varjú et al. (Eds.): MFCS 2014, Part I, LNCS 8634, pp. 486–498, 2014.

technique for Streett automata in the next. Streett determinisation, however, is significantly more involved and expensive [13,9].

In this paper, we introduce determinisation procedures for nondeterministic parity automata (NPA) to deterministic Rabin and parity automata (DRA and DPA). Using an algorithmic representation that extends the determinisation procedures from [15], we show that the number of states used in the determinisation of nondeterministic Büchi automata cannot be reduced by a single state, while we establish the tightness of our parity determinisation procedure to below an approximation factor of 1.5, even if we allow for Streett acceptance. This also shows that determinising parity automata to Rabin automata leads to a smaller blow-up than the determinisation to parity or Streett. As a special case, this holds in particular for Büchi automata.

Transition-Based Acceptance. We use a transition based acceptance mechanism for various reasons, but most importantly, cleaner results. Transition-based acceptance mechanisms have proven to be a more natural target of automata transformations. Indeed, all determinisation procedures quoted above have a natural representation with an acceptance condition on transitions, and their translation to state-based acceptance is by multiplying the acceptance from the last transition to the statespace. A similar observation can be made for other automata transformations, like the removal of ε-transitions from translations of μ-calculi [20,16] and the treatment of asynchronous systems [17], where the statespace grows by multiplication with the acceptance information (e.g., maximal priority on a finite sequence of transitions), while it can only shrink in case of transition based acceptance. Similarly, tools like SPOT [4] translate LTL formulas to more concise automata with a transition-based acceptance mechanism.

Related Work. Besides the work on complementing [19,21,14] and determinising [11,12,9,15,3] Büchi automata, tight bounds have been obtained for generalised Büchi automata [18], and specialised algorithms for complementing [2] and determinising Streett [13,9] automata have been studied. The construction of deterministic CoBüchi automata with a one-sided error, which is correct for CoBüchi recognisable languages [1], and decision procedures that use emptiness equivalent Büchi [8] or safety [6] automata have also been studied.

2   Preliminaries

We denote the set of non-negative integers by ω, i.e., $\omega = \{0, 1, 2, 3, ...\}$. For a finite alphabet Σ, we use Σ^*, Σ^+, and Σ^ω to denote the set of finite, non-empty finite, and infinite sequences over Σ, respectively. An infinite *word* $\alpha : \omega \to \Sigma$ is an infinite sequence of letters $\alpha_0 \alpha_1 \alpha_2 \cdots$ from Σ. We use $[k]$ to represent $\{1, 2, \ldots, k\}$.

ω-automata are finite automata that are interpreted over infinite words and recognise ω-regular languages $L \subseteq \Sigma^\omega$. Nondeterministic ω-automata are quintuples $\mathcal{N} = (Q, \Sigma, I, T, \mathcal{F})$, where Q is a finite set of states with a non-empty subset $I \subseteq Q$ of initial states, Σ is a finite alphabet, $T \subseteq Q \times \Sigma \times Q$ is a transition relation that maps states and input letters to sets of successor states, and $\mathcal{F}$ is an acceptance condition. A *run* ρ of a nondeterministic ω-automaton $\mathcal{N}$ on an input word α is an infinite sequence $\rho : \omega \to Q$ of states of $\mathcal{N}$, also denoted $\rho = q_0 q_1 q_2 \cdots \in Q^{\omega}$, such that the first symbol of ρ is an initial state $q_0 \in I$ and, for all $i \in \omega$, $(q_i, \alpha_i, q_{i+1}) \in T$ is a valid transition. For a run ρ on a word α, we denote with $\overline{\rho} : i \mapsto \big(\rho(i), \alpha(i), \rho(i+1) \big)$ the transitions of ρ. Let $\mathsf{infin}(\rho) = \{ q \in Q \mid \forall i \in \omega \; \exists j > i \text{ such that } \rho(j) = q \}$ denote the set of all states that occur infinitely often during the run ρ. Likewise, let $\mathsf{infin}(\overline{\rho}) = \{ t \in T \mid \forall i \in \omega \; \exists j > i \text{ such that } \overline{\rho}(j) = t \}$ denote the set of all transitions that are taken infinitely many times in ρ.

In this paper, we use acceptance conditions over transitions. Acceptance mechanisms over states can be defined accordingly. *Rabin* automata are ω-automata, whose acceptance is defined by a family of pairs $\{ (A_i, R_i) \mid i \in J \}$, with $A_i, R_i \subseteq T$, of accepting and rejecting transitions for all indices i of some index set J. A run ρ of a Rabin automaton is *accepting* if there is an index $i \in J$, such that infinitely many accepting transitions $t \in A_i$, but only finitely many rejecting transitions $t \in R_j$ occur in $\overline{\rho}$. That is, if there is an $i \in J$ such that $\mathsf{infin}(\overline{\rho}) \cap A_i \neq \emptyset = \mathsf{infin}(\overline{\rho}) \cap R_i$. *Streett* automata are ω-automata, whose acceptance is defined by a family of pairs $\{ (G_i, B_i) \mid i \in J \}$, with $G_i, B_i \subseteq T$, of good and bad transitions for all indices i of some index set J. A run ρ of a Streett automaton is *accepting* if, for all indices $i \in J$, some good transition $t \in G_i$ or no bad transition $t \in B_j$ occur infinitely often in $\overline{\rho}$. That is, if, for all $i \in J$, $\mathsf{infin}(\overline{\rho}) \cap G_i \neq \emptyset$ or $\mathsf{infin}(\overline{\rho}) \cap B_i = \emptyset$ holds.

Parity automata are ω-automata, whose acceptance is defined by a priority function $\mathsf{pri} : T \to [c]$ for some $c \in \mathbb{N}$. A run ρ of a parity automaton is *accepting* if $\limsup_{n \to \infty} \mathsf{pri}\big(\overline{\rho}(n) \big)$ is even, that is, if the highest priority that occurs infinitely often is even. Parity automata can be viewed as special Rabin or Streett automata. In older works, the parity condition was referred to as Rabin chain: one can represent it by choosing A_i and R_i as the set of transitions with priority $\leq 2i$ and $\leq 2i - 1$, respectively. This results in a chain $A_i \subseteq R_i \subseteq A_{i+1} \subseteq \ldots$.

One-pair Rabin automata $\mathcal{R}_1 = \big(Q, \Sigma, I, T, (A, R) \big)$ are of special technical interest in this paper. They are Rabin automata with a singleton index set, such that we directly refer to the only pair (A, R). *Büchi* automata can be viewed as one-pair Rabin automata with an empty set of rejecting states $R = \emptyset$.

For all types of automata, a word α is accepted by an automaton $\mathcal{A}$ iff it has an accepting run, and its language $\mathcal{L}(\mathcal{A})$ is the set of words it accepts.

We call an automaton $(Q, \Sigma, I, T, \mathcal{F})$ *deterministic* if I is singleton and T contains at most one target node for all pairs of states and input letters. Deterministic automata are denoted $(Q, \Sigma, q_0, \delta, \mathcal{F})$, where q_0 is the only initial state and δ is the partial function with $\delta : (q, \alpha) \mapsto r \Leftrightarrow (q, \alpha, r) \in T$.

3   Determinisation

We will tackle the determinisation of parity automata in three steps. Firstly, we will recall history trees, the data structure for determinising Büchi automata. Secondly, we will adjust this data structure and adapt the Büchi determinisation procedure to determinise one-pair Rabin automata. Finally, we will show that this data structure can be nested for the determinisation of parity automata.

In [15,18], we use ordered labelled trees to depict the states of the deterministic automaton. The ordered labelled trees are called *history trees* in [15,18].

A *history tree* is an ordered labelled tree $(\mathcal{T}, l)$, where $\mathcal{T}$ is a finite, prefix closed subset of finite sequences of natural numbers ω. Every element $v \in \mathcal{T}$ is called a *node*. Prefix closedness implies that, if a node $v = n_1 \ldots n_j n_{j+1} \in \mathcal{T}$ is in $\mathcal{T}$, then $v' = n_1 \ldots n_j$ is also in $\mathcal{T}$. We call v' the predecessor of v, denoted $\mathsf{pred}(v)$. The empty sequence $\epsilon \in \mathcal{T}$ is called the *root* of the ordered tree $\mathcal{T}$. ϵ has no predecessor. We further require $\mathcal{T}$ to be *order closed* with respect to siblings: if a node $v = n_1 \ldots n_j$ is in $\mathcal{T}$, then $v' = n_1 \ldots n_{j-1} i$ is also in $\mathcal{T}$ for all $i \in \omega$ with $i < n_j$. In this case, we call v' an *older sibling* of v (and v a *younger sibling* of v'). We denote the set of older siblings of v by $\mathsf{os}(v)$.

A history tree is a tree labelled with sets of states. That is, $l : \mathcal{T} \to 2^Q \setminus \{\emptyset\}$ is a labelling function, which maps nodes of $\mathcal{T}$ to non-empty sets of automata states. For Büchi automata, the labelling is subject to the following criteria.

1. The label of each node is a subset of the label of its predecessor:
   $l(v) \subseteq l(\mathsf{pred}(v))$ holds for all $\varepsilon \neq v \in \mathcal{T}$.
2. The intersection of the labels of two siblings is disjoint:
   $\forall v, v' \in \mathcal{T}. \ v \neq v' \wedge \mathsf{pred}(v) = \mathsf{pred}(v') \Rightarrow l(v) \cap l(v') = \emptyset$.
3. The union of the labels of all siblings is *strictly* contained in the label of their predecessor: $\forall v \in \mathcal{T} \ \exists q \in l(v) \ \forall v' \in \mathcal{T}. \ v = \mathsf{pred}(v') \Rightarrow q \notin l(v')$.

Determinising One-Pair Rabin Automata. For one-pair Rabin automata, it suffices to adjust this data structure slightly. A *root history tree* (RHT) satisfies (1) and (2) from the definition of history trees, and a relaxed version of (3) that allows for non-strict containment of the label of the root, $\forall v \in \mathcal{T} \setminus \{\varepsilon\} \ \exists q \in l(v) \ \forall v' \in \mathcal{T}. \ v = \mathsf{pred}(v') \Rightarrow q \notin l(v')$, and the label of the root ε *equals* the union of its children's labels, $l(\varepsilon) = \bigcup \{l(v) \mid v \in \mathcal{T} \cap \omega\}$.

Let $\mathcal{R}_1 = (Q, \Sigma, I, T, (A, R))$ be a nondeterministic one-pair Rabin automaton with $|Q| = n$ states. We first construct a language equivalent deterministic Rabin automaton $\mathcal{D}_1 = (D, \Sigma, d_0, \Delta, \{(A_i, R_i) \mid i \in J\})$ where, D is the set of RHTs over Q, d_0 is the history tree $(\{\varepsilon, 0\}, l : \varepsilon \mapsto I, l : 0 \mapsto I)$, J is the set of nodes $\neq \varepsilon$ that occur in some RHT of size $n + 1$ (due to the definition of RHTs, an RHT can contain at most $n + 1$ nodes), and for every tree $d \in D$ and letter $\sigma \in \Sigma$, the transition $d' = \Delta(d, \sigma)$ is the result of the sequence of the transition mechanism described below. The index set is the set of nodes, and, for each index, the accepting and rejecting sets (defined later) refer to this node.

Transition Mechanism. We determine $\Delta : ((\mathcal{T}, l), \sigma) \mapsto (\mathcal{T}', l')$ as follows:

1. *Update of node labels (subset constructions).* The root of a history tree d collects the momentarily reachable states $Q_r \subseteq Q$ of the automaton $\mathcal{R}_1$. In the first step of the construction, we update the label of the root to the set of reachable states upon reading a letter $\sigma \in \Sigma$, using the classical subset construction. We update the label of every other node of the RHT d to

reflect the successors reachable through accepting or neutral transitions. For ε, we update l to the function l_1 by assigning $l_1 : \varepsilon \mapsto \{q' \in Q \mid \exists q \in l(\varepsilon). (q, \sigma, q') \in T\}$, and for all $\varepsilon \neq v \in \mathcal{T}$, we update l to the function l_1 by assigning $l_1 : v \mapsto \{q' \in Q \mid \exists q \in l(v). (q, \sigma, q') \in T \smallsetminus R\}$.

2. *Splitting of run threads / spawning new children.* We spawn new children for every node in the RHT. For nodes other than the root ε, we spawn a child labelled with the set of states reachable through accepting transitions; for the root ε, we spawn a child labelled like the root. Thus, for every node $\varepsilon \neq v \in d$ with c children, we spawn a new child vc and expand l_1 to vc by assigning $l_1 : vc \mapsto \{q' \in Q \mid \exists q \in l(v). (q, \sigma, q') \in A\}$. If ε has c children, we spawn a new child c of the root ε and expand l_1 to c by assigning $l_1 : c \mapsto l_1(\varepsilon)$. We use $\mathcal{T}_n$ to denote the extended tree that includes the new children.

3. *Removing states from labels – horizontal pruning.* We obtain a function l_2 from l_1 by removing, for every node v with label $l(v) = Q'$ and all states $q \in Q'$, q from the labels of all younger siblings of v and all of their descendants.

4. *Identifying breakpoints – vertical pruning.* We denote with $\mathcal{T}_e \subseteq \mathcal{T}_n$ the set of all nodes $v \neq \varepsilon$ whose label $l_2(v)$ is now equal to the union of the labels of its children. We obtain $\mathcal{T}_v$ from $\mathcal{T}_n$ by removing all descendants of nodes in $\mathcal{T}_e$, and restrict the domain of l_2 accordingly. Nodes in $\mathcal{T}_v \cap \mathcal{T}_e$ represent the breakpoints reached during the infinite run ρ and are called *accepting*, that is, the transition of $\mathcal{D}_1$ will be in A_v for exactly the $v \in \mathcal{T}_v \cap \mathcal{T}_e$. Note that the root cannot be accepting.

5. *Removing nodes with empty label.* We denote with $\mathcal{T}_r = \{v \in \mathcal{T}_v \mid l_2(v) \neq \emptyset\}$ the subtree of $\mathcal{T}_v$ that consists of the nodes with non-empty label and restrict the domain of l_2 accordingly.

6. *Reordering.* To repair the orderedness, we call $\|v\| = |\mathsf{os}(v) \cap \mathcal{T}_r|$ the number of (still existing) older siblings of v, and map $v = n_1 \ldots n_j$ to $v' = \|n_1\| \, \|n_1 n_2\| \, \|n_1 n_2 n_3\| \ldots \|v\|$, denoted $\mathsf{rename}(v)$. For $\mathcal{T}' = \mathsf{rename}(\mathcal{T}_r)$, we update a pair $(\mathcal{T}_r, l_2)$ from Step 5 to $d' = (\mathcal{T}', l')$ with $l' : \mathsf{rename}(v) \mapsto l_2(v)$. We call a node $v \in \mathcal{T}' \cap \mathcal{T}$ *stable* if $v = \mathsf{rename}(v)$, and we call all nodes in J *rejecting* if they are not stable. That is, the transition will be in R_v exactly for those $v \in J$, such that v is not a stable node in $\mathcal{T} \cap \mathcal{T}'$.

Note that this construction is a generalisation of the same construction for Büchi automata: if $R = \emptyset$, then the label of 0 is always the label of ε in this construction, and the node 1 is not part of any reachable RHT. (We would merely write 0 in front of every node of a history tree.) The correctness proof of this construction follows the same lines as the proof of the Büchi construction.

Lemma 1. $L(\mathcal{R}_1) \subseteq L(\mathcal{D}_1)$

Notation. For a state q of $\mathcal{R}_1$ and an RHT $d = (\mathcal{T}, l)$, we call a node v the *host node* of q, denoted $\mathsf{host}(q, d)$, if $q \in l(v)$, but not in $l(vc)$ for any child vc of v.

The *proof idea* is the same as for Büchi determinisation [15]: the state of each accepting run is eventually 'trapped' in the same node of the RHT, and this node must be accepting infinitely often. Let $d_0, d_1 \ldots$ be the run of $\mathcal{D}_1$ on α and $q_0, q_1, \ldots$ an accepting run of $\mathcal{R}_1$ on α. We then define the sequence $v_0, v_1, \ldots$ with $v_i = \mathsf{host}(q_i, d_i)$, which contains a longest eventually stable prefix v.

An inductive argument can then be exploited to show that, once this prefix v is henceforth stable, the index v cannot be rejecting. The assumption that v is eventually stable but never again accepting leads to a contradiction. Once the transition $(q_i, \alpha(i), q_{i+1})$ is accepting, $q_{i+1} \in l_{i+1}(vc)$ for some $c \in \omega$ and for $d_{i+1} = (\mathcal{T}_{i+1}, l_{i+1})$. As v is never again accepting or rejecting, we can show for all $j > i$ that, if $q_j \in l_j(vc_j)$, then $q_{j+1} \in l_{j+1}(vc_{j+1})$ for some $c_{j+1} \le c_j$. This monotonicity contradicts the assumption that v is the $longest$ stable prefix.

Lemma 2. $L(\mathcal{D}_1) \subseteq L(\mathcal{R}_1)$

For the run $d_0 d_1 \dots$ of $\mathcal{D}_1$ on α, we fix an ascending chain $i_0 < i_1 < i_2 \dots$ of indices, such that v is not rejecting in any transition $(d_{j-1}, \alpha(j-1), d_j)$ for $j \ge i_0$ and such that $(d_{i_j-1}, \alpha(i_j - 1), d_{i_j}) \in A_v$ for all $j \ge 0$. The proof idea is the usual way of building a tree of initial sequences of runs: we build a tree of initial sequences of runs of $\mathcal{R}_1$ that contains a sequence $q_0 q_1 q_2 \dots q_{i_j}$ for any $j \in \omega$ iff

- $(q_i, \alpha(i), q_{i+1}) \in T$ is a transition of $\mathcal{R}_1$ for all $i < i_j$,
- $(q_i, \alpha(i), q_{i+1}) \notin R$ is not rejecting for all $i \ge i_0 - 1$, and
- for all $k < j$ there is an $i \in [i_k, i_{k+1}[$ s.t. $(q_i, \alpha(i), q_{i+1}) \in A$ is accepting.

This infinite tree has an infinite branch by König's Lemma. By construction, this branch is an accepting run of $\mathcal{R}_1$ on α.

Corollary 1. $L(\mathcal{R}_1) = L(\mathcal{D}_1)$.

Let $\#\mathsf{ht}(n)$ and $\#\mathsf{rht}(n)$ be the number of history trees and RHTs, respectively, over sets with n states. First, $\#\mathsf{rht}(n) \ge \#\mathsf{ht}(n)$ holds, because the subtree rooted in 0 of an RHT is a history tree. Second, $\#\mathsf{ht}(n+1) \ge \#\mathsf{rht}(n)$, because adding the additional state to $l(\varepsilon)$ turns an RHT into a history tree. With an estimation similar to that of history trees [15], we get:

Theorem 1. $\inf \{ c \mid \#\mathsf{rht}(n) \in O((cn)^n) \} = \inf \{ c \mid \#\mathsf{ht}(n) \in O((cn)^n) \} \approx 1.65$.

The full version shows that $\#\mathsf{rht}(n)$ is only a constant factor bigger than $\#\mathsf{ht}(n)$.

Determinising Parity Automata. Having outlined a determinisation construction for one-pair Rabin automata using root history trees, we proceed to define *nested history trees* (NHTs), the data structure we use for determinising parity automata. We assume that we have a parity automaton $\mathcal{P} = (Q, \Sigma, I, T, \mathsf{pri} : T \to [c])$, and we select $e = 2\lfloor 0.5c \rfloor$.

A *nested history tree* is a triple $(\mathcal{T}, l, \lambda)$, where $\mathcal{T}$ is a finite, prefix closed subset of finite sequences of natural numbers and a special symbol $\mathfrak{s}$ (for *stepchild*), $\omega \cup \{\mathfrak{s}\}$. We refer to all other children vc, $c \in \omega$ of a node v as its *natural children*. We call $l(v)$ the label of the node $v \in \mathcal{T}$, and $\lambda(v)$ its *level*.

A node $v \ne \varepsilon$ is called a *Rabin root*, iff it ends in $\mathfrak{s}$. The root ε is called a Rabin root iff $c > e$. A node $v \in \mathcal{T}$ is called a *base node* iff it is not a Rabin root and $\lambda(v) = 2$. The set of base nodes is denoted $\mathsf{base}(\mathcal{T})$.

- The label $l(v)$ of each node $v \ne \varepsilon$ is a subset of the label of its predecessor: $l(v) \subseteq l(\mathsf{pred}(v))$ holds for all $\varepsilon \ne v \in \mathcal{T}$.
- The intersection of the labels of two siblings is disjoint:
  $\forall v, v' \in \mathcal{T}. \ v \ne v' \wedge \mathsf{pred}(v) = \mathsf{pred}(v') \Rightarrow l(v) \cap l(v') = \emptyset$.

- For all *base nodes*, the union of the labels of all siblings is *strictly* contained in the label of their predecessor:
  $\forall v \in \mathsf{base}(\mathcal{T}) \ \exists q \in l(v) \ \forall v' \in \mathcal{T}. \ v = \mathsf{pred}(v') \Rightarrow q \notin l(v')$.
- A node $v \in \mathcal{T}$ has a stepchild iff v is neither a base-node, nor a Rabin root.
- The union of the labels of all siblings of a non-base node *equals* the union of its children's labels: $\hfill \forall v \in \mathcal{T} \smallsetminus \mathsf{base}(\mathcal{T})$
  $l(v) = \{q \in l(v') \mid v' \in \mathcal{T} \text{ and } v = \mathsf{pred}(v')\}$ holds.
- The level of the root is $\lambda(\varepsilon) = e$.
- The level of a stepchild is 2 smaller than the level of its parent:
  for all $v\mathsf{s} \in \mathcal{T}$, $\lambda(v\mathsf{s}) = \lambda(v) - 2$ holds.
- The level of all other children equals the level of its parent:
  for all $i \in \omega$ and $vi \in \mathcal{T}$, $\lambda(vi) = \lambda(v)$ holds.

While the definition sounds rather involved, it is (for odd c) a nesting of RHTs. Indeed, for $c = 3$, we simply get the RHTs, and λ is the constant function with domain $\{2\}$. For odd $c > 3$, removing all nodes that contain an s somewhere in the sequence again resemble RHTs, while the sub-trees rooted in a node $v\mathsf{s}$ such that v does not contain a s resemble NHTs whose root has level $c - 3$.

The transition mechanism from the previous subsection is adjusted accordingly. For each level a (note that levels are always even), we define three sets of transitions for the parity automaton $\mathcal{P}$: the rejecting transitions $R_a = \{t \in T \mid \mathsf{pri}(t) > a \text{ and } \mathsf{pri}(t) \text{ is odd}\}$, the accepting transitions $A_a = \{t \in T \mid \mathsf{pri}(t) \geq a \text{ and } \mathsf{pri}(t) \text{ is even}\}$, and the (at least) neutral transitions, $N_a = T \smallsetminus R_a$.

Construction. Let $\mathcal{P} = (P, \Sigma, I, T, \{\mathsf{pri} : P \to [c]\})$ be a nondeterministic parity automaton with $|P| = n$ states. We construct a language equivalent deterministic Rabin automaton $\mathcal{DR} = (D, \Sigma, d_0, \Delta, \{(A_i, R_i) \mid i \in J\})$ where,

- D is the set of NHTs over P (i.e., with $l(\varepsilon) \subseteq P$) whose root has level e, where $e = c$ if c is even, and $e = c - 1$ if c is odd,
- d_0 is the NHT we obtain by starting with $(\{\varepsilon\}, \ l : \varepsilon \mapsto I, \ \lambda : \varepsilon \mapsto e)$, and performing Step 7 from the transition construction until we obtain an NHT.
- J is the set of nodes v that occur in some NHT of level e over P, and
- for every tree $d \in D$ and letter $\sigma \in \Sigma$, the transition $d' = \Delta(d, \sigma)$ is the result of the sequence of transformations described below.

Transition Mechanism. Note that we do not define the update of λ, but use λ. This can be done because the level of the root always remains $\lambda(\varepsilon) = e$; the level $\lambda(v)$ of all nodes v is therefore defined by the number of s occurring in v. Likewise, the property of v being a base-node or a Rabin root is, for a given c, a property of v and independent of the labelling function.

Starting from an NHT $d = (\mathcal{T}, l, \lambda)$, we define the transitions $\Delta : (d, \sigma) \mapsto d'$ as follows:

1. *Update of node labels (subset constructions):* For the root, we continue to use $l_1(\varepsilon) = \{q' \in Q \mid \exists q \in l(\varepsilon). \ (q, \sigma, q') \in T\}$.
   For other nodes $v \in \mathcal{T}$ that are *no Rabin roots*, we use $l_1(v) = \{q' \in Q \mid \exists q \in l(v). \ (q, \sigma, q') \in N_{\lambda(v)}\}$.

For the remaining Rabin roots $v\mathfrak{s} \in \mathcal{T}$, we use $l_1(v\mathfrak{s}) = \{q' \in Q \mid \exists q \in l(v\mathfrak{s}). (q, \sigma, q') \in N_{\lambda(v)}\}$. That is, we use the neutral transition of the higher level of the *parent* of the Rabin node.

2. *Splitting of run threads / spawning new children.* We spawn new children for every node in the NHT. For nodes $v \in \mathcal{T}$ that are no Rabin roots, we spawn a child labelled with the set of states reachable through accepting transitions. For a Rabin root $v \in \mathcal{T}$, we spawn a new child labelled like the root. Thus, for every node $v \in \mathcal{T}$ which is no Rabin root and has c *natural* children, we spawn a new child vc and expand l_1 to vc by assigning $l_1 : vc \mapsto \{q \in Q \mid \exists q' \in l(v). (q', \sigma, q) \in A_{\lambda(v)}\}$. If a Rabin root v has c natural children, we spawn a new child vc of the Rabin root v and expand l_1 to vc by assigning $l_1 : vc \mapsto l_1(v)$. We use $\mathcal{T}_n$ to denote the extended tree that includes the new children.

3. *Removing states from labels – horizontal pruning.* We obtain a function l_2 from l_1 by removing, for every node v with label $l(v) = Q'$ and all states $q \in Q'$, q from the labels of all younger siblings of v and all of their descendants. Stepchildren are always treated as the *youngest* sibling, irrespective of the order of creation.

4. *Identifying breakpoints – vertical pruning.* We denote with $\mathcal{T}_e \subseteq \mathcal{T}_n$ the set of all nodes $v \neq \varepsilon$ whose label $l_2(v)$ is now equal to the union of the labels of its *natural* children. We obtain $\mathcal{T}_v$ from $\mathcal{T}_n$ by removing all descendants of nodes in $\mathcal{T}_e$, and restrict the domain of l_2 accordingly.

   Nodes in $\mathcal{T}_v \cap \mathcal{T}_e$ represent the breakpoints reached during the infinite run ρ and are called *accepting*. That is, the transition of $\mathcal{DR}$ will be in A_v for exactly the $v \in \mathcal{T}_v \cap \mathcal{T}_e$. Note that Rabin roots cannot be accepting.

5. *Removing nodes with empty label.* We denote with $\mathcal{T}_r = \{v \in \mathcal{T}_v \mid l_2(v) \neq \emptyset\}$ the subtree of $\mathcal{T}_v$ that consists of the nodes with non-empty label and restrict the domain of l_2 accordingly.

6. *Reordering.* To repair the orderedness, we call $\|v\| = |\mathsf{os}(v) \cap \mathcal{T}_r|$ the number of (still existing) older siblings of v, and map $v = n_1 \ldots n_j$ to $v' = \|n_1\| \, \|n_1 n_2\| \, \|n_1 n_2 n_3\| \ldots \|v\|$, denoted $\mathsf{rename}(v)$.

   For $\mathcal{T}_o = \mathsf{rename}(\mathcal{T}_r)$, we update a pair $(\mathcal{T}_r, l_2)$ from Step 5 to $d' = (\mathcal{T}_o, l')$ with $l' : \mathsf{rename}(v) \mapsto l_2(v)$.

   We call a node $v \in \mathcal{T}_o \cap \mathcal{T}$ *stable* if $v = \mathsf{rename}(v)$, and we call all nodes in J *rejecting* if they are not stable. That is, the transition will be in R_v exactly for those $v \in J$, such that v is not a stable node in $\mathcal{T} \cap \mathcal{T}'$.

7. *Repairing nestedness.* We initialise $\mathcal{T}'$ to $\mathcal{T}_o$ and then add recursively for
   - Rabin roots v without children a child $v0$ to $\mathcal{T}'$ and expand l' by assigning $l' : v0 \mapsto l'(v)$, and for
   - nodes v, which are neither Rabin roots nor base-nodes, without children a child $v\mathfrak{s}$ to $\mathcal{T}'$ and expand l' by assigning $l' : v\mathfrak{s} \mapsto l'(v)$

   until we have constructed an NHT $d' = (\mathcal{T}', l', \lambda')$.

Lemma 3. $L(\mathcal{P}) \subseteq L(\mathcal{DR})$

Notation. For a state q of $\mathcal{P}$, an NHT $d = (\mathcal{T}, l, \lambda)$ and an even number $a \leq e$, we call a node v' the a *host node of* q, denoted $\mathsf{host}_a(q, d)$, if $q \in l(v')$, but not in $l(v'c)$ for any natural child $v'c$ of v', and $\lambda(v') = a$.

Let $\rho = q_0, q_1, q_2 \ldots$ be an accepting run of the NPA $\mathcal{P}$ with even $a = \liminf_{i \to \infty} \mathsf{pri}(q_i, \alpha(i), q_{i+1})$ on an ω-word α, let $d_0 d_1 d_2 \ldots$ be the run of $\mathcal{DR}$ on α, and let $v_i = \mathsf{host}_a(q_i, d_i)$ for all $i \in \omega$.

The core idea of the proof is again that the state of each accepting run is eventually 'trapped' in a maximal initial sequence v of a-hosts, with the additional constraint that neither v nor any of its ancestors are infinitely often rejecting, and the transitions of the run of $\mathcal{P}$ are henceforth in N_a.

We show by contradiction that v is accepting infinitely often. For $\lambda(v) = a$, the proof is essentially the same as for one-Rabin determinisation. For $\lambda(v) > a$, the proof is altered by a case distinction, where one case assumes that, for some index $i > 0$ such that, for all $j \geq i$, v is a prefix of all v_j, $(q_{j-1}, \alpha(j-1), q_j) \in N_a$, and $(d_{j-1}, \alpha(j-1), d_j) \notin R_v \cup A_v$, q_i is in the label of a natural child vc of v. This provides the induction basis – in the one-pair Rabin case, the basis is provided through the accepting transition of the one-pair Rabin automaton, and we have no corresponding transition with even priority $\geq \lambda(v)$ – by definition. If no such i exists, we choose an i that satisfies the above requirements except that q_i is in the label of a natural child vc of v. We can then infer that the label of $v\mathfrak{s}$ also henceforth contains q_i. As a Rabin root whose parent is not accepting or rejecting, $v\mathfrak{s}$ is not rejecting either.

Lemma 4. $L(\mathcal{DR}) \subseteq L(\mathcal{P})$

The proof of this lemma is essentially the proof of Lemma 2 where, for the priority $a = \lambda(v)$ chosen to be the level of the accepting index v, A_a takes the role of the accepting set A from the one-pair Rabin automaton.

Corollary 2. $L(\mathcal{P}) = L(\mathcal{DR})$.

Determinising to a Deterministic Parity Automata $\mathcal{D}$. Deterministic parity automata seem to be a nice target when determinising ω-automata [9,15,18] given that algorithms that solve parity games (e.g., for acceptance games of alternating and emptiness games of nondeterministic tree automata) have a lower complexity when compared to solving Rabin games. For applications that involve co-determinisation, the parity condition also avoids the Streett condition.

Safra's determinisation construction (and younger variants) intuitively enforces a parity-like order on the nodes of history trees. By storing the order in which nodes are introduced during the construction, we can capture the Index Appearance Records construction that is used to convert Rabin or Streett automata to parity automata. To achieve this, we augment the states of the deterministic automaton (RHTs or NHTs) with a *later introduction record* (LIR), an abstraction of the order in which the non-Rabin roots of the ordered trees are introduced. (Rabin roots provide redundant information and are omitted.)

For an ordered tree $\mathcal{T}$ with m nodes that are no Rabin roots, an LIR is a sequence $v_1, v_2, \ldots v_m$ that contains the nodes of $\mathcal{T}$ that are no Rabin roots nodes, such that, each node appears after its ancestors and older siblings. For convenience in the lower bound proof, we represent a node $v \in \mathcal{T}$ of an NHT $d = (\mathcal{T}, l, \lambda)$ in the LIR by a triple (S_v, c_v, P_v) where $S_v = l(v)$, is the label of v, $c_v = \lambda(v)$ the level of v, and $P_v = \{q \in Q \mid v = \mathsf{host}_{c_v}(q, d)\}$ is the set of states c_v hosted by v. The v can be reconstructed by the order and level. We call the possible sequences of these triples *LIR-NHTs*. Obviously, each LIR-NHT defines an NHT, but not the other way round.

A finite sequence $(S_1, c_1, P_1)(S_2, c_2, P_2)(S_3, c_3, P_3) \ldots (S_k, c_k, P_k)$ of triples is a LIR-NHT if it satisfies the following requirements for all $i \in [k]$.

1. $P_i \subseteq S_i$,
2. $\{P_i\} \cup \{S_j \mid j > i,\ c_i = c_j,\ \text{and}\ S_j \cap S_i \neq \emptyset\}$ partitions S_i.
3. $\{S_j \mid j > i,\ c_i = c_j + 2,\ \text{and}\ S_j \cap P_i \neq \emptyset\}$ partition P_i.
4. If the highest priority of $\mathcal{P}$ is even, then $c_i = e$ implies $S_i \subseteq S_1$. (Then, the lowest level construction is Büchi and the first triple refers to the root.)
5. For $c_i < e$, there is a $j < i$ with $S_i \subseteq P_j$.

To define the transitions of $\mathcal{D}$, we can work in two steps. First, we identify, for each position i of a state $N = (S_1, c_1, P_1)(S_2, c_2, P_2)(S_3, c_3, P_3) \ldots$ of $\mathcal{D}$, the node v_i of the NHT $d = (\mathcal{T}, l, \lambda)$ for the same input letter. We then perform the transition $(d, \sigma, (\mathcal{T}', l', \lambda'))$ on this Rabin automaton. We are then first interested in the set of non-rejecting nodes from this transition and their indices. These indices are moved to the left, otherwise maintaining their order. All remaining vertices of $\mathcal{T}'$ are added at the right, maintaining orderedness.

The priority of the transition is determined by the smallest position i in the sequence, where the related node in the underlying tree is accepting or rejecting. It is therefore more convenient to use a min-parity condition, where the parity of $\liminf_{n \to \infty} \mathsf{pri}(\bar{\rho})$ determines acceptance of a run ρ. As this means smaller numbers have higher priority, pri is representing the opposite of a priority function, and we refer to the priority as the *co-priority* for clear distinction.

If the smallest node is rejecting, the transition has co-priority $2i - 1$, if it is accepting (and not rejecting), then the transition has co-priority $2i$, and if no such node exists, then the transition has co-priority $ne + 1$.

Lemma 5. *Given an NPA $\mathcal{P}$ with n states and maximal priority c, we can construct a language equivalent deterministic parity automaton $\mathcal{D}$ with $ne + 1$ priorities for $e = 2\lfloor 0.5c \rfloor$, whose states are the LIR-NHTs described above.*

Lemma 6. *The DPA resulting from determinising a one-pair Rabin automaton $\mathcal{R}_1$ has $O(n!^2)$ states, and $O\big(n!(n-1)!\big)$ if $\mathcal{R}_1$ is Büchi.*

Lower Bound. We finally show that our determinisation to Rabin automata is optimal, and that our determinisation to parity automata is optimal up to a small constant factor. What is more, this lower bound extends to the more liberal Streett acceptance condition. The technique we employ to establish the lower bound for determinisation to Rabin is similar to [3,18], in that we use the states of the Rabin automaton in a language game. Such a language game is an initialised two player game $G = (V, E, v_0, \mathcal{L})$, which is played between a verifier and a spoiler on a star-shaped directed labelled simple multi-graph (V, E). It has a finite set V of vertices, but a potentially infinite set of edges.

The centre of the star, which we refer to by $c \in V$, is the only vertex of the verifier, while all other vertices are owned by the spoiler. Besides the centre, the game has a second distinguished vertex, the initial vertex v_0, where a play of the game starts. The remaining vertices $W = V \smallsetminus \{v_0, c\}$ are called the working vertices. Like v_0, they are owned by the spoiler.

The edges are labelled by finite words over an alphabet Σ. Edges leaving the centre vertex are labelled by the empty word ε, and there is exactly one edge leaving from the edge to each working vertex, and no outgoing edge to the initial vertex. The set of these outgoing edges is thus $\{(c, \varepsilon, v) \mid v \in W\}$. The edges that lead to the centre vertex are labelled with non-empty words.

The players play out a run of the game in the usual way by placing a pebble on the initial vertex v_0, letting the owner of that vertex select an outgoing edge, moving the pebble along it, and so forth. This way, an infinite sequence of edges is produced, and concatenating the finite words by which they are labelled provides an infinite word w over Σ. The verifier has the objective to construct a word in $\mathcal{L}$, while the spoiler has the antagonistic objective to construct a word in $\Sigma^\omega \smallsetminus \mathcal{L}$.

Theorem 2. *[3] If the verifier wins a language game for a language recognised by a DRA $\mathcal{R}$ with r states, then he wins it using a strategy with memory r.*

To prove optimality of determinising to Rabin, we expand the hardness proofs [3,18] based on *full automata* to the determinisation of parity automata.

The words from each vertex is the language for the Rabin automaton where we have some well defined progress, either some index which occurs accepting but not rejecting, or an index which occurs not rejecting with the target NHT either growing or shrinking.

It is easy to establish that the verifier wins these games. To prove that the minimal DRA that recognises the language of a full parity automaton cannot be smaller than a DRA obtained with our construction, we show that the verifier needs all edges to win these games. The techniques to establish this are similar to those for previous hardness proofs [3,18].

The surprising result is that this technique can be extended to establish optimality for determinisation *to* parity automata. This can be achieved by aiming for Streett automata that accept the complement language. This technique is successful, because the parity construction proves to be optimal (up to a factor of 1.5) for determinisation to Streett automata.

Theorem 3. *The DPA that we construct has less than 1.5 times as many states as the smallest deterministic Streett (or parity) automaton that recognises the language of an NPA recognising the hardest possible language.*

State sizes for two parameters are usually not crisp to represent. But for the simple base cases, Büchi and one pair Rabin automata, we get very nice results:

the known upper bound for determinising Büchi to parity automata [15] are tight and Piterman's algorithm for it [9] is **optimal** modulo a factor of $3n$, where $2n$ stem from the fact that [9] uses state based acceptance. With Lemma 6 we get:

Corollary 3. *The determinisation of Büchi automata to Streett or parity automata leads to $\theta(n!(n-1)!)$ states, and the determinisation of one-pair Rabin automata to Streett or parity automata leads to $\theta(n!^2)$ states.*

References

1. Boker, U., Kupferman, O.: Co-ing Büchi made tight and useful. In: Proc. of LICS, pp. 245–254 (2009)
2. Cai, Y., Zhang, T.: Tight upper bounds for Streett and parity complementation. In: Proc. of CSL, pp. 112–128 (2011)
3. Colcombet, T., Zdanowski, K.: A tight lower bound for determinization of Büchi automata. In: Albers, S., Marchetti-Spaccamela, A., Matias, Y., Nikoletseas, S., Thomas, W. (eds.) ICALP 2009, Part II. LNCS, vol. 5556, pp. 151–162. Springer, Heidelberg (2009)
4. Duret-Lutz, A.: LTL translation improvements in SPOT. In: VECoS, pp. 72–83. BCS (2011)
5. Finkbeiner, B., Schewe, S.: Uniform distributed synthesis. In: Proc. of LICS 2005, pp. 321–330 (2005)
6. Finkbeiner, B., Schewe, S.: Bounded synthesis. International Journal on Software Tools for Technology Transfer, online-first: 1–12 (2012)
7. Kupferman, O., Vardi, M.Y.: Synthesizing distributed systems. In: Proc. of LICS 2001, pp. 389–398 (2001)
8. Kupferman, O., Vardi, M.Y.: Safraless decision procedures. In: Proc. of FOCS 2005, pp. 531–540 (2005)
9. Piterman, N.: From nondeterministic Büchi and Streett automata to deterministic parity automata. Journal of LMCS 3(3:5) (2007)
10. Pnueli, A., Rosner, R.: Distributed reactive systems are hard to synthesize. In: Proc. of FOCS 1990, pp. 746–757 (1990)
11. Rabin, M.O.: Decidability of second order theories and automata on infinite trees. Transaction of the AMS 141, 1–35 (1969)
12. Safra, S.: On the complexity of ω-automata. In: Proc. of FOCS 1988, pp. 319–327 (1988)
13. Safra, S.: Exponential determinization for omega-automata with strong-fairness acceptance condition. In: Proc. of STOC 1992, pp. 275–282 (1992)
14. Schewe, S.: Büchi complementation made tight. In: Proc. of STACS 2009, pp. 661–672 (2009)
15. Schewe, S.: Tighter bounds for the determinisation of büchi automata. In: de Alfaro, L. (ed.) FOSSACS 2009. LNCS, vol. 5504, pp. 167–181. Springer, Heidelberg (2009)
16. Schewe, S., Finkbeiner, B.: Satisfiability and finite model property for the alternating-time μ-calculus. In: Ésik, Z. (ed.) CSL 2006. LNCS, vol. 4207, pp. 591–605. Springer, Heidelberg (2006)
17. Schewe, S., Finkbeiner, B.: Synthesis of asynchronous systems. In: Puebla, G. (ed.) LOPSTR 2006. LNCS, vol. 4407, pp. 127–142. Springer, Heidelberg (2007)
18. Schewe, S., Varghese, T.: Tight bounds for the determinisation and complementation of generalised büchi automata. In: Chakraborty, S., Mukund, M. (eds.) ATVA 2012. LNCS, vol. 7561, pp. 42–56. Springer, Heidelberg (2012)

19. Vardi, M.Y.: The Büchi complementation saga. In: Thomas, W., Weil, P. (eds.) STACS 2007. LNCS, vol. 4393, pp. 12–22. Springer, Heidelberg (2007)
20. Wilke, T.: Alternating tree automata, parity games, and modal μ-calculus. Bull. Soc. Math. Belg. 8(2) (May 2001)
21. Yan, Q.: Lower bounds for complementation of omega-automata via the full automata technique. Journal of LMCS, 4(1:5) (2008)

Tight Bounds for Complementing Parity Automata[*]

Sven Schewe and Thomas Varghese

University of Liverpool

Abstract. We follow a connection between tight determinisation and complementation and establish a complementation procedure from transition labelled parity automata to transition labelled nondeterministic Büchi automata. We prove it to be tight up to an $O(n)$ factor, where n is the size of the nondeterministic parity automaton. This factor does not depend on the number of priorities.

1 Introduction

The precise complexity of complementing ω-automata is an intriguing problem for two reasons: first, the quest for optimal algorithms is a much researched problem [1,16,11,23,14,10,24,9,7,6,12,26,28], and second, complementation is a valuable tool in formal verification (c.f., [8]), in particular when studying language inclusion problems of ω-regular languages. Complementation is also useful to check the correctness of translation techniques [26,25]. The GOAL tool [25], for example, provides such a test suite and incorporates recent algorithms [14,24,7,12] for Büchi complementation.

While devising optimal algorithms for complementing nondeterministic finite automata is simple—nondeterministic *finite* automata can be determinised using a simple subset construction, and deterministic finite automata can be complemented by complementing the set of final states [13,16]—devising optimal algorithms for complementing nondeterministic ω-automata is hard, because simple subset constructions are insufficient to determinise or complement them [10,9].

Given the hardness and importance of the problem, complementation of ω-automata enjoyed much attention. The initial focus was on the complementation of Büchi automata with state-based acceptance [1,11,23,10,14,9,24,7,6,26,28,17,25], resulting in a continuous improvement of its upper and lower bounds.

The first complementation algorithm dates back to the introduction of Büchi automata in 1962. In his seminal paper "On a decision method in restricted second order arithmetic" [1], Büchi develops a doubly exponential complementation procedure. While Büchi's result shows that nondeterministic Büchi automata (and thus ω-regular expressions) are closed under complementation, complementing an automaton with n states may, when using Büchi's complementation procedure, result in an automaton with $2^{2^{O(n)}}$ states, while an $\Omega(2^n)$ lower bound [16] is inherited from finite automata.

---

[*] Extended version with omitted proofs at `http://arxiv.org/abs/1406.1090`.

E. Csuhaj-Varjú et al. (Eds.): MFCS 2014, Part I, LNCS 8634, pp. 499–510, 2014.

In the late 80s, these bounds were improved in a sequence of results, starting with establishing an EXPTIME upper bound [11,23], matching the EXPTIME lower bound [16] inherited from finite automata. However, the early EXPTIME complementation techniques produce automata with up to $2^{O(n^2)}$ states [11,23]; thus, these upper bounds were still exponential in the lower bounds.

This situation changed in 1988, when Safra introduced his famous determinisation procedure for nondeterministic Büchi automata [14], resulting in an $n^{O(n)}$ bound for Büchi complementation, while Michel [10] established a seemingly matching $\Omega(n!)$ lower bound shortly. Together, these results imply that Büchi complementation is in $n^{\theta(n)}$, leaving again the impression of a tight bound.

As pointed out by Vardi [26], this impression is misleading, because the $O()$ notation hides an $n^{\theta(n)}$ gap between both bounds. This gap was narrowed down in 2001 to $2^{\theta(n)}$ by the introduction of an alternative technique that builds on level rankings and a cut-point construction [7]. The complexity of the plain method is approximately $(6n)^n$, leaving a $(6e)^n$ gap to Michel's lower bound [10].

Subseqently, tight level rankings [6,28] were exploited by Friedgut, Kupferman, and Vardi [6] to improve the upper complexity bound to $O((0.96n)^n)$, and by Yan [28] to improve the lower complexity bound to $\Omega((0.76n)^n)$. Schewe [17] provided a matching upper bound, showing tightness up to an $O(n^2)$ factor.

In recent works, more succinct acceptance mechanism have been studied, where the most important ones are parity and generalised Büchi automata, as they occur naturally in the translation of μ-calculi and LTL specifications, respectively. In [21], we gave tight bounds for the determinisation and complementation of generalised Büchi automata. For Rabin, Streett, and parity automata, there has been much progress [4,3,2], in particular establishing an $n^{\theta(n)}$ bound for parity complementation with state-based acceptance, which has been a great improvement and pushed tightness of parity complementation to the level known from Büchi complementation since the late 80s [14,10].

Contribution. In this paper, we establish tight bounds for the complementation of parity automata with transition-based acceptance. A generalisation of the ranking-based complementation procedures quoted above to transition-based acceptance is straight forward, and the Safra-style determinisation procedures from the literature [14,15,12,18,21] have a natural representation with an acceptance condition on transitions. Their translation to state-based acceptance is by multiplying the acceptance from the last transition to the state space.

A similar observation can be made for other automata transformations, like the removal of ε-transitions from translations of μ-calculi [27,19] and the treatment of asynchronous systems [20], where the state-space grows by multiplication with the acceptance information (e.g., maximal priority on a finite sequence of transitions), while it cannot grow in case of transition-based acceptance. Similarly, tools like SPOT [5] offer more concise automata with transition-based acceptance mechanism as a translation from LTL. Using state-based acceptance in the automaton that we want to complement would also complicate the presentation of the complementation procedure. But first and foremost, using transition-based acceptance provides cleaner results. This is the case because in state-based acceptance, the role of the states is overloaded. In finite automata over infinite structures, each state represents the class of tails of the word that can be accepted from this state. In state-based acceptance, they must account for the acceptance mechanism itself, while they are relieved from this burden in transition-based acceptance. In ranking-based complementation techniques, this results in a situation where states with certain properties, such as final states for Büchi automata, can only occur with some ranks, but not with all.

As transition-based acceptance separates these concerns, the presentation becomes cleaner. The natural downside is that we lose the $n^{O(n)}$ bound [3] for parity complementation, as the number of priorities in a parity automaton with transition-based acceptance can grow arbitrarily. But in return, we do get a clean and simple complementation procedure based on a data structure we call flattened nested history trees (FNHTs), which is inspired by a generalisation of history trees [18] to multiple levels, one for each even priority ≥ 2.

In [21], we show a connection between optimal determinisation and complementation for generalised Büchi automata, where we exploit the nondeterministic power of a Büchi automaton to devise a tight complementation procedure. In this paper, we follow this connection between tight determinisation [22] and complementation to devise a tight complementation construction from parity to nondeterministic Büchi automata.

We show that any procedure that complements full parity automata with states Q and maximal priority π has at least $|\mathsf{fnht}(Q, \pi)|/2$ states, where $\mathsf{fnht}(Q, \pi)$ is the set of FNHTs for a given set Q of states and maximal priority π of the parity automaton that is to be complemented. Our complementation construction uses a marker in addition for its acceptance mechanism. Essentially, it is used to mark some position of interest in an FNHT. It accounts for the $O(n)$ gap between the upper and lower bound. We show that, for $\pi \geq 2$ (and hence for Büchi automata upwards) the number of states of our complementation construction is bounded by $4n + 1$ times the lower bound.

2   Preliminaries

We denote the non-negative integers by $\omega = \{0, 1, 2, 3, ...\}$. For a finite alphabet Σ, an infinite *word* α is an infinite sequence $\alpha_0 \alpha_1 \alpha_2 \cdots$ of letters from Σ. We sometimes interpret ω-words as functions $\alpha : i \mapsto \alpha_i$, and use Σ^ω to denote the ω-words over Σ.

ω-automata are finite automata that are interpreted over infinite words and recognise ω-regular languages $L \subseteq \Sigma^\omega$. Nondeterministic parity automata are quintuples $\mathcal{P} = (Q, \Sigma, I, T, \mathsf{pri} : T \to \Pi)$, where Q is a finite set of states with a non-empty subset $I \subseteq Q$ of initial states, Σ is a finite alphabet, $T \subseteq Q \times \Sigma \times Q$ is a transition relation that maps states and input letters to sets of successor states, and pri is a priority function that maps transitions to a finite set $\Pi \subset \omega$.

A *run* ρ of a nondeterministic parity automaton $\mathcal{P}$ on an input word α is an infinite sequence $\rho : \omega \to Q$ of states of $\mathcal{P}$, also denoted $\rho = q_0 q_1 q_2 \cdots \in Q^\omega$, such that the first symbol of ρ is an initial state $q_0 \in I$ and, for all $i \in \omega$,

$(q_i, \alpha_i, q_{i+1}) \in T$ is a valid transition. For a run ρ on a word α, we denote with $\overline{\rho} : i \mapsto (\rho(i), \alpha(i), \rho(i+1))$ the transitions of ρ. Let $\mathsf{infin}(\rho) = \{q \in Q \mid \forall i \in \omega \; \exists j > i \text{ such that } \rho(j) = q\}$ denote the set of all states that occur infinitely often during the run ρ. Likewise, let $\mathsf{infin}(\overline{\rho}) = \{t \in T \mid \forall i \in \omega \; \exists j > i \text{ such that } \overline{\rho}(j) = t\}$ denote the set of all transitions that are taken infinitely many times in $\overline{\rho}$. Acceptance of a run is defined through the priority function pri. A run ρ of a parity automaton is *accepting* if $\limsup_{n \to \infty} \mathsf{pri}(\overline{\rho}(n))$ is even, that is, if the highest priority that occurs infinitely often in the transitions of ρ is even. A word α is accepted by a parity automaton $\mathcal{P}$ iff it has an accepting run, and its language $\mathcal{L}(\mathcal{P})$ is the set of words it accepts.

Parity automata with $\Pi \subseteq \{1, 2\}$ are called *Büchi* automata. Büchi automata are denoted $\mathcal{B} = (Q, \Sigma, I, T, F)$, where $F \subseteq T$ are called the final or accepting transitions. A run is accepting if it contains infinitely many accepting transitions.

We assume w.l.o.g. that the set Π of priorities satisfies that $\min \Pi \in \{0, 1\}$. If this is not the case, we can simply change pri accordingly to $\mathsf{pri}' : t \mapsto \mathsf{pri}(t) - 2$ several times until this constraint is satisfied. We likewise assume that Π has no holes, that is, $\Pi = \{i \in \omega \mid \max \Pi \geq i \geq \min \Pi\}$. If there is a hole $h \notin \Pi$ with $\max \Pi > h > \min \Pi$, we can change pri to $\mathsf{pri}' : t \mapsto \mathsf{pri}(t)$ if $\mathsf{pri}(t) < h$ and $\mathsf{pri}' : t \mapsto \mathsf{pri}(t) - 2$ if $\mathsf{pri}(t) > h$. These changes do not affect the acceptance of any run, and applying finitely many of these changes brings Π into this form.

The different priorities have a natural order $\succcurlyeq$, where $i \succ j$ if i is even and j is odd; i is even and $i > j$; or j is odd and $i < j$. For a non-empty set $\Pi' \subseteq \Pi$ of priorities, $\mathsf{opt}\Pi' = \{i \in \Pi' \mid \forall j \in \Pi'. \; i \succcurlyeq j\}$ is the optimal priority for acceptance.

The complexity of a parity automaton $\mathcal{P} = (Q, \Sigma, I, T, \mathsf{pri} : T \to \Pi)$ is measured by its size $n = |Q|$ and its set of priorities Π. For a given size n and set of priorities Π, there is an automaton that recognises a hardest language. This automaton is referred to as the full automaton $\mathcal{P}_n^\Pi = (Q, \Sigma, I, T, \mathsf{pri} : T \to \Pi)$, with $|Q| = n$, $I = Q$, $\Sigma = Q \times Q \to 2^\Pi$, $T = \{q, \sigma, q') \mid \sigma(q, q') \neq \emptyset\}$, and $\mathsf{pri}(q, \sigma, q') = \mathsf{opt}\sigma(q, q')$. Note that partial functions from $Q \times Q$ to Π would work as well as the alphabet. The larger alphabet is chosen for technical convenience in the proofs. Any other language recognised by a nondeterministic parity automaton $\mathcal{P}$ with n states and priorities Π can essentially be obtained by a language restriction via alphabet restriction from $\mathcal{P}_n^\Pi$.

3   Complementing Parity Automata

The construction described in this section draws from two main sources of inspiration. One source is the introduction of efficient techniques for the determinisation of parity automata in [22]. The nested history trees used there have been our inspiration for the *flattened nested history trees* that form the core data structure in the complementation from Subsection 3.2 and are the backbone of the lower bound proof from Subsection 3.4. The second source of inspiration is the connection [21] between the efficient determinisation based on history trees [18] for Büchi automata and generalised Büchi automata [21] and their level ranking based complementation [7,6,17,21].

The intuition for the complementation is to use the nondeterministic power of a Büchi automaton to reduce the size of the data stored for determinisation. As usual, this nondeterministic power is intuitively used to guess a point in time, where all nodes of the nested history trees from parity determinisation [22], which are eventually always stable, are henceforth stable. Alongside, the set of stable nodes can be guessed. Like in the construction for Büchi automata, the structure can then be flattened, preserving the 'nicking order', the order in which older nodes and descendants take preference in taking states of the nondeterministic parity automaton that is determinised. The complement automaton runs in two phases: a first phase before this guessed point in time, and a second phase after this point, where the run starts in such a flattened tree.

We first introduce *flattened nested history trees* as our main data structure. While we take inspiration from nested history trees [22], the construction is self-contained. In the second subsection, we show that Büchi automata recognising the complement language of the full nondeterministic parity automaton $\mathcal{P}_n^{\Pi}$ need to be large by showing disjointness properties of accepting runs for a large class of words, one for each full flattened nested history tree introduced in Subsection 3.1. This language is also instructive in how the data structure is exploited.

We extend our data structure by markers, resulting in *marked flattened trees*, which are then used as the main part of the state space of the natural complementation construction introduced in Subsection 3.2. We show correctness of our complementation construction in Subsection 3.3 and tightness up to an $O(n)$ factor in Subsection 3.4. Note that all our constructions assume max $\Pi \geq 2$, and therefore do not cover the less expressive CoBüchi automata.

3.1 Flattened Nested History Trees and Marked Flattened Trees

Flattened nested history trees (FNHTs) are the main data structure used in our complementation algorithm. For a given parity automaton $\mathcal{P} = (Q, \Sigma, I, T, \text{pri} : T \to \Pi)$, an FNHT over the set Q of states, maximal priority $\pi_m = \max \Pi$ and maximal even priority $\pi_e = \text{opt}\Pi$, is a tuple $(\mathcal{T}, l_s : \mathcal{T} \to 2^Q, l_l : \mathcal{T} \to 2\mathbb{N}, l_p : \mathcal{T} \to 2^Q, l_r : \mathcal{T} \to 2^Q)$, where $\mathcal{T}$ (an ordered, labelled tree) is a non-empty, finite, and prefix closed subset of finite sequences of natural numbers and a special symbol $\mathfrak{s}$ (for *stepchild*), $\omega \cup \{\mathfrak{s}\}$, that satisfies the constraints given below. We call a node $v\mathfrak{s} \in \mathcal{T}$ a *stepchild* of v, and refer to all other nodes vc with $c \in \omega$ as the *natural children* of v. $\text{nc}(v) = \{vc \mid c \in \omega \text{ and } vc \in \mathcal{T}\}$ is the set of natural children of v. The root is a stepchild.

The **constraints** an FNHT quintuple has to satisfy are as follows:

1. Stepchildren have only natural children and natural children have only stepchildren.
2. Only natural children and, when the highest priority π is odd, the root may be leafs.
3. $\mathcal{T}$ is order closed: for all $c, c' \in \omega$ with $c < c'$, $vc' \in \mathcal{T}$ implies $vc \in \mathcal{T}$.
4. For all $v \in \mathcal{T}$, $l_s(v) \neq \emptyset$.
5. If v is a stepchild, then $l_p(v) = \emptyset$.

6. If v is a stepchild, then $l_s(v) = l_r(v) \cup \bigcup_{v' \in nc(v)} l_s(v')$.
   The sets $l_s(v')$ and $l_s(v'')$ are disjoint for all $v', v'' \in nc(v)$ with $v' \neq v''$, and $l_r(v)$ is disjoint with $\bigcup_{v' \in nc(v)} l_s(v')$.
7. If v is a natural child, then $l_p(v) \neq \emptyset$, $l_s(v) = l_p(v) \cup l_r(v)$, and $l_p(v) \cap l_r(v) = \emptyset$.
8. If a natural child v is not a leaf, then $l_s(v\mathbf{s}) = l_p(v)$.
9. $l_l(\varepsilon) = \pi_e$ and, for all $v \in \mathcal{T}$, $l_l(v) \geq 2$.
10. If $v\mathbf{s} \in \mathcal{T}$, then $l_l(v\mathbf{s}) = l_l(v) - 2$, and if $vc \in \mathcal{T}$ for $c \in \omega$, then $l_l(vc) = l_l(v)$.

The elements in $l_s(v)$ are called the states, $l_p(v)$ the pure states, and $l_r(v)$ the recurrent states of a node v, and $l_l(v)$ is called its level. Note that the level follows a simple pattern: the root is labelled with the maximal even priority, $l_l(\varepsilon) = \pi_e$, the level of natural children is the same as the level of their parents, and the level of a stepchild $v\mathbf{s}$ of a node v is two less than the level of v. For a given maximal even priority π_e, the level is therefore redundant information that can be reconstructed from the node and π_i. For a given set Q and maximal priority π, $\mathsf{fnht}(Q, \pi)$ denotes the flattened nested history trees over Q. An FNHT is called *full* if the states $l_s(\varepsilon) = Q$ of the root is the full set Q.

To include an acceptance mechanism, we enrich FNHTs to *marked flattened tress* (MFTs) that additionally contain a *marker* v_m and a marking set Q_m, s.t.
- either $v_m = (\bar{v}, r)$ with $\bar{v} \in \mathcal{T}$ is used to mark that we follow a breakpoint construction on the recurrent states, in this case $l_r(\bar{v}) \supseteq Q_m \neq \emptyset$,
- or $v_m = (\bar{v}, p)$ with $\bar{v}$ is a leaf in $\mathcal{T}$ is used to mark that we follow a breakpoint construction on the pure states of a leaf $\bar{v}$, in this case $l_p(\bar{v}) \supseteq Q_m \neq \emptyset$.

The marker is used to mark a property to be checked. For markers $v_m = (\bar{v}, r)$, the property is that a particular node would not spawn stable children in a nested history tree [22]. As usual in Safra like constructions, this is checked with a breakpoint, where a breakpoint is reached when all children of a node spawned prior to the last breakpoint die. For markers $v_m = (\bar{v}, p)$, the property is that all runs that are henceforth trapped in the pure nodes of v must eventually encounter a priority $l_l(v) - 1$. This priority is then dominating, and implies rejection as an odd priority. We check these properties round robin for all nodes in $\mathcal{T}$, skipping over nodes, where the respective sets $l_r(\bar{v})$ or $l_p(\bar{v})$ are empty, as the breakpoint there is trivially reached immediately.

For a given FNHT $(\mathcal{T}, l_s, l_l, l_p, l_r)$, $\mathsf{next}(v_m)$ is a mapping from a marker v_m to a marker/marking set pair $(\bar{v}, r), l_r(\bar{v})$ or $(\bar{v}, p), l_p(\bar{v})$. The new marker is the first marker after v_m in some round robin order such that the set $l_r(\bar{v})$ or $l_p(\bar{v})$, resp., is non-empty.

If $(\mathcal{T}, l_s, l_l, l_p, l_r)$ is an FNHT and v_m and Q_m satisfy the constraints for markers and marking sets from above, then $(\mathcal{T}, l_s, l_l, l_p, l_r; v_m, Q_m)$ is a marked flattened tree. For a given set Q and priorities Π with maximal priority $\pi = \max \Pi$, $\mathsf{mft}(Q, \pi)$ denotes the marked flattened trees over Q. A marking is called *full* if either $v_m = (\bar{v}, r)$ and $Q_m = l_r(\bar{v})$, or $v_m = (\bar{v}, p)$ and $Q_m = l_p(\bar{v})$.

3.2 Construction

For a given nondeterministic parity automaton $\mathcal{P} = (Q, \Sigma, I, T, \mathsf{pri}: T \to \Pi)$ with maximal even priority $\pi_e > 1$, we construct a nondeterministic Büchi automaton

$\mathcal{C} = (Q', \Sigma, \{I\}, T', F)$ that recognises the complement language of $\mathcal{P}$ as follows. First we set $Q' = Q_1 \cup Q_2$ with $Q_1 = 2^Q$ and $Q_2 = \mathsf{mft}(Q, \pi)$, and $T' = T_1 \cup T_t \cup T_2$, where

- $T_1 \subseteq Q_1 \times \Sigma \times Q_1$ are transitions in an initial part Q_1 of the states of $\mathcal{C}$,
- $T_t \subseteq Q_1 \times \Sigma \times Q_2$ are transfer transitions that can be taken only once in a run, and
- $T_2 \subseteq Q_2 \times \Sigma \times Q_2$, are transitions in a final part Q_2 of the states of $\mathcal{C}$,

where T_1 and T_2 are deterministic. We first define a transition function δ for the subset construction and functions δ_i for all priorities $i \in \Pi$, and then the sets T_1, T_t, and T_2.

- $\delta : (S, \sigma) \mapsto \{q \in Q \mid \exists s \in S.\ (s, \sigma, q) \in T\}$,
- $\delta_i : (S, \sigma) \mapsto \{q \in Q \mid \exists s \in S.\ (s, \sigma, q) \in T \text{ and } \mathsf{pri}\big((s, \sigma, q)\big) \succcurlyeq i\}$,
- $T_1 = \{(S, \sigma, S') \in Q_1 \times \Sigma \times Q_1 \mid S' = \delta(S, \sigma)\}$,
  where only transitions $(\emptyset, \sigma, \emptyset)$ are accepting.
- $T_t = \{(S, \sigma, (\mathcal{T}, l_s, l_l, l_p, l_r; v_m, Q_m)) \in Q_1 \times \Sigma \times Q_2 \mid l_s(\varepsilon) = \delta(S, \sigma)\}$ and $(\mathcal{T}, l_s, l_l, l_p, l_r; v_m, Q_m)$ is a marked flattened tree.
- $T_2 = \{((\mathcal{T}, l_s, l_l, l_p, l_r; v_m, Q_m), \sigma, s) \in Q_2 \times \Sigma \times Q_2 \mid$
  - if v is a stepchild, then $l''_s(v) = \delta_{l_l(v)+1}\big(l_s(v), \sigma\big)$
  - if v is a natural child, then $l''_s(v) = \delta_{l_l(v)-1}\big(l_s(v), \sigma\big)$
  - if v is a natural child, then $l''_r(v) = \delta_{l_l(v)-1}\big(l_r(v), \sigma\big) \cup \delta_{l_l(v)}\big(l_s(v), \sigma\big)$,
  - starting at the root, we then define inductively:
    * $l'_s(\varepsilon) = l''_s(\varepsilon)$,
    * if vc is a natural child, then $l'_s(vc) = \big(l''_s(vc) \cap l'_s(v)\big) \setminus \bigcup_{c' < c} l''_s(vc')$, $l'_r(vc) = l''_r(vc) \cap l'_s(vc)$, and $l'_p(vc) = l'_s(vc) \setminus l'_r(vc)$, and
    * if vs is a stepchild, then $l'_s(vs) = l'_p(v)$.
  - if there exists one, we extend the functions to obtain the unique FNHT $(\mathcal{T}, l'_s, l_l, l'_p, l'_r)$ (otherwise $\mathcal{C}$ blocks)
  - if $v_m = (\overline{v}, r)$ then $Q'_m = \delta_{l_l(\overline{v})-1}(Q_m, \sigma) \cap l'_r(\overline{v})$, and
    if $v_m = (\overline{v}, p)$ then $Q'_m = \delta_{l_l(\overline{v})-3}(Q_m, \sigma) \cap l'_p(\overline{v})$,
  - if $Q'_m = \emptyset$, then the transition is accepting and $s = \big(\mathcal{T}, l'_s, l_l, l'_p, l'_r; \mathsf{next}(v_m)\big)$,
  - if $Q'_m \neq \emptyset$, then the transition is not accepting and we have that $s = (\mathcal{T}, l'_s, l_l, l'_p, l'_r; v_m, Q'_m)$.

3.3   Correctness

To show that $\mathcal{L}(\mathcal{C})$ is the complement of $\mathcal{L}(\mathcal{P})$, we first show that a word accepted by $\mathcal{C}$ is rejected by $\mathcal{P}$ and then that a word accepted by $\mathcal{P}$ is rejected by $\mathcal{C}$.

Lemma 1. *If $\mathcal{C}$ has an accepting run on α, then $\mathcal{P}$ rejects α.*

Proof. Let $\rho = S_0 S_1 \ldots$ be an accepting run of $\mathcal{C}$ on α that stays in Q_1. Thus, there is an $i \in \omega$ such that, for all $j \geq i$, $S_j = \emptyset$. But if we consider any run $\rho' = q_0 q_1 q_2 \ldots$ of $\mathcal{P}$ on α, then it is easy to show by induction that $q_k \in S_k$ holds for all $k \in \omega$, which contradicts $S_i = \emptyset$; that is, in this case $\mathcal{P}$ has no run on α.

Let us now assume that $\rho = S_0 S_1 \ldots S_i s_{i+1} s_{i+2} \ldots$ is an accepting run of $\mathcal{C}$ on α, where $(S_i, \alpha_i, s_{i+1}) \in T_t$ is the transfer transition taken. (Recall that runs of $\mathcal{C}$ must either stay in Q_1 or contain exactly one transfer transition.)

Let us assume for contradiction that $\mathcal{P}$ has an accepting run $\rho' = q_0 q_1 q_2 \ldots$ with even dominating priority $e = \limsup_{j \to \infty} \mathrm{pri}((q_j, \alpha_j, q_{j+1}))$. Let, for all $j > i$, $s_j = (\mathcal{T}, l_s^j, l_l, l_p^j, l_r^j; v_m^j, Q_m^j)$ and $S_j = l_s^j(\varepsilon)$. It is again easy to show by induction that $q_j \in S_j$ for all $j \in \omega$. Let now $v_j \in \mathcal{T}$ be the longest node with $l_l^j(v_j) \geq e$ and $q_j \in l_s^j(v_j)$. Note that such a node exists, as $q_j \in S_j = l_s^j(\varepsilon)$ holds. We now distinguish the two cases that the v_j do and do not stabilise eventually.

First case. Assume that there are an $i' > i$ and a $v \in \mathcal{T}$ such that, for all $j \geq i'$, $v_j = v$. We choose i' big enough that $\mathrm{pri}(q_{j-1}, \alpha_{j-1}, q_j) \succ e + 1$ holds for all $j \geq i'$.

If v **is a stepchild**, then $q_j \in l_r^j(v)$ for all $j \geq i'$. Using the assumption that ρ is accepting, there is an $i'' > i'$ such that $(s_{i''-1}, \alpha_{i''-1}, s_{i''})$ is accepting, and $v_m^{i''} = (v, r)$. (Note that $q_{i''} \in l_r^{i''}(v)$ implies $l_r^{i''}(v) \neq \emptyset$.) But then we have $q_{i''} \in Q_m^{i''} = l_r^{i''}(v)$, and an inductive argument provides $(s_j, \alpha_j, s_{j+1}) \notin F$ and $q_j \in Q_m^j$ for all $j \geq i''$. This contradicts that ρ is accepting.

If v **is a natural child**, then we distinguish three cases. The first one is that there is a $j' \geq i'$ such that $q_{j'} \in l_r^{j'}(v)$. Then we can show by induction that $q_j \in l_r^j(v)$ for all $j \geq j'$ and follow the same argument as for stepchildren, using $i'' > j'$.

The second is that $q_j \in l_p^j(v)$ holds for all $j \geq i'$. There are now again a few sub-cases that each lead to contradiction. The first is that $l_l(v) = e$. But in this case, we can choose a $j > i'$ with $\mathrm{pri}((q_j, \alpha_j, q_{j+1})) = e$ and get $q_{j+1} \in l_r^{j+1}(v)$ (contradiction). The second is that $l_l(v) > e$ and v is not a leaf. But in that case, $l_l(vs) \geq e$ holds and $q_j \in l_p^j(v)$ implies $q_j \in l_p^j(vs)$, which contradicts the maximality of v. Finally, if $l_l(v) > e$ and v is a leaf of $\mathcal{T}$, we get a similar argument as for stepchildren: Using the assumption that ρ is accepting, there is an $i'' > i'$ such that $(s_{i''-1}, \alpha_{i''-1}, s_{i''})$ is accepting, and $v_m^{i''} = (v, p)$. (Note that $q_{i''} \in l_p^{i''}(v)$ implies $l_p^{i''}(v) \neq \emptyset$.) But then we have $q_{i''} \in Q_m^{i''} = l_p^{i''}(v)$, and an inductive argument provides $(s_j, \alpha_j, s_{j+1}) \notin F$ and $q_j \in Q_m^j$ for all $j \geq i''$. This contradicts that ρ is accepting.

Second case. Assume that the v_j do not stabilise. Let v be the longest sequence such that v is an initial sequence of almost all v_j, and let $i' > i$ be an index such that v is an initial sequence of v_j for all $j \geq i'$. Note that q_j is in $l_s(v_j')$ for all ancestors v_j' of v_j.

First, we assume for contradiction that there exists some $j > i'$ with $\mathrm{pri}((q_j, \alpha_j, q_{j+1})) = e' \succ l_l(v)$ (note that the 'better than' relation implies that $e' > l_l(v)$ is even). Then we select a maximal ancestor v' of v with $l_l(v') = e'$; note that such an ancestor is a natural child, as a stepchild has only natural children, and all of them have the same level.

As v' is an ancestor of v_j and v_{j+1}, $q_j \in l_s^j(v')$ and $q_{j+1} \in l_s^{j+1}(v')$ hold, and by the transition rules thus imply $q_{j+1} \in l_r^{j+1}(v')$, which contradicts $q_{j+1} \in l_s^{j+1}(v_{j+1})$. (Note that $l_l(v') > l_l(v) \geq l_l(v_{j+1})$ holds.)

Second, we show that $\mathrm{pri}\big((q_j, \alpha_j, q_{j+1})\big) \preceq l_l(v)+1$ holds infinitely many times. For this, we first note that the non-stability of the sequence of v_j-s implies that at least one of the following three events happen for infinitely many $j > i'$.

1. v is a stepchild, $q_j \in l_s^j(vc)$ for some child vc of v, but, for all children vc' of v, $q_{j+1} \notin l_s^{j+1}(vc')$,
2. v is a stepchild, $q_j \in l_s^j(vc)$ for some child vc of v, and $q_{j+1} \in l_s^{j+1}(vc')$ for some older sibling vc' of vc, that is, for $c' > c$, or
3. v is a natural child, $q_j \notin l_s^j(v\mathfrak{s})$, but $q_{j+1} \in l_s^{j+1}(v\mathfrak{s})$.

This is just the counter position to "v_j stabilises or v is not maximal". In all three cases, the definition of T_2 requires that $\mathrm{pri}\big((q_j, \alpha_j, q_{j+1})\big) \preceq l_l(v) + 1$.

As the first observation implies that there may only be finitely many transitions with even priority $> l_l(v)$ and the second observation implies that there are infinitely many transitions in ρ' with odd priority $> l_l(v)$, they together imply that $\limsup_{j \to \infty} \mathrm{pri}\big((q_j, \alpha_j, q_{j+1})\big)$ is odd, which leads to a contradiction. □

Lemma 2. *If $\mathcal{P}$ has an accepting run on α, then $\mathcal{C}$ rejects α.* □

We prove this with a contradiction. We consider the runs of $\mathcal{C}$ in two phases. We assume for contradiction that $\mathcal{C}$ has an accepting run $\rho' = S_0 S_1 \ldots$ before the transfer transition is taken. It is easy to show by induction that no transition of (S_i, α_i, S_{i+1}) is accepting. We then guess a point in time where the transfer transition is taken and we assume for contradiction that $\mathcal{C}$ has an accepting run $\rho' = S_0 S_1 \ldots S_i s_{i+1} s_{i+2} \ldots$, where $(S_i, \alpha_i, s_{i+1}) \in T_t$ is the transfer transition taken. It is similarly easy to now show by induction that for all $j \in \omega$, $q_j \in S_j$ holds. We then choose an $i_\varepsilon > i$ such that, for all $k \geq i_\varepsilon$, $\mathrm{pri}\big((q_{k-1}, \alpha_{k-1}, q_k)\big) \leq e$ holds.

We now look at the nodes $v \in \mathcal{T}$, such that $q_j \in l_s^j(v)$, where $j \geq i_\varepsilon$. We exploit an inductive argument based on our construction for two possible cases and we reach a point in the run which is stable but not accepting. This leads to a contradiction.

Corollary 1. *$\mathcal{C}$ recognises the complement language of $\mathcal{P}$.* □

3.4 Lower Bound and Tightness

In order to establish a lower bound, we use a sub-language of the full automaton $\mathcal{P}_n^{\Pi}$, and show that an automaton that recognises it must have at least as many states as there are full FNHTs in $\mathrm{fnht}(Q, \pi)$ for $n = |Q|$ and $\pi = \max \Pi$.

To show this, we define two letters for each full FNHT $t = (\mathcal{T}, l_s, l_l, l_p, l_r) \in \mathrm{fnht}(Q, \pi)$. $\beta_t : Q \times Q \to 2^{\Pi}$ is the letter where:

- if v is a stepchild and $p, q \in l_s(v)$, then $l_l(v)+1 \in \beta_t(p, q)$ (provided $l_l(v)+1 \in \Pi$),
- if v is a stepchild, $p \in l_r(v)$, and $q \in l_s(vc)$ for some $c \in \omega$, then $l_l(v) \in \beta_t(p, q)$,
- if v is a stepchild, $c, c' \in \omega$, $c < c'$, $vc' \in \mathcal{T}$, $p \in l_s(vc')$, and $q \in l_s(vc)$, then $l_l(v) \in \beta_t(p, q)$,

- if v is a natural child, $p \in l_p(v)$, and $q \in l_r(v)$ then $l_l(v) \in \beta_t(p,q)$.
- if v is a natural child and $p, q \in l_r(v)$, then $l_l(v) - 1 \in \beta_t(p,q)$, and
- if v is a natural child and $p, q \in l_p(v)$, then $l_l(v) - 1 \in \beta_t(p,q)$.

$\gamma_t : Q \times Q \to 2^\Pi$ is the letter where $i \in \gamma_t(p,q)$ if $i \in \beta_t(p,q)$ and additionally:
- if v is a natural child, $l_l(v) - 2 \in \Pi$, and $p, q \in l_r(v)$, then $l_l(v) - 2 \in \gamma_t(p,q)$,
- if v is a stepchild and $p, q \in l_r(v)$, then $l_l(v) \in \gamma_t(p,q)$, and
- if v is a natural child, $l_l(v) - 2 \in \Pi$, and $p, q \in l_p(v)$, then $l_l(v) - 2 \in \gamma_t(p,q)$.

For a high integer $h > |\text{fnht}(Q, \pi)|$, we now define the ω-word $\alpha^t = (\beta_t \gamma_t^{h-1})^\omega$, which consists of infinitely many sequences of length h that start with a letter β_t and continue with $h - 1$ repetitions of the letter γ_t, for each full FNHT $t \in \text{fnht}(Q, \pi)$.

We first observe that α^t is rejected by $\mathcal{P}_n^\Pi$.

Lemma 3. $\alpha^t \notin \mathcal{L}(\mathcal{P}_n^\Pi)$.

Proof. By Lemma 1, it suffices to show that the complement automaton $\mathcal{C}$ of $\mathcal{P}_n^\Pi$, as defined in Section 3.2 accepts α^t. The language is constructed such that $\mathcal{C}$ has a run $\rho = Q(t; v_m^1, Q_m^1)(t; v_m^2, Q_m^2)(t; v_m^3, Q_m^3) \ldots$, such that the transition $\left((t; v_m^i, Q_m^i), \alpha_t^i, (t; v_m^{i+1}, Q_m^{i+1})\right)$ is accepting for $i > 0$ if $i \mod h = 0$. $\square$

Let $\mathcal{B}$ be some automaton with states S that recognises the complement language of $\mathcal{P}_n^\Pi$. We now fix an accepting run $\rho_t = s_0 s_1 s_2 \ldots$ for each word α^t and define the set A_t of states in an 'accepting cycle' as $A_t = \left\{ s \in S \mid \exists i, j, k \in \omega \text{ with } 1 \leq j < k \leq h \text{ such that } s = s_{ih+j} = s_{ih+k} \right\}$ holds, and define the interesting states $I_t = A_t \cap \text{infin}(\rho_t)$.

Lemma 4. For $t \neq t'$, I_t and $I_{t'}$ are disjoint $(I_t \cap I_{t'} = \emptyset)$. $\square$

The proof idea is to assume that a state $s \in I_t \cap I_{t'}$, and use it to construct a word from α_t and $\alpha_{t'}$ and an accepting run of $\mathcal{B}$ on the resulting word from ρ_t and $\rho_{t'}$, and then show that it is also accepted by $\mathcal{P}_n^\Pi$.

Theorem 1. $\mathcal{B}$ has at least one state for each full FNHTs in $\text{fnht}(Q, \max \Pi)$.

Proof. We prove the claim with a case distinction. The first case is that $I_t \neq \emptyset$ holds for all full FNHT $t \in \text{fnht}(Q, \max \Pi)$. Lemma 4 shows that the sets of interesting states are pairwise disjoint for different trees $t \neq t'$. As none of them is empty, $\mathcal{B}$ has at least as many states as $\text{fnht}(Q, \max \Pi)$ contains full FNHTs.

The second case is there is a full FNHT $t \in \text{fnht}(Q, \max \Pi)$ such that $I_t = \emptyset$. By Lemma 3, each $\rho_t = s_0 s_1 s_2 \ldots$ is an accepting run. Let now $i \in \omega$ be an index, such that, for all $j \geq i$, $s_j \in \text{infin}(\rho_t)$, and $k \geq i$ an integer with $k \mod h = 0$. $I_t = \emptyset$ implies that $s_{k+j} \neq s_{k+j'}$ for all j, j' with $1 \leq j < j' \leq h$. Then $\mathcal{B}$, and even $\text{infin}(\rho_t)$, has at least $h - 1$ different states, and the claim follows with $h > |\text{fnht}(Q, \max \Pi)|$. $\square$

To show tightness, we proceed in three steps. In a first step, we provide an injection from MFTs with non-full marking to MFTs with full marking.

Next, we argue that the majority of FNHTs is full. Taking into account that there are at most $|Q|$ different markers makes it simple to infer that the states of our complementation construction divided by the lower bound from Theorem 1 is in $O(n)$.

Lemma 5. *There is an injection from MFTs with non-full marking to MFTs with full marking in* $\mathsf{mft}(Q, \pi)$. □

In Lemma 5, we have shown that the majority of MFTs have a full marking. Next we show that the majority of FNHTs is full. (Neither mapping is surjective.)

Lemma 6. *There is an injection from non-full to full FNHTs in* $\mathsf{fnht}(Q, \pi)$. □

Theorem 2. *The complementation construction is tight up to a factor of* $4n+1$, *where* $n = |Q|$ *is the number of states of the complemented parity automaton.*

Proof. For the number of MFTs, Lemma 5 shows that they are at most twice the number of MFTs with full marking. Note that the marker (v_m, p) can only refer to leafs where $l_p(v_m)$ is non-empty and markers (v_m, r) can only refer to nodes where $l_r(v_m)$ is non-empty. It is easy to see that all sets described in this way are pairwise disjoint. Consequently, there are at most $|Q|$ such markers. Thus, the number of MFTs with full marking is at most n times the number of FNHTs.

By Lemma 6, the number of FNHTs is in turn at most twice as high as the number of all full FNHTs. Thus we have bounded the number of MFTs by $4n$ times the number of full FNHTs used to estimate the lower bound in Theorem 1, irrespective of the priorities.

What remains is the trivial observation that the second part of the state-space, the subset construction, is dwarfed by the number of MFTs. Consequently, we can estimate the state-space of the complement automaton divided by the lower bound from Theorem 1 by $4n + 1$. □

References

1. Büchi, J.R.: On a decision method in restricted second order arithmetic. In: Proc. of the Int. Congress on Logic, Methodology, and Philosophy of Science 1960, pp. 1–11 (1962)
2. Cai, Y., Zhang, T.: A tight lower bound for streett complementation. In: Proc. of FSTTCS 2011. LIPIcs, vol. 13, pp. 339–350 (2011)
3. Cai, Y., Zhang, T.: Tight upper bounds for streett and parity complementation. In: Proc. of CSL 2011. LIPIcs, vol. 12, pp. 112–128 (2011)
4. Cai, Y., Zhang, T., Luo, H.: An improved lower bound for the complementation of rabin automata. In: Proc. of LICS 2009 (2009)
5. Duret-Lutz, A.: Ltl translation improvements in spot. In: Proc. of VECoS 2011, pp. 72–83. British Computer Society (2011)
6. Friedgut, E., Kupferman, O., Vardi, M.Y.: Büchi complementation made tighter. International Journal of Foundations of Computer Science 17(4), 851–867 (2006)
7. Kupferman, O., Vardi, M.Y.: Weak alternating automata are not that weak. ACM Transactions on Computational Logic 2(2), 408–429 (2001)

8. Kurshan, R.P.: Computer-aided verification of coordinating processes: the automata-theoretic approach. Princeton University Press (1994)
9. Löding, C.: Optimal bounds for transformations of ω-automata. In: Pandu Rangan, C., Raman, V., Sarukkai, S. (eds.) FST TCS 1999. LNCS, vol. 1738, pp. 97–121. Springer, Heidelberg (1999)
10. Michel, M.: Complementation is more difficult with automata on infinite words. Technical report, CNET, Paris (1988) (manuscript)
11. Pécuchet, J.-P.: On the complementation of Büchi automata. TCS 47(3), 95–98 (1986)
12. Piterman, N.: From nondeterministic Büchi and Streett automata to deterministic parity automata. Journal of Logical Methods in Computer Science 3(3:5) (2007)
13. Rabin, M.O., Scott, D.S.: Finite automata and their decision problems. IBM Journal of Research and Development 3, 115–125 (1959)
14. Safra, S.: On the complexity of ω-automata. In: Proc. of FOCS 1988, pp. 319–327 (1988)
15. Safra, S.: Exponential determinization for omega-automata with strong-fairness acceptance condition (extended abstract). In: Proc. of STOC 1992, pp. 275–282 (1992)
16. Sakoda, W.J., Sipser, M.: Non-determinism and the size of two-way automata. In: Proc. of STOC 1978, pp. 274–286. ACM Press (1978)
17. Schewe, S.: Büchi complementation made tight. In: Proc. of STACS 2009. LIPIcs, vol. 3, pp. 661–672 (2009)
18. Schewe, S.: Tighter bounds for the determinisation of Büchi automata. In: de Alfaro, L. (ed.) FOSSACS 2009. LNCS, vol. 5504, pp. 167–181. Springer, Heidelberg (2009)
19. Schewe, S., Finkbeiner, B.: Satisfiability and finite model property for the alternating-time μ-calculus. In: Ésik, Z. (ed.) CSL 2006. LNCS, vol. 4207, pp. 591–605. Springer, Heidelberg (2006)
20. Schewe, S., Finkbeiner, B.: Synthesis of asynchronous systems. In: Puebla, G. (ed.) LOPSTR 2006. LNCS, vol. 4407, pp. 127–142. Springer, Heidelberg (2007)
21. Schewe, S., Varghese, T.: Tight bounds for the determinisation and complementation of generalised Büchi automata. In: Chakraborty, S., Mukund, M. (eds.) ATVA 2012. LNCS, vol. 7561, pp. 42–56. Springer, Heidelberg (2012)
22. Schewe, S., Varghese, T.: Determinising parity automata. In: Csuhaj-Varjú, E., et al. (eds.) MFCS 2014, Part I. LNCS, vol. 8634, pp. 486–498. Springer, Heidelberg (2014)
23. Sistla, A.P., Vardi, M.Y., Wolper, P.: The complementation problem for Büchi automata with applications to temporal logic. Theoretical Computer Science 49(3), 217–239 (1987)
24. Thomas, W.: Complementation of Büchi automata revisited. In: Jewels are Forever, Contributions on Theoretical Computer Science in Honor of Arto Salomaa, pp. 109–122 (1999)
25. Tsai, M.-H., Tsay, Y.-K., Hwang, Y.-S.: GOAL for games, omega-automata, and logics. In: Sharygina, N., Veith, H. (eds.) CAV 2013. LNCS, vol. 8044, pp. 883–889. Springer, Heidelberg (2013)
26. Vardi, M.Y.: The Büchi complementation saga. In: Thomas, W., Weil, P. (eds.) STACS 2007. LNCS, vol. 4393, pp. 12–22. Springer, Heidelberg (2007)
27. Wilke, T.: Alternating tree automata, parity games, and modal μ-calculus. Bulletin of the Belgian Mathematical Society 8(2) (May 2001)
28. Yan, Q.: Lower bounds for complementation of ω-automata via the full automata technique. Journal of Logical Methods in Computer Science 4(1:5) (2008)

On Infinite Words Determined by Indexed Languages*

Tim Smith

Northeastern University
Boston, MA, USA
smithtim@ccs.neu.edu

Abstract. We characterize the infinite words determined by indexed languages. An infinite language L determines an infinite word α if every string in L is a prefix of α. If L is regular or context-free, it is known that α must be ultimately periodic. We show that if L is an indexed language, then α is a morphic word, i.e., α can be generated by iterating a morphism under a coding. Since the other direction, that every morphic word is determined by some indexed language, also holds, this implies that the infinite words determined by indexed languages are exactly the morphic words. To obtain this result, we prove a new pumping lemma for the indexed languages, which may be of independent interest.

1 Introduction

Formal languages and infinite words can be related to each other in various ways. One natural connection is via the notion of a prefix language. A prefix language is a language L such that for all $x, y \in L$, x is a prefix of y or y is a prefix of x. Every infinite prefix language determines an infinite word. Prefix languages were introduced by Book [5] in an attempt to study the complexity of infinite words in terms of acceptance and generation by automata. Book used the pumping lemma for context-free languages to show that every context-free prefix language is regular, implying that any infinite word determined by such a language is ultimately periodic. Recent work has continued and expanded Book's project, classifying the infinite words determined by various classes of automata [19] and parallel rewriting systems [18].

In this paper we characterize the infinite words determined by indexed languages. The indexed languages, introduced in 1968 by Alfred Aho [1], fall between the context-free and context-sensitive languages in the Chomsky hierarchy. More powerful than the former class and more tractable than the latter, the indexed languages have been applied to the study of natural languages [9] in computational linguistics. Indexed languages are generated by indexed grammars, in which nonterminals are augmented with stacks which can be pushed,

---

* Due to space constraints, some proofs are omitted or only sketched. The full version is available at http://arxiv.org/abs/1406.3373.

E. Csuhaj-Varjú et al. (Eds.): MFCS 2014, Part I, LNCS 8634, pp. 511–522, 2014.
© Springer-Verlag Berlin Heidelberg 2014

popped, and copied to other nonterminals as the derivation proceeds. Two automaton characterizations are the nested stack automata of [2], and the order-2 pushdown automata within the Maslov pushdown hierarchy [15].

The class of indexed languages IL includes all of the stack automata classes whose infinite words are characterized in [19], as well as all of the rewriting system classes whose infinite words are characterized in [18]. In particular, IL properly includes ET0L [8], a broad class within the hierarchy of parallel rewriting systems known as L systems. L systems have close connections with a class of infinite words called morphic words, which are generated by repeated application of a morphism to an initial symbol, under a coding [3]. In [18] it is shown that every infinite word determined by an ET0L language is morphic. This raises the question of whether the indexed languages too determine only morphic words, or whether indexed languages can determine infinite words which are not morphic.

To answer this question, we employ a new pumping lemma for IL. In Book's paper, as well as in [18] and [19], pumping lemmas played a prominent role in characterizing the infinite words determined by various language classes. A pumping lemma for a language class C is a powerful tool for proving that certain languages do not belong to C, and thereby for proving that certain infinite words cannot be determined by any language in C. For the indexed languages, a pumping lemma exists due to Hayashi [11], as well as a "shrinking lemma" due to Gilman [10]. We were not successful in using these lemmas to characterize the infinite words determined by IL, so instead have proved a new pumping lemma for this class (Theorem 6), which may be of independent interest.

Our lemma generalizes a pumping lemma recently proved for ET0L languages [16]. Roughly, it states that for any indexed language L, any sufficiently long word $w \in L$ may be written as $u_1 \cdots u_n$, each u_i may be written as $v_{i,1} \cdots v_{i,n_i}$, and the $v_{i,j}$s may be replaced with u_is to obtain new words in L. Using this lemma, we extend to IL a theorem about frequent and rare symbols proved in [16] for ET0L, which can be used to prove that certain languages are not indexed. We also use the lemma to obtain the new result that every infinite indexed language has an infinite subset in a smaller class of L systems called CD0L. This implies that every infinite word determined in IL can also be determined in CD0L, and thus that every such word is morphic. Since every morphic word can be determined by some CD0L language [18], we therefore obtain a complete characterization of the infinite words determined by indexed languages: they are exactly the morphic words.

1.1   Proof Techniques

Our pumping lemma for IL generalizes the one proved in [16] for ET0L. Derivations in an ET0L system, like those in an indexed grammar, can be viewed as having a tree structure, but with certain differences. In ET0L, symbols are rewritten in parallel, and the tree is organized into levels corresponding to the steps of the derivation. Further, each node in the tree has one of a finite set of possible labels, corresponding to the symbols in the ET0L system. The proof in [16] classifies each level of the tree according to the set of symbols which appear at that level, and then finds two levels with the same symbol set, which are used to construct the pumping operation. By contrast, the derivation tree of an indexed grammar is not organized into levels in this way, and there is no bound on the number of possible labels for the nodes, since each nonterminal can have an arbitrarily large stack. We deal with these differences by assigning each node a "type" based on the set of nonterminals which appear among its descendants immediately before its stack is popped. These types then play a role analogous to the symbol sets of [16] in our construction of the pumping operation.

1.2   Related Work

The model used in this paper, in which infinite words are determined by languages of their prefixes, originates in Book's 1977 paper [5]. Book formulated the "prefix property" in order to allow languages to "approximate" infinite sequences, and showed that for certain classes of languages, if a language in the class has the prefix property, then it is regular. A follow-up by Latteux [14] gives a necessary and sufficient condition for a prefix language to be regular. Languages whose complement is a prefix language, called "coprefix languages", have also been studied; see Berstel [4] for a survey of results on infinite words whose coprefix language is context-free. In Smith [18], prefix languages are used to categorize the infinite words determined by a hierarchy of L system classes. In Smith [19], they are used to characterize the infinite words determined by several classes of one-way stack automata, and also studied in connection with multihead deterministic finite automata.

Hayashi's 1973 pumping lemma for indexed languages is proved in a dense thirty-page paper [11]. The main theorem states that if a given terminal derivation tree is big enough, new terminal derivation trees can be generated by the insertion of other trees into the given one. Hayashi applies his theorem to give a new proof that the finiteness problem for indexed languages is solvable and to show that certain languages are not indexed. Gilman's 1996 "shrinking lemma" for indexed languages [10] is intended to be easier to employ, and operates directly on terminal strings rather than on derivation trees. Our lemma generalizes the recent ET0L pumping lemma of Rabkin [16]. Like Gilman's lemma, it is stated in terms of strings rather than derivation trees, making it easier to employ, while like Hayashi's lemma and unlike Gilman's, it provides a pumping operation which yields an infinity of new strings in the language.

Another connection between indexed languages and morphic words comes from Braud and Carayol[6], in which morphic words are related to a class of graphs at level 2 of the pushdown hierarchy. The string languages at this level of the hierarchy are the indexed languages.

1.3   Outline of Paper

The paper is organized as follows. Section 2 gives preliminary definitions and propositions. Section 3 gives our pumping lemma for indexed languages. Section 4 gives applications for the lemma, in particular characterizing the infinite words determined by indexed languages. Section 5 gives our conclusions.

2   Preliminaries

An **alphabet** A is a finite set of symbols. A **word** is a concatenation of symbols from A. We denote the set of finite words by A^* and the set of infinite words by A^ω. A **string** x is an element of A^*. The length of x is denoted by $|x|$. We denote the empty string by λ. For a symbol c, $\#_c(x)$ denotes the number of appearances of c in x, and for an alphabet B, $\#_B(x)$ denotes $\sum_{c \in B} \#_c(x)$. A **language** is a subset of A^*. A (symbolic) **sequence** S is an element of $A^* \cup A^\omega$. A **prefix** of S is a string x such that $S = xS'$ for some sequence S'. A **subword** (or factor) of S is a string x such that $S = wxS'$ for some string w and sequence S'. For $i \geq 1$, $S[i]$ denotes the ith symbol of S. For a string $x \neq \lambda$, x^ω denotes the infinite word $xxx \cdots$. An infinite word of the form xy^ω, where x and y are strings and $y \neq \lambda$, is called **ultimately periodic**.

2.1   Prefix Languages

A **prefix language** is a language L such that for all $x, y \in L$, x is a prefix of y or y is a prefix of x. A language L **determines** an infinite word α iff L is infinite and every $x \in L$ is a prefix of α. For example, the infinite prefix language $\{\lambda$, ab, abab, ababab, $\dots \}$ determines the infinite word $(\mathsf{ab})^\omega$. For a language class C, let $\omega(C) = \{\alpha \mid \alpha$ is an infinite word determined by some $L \in C\}$. The following propositions are basic consequences of the definitions.

Remark 1. A language determines at most one infinite word.

Remark 2. A language L determines an infinite word iff L is an infinite prefix language.

Remark 3. If a language L determines an infinite word α and L' is an infinite subset of L, then L' determines α.

2.2   Morphic Words

A **morphism** on an alphabet A is a map h from A^* to A^* such that for all $x, y \in A^*$, $h(xy) = h(x)h(y)$. Notice that $h(\lambda) = \lambda$. The morphism h is a **coding** if for all $a \in A$, $|h(a)| = 1$. A string $x \in A^*$ is **mortal** (for h) if there is an $m \geq 0$ such that $h^m(x) = \lambda$. The morphism h is **prolongable** on a symbol a if $h(a) = ax$ for some $x \in A^*$, and x is not mortal. If h is prolongable on a, $h^\omega(a)$ denotes the infinite word $a \; x \; h(x) \; h^2(x) \; \cdots$. An infinite word α is **morphic** if there is a morphism h, coding e, and symbol a such that h is prolongable on a and $\alpha = e(h^\omega(a))$. For example, let:

$$h(\mathsf{s}) = \mathsf{sbaa} \quad e(\mathsf{s}) = \mathsf{a}$$
$$h(\mathsf{a}) = \mathsf{aa} \quad\;\; e(\mathsf{a}) = \mathsf{a}$$
$$h(\mathsf{b}) = \mathsf{b} \quad\quad\; e(\mathsf{b}) = \mathsf{b}$$

Then $e(h^\omega(\mathsf{s})) = \mathsf{a}^1\mathsf{ba}^2\mathsf{ba}^4\mathsf{ba}^8\mathsf{ba}^{16}\mathsf{b} \cdots$ is a morphic word. See [3] for more on morphic words. Morphic words have close connections with the parallel rewriting systems known as L systems. Many classes of L systems appear in the literature; here we define only HD0L and CD0L. For more on L systems, including the class ET0L, see [13] and [17]. An **HD0L system** is a tuple $G = (A, h, w, g)$ where A is an alphabet, h and g are morphisms on A, and w is in A^*. The language of G is $L(G) = \{g(h^i(w)) \mid i \geq 0\}$. If g is a coding, G is a **CD0L system**. **HD0L** and **CD0L** are the sets of HD0L and CD0L languages, respectively. From [13] and [8] we have CD0L $\subset$ HD0L $\subset$ ET0L $\subset$ IL. In [18] it is shown that $\omega(\text{CD0L})$ $= \omega(\text{HD0L}) = \omega(\text{ET0L})$, and α is in this class of infinite words iff α is morphic.

2.3   Indexed Languages

The class of indexed languages IL consists of the languages generated by indexed grammars. These grammars extend context-free grammars by giving each non-terminal its own stack of symbols, which can be pushed, popped, and copied to other nonterminals as the derivation proceeds. Indexed grammars come in several forms [1,9,12], all generating the same class of languages, but varying with respect to notation and which productions are allowed. The following definition follows the form of [12].

An **indexed grammar** is a tuple $G = (N, T, F, P, S)$ in which N is the nonterminal alphabet, T is the terminal alphabet, F is the stack alphabet, $S \in N$ is the start symbol, and P is the set of productions of the forms

$$A \to r \qquad A \to Bf \qquad Af \to r$$

with $A, B \in N$, $f \in F$, and $r \in (N \cup T)^*$. In an expression of the form $Af_1 \cdots f_n$ with $A \in N$ and $f_1, \ldots, f_n \in F$, the string $f_1 \cdots f_n$ can be viewed as a stack joined to the nonterminal A, with f_1 denoting the top of the stack and f_n the bottom. For $r \in (N \cup T)^*$ and $x \in F^*$, we write $r\{x\}$ to denote r with every $A \in N$ replaced by Ax. For example, with $A, B \in N$ and $\mathtt{c}, \mathtt{d} \in T$, $\mathtt{cd}AB\{f\} = \mathtt{cd}AfBf$. For $q, r \in (NF^* \cup T)^*$, we write $q \longrightarrow r$ if there are $q_1, q_2 \in (NF^* \cup T)^*$, $A \in N$, $p \in (N \cup T)^*$, and $x, y \in F^*$ such that $q = q_1 \, Ax \, q_2$, $r = q_1 \, p\{y\} \, q_2$, and one of the following is true: (1) $A \to p$ is in P and $y = x$, (2) $A \to pf$ is in P and $y = fx$, or (3) $Af \to p$ is in P and $x = fy$. Let $\overset{*}{\longrightarrow}$ be the reflexive, transitive closure of $\longrightarrow$. For $A \in N$ and $x \in F^*$, let $L(Ax) = \{s \in T^* \mid Ax \overset{*}{\longrightarrow} s\}$. The language of G, denoted $L(G)$, is $L(S)$. The class IL of indexed languages is $\{L(G) \mid G \text{ is an indexed grammar}\}$.

See Example 7 and Figure 1 for a sample indexed grammar and derivation tree. For convenience, we will work with a form of indexed grammar which we call "grounded", in which terminal strings are produced only at the bottom of the stack. G is **grounded** if there is a symbol $\$ \in F$ (called the bottom-of-stack symbol) such that every production has one of the forms

$$S \to A\$ \qquad A \to r \qquad A \to Bf \qquad Af \to r \qquad A\$ \to s$$

with $A, B \in N \setminus S$, $f \in F \setminus \$$, $r \in (N \setminus S)^+$, and $s \in T^*$. It is not difficult to verify the following proposition.

Proposition 4. *For every indexed grammar G, there is a grounded indexed grammar G' such that $L(G') = L(G)$.*

3   Pumping Lemma for Indexed Languages

In this section we present our pumping lemma for indexed languages (Theorem 6) and give an example of its use. Our pumping lemma generalizes the ET0L pumping lemma of [16]. Like that lemma, it allows positions in a word to be designated as "marked", and then provides guarantees about the marked positions during the pumping operation.

Let $G = (N, T, F, P, S)$ be a grounded indexed grammar. To prove Theorem 6, we will first prove a lemma about paths in derivation trees of G. A derivation tree D of a string s has the following structure. Each internal node of D has a label in NF^* (a nonterminal with a stack), and each leaf has a label in T^* (a terminal string). Each internal node has either a single leaf node as a child, or one or more internal children. The root of D is labelled by the start symbol S, and the terminal yield of D is the string s.

If the string s contains marked positions, then we will take D to be marked in the following way. Mark every leaf whose label contains a marked position of s, and then mark every internal node which has a marked descendant. Call any node with more than one marked child a **branch node**.

A **path** H in D is a list of nodes $(v_0, \ldots, v_m)$ with $m \geq 0$ such that for each $1 \leq i \leq m$, v_i is a child of v_{i-1}. For convenience, we will sometimes refer to nodes in H by their indices; e.g. node i in the context of H means v_i. When we say that there is a branch node between i and j we mean that the branch node is between v_i (inclusive) and v_j (exclusive).

We define several operations on internal nodes of D. Each such node v has the label Ax for some $A \in N$ and $x \in F^*$. Let $\sigma(v) = A$ and $\eta(v) = |x|$. $\sigma(v)$ gives the nonterminal symbol of v and $\eta(v)$ gives the height of v's stack. We say that a node v' is in the scope of v iff v' is an internal node and there is a path in D from v to v' such that for every node v'' on the path (including v'), $\eta(v'') \geq \eta(v)$. Let $\beta(v)$ be the set of nodes v' such that v' is in the scope of v but no child of v' is in the scope of v. The set $\beta(v)$ can be viewed as the "last" nodes in the scope of v. Notice that for all $v' \in \beta(v)$, $\eta(v') = \eta(v)$. Finally, we give v a "type" $\tau(v)$ based on which nonterminal symbols appear in $\beta(v)$. Let $\tau(v)$ be a 3-tuple such that:

- $\tau(v)[1] = \{A \in N \mid \text{for all } v' \in \beta(v), \sigma(v') \neq A\}$
- $\tau(v)[2] = \{A \in N \mid \text{for some } v' \in \beta(v), \sigma(v') = A, \text{ and for all marked } v' \in \beta(v), \sigma(v') \neq A\}$
- $\tau(v)[3] = \{A \in N \mid \text{for some marked } v' \in \beta(v), \sigma(v') = A\}$

Notice that for each v, $\tau(v)$ partitions N: every $A \in N$ occurs in exactly one of $\tau(v)[1]$, $\tau(v)[2]$, and $\tau(v)[3]$. So there are $3^{|N|}$ possible values for $\tau(v)$.

Lemma 5. Let $H = (v_0, \ldots, v_m)$ be a path in a derivation tree D from the root to a leaf (excluding the leaf) with more than $(|N| \cdot 3^{|N|})^{|N|^2 \cdot 3^{|N|} + 1}$ branch nodes. Then there are $0 \leq b_1 < t_1 < t_2 \leq b_2 \leq m$ such that

- $\sigma(b_1) = \sigma(t_1)$ and $\sigma(t_2) = \sigma(b_2)$,
- b_2 is in $\beta(b_1)$ and t_2 is in $\beta(t_1)$,

- $\tau(b_1) = \tau(t_1)$, and
- there is a branch node between b_1 and t_1 or between t_2 and b_2.

Proof (Sketch). If H is flat, i.e. if all of the nodes in H have the same stack, then after $|N| \cdot 3^{|N|}$ branch nodes, there will have been two nodes with the same σ and τ, with a branch node between them. Then we can set b_1 and t_1 to these two nodes and set $t_2 = b_2 = m$, since m will be in $\beta(v)$ for every node v on the path, because H is flat. If H is not flat, then consider just the "base" of H, i.e. the nodes in H with the smallest stack. These nodes are separated by "hills" in which the stack is bigger. The base of H can be viewed as a flat path with gaps corresponding to the hills. Then at most $|N| \cdot 3^{|N|}$ of the hills can contain branch nodes. We can then use an inductive argument to bound the number of branch nodes in each hill. In this argument, each hill is itself treated as a path, which is shorter than the original path H and so subject to the induction. Since node 0 and node m in H can serve as a potential b_1 and b_2 for any of the hills, each hill has fewer configurations of σ and τ to "choose from" if it is to avoid containing nodes which could serve as t_1 and t_2. Working out the details of the induction gives the bound stated in the lemma. □

We are now ready to state our pumping lemma for indexed languages, which generalizes the pumping lemma for ET0L languages of [16]. As noted, this lemma allows arbitrary positions in a word to be designated as "marked", and then provides guarantees about the marked positions during the pumping operation. The only difference between our pumping operation and that of Theorem 15 of [16] is that in the latter, there are guaranteed to be at least two marked positions in the $v_{i,j}$ of part 4, whereas in our lemma, this $v_{i,j}$ might not contain any marked positions and could even be an empty string (which nonetheless maps under ϕ to u_i, which does contain a marked position).

Theorem 6. *Let L be an indexed language. Then there is an $l \geq 0$ (which we will call a threshold for L) such that for any $w \in L$ with at least l marked positions,*

1. *w can be written as $w = u_1 u_2 \cdots u_n$ and each u_i can be written $u_i = v_{i,1} v_{i,2} \cdots v_{i,n_i}$ (we will denote the set of subscripts of v, i.e. $\{(i,j) \mid 1 \leq i \leq n \text{ and } 1 \leq j \leq n_i\}$, by I);*
2. *there is a map $\phi : I \to \{1, \ldots, n\}$ such that if each $v_{i,j}$ is replaced with $u_{\phi(i,j)}$, then the resulting word is still in L, and this process can be applied iteratively to always yield a word in L;*
3. *if $v_{i,j}$ contains a marked position then so does $u_{\phi(i,j)}$;*
4. *there is an $(\mathbf{i}, \mathbf{j}) \in I$ such that $\phi(\mathbf{i}, \mathbf{j}) = \mathbf{i}$, and there is at least one marked position in $u_\mathbf{i}$ but outside of $v_{\mathbf{i},\mathbf{j}}$.*

Proof (Sketch). We take a grounded indexed grammar G with language L and set the threshold l using the bound from Lemma 5 together with some properties of the productions of G. Then we take any $w \in L$ with at least l marked positions and take a derivation tree D for w. Some path in D from the root to a leaf then has enough branch nodes to give us the b_1, t_1, t_2, and b_2 from Lemma 5. We then need to construct the map ϕ and the factors u_i and $v_{i,j}$. To do this, we use the nodes in $\beta(b_1)$ and $\beta(t_1)$. The nodes in $\beta(b_1)$ will correspond to $v_{i,j}$s and those in $\beta(t_1)$ will correspond to u_is. The operation ϕ will then map each node in $\beta(b_1)$ to a node in $\beta(t_1)$ with the same σ, and which is marked if the node being mapped is marked. This is possible because $\tau(b_1) = \tau(t_1)$. The justification for this construction is that in D, between b_1 and t_1 the stack grows from x to yx for some $x, y \in F^*$, and then shrinks back to x between $\beta(t_1)$ and $\beta(b_1)$. Since $\sigma(b_1) = \sigma(t_1)$, the steps between b_1 and t_1 can be repeated, growing the stack from x to yx to yyx to $yyyx$, and so on. Then the ys can be popped back off by repeating the steps between the nodes in $\beta(t_1)$ and $\beta(b_1)$. This construction gives us parts 1, 2, and 3 of the theorem. Part 4 follows from the fact that there is a branch node between b_1 and t_1 or between t_2 and b_2. In the former case, the u_i in part 4 corresponds to the yield produced between b_1 and t_1 involving the branch node, and the $v_{i,j}$ is specially constructed as an empty factor which maps to u_i. In the latter case, since t_2 is in $\beta(t_1)$, b_2 is in $\beta(b_1)$, and $\sigma(t_2) = \sigma(b_2)$, the u_i in part 4 corresponds to t_2 and the $v_{i,j}$ corresponds to b_2. □

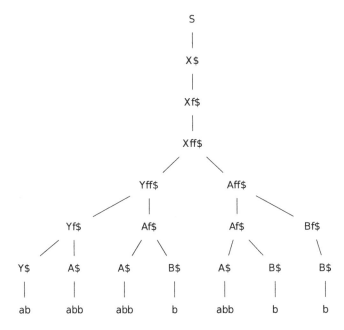

Fig. 1. A derivation tree for the string $ab^1ab^2ab^3ab^4$

Example 7. We now give an example of how our pumping operation works on a derivation tree of an indexed grammar. Let the nonterminal alphabet N be $\{S, X, Y, A, B\}$, the terminal alphabet T be $\{a, b\}$, the stack alphabet F be $\{f\}$, and the set of productions P be $\{S \rightarrow X\$, X \rightarrow Xf, X \rightarrow YA, Yf \rightarrow YA, Y\$ \rightarrow ab, Af \rightarrow AB, A\$ \rightarrow abb, Bf \rightarrow B, B\$ \rightarrow b\}$. Let G be the indexed grammar (N, T, F, P, S). Notice that G is grounded and that $L(G)$ determines the infinite word $ab^1ab^2ab^3\cdots$. Figure 1 depicts a derivation tree of G with terminal yield $ab^1ab^2ab^3ab^4$. Notice how stacks are copied as the derivation proceeds; for example, the production $X \to YA$ applied to $Xff\$$ copies X's stack $ff\$$ to both Y and A, yielding $Yff\$$ $Aff\$$.

Take every position in the terminal yield of the tree to be marked. Let b_1 and t_1 be the nodes labelled $X\$$ and $Xf\$$, respectively, and let t_2 and b_2 be the nodes labelled $Af\$$ and $A\$$, respectively, in the $Yff\$$ subtree. We have $\sigma(b_1) = \sigma(t_1) = X$, $\sigma(t_2) = \sigma(b_2) = A$, and $\tau(b_1) = \tau(t_1) = [\{S, X\}, \{\}, \{Y, A, B\}]$. Additionally, b_2 is in $\beta(b_1)$, which consists of all the nodes labelled $A\$$, $B\$$, or $Y\$$, and t_2 is in $\beta(t_1)$, which consists of all the nodes labelled $Af\$$, $Bf\$$, or $Yf\$$. Also, there is a branch node between t_2 and b_2, namely t_2 itself. This satisfies the conditions of Lemma 5. We break up the string as $\overline{ab}\,\overline{abb}\,\overline{abbb}\,\overline{abbb}\,\overline{b}$, where the outer brackets delimit the u_is and the inner brackets delimit the $v_{i,j}$s. Set $\phi(1,1) = 1$, $\phi(1,2) = 2$, $\phi(2,1) = 2$, $\phi(2,2) = 4$, $\phi(3,1) = 2$, $\phi(3,2) = 4$, and $\phi(4,1) = 4$. Applying the pumping operation yields $\overline{ab}\,\overline{abb}\,\overline{abbb}\,\overline{abbb}\,\overline{b}\,\overline{abbb}\,\overline{b}\,\overline{b} = ab^1ab^2ab^3ab^4ab^5$, which is indeed in $L(G)$. Applying it again yields $ab^1ab^2ab^3ab^4ab^5ab^6$, and so on.

We now follow [16] in giving a more formal description of the replacement operation in part 2 of Theorem 6. This operation produces the words $w^{(t)}$ for all $t \geq 0$, where

$$v_{i,j}^{(0)} = v_{i,j}$$
$$u_i^{(t)} = v_{i,1}^{(t)} v_{i,2}^{(t)} \cdots v_{i,n_i}^{(t)}$$
$$v_{i,j}^{(t+1)} = u_{\phi(i,j)}^{(t)}$$
$$w^{(t)} = u_1^{(t)} u_2^{(t)} \cdots u_n^{(t)}$$

Notice that $w^{(0)} = w$. The following lemma states that the number of marked symbols tends to infinity as the replacement operation is repeatedly applied.

Lemma 8. *If L is an indexed language with threshold l, and $w \in L$ has at least l marked symbols, then for all $t \geq 0$, $w^{(t)}$ has at least t marked symbols.*

Proof. Call each $v_{i,j}$ a v-word and call a v-word marked if it contains a marked position. We will show by induction on t that for all $t \geq 0$, $w^{(t)}$ contains at least t occurrences of marked v-words. Obviously the statement holds for $t = 0$. So say $t \geq 1$ and suppose for induction that $w^{(t-1)}$ contains at least $t - 1$ occurrences of marked v-words. By part 3 of Theorem 6, for every marked $v_{i,j}$ in $w^{(t-1)}$, $v_{i,j}^{(1)}$ contains a marked position. Then $w^{(t)}$ contains at least $t - 1$ occurrences of marked v-words. Now by part 4 of the theorem, there is an $(i, j) \in I$ such that $\phi(i,j) = i$ and there is at least one marked position in u_i but outside of $v_{i,j}$. Then $v_{i,j}^{(1)} = u_i$ contains at least one more occurrence of a marked v-word than $v_{i,j}$. Now since w contains $v_{i,j}$, $w^{(t-1)}$ contains $v_{i,j}$. Then $w^{(t)}$ contains at least one more occurrence of a marked v-word than $w^{(t-1)}$. So $w^{(t)}$ contains at least t occurrences of marked v-words, completing the induction. Thus for all $t \geq 0$, $w^{(t)}$ contains at least t occurrences of marked v-words, hence $w^{(t)}$ contains at least t marked positions.   □

4 Applications

In this section we give some applications of our pumping lemma for indexed languages (Theorem 6). We prove for IL a theorem about frequent and rare symbols which is proved in [16] for ET0L languages. Then we characterize the infinite words determined by indexed languages.

4.1 Frequent and Rare Symbols

Let L be a language over an alphabet A, and $B \subseteq A$. B is **nonfrequent** if there is a constant c_B such that $\#_B(w) \leq c_B$ for all $w \in L$. Otherwise it is called **frequent**. B is called **rare** if for every $k \geq 1$, there is an $n_k \geq 1$ such that for all $w \in L$, if $\#_B(w) \geq n_k$ then the distance between any two appearances in w of symbols from B is at least k.

Theorem 9. *Let L be an indexed language over an alphabet A, and $B \subseteq A$. If B is rare in L, then B is nonfrequent in L.*

Proof. Suppose B is rare and frequent in L. By Theorem 6, L has a threshold $l \geq 0$. Since B is frequent in L, there is a $w \in L$ with more than l symbols from B. If we mark them all, parts 1 to 4 of the theorem apply. By part 3 of the theorem, if $v_{i,j}$ contains a marked position then so does $u_{\phi(i,j)}$, and by part 4 of the theorem, there is an $(i,j) \in I$ such that $\phi(i,j) = i$ and there is at least one marked position in u_i but outside of $v_{i,j}$. Then $u_i^{(1)}$ contains at least two marked positions. So take any two marked positions in $u_i^{(1)}$ and let d be the distance between them. Let $k = d + 1$. Since B is rare in L, there is an $n_k \geq 1$ such that for all $w' \in L$, if $\#_B(w') \geq n_k$ then the distance between any two appearances in w' of symbols from B is at least k. By Lemma 8, $w^{(n_k)}$ contains at least n_k marked symbols, so $\#_B(w^{(n_k)}) \geq n_k$. Then the distance between any two appearances in $w^{(n_k)}$ of symbols from B is at least k. Since u_i appears in w and $u_i^{(1)}$ contains u_i, $u_i^{(1)}$ appears in $w^{(t)}$ for all $t \geq 1$. Then $u_i^{(1)}$ appears in $w^{(n_k)}$ and contains two symbols from B separated by $d < k$, a contradiction. So if B is rare in L, then B is nonfrequent in L.   □

Theorem 9 gives us an alternative proof of the result of Hayashi [11] and Gilman [10] that the language L below is not indexed.

Corollary 10 ([11] Theorem 5.3; [10] Corollary 4). *The language $L = \{(\mathsf{ab}^n)^n \mid n \geq 1\}$ is not indexed.*

Proof. The subset $\{\mathsf{a}\}$ of $\{\mathsf{a}, \mathsf{b}\}$ is rare and frequent in L. So by Theorem 9, L is not indexed.   □

4.2 CD0L and Morphic Words

Next, we turn to characterizing the infinite words determined by indexed languages. We show that every infinite indexed language has an infinite CD0L subset, which then implies that $\omega(IL)$ contains exactly the morphic words. This is a new result which we were not able to obtain using the pumping lemmas of [11] or [10].

Theorem 11. *Let L be an infinite indexed language. Then L has an infinite CD0L subset.*

Proof. By Theorem 6, L has a threshold $l \geq 0$. Take any $w \in L$ such that $|w| \geq l$, and mark every position in w. Then parts 1 to 4 of the theorem apply. Now for each $(i, j) \in I$, create a new symbol $x_{i,j}$. Let X be the set of these symbols. For i from 1 to n, let $x_i = x_{i,1}x_{i,2} \cdots x_{i,n_i}$. Let $x = x_1 x_2 \cdots x_n$. Let h be a morphism such that $h(x_{i,j}) = x_{\phi(i,j)}$ for all $(i, j) \in I$. Let g be a morphism such that $g(x_{i,j}) = v_{i,j}$ for all $(i, j) \in I$. Let A be the alphabet of L. Let G be the HD0L system $(X \cup A, h, x, g)$. Then for all $t \geq 0$, $g(h^t(x)) = w^{(t)}$. By Lemma 8, for all $t \geq 0, |w^{(t)}| \geq t$. Then $L(G)$ is an infinite HD0L subset of L. By Theorem 18 of [18], every infinite HD0L language has an infinite CD0L subset. Therefore L has an infinite CD0L subset. □

Theorem 12. $\omega(IL)$ *contains exactly the morphic words.*

Proof. For any infinite word $\alpha \in \omega(IL)$, some $L \in IL$ determines α. Then L is an infinite indexed language, so by Theorem 11, L has an infinite CD0L subset L'. Then L' determines α, so α is in $\omega(CD0L)$. Then by Theorem 23 of [18], α is morphic. For the other direction, by Theorem 23 of [18], every morphic word is in $\omega(CD0L)$, so since CD0L $\subset$ IL, every morphic word is in $\omega(IL)$. □

Theorem 12 lets us use existing results about morphic words to show that certain languages are not indexed, as the following example shows.

Corollary 13. *Let $L = \{$0, 0:1, 0:1:01, 0:1:01:11, 0:1:01:11:001, ... $\}$, the language containing for each $n \geq 0$ a word with the natural numbers up to n written in backwards binary and colon-separated. Then L is not indexed.*

Proof. L determines the infinite word $\alpha = $ 0:1:01:11:001:101:011:111:$\cdots$. By Theorem 3 of [7], α is not morphic. Then by our Theorem 12, α is not in $\omega(IL)$, so no language in IL determines α, hence L is not indexed. □

5 Conclusion

In this paper we have characterized the infinite words determined by indexed languages, showing that they are exactly the morphic words. In doing so, we proved a new pumping lemma for the indexed languages, which may be of independent interest and which we hope will have further applications. One direction for future work is to look for more connections between formal languages and infinite words via the notion of prefix languages. It would be interesting to see what other language classes determine the morphic words, and what language classes are required to determine infinite words that are not morphic. More generally, for any language class, we can ask what class of infinite words it determines, and for any infinite word, we can ask in what language classes it can be determined, yielding many opportunities for future research. It is hoped that work in this area will help to build up a theory of the complexity of infinite words with respect to what language classes can determine them.

Acknowledgments. I want to thank my advisor, Rajmohan Rajaraman, for supporting this work, encouraging me, and offering many helpful comments and suggestions.

References

1. Aho, A.V.: Indexed grammars - an extension of context-free grammars. J. ACM 15(4), 647–671 (1968)
2. Aho, A.V.: Nested stack automata. J. ACM 16(3), 383–406 (1969)
3. Allouche, J.P., Shallit, J.: Automatic Sequences: Theory, Applications, Generalizations. Cambridge University Press, New York (2003)
4. Berstel, J.: Properties of infinite words: Recent results. In: Cori, R., Monien, B. (eds.) STACS 1989. LNCS, vol. 349, pp. 36–46. Springer, Heidelberg (1989)
5. Book, R.V.: On languages with a certain prefix property. Mathematical Systems Theory 10, 229–237 (1977)
6. Braud, L., Carayol, A.: Linear orders in the pushdown hierarchy. In: Abramsky, S., Gavoille, C., Kirchner, C., Meyer auf der Heide, F., Spirakis, P.G. (eds.) ICALP 2010. LNCS, vol. 6199, pp. 88–99. Springer, Heidelberg (2010)
7. Culik, K., Karhumäki, J.: Iterative devices generating infinite words. Int. J. Found. Comput. Sci. 5(1), 69–97 (1994)
8. Ehrenfeucht, A., Rozenberg, G., Skyum, S.: A relationship between ET0L and EDT0L languages. Theoretical Computer Science 1(4), 325–330 (1976)
9. Gazdar, G.: Applicability of indexed grammars to natural languages. In: Reyle, U., Rohrer, C. (eds.) Natural Language Parsing and Linguistic Theories. Studies in Linguistics and Philosophy, vol. 35, pp. 69–94. Springer, Netherlands (1988)
10. Gilman, R.H.: A shrinking lemma for indexed languages. Theor. Comput. Sci. 163(1-2), 277–281 (1996)
11. Hayashi, T.: On derivation trees of indexed grammars: An extension of the $uvwxy$-theorem. Publications of the Research Institute for Mathematical Sciences 9, 61–92 (1973)
12. Hopcroft, J., Ullman, J.: Introduction to automata theory, languages, and computation. Addison-Wesley series in computer science. Addison-Wesley (1979)
13. Kari, L., Rozenberg, G., Salomaa, A.: L systems. In: Rozenberg, G., Salomaa, A. (eds.) Handbook of Formal Languages, vol. 1, pp. 253–328. Springer-Verlag New York, Inc. (1997)
14. Latteux, M.: Une note sur la propriété de prefixe. Mathematical Systems Theory 11, 235–238 (1978)
15. Maslov, A.N.: Multilevel stack automata. Problems of Information Transmission 12, 38–43 (1976)
16. Rabkin, M.: Ogden's lemma for ET0L languages. In: Dediu, A.-H., Martín-Vide, C. (eds.) LATA 2012. LNCS, vol. 7183, pp. 458–467. Springer, Heidelberg (2012)
17. Rozenberg, G., Salomaa, A.: Mathematical Theory of L Systems. Academic Press, Inc., Orlando (1980)
18. Smith, T.: On infinite words determined by L systems. In: Karhumäki, J., Lepistö, A., Zamboni, L. (eds.) WORDS 2013. LNCS, vol. 8079, pp. 238–249. Springer, Heidelberg (2013)
19. Smith, T.: On Infinite Words Determined by Stack Automata. In: FSTTCS 2013. Leibniz International Proceedings in Informatics (LIPIcs), vol. 24, pp. 413–424. Schloss Dagstuhl–Leibniz-Zentrum fuer Informatik, Dagstuhl (2013)

A Pumping Lemma for Two-Way Finite Transducers

Tim Smith

Northeastern University
Boston, MA, USA
smithtim@ccs.neu.edu

Abstract. A two-way nondeterministic finite transducer (2-NFT) is a finite automaton with a two-way input tape and a one-way output tape. The generated language of a 2-NFT is the set of all strings it can output (across all inputs). Whereas two-way nondeterministic finite acceptors (2-NFAs) accept only regular languages, 2-NFTs can generate languages which are not even context-free, e.g. $\{a^n b^n c^n \mid n \geq 0\}$. We prove a pumping lemma for 2-NFT languages which strengthens and generalizes previous results. Our pumping lemma states that every 2-NFT language L is k-iterative for some $k \geq 1$. That is, every string in L above a certain length can be expressed in the form $x_1 y_1 x_2 y_2 \cdots x_k y_k x_{k+1}$, where the ys can be "pumped" to produce new strings in L of the form $x_1 y_1^i x_2 y_2^i \cdots x_k y_k^i x_{k+1}$.

1 Introduction

A pumping lemma for a language class C is a powerful tool for proving that certain languages do not belong to C, and thus for separating one language class from another. The pumping lemmas for regular and context-free languages are well-known, and pumping lemmas have been proved for other language classes as well. These lemmas differ in their specifics, but have in common that they subject elements of the language to an iterated pumping operation which yields new elements in the language. Pumping lemmas come in two strengths: universal and existential. A universal pumping lemma states that all but finitely many elements of the language can be pumped, whereas an existential pumping lemma guarantees only that some element in the language can be pumped. The broader the class of languages, the harder it is to provide it with a universal pumping lemma, and the more intricate the pumping operation becomes.

In this paper we prove a universal pumping lemma for two-way finite transducers. A two-way nondeterministic finite transducer (2-NFT) is a finite automaton with a two-way input tape and a one-way output tape. At each step, the machine can read the symbol under its input head, move its input head left or right, change state, and append a string to the output tape. The final content of the output tape is the output of the computation. Early studies of two-way finite transducers include [1] and [12]; for a survey of early results, see [2].

E. Csuhaj-Varjú et al. (Eds.): MFCS 2014, Part I, LNCS 8634, pp. 523–534, 2014.

More recent work connects two-way finite transducers to monadic second-order (MSO) logic [3,4] and explores the relationship between nondeterministic and sequential transducers [15] and between two-way and one-way transducers [5].

As a transducer, a 2-NFT M can be viewed as a relation of input strings to output strings, and thus as an operation which maps languages to other languages. In addition to its role as a transducer, M can also be treated as a language generator, generating a single language $L(M)$, consisting of all strings it can output (across all inputs). We call $L(M)$ the generated language or range of M. We denote the class of 2-NFT generated languages by $\mathscr{L}(2\text{-NFT})$, and the deterministic restriction by $\mathscr{L}(2\text{-DFT})$.

It is well known that for finite-state acceptors, neither nondeterminism nor two-way input allows recognition of any additional languages, so that we have:

$$1\text{-DFA} = 1\text{-NFA} = 2\text{-DFA} = 2\text{-NFA} = \text{REG}.$$

For transducers the case is different: 2-DFTs and 2-NFTs can generate languages which are not even context-free, e.g. $\{a^n b^n c^n \mid n \geq 0\}$. In fact, the class $\mathscr{L}(2\text{-DFT})$ equals APL, the languages generated by the absolutely parallel grammars of [12], while $\mathscr{L}(2\text{-NFT})$ equals 1-NCSA, the languages recognized by one-way nondeterministic checking stack automata [12]. $\mathscr{L}(2\text{-NFT})$ also equals 2-NFT(REG), the images of regular languages under 2-NFT transductions. Our pumping lemma, which we prove for $\mathscr{L}(2\text{-NFT})$, therefore holds for all of the classes $\mathscr{L}(2\text{-NFT}) = 1\text{-NCSA} = 2\text{-NFT(REG)}$ and $\mathscr{L}(2\text{-DFT}) = \text{APL}$.

The pumping operation we use involves the notion of k-iterativity [6]. For $k \geq 1$, a string s is k-**iterative for a language** L if $s = x_1 y_1 x_2 y_2 \cdots x_k y_k x_{k+1}$ for some strings $x_1, \ldots, x_{k+1}, y_1, \ldots, y_k$, and $\{x_1 y_1^i x_2 y_2^i \cdots x_k y_k^i x_{k+1} \mid i \geq 0\}$ is an infinite subset of L. (The condition that the subset be infinite is equivalent to requiring that $y_1 \cdots y_k$ is not empty.) Notice that if s is k-iterative for L, then s is i-iterative for L for all $i \geq k$. For $k \geq 1$, a language L is k-**iterative** if there is a $c \geq 0$ such that for every $s \in L$ where $|s| > c$, s is k-iterative for L. L is **weakly** k-**iterative** if either L is finite, or some string is k-iterative for L. Notice that every regular language is 1-iterative and every context-free language is 2-iterative, due to the pumping lemmas for those classes.

In this paper we show that every language in $\mathscr{L}(2\text{-NFT})$ is k-iterative for some $k \geq 1$. Our work strengthens and generalizes the following results, which were proved for checking stack automata or related models but which apply to two-way transducers through the equivalence $\mathscr{L}(2\text{-NFT}) = 1\text{-NCSA}$.

(a) Greibach [8, Lemma 2.1] showed that every language in 1-NCSA over a single-letter alphabet is 1-iterative.
(b) Rodriguez [13] showed that every reversal-bounded 1-NCSA language is k-iterative for some $k \geq 1$.
(c) As observed in [14, Lemma 3], the results of Greibach [6] imply that every language in 1-NCSA is weakly k-iterative for some $k \geq 1$.

Our result generalizes (a) to alphabets with multiple letters, extends (b) to automata which are not reversal-bounded, and strengthens (c) to k-iterativity instead of weak k-iterativity.

1.1   Proof Techniques

A useful tool for analyzing the behavior of an automaton on a two-way tape is the notion of a "visiting sequence". For each square of the tape, the visiting sequence at that square is the sequence of states in which the square is visited during the computation. In a pumping argument, one shows that either the machine visits the same input square twice in the same state, in which case the intervening computation can be repeated, or else two squares have the same visiting sequence, in which case the region between them can be pumped.

We extend this argument to work with two-way finite transducers. Here, it is not enough to show that the input can be pumped to yield new accepting computations; it is also necessary that the outputs of those computations exhibit a k-iterative pattern. In particular, it is necessary to deal with zigzags (repeated changes of direction) in the computation path, which tend to fragment the pumped output. We do so by finding regions of the computation with small zigzags, occurring near positions of the input string with matching "neighborhoods" of surrounding input symbols. Zigzags at these positions stay within their neighborhoods, allowing pumping to proceed with a k-iterative output pattern.

1.2   Related Work

The classic paper of Rabin and Scott [10] presented a technique of zigzag elimination to show the equivalence of two-way and one-way finite acceptors. From an original two-way automaton they define a new derived automaton which performs fewer zigzags, and then repeat this derivation operation until a one-way automaton is obtained. Recent work of Filiot et al. [5] extends Rabin and Scott's proof to a subset of nondeterministic transducers called functional transducers, in order to build a one-way functional transducer from a two-way functional transducer whenever one exists. In the present work we take a different approach: instead of eliminating zigzags, we locate regions of the computation with small zigzags and identical neighborhoods of surrounding input symbols, so that zigzags can occur within these neighborhoods without disrupting our pumping operation.

Greibach [7] defines a notion of strong k-iterativity, which goes beyond k-iterativity by allowing certain positions of a string to be designated as distinguished in the pumping operation. Greibach shows that a certain language L in 2-DFT(REG) is not strongly k-iterative for any $k \geq 1$ (Lemma 5.4 of [7] and its corollary). It is not difficult to show that this particular language L is nonetheless k-iterative for some $k \geq 1$, in accordance with our main result.

The class of languages MCFL generated by multiple context-free grammars, a generalization of context-free grammars, has been studied in connection with k-iterativity. As with $\mathscr{L}(2\text{-NFT})$, it was known that every MCFL is weakly k-iterative for some $k \geq 1$, but whether k-iterativity held was not known. Recent work resolves this question, showing that in fact there is an MCFL which is not k-iterative for any k [9].

1.3   Outline of Paper

The paper is organized as follows. Section 2 gives preliminary definitions concerning two-way finite transducers. Section 3 gives a framework for a pumping argument and explains the challenges to be overcome in applying it to transducers. Section 4 proves our pumping lemma for $\mathscr{L}(\text{2-NFT})$. Section 5 provides an application of the pumping lemma. Section 6 gives our conclusions.

2   Preliminaries

An **alphabet** A is a finite set of symbols. A **string** x is an element of A^*. The length of x is denoted by $|x|$. We denote the empty string by λ. For $1 \leq i \leq |x|$, $x[i]$ denotes the ith symbol of x. A **language** is a subset of A^*.

A **two-way nondeterministic finite transducer (2-NFT)** is a tuple $M = (Q, A, B, P, q_{in}, q_{out})$ where Q is a finite set of states, A is the input alphabet, B is the output alphabet, $q_{in}, q_{out} \in Q$ are the initial and final states, respectively, and P is a finite subset of $(Q - \{q_{out}\}) \times (A \cup \{\triangleright, \triangleleft\}) \times B^* \times Q \times \{-1, 0, 1\}$. The symbols $\triangleright$ and $\triangleleft$ are the left and right endmarkers, respectively.

A **step** I of M is a tuple $(q, \triangleright x \triangleleft, y, i)$ for $q \in Q$, $x \in A^*$, $y \in B^*$, and i an integer. We call I a **visit** of i. We write $(q, \triangleright x \triangleleft, y, i) \vdash (q', \triangleright x \triangleleft, ys, i+j)$ if $1 \leq i \leq |\triangleright x \triangleleft|$ and $(q, (\triangleright x \triangleleft)[i], s, q', j)$ is in P. For $n \geq 1$, an **accepting computation** C with **input** x and **output** y is a sequence of steps $I_1 \vdash I_2 \vdash \cdots \vdash I_n$ where $I_1 = (q_{in}, \triangleright x \triangleleft, \lambda, 1)$ and $I_n = (q_{out}, \triangleright x \triangleleft, y, i)$ for some i. By $|C|$ we mean the number of steps in C and by $C[i]$ we mean the ith step of C.

We call M **returning** if for every accepting computation C, the last step of C has the form $(q_{out}, \triangleright x \triangleleft, y, 1)$. Clearly for every 2-NFT M, there is a returning 2-NFT M' such that $L(M) = L(M')$. (Whenever M would enter q_{out}, M' first moves to the left endmarker, and then enters q_{out}.)

We define 2-NFT transductions over strings, languages, and families of languages. For a string x, let $M(x) = \{y \mid M$ has an accepting computation with input x and output $y\}$. For a language L, let $M(L) = \{y \in M(x) \mid x$ is in $L\}$. For a family of languages $\mathscr{L}$, let $\text{2-NFT}(\mathscr{L}) = \{M(L) \mid M$ is a 2-NFT and L is in $\mathscr{L}\}$.

2-NFTs can also be viewed as language generators. Let $L(M) = M(A^*)$. $L(M)$ is called the "generated language", or "range" of M. Let $\mathscr{L}(\text{2-NFT}) = \{L(M) \mid M$ is a 2-NFT$\}$. Clearly $\mathscr{L}(\text{2-NFT}) = \text{2-NFT}(\text{REG})$, since the finite-state control of a 2-NFT can be used to check whether or not an input word in A^* is in some particular regular language L.

3   Pumping Framework

In this section we give a framework for a pumping argument on a two-way tape and explain the challenges to be overcome in applying it to transducers. We keep the discussion at a high level, giving a more formal treatment in Section 4.

A useful tool for analyzing the behavior of an automaton on a two-way tape is the notion of a visiting sequence. For each square of the tape, the visiting sequence at that square is the sequence of states in which the square is visited during the computation. In a pumping argument, one shows that either the machine visits the same input square twice in the same state, in which case the intervening computation can be repeated, or else two squares have the same visiting sequence, in which case the region between them can be pumped. By choosing the original computation to be a shortest computation for its output string, we ensure that the pumped portion of the path has non-empty output, and therefore that the pumping operation produces an infinity of new strings.

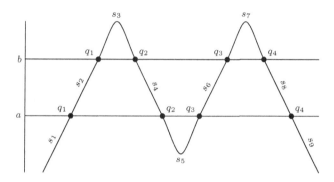

Fig. 1.

Let us apply this basic idea to a 2-NFT M. Consider a shortest accepting computation C (i.e., one with fewest steps, and in the case of a tie, with shortest input) for an output string s. Suppose C has the form shown in Figure 1. The input tape is represented as a vertical line whose bottom corresponds to the left end of the tape and whose top corresponds to the right end. The steps of the computation are depicted as a path winding back and forth along the input tape. Each s_i designates the output of the machine during some portion of the computation, and each q_i designates the state of the machine at a particular point. Thus the output of this computation is the string $s = s_1 s_2 s_3 s_4 s_5 s_6 s_7 s_8 s_9$. The two input positions a and b can be seen to have the same visiting sequence (q_1, q_2, q_3, q_4). This means that the input region r which separates a and b can be removed, yielding a computation C_0 with output $s_1 s_3 s_5 s_7 s_9$. Alternatively, r can be duplicated, yielding a computation C_2 with output $s_1 s_2^2 s_3 s_4^2 s_5 s_6^2 s_7 s_8^2 s_9$. In general, with $i \geq 0$ copies of region r, we can obtain a computation C_i with output $s_1 s_2^i s_3 s_4^i s_5 s_6^i s_7 s_8^i s_9$. Suppose $s_2 s_4 s_6 s_8 = \lambda$. Then $s = s_1 s_3 s_5 s_7 s_9$. But then C_0 is a shorter computation for s than C, a contradiction. Therefore $|s_2 s_4 s_6 s_8| \geq 1$, making s 4-iterative for $L(M)$.

Problems arise, however, if C instead has a form which zigzags through crossings of r, as in Figure 2. Here, a and b still have the same visiting sequence, so we can still remove or duplicate r and complete the computation, but the new output strings will not have the form needed to make the original string s k-iterative for $L(M)$. For example, if we remove r, we can complete the computation by skipping

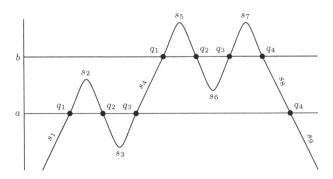

Fig. 2.

from q_1 at a to q_1 at b, from q_2 at b back to q_2 at a, from q_3 at a to q_3 at b, and finally from q_4 at b to q_4 at a. We thereby obtain the output string $s_1 s_5 s_3 s_7 s_9$. But in this string, s_5 precedes s_3, whereas in s, s_3 precedes s_5. The new string therefore does not have the right form for showing that s is k-iterative for $L(M)$. Thus the pumping argument does not go through as it stands.

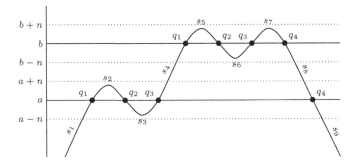

Fig. 3.

To resolve this issue, we will find input positions a and b which not only have appropriate visiting sequences, but also have matching "neighborhoods" of surrounding input symbols large enough to encompass any problematic zigzags. The zigzags in these neighborhoods can then be retained for the pumping and shrinking operations. For example, in Figure 3, suppose the input region from $a-n$ to $a+n$ is identical to the input region from $b-n$ to $b+n$; that is, a and b have the same neighborhood out to n symbols above and below. Now when we remove r, we do not have to skip from q_2 at b back to q_2 at a as before, but can proceed from q_2 at b to q_3 at b, outputting s_6. This is possible because the s_6 portion of the computation never leaves the lower neighborhood of b, so with r removed, it will never leave the lower neighborhood of a, and these two neighborhoods are the same. We thus obtain a computation C_0 with output $s_1 s_5 s_6 s_7 s_9$. If we duplicate r, then since the s_3 portion of the computation never leaves the lower neighborhood of a, we can repeat the segments from q_1 at a to q_1 at b and from q_4 at b to q_4 at a, obtaining the output $s_1 (s_2 s_3 s_4)^2 s_5 s_6 s_7 s_8^2 s_9$. In general, with $i \geq 0$ copies of region r, we can obtain the output $s_1(s_2s_3s_4)^i s_5 s_6 s_7 s_8^i s_9$. Now suppose $s_2s_3s_4s_8 = \lambda$. Then $s = s_1s_5s_6s_7s_9$ and C_0 is a shorter computation for s than C, a contradiction. Hence $|s_2s_3s_4s_8| \geq 1$, making s 2-iterative for $L(M)$.

In Section 4 we give the details of this argument, showing that if the original output string is sufficiently long, then there is always a region with visiting sequences and surrounding neighborhoods fit for shrinking and pumping in a form which yields k-iterativity. This will allow us to prove our pumping lemma for $\mathscr{L}$(2-NFT).

4 Pumping Lemma

In this section we prove our pumping lemma for $\mathscr{L}$(2-NFT), formalizing the high-level approach outlined in Section 3.

Theorem 1. *Suppose L is in $\mathscr{L}$(2-NFT). Then L is k-iterative for some $k \geq 1$.*

Proof. We begin with some definitions. Take $M = (Q, A, B, P, q_{in}, q_{out})$ to be a returning 2-NFT such that $L(M) = L$. We define a function f over the integers from 1 to $\lfloor \frac{|Q|}{2} \rfloor + 1$, as follows.

$$f(i) = \begin{cases} 1 & : i = \lfloor \frac{|Q|}{2} \rfloor + 1 \\ |A|^{2f(i+1)+1} \cdot |Q|^{2i} + 2f(i+1) + 2 : 1 \leq i \leq \lfloor \frac{|Q|}{2} \rfloor \end{cases}$$

Notice that for $1 \leq i \leq \lfloor \frac{|Q|}{2} \rfloor$, $f(i) > f(i+1)$. Let $k = |Q|$. If L is finite, then trivially L is k-iterative. So say L is infinite. Let r be the highest i such that for some $(q, d, s, q', j) \in P$, $|s| = i$. This is the length of the longest string that M can add to its output in a single step. Let $c = (f(1) + 2) \cdot |Q| \cdot r + r$. Take any $y \in L$ such that $|y| > c$. We will show that y is k-iterative for L.

Let a shortest computation for y be an accepting computation C with output y such that for every accepting computation C' with output y, $|C| \leq |C'|$ and if $|C| = |C'|$, then $|x| \leq |x'|$, where x is the input of C and x' is the input of C'. Take any shortest computation C for y. Let x be the input of C.

Each step i of the computation C has the form $(q_i, \triangleright x \triangleleft, s_i, v_i)$. We will view C as a path and each step of C as a **node** on the path. We call each position on the input tape from 1 to $|\triangleright x \triangleleft|$ a **level** and we call v_i the level at node i. We have $v_1 = v_{|C|} = 1$; that is, the path starts at level 1 (the left endmarker) and also ends at level 1 (since M is returning). The level of each node differs from that of its predecessor by at most 1.

For $1 \leq i \leq j \leq |C|$, we call the sequence $C[i], \ldots, C[j]$ a subpath from i to j. A **hill** h **at level** l is a subpath from i to j of length > 2 such that $v_i = v_j = l$ and for all m such that $i < m < j$, $v_m > l$. The **top** of h is $\max(v_i, \ldots, v_j)$ and the **height** of h is $\max(v_i, \ldots, v_j) - l$. A **valley** v **at level** l is a subpath from i to j of length > 2 such that $v_i = v_j = l$ and for all m such that $i < m < j$, $v_m < l$. The **bottom** of v is $\min(v_i, .., v_j)$ and the **depth** of v is $l - \min(v_i, .., v_j)$.

With these definitions in place, the proof idea is as follows. If some level has more than $|Q|$ visits, we will see that y is 1-iterative for L. Otherwise, the input string x is long enough that it contains a "smooth" region (l to $l + f(n)$ below), one with small hills and valleys. Within this region we find two levels a and b with similar neighborhoods and visiting sequences. The smoothness of the region then permits a pumping argument applied to the area between a and b to show that y is k-iterative for L.

So suppose some level has more than $|Q|$ visits. Then M visits the same input position twice in the same state. Suppose M produces no output between these two visits. Then C could be shortened, a contradiction. So between the two visits M produces some non-empty output w. Then we can "pump" (repeat) the intervening computation. Hence $y = uwz$ for some strings u and z, and $\{uw^i z \mid i \geq 0\}$ is an infinite subset of L. Then y is 1-iterative for L, hence k-iterative for L. So say no level has more than $|Q|$ visits. Notice that we now have $|Q| \geq 2$, since if $|Q| = 1$, then M can only visit the left endmarker once, so since C must begin and end at the left endmarker, we have $|C| = 1$ and $y = \lambda$, which contradicts the fact that $|y| > c$.

Since no level has more than $|Q|$ visits, $|\triangleright x \triangleleft| \geq \frac{|y|}{|Q| \cdot r}$. Hence $|x| \geq \frac{|y|}{|Q| \cdot r} - 2 > \frac{c}{|Q| \cdot r} - 2 = \frac{(f(1)+2) \cdot |Q| \cdot r + r}{|Q| \cdot r} - 2 > f(1)$. Every position of x is visited at least once, otherwise x could be shortened and C would still output y, a contradiction. Therefore level 1 has a hill of height $> f(1)$. So take the highest $n \leq \lfloor \frac{|Q|}{2} \rfloor$ such that some level has $\geq n$ hills of height $> f(n)$. We have $1 \leq n \leq \lfloor \frac{|Q|}{2} \rfloor$ and $f(n) > f(n+1) \geq 1$. Notice that no level i has $\geq n + 1$ hills of height $> f(n+1)$, since if $n < \lfloor \frac{|Q|}{2} \rfloor$, then this would contradict the construction of n, and if $n = \lfloor \frac{|Q|}{2} \rfloor$, then since level $i + 1$ has at least two visits for each hill of height > 1 at level i, level $i+1$ would have more than $|Q|$ visits, a contradiction. So take any level l with $\geq n$ hills of height $> f(n)$. Then since $f(n) > f(n+1)$, l must have exactly n hills of height $> f(n)$. For the same reason, level l cannot have a hill whose height is $> f(n+1)$ but $\leq f(n)$.

Further, suppose some level $i \leq l + f(n)$ has a valley v of depth $> f(n+1)$ whose bottom is above l. Consider the n hills of height $> f(n)$ at level l. Since the bottom of v is above l, either v is contained completely by one of these n hills, or it is not part of any of them. If it is not part of any of them, then it is part of another hill at level l, but then level l has $\geq n + 1$ hills of height $> f(n+1)$, a contradiction. If v is contained completely by one of the n hills of height $> f(n)$ at level l, then this hill contains two hills of height $> f(n+1)$ at level $i - f(n+1)$ (one on each side of v). But then level $i - f(n+1)$ contains $\geq n + 1$ hills of height $> f(n+1)$, a contradiction. So there is no such level i.

We will refer to the n hills at level l which are of height $> f(n)$ as hill $1, \ldots,$ hill n. For any level i from $l + f(n+1)$ to $l + f(n) - f(n+1)$, call $x[i - f(n+1)] \cdots x[i + f(n+1)]$ the **neighborhood** of i. For j from 1 to n, let $\mathrm{in}(i, j)$ be the first node at level i in hill j, and let $\mathrm{out}(i, j)$ be the last node at level i in hill j. Recall that q_m is the state of M at step m of C. Let the **pair list** of i be a list of n pairs of states such that for $1 \leq j \leq n$, the jth pair is $(q_{\mathrm{in}(i,j)}, q_{\mathrm{out}(i,j)})$.

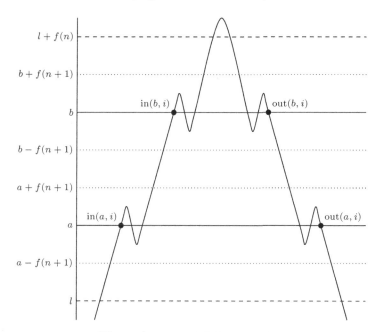

Fig. 4. An example hill i at level l

There are at most $p_1 = |A|^{2f(n+1)+1}$ distinct neighborhoods, and at most $p_2 = |Q|^{2n}$ distinct pair lists. From $l + f(n+1) + 1$ to $l + f(n) - f(n+1) - 1$ there are $f(n) - 2f(n+1) - 1$ levels. Then since $f(n) - 2f(n+1) - 1 > p_1 p_2$, there are levels a, b such that $l + f(n+1) < a < b < l + f(n) - f(n+1)$ and a and b have the same neighborhood and pair list. See Figure 4 for an example hill i passing through levels a and b. We make some observations O1, O2, O3 for use below.

O1. For any hill i such that $1 \leq i < n$, the portion of C between $\mathrm{out}(a, i)$ and $\mathrm{in}(a, i+1)$ never reaches level a, since if it did so as part of hill i, then $\mathrm{out}(a, i)$ would not be the last node at level a in hill i, if it did so as part of hill $i + 1$, then $\mathrm{in}(a, i+1)$ would not be the first node at level a in hill $i + 1$, and if it did so between hills i and $i + 1$, then level l would have a hill whose height is $> f(n+1)$ but $\leq f(n)$, which we observed to be impossible. Similarly, the portion of C before $\mathrm{in}(a, 1)$ never reaches level a, and the portion of C after $\mathrm{out}(a, n)$ never reaches level a.

O2. For any hill i, the portion of C from $\mathrm{in}(a, i)$ to $\mathrm{out}(a, i)$ never goes below the neighborhood of a, since if it did, then a would have a valley of depth $> f(n+1)$ whose bottom is above l (since this portion of the path is in a hill of l), which we observed to be impossible.

O3. Similarly, for any hill i, the portion of C from $\mathrm{in}(b, i)$ to $\mathrm{out}(b, i)$ never goes below the neighborhood of b.

We now break up y into substrings, where each substring designates the output produced during a range of steps of C. Recall that s_i is the output produced from step i of C. Let $h_0 = s_1 \cdots s_{\mathrm{in}(a,1)-1}$ and for i from 1 to n, let:

$$e_i = s_{\text{in}(a,i)} \cdots s_{\text{in}(b,i)-1}$$
$$f_i = s_{\text{in}(b,i)} \cdots s_{\text{out}(b,i)-1} \qquad\qquad h_i = \begin{cases} s_{\text{out}(a,i)} \cdots s_{\text{in}(a,i+1)-1} & \text{if } 1 \le i < n \\ s_{\text{out}(a,i)} \cdots s_{|C|} & \text{if } i = n \end{cases}$$
$$g_i = s_{\text{out}(b,i)} \cdots s_{\text{out}(a,i)-1}$$

We have $y = h_0\, e_1 f_1 g_1 h_1 \cdots e_n f_n g_n h_n$. For $i \ge 0$, let $y_i = h_0\, e_1^i f_1 g_1^i h_1 \cdots e_n^i f_n g_n^i h_n$. We will show that $\{y_i \mid i \ge 0\}$ is an infinite subset of L, and thus that $y\ (= y_1)$ is $2n$-iterative for L.

First we show that y can be "shrunk", i.e. that y_0 is in L. We construct a computation C_0 with input $x[1] \cdots x[a-1]\, x[b] \cdots x[|x|]$ which will follow C, but skip some steps. C_0 proceeds as follows. Follow C until $\text{in}(a,1)$, outputting h_0. This is possible due to O1. For i from 1 to n, continue as follows. Skip to $\text{in}(b,i)$, which is possible because $q_{\text{in}(a,i)} = q_{\text{in}(b,i)}$. Proceed from $\text{in}(b,i)$ to $\text{out}(b,i)$, outputting f_i. This is possible because due to O3, in C this portion never went below the neighborhood of b, so in C_0 it will never go below the neighborhood of a, and these two neighborhoods are equal. Skip from $\text{out}(b,i)$ to $\text{out}(a,i)$. This is possible because $q_{\text{out}(b,i)} = q_{\text{out}(a,i)}$. If $i < n$, proceed from $\text{out}(a,i)$ to $\text{in}(a,i+1)$, outputting h_i (possible due to O1). If $i = n$, proceed from $\text{out}(a,n)$ to the end of C, outputting h_n and finishing in the final state q_{out} (possible due to O1). C_0 is now an accepting computation with output y_0, which is therefore in L.

Next we show that y can be "pumped", i.e. that y_m is in L for each $m \ge 2$. So take any $m \ge 2$. We show how to construct a computation C_m with input $x[1] \cdots x[a-1]\, (x[a] \cdots x[b-1])^m\, x[b] \cdots x[|x|]$ which will follow C, but repeat some steps. C_m proceeds as follows. Follow C until $\text{in}(a,1)$, outputting h_0. This is possible due to O1. For i from 1 to n, continue as follows. Perform the portion of C from $\text{in}(a,i)$ to $\text{in}(b,i)$ m times, each time outputting e_i. This is possible because $q_{\text{in}(a,i)} = q_{\text{in}(b,i)}$, and because due to O2, in C this portion never goes below the neighborhood of a (or above level b, since $\text{in}(b,i)$ is the first node at level b in hill i), and in C_m, for j from 1 to m, the neighborhood of $a + j(b-a)$ equals the neighborhood of a. Now C_m is at level $a + m(b-a)$. Proceed from $\text{in}(b,i)$ to $\text{out}(b,i)$, outputting f_i. This is possible due to O3. Next, perform the portion of C from $\text{out}(b,i)$ to $\text{out}(a,i)$ m times, each time outputting g_i. This is possible because $q_{\text{out}(b,i)} = q_{\text{out}(a,i)}$, and because in C this portion never goes above level b (since $\text{out}(b,i)$ is the last node at level b in hill i) or below the neighborhood of a (due to O2). Now if $i < n$, proceed from $\text{out}(a,i)$ to $\text{in}(a,i+1)$, outputting h_i (possible due to O1). If $i = n$, proceed from $\text{out}(a,n)$ to the end of C, outputting h_n and finishing in the final state q_{out} (possible due to O1). C_m is now an accepting computation with output y_m, which is therefore in L. Hence $\{y_i \mid i \ge 2\}$ is a subset of L.

Finally, suppose $e_1 g_1 \cdots e_n g_n = \lambda$. Then $y_0 = y$ and C_0 is an accepting computation with output y, a contradiction, since $|C_0| < |C|$ and C is a shortest computation for y. So $e_1 g_1 \cdots e_n g_n \ne \lambda$. Then $\{y_i \mid i \ge 0\}$ is an infinite language. Hence $\{y_i \mid i \ge 0\}$ is an infinite subset of L. Therefore y is $2n$-iterative for L. Then since $n \le \lfloor \frac{|Q|}{2} \rfloor$ and $k = |Q|$, we have $2n \le k$, so y is k-iterative for L. So for any $y \in L$ such that $|y| > c$, y is k-iterative for L. Hence L is k-iterative, which was to be shown. $\qquad\square$

5   Application

In this section we apply our pumping lemma to show that a particular language of interest does not belong to $\mathscr{L}(2\text{-NFT})$. Addressing the question of whether a certain type of "mildly context-sensitive" grammars can generate the class of natural languages, Radzinksi [11] considers the system of Chinese number-names. In particular, he examines the set L consisting of number-names composed only of instances of *wu* (five) and *zhao* (trillion):

$$L = \{\mathtt{wu}\,(\mathtt{zhao})^{k_1}\,\mathtt{wu}\,(\mathtt{zhao})^{k_2}\,\cdots\,\mathtt{wu}\,(\mathtt{zhao})^{k_n} \mid k_1 > k_2 > \cdots k_n > 0\}$$

Radzinski shows that L cannot be generated by the class TAG of tree adjoining grammars; we will show that it also cannot be generated by a 2-NFT. The string wu zhao zhao is 1-iterative for L, since $\{\mathtt{wu}\,\mathtt{zhao}\,(\mathtt{zhao})^i \mid i \geq 0\}$ is an infinite subset of L. The language L is therefore weakly 1-iterative. At first glance it might seem that L is also 1-iterative, since we can pump the first zhao in any string. For example, for the string wu zhao zhao wu zhao we have $\{\mathtt{wu}\,(\mathtt{zhao})^i\,\mathtt{zhao}\,\mathtt{wu}\,\mathtt{zhao} \mid i \geq 1\}$, which is an infinite subset of L. But recall that k-iterativity requires the pumping index i to start at 0, not 1, and $\{\mathtt{wu}\,(\mathtt{zhao})^i\,\mathtt{zhao}\,\mathtt{wu}\,\mathtt{zhao} \mid i \geq 0\}$ is not a subset of L. In fact, Radzinski shows that a related language (K below) is not k-iterative for any $k \geq 1$. Our pumping lemma then gives the following result.

Theorem 2. *L is not in $\mathscr{L}(2\text{-NFT})$.*

Proof. Let K be the language $\{\mathtt{a}\,\mathtt{b}^{k_1}\,\mathtt{a}\,\mathtt{b}^{k_2}\,\cdots\,\mathtt{a}\,\mathtt{b}^{k_n} \mid k_1 > k_2 > \cdots k_n > 0\}$. The proof of Lemma 2 of Radzinski [11] shows that K is not k-iterative for any $k \geq 1$. Then by our Theorem 1, K is not in $\mathscr{L}(2\text{-NFT})$. Suppose L is in $\mathscr{L}(2\text{-NFT})$. Let h be a homomorphism from $\{\mathtt{w},\mathtt{u},\mathtt{z},\mathtt{h},\mathtt{a},\mathtt{o}\}$ to $\{\mathtt{a},\mathtt{b}\}$ such that $h(\mathtt{wu}) = \mathtt{a}$ and $h(\mathtt{zhao}) = \mathtt{b}$. Then K is the image of L under h. By Lemma 1.1 of [8], $\mathscr{L}(2\text{-NFT})$ is closed under substitution. Then K is in $\mathscr{L}(2\text{-NFT})$, a contradiction. So L is not in $\mathscr{L}(2\text{-NFT})$.   □

6   Conclusion

In this paper we have proved a pumping lemma for the class $\mathscr{L}(2\text{-NFT})$ of languages generated by two-way nondeterministic finite transducers. Our pumping lemma strengthens and generalizes previous results for this class. We gave an example of a language of interest which can be shown using our pumping lemma not to belong to $\mathscr{L}(2\text{-NFT})$, and we hope that our lemma will help to obtain similar results in other cases. One direction for further research would be to generalize our lemma to broader classes of languages. For example, recall that $\mathscr{L}(2\text{-NFT}) = 2\text{-NFT}(\text{REG})$. Where CFL denotes the context-free languages, the class $2\text{-NFT}(\text{CFL})$ properly contains $2\text{-NFT}(\text{REG})$, since $2\text{-NFT}(\text{REG})$ does not contain CFL [6, Theorem 4.26]. We can then ask whether our pumping lemma can be generalized to apply to $2\text{-NFT}(\text{CFL})$. More broadly, it would be interesting to know whether such a lemma holds for all $2\text{-NFT}(\mathscr{L})$ where $\mathscr{L}$ is a language class such that every language in $\mathscr{L}$ is k-iterative.

Acknowledgments. I want to thank my advisor, Rajmohan Rajaraman, for supporting this work, encouraging me, and offering many helpful comments and suggestions.

References

1. Ehrich, R., Yau, S.: Two-way sequential transductions and stack automata. Information and Control 18(5), 404–446 (1971)
2. Engelfriet, J.: Two-way automata and checking automata. In: de Bakker, J.W., van Leeuwen, J. (eds.) Foundations of Computer Science III Part 1. Mathematical Centre Tracts, vol. 108, pp. 1–69. Mathematisch Centrum, Amsterdam (1979)
3. Engelfriet, J., Hoogeboom, H.J.: MSO definable string transductions and two-way finite-state transducers. ACM Trans. Comput. Logic 2(2), 216–254 (2001)
4. Engelfriet, J., Hoogeboom, H.J.: Finitary compositions of two-way finite-state transductions. Fundam. Inf. 80(1-3), 111–123 (2007)
5. Filiot, E., Gauwin, O., Reynier, P.A., Servais, F.: From two-way to one-way finite state transducers. In: LICS 2013, pp. 468–477. IEEE Computer Society (2013)
6. Greibach, S.A.: One way finite visit automata. Theoretical Computer Science 6(2), 175–221 (1978)
7. Greibach, S.A.: The strong independence of substitution and homomorphic replication. RAIRO - Theoretical Informatics and Applications 12(3), 213–234 (1978)
8. Greibach, S.: Checking automata and one-way stack languages. J. Comput. Syst. Sci. 3(2), 196–217 (1969)
9. Kanazawa, M., Kobele, G., Michaelis, J., Salvati, S., Yoshinaka, R.: The failure of the strong pumping lemma for multiple context-free languages. Theory of Computing Systems, 1–29 (2014)
10. Rabin, M.O., Scott, D.: Finite automata and their decision problems. IBM J. Res. Dev. 3(2), 114–125 (1959)
11. Radzinski, D.: Chinese number-names, tree adjoining languages, and mild context-sensitivity. Comput. Linguist. 17(3), 277–299 (1991)
12. Rajlich, V.: Absolutely parallel grammars and two-way finite-state transducers. J. Comput. Syst. Sci. 6(4), 324–342 (1972)
13. Rodriguez, F.: Une double hiérarchie infinie de langages vérifiables. RAIRO - Theoretical Informatics and Applications 9(R1), 5–19 (1975)
14. Smith, T.: On Infinite Words Determined by Stack Automata. In: FSTTCS 2013. Leibniz International Proceedings in Informatics (LIPIcs), vol. 24, pp. 413–424. Schloss Dagstuhl–Leibniz-Zentrum fuer Informatik, Dagstuhl (2013)
15. de Souza, R.: Uniformisation of two-way transducers. In: Dediu, A.-H., Martín-Vide, C., Truthe, B. (eds.) LATA 2013. LNCS, vol. 7810, pp. 547–558. Springer, Heidelberg (2013)

Tractability Frontier for Dually-Closed Ord-Horn Quantified Constraint Satisfaction Problems

Michał Wrona*

Department of Computer and Information Science
University of Linköping
Linköping, Sweden
michal.wrona@liu.se

Abstract. A temporal constraint language is a relational structure with a first-order definition in the rational numbers with the order. We study here the complexity of the Quantified Constraint Satisfaction Problem (QCSP) for Ord-Horn languages: probably the most widely studied family of all temporal constraint languages.

We restrict ourselves to a natural subclass that we call dually-closed Ord-Horn languages. The main result of the paper states that the QCSP for a dually-closed Ord-Horn language is either in P or it is coNP-hard.

1 Introduction

One of the most widely studied computational problems in Computer Science is the Constraint Satisfaction Problem (CSP). An instance of the CSP is a primitive-positive (pp) sentence of the form: $(\exists v_1 \ldots \exists v_n (R(v_{i_1}, \ldots, v_{i_k}) \wedge \ldots))$, where the inner quantifier-free part is the conjunction of relational symbols with variables. The CSP is NP-hard and therefore one often looks at various restrictions of the problem. The one that appears often in the literature and is considered in this paper is the parametrization $\mathrm{CSP}(\Gamma)$ of the CSP with a relational structure Γ. An instance of the $\mathrm{CSP}(\Gamma)$ is a pp-sentence in which all relational symbols comes from a signature of Γ.

This framework on one hand allows to define many problems arising naturally in different branches of computer science such as Artificial Intelligence. On the other hand, complexity classifications of the flavour of Schaefer's theorem [1] these problems tend to display are of theoretical interest. Directly relevant to this paper are temporal CSPs that are $\mathrm{CSP}(\Gamma)$ for Γ with a first-order definition in $(\mathbb{Q}; <)$, called *temporal (constraint) languages.* In [2], nine large classes of tractable (solvable in polynomial time) temporal CSPs has been identified and it was proved that all other such problems are NP-complete. The strength of every such classification can be naturally measured by the number of natural problems

* The author is supported by the *Swedish Research Council* (VR) under grant 621-2012-3239.

E. Csuhaj-Varjú et al. (Eds.): MFCS 2014, Part I, LNCS 8634, pp. 535–546, 2014.

expressible within the framework under consideration existing in the literature prior to the classification itself. In this case, there is a great deal of such problems, e.g., the Betweenness and the Cyclic Ordering Problem mentioned in [3], and network satisfaction problem for Point Algebra [4]. Furthermore, temporal CSP(Γ) has been studied in the literature for so-called Ord-Horn languages [5], and AND/OR precedence constraints in scheduling [6].

A natural generalization of the CSP is the Quantified Constraint Satisfaction Problem (QCSP) for a relational structure Γ, denoted by QCSP(Γ) that next to existential allows also universal quantifiers in the input sentence. Similarly as the CSP, this problem has been studied widely in the literature, see e.g. [7,8]. In this paper, we study QCSP(Γ) for temporal languages. Although a number of partial results has been obtained [9,10,11,12,13], these efforts did not lead to a full complexity classification of all temporal QCSP(Γ). One of the reasons is that QCSPs are usually harder to classify than CSPs. This also holds in our case. The hardest temporal QCSPs are PSPACE-complete, as opposed to temporal CSPs that are all in NP. Furthermore, already QCSPs for so-called positive temporal languages (that are all closed under a constant operation and therefore the corresponding CSPs are trivially tractable) classified in [11,12] may be in LOGSPACE, be NLOGSPACE-complete, be P-complete, be NP-complete, or be PSPACE-complete. The ultimate goal for temporal QCSPs is to provide a similar classification for the entire class. In this paper we look at the complexity of QCSPs for *dually-closed Ord-Horn* languages. Ord-Horn languages lie at the intersection of two large CSP-tractable classes (languages preserved by operations ll and dual-ll, see [2] for details), whereas dually-closed languages (as explained in Section 2) contain equality languages (with a first-order definition over $(\mathbb{N}; =)$) whose QCSPs has been studied in [9,10]. The main result of the paper is a dichotomy theorem that reveals the tractability frontier for the complexity of QCSPs for dually-closed Ord-Horn languages: we show that QCSP(Γ) for such languages Γ is either coNP-hard or it is in P. In the second case Γ is a Guarded Ord-Horn language and the corresponding QCSP can be solved by an algorithm based on establishing local consistency presented in [13]. In the first case Γ simulates one of three relations whose QCSP is coNP-hard, defined by formulas: $(1)(x = y \rightarrow y = z)$, $(2)(x = y \rightarrow z < v) \wedge (x < z) \wedge (x < v) \wedge (y < z) \wedge (y < v)$, and $(3)(x = y \rightarrow z > v) \wedge (x > z) \wedge (x > v) \wedge (y > z) \wedge (y > v)$. While the first result comes from [9], the other two are new. The exact complexity in all three cases remains open.

2   Preliminaries

Formulas and definability. We consider two restricted forms of first-order (fo)-formulas. Let τ be a signature. A first-order τ-formula is a $\forall\exists\wedge$-*formula* if it has the form $Q_1 v_1 \ldots Q_n v_n (\psi_1 \wedge \cdots \wedge \psi_m)$, where each Q_i is a quantifier from $\{\forall, \exists\}$, and each ψ_i is an atomic τ-formula of the form $R(x_1, \ldots, x_k)$ where $R \in \tau$. A *primitive positive (pp)-formula* is a $\forall\exists\wedge$-formula where all quantifiers are existential. A $\forall\exists\wedge$-*sentence (a pp-sentence)* is a $\forall\exists\wedge$-formula (pp-formula) without free variables.

We say that a relation R is (fo)-definable ($\forall\exists\wedge$-, or pp-definable) in a relational structure Γ if R has the same domain as Γ, and there is a fo-formula ($\forall\exists\wedge$-, or pp-formula) ϕ in the signature of Γ such that ϕ holds exactly on those tuples that are contained in R. A relational structure Δ is fo-definable in Γ if every relation in Δ is fo-definable in Γ.

Temporal languages and formulas. In this paper a *temporal formula* is a fo-formula built from quantifiers, logical connectivities and relational symbols: $<, \leq$ $, \neq, =$. A *temporal relation* is a relation with a fo-definition in $(\mathbb{Q}; <, \leq, \neq, =)$ and a *temporal (constraint) language* is a relational structure over a finite signature consisting of the domain $\mathbb{Q}$ and a finite number of temporal relations.

Entailment. For a temporal formula φ we write $\mathrm{Var}(\varphi)$ to denote the set of free variables in φ. Let φ_1, φ_2 be temporal formulas with $\mathrm{Var}(\varphi_1), \mathrm{Var}(\varphi_2) \subseteq \{v_1, \ldots, v_n\}$. We say that φ_1 *entails* φ_2 if $(\forall v_1 \cdots \forall v_n(\varphi_1 \to \varphi_2))$ is true in $(\mathbb{Q}; <, \leq, =, \neq)$. We say that an n-ary relation R entails a formula φ with $\mathrm{Var}(\varphi) \subseteq \{v_1, \ldots, v_n\}$ if a formula φ_R defining R over variables $\{v_1, \ldots, v_n\}$ entails φ.

QCSPs. Let Γ be a relational structure over a finite signature. The *Quantified Constraint Satisfaction Problem* for Γ, denoted by $\mathrm{QCSP}(\Gamma)$, is the problem to decide if a given $\forall\exists\wedge$-sentence over a signature of Γ is true in Γ.

Lemma 1. *([9,7]) Let Γ_1, Γ_2 be constraint languages. If Γ_1 has a $\forall\exists\wedge$-definition in Γ_2, then $\mathrm{QCSP}(\Gamma_1)$ is logarithmic space reducible to $\mathrm{QCSP}(\Gamma_2)$.*

Polymorphisms. The set of relations with a pp-definition in Γ will be denoted by $[\Gamma]_{\mathrm{pp}}$ A polymorphism of a temporal language Γ is a homomorphism from Γ^k to Γ. The set of all polymorphisms of Γ is denoted by $Pol(\Gamma)$.

For temporal languages we have Galois correspondence [14].

Theorem 1. *Let Γ be a temporal language, then $R \in [\Gamma]_{pp}$ if and only if $Pol(\Gamma) \subseteq Pol(\mathbb{Q}; R)$.*

Dually-Closed Temporal Languages. Let $- : \mathbb{Q} \to \mathbb{Q}$ be the unary operation such that for every $q \in \mathbb{Q}$ we have $-(q) = -q$. Let $f : \mathbb{Q}^n \to \mathbb{Q}$. The *dual* of f is the operation $-f(-x_1, \ldots, -x_n)$ denoted by $\overline{f}$. We say that a temporal language Γ is *dually-closed* if $f \in Pol(\Gamma)$ if and only if $\overline{f} \in Pol(\Gamma)$. We can now elaborate on the statement from the introduction that every equality language [9] is in particular dually-closed. Indeed, every such language has as polymorphisms all injective operation from $\mathbb{Q}$ to $\mathbb{Q}$, hence in particular $-$. Observe that in general a dually-closed language does not have to be closed under $-$. As an example consider $(\mathbb{Q}; <)$.

Let $R \subseteq \mathbb{Q}^n$. The dual of R is the relation $\{(-t[1], \ldots, -t[n]) \mid t \in R\}$ denoted by $\overline{R}$.

Proposition 1. *Let Γ be a dually closed temporal language. Then $R \in [\Gamma]_{pp}$ if and only if $\overline{R} \in [\Gamma]_{pp}$.*

3   Ord-Horn Languages and the Main Result

Definition 1. *An* Ord-Horn (OH) clause *is a disjunction in one of the following forms:*

1. $(x_1 \neq y_1 \vee \cdots \vee x_k \neq y_k)$ *(called* negative*),*
2. $(x_1 \neq y_1 \vee \cdots \vee x_k \neq y_k \vee x < y)$ *(called* strict*), or*
3. $(x_1 \neq y_1 \vee \cdots \vee x_k \neq y_k \vee x \leq y)$ *(called* positive*).*

We will often write $\lhd$ to denote any element of the set $\{<, \leq\}$.

Definition 2. *A conjunction of OH clauses will be called an* OH formula. *A relation is OH if it is definable by an OH formula. A temporal language Γ is OH if all relations in Γ are OH.*

When writing about OH clauses, formulas, relations or languages, we will often omit 'OH' if it is clear from the context what a clause, formula, etc., we have in mind.

Definition 3. *We say that an OH clause is* basic *if it is in one of the following forms:*

- $x = y$, $x \leq y$,
- $(x_1 \neq y_1 \vee \cdots \vee x_p \neq y_p)$, *or*
- $(x_1 \neq x_2 \vee \cdots \vee x_1 \neq x_q) \vee (x_1 < y_1) \vee (y_1 \neq y_2 \vee \cdots \vee y_1 \neq y_{q'})$.

Definition 4. Guarded Ord-Horn (GOH) *formulas are defined inductively as follows.*

1. *Basic OH formulas are GOH.*
2. *If ψ_1 and ψ_2 are GOH formulas, then $\psi_1 \wedge \psi_2$ is a GOH formula.*
3. *If ψ is a GOH formula, then*

$$(x_1 \leq y_1) \wedge \ldots \wedge (x_m \leq y_m) \wedge$$
$$(x_1 \neq y_1 \vee \ldots \vee x_m \neq y_m \vee \psi)$$

is a GOH formula.

Definition 5. *We say that an OH relation R is* GOH *if it can be defined by a GOH formula and that an OH language Γ is GOH if every R in Γ is GOH.*

The complexity of QCSP(Γ) for GOH Γ is known.

Theorem 2. *([13]) Let Γ be a GOH structure. Then QCSP(Γ) is solvable in polynomial time.*

On the other hand, dually-closed languages containing one of the following relations give raise to coNP-hard problems.

Definition 6. *We define the following relations.*

- $I = \{(x, y, z) \in \mathbb{Q}^3 \mid (x = y \to y = z)\}$,
- $LeftPointer = \{(x, y, z) \in \mathbb{Q}^3 \mid (x = y \to y \leq z)\}$,
- $RightPointer = \{(x, y, z) \in \mathbb{Q}^3 \mid (x = y \to y \geq z)\}$,
- $WindMillLeft = \{(x, y, z, v) \in \mathbb{Q}^4 \mid (x = y \to z < v) \wedge (x < z) \wedge (x < v) \wedge (y < z) \wedge (y < v)\}$,
- $WindMillRight = \{(x, y, z, v) \in \mathbb{Q}^4 \mid (x = y \to z > v) \wedge (x > z) \wedge (x > v) \wedge (y > z) \wedge (y > v)\}$

The hardness results we use explicitly in our classification are as follows.

Theorem 3. *The problems:* $\mathrm{QCSP}(\mathbb{Q}; I)$ *(proved in [9]),* $\mathrm{QCSP}(\mathbb{Q}; LeftPointer, RightPointer)$, $\mathrm{QCSP}(\mathbb{Q}; WindMillLeft)$, *and* $\mathrm{QCSP}(\mathbb{Q}; WindMillRight)$ *are coNP-hard.*

We are now ready to state the main result of the paper.

Theorem 4. *Let* Γ *be a dually-closed OH constraint language. Then* Γ *is either GOH and then the problem* $\mathrm{QCSP}(\Gamma)$ *is in P or* $\mathrm{QCSP}(\Gamma)$ *is coNP-hard.*

Proof. In Section 5 in Definition 11 we provide a definition of a blocked OH structure. In Theorem 6 we show that every blocked OH structure is GOH. By Theorem 2, in this case $\mathrm{QCSP}(\Gamma)$ is in P. On the other hand, if Γ is not blocked, then by Theorem 7 either Γ $\forall\exists\wedge$-defines one of the following: WindMillLeft, WindMillRight, I, or Γ pp-defines LeftPointer or RightPointer. In the first case by Lemma 1 and Theorem 3, the problem $\mathrm{QCSP}(\Gamma)$ is coNP-hard. In the second case since LeftPointer and RightPointer are the duals of each other and the language Γ is dually-closed, it follows by Proposition 1 that Γ pp-defines both LeftPointer and RightPointer and thus I. Again, by Lemma 1 and Theorem 3, $\mathrm{QCSP}(\Gamma)$ is coNP-hard. ∎

4   Representation of Ord-Horn Relations

The following is easy to observe.

Proposition 2. *Let* R *be an* n-*ary OH relation then* R *can be defined by a conjunction of all clauses over variables* $\{v_1, \ldots, v_n\}$ *entailed by* R.

Indeed, an n-ary OH relation may be defined by a conjunction φ of Ord-Horn clauses over $\{v_1, \ldots, v_n\}$. Adding clauses entailed by φ does not change the defined relation. We will represent OH relations by sets of so-called OH types entailed by R defined in what follows. We need first some auxiliary definitions.

Partitions of variables. Let V be a set of variables. By $\mathcal{P}_V$ we denote the set of partitions of V and $\mathcal{L}_V = (\mathcal{P}_V, \preceq_V)$ the lattice of partitions of V. The least element in $\mathcal{L}_V$, that is $(\{\{v_1\}, \ldots, \{v_n\}\})$ will be denoted by $\perp_V$, the join and the meet in $\mathcal{L}_V$ by $\sqcup_V$ and $\sqcap_V$, respectively. We write $P_1 \prec_V P_2$ for $P_1, P_2 \in \mathcal{P}_V$ if $P_1 \preceq_V P_2$ and $P_1 \neq P_2$. An element $S \in P$ will be called a component of P, trivial if $|S| = 1$ and non-trivial otherwise. Let $P \in \mathcal{P}_V$, T a non-trivial component of P and $\{T_1, T_2\}$ be a partition of T. By $P \downarrow^V_{\{T_1, T_2\}}$ we denote $P' := (P \setminus \{T\}) \cup \{T_1, T_2\}$. Observe that P' is a predecessor of P in $\mathcal{L}_V$.

Types of OH clauses. Let ψ be an OH clause over variables V. Then ψ imposes a partition P of V defined as follows. We say that two variables $x, y \in V$ are linked in ψ if ψ contains a disjunct $(x \neq y)$ or there is $z \in V$ such that both pairs x, z and z, y are linked in ψ. Now two variables x and y are in the same $S \in P$ if and only if they are linked in ψ.

Example 1. Let $\psi := (v_1 \neq v_2 \vee v_1 \neq v_3 \vee v_4 \neq v_5 \vee v_4 < v_6)$ and $V = \{v_1, \ldots, v_7\}$. Then $P = \{\{v_1, v_2, v_3\}, \{v_4, v_5\}, \{v_6\}, \{v_7\}\}$ is a partition of V imposed by ψ.

We are now ready to define an OH type. Every OH type will represent certain sets of clauses that differ only syntactically.

Definition 7. *Let ψ be an OH clause with* $\mathrm{Var}(\psi) \subseteq V$. *We define a type of an OH clause ψ over V in one of the following ways.*

1. *If ψ is negative, the type of ψ over V is graphically represented by $[P]$ where P is a partition of V imposed by ψ.*
2. *If ψ is strict or positive, that is, of the form $(x_1 \neq y_1 \vee \cdots \vee x_k \neq y_k \vee x \triangleleft y)$ with $\triangleleft$ equal to $<$ in the first and to $\leq$ in the second case, then the type of ψ over V is a triple graphically represented by $[P, S_1, \triangleleft, S_2]$ where P is a partition of V imposed by ψ, S_1 a component of P containing x and S_2 a component of P containing y.*

We say that $\mathtt{Typ}$ is an OH type (or simply a type) over V if it is a type of some OH clause over V. We will omit V if it is clear from the context. We say that an OH type $\mathtt{Typ}$ is negative, positive, strict, or basic if there is an OH clause of type $\mathtt{Typ}$ that is negative, positive, strict, or basic.

Example 2. As an example consider OH clauses $(x_1 \neq x_2 \vee x_1 \neq x_3 \vee y_1 \leq x_1)$ and $(x_2 \neq x_3 \vee x_1 \neq x_3 \vee y_1 \leq x_2)$. Observe that they are both of the same OH type $[\{\{x_1, x_2, x_3\}, \{y_1\}\}, \{y_1\}, \leq, \{x_1, x_2, x_3\}]$ over $V = \{x_1, x_2, x_3, y_1\}$.

Observe that all OH clauses of the same type entail each other. Thus it makes sense to say that a formula or a relation entails a type. We say that an OH formula φ over variables V entails an OH type $\mathtt{Typ}$ over V if φ entails an OH clause, or equivalently all OH clauses of the type $\mathtt{Typ}$ over V. We say that an OH relation entails an OH type $\mathtt{Typ}$ over $\{v_1, \ldots, v_n\}$ if R entails every clause of the type $\mathtt{Typ}$ over $\{v_1, \ldots, v_n\}$. We say that an OH type $\mathtt{Typ}_1$ (set of OH types $\mathtt{Types}_1$) entails an OH type $\mathtt{Typ}_2$ if all OH clauses of the type $\mathtt{Typ}_1$ (conjunction of OH clauses of types in $\mathtt{Types}_1$) entail an OH clause of the type $\mathtt{Typ}_2$. We say that a formula, a relation, a type or a set of types entails a set of types $\mathtt{Types}_2$ if it entails every type in $\mathtt{Types}_2$.

Let R be an n-ary OH relation. By V_R we will denote $\{v_1, \ldots, v_n\}$. The set of all OH types over variables V_R entailed by an n-ary relation R will be denoted by $\mathrm{Ent}(R)$.

Proposition 3. *Let R be an n-ary OH relation. Then the conjunction of all OH clauses of types in $\mathrm{Ent}(R)$ defines R.*

We will now show that for every OH relation R there is a subset $\mathrm{Ent}_{\mathrm{Pr}}(R)$ of $\mathrm{Ent}(R)$ with some desirable properties such that the conjunction of clauses of types in $\mathrm{Ent}_{\mathrm{Pr}}(R)$ also defines R. We first mention properties.

Definition 8. *Let R be an n-ary OH relation. We say that a non-basic OH type $\mathit{Typ} \in \mathrm{Ent}(R)$ is*

- falsely non-negative *if Typ is of the form $[P, S_1, \lhd, S_2]$ and $\mathrm{Ent}(R)$ contains also $[P]$;*
- subsumed *if Typ is of the form $[P, S_1, \lhd, S_2]$ and there is $[P', S'_1, \lhd, S'_2]$ in $\mathrm{Ent}(R)$ for some $P' \prec_{V_R} P$, $S'_1 \subseteq S_1$, and $S'_2 \subseteq S_2$;*
- falsely strict *if Typ is of the form $[P, S_1, <, S_2]$, there are $P' \prec_{V_R} P$, and $S'_i \subseteq S_i$ for $i \in [2]$ such that $\mathrm{Ent}(R)$ contains also $[P', S'_1, \leq, S'_2]$;*
- falsely positive *if Typ is of the form $[P, S_1, \leq, S_2]$ and $\mathrm{Ent}(R)$ contains also $[P, S_1, <, S_2]$;*
- strictly sandwiched *if Typ is of the form $[P, S_1, <, S_2]$ and $\mathrm{Ent}(R)$ contains also both $[P, S_1, <, S_3]$ and $[P, S_3, <, S_2]$ for some $S_3 \in P$;*
- positively sandwiched *if Typ is of the form $[P, S_1, \leq, S_2]$ and $\mathrm{Ent}(R)$ contains also $[P \downarrow^{V_R}_{\{T_1, T_2\}}, S_1, \leq, T_1]$ and $[P \downarrow^{V_R}_{\{T_1, T_2\}}, T_2, \leq, S_2]$ for some predecessor $P \downarrow^{V_R}_{\{T_1, T_2\}}$ of P;*
- falsely non-basic *if Typ is of the form $[P, S_1, <, S_2]$ and for every non-trivial component S in P there is $i \in [2]$ such that $\mathrm{Ent}(R)$ contains both $[P, S_i, \leq, S]$ and $[P, S, \leq, S_i]$.*

We now prove that the desirable set of types $\mathrm{Ent}_{\mathrm{Pr}}(R)$ exists.

Theorem 5. *Let R be an OH relation. Then there is a subset $\mathrm{Ent}_{Pr}(R)$ of $\mathrm{Ent}(R)$ that contains no falsely non-negative, no subsumed, no falsely strict, no falsely positive, no strictly sandwiched, no positively sandwiched, and no falsely non-basic OH types such that the conjunction of clauses of types in $\mathrm{Ent}_{Pr}(R)$ defines R.*

5   Classification

We will here define a notion of a *blocked* OH relation and a *blocked* OH language. We will then show that every blocked OH relation is GOH and if an OH language Γ contains R that is not blocked, then $\mathrm{QCSP}(\Gamma)$ is coNP-hard.

To that end we set an n-ary OH relation R, the set of variables $V_R = \{v_1, \ldots, v_n\}$, the set $\mathrm{Ent}(R)$ of OH types over V_R entailed by R and the set $\mathrm{Ent}_{\mathrm{Pr}}(R)$ from Theorem 5. Beforehand, we need two more auxiliary definitions.

Definition 9. *Let $\mathit{Typ} := [P, S_1, \lhd, S_2]$ be an OH type over variables V_R. If $\lhd$ is equal to $\leq$, then all non-trivial components in P are called free in Typ. If $\lhd$ is $<$, then a non-trivial $S \in P$ is called free in Typ if neither both $[P, S_1, \leq, S]$ and $[P, S, \leq, S_1]$ nor both $[P, S_2, \leq, S]$ and $[P, S, \leq, S_2]$ are in $\mathrm{Ent}(R)$.*

Definition 10. *We say that a type* $\text{Typ} := [P, S_1, \lhd, S_2]$ *over* V_R *is blocked if for every non-trivial component* $S \in P$ *free in* Typ *and every partition* $\{T_1, T_2\}$ *of* S *the set* $\text{Ent}(R)$ *contains* $[P \downarrow^{V_R}_{\{T_1,T_2\}}, T_1, \leq, T_2]$ *or* $[P \downarrow^{V_R}_{\{T_1,T_2\}}, T_2, \leq, T_1]$.

We will now say what is a blocked OH relation and a blocked OH language.

Definition 11. *We say that an OH relation* R *is blocked if every non-basic OH type in* $\text{Ent}_{Pr}(R)$ *is blocked and that OH language* Γ *is blocked if every relation in* Γ *is blocked.*

As we show every blocked OH structure is GOH.

Theorem 6. *Let* Γ *be a blocked OH structure, then* Γ *is GOH.*

We will now turn to the case where Γ contains a relation R that is not blocked. In this case $\text{Ent}_{Pr}(R)$ contains a type Typ which is not blocked. We use R to $\forall\exists\wedge$-define a relation of arity three or four. Such a relation is again OH. We show that every time we can reduce from a problem from Theorem 3.

Lemma 2. *([15]) The set of OH relations is closed under pp-definitions.*

We now give a sufficient and a necessary condition for a tuple with pairwise different entries to be in an OH relation.

Lemma 3. *([15]) Let* R *be an* n-*ary OH relation. The relation* R *contains a tuple* t *satisfying* $t[\pi(1)] < \cdots < t[\pi(n)]$ *for some permutation* π *of* $[n]$ *if and only if it contains a tuple* t' *satisfying* $t'[\pi[j]] < t'[\pi[k]]$ *for all* j, k *in* $[n]$ *such that* $t[j] < t[k]$.

The next two lemmas provide some information about ternary, four-ary OH relations we intend to define from non-blocked OH relations of arbitrary arity.

Lemma 4. *Let* R *be an* n-*ary OH relation. Let* $V_R = \{v_1, \ldots, v_n\}$, *and* $P \in \mathcal{P}_{V_R}$, *and* $T_0, T_3, T_4 \in P$ *be pairwise different, and* $\{T_1, T_2\}$ *a partition of* T_0. *Let*

$$R_1 = \{(w_1, w_2, w_3, w_4) \mid \exists v_1 \cdots \exists v_n \ R(v_1, \ldots, v_n) \wedge \bigwedge_{S \in P \setminus \{T_0, T_3, T_4\}} \bigwedge_{x,y \in S} x = y \wedge$$

$$\bigwedge_{x \in T_1} x = w_1 \wedge \bigwedge_{x \in T_2} x = w_2 \wedge \bigwedge_{x \in T_3} x = w_3 \wedge \bigwedge_{x \in T_4} x = w_4\}$$

We have both of the following:

1. *The relation* R *entails* $[P, T_{i_1}, \lhd, T_{i_2}]$ *where* $i_1, i_2 \in \{0, 3, 4\}$ *if and only if* R_1 *entails* $(v_1 \neq v_2 \vee v_{j_1} \lhd v_{j_2})$ *where for* $k \in [2]$ *we have that* $(j_k = 1)$ *if* $(i_k = 0)$ *and* $j_k = i_k$ *if* $i_k \in \{3, 4\}$.
2. *The relation* R *entails* $[P \downarrow^{V_R}_{\{T_1,T_2\}}, T_i, \lhd, T_j]$ *where* $i, j \in [4]$ *if and only if* R_1 *entails* $(v_i \lhd v_j)$.

Lemma 5. *Let R be an n-ary OH relation. Let $V_R = \{v_1, \ldots, v_n\}$, and $P \in \mathcal{P}_{V_R}$, and $T_0, T_3 \in P$ be different, and $\{T_1, T_2\}$ a partition of T_0. Let*

$$R_1 = \{(w_1, w_2, w_3) \mid \exists v_1 \cdots \exists v_n \ R(v_1, \ldots, v_n) \wedge \bigwedge_{S \in P \setminus \{T_0, T_3\}} \bigwedge_{x, y \in S} x = y \wedge$$

$$\bigwedge_{x \in T_1} x = w_1 \wedge \bigwedge_{x \in T_2} x = w_2 \wedge \bigwedge_{x \in T_3} x = w_3$$

We have both of the following:

1. *The relation R entails $[P, T_{i_1}, \lhd, T_{i_2}]$ where $i_1, i_2 \in \{0, 3\}$ if and only if R_1 entails $(v_1 \neq v_2 \vee v_{j_1} \lhd v_{j_2})$ where for $k \in [2]$ we have that $(j_k = 1)$ if $(i_k = 0)$ and $j_k = i_k$ if $i_k = 3$.*
2. *The relation R entails $[P \downarrow_{\{T_1, T_2\}}^{V_R}, T_i, \lhd, T_j]$ where $i, j \in [3]$ if and only if R_1 entails $(v_i \lhd v_j)$.*

We are now ready to prove that a non-blocked OH structure Γ defines one of the relations from Section 3.

Theorem 7. *Let R be an OH relation that is not blocked. Then one of the following holds:*

- *$R \ \forall\exists\wedge$-defines WindMillLeft, WindMillRight, or I;*
- *R pp-defines LeftPointer or RightPointer.*

Proof. If R is not blocked, then $\mathrm{Ent}_{\mathrm{Pr}}(R)$ contains a type Typ which is neither basic nor blocked. The proof consists of three parts in each of which we consider a different form of Typ. Before we move on to the first part, we present an auxiliary lemma.

Lemma 6. *Let R be a four-ary OH relation that entails $(v_1 \neq v_2 \vee v_3 \leq v_4)$ and contains tuples: t_d such that $(t_d[1] = t_d[2] < t_d[3] \leq t_d[4])$, t_l such that $(t_l[1] < t_l[2] < t_l[4] < t_l[3])$, and t_g such that $(t_g[2] < t_g[1] < t_g[4] < t_g[3])$. Then $R \ \forall\exists\wedge$-defines I or WindMillLeft.*

Part One. We look at a type Typ of the form $[P, T_3, <, T_4]$. Since it is not blocked there is a free component $T_0 \in P$ and disjoint T_1, T_2 with $T_0 = T_1 \cup T_2$ such that neither $[P \downarrow_{\{T_1, T_2\}}^{V_R}, T_1, \leq, T_2]$ nor $[P \downarrow_{\{T_1, T_2\}}^{V_R}, T_2, \leq, T_1]$ is in $\mathrm{Ent}(R)$. We consider the relation R_1 defined as in Lemma 4. Since Typ is entailed by R, we have by Lemma 4 that R_1 entails $(v_1 \neq v_2 \vee v_3 < v_4)$. Since Typ is not blocked, neither $[P \downarrow_{\{T_1, T_2\}}^{V_R}, T_1, \leq, T_2]$ nor $[P \downarrow_{\{T_1, T_2\}}^{V_R}, T_2, \leq, T_1]$ is in $\mathrm{Ent}(R)$. By Lemma 4, it follows that R_1 entails neither $(v_1 \leq v_2)$ nor $(v_2 \leq v_1)$. Thus, R_1 has tuples t_{BG} such that $(t_{\mathrm{BG}}[1] > t_{\mathrm{BG}}[2])$ and t_{BL} such that $(t_{\mathrm{BL}}[1] < t_{\mathrm{BL}}[2])$.

The type Typ is not strictly sandwiched and therefore $\mathrm{Ent}_{\mathrm{Pr}}(R)$ does not contain either $[P, T_3, <, T_0]$ or $[P, T_0, <, T_4]$. We consider the first situation and we will show that in this case $R \ \forall\exists\wedge$-defines WindMillLeft. The other case may be treated analogously and it can be shown that WindMillRight has a $\forall\exists\wedge$-definition by R. The type $\mathtt{Typ}$ is not falsely non-negative and hence $\mathrm{Ent}(R)$ does not contain $[P]$. By Lemma 4, it follows that R_1 does not entail $(v_1 \neq v_2)$. Thus it has a tuple t_0 such that $t_0[1] = t_0[2]$. Since $[P, T_3, <, T_0]$ is not in $\mathrm{Ent}(R)$, by Lemma 4, the relation R_1 does not entail $(v_1 = v_2 \rightarrow v_1 > v_3)$. Thus, R_1 has a tuple t_1 such that $(t_1[1] = t_1[2] \leq t_1[3] < t_1[4])$. Furthermore, since T_0 is free in $\mathtt{Typ}$, the relation R_1 does not entail either $[P, T_0, \leq, T_3]$ or $[P, T_3, \leq, T_0]$. Hence R_1 contains a tuple t_2 such that $(t_2[1] = t_2[2] \neq t_2[3])$. Let $\mathrm{ll} : \mathbb{Q}^2 \rightarrow \mathbb{Q}$ be a binary operation defined in [2]. Then $t_{\mathrm{d}} = \mathrm{ll}(t_1, t_2)$, when ll is applied to t_1, t_2 coordinatewise, satisfies $(t_{\mathrm{d}}[1] = t_{\mathrm{d}}[2] < t_{\mathrm{d}}[3] < t_{\mathrm{d}}[4])$.

The type $\mathtt{Typ}$ is not falsely strict in $\mathrm{Ent}(R)$ and therefore $\mathrm{Ent}(R)$ does not contain $[P \downarrow_{\{T_1,T_2\}}^{V_R}, T_3, \leq, T_4]$. Thus, the relation R_1 contains a tuple t_3 such that $(t_3[3] > t_3[4])$. Now, by applying Lemma 3 to tuples $t_{\mathrm{BL}}, t_{\mathrm{BG}}, t_{\mathrm{d}}, t_3$, we obtain that R contains tuples t_l such that $(t_l[1] < t_l[2] < t_l[4] < t_l[3])$ and t_{g} satisfying $(t_{\mathrm{g}}[2] < t_{\mathrm{g}}[1] < t_{\mathrm{g}}[4] < t_{\mathrm{g}}[3])$.

Now, since R_1 entails $(v_1 \neq v_2 \vee v_3 < v_4)$ (and also $(v_1 \neq v_2 \vee v_3 \leq v_4)$) and contains t_d, t_l, and t_g, by Lemma 6, the relation R_1 $\forall\exists\wedge$-defines either I or WindMillLeft.

Before we move on to the second case, we present another auxiliary lemma.

Lemma 7. *Let R be a ternary OH relation that entails LeftPointer and contains tuples:* (1) *t_l such that $t_l(v_3) < t_l(v_1) < t_l(v_2)$,* (2) *$t_g$ such that $t_g(v_3) < t_g(v_2) < t_g(v_1)$, and* (3) *$t_e$ such that $t_e(v_3) = t_e(v_2) = t_e(v_1)$. Then R pp-defines LeftPointer or I.*

Part Two. Here, we consider the situation where a non-blocked $\mathtt{Typ}$ is of the form $[P, S, \leq, T_3]$ and $T_0 \in P$, for which there is a partition $\{T_1, T_2\}$ such that neither $[P \downarrow_{\{T_1,T_2\}}^{V_R}, T_1, \leq, T_2]$ nor $[P \downarrow_{\{T_1,T_2\}}^{V_R}, T_2, \leq, T_1]$ is in $\mathrm{Ent}(R)$, is equal to S or T_3. We assume that T_0 is equal to S and show that in this case R pp-defines either LeftPointer or I. The other case is analogous and one can show that then R pp-defines either RightPointer or I. We have therefore that R entails a type $[P, T_0, \leq, T_3]$. Consider the relation R_1 defined as in Lemma 5. Since neither $[P \downarrow_{\{T_1,T_2\}}^{V_R}, T_1, \leq, T_2]$ nor $[P \downarrow_{\{T_1,T_2\}}^{V_R}, T_2, \leq, T_1]$ is entailed by R, it follows by the lemma that neither $(v_1 \leq v_2)$ nor $(v_1 \geq v_2)$ is entailed by R_1. Thus R_1 contains tuples t_{BG} such that $(t_{\mathrm{BG}}[1] > t_{\mathrm{BG}}[2])$ and t_{BL} such that $(t_{\mathrm{BL}}[1] < t_{\mathrm{BL}}[2])$. Further, the type $\mathtt{Typ}$ is not subsumed in $\mathrm{Ent}(R)$. Thus, R entails neither $[P \downarrow_{\{T_1,T_2\}}^{V_R}, T_1, \leq, T_3]$ nor $[P \downarrow_{\{T_1,T_2\}}^{V_R}, T_2, \leq, T_3]$. By Lemma 5, the relation R_1 entails neither $(v_1 \leq v_3)$ nor $(v_2 \leq v_3)$. Thus R_2 has tuples t_1 such that $(t_1[3] < t_1[1])$ and t_2 such that $(t_2[3] < t_2[2])$. Now by applying Lemma 3 to tuples $t_{\mathrm{BG}}, t_{\mathrm{BL}}, t_1$, and t_2, we obtain that R_2 contains tuples t_l such that $(t_l[3] < t_l[1] < t_l[2])$ and t_g such that $(t_g[3] < t_g[2] < t_g[1])$. To complete this part of the proof of the theorem we intend to use Lemma 7. To this end, we first use Lemma 5 to show that R_2 entails $(v_1 \neq v_2 \vee v_1 \leq v_3)$. Indeed, it holds since R entails $\mathtt{Typ}$. Finally, we argue that a tuple t_e with all entries equal is contained in R_1. Since $\mathtt{Typ}$ is not falsely non-negative, the set $\mathrm{Ent}_{\mathrm{Pr}}(R)$ does not contain $[P]$ and hence by Lemma 5 the relation R_1 does not entail $(v_1 \neq v_2)$. Thus R_1 contains a tuple t_0 such that $t_0[1] = t_0[2]$. The type Typ is not falsely positive and hence R does not entail $[P, T_2, <, T_3]$. By Lemma 5, it follows that R_1 does not entail $(v_1 \neq v_2 \vee v_1 < v_3)$. Thus, the relation R_1 contains a tuple t such that $(t[1] = t[2] \geq t[3])$. Since R_1 entails $(v_1 \neq v_2 \vee v_1 \leq v_3)$, it follows that R_2 has a tuple t_e with all entries equal. Now we can use Lemma 7 and complete the second part of the proof.

Part Three. Here a non-blocked type from $\mathrm{Ent}_{Pr}(R)$ is of the form $[P, T_3, \leq, T_4]$ and a component $T_0 \in P$ for which there is a partition $\{T_1, T_2\}$ of T_0 such that neither $[P \downarrow^{V_R}_{\{T_1, T_2\}}, T_1, \leq, T_2]$ nor $[P \downarrow^{V_R}_{\{T_1, T_2\}}, T_2, \leq, T_1]$ belongs to the set $\mathrm{Ent}(R)$, is equal neither to T_3 nor to T_4. We consider the relation R_1 defined as in Lemma 4. Since neither $[P \downarrow^{V_R}_{\{T_1, T_2\}}, T_1, \leq, T_2]$ nor $[P \downarrow^{V_R}_{\{T_1, T_2\}}, T_2, \leq, T_1]$ is entailed by R, it follows by the lemma that neither $(v_1 \leq v_2)$ nor $(v_1 \geq v_2)$ is entailed by R_1. Thus R_2 contains tuples t_{BG} such that $(t_{\mathrm{BG}}[1] > t_{\mathrm{BG}}[2])$ and t_{BL} such that $(t_{\mathrm{BL}}[1] < t_{\mathrm{BL}}[2])$. Furthermore, since R entails Typ, it follows, by Lemma 4 that R_3 entails $(v_1 \neq v_2 \vee v_3 \leq v_4)$.

We first assume that R_1 contains a tuple t_d such that $(t_d(v_1) = t_d(v_2) < t_d(v_3) \leq t_d(v_4))$ or $(t_d(v_3) \leq t_d(v_4) < t_d(v_1) = t_d(v_2))$. We consider only the first case in which we show that R_3 $\forall\exists\wedge$-define WindMillLeft. (In the other case one can analogously show that R_3 $\forall\exists\wedge$-defines WindMillRight.) Since Typ is not subsumed, the relation R does not entail $[P \downarrow^{V_R}_{\{T_1, T_2\}}, T_3, \leq, T_4]$. By Lemma 4, the relation R does not entail $(v_3 \leq v_4)$. Thus R_1 has a tuple t_h satisfying $(t_h[3] > t_h[4])$. Now by applying Lemma 3 to tuples $t_d, t_h, t_{\mathrm{BG}}, t_{\mathrm{BL}}$, we obtain that R_3 has tuples t_l such that $(t_l[1] < t_l[2] < t_l[4] < t_l[3])$, and t_g such that $(t_g[2] < t_g[1] < t_g[4] < t_g[3])$. To complete this case, we take use of Lemma 6.

We now assume that R_3 entails $(v_1 = v_2 \rightarrow v_3 \leq v_1 = v_2 \leq v_4)$. We will now show that either R_3 contains a tuple t_a satisfying $t_a(v_1) < t_a(v_3)$ and $t_a(v_2) < t_a(v_3)$ or a tuple t_b satisfying $t_b(v_4) < t_b(v_1)$ and $t_b(v_4) < t_b(v_2)$. Suppose it is not the case. Then R_3 entails $(v_1 \geq v_3 \vee v_2 \geq v_3) \wedge (v_4 \geq v_1 \vee v_4 \geq v_2)$. It follows that R_3 entails $(v_3 \leq v_4)$, $(v_1 \geq v_3 \wedge v_4 \geq v_2)$, $(v_4 \geq v_1 \wedge v_2 \geq v_3)$. But, the first case contradicts the fact that Typ is not subsumed, wherever the other two that Typ is not positively sandwiched. To see it, we use Lemma 4. Indeed, if R_1 entails $(v_3 \leq v_4)$, then R entails $[P \downarrow^{V_R}_{\{T_1, T_2\}}, T_3, \leq, T_4]$ which implies that Typ is subsumed. If R_1 entails $(v_1 \geq v_3 \wedge v_4 \geq v_2)$, then R entails both $[P \downarrow^{V_R}_{\{T_1, T_2\}}, T_3, \leq, T_1]$ and $[P \downarrow^{V_R}_{\{T_1, T_2\}}, T_2, \leq, T_4]$ which implies that Typ is positively sandwiched. We treat the third case in the similar way.

By the previous paragraph, we can assume that R_3 contains either t_a or t_b. We consider the second case only and using Lemma 7, we will show that then R_1 pp-defines either LeftPointer or I. In the other case one can analogously show that R_1 pp-defines RightPointer or I. Let R_2 be the relation $\{(v_1, v_2, v_4) \mid (\exists v_3 \ R_1(v_1, v_2, v_3, v_4))\}$. We intend to use Lemma 7 in order to show that R_4 is LeftPointer or I. Observe that R_2 entails $(v_1 \neq v_2 \vee v_2 \leq v_3)$. Furthermore, Typ is not falsely positive in $\mathrm{Ent}(R)$, and hence R does not entail $[P, T_3, <, T_4]$.

Thus, by Lemma 4, the relation R_2 does not entail $(v_1 \neq v_2 \vee v_3 < v_4)$. It implies that R_1 contains a tuple t_e satisfying $t_e[3] = t_e[1] = t_e[2] = t_e[4]$ and R_2 a tuple t'_e satisfying $t'_e[1] = t'_e[2] = t'_e[3]$. Since R_1 contains $t_a, t_{\mathrm{BL}}, t_{\mathrm{BG}}$, the relation R_2 contains: t'_a such that $t'_a[3] < t'_a[1]$ and $t'_a[3] < t'_a[2]$, t'_{BL} such that $t'_{\mathrm{BL}}[1] < t'_{\mathrm{BL}}[2]$ and t'_{BG} satisfying $t'_{\mathrm{BG}}[1] > t'_{\mathrm{BG}}[2]$. We first use Lemma 3 to argue that R_2 contains t'_l such that $t'_l(v_3) < t'_l(v_1) < t'_l(v_2)$ and t'_g such that $t'_g(v_3) < t'_g(v_2) < t'_g(v_1)$. Now, we use Lemma 7 to argue that R_2 pp-defines either LeftPointer or I.

References

1. Schaefer, T.J.: The complexity of satisfiability problems. In: Proceedings of the Symposium on Theory of Computing (STOC), pp. 216–226 (1978)
2. Bodirsky, M., Kára, J.: The complexity of temporal constraint satisfaction problems. Journal of the ACM 57(2), 1–41 (2009); An extended abstract appeared in the Proceedings of the Symposium on Theory of Computing (STOC 2008)
3. Garey, M., Johnson, D.: A guide to NP-completeness. CSLI Press, Stanford (1978)
4. Vilain, M., Kautz, H., van Beek, P.: Constraint propagation algorithms for temporal reasoning: A revised report. Reading in Qualitative Reasoning about Physical Systems, 373–381 (1989)
5. Nebel, B., Bürckert, H.J.: Reasoning about temporal relations: A maximal tractable subclass of Allen's interval algebra. Journal of the ACM 42(1), 43–66 (1995)
6. Möhring, R.H., Skutella, M., Stork, F.: Scheduling with and/or precedence constraints. SIAM Journal on Computing 33(2), 393–415 (2004)
7. Börner, F., Bulatov, A.A., Jeavons, P., Krokhin, A.A.: Quantified constraints: Algorithms and complexity. In: CSL, pp. 58–70 (2003)
8. Chen, H.: Meditations on quantified constraint satisfaction. In: Logic and Program Semantics, pp. 35–49 (2012)
9. Bodirsky, M., Chen, H.: Quantified equality constraints. SIAM Journal on Computing 39(8), 3682–3699 (2010); A preliminary version of the paper appeared in the proceedings of LICS 2007
10. Chen, H., Müller, M.: An algebraic preservation theorem for aleph-zero categorical quantified constraint satisfaction. In: LICS, pp. 215–224 (2012)
11. Charatonik, W., Wrona, M.: Tractable quantified constraint satisfaction problems over positive temporal templates. In: Cervesato, I., Veith, H., Voronkov, A. (eds.) LPAR 2008. LNCS (LNAI), vol. 5330, pp. 543–557. Springer, Heidelberg (2008)
12. Charatonik, W., Wrona, M.: Quantified positive temporal constraints. In: Kaminski, M., Martini, S. (eds.) CSL 2008. LNCS, vol. 5213, pp. 94–108. Springer, Heidelberg (2008)
13. Chen, H., Wrona, M.: Guarded ord-horn: A tractable fragment of quantified constraint satisfaction. In: TIME, pp. 99–106 (2012)
14. Bodirsky, M., Nešetřil, J.: Constraint satisfaction with countable homogeneous templates. Journal of Logic and Computation 16(3), 359–373 (2006)
15. Wrona, M.: Syntactically characterizing local-to-global consistency in ORD-horn. In: Milano, M. (ed.) CP 2012. LNCS, vol. 7514, pp. 704–719. Springer, Heidelberg (2012)

The Dynamic Descriptive Complexity
of k-Clique

Thomas Zeume*

TU Dortmund University, Dortmund, Germany
thomas.zeume@cs.tu-dortmund.de

Abstract. In this work the dynamic descriptive complexity of the k-clique query is studied in a framework introduced by Patnaik and Immerman. It is shown that when edges may only be inserted then k-clique can be maintained by a quantifier-free update program of arity $k - 1$, but it cannot be maintained by a quantifier-free update program of arity $k-2$ (even in the presence of unary auxiliary functions). This establishes an arity hierarchy for graph queries for quantifier-free update programs under insertions. The proof of the lower bound uses upper and lower bounds for Ramsey numbers.

1 Introduction

The k-clique query — does a given graph contain a k-clique? — can be expressed by an existential first-order formula with k quantifiers. In this work we study the descriptive complexity of the k-clique query in a setting where edges may be inserted dynamically into a graph. In particular we are interested in lower bounds for the resources necessary to answer this query dynamically.

The dynamic descriptive complexity framework (short: dynamic complexity) introduced by Patnaik and Immerman [14] models the setting of dynamically changing graphs. For a graph subject to changes, auxiliary relations are maintained with the intention to help answering a query $\mathcal{Q}$. When an insertion (or, in the general setting, a deletion) of an edge occurs, every auxiliary relation is updated through a first-order query that can refer to both the graph itself and the auxiliary relations. The query $\mathcal{Q}$ is maintained by such a program, if one designated auxiliary relation always stores the current query result. The class of all queries maintainable by first-order update programs is called DYNFO.

Since k-clique can be expressed in existential first-order logic, it can be trivially maintained by a first-order update program. Therefore for characterizing the precise dynamic complexity of this query we need to look at fragments of DYNFO. It turns out that k-clique can still be maintained when the update formulas are not allowed to use quantifiers at all and auxiliary relations may only have restricted arity. We obtain the following characterization.

Main result: When only edge insertions are allowed then k-clique ($k \geq 3$) can be maintained by a quantifier-free update program of arity $k - 1$, but it cannot be maintained by a quantifier-free update program of arity $k - 2$.

* The author acknowledges the financial support by DFG grant SCHW 678/6-1.

E. Csuhaj-Varjú et al. (Eds.): MFCS 2014, Part I, LNCS 8634, pp. 547–558, 2014.

Actually we prove that every property expressible by a positive existential first-order formula with k quantifiers and, possibly, negated equality atoms can be maintained by a $(k-1)$-ary quantifier-free program under insertions.

In order to understand why the lower bound contained in the above result is interesting, we shortly discuss the status quo of lower bound methods for the dynamic complexity framework. Up to now very few lower bounds are known; all of them for fragments of DYNFO obtained by either bounding the arity of the auxiliary relations or by restricting the usage of quantifiers (or by restricting both). Usually those bounds have been stated only for the setting where both insertions and deletions are allowed. We emphasize that our lower bound for the insertion-only setting immediately transfers to this more general setting.

The study of bounded arity auxiliary relations was started by Dong and Su [4]. They exhibited concrete graph queries that cannot be maintained in unary DYNFO, and they showed that DYNFO has an arity hierarchy for general (that is non-graph) queries. Both results rely on previously obtained static lower bounds.

Hesse started the study of the quantifier-free fragment of DYNFO (short: DYNPROP) in [13]. Although this fragment appears to be rather weak at first glance, deterministic reachability [13] and regular languages [9] can be maintained in DYNPROP. In [9], Gelade et al. also provided first lower bounds. They proved that non-regular languages as well as the alternating reachability problem cannot be maintained in this fragment. The use of very restricted graphs in the proof of the latter result implies that there is a $\exists^*\forall\exists^*$FO-definable query that cannot be maintained in DYNPROP. In [16] it was shown that reachability and 3-clique cannot be maintained in the binary quantifier-free fragment of DYNFO.

In general, it is a difficult task to prove lower bounds in the dynamic complexity setting; even when update formulas are restricted to the quantifier-free fragment of first-order logic. We are not at the point where we can, when given a query, apply a set of tools in order to prove that the query cannot be maintained in DYNPROP. Finding more queries that cannot be maintained in DYNPROP seems to be a reasonable approach towards finding more generic proof methods.

The lower bound provided by the main result follows this approach and is interesting in two ways. First, it exhibits, for every k, a query in $\exists^k$FO that cannot be maintained in $(k-2)$-ary DYNPROP, even when only insertions are allowed. We believe that finding simple queries that cannot be maintained will advance the understanding of dynamic complexity. Second, the main result establishes the first arity hierarchy for graph queries, although for a weak fragment of DYNFO and for insertions only.

The proof of the lower bound uses upper and lower bounds for Ramsey numbers. This has been quite curious for us.

A natural question is how far this method to prove lower bounds can be pushed. As an intermediate step between the quantifier-free fragment and DYNFO itself, Hesse suggested the study of quantifier-free update programs with auxiliary functions [13]. The main result can be extended as follows.

Extension of the main result: k-clique ($k \geq 3$) cannot be maintained by a quantifier-free update program of arity $k-2$ with unary auxiliary functions.

So far there have been only two lower bounds for dynamic classes with auxiliary functions. Alternating reachability was actually shown to be not maintainable in the quantifier-free fragment of DYNFO even in the presence of a successor and a predecessor function [9]. Further, in [16], it was shown that reachability cannot be maintained in unary DYNPROP with unary auxiliary functions. Thus our extension is a first lower bound for arbitrary unary auxiliary functions and k-ary auxiliary relations, for every fixed k. We also explain why the lower bound technique does not extend to binary auxiliary functions. To this end we show that binary DYNQF can maintain every boolean graph property when the domain is large with respect to the actually used domain.

Related work. Up to now we mentioned only work immediately relevant for this work. For the interested reader we give a short list of further related work.

Further lower bounds have been shown in [1, 2, 10]. Further upper bounds have been shown in [8, 12, 15, 10]. Many other aspects such as whether the auxiliary relations are determined by the current structure (see e.g. [14, 3, 10]) and the presence of an order (see e.g. [10]) have been studied.

Outline. In Section 2 we fix some of our notations and in Section 3 we recapitulate the formal dynamic complexity framework. In Sections 4 and 5 we prove the upper and lower bound of the main result, respectively. In Section 6 we study the extension of DYNPROP by auxiliary functions.

2 Preliminaries

We fix some of our notations. Most notations are reused from [16]. The reader can feel free to skip this section and return when encountering unknown notations.

A *domain* D is a finite set. A (relational) *schema* τ consists of a set τ of relation symbols together with an arity function $\mathrm{Ar} : \tau \to \mathbb{N}$. A *database* $\mathcal{D}$ of schema τ with domain D is a mapping that assigns to every relation symbol $R \in \tau$ a relation of arity $\mathrm{Ar}(R)$ over D. A τ-*structure* $\mathcal{S}$ is a pair $(D, \mathcal{D})$ where $\mathcal{D}$ is a database with schema τ and domain D. If $\mathcal{S}$ is a structure over domain D and D' is a subset of D, then the substructure of $\mathcal{S}$ induced by D' is denoted by $\mathcal{S} \restriction D'$.

A tuple $\vec{a} = (a_1, \ldots, a_k)$ is $\prec$-*ordered* with respect to a linear order[1] $\prec$ of the domain, if $a_1 \prec \ldots \prec a_k$. The k-*ary atomic type* $\langle \mathcal{S}, \vec{a} \rangle$ of $\vec{a}$ over D with respect to $\mathcal{S}$ is the set of all atomic formulas $\varphi(\vec{x})$ with $\vec{x} = (x_1, \ldots, x_k)$ for which $\varphi(\vec{a})$ holds in $\mathcal{S}$, where $\varphi(\vec{a})$ is short for the substitution of $\vec{x}$ by $\vec{a}$ in φ. As we only consider atomic types here, we will often simply say type instead of atomic type.

For a set A, denote by A^k the set of all k-tuples over A and, following [11], by $[A]^k$ the set of all k-element subsets of A. A k-*hypergraph* G is a pair (V, E) where V is a set and E is a subset of $[V]^k$. If $E = [V]^k$ then G is called *complete*. An r-*coloring col* of G is a mapping that assigns to every edge in E a color from $\{1, \ldots, r\}$. A r-*colored k-hypergraph* is a pair (G, col) where G is a k-hypergraph and *col* is a r-coloring of G. If the name of the r-coloring is not important we also say G *is r-colored*.

---

[1] All linear orders in this work are strict.

3   Dynamic Setting

The following introduction to dynamic descriptive complexity is borrowed from previous work [16, 17]. Although the focus of this work is on maintaining the k-clique query under insertions, we introduce the general dynamic complexity framework in order to be able to give a broader discussion of concrete results.

A *dynamic instance* of a query $\mathcal{Q}$ is a pair $(\mathcal{D}, \alpha)$, where $\mathcal{D}$ is a database over a domain D and α is a sequence of modifications to $\mathcal{D}$, i.e. a sequence of insertions and deletions of tuples over D. The dynamic query $\mathrm{DYN}(\mathcal{Q})$ yields as result the relation that is obtained by first applying the modifications from α to $\mathcal{D}$ and then evaluating the query $\mathcal{Q}$ on the resulting database. The database resulting from applying a modification δ to a database $\mathcal{D}$ is denoted by $\delta(\mathcal{D})$. The result $\alpha(\mathcal{D})$ of applying a sequence of modifications $\alpha = \delta_1 \dots \delta_m$ to a database $\mathcal{D}$ is defined by $\alpha(\mathcal{D}) \stackrel{\text{def}}{=} \delta_m(\dots (\delta_1(\mathcal{D})) \dots)$.

Dynamic programs, to be defined next, consist of an initialization mechanism and an update program. The former yields, for every (input) database $\mathcal{D}$, an initial state with initial auxiliary data. The latter defines the new state of the dynamic program for each possible modification δ.

A *dynamic schema* is a tuple where τ_{in} and τ_{aux} are the schemas of the input database and the auxiliary database, respectively. We always let $\tau \stackrel{\text{def}}{=} \tau_{\text{in}} \cup \tau_{\text{aux}}$.

Definition 1. *(Update program) An* update program $\mathcal{P}$ *over a dynamic schema* (τ_{in}, τ_{aux}) *is a set of first-order formulas (called* update formulas *in the following) that contains, for every* $R \in \tau_{aux}$ *and every* $\delta \in \{\mathrm{INS}_S, \mathrm{DEL}_S\}$ *with* $S \in \tau_{in}$, *an* update formula $\phi_\delta^R(\vec{x}; \vec{y})$ *over the schema* τ *where* $\vec{x}$ *and* $\vec{y}$ *have the same arity as* S *and* R, *respectively.*

A *program state* $\mathcal{S}$ over dynamic schema $(\tau_{\text{in}}, \tau_{\text{aux}})$ is a structure $(D, \mathcal{I}, \mathcal{A})$ where D is a finite domain, $\mathcal{I}$ is a database over the input schema (the *current database*) and $\mathcal{A}$ is a database over the auxiliary schema (the *auxiliary database*). The *semantics of update programs* is as follows. For a modification $\delta(\vec{a})$, where $\vec{a}$ is a tuple over D, and program state $\mathcal{S} = (D, \mathcal{I}, \mathcal{A})$ we denote by $P_\delta(\mathcal{S})$ the state $(D, \delta(\mathcal{I}), \mathcal{A}')$, where $\mathcal{A}'$ consists of relations $R' \stackrel{\text{def}}{=} \{\vec{b} \mid \mathcal{S} \models \phi_\delta^R(\vec{a}; \vec{b})\}$. The effect $P_\alpha(\mathcal{S})$ of a modification sequence $\alpha = \delta_1 \dots \delta_m$ to a state $\mathcal{S}$ is the state $P_{\delta_m}(\dots (P_{\delta_1}(\mathcal{S})) \dots)$.

Definition 2. *(Dynamic program) A* dynamic program *is a triple* (P, INIT, Q), *where*

- *P is an update program over some dynamic schema* (τ_{in}, τ_{aux}),
- INIT *is a mapping that maps* τ_{in}-*databases to* τ_{aux}-*databases, and*
- $Q \in \tau_{aux}$ *is a designated* query symbol.

A dynamic program $\mathcal{P} = (P, \mathrm{INIT}, Q)$ *maintains* a dynamic query $\mathrm{DYN}(\mathcal{Q})$ if, for every dynamic instance $(\mathcal{D}, \alpha)$, the relation $\mathcal{Q}(\alpha(\mathcal{D}))$ coincides with the query relation $Q^{\mathcal{S}}$ in the state $\mathcal{S} = P_\alpha(\mathcal{S}_{\mathrm{INIT}}(\mathcal{D}))$, where $\mathcal{S}_{\mathrm{INIT}}(\mathcal{D})$ is the initial state for $\mathcal{D}$, i.e. $\mathcal{S}_{\mathrm{INIT}}(\mathcal{D}) \stackrel{\text{def}}{=} (D, \mathcal{D}, \mathrm{INIT}_{\mathrm{aux}}(\mathcal{D}))$.

Definition 3. *(DynFO and DynProp) DynFO is the class of all dynamic queries that can be maintained by dynamic programs with first-order update formulas and arbitrary initialization mappings. DynProp is the subclass of DynFO, where update formulas are not allowed to use quantifiers. A dynamic program is k-ary if the arity of its auxiliary relation symbols is at most k. By k-ary DynProp (resp. DynFO) we refer to dynamic queries that can be maintained with k-ary dynamic programs.*

In the literature, classes with restricted initialization mappings have been studied as well, see [17] for a discussion. The choice made here is not a real restriction as lower bounds proved for arbitrary initialization hold for restricted initialization as well. On the other hand, our upper bounds also hold for other settings of initialization; with the single exception of Theorem 6, which requires arbitrary initialization. Furthermore our results also hold in the related setting where domains can be infinite.

4 k-Clique Can Be Maintained under Insertions with Arity $k - 1$

In this section we prove that the k-clique query can be maintained in $(k-1)$-ary DynProp when only edge insertions are allowed. Instead of proving this result directly, we show that the class of all semi-positive existential first-order queries can be maintained in DynProp under insertions.

A *positive existential first-order query* over schema τ is a query that can be expressed by a first-order formula of the form $\varphi(\vec{y}) = \exists \vec{x} \psi(\vec{x}, \vec{y})$ where ψ is a quantifier-free formula that contains only literals of the form $z_i = z_j$ and $R(\vec{z})$ with $R \in \tau$. For *semi-positive existential first-order queries* literals of the form $z_i \neq z_j$ are allowed as well.

We will prove that every semi-positive existential first-order query can be maintained in DynProp when only insertions are allowed. More precisely, it will be shown that $(k-1)$-ary DynProp is sufficient for boolean queries with k existential quantifiers. In particular k-Clique can be maintained in $(k-1)$-ary DynProp. Before turning to the proof we give some intuition.

Example 1. We show how to maintain 3-Clique in binary DynProp under insertions. The very simple idea is to use an additional binary auxiliary relation R that stores all edges whose insertion would complete a triangle. Hence a tuple (a, b) is inserted into R as soon as deciding whether there is a 3-clique containing the nodes a and b only depends on those two nodes. More precisely (a, b) is added to R, if an edge (c, a) is inserted to the input graph and the edge (c, b) is already present (or vice versa).

Thus the update formula for R is

$$\phi_{\text{INSE}}^{R}(u, v; x, y) \stackrel{\text{def}}{=} u \neq v \wedge x \neq y \wedge \Big(\big(E\{u, y\} \wedge v = x \big) \vee \big(E\{u, x\} \wedge v = y \big)$$

$$\vee \big(E\{v, y\} \wedge u = x \big) \vee \big(E\{v, x\} \wedge u = y \big) \Big)$$

where $E\{x, y\}$ is an abbreviation for $E(x, y) \vee E(y, x)$.

The update formula for the query symbol Q is

$$\phi^Q_{\text{INS}E}(u, v; x, y) = Q \vee R(u, v)$$

$\square$

The general proof for arbitrary semi-positive existential first-order properties extends the approach taken in the example.

Theorem 1. *An ℓ-ary query expressible by a semi-positive existential first-order formula with k quantifiers can be maintained under insertions in $(\ell + k - 1)$-ary* DynProp.

Proof sketch.   For simplicity we restrict the sketch to boolean graph queries. The proof easily carries over to arbitrary semi-positive existential queries.

Basically a semi-positive existential sentence with k quantifiers can state which subgraphs with k nodes shall occur in a graph. Therefore it is sufficient to construct a dynamic quantifier-free program that maintains whether the input graph contains a subgraph H. Such a program can work as follows. For every induced, proper subgraph $H' = \{u_1, \ldots, u_m\}$ of H, the program maintains an auxiliary relation that stores all tuples $\vec{a} = (a_1, \ldots, a_m)$ such that inserting H' into $\{a_1, \ldots, a_m\}$ (with a_i corresponding to u_i) yields a graph that contains H.

In particular, auxiliary relations have arity of at most $k - 1$ (as only proper subgraphs of H have a corresponding auxiliary relation). Furthermore the graph H is contained in the input graph whenever the value of the 0-ary relation corresponding to the empty subgraph of H is true. In the example above, the relation R is the relation for the subgraph of the 3-clique graph that consists of a single edge, and the designated query relation is the 0-ary relation for the empty subgraph.

Those auxiliary relations can be updated as follows. Assume that a tuple $\vec{a} = (a_1, \ldots, a_m)$ is contained in the relation corresponding to H'. If, after the insertion of an edge with end point a_m, every edge from u_m in H' has a corresponding edge from a_m in the graph induced by $\{a_1, \ldots, a_m\}$, then the tuple $\vec{a}' = (a_1, \ldots, a_{m-1})$ has to be inserted into the auxiliary relation for the subgraph $H' \upharpoonright \{u_1, \ldots, u_{m-1}\}$. This is because inserting the graph $H' \upharpoonright \{u_1, \ldots, u_{m-1}\}$ into $\{a_1, \ldots, a_{m-1}\}$ will now yield a graph that contains H. Observe that for those updates no quantifiers are needed. $\square$

5 k-Clique Cannot Be Maintained with Arity $k - 2$

In this section we prove that the k-clique query cannot be maintained by a $(k-2)$-ary quantifier-free update program when $k \geq 3$. The proof uses two main ingredients; the substructure lemma from [9, 16] and a new Ramsey-like lemma. We state those lemmas next.

For stating the substructure lemma we need the following notion of corresponding modifications in isomorphic structures. Let π be an isomorphism from a structure $\mathcal{A}$ to a structure $\mathcal{B}$. Two modifications $\delta(\vec{a})$ on $\mathcal{A}$ and $\delta'(\vec{b})$ on $\mathcal{B}$ are said to be π-*respecting* if $\delta = \delta'$ and $\vec{b} = \pi(\vec{a})$. Two sequences $\alpha = \delta_1 \cdots \delta_m$ and $\beta = \delta'_1 \cdots \delta'_m$ of modifications respect π if δ_i and δ'_i are π-respecting for every $i \leq m$. For a discussion of the lemma we refer the reader to the long version of [16]. Recall that $P_\alpha(\mathcal{S})$ denotes the state obtained by executing the dynamic program $\mathcal{P}$ for the modification sequence α from state $\mathcal{S}$.

Lemma 1 (Substructure Lemma [9, 16]). *Let $\mathcal{P}$ be a* DynProp-*program with designated Boolean query symbol Q, and let $\mathcal{S}$ and $\mathcal{T}$ be states of $\mathcal{P}$ with domains S and T. Further let $A \subseteq S$ and $B \subseteq T$ such that $\mathcal{S} \restriction A$ and $\mathcal{T} \restriction B$ are isomorphic via π. Then Q has the same value in $P_\alpha(\mathcal{S})$ and $P_\beta(\mathcal{T})$ for all π-respecting sequences α, β of modifications on A and B.*

The second ingredient exhibits a disparity between upper bounds for Ramsey numbers in k-ary structures and lower bounds for Ramsey numbers in $(k+1)$-dimensional hypergraphs. While the first condition in the following lemma guarantees the existence of a Ramsey clique of size $f(|A|)$ in k-ary structures over A, the second condition states that there is a 2-coloring of the complete $(k+1)$-hypergraph over A that does not contain a Ramsey clique of size $f(|A|)$. This disparity is the key to the lower bound proof.

Lemma 2. *Let $k \in \mathbb{N}$ be arbitrary and τ a k-ary schema. Then there is a function $f : \mathbb{N} \to \mathbb{N}$ and an $n \in \mathbb{N}$ such that for every domain A larger than n the following conditions are satisfied:*

(S1) For every τ-structure $\mathcal{S}$ over A and every linear order $\prec$ on A there is a subset A' of A of size $|A'| \geq f(|A|)$ such that all $\prec$-ordered k-tuples over A' have the same type in $\mathcal{S}$.

(S2) The set $[A]^{k+1}$ of all $(k+1)$-hyperedges over A can be partitioned into two sets B and B' such that for every set $A' \subseteq A$ of size $|A'| \geq f(|A|)$ there are $(k+1)$-hyperedges $b, b' \subseteq A'$ with $b \in B$ and $b' \in B'$.

The two lemmas above can be used to obtain the lower bound for the k-clique query as follows. The proof of Lemma 2 will be sketched afterwards.

Theorem 2. $(k+2)$-Clique *($k \geq 1$) cannot be maintained under insertions by a k-ary* DynProp-*program.*

Proof.   Towards a contradiction assume that there is a k-ary DynProp-program $\mathcal{P}$ over k-ary schema τ that maintains $(k+2)$-Clique. Let n and f be as in Lemma 2. For a set A larger than n let $\prec$ be an arbitrary order on A and let $D \stackrel{\text{def}}{=} A \uplus C$ be a domain with $C \stackrel{\text{def}}{=} [A]^{k+1}$. Further let B, B' be the partition of $[A]^{k+1}$ guaranteed to exist by (S2) in Lemma 2.

We consider a state $\mathcal{S}$ over domain D where the input graph G contains the edges

$$\{(b_1, b), (b_2, b), \ldots, (b_{k+1}, b) \mid b = \{b_1, b_2, \ldots, b_{k+1}\} \in B\}$$

See Figure 1 for an illustration.

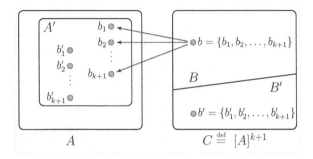

Fig. 1. The construction from the proof of Theorem 2

By Condition (S1) there is a subset $A' \subseteq A$ of size $|A'| \geq f(|D|)$ such that all ordered k-tuples over A' have the same τ-type in $\mathcal{S}$. Then by (S2) there are $(k+1)$-hyperedges $b, b' \subseteq A'$ with $b \in B$ and $b' \in B'$. Without loss of generality $b = \{b_1, b_2, \ldots, b_{k+1}\}$ with $b_1 \prec \ldots \prec b_{k+1}$ and $b' = \{b'_1, b'_2, \ldots, b'_{k+1}\}$ with $b'_1 \prec \ldots \prec b'_{k+1}$. By construction of the graph G, all elements in b are connected to the node $b \in C$ while there is no node in C connected to all elements of b'. Thus applying the modification sequences

(α) Insert the edges (b_i, b_j) in lexicographic order with respect to $\prec$.
(β) Insert the edges (b'_i, b'_j) in lexicographic order with respect to $\prec$.

yields one graph with a $(k+2)$-clique and one graph without a $(k+2)$-clique, respectively. However, by the substructure lemma, the program $\mathcal{P}$ yields the same result since the substructures induced by $\vec{b} = (b_1, \ldots, b_{k+1})$ and $\vec{b}' = (b'_1, \ldots, b'_{k+1})$ are isomorphic. This is the desired contradiction. $\square$

In the following we give a rough sketch of the proof of Lemma 2. The k-*dimensional Ramsey* number for r colors and clique-size l, denoted by $R_k(l; r)$, is the smallest number n such that every r-coloring of a complete k-hypergraph with n nodes has a monochromatic clique of size l. The *tower function* $\mathrm{tow}_k(n)$ is defined by $\mathrm{tow}_k(n) \stackrel{\mathrm{def}}{=} 2^{2^{\cdot^{\cdot^{\cdot^{2^n}}}}}$ with $(k-1)$ many 2's. The following classical result for asymptotic bounds on Ramsey numbers due to Erdős, Hajnal and Rado is the key to prove Lemma 2. The concrete formulation is from [5].

Theorem 3. *[6, 7] Let k, l and r be positive integers. Then there exist positive constants c_k, $c_{k,r}$ and l_k such that*

(a) $R_k(l; r) \leq \mathrm{tow}_k(c_{k,r}l)$
(b) $R_k(l; 2) \geq \mathrm{tow}_{k-1}(c_k l^2)$ for all $l \geq l_k$

The theorem immediately implies that (T1) Ramsey cliques in r-colored k-dimensional complete hypergraphs are of size at least $\Omega(\log^{(k-1)}(n))$; and that (T2) there are 2-colorings of the $(k+1)$-dimensional complete hypergraphs such that monochromatic cliques are of size $O((\log^{(k-1)}(n))^{\frac{1}{2}})$. Here $\log^{(k)}(n)$ denotes $\log(\log(\ldots(\log n)\ldots))$ with k many log's. Those conditions are already quite similar to the conditions (S1) and (S2). The major difference is that (T1) is about hypergraphs and not about structures with a k-ary schema.

Fortunately the upper bound from Theorem 3 can be generalized to Ramsey numbers for structures. To this end some notions need to be transferred from hypergraphs to structures. Let τ be a k-ary schema, let $\mathcal{S}$ be a τ-structure over domain D and $\prec$ an order on D. A subset $D' \subseteq D$ of $\mathcal{S}$ is called an $\prec$-*ordered* τ-*clique* if all $\prec$-ordered k-tuples $\vec{a} \in D'^k$ have the same τ-type. Denote by $R(l; \tau)$ the smallest number n such that every τ-structure with n elements contains an $\prec$-ordered τ-clique of size l, for every order $\prec$ of the domain.

Theorem 4. *Let τ be a schema with maximal arity k and let l be a positive integer. Then there exists a constant c such that $R(l; \tau) \leq tow_k(cl)$.*

Proof. The proof of Observation 1' in [9, p. 11] yields this bound. For the sake of completeness we repeat the construction.

Let $\mathcal{S}$ be a τ-structure over domain D of size $tow_k(cl)$. Further let $\prec$ be an arbitrary order on D. Define a coloring col of the complete k-dimensional hypergraph with nodes D as follows. An edge $\{e_1, \ldots, e_k\}$ with $e_1 \prec \ldots \prec e_k$ is colored by the type $\langle \mathcal{S}, e_1, \ldots, e_k \rangle$. By Theorem 3 there is an induced monochromatic sub-k-hypergraph with domain $D' \subseteq D$ with $|D'| \geq l$. By the definition of the coloring col, two $\prec$-ordered k-tuples over D' have the same type and therefore D' is a $\prec$-ordered τ-clique in $\mathcal{S}$ as well.   □

The previous theorem implies that (T1') Ramsey cliques in k-ary structures are of size at least $\Omega(\log^{(k-1)}(n))$. Then Lemma 2 follows from the facts (T1') and (T2) by choosing f as the function in $\Omega(\log^{(k-1)}(n))$ guaranteed to exist by (T1'). The function f satisfies (S1) and (S2) due to (T1') and (T2). This completes the lower bound proof.

6   Adding Auxiliary Functions

In quantifier-free update programs, as considered up to here, only the modified and the updated tuple can be accessed while updating an auxiliary tuple. Since lower bounds for first-order update programs where arbitrary elements can be accessed in updates seem to be out of reach for the moment, it seems natural to look for extensions of quantifier-free update programs that allow for accessing more elements in some restricted way.

Here we study DYNPROP programs extended by auxiliary functions, an extension proposed by Hesse [13]. Auxiliary functions are updated by update terms that may use function symbols and a special if-then-else construct. For a discussion of previous work on this extension we refer to the introduction. A formal treatment of this extension can be found in [16, 9].

The lower bound from the previous section can be generalized to quantifier-free programs that may use unary functions.

Theorem 5. $(k+2)$-CLIQUE *(k ≥ 1) cannot be maintained under insertions by a k-ary* DYNPROP-*program with unary auxiliary functions.*

The proof is along the same lines as the proof of Theorem 2. Instead of the substructure lemma for DYNPROP a corresponding lemma for DYNQF from [9, 16] is used. However, this substructure lemma for DYNQF requires to exhibit isomorphic substructures that, additionally, have similar neighbourhoods.

A natural question is whether the lower bounds transfer to k-ary auxiliary functions. We conjecture that they do, but we will argue that the techniques used so far are not sufficient for proving lower bounds for binary auxiliary functions.

The fundamental difference between unary and binary auxiliary functions is that, on the one hand, unary functions can access elements that depend either on the tuple that has been modified in the input structure or on the auxiliary tuple under consideration but not on both. On the other hand binary functions can access elements that depend on both tuples.

A consequence is that binary DYNQF can maintain every boolean graph property when the domain is large with respect to the actually used domain. We make this more precise. In the following we assume that all domains D are a disjoint union of a modifiable domain D^+ and a non-modifiable domain D^-, and that modifications may only involve tuples over D^+. Auxiliary data, however, may use the full domain. A dynamic complexity class C *profits from padding* if every boolean graph property can be maintained whenever the non-modifiable domain is sufficiently large in comparison to the modifiable domain[2].

Above we have seen that DYNPROP with unary auxiliary functions does not profit from padding.

Theorem 6. *Binary* DYNQF *profits from padding.*

Proof sketch.   We only show how ternary DYNQF profits from padding, the adaption to binary DYNQF is not difficult. Let Q be an arbitrary boolean graph property. In the following we construct a ternary DYNQF program $\mathcal{P}$ which maintains Q if $2^{|D^+|^2} = |D^-|$. The idea is to identify D^- with the set of all graphs over D^+, that is D^- contains an element c_G for every graph G over D^+. A unary relation R_Q stores those elements of D^- that correspond to graphs with the property Q. Finally the program maintains a pointer p to the element in D^- corresponding to the graph stored in D^+. The pointer is updated upon edge modification by using ternary functions f_{INS} and f_{DEL} initialized by the initialization mapping in a suitable way.

The program $\mathcal{P}$ is over schema $\tau = \{Q, p, f_{\text{INS}}, f_{\text{DEL}}, R_Q\}$ where p is a constant, f_{INS} and f_{DEL} are ternary function symbols, R_Q is a unary relation symbol and Q is the designated query symbol.

We present the initialization mapping of $\mathcal{P}$ first. The initial state $\mathcal{S}$ for a graph H is defined as follows. The functions f_{INS} and f_{DEL} are independent of H and defined via

$$f_{\text{INS}}^{\mathcal{S}}(a, b, c_G) = c_{G+(a,b)}$$

$$f_{\text{DEL}}^{\mathcal{S}}(a, b, c_G) = c_{G-(a,b)}$$

---

[2] Note that this type of padding is different from the padding technique used by Patnaik and Immerman for maintaining a PTIME-complete problem in DYNFO [14].

for $a, b \in D^+$ and $c_G \in D^-$. For all other arguments the value of the functions is arbitrary. Here $G + (a, b)$ and $G - (a, b)$ denote the graphs obtained by adding the edge (a, b) to G and removing the edge (a, b) from G, respectively. The relation R_Q^S contains all c_G with $G \in Q$. Finally the constant p^S points to c_H.

It remains to exhibit the update formulas. After a modification, the pointer p is moved to the node corresponding to the modified graph, and the query bit is updated accordingly:

$$t_{\text{INS}}^p(u, v) = f_{\text{INS}}(u, v, p) \qquad\qquad t_{\text{INS}}^Q(u, v) = R_Q(f_{\text{INS}}(u, v, p))$$
$$t_{\text{DEL}}^p(u, v) = f_{\text{DEL}}(u, v, p) \qquad\qquad t_{\text{DEL}}^Q(u, v) = R_Q(f_{\text{DEL}}(u, v, p))$$

By further extending the non-modifiable domain, this construction can be extended to binary DYNQF. $\qquad\qquad\qquad\qquad\qquad\qquad\qquad\qquad\qquad\qquad\square$

Hence the ability to profit from padding distinguishes binary DYNQF and DYNPROP extended by unary functions. Although the proof of the preceding theorem requires the non-modifiable domain to be of exponential size with respect to the modifiable domain, the construction also explains why the lower bound technique from the previous sections cannot be immediately applied to binary DYNQF. In the lower bound construction only tuples over the set A are modified, while tuples containing elements from $C = [a]^k$ are not modified. Thus, by treating C as a non-modifiable domain, it can be used to store information as in the proof above. As the modification sequences used in the lower bounds are of length k^2, finding similar substructures in structures with binary auxiliary functions becomes much harder.

7   Conclusion and Future Work

In this work we exhibited a precise dynamic descriptive complexity characterization of the k-clique query when only insertions are allowed. The characterization implies an arity hierarchy for graph queries for DYNPROP under insertions. We also discussed the limit of our proof methods.

While proving lower bounds for full DYNFO — a major long-term goals in dynamic descriptive complexity — might be really hard to achieve, we believe that the following goals are suitable for both developing new lower bound methods and for further improving the current methods.

Goal 1. *Prove general quantifier-free lower bounds for insertions and deletions for the reachability query and the k-clique query.*

It is known that both queries cannot be maintained in binary DYNPROP. We conjecture that 3-clique cannot be maintained in DYNPROP under deletions.

Goal 2. *Find a general framework for proving quantifier-free lower bounds.*

Goal 3. *Find a natural query that cannot be maintained in binary DYNQF.*

Acknowledgement. I am grateful to Thomas Schwentick for encouraging discussions and many suggestions for improving a draft of this work. Further I thank Nils Vortmeier for proofreading.

References

[1] Dong, G., Libkin, L., Wong, L.: On impossibility of decremental recomputation of recursive queries in relational calculus and SQL. In: DBPL 1995, p. 7 (1995)

[2] Dong, G., Libkin, L., Wong, L.: Incremental recomputation in local languages. Inf. Comput. 181(2), 88–98 (2003)

[3] Dong, G., Su, J.: Deterministic FOIES are strictly weaker. Ann. Math. Artif. Intell. 19(1-2), 127–146 (1997)

[4] Dong, G., Su, J.: Arity bounds in first-order incremental evaluation and definition of polynomial time database queries. J. Comput. Syst. Sci. 57(3), 289–308 (1998)

[5] Duffus, D., Lefmann, H., Rödl, V.: Shift graphs and lower bounds on Ramsey numbers $r_k(l; r)$. Discrete Mathematics 137(1-3), 177–187 (1995)

[6] Erdös, P., Rado, R.: Combinatorial theorems on classifications of subsets of a given set. Proc. London Math. Soc. 2(3), 417–439 (1952)

[7] Erdős, P., Hajnal, A., Rado, R.: Partition relations for cardinal numbers. Acta Mathematica Hungarica 16(1), 93–196 (1965)

[8] Etessami, K.: Dynamic tree isomorphism via first-order updates. In: PODS 1998, pp. 235–243 (1998)

[9] Gelade, W., Marquardt, M., Schwentick, T.: The dynamic complexity of formal languages. ACM Trans. Comput. Log. 13(3), 19 (2012)

[10] Grädel, E., Siebertz, S.: Dynamic definability. In: ICDT 2012, pp. 236–248 (2012)

[11] Graham, R.L., Rothschild, B.L., Spencer, J.H.: Ramsey Theory. Wiley Series in Discrete Mathematics and Optimization. Wiley (1990)

[12] Hesse, W.: The dynamic complexity of transitive closure is in $DynTC^0$. Theor. Comput. Sci. 296(3), 473–485 (2003)

[13] Hesse, W.: Dynamic Computational Complexity. PhD thesis, University of Massachusetts Amherst (2003)

[14] Patnaik, S., Immerman, N.: Dyn-FO: A parallel, dynamic complexity class. J. Comput. Syst. Sci. 55(2), 199–209 (1997)

[15] Weber, V., Schwentick, T.: Dynamic complexity theory revisited. Theory Comput. Syst. 40(4), 355–377 (2007)

[16] Zeume, T., Schwentick, T.: On the quantifier-free dynamic complexity of reachability. In: Chatterjee, K., Sgall, J. (eds.) MFCS 2013. LNCS, vol. 8087, pp. 837–848. Springer, Heidelberg (2013), Full version available at http://arxiv.org/abs/1306.3056

[17] Zeume, T., Schwentick, T.: Dynamic conjunctive queries. In: ICDT 2014, pp. 38–49 (2014)

Author Index